Water Resources Engineering

Ralph A. Wurbs
Department of Civil Engineering
Texas A&M University
College Station, Texas

Wesley P. James
H2WR, Inc.
College Station, Texas

PHI Learning Private Limited
Delhi-110092
2014

This Indian Reprint—₹ 595.00
(Original U.S. Edition—₹ 11293.00)

WATER RESOURCES ENGINEERING
by Ralph A. Wurbs and Wesley P. James

Original edition, entitled *Water Resources Engineering*, by Ralph A. Wurbs and Wesley P. James, published by Pearson Education, Inc., publishing as Pearson Prentice Hall.

ISBN-978-81-203-2151-9

Indian edition published by PHI Learning Private Limited.

This edition is manufactured in India, and is authorized for sale in India, Pakistan, Sri Lanka, Bhutan, Bangladesh, Nepal and the Maldives only.

Published by Asoke K. Ghosh, PHI Learning Private Limited, Rimjhim House, 111, Patparganj Industrial Estate, Delhi-110092 and Printed by V.K. Batra at Pearl Offset Press Private Limited, New Delhi-110015.

Contents

Preface

Effective management of water resources, locally, regionally, and globally, is crucial for human welfare, economic prosperity, and environmental vitality. The professional field of water resources engineering is concerned with solving problems and meeting needs associated with municipal, industrial, and agricultural water supply and use, water quality in streams and aquifers, erosion and sedimentation, protection of ecosystems and other natural resources, recreation, navigation, hydroelectric power generation, stormwater drainage, and flood damage mitigation.

This textbook is designed for a basic course in water resources engineering focusing on fundamental topics of hydraulics, hydrology, and water management. It is also appropriate for advanced undergraduate and graduate courses and as a reference for practicing engineers. Water resources engineering concepts and methods are addressed from the perspective of practical applications in water management and associated environmental and infrastructure management. The focus is on mathematical modeling and analysis using state-of-the-art computational techniques and computer software.

The book is based largely on the authors' combined total of more than 40 years experience in teaching the required undergraduate civil engineering course at Texas A&M University entitled Water Resources Engineering and other undergraduate and graduate courses that build on this course. The book reflects ideas formulated by the authors and their colleagues and students for updating and improving these courses. It is designed to provide broad coverage of pertinent topics with flexibility for adaptation to the spectrum of ways that individual courses and sequences of undergraduate and graduate courses are organized at various universities.

The first three chapters are introductory overviews of the professional practice of water resources engineering (Chapter 1), the science of hydrology (Chapter 2), and fundamentals of fluid mechanics (Chapter 3). Chapters 4 through 9 each focus on specific aspects of water resources engineering design and analysis, including pressure conduit hydraulics, open channel hydraulics, hydraulic and hydrologic routing, frequency analysis, watershed modeling, and groundwater engineering. In Chapter 10, selected methods from previous chapters are applied to urban stormwater management. Chapter 11 is an introduction to the application of systems simulation, optimization, and economic evaluation techniques in water management decision-making. Chapter 12 introduces practices and issues of comprehensive river basin management.

Chapters 1, 2, and 12 provide a broad qualitative overview of hydrology and water resources management. Chapter 3 reviews basic principles of fluid mechanics. The hydrologic and hydraulic analysis concepts and computational methods presented in Chapters 4 through 10 are applied in professional practice as components of computer models that simulate natural and constructed water systems. Chapter 11 integrates economic and systems analysis methods, along with hydrologic and hydraulic engineering methods, in a broader systems view of water resources planning and management.

The chapters can be covered in essentially any order, subject to the following considerations. Students using the text will likely have already completed a course in fluid mechanics and may not need Chapter 3. However, because hydraulics is built on fluid mechanics, a brief review of basic fluid mechanics concepts is presented in Chapter 3 prior to addressing hydraulics in depth in Chapters 4–6 and 9–10. The basic overview of hydrology presented in Chapter 2 is prerequisite for Chapter 7, "Hydrologic Frequency Analysis," and Chapter 8, "Modeling Watershed Hydrology," and also provides an introduction for Chapter 9, "Groundwater Engineering." Chapter 10, "Urban Stormwater Management," applies methods from all of the hydrology and hydraulics chapters. Chapter 1 is the only prerequisite chapter for Chapters 11 and 12. Although Chapters 11 and 12 complement each other, either may be covered without the other.

Sufficient material is provided for multiple courses, particularly if supplemented by the computer models discussed in the book. The text may be used for courses in hydrology, hydraulics, urban stormwater management, and water systems planning and management, as well as for a fundamental first course in water resources engineering. A set of computer programs developed in conjunction with the text enhances understanding and application of computational methods. Modeling capabilities provided by generalized simulation models developed by federal water agencies and other entities greatly contribute to water resources engineering practice and play an important role in education.

The authors gratefully acknowledge the contributions of our students and colleagues at Texas A&M University in shaping our perspectives on the subject matter of this book. We join the publisher in thanking the following reviewers for their thoughtful comments during the development of the manuscript: Paul C. Chan, New Jersey Institute of Technology, Robert D. Kersten, University of Central Florida, and Thomas C. Piechota, University of Nevada, Las Vegas. Mrs. Joyce Hyden typed much of the manuscript, proficiently as always. Finally, we thank our wives, Keri and Karen, for their enduring patience and support during the project.

Ralph A. Wurbs
Wesley P. James

1 Introduction

Water resources engineering involves people, natural resources, and constructed facilities. In meeting the water-related needs of society, water resources engineers both (1) formulate and implement resource management strategies, and (2) plan, design, construct, and operate structures and facilities. Development and management of the natural resource *water* are essential for human survival and prosperity. Water management is also integrally linked to stewardship of land and environmental resources. The water-related infrastructure of a city, region, or nation includes river regulation structures, wells for pumping groundwater, storage and conveyance facilities, treatment plants, water distribution networks, wastewater management systems, flood damage reduction measures, erosion mitigation practices, stormwater drainage systems, bridges, hydroelectric power plants, and various other constructed facilities.

1.1 WATER RESOURCES ENGINEERING DISCIPLINES

Water resources engineering is one of the several major subdisciplines of civil engineering. Civil engineers serve the public by solving problems and addressing needs related to developing/maintaining the physical infrastructure and protecting/restoring the environment. Most civil engineers deal with water problems to at least some extent throughout their professional careers. Many civil engineers specialize specifically in water resources engineering.

This book covers water management practices; hydrologic and hydraulic engineering principles; and hydrosystems planning, design, and analysis. Basic concepts of hydrology, hydraulics, and water resources systems engineering are applied to practical problems of managing water resources.

Hydrology is the study of the occurrence, distribution, movement, and properties of the waters of the earth. Engineering hydrology involves understanding hydrologic processes, data collection and analysis, mathematical modeling, and hydrologic design. Although the book devotes more attention to water quantity than water quality, both of these closely interrelated aspects of hydrology are covered.

Hydraulics is the study of the mechanical behavior of water in physical systems and processes. Hydraulics is the practical application of the principles of fluid mechanics in water resources engineering. Hydraulic analysis of natural and man-made systems and/or design of constructed facilities may involve flow through open channels, pressure conduits, and/or porous media. Sediments and other contaminants may be transported with the water.

Water resources systems engineering may be defined as the formulation and evaluation of alternative plans to determine that particular system configuration or set of actions that will best accomplish public objectives within the constraints of governing natural laws, engineering principles, economics, environmental protection objectives, social and political pressures, legal restrictions, and institutional and financial capabilities. Hydrologic and hydraulic engineering are combined with other disciplines to support analysis and decision-making processes involved in planning, designing, maintaining, and operating water management systems.

The term *hydrosystems engineering* has been used synonymously with *water resources engineering* to highlight the systems analysis perspective. The broader term *water resources planning and management* encompasses water resources engineering, but emphasizes the interdisciplinary institutional, political, and socio-economic, as well as technical engineering, aspects of the processes by which society addresses its water-related problems and needs.

Environmental engineering and *water resources engineering* are integrally connected and greatly overlapping. Both are specialty fields within civil engineering. The environmental/water resources interconnections have grown in importance over the past several decades, as the fields of *hydraulic engineering* and *sanitary engineering* grew into *water resources engineering* and *environmental engineering*, respectively. Environmental engineering is concerned with provision of safe, palatable, and ample water supplies; disposal or recycling of wastewater and solid wastes; control of water, soil, and atmospheric pollution; mitigation of the adverse social and environmental impacts of human activities; and engineering aspects of the public health field, such as sanitation, control of arthropod-borne diseases, and elimination of industrial health hazards. This book omits water-related subjects such as water and wastewater treatment that are covered in depth by environmental engineering textbooks. However, all of the material presented in this book is relevant to environmental engineers.

The book is written by civil engineers for civil engineers. However, water resources development and management are interdisciplinary in nature. Water-related needs and issues are broad, complex, and crucial to economic/social development and environmental protection. Consequently, water resources engineers work with political officials, economists, lawyers, urban planners, agricultural scientists, chemists, biologists, geologists, meteorologists, computer analysts, and professionals from various other scientific and engineering disciplines, as well as water users and the public.

1.2 WATER MANAGEMENT SECTORS

Water management involves the development, control, regulation, protection, and beneficial use of surface and groundwater resources. Water management activities include policy formulation; national, regional, and local resource assessments; regulatory and permitting functions; formulation and implementation of resource management strategies; planning, design, construction, maintenance, rehabilitation, and operation of structures and facilities; scientific and engineering research; and education and training.

Water is essential to all of us. Human health and socioeconomic welfare is dependent on adequate supplies of suitable quality water. Conversely, too much water results in socioeconomic damages and loss of life due to flooding. Flood mitigation, stormwater management, and erosion control are important concerns in water resources engineering. The vitality of natural ecological systems is dependent on mankind's stewardship of water resources. Water management has played a key

Figure 1.1 The Seine River flowing through Paris is illustrative of the propensity for cities to develop adjacent to streams and rivers. (Photo by R. A. Wurbs, June 1996)

Figure 1.2 The London docks extend along 56 km of the 340-km Thames River. The river is the main water supply source for London. (Photo by R. A. Wurbs, May 1998)

role in the development of civilizations since ancient times. Population growth, urban and industrial development, and expansion of irrigated agriculture have resulted in drastically increased demands and problems during the past century.

The 1940's–1980's were the construction era of water resources development in the United States. Numerous multipurpose dams, hydropower plants, flood control improvements, navigation facilities, and irrigation projects were designed and constructed during the 1940's–1970's. The 1970's–1980's were a peak period for construction of wastewater treatment plants. Construction of new facilities continues to be important today. However, water resources management policy and practice shifted during the 1970's–1980's to major emphases on (1) maintenance, rehabilitation, and operation of the massive inventory of existing facilities, and (2) nonstructural strategies for managing water and related land resources. Protection and restoration of water quality and environmental resources have also been driving concerns since the 1970's.

The scale and complexity of water resources engineering projects vary greatly. For example, an engineer may design a culvert for a road construction project to convey rainfall runoff from a watershed with a drainage area of less than a square kilometer. Other engineers may be conducting studies to optimize the operations of a major multiple-purpose, multiple-reservoir system regulating stream flows of an interstate river basin with a watershed area of many thousand square kilometers. From the perspective of the citizens of a small community, a local groundwater aquifer contained completely within their county is of the utmost importance because it provides their domestic water supply. However, the pumpage from the few wells in this relatively small aquifer is a minute fraction of the volume of water pumped from the High Plains Ogallala Aquifer that extends across seven states, from South Dakota to northern Texas, with economies highly dependent on irrigated agriculture. The scale and complexity of construction projects range from a small pipeline extending water supply services to several new residential homes to dams and appurtenant structures that are among the largest engineering projects ever constructed.

Although water resources engineering projects often focus on a single purpose, multiple-purpose development is also common. The important concept of multiple-purpose water management is illustrated by a reservoir system that stores flood waters that are later released through hydroelectric power turbines and diverted further downstream for municipal and industrial water supply while ensuring adequate instream flows to maintain fisheries and wildlife habitat. An athletic field in a neighborhood park may be used as a stormwater detention basin. Detention storage may provide sediment and water quality control functions, as well as reduce flood peaks.

Purposes achieved by water projects and programs are outlined in Table 1.1. Human use of water is categorized in Table 1.1 by whether the water is withdrawn/diverted from its source or used in the stream without being withdrawn. Hydroelectric power generation, navigation, and recreation are major ways that people use water without removing it from stream/reservoir systems. Environmental management includes protecting or restoring water quality and maintaining stream flow quantities required for healthy ecosystems. Stormwater management and flood mitigation deal with problems of too much water.

TABLE 1.1 WATER RESOURCES DEVELOPMENT AND MANAGEMENT PURPOSES

- Water Supply Diversions with Consumptive Use
  - -Municipal supply and use
  - -Industrial supply and use
  - -Agricultural supply and use
- Instream Water Use
  - -Hydroelectric power generation
  - -Inland navigation
  - -Water-based recreation
- Environmental Management
  - -Wastewater collection, treatment, and disposal
  - -Water quality management
  - -Protection/restoration/enhancement of biological resources
- Stormwater Management and Flood Mitigation
  - -Stormwater drainage and management
  - -Flood damage reduction
  - -Erosion and sedimentation control
- Multiple-Purpose Development and Management

1.2.1 Municipal, Industrial, and Agricultural Water Supply

We use water in a myriad of ways. Water is diverted from rivers, lakes, and aquifers to supply municipal (domestic and public), rural domestic, livestock, agricultural irrigation, and various industrial uses. Consumptive use is water that is withdrawn and not directly returned to a stream or aquifer. Depending on the proportion of municipal water use devoted to lawn watering, flows returned to a river as wastewater treatment plant effluent may be about half the amount withdrawn. In terms of total withdrawals, the largest water use in the U.S. is cooling water for thermal-electric power plants. However, after circulating through the power plant cooling system, most of this water, now warmer, is returned to the river. Typical irrigation return flows range from none to almost half of the water withdrawn. Irrigation accounts for more consumptive use than any other water use sector.

Water use in the U.S. is summarized in Table 1.2 using data for the year 1995. The 554 billion cubic meters (m^3) of water withdrawn divided by the 1995 U.S. population of 267 million people results in a mean per capita water use of 2,070 m^3/year/person, which is equivalent to 5.68 m^3/day/person or 1,500 gallons/day/person. The consumptive use is 24.9 percent of the amount withdrawn; the other

Figure 1.3 This 150 m (490 ft) well being drilled into the Ogallala Aquifer in Nebraska will supply a center pivot irrigation system. (Photo by R. A. Wurbs, June 2001)

TABLE 1.2 WATER USE IN THE UNITED STATES IN 1995 (GLEICK, 2000)

	Billion m^3/yr	Percent
1995 Withdrawals in the United States		
Agricultural irrigation	185	33.4
Rural domestic and livestock use	12	2.2
Municipal domestic and public use	56	10.1
Thermal-electric power plants	262	47.3
Other industrial uses	39	7.0
Total 1995 Withdrawals	554	100
Total 1995 Consumptive Use	138	24.9
Source of supply		
Surface water	447	80.7
Groundwater	107	19.3

75.1 percent is returned to its source. Surface water rivers and lakes and groundwater aquifers supply 80.7 and 19.3 percent, respectively, of the water withdrawn.

1.2.1.1 Irrigation Agricultural irrigation accounts for a third of total withdrawals in the U.S., 65 percent worldwide, and about 75 percent in developing countries (Chaturvedi, 2000). The 16 percent of cultivated land that is irrigated worldwide contributes 36 percent of total food production. The land areas being irrigated throughout the world in 1940, 1970, and 1995, respectively, comprised about 76, 242, and 256 million hectares. Fifty-four percent of the irrigated land is in four countries: China (20%), India (19%), the United States (8%), and Pakistan (7%). Irrigation increases crop yields and the amount of land that can be productively farmed, stabilizes productivity, facilitates a greater diversity of crops, increases farm income and employment, helps alleviate poverty, and contributes to regional development.

The earliest civilizations were developed in the Middle East and Asia along major rivers such as the Tigris, Euphrates, Nile, and Indus, which supplied water for irrigation. Increased demands for food accompanying population growth have resulted in irrigation projects worldwide over the past century. The Green Revolution of the 20th century relied heavily on water resources development and irrigation technology. Irrigation has been essential for the settlement and economic development of the dry western half of the United States. In recent years, in regions of the United States and elsewhere, use of water for irrigation has leveled off and even decreased due largely to depleting groundwater reserves and competition from cities for limited water resources.

1.2.1.2 Domestic water supply and sanitation worldwide. Basic domestic water supply and sanitation are fundamental to economic and social development and prevention of a multitude of diseases. The World Health Organization, the World Bank, and the U.S. Agency for International Development (USAID) estimates the minimum amount of water required for drinking, cooking,

Figure 1.4 Low elevation spray application center pivot irrigation systems are used in the Winter Garden Region of South Texas shown here and elsewhere (New and Fipps, 2001).

cleaning, and sanitation range from 20 to 40 liters/day (5.3–10.6 gallons/day) per person. Per capita water use for household purposes, including lawn watering as well as indoor uses, varies greatly regionally throughout the United States, but about 300 liters/day (80 gallons/day) is a typical average per capita use rate. More than a billion people in the developing world lack safe drinking water that those in the developed nations take for granted. About half of the six billion people in the world live without access to adequate sanitation systems necessary for reducing exposure to water-related diseases. Construction of water supply and wastewater treatment facilities during the past century have dramatically reduced or eliminated typhoid fever, cholera, dysentery, and many other water-borne diseases in the developed nations. In the developing world, an estimated 14,000–30,000 people, mostly young children and the elderly, die every day from disease caused by drinking contaminated water or eating contaminated food (Gleick, 2000).

An airplane crash killing 300 people would be in the headlines of all major newspapers. The death toll attributable to inadequate water supply and sanitation is equivalent to such airplane crashes occurring every 15–30 minutes, 24 hours a day, 365 days a year. Those of us fortunate enough to live in nations with reliable, safe water supply systems, which we often take for granted, owe a great debt of gratitude to the engineering profession. Water resources engineers face a staggering challenge in responding to the needs of the developing world.

1.2.1.3 Water supply systems. Municipal, industrial, and agricultural water supply systems include sources of supply and facilities for storing, transporting, and distributing the water. Municipal and industrial systems also include treatment plants. Supply sources are usually river/reservoir systems and/or groundwater aquifers. Treatment of seawater, though very expensive, provides an alternative source for cities in coastal regions without access to rivers or aquifers. Farmers' crops are watered directly by precipitation and by irrigation. Wells are required to develop groundwater sources. Pipelines, pumps, and open channels serve to transport water between components of the supply system and distribute the water to consumers. Water distribution systems for large cities are typically comprised of thousands of pipelines and numerous pumps and storage facilities.

Dams, reservoirs, and appurtenant structures play a key role in water supply and multiple-purpose water management. Most of the rivers of the world are characterized by flows that are highly variable, random, and subject to extremes. In addition to cyclic seasonal fluctuations within each year, severe droughts with durations of several years and extreme flood events are major concerns. Reservoir storage is necessary to regulate stream flow fluctuations and develop reliable water supplies.

With increased hydrologic, environmental, and economic constraints on developing additional water supplies, demand management has become a major focus in the U.S. and elsewhere since the 1970's. Many cities waste much of their water through undetected pipeline leaks in aging water distribution systems. Agricultural irrigation delivery systems are notorious for losses due to seepage and evaporation. People use more water than they really need, with water inefficient landscaping and plumbing. Demand management measures may be categorized as long-term and short-term. Leak detection and repair, reuse of treated wastewater, pricing incentives, and water efficient plumbing, irrigation equipment, and landscaping achieve long-term reductions in water use. Rationing restrictions placed on water use during drought conditions is an emergency short-term measure.

1.2.2 Hydroelectric Power Generation

Electrical energy is produced by two types of plants, thermal and hydro. Most thermal plants use steam turbines and coal, natural gas, or nuclear fuel. As previously noted, circulation of water through the cooling systems of thermal-electric plants accounts for almost half of the water diverted from streams and aquifers in the U.S., although most of this water is returned to the streams. Hydroelectric generators are driven by water turbines. Many large electric power production systems include interconnected thermal and hydroelectric plants. Hydroelectric power generation is limited by water availability, but during wet periods, reliance on hydropower reduces fuel costs associated with the thermal plants. Thermal plants are often used to meet base-load requirements, supplemented by hydropower during times of peak energy demands.

Two-thirds of the 2,445,000 gigawatt-hours of electric energy generated by hydropower plants worldwide in 1996 were produced in the 10 countries listed in

TABLE 1.3 HYDROELECTRIC CAPACITY AND PRODUCTION IN 1996 (GLEICK, 2000)

Country	Installed capacity (megawatts)	Hydropower production (gigawatt-hours/year)	Percentage of country's total electrical energy
United States	74,860	296,380	10
Canada	64,770	330,690	62
China	52,180	166,800	18
Brazil	51,100	250,000	97
Russian Federation	39,990	162,800	27
Norway	26,000	112,680	99
France	23,100	65,500	15
Japan	21,170	91,300	9
India	20,580	72,280	25
Sweden	16,540	63,500	52
Total for 10 countries listed above	390,290	1,611,930	22
Worldwide total	633,730	2,445,390	20

Table 1.3. The 296,380 gigawatt-hours produced by hydroelectric plants in the U.S. represents about 10 percent of the total U.S. electric energy production. Hydroelectric plants account for about 20 percent of the total electric energy production worldwide. Hydropower supplies over 50 percent of the total electric energy supplied in each of 63 countries and over 90 percent of the electricity generated in 23 countries (Gleick, 2000).

Each hydroelectric power plant is unique with its own design. Hydropower projects normally include a dam, turbines, intake and conduit (called a penstock) to convey water to the turbines, generators, control mechanisms, housing for the equipment, transformers, and transmission lines. Trash racks, gates, forebay, surge tanks, and other appurtenant hydraulic structures may also be required. A tailrace, or channel, from the powerhouse back to the river is provided if the location of the turbines prevents discharge directly into the river. Hydropower plants may be classified as run-of-river, storage, or pumped storage. A storage-type plant has sufficient storage capacity to carry-over water from a wet season to a dry season or from year to year. Run-of-river plants have little storage and must use stream flow as it occurs. A pumped-storage plant generates energy during periods of peak demand, but during off-peak periods water is pumped from the tailwater pool back to the headwater pool.

1.2.3 Navigation

River navigation is another major instream use of water resources. Waterways have been important avenues of commerce in world history. Rivers offered comparatively easy routes through unmapped wilderness for the exploration of new lands. However, navigation on natural rivers is limited by sandbars, debris, and other obstructions, turbulent water, low flows, seasonal variations in flows, and floods. Consequently, extensive systems of channel improvements, canals, reservoirs, dams and locks, and bank stabilization measures have been constructed to facilitate navigation.

Petroleum, coal, construction materials, iron and steel products, grain, and other heavy bulky commodities are transported by barge on waterways relatively inexpensively compared with other alternative means of transport. However, river navigation has the disadvantage of being much slower than transport by rail or truck, and it is limited to serving cities located along the waterway. Other modes of transportation must be used to complete delivery of goods to locations away from the waterway system.

The navigation system of the United States consists of seven major groups of waterway routes with a total of 42,000 km of navigable channels that enable commercial water transportation to serve 38 states. This system includes the Atlantic Coast Waterways, Atlantic Intracoastal Waterway, Gulf Coast Waterways, Gulf Intracoastal Waterway, Mississippi River System, Pacific Coast Waterways, and Great Lakes System.

Navigation systems consist of natural and improved channels, man-made canals, contraction works, bank stabilization, and reservoir/dam/lock facilities. Where river flows are otherwise too shallow for navigation, dams are constructed to create sufficient depth. Lock structures allow boats and barges to pass through the dams. Canals are constructed to connect existing water bodies, to connect an inland city with a water body, and to circumvent unnavigable portions of a river such as rapids or falls.

Dredging consists of deepening and/or widening a channel by removing channel-bed material. Dredges are machines equipped with scooping or suction devices for removing debris and soil. Dredging is used to initially improve a channel where navigation is impeded by sandbars or deposits of silt. Annual maintenance dredging is commonly required to maintain adequate depths for navigation.

1.2.4 Environmental Management

Environmental management is concerned with minimizing the adverse effects of human activity on the environment. Environmental problems stem largely from population growth and the rising standard of living. The world population is over 6 billion and growing. Increased affluence in many nations, including the U.S. over the past century, has generated greater consumption of water, energy, and other resources and more pollution. Environmental management aspects of water resources engineering include wastewater collection, treatment, and disposal; protection and restoration of water quality in riverine and groundwater systems; and protection and enhancement of ecological systems and biological resources.

1.2.4.1 Wastewater collection, treatment, and disposal. Wastewater and sewage from homes, apartments, businesses, and industries are collected; pollutants are removed from the water; and the treated effluent is returned to a stream system. The elements of a municipal wastewater management system include: (1) individual sources of wastewater, (2) on-site processing facilities, (3) collection network, (4) conveyance to treatment plants, (5) treatment facilities, and (6) disposal facilities. Conventional municipal wastewater treatment plants use a multiple-step process: (1) removal of materials that will interfere with pumping and later treatment steps, (2) removal of the solid materials that will settle by gravity under quiescent conditions, (3) conversion of the remaining soluble and colloidal material into microbial solids, (4) removal of the remaining pollutant materials in a second sedimentation, and (5) treatment and disposal of the residual solids and sludges generated in the treatment process. Industrial facilities must either treat wastewater prior to discharge into a receiving water, or send it to a publicly owned municipal treatment plant, or in some cases, do both. Since industrial wastewater composition is highly variable, its treatment is very industry- and site-specific.

Wastewater treatment facilities in the United States are regulated by state environmental agencies in collaboration with the U.S. Environmental Protection Agency (EPA) through the National Pollutant Discharge Elimination System (NPDES). The Federal Water Pollution Control Act of 1948 provided a program for federal grants to municipalities for construction of wastewater treatment facilities, but funding was minimal relative to needs. In 1972, Congress amended the Water Pollution Control Act with Public Law 92-500 (PL 92-500), called the Clean Water Act, establishing programs that have greatly impacted water management, including the NPDES and federal grants for construction of publicly owned treatment works. Under the NPDES, the EPA requires each state to establish and enforce effluent limitations and performance standards for sources of water pollution, including wastewater treatment plants, industries, power plants, confined agricultural operations, and urban stormwater. PL 92-500 also established a massive construction grants program, providing several billion dollars of federal funds each year for the construction of municipal wastewater treatment plants that continued through the 1980's. The Water Quality Act of 1987 converted the construction grants to state-revolving loan programs, with state agencies administering loans to municipalities for construction of wastewater treatment plants and other water quality control projects.

1.2.4.2 Water quality management. Water quality management is concerned with the control of pollution from human activity so that the water is not degraded to the point that it is no longer suitable for intended uses. The uses of streams, rivers, lakes, and groundwater are greatly influenced by water quality. Activities such as hydropower generation, navigation, fishing, swimming, and potable water supply have different requirements for water quality. Particularly high quality is required for potable water supplies. In many parts of the world, municipal and industrial wastewater discharges and other human activities have transformed pristine natural streams with abundant fish and diverse ecological systems into foul open sewers with few life forms and fewer beneficial uses.

Domestic sewage and industrial wastes are called point sources because they are generally collected by a system of pipes or channels and conveyed to definite points of discharge into receiving water. Urban and agricultural runoff characterized by multiple disperse discharge sites are called nonpoint sources. Major categories of pollutants include oxygen-demanding material, typically biodegradable organic matter; nutrients, primarily nitrogen and phosphorous; pathogenic organisms; suspended solids; salts; toxic metals and toxic organic compounds; and heat.

Surface water quality in the United States has improved more since the 1970's than perhaps any other area of the environment. The Clean Water Act of 1972 (PL 92-500) established national water quality goals and created programs to implement the goals, including the previously noted NPDES permit system and federal funding of municipal wastewater treatment facilities. Water quality in the U.S. was dreadful prior to the 1970's. Cities and industries dumped large quantities of poorly treated or raw wastewater directly into many of our rivers and lakes. Today most, although not all, surface waters are in much better condition. Although much has been accomplished, more work is required. Some municipalities and industries have still not met the standards. Nonpoint pollution, such as agricultural fertilizers and pesticides and urban stormwater runoff, account for much of the present pollution.

Groundwater is a major source of domestic and public water supply. Treatment usually consists of chlorination and perhaps removal of iron and manganese or other specific constituents. Groundwater is typically viewed as being protected from pollution by overlying formations. However, since the 1970's, many aquifers in the U.S. have been found to be contaminated by compounds other than those present in the natural environment. Many of the compounds that have been found in groundwater are known to be carcinogenic and/or mutagenic. Sources of groundwater contamination include leachates from municipal and industrial solid waste landfills; land disposal of sludges from municipal and industrial wastewater treatment plants; industrial waste storage ponds and lagoons; septic tanks; underground storage tanks; hazardous waste disposal sites; deep-well injection of wastes; deicing of roads; accidental spills; agricultural pesticides, herbicides, and fertilizers; animal feedlots; mining; and saltwater intrusion. Groundwater aquifers are somewhat protected from surface contamination by overlying earth formations. However, being located at a depth into the earth also makes clean up of contaminated aquifers extremely difficult.

1.2.4.3 Protection and restoration of biological resources. Environmental management is concerned with water quantity as well as water quality. Agricultural, industrial, and municipal water supply diversions often conflict with environmental instream flow needs. Instream flow requirements include suitable stream flow amounts and seasonal variations to support fish, wildlife, and diverse ecological systems. Freshwater inflows into bays and estuaries are required to maintain proper salinity gradients for estuarine ecosystems. Environmental instream flow requirements are important considerations in water rights permitting programs and reservoir system operations.

Protection and restoration of wetlands is another important issue in water management. Wetlands occur in many forms, including swamps, bogs, marshes, shallow lakes, sloughs, flood plains, and estuaries. Water inundates or is near the surface of the ground for much of the year. Wetland ecosystems represent the transition between terrestrial and aquatic systems. Wetlands play important roles in the natural environment, including supporting ecosystems, naturally removing pollutants from water, and mitigating flooding. About half of the original wetlands in the U.S. have been lost through agricultural and urban development. In the 1980's, the federal government adopted policies and programs to prevent further loss of wetlands. In addition to natural wetlands, there are a growing number of restored, created or artificial, and constructed wetlands. Restored wetlands are those that existed as wetlands previously, but have had to be restored to their original conditions after being converted to dry land. Created or artificial wetlands are those that are built in previously dry areas to emulate natural wetlands. They are often created to mitigate the loss of wetlands at other locations due to construction or land development activities. Constructed wetlands may also be created specifically to serve as water treatment systems using wetland processes.

1.2.5 Stormwater Management, Flood Mitigation, and Erosion Control

Stormwater management, drainage, flood mitigation, and mitigation of erosion and sedimentation deal with problems caused by precipitation and runoff, particularly during periods of excessively high rainfall and/or snowmelt. Stormwater drainage refers to the runoff of precipitation from a watershed to a major stream. Flood mitigation addresses problems associated with the overflow of major streams. Although also occurring during nonflooding periods, erosion and sedimentation problems are generally associated with significant storms.

1.2.5.1 Stormwater management. Stormwater management is a major consideration for essentially all municipalities, small and large. Urban drainage systems include streets, curbs and gutters, ditches, channels, streams, culverts and bridges, and storm sewer networks, including pipes, inlets, and other appurtenant structures. The purpose of such systems is to collect and convey rainfall to a stream system. Drainage improvements are necessary to prevent ponding of water in streets, homes, and properties. However, efficient drainage also results in higher peak flows occurring downstream. Stormwater management entails a more comprehensive consideration of detention storage and other strategies for detaining runoff, as well as drainage improvements for expediting runoff.

Urbanization often results in significant increases in storm runoff as buildings, parking lots, and streets replace pastures and woods. Stormwater management strategies are often based on mitigating the increases in runoff to protect downstream properties. Detention basins of various configurations represent the most common approach for limiting postdevelopment peak flows to predevelopment conditions.

Precipitation runoff from urban areas washes off soil, lawn fertilizers, oil from streets, debris, and other contaminants and transports them to streams. The Water Quality Act of 1987 mandated incorporation of urban stormwater in the NPDES permit process. Stormwater permits became a major consideration for cities during the 1990's.

Drainage is also a major consideration for transportation engineers. Streets, highways, and railroads cross rivers, streams, and drainage channels. Bridges and culverts are constructed for these crossings. Provisions are also required to drain rainfall from pavements and adjacent right-of-ways. Drainage is an important component of airport design. Stormwater management is also important in agriculture, mining, and other industrial sectors.

1.2.5.2 Flood mitigation. Floodplain lands near rivers and other water bodies offer significant advantages for locating cities and agricultural development. Proximity to rivers facilitates water supply and wastewater disposal, recreation, and provision of other water-related services. The aesthetics of tree-lined streams attract residential development. Floodplain fertility encourages agriculture. However, human activities result in susceptibility to damage as rivers naturally overflow their banks periodically. Floodplains are a natural part of a stream system. Coastal areas are subject to flooding associated with tropical storms and hurricanes.

River and coastal flooding are major problems throughout the world. The Flood Control Act of 1936 and subsequent legislation initiated major federal programs for constructing flood control projects throughout the United States. The U.S.

Figure 1.5 The April 2001 flood on the Mississippi River inundated these homes in Campbell's Island, Illinois. (Courtesy of Federal Emergency Management Agency.)

Figure 1.6 Rescue workers patrol flooded streets in Davenport, Iowa, in April 2001. (Courtesy of Federal Emergency Management Agency.)

Army Corps of Engineers (USACE) constructed numerous dams, levees, floodwalls, and channel improvements on the major rivers of the nation during the 1930's–1970's. The Natural Resource Conservation Service (NRCS) has constructed many thousands of flood-retarding structures controlling flood flows from smaller watersheds. Nonfederal entities also construct and maintain flood control facilities.

These flood control structures, with few exceptions, have functioned as designed and greatly reduced flood damages. However, as illustrated by Figs. 1.7 and 1.8, flood losses have continued to increase nationwide, despite the large investments in flood control structures, as a result of intensified development of floodplain lands. People tend to build their homes and businesses near streams. Essentially all cities and towns are superimposed on a system of natural and man-altered streams.

The National Flood Insurance Program (NFIP) was established pursuant to the National Flood Insurance Act of 1968 as amended by the Federal Flood Disaster Protection Act of 1973 to supplement structural measures with nonstructural approaches. About 20,000 communities participate in the NFIP. Flood insurance is available to the citizens of participating communities. Participating local communities are required to enact and enforce floodplain management regulations. Hydrologic and hydraulic studies are performed to delineate floodplains. The

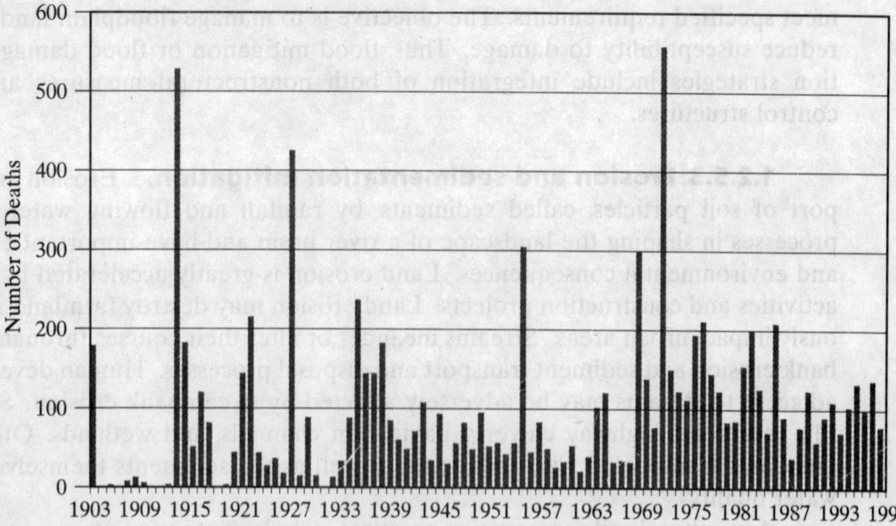

Figure 1.7 Flood-related deaths in the United States each year from 1903 to 1999. (National Weather Service.)

floodplain delineations continue to be updated to reflect watershed land use changes, floodplain encroachments, and construction of stormwater management and flood control improvements. Building of homes and businesses within the 1.0 percent annual exceedance frequency floodplain is either prohibited or must

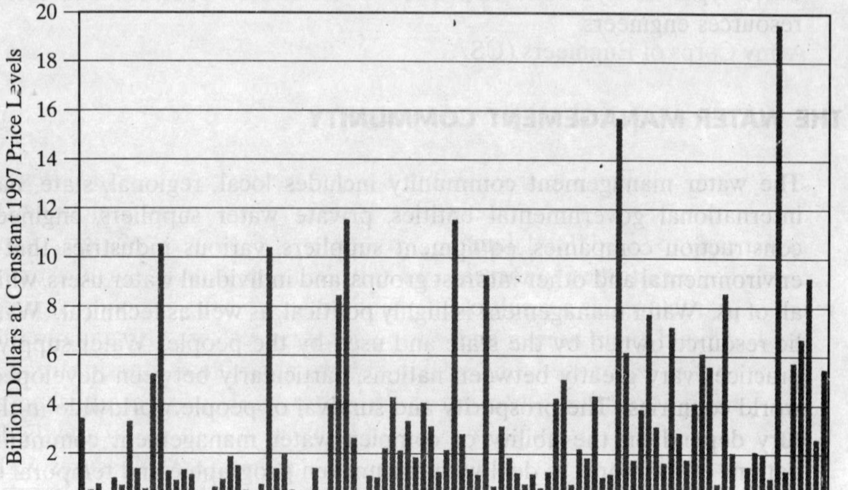

Figure 1.8 Flood damages in the United States in constant 1997 dollars. (National Weather Service.)

meet specified requirements. The objective is to manage floodplain land use and reduce susceptibility to damage. Thus, flood mitigation or flood damage reduction strategies include integration of both nonstructural measures and flood control structures.

1.2.5.3 Erosion and sedimentation mitigation. Erosion and transport of soil particles, called sediments, by rainfall and flowing water are key processes in shaping the landscape of a river basin and have important economic and environmental consequences. Land erosion is greatly accelerated by farming activities and construction projects. Land erosion may destroy farmland and seriously impact urban areas. Streams meander or alter their courses through streambank erosion and sediment transport and disposal processes. Human development adjacent to streams may be adversely affected by streambank erosion. Sediment fills reservoirs, highway culverts, navigation channels, and wetlands. Other contaminants transported with sediments, as well as the sediments themselves, affect water quality.

Agricultural soil conservation practices include contouring, strip cropping, and terracing. Streambank erosion control measures include revetments and lining materials, such as vegetation, rock riprap, and concrete. A variety of measures are applied in construction activities, such as highway projects, to prevent bare soil from being washed off the construction site.

Hydraulic structures often drastically increase flow velocities and require stilling basins and other energy dissipation measures. Erosion control below dam outlet structures and highway culverts, around bridge piers, and in association with other types of hydraulic structures is a common problem addressed by water resources engineers.

1.3 THE WATER MANAGEMENT COMMUNITY

The water management community includes local, regional, state, national, and international governmental entities, private water suppliers, engineering firms, construction companies, equipment suppliers, various industries that use water, environmental and other interest groups, and individual water users, which includes all of us. Water management is highly political, as well as technical. Water is a public resource owned by the state and used by the people. Water supply needs and practices vary greatly between nations, particularly between developed and third world countries. The prosperity and survival of people worldwide in the 21st century depend on the ability of complex water management communities in the nations of the world to deal with the uneven geographic and temporal distribution of water resources and to protect the water from pollution.

Water resources engineers in the United States work within the organizational setting outlined in Table 1.4. The entities listed in the table are interconnected, working together and interacting in a variety of ways. Thus, an engineer employed

TABLE 1.4 THE WATER MANAGEMENT COMMUNITY IN THE UNITED STATES

Private Sector
Engineering consulting firms
Construction contractors, Equipment suppliers
Private water suppliers
Industrial water users, Farmers, Developers

Local Public Agencies	*Interstate Regional Agencies*
Cities	River basin commissions
Water districts	Tennessee Valley Authority http://www.tva.com

State Agencies
California Department of Water Resources http://www.dwr.water.ca.gov
California Water Resources Control Board http://www.swrcb.ca.gov
Texas Water Development Board http://www.twdb.state.tx.us
Texas Natural Resource Conservation Commission http://www.tnrcc.state.tx.us
Florida Department of Environmental Protection http://www.dep.state.fl.us/org/
Illinois State Water Survey http://www.sws.uiuc.edu
Numerous other water agencies in the 50 states

Federal Agencies
Army Corps of Engineers (USACE) http://www.usace.army.mil
Bureau of Reclamation (USBR) http://www.usbr.gov
Natural Resource Conservation Service (NRCS) http://www.nrcs.usda.gov
Geological Survey (USGS) http://www.usgs.gov
National Weather Service (NWS) http://www.nws.noaa.gov
Federal Emergency Management Agency (FEMA) http://www.fema.gov
Federal Energy Regulatory Commission (FERC) http://www.ferc.fed.us
Fish and Wildlife Service (USFWS) http://www.fws.gov
Environmental Protection Agency (EPA) http://www.epa.gov

International Agencies
United Nations System agencies and programs http://www.un.org
U.S. Agency for International Development http://www.usaid.gov
International Boundary and Water Commission http://www.ibwc.state.gov

Professional Societies and Associations
Environmental and Water Resources Institute (EWRI) http://www.ewrinstitute.org
of the American Society of Civil Engineers (ASCE) http://www.asce.org
American Water Resources Association (AWRA) http://www.awra.org
International Water Resources Association (IWRA) http://www.iwra.siu.edu
American Water Works Association (AWWA) http://www.awwa.org
Water Environment Federation (WEF) http://www.wef.org
American Institute of Hydrology (AIH) http://www.aihydro.org
National Groundwater Association (NGWA) http://www.ngwa.org
Association of State Floodplain Managers (ASFPM) http://www.floods.org
Universities Council on Water Resources (UCOWR)
http://www.uwin.siu.edu/ucowr/

Universities

by any one of them should also have a general understanding of the activities of the others.

The water agencies maintain web sites, with addresses shown in Table 1.4, that describe their missions and programs and provide access to a wealth of information, including announcements of current activities and employment opportunities, directories of people to contact for various types of assistance, technical publications, computer software, and databases. The addresses shown provide connections to numerous other sites. For example, the centralized federal agency web sites provide connections to sites for their regional and field offices and research organizations.

Table 1.4 highlights key entities with water management responsibilities, but certainly is not a comprehensive listing of all organizations with interests in water resources engineering. For example, hydrologic and hydraulic design of bridges and drainage facilities is a major concern for engineers in state transportation departments that are not included in the table. Many agencies responsible for managing public lands and other related resources are not listed. The water-related work of the organizations listed in Table 1.4 is performed by water resources engineers and other water professionals. Many volunteer groups, not listed, also play important roles in the water management community. Environmental and natural resource conservation organizations encourage wise stewardship of water resources. Other organizations have been created to lobby for political support for various types of water development projects. A myriad of local and regional citizen advisory groups guide water management in their local areas.

1.3.1 Private Industry

Engineering consulting firms perform feasibility and design studies, construction supervision, and other professional services for the other water management entities listed in Table 1.4. A majority of the thousands of civil engineering consulting firms in the U.S. include water resources engineering in their repertoire of expertise. Many specialize exclusively in water resources engineering. Consulting firms range in size from one professional engineer working alone to firms with several thousand employees in many offices located throughout the nation and world. The public agencies listed in Table 1.4 accomplish their water resources engineering work with various combinations of in-house expertise and contracts with consulting firms.

Construction contractors are hired by many of the entities listed in Table 1.4 to construct water projects. Equipment suppliers manufacture and sell pumps, pipes, irrigation equipment, and the various other types of equipment required for water management systems.

Of the myriad industrial water users, electric power utilities divert the most water. Farmers irrigating their crops account for the largest consumptive use of water. Water supply, sewage collection, drainage, and stormwater management are major concerns to developers who build residential neighborhoods, commercial facilities, and other land development projects.

1.3.2 Local Public Agencies

Cities provide water supply and wastewater management services to their citizens and industrial customers. Drainage, stormwater management, and flood mitigation are also important functions of municipal government. The professional staff of public works, development services, planning, and engineering departments of cities perform some of their water resources engineering work in-house and contract with consulting firms to perform other work.

Municipal water districts are created to develop and operate regional water supply and wastewater management facilities for multiple member cities. The regional approach is often more efficient than each city owning and operating its own individual facilities. Likewise, urban drainage and flood control districts encompass geographic areas larger than a single city. Agricultural levee districts allow farmers to pool their resources to protect their land from flooding. Soil conservation districts work to mitigate agricultural land erosion and manage land and water resources. Groundwater conservation districts are created to manage the use of particular aquifers. River authorities have comprehensive river basin management responsibilities.

1.3.3 Interstate Regional Agencies

Water agencies in the U.S. are created primarily at the local, state, and federal levels. Various interstate river basin commissions, compact commissions, and other governmental bodies have been established to allocate the water resources of interstate river basins between the affected states and to coordinate water management. However, the Tennessee Valley Authority (TVA) is unique in its comprehensive role in developing and managing the water resources of a multiple-state region.

The TVA was created by the federal government in 1933 as a regional agency to promote economic development and social betterment of a depressed area of the nation. In 1933, land in the Tennessee River Valley was underdeveloped and neglected; heavy rainfall had eroded the soil; and the forests had been cut over and burned. Almost every community along the major streams was subject to flood damage. Heavy flows from the Tennessee River also contributed to flooding on the Ohio and Mississippi Rivers. The TVA was charged with planning for the proper use, conservation, and development of the natural resources of the Tennessee River Basin. This was to be accomplished through flood control, power production, navigation, reduction of soil erosion, afforestation, elimination of agricultural use of marginal lands, industrial development and diversification, and community development.

The Tennessee River Basin encompasses portions of Alabama, Georgia, Kentucky, Mississippi, North Carolina, Tennessee, and Virginia. The Tennessee River flows into the Ohio River just above its confluence with the Mississippi River. The TVA is the largest public power company in the U.S., with 28,500 megawatts of

Figure 1.9 Norris Dam on the Clinch River in Tennessee was the first dam constructed by the TVA and is operated by the TVA for flood control, hydropower, water supply, and recreation. (Courtesy of Tennessee Valley Authority.)

dependable generating capacity. TVA's power facilities include 11 fossil-fuel plants, 3 nuclear plants, 29 hydroelectric plants, and 27,000 km of transmission lines. The TVA operates a multipurpose reservoir system with 50 dams and appurtenant structures for flood control, hydroelectric power, water supply, navigation, fish and wildlife, and recreation. The 10 dams on the main stem of the Tennessee River have navigation locks facilitating barge traffic on the 1,000-km reach from the Ohio River to the city of Knoxville, Tennessee.

1.3.4 State Agencies

State water agencies are different in the different states, having evolved historically in response to varying needs and conditions. Several of the larger state agencies are noted here to illustrate the types of state organizations functioning across the U.S.

The California Department of Water Resources was created by the California Legislature in 1956 to plan and guide the development of the state's water resources. The agency now has a staff of 2,700 people. Responsibilities include preparing and updating the state water plan, providing technical and financial assistance to local communities, administering a dam safety program, collecting and distributing data, and operating the State Water Project—one of the largest water development and distribution systems in the nation. The State Water Resources Control Board, created by the California Legislature in 1967, administers regulatory programs to protect water quality and allocate water resources. Nine regional Water Quality Control Boards develop and implement plans for protecting water quality.

The Texas Natural Resource Conservation Commission, with 3,000 employees, is another of the larger and more comprehensive state environmental regulatory agencies in the nation. The agency administers an assortment of regulatory programs, including water quality and water rights permit programs. The Texas Water Development Board prepares and updates the state water plan, administers a variety of grant and loan programs, and maintains a centralized data bank called the Texas Natural Resources Information System. Nineteen river authorities in Texas serve as nonfederal sponsors for federal projects; construct and operate multipurpose reservoir projects, regional water supply systems, and hydropower plants; and perform other basinwide planning and management responsibilities.

Florida is divided into five regions, with a water management district responsible for multipurpose water management for each region, including planning, project development, financing, and regulatory functions. The five water management districts have a total of about 3,200 employees. The Florida Department of Environmental Protection, with 3,000 employees, exercises general supervisory authority over the water management districts, administers programs statewide to protect air and water quality, ensures proper waste management, and manages state parks and other lands.

The Illinois State Water Survey is a division of the Office of Scientific Research and Analysis of the Illinois Department of Natural Resources and is affiliated with the University of Illinois. The State Water Survey conducts investigations, performs applied research, and disseminates information regarding groundwater and surface water resources and atmospheric science.

1.3.5 Federal Agencies

Each of the federal agencies listed in Table 1.4 has multiple missions. The three agencies that construct and operate water resources development projects, along with other water management responsibilities, are the USACE, U.S. Bureau of Reclamation (USBR), and the Natural Resource Conservation Service (NRCS). The responsibilities of the U.S. Geological Survey (USGS) and the National Weather Service (NWS) deal primarily with collecting and managing data and disseminating information used throughout the water management community. The Federal Emergency Management Agency (FEMA) administers the NFIP and emergency relief programs for responding to the full spectrum of disasters, including floods, droughts, and hurricanes. The EPA and the Federal Energy Regulatory Commission (FERC) are regulatory agencies.

The USACE is the nation's oldest and largest water resources development and management agency. The Corps traces its military roots to the founding of the nation, with its water development program beginning with the construction of navigation improvements in the 1800's. The agency employs 34,600 civilian and 650 military men and women. The Corps is organized geographically into 8 divisions that are further subdivided by river basins into 40 districts in the U.S. Other offices are located throughout Europe and Asia primarily to support military functions. USACE water research and development organizations include the Engineering Research and Development Center (formerly called the Waterways Experiment Station) in Vicksburg, Mississippi; Hydrologic Engineering Center in Davis, California; and Institute for Water Resources in Alexandria, Virginia. In addition to its extensive water resources development and management activities, the USACE is responsible for designing and managing construction of military facilities for the Army and Air Force and providing similar support for other federal agencies. Water resources engineering expertise is important in accomplishing military construction as well as water management responsibilities.

The USACE civil works program includes planning, design, construction, and operation of water resources development projects, as well as other water resources planning, research, and regulatory activities. The Corps has constructed and now maintains and operates most of the inland navigation facilities in the U.S. and over 500 major multipurpose reservoir projects nationwide. The agency has constructed numerous flood control levee and channel improvement projects that have been turned over to local project sponsors for maintenance. Under authority of Section 404 of the Clean Water Act of 1977, the USACE administers a permit program regulating construction and other activities involving dredging and/or filling in rivers, streams, and wetlands. The Section 404 permit process ensures that all construction projects and other filling or dredging activities are in compliance with all federal environmental protection laws and requirements. Obtaining a Section 404 permit from the USACE is an important aspect of private and public development projects.

The USBR of the Department of the Interior was created by the Reclamation Act of 1902 to plan and implement water projects needed to support population and economic growth in the arid West. The USBR is responsible for development and management of water resources in the 17 western states for multiple purposes, including irrigation, hydroelectric power, municipal and industrial water supply, pollution abatement, propagation of fish and wildlife, recreation, erosion control, and flood control. The Bureau has constructed more than 230 dams, including some of the largest in the nation. Bureau projects supply wholesale municipal water to 31 million people and irrigation water to 140,000 farmers for irrigating 4 million hectares of farmland in the dry West that produce 60 percent of the nation's vegetables and 25 percent of its fruits and nuts. USBR's 58 hydroelectric power plants produce 40 billion kilowatts of electricity per year to serve 6 million homes.

The NRCS (formerly called the Soil Conservation Service) of the Department of Agriculture, dates back to the 1930's. The NRCS conducts national programs

Figure 1.10 The USBR operates Hoover Dam to regulate the flow of the Colorado River to supply water for Las Vegas, Los Angeles, other cities, agricultural irrigators, and environmental instream flow needs; meet flow commitments to Mexico; generate hydroelectric power; and control floods. (Photo by W. P. James, June 1998)

dealing with soil and water conservation and small watershed flood protection. The NRCS works closely with farmers, ranchers, landowners, environmental groups, and state and local organizations in carrying out its programs. The NRCS is linked with 3,000 conservation districts, essentially one in every county nationwide, organized by local citizens under state law. The NRCS has 10,800 full-time and 1,650 part-time and temporary employees. Partners include 17,000 unpaid conservation district officials, 8,000 permanent conservation district employees, and many thousands of volunteers associated with environmental conservation groups. The NRCS has constructed many thousands of relatively small flood control dams and other structures on private and public lands and provides technical water resources engineering assistance for its local collaborators.

The mission of the Water Resources Division (WRD) of the USGS, of the Department of the Interior, is to provide the hydrologic information and understanding needed to support management and use of the nation's water resources. The USGS organizational structure includes 48 district offices generally located in state capitals and four regional offices. The USGS WRD conducts three major types of interrelated activities: (1) data collection and dissemination, (2) problem-oriented water resources appraisals and interpretive studies, and (3) research. Data collection programs provide water quality and quantity data for stream flow, reservoir storage, and groundwater. Data are stored in computer-based data management systems accessible to the public, as well as printed in published reports. The water management community relies heavily on USGS water data.

Other divisions of the USGS provide mapping products and services and information regarding geological and biological resources. Topographic quadrangle maps with nationwide coverage and related mapping products developed by the USGS are used extensively by water resources engineers. Maps are in both paper and digital formats.

The NWS is responsible for the hydrologic services programs of the National Oceanic and Atmospheric Administration of the Department of Commerce. The NWS provides weather, hydrologic, and climate forecasts and warnings for the U.S., its territories, and adjacent waters and ocean areas. The NWS also collects and disseminates weather and climatic data, which is widely used throughout the water management community.

The mission of FEMA, created in 1979, is to reduce loss of life and property and protect the nation's infrastructure from all types of hazards through a comprehensive, risk-based, emergency management program of mitigation, preparedness, response, and recovery. FEMA has 2,500 employees supplemented by over 5,000 stand-by disaster reservists from other agencies. The Flood Insurance Administration of FEMA administers the flood insurance component of the previously noted NFIP, and the Mitigation Directorate of FEMA oversees the floodplain management aspect of the program.

FERC is the agency within the Department of Energy that regulates the transmission and sale of natural gas, oil, and electricity in interstate commerce and licenses and inspects private, municipal, and state hydroelectric projects.

Hydroelectric power regulation includes issuing preliminary permits, project licenses, and exemptions from licensing; ensuring dam safety; performing project compliance activities; and coordinating with other agencies.

The U.S. Fish and Wildlife Service (USFWS) is responsible for enforcing wildlife laws, protecting endangered species, and conserving and protecting wildlife and wildlife habitat on public and private lands nationwide. These responsibilities significantly affect the activities of other agencies in developing and managing water projects. The USFWS also manages a system of 520 national wildlife refuges and thousands of wetlands and other special management areas and operates 66 national fish hatcheries, 64 fishery resources offices, and 78 ecological services offices. The USFWS employs 7,500 people in facilities across the nation.

The EPA was established in 1970 to consolidate in one agency a variety of federal research, monitoring, standard-setting, and enforcement activities to ensure environmental protection. EPA's mission is to protect human health and to safeguard the natural environment (air, water, and land) on which life depends. The EPA has grown from 4,084 employees in 1970 to more than 18,000 since 1998. EPA headquarters is in Washington, D.C. Regional offices in each of 10 regions of the U.S. are responsible for execution of EPA's programs within the states in that region. The EPA Office of Research and Development operates 14 national laboratories. The EPA Office of Water, working through the EPA regional offices in collaboration with state environmental agencies, is responsible for implementing the Clean Water Act, Safe Drinking Water Act, and other federal laws pertaining to protection of water quality, public health, and environmental resources. The NPDES, safe drinking water programs, and an array of other regulatory activities are accomplished largely through issuance and enforcement of permits by state regulatory agencies that meet requirements outlined by the federal EPA.

1.3.6 International Agencies

The United Nations (UN) System, consisting of the UN itself and over 30 affiliated organizations, is central to global efforts to solve a broad diversity of problems that challenge humanity. The UN was established in 1945 with the mission of preserving peace through international cooperation and collective security. The 189 member countries account for most of the nations in the world. The United States (25%), Japan (18%), Germany (9.6%), Italy (5.4%), the United Kingdom (5.1%), and Russia (2.9%) contribute over 72 percent of the regular UN budget. About 52,100 people work in the UN System.

The International Monetary Fund, the World Bank group, and 12 other independent organizations known as *specialized agencies* are linked to the UN through cooperative agreements. These agencies are autonomous bodies created by intergovernmental agreement. In addition, a number of other UN programs, funds, and offices report to the UN General Assembly or the Economic and Social Council. These organizations have their own governing bodies, budgets, and secretariats. These agencies and programs along with the UN are collectively known as the UN

System or UN family. They provide coordinated yet diverse programs to improve the economic and social conditions of people around the world.

The UN System organizations for which water is a key concern include the World Bank, the World Health Organization, the World Meteorological Organization, the International Fund for Agricultural Development, and the UN Development Program (UNDP). For example, the World Bank provides funds and technical assistance to developing countries to reduce poverty and advance sustainable economic growth. Likewise, the UNDP supports technical support of development projects for developing nations. Many of the World Bank and UNDP projects involve multipurpose water resources development projects and/or water supply and sanitation services.

In addition to participating in the UN System, the United States also has its own programs for helping people in other countries. USAID is the principal U.S. agency responsible for extending assistance to countries recovering from disaster or trying to escape poverty. Water is a key concern. USAID has headquarters in Washington, D.C., and field offices throughout the world.

Worldwide, two or more countries share each of 261 international river basins. Water allocation, cooperation, and conflict mitigation are important aspects of water management in these shared river basins. For example, the United States and Mexico share the waters of the Rio Grande and Colorado River Basins. The International Boundary and Water Commission (IBWC) consists of a Mexican Section and a U.S. Section, with headquarters in the adjoining cities of El Paso, Texas, and Ciudad Juarez, Chihuahua.

The international boundary between the U.S. and Mexico follows the middle of the Rio Grande from its mouth on the Gulf of Mexico 2,019 km to a point just upstream of El Paso, Texas. From there, the boundary follows an alignment westward overland for 858 km to the Colorado River, follows the middle of that river for 38 km, and then extends overland for 226 km to the Pacific Ocean. By the Convention of 1889, the governments of the two countries established the International Boundary Commission to settle questions regarding the location of the boundary when the rivers changed their course. A 1944 treaty allocating the waters of the Rio Grande and Colorado River between the two nations also changed the name of the International Boundary Commission (IBC) to the IBWC. The IBWC administers the allocation of the waters of the two river basins between the two nations and operates a multipurpose reservoir system on the Rio Grande for water supply, flood control, hydroelectric power, and recreation. The IBWC also conducts planning studies and implements projects for border water supply and sanitation, salinity mitigation, local flood control, and stream bank stabilization.

1.3.7 Professional Societies and Associations

Several of the many water-related professional societies and associations are listed in Table 1.4. These organizations facilitate sharing of information, foster professional growth of its members, influence political processes, and provide leadership in continually improving the effectiveness of the professional community in serving the public. The professional societies are perhaps best known for their journals and other publications and technical conferences and meetings.

The 123,000 member American Society of Civil Engineers (ASCE) founded in 1852 is the oldest engineering society in the nation. Technical specialty area activities within ASCE, as reorganized in the late 1990's, are focused in the following institutes: Geo-Institute; Structural Engineering Institute; Architectural Engineering Institute; Construction and Materials Institute; Coasts, Oceans, Ports, and Rivers Institute; and Environmental and Water Resources Institute (EWRI). The ASCE EWRI has 20,000 members and over 100 active task committees working on manuals of practice and technical standards, as well as providing commentary on public policy issues. The EWRI holds several annual technical conferences and publishes the following journals.

- *Journal of Environmental Engineering*
- *Journal of Hydraulic Engineering*
- *Journal of Hydrologic Engineering*
- *Journal of Irrigation and Drainage Engineering*
- *Journal of Water Resources Planning and Management*

1.3.8 Universities

The important roles of universities in the water management community include educating undergraduate and graduate students, providing continuing education opportunities for practicing engineers, research, publications, and service activities. Water resources engineering is a subject or option program in undergraduate civil, environmental, and agricultural engineering curricula and a major specialty area for graduate study. Basic and applied research contributes to the knowledge base of the array of engineering and scientific fields related to water. Many universities have water resources institutes or centers to support interdisciplinary research in water resources.

1.4 COMPUTER MODELS IN WATER RESOURCES ENGINEERING

The term *model* may be defined as any simplified representation of a real-world system. The discussion here deals with representing water-related systems with mathematical formulations that are solved using a computer. Computer modeling of natural and man-made water resources systems is a central focus in water resources engineering. Most of the analysis techniques presented in this textbook are implemented, in professional practice, using computers.

TABLE 1.5 CATEGORIES OF COMPUTER SOFTWARE

General-Purpose Commercial Software	
Spreadsheet-based packages	Excel by Microsoft Corporation Lotus 1-2-3 by Lotus Development Corporation Quattro Pro by Borland International, Inc.
Mathematical modeling environments	MATLAB by The MatWorks, Inc. MathCAD by MathSoft, Inc. Mathematica by Wolfram Research, Inc.
Geographical information systems	ArcInfo, ArcView, and ArcGIS by ESRI
Computer-aided drafting and design	AutoCAD by Autodesk, Inc. MicroStation by Intergraph Corporation
High-Level Programming Languages	
Fortran	Compaq Computer Corporation, Lahey Computer Systems, Inc., and other companies
C, C++, BASIC	Microsoft Corporation and other companies
Generalized Water Resources Engineering Models	
See Table 1.6	

Different types of software tools available for building models are outlined in Table 1.5. Categories of software include (1) general-purpose commercial products, (2) conventional programming languages, and (3) generalized water resources engineering models. Several alternative types of software may be used for a particular application. Various options may be adopted for solving the examples and end-of-chapter problems in this textbook. Different tools have certain advantages and disadvantages in various situations. Choice of software is also largely dependent on personal preferences.

1.4.1 General-Purpose Commercial Software

Much of the proliferation of software on the commercial market, which is widely used in various business, scientific, and engineering fields, is also extensively used in water resources engineering. Computer programs are dynamic with new versions being released periodically. Water resources engineers continue to discover new uses for old software, as well as for new products being marketed. Several particularly useful categories of general-purpose software packages are noted in Table 1.5, along with examples of popular products in each category.

Spreadsheet programs are used routinely by millions of businesses, professional offices, universities, and homes throughout the world. Water resources engineers have recognized the usefulness of electronic spreadsheets since the early 1980's when they were first marketed. Spreadsheet programs may be conveniently applied in solving many of the problems covered in this book. They have the advantage of applying the same familiar software to many different applications. For relatively simple applications, they are used to develop complete models. Spreadsheet packages also are commonly used as preprocessors for preparing and organizing input data for complex models and as postprocessors for summarizing, plotting, and analyzing simulation results.

Water resources engineering models are based on sets of algebraic or differential equations representing governing principles, such as conservation of mass, momentum, and energy. For fairly simple applications, model development may consist of formulating the appropriate equations and then solving them using mathematical modeling environments, such as those cited in Table 1.5. These software products provide capabilities for solving algebraic and differential equations; performing differentiation and integration, matrix operations, and statistical computations; and displaying results in numbers, tables, symbols, and graphs. The packages provide built-in mathematical and statistical functions and programming capabilities.

A geographic information system (GIS) is a set of computer-based tools for storing, processing, combining, manipulating, analyzing, and displaying data that are spatially referenced to the earth. Many types of water resources engineering models are developed within GIS environments or interconnected with GIS. A complete model may be constructed with a GIS software package, but more often the GIS serves to manage voluminous spatial input and output data for other models. Water-related spatial information managed with GIS include topographic maps, watershed characteristics (land use, vegetation, and soil types), stream configurations, floodplain delineations, geologic formations, water distribution and other utility system layouts and components, demographic data, precipitation and stream gage information, and other climatic information.

Computer-aided drafting and design (CADD) software is used for a variety of graphics applications in various fields, including water resources engineering. CADD programs provide drawing and specialized graphics capabilities used for mapping and general technical illustration purposes, as well as traditional drafting and design functions. Because CADD programs store and display spatial data, they often serve as components of GISs. An example of a CADD application is to draw complex water distribution system pipe networks.

1.4.2 Programming Languages

Many computer language translation software packages are marketed for developing application programs. Although water resources engineering models have been written in a variety of high-level languages, Fortran has dominated. Fortran (*For*mula *Tran*slator) is the oldest high-level programming language, dating back to the 1950's, and continues to be improved with new versions. It is widely used in engineering and science. The C and object-oriented C++ programming languages are also popular and provide excellent graphics capabilities, as well as optimizing computational efficiency. BASIC (*B*eginner's *A*ll-Purpose *S*ymbolic *I*nstruction *C*ode) is an example of other programming languages used in water resources engineering. Different languages are often used in combination. Visual BASIC or C may be used to develop graphical user interfaces for models with computational routines written in Fortran. Water resources engineers develop models in various versions of these languages using programming environment software products sold by a variety of companies, including those cited in Table 1.5.

Computer programs can be written fairly easily for relatively simple models. However for complex models, formulating computational algorithms, devising data management schemes, writing and debugging code, and testing new programs are very time consuming. Developing new computer programs, written in Fortran, C, C++, and/or other languages, provides flexibility in situations where other simpler options are not readily available. However, as discussed next, generalized programs have been developed for a broad spectrum of water resources engineering applications. These generalized models are extensively applied, greatly expanding the modeling and analysis capabilities of the professional water resources engineering community.

1.4.3 Generalized Water Resources Engineering Models

Most of the modeling and analysis methods covered in this textbook are incorporated in the several generalized water resources engineering models listed in Table 1.6. Sections of the book that discuss these and other similar models are cited in Table 1.6. These models are examples representative of many other similar models, also reflecting the methods presented in this book, available from the organizations listed in Table 1.7 (Wurbs, 1995, 1998). These widely applied generalized models continue to be expanded and updated, with new versions periodically being released.

TABLE 1.6 GENERALIZED WATER RESOURCES ENGINEERING MODELS

Generalized model	Model developer	Textbook section
KYPIPE Pipe Network Analysis	University of Kentucky	4.7
EPANET Pipe Network Water Quality Analysis	Environmental Protection Agency	4.7
HEC-RAS River Analysis System	Hydrologic Engineering Center	5.10
FLDWAV Flood Wave Model	National Weather Service	6.6
HEC-FFA Flood Frequency Analysis	Hydrologic Engineering Center	7.7.1
HEC-HMS Hydrologic Modeling System	Hydrologic Engineering Center	8.9.1
SWMM Stormwater Management Model	Environmental Protection Agency	8.9.2
SWAT Soil and Water Assessment Tool	Agricultural Research Service	8.9.3
BASINS Better Assessment Science Integrating Point and Nonpoint Sources	Environmental Protection Agency	8.9
MODFLOW Modular Groundwater Flow Model	U.S. Geological Survey	9.7
HEC-FDA Flood Damage Analysis	Hydrologic Engineering Center	11.3.5
HEC-RESSIM Reservoir System Simulation	Hydrologic Engineering Center	12.8.2
HEC-5 Flood Control and Conservation Systems	Hydrologic Engineering Center	12.8.2
WRAP Water Rights Analysis Package	Texas A&M University	12.8.2
MODSIM River Basin Network Flow Model	Colorado State University	12.8.2
RiverWare Reservoir and River Operations	USBR, TVA, CADSWES	12.8.2

TABLE 1.7 MODEL DEVELOPMENT AND DISTRIBUTION ORGANIZATIONS

Hydrologic Engineering Center U.S. Army Corps of Engineers Davis, California 95616 http://www.hec.usace.army.mil/	Engineering Research and Development Center U.S. Army Corps of Engineers Vicksburg, Mississippi 39180-6199 http://www.erdc.army.mil/
Water Resources Division U.S. Geological Survey Reston, Virginia 22092 http://www.usgs.gov/	Office of Hydrology National Weather Service, NOAA Silver Spring, Maryland 20910 http://www.nws.noaa.gov/
Center for Exposure Assessment Modeling National Exposure Research Laboratory U.S. Environmental Protection Agency Athens, Georgia 30613-0801 http://www.epa.gov/ceampubl/ceamhome.htm	Center for Subsurface Modeling Support Kerr Environmental Research Laboratory U.S. Environmental Protection Agency Ada, Oklahoma 74820 http://www.epa.gov/ada/csmos.html
Hydrologic Modeling Inventory U.S. Bureau of Reclamation Denver, Colorado 80225-0007 http://www.usbr.gov/hmi	National Technical Information Service Federal Computer Products Center Springfield, Virginia 22161 http://www.ntis.gov/fcpc/opd.htm
International Ground Water Modeling Center Colorado School of Mines Golden, Colorado 80401-1887 http://www.mines.edu/igwmc/	Hydrology and Water Resources Programme World Meteorological Organization Geneva 2, Switzerland http://www.wmo.ch/web/homs/hwrphome.html
Civil Engineering Software Center University of Kentucky Lexington, Kentucky 40506-0281 http://www.kypipe.com/	Haestad Methods 37 Brookside Road Waterbury, Connecticut 06708 http://www.haestad.com/

Generalized means the computer model is designed for application to a range of problems dealing with systems of various configurations and locations, rather than being developed to address a particular problem at a specific site. In applying the software package, the model user develops input for the system of concern. For example, the Hydrologic Modeling System (HEC-HMS) and River Analysis System (HEC-RAS) developed by the Hydrologic Engineering Center (HEC) may be applied to essentially any watershed and river system. HEC-HMS and HEC-RAS and their predecessors, called HEC-1 and HEC-2, have been applied by cities, agencies, and consulting firms to delineate floodplains in most of the 20,000 cities participating in the NFIP, as well as in many other types of applications throughout the U.S. and abroad. The EPA's Stormwater Management Model has been applied in investigating both water quality and quantity aspects of stormwater management in numerous cities. KYPIPE and EPANET have been applied to model the hydraulics of pipe networks in numerous cities. Likewise, the widely used MODFLOW is generalized for application to essentially any groundwater aquifer.

The widely used software packages developed and distributed by the federal agencies listed in Table 1.7 are in the public domain. These generalized models required large amounts of public funds to develop, but are available to all interested users, either free-of-charge or with only nominal handling charges, and may be freely copied. Instructions for obtaining numerous software packages, documentation, and supporting information are available at the web sites listed in Table 1.7. Many of the models may be directly downloaded through the internet. Proprietary commercially marketed pipe network analysis models developed without federal support are a notable example of generalized models that must be purchased.

Computer models play important roles in all aspects of water resources engineering. A tremendous amount of work has been accomplished during the past several decades in developing powerful generalized software packages. With advances in computer technology since the 1980's, everyone working in the water resources engineering field has convenient access to desktop computers, providing all the hardware capabilities needed to execute a mighty arsenal of available models. The principles and methods presented throughout this textbook are embedded within the computer models. Application of the generalized models requires a thorough knowledge of the hydraulics, hydrology, and systems analysis concepts and techniques on which the models are built.

A word of caution is warranted regarding the use of generalized models. The danger always exists of providing the novice a weapon with which to shoot himself/herself in the foot. Easy access to computer software does not diminish the necessity for high levels of technical knowledge and expertise. The user of off-the-shelf software must still have a thorough understanding of the computations performed by the model, and the capabilities and limitations of the model in representing real-world processes. Models must be carefully and meticulously applied with professional judgment and good common sense. Although the effectiveness and efficiency of engineering work can be greatly enhanced by exploiting the capabilities provided by readily available software, modeling still requires significant time and effort, as well as high levels of technical expertise.

Regardless of the sophistication of a generalized computer program, the quality of the modeling results for a particular application can be no better than the input data for that application. Parameter values are required for the governing equations representing the processes being modeled. A variety of other input data are also required. Calibration and verification are crucial to modeling studies.

Generalized models are used to manage data, perform computations to simulate real-world processes, and display results for various analysis applications. Design processes are typically based on iterative executions of simulation models. Computer models also play an important role in documenting and transferring knowledge. Much like textbooks and other reference books and manuals, generalized modeling packages serve to organize, record, and pass on state-of-the-art knowledge.

1.4.4 Modeling Systems

A water resources engineering modeling application often involves several software packages used in combination. For example, a river basin management application might involve a modeling system that includes a watershed runoff model used to develop runoff hydrographs and pollutant loadings for input to a reservoir/river system operation model and/or water quality model which, in turn, determines discharges and contaminant concentrations at pertinent locations. The example modeling system could also include a hydraulics model to compute flow depths and velocities. A GIS, spreadsheet program, and other data management programs are included in the modeling system to develop and manage voluminous input data; perform various statistical and graphical analyses of simulation output; and display and communicate modeling results.

The software incorporated into modeling systems can be categorized as follows:

- models that simulate real-world systems and subsystems
- user interfaces
- preprocessor programs for acquiring, preparing, checking, manipulating, managing, and analyzing model input data
- postprocessor programs for managing, analyzing, interpreting, summarizing, displaying, and communicating modeling results.

The concept of decision support systems is popular in the water management community, as well as in business, engineering, and other professional fields in general. A decision support system is a user-oriented, computer-based system that supports decision-makers in addressing unstructured problems. The general concept emphasizes:

- solving unstructured problems that require combining the judgment of human decision-makers with quantitative information
- capabilities to answer *"what if"* questions quickly and conveniently by making multiple runs of one or more models
- use of enhanced user–machine interfaces and graphical displays
- efficient management of large quantities of spatial, time series, and other data.

Decision support systems include a collection of software packages and hardware. For example, decision support systems are used for real-time flood control operations of reservoir systems. Making release decisions during a flood event is a highly unstructured problem because reservoir operations are highly dependent on operator judgment, as well as prespecified operating rules and current and forecasted stream flow, reservoir storage level, and other available data. The decision support system includes data management software; watershed runoff, stream hydraulics, and reservoir/river system operation models; a computer platform with various peripheral hardware devices; and an automated real-time hydrologic data collection system.

TABLE 1.8 HYDRAULIC MODELING PACKAGE

Program	Description	Examples
	Chapter 4: Hydraulics of Pipelines and Pipe Networks	
SIMEQ	solution of linear equations	4.13
NETWORK	linear method pipe network analysis	4.15–4.18
NETOPSIM	pipe network operation simulation	4.19(a)
NETOPCON	concentration of contaminant in water distribution system	4.19(b)
PNETUNS1	initial conditions for unsteady flow	4.22–4.23
PNETUNS2	unsteady flow in a pipe network using method of characteristics	4.22–4.23
HYRAM	simulation of hydraulic ram	4.24–4.25
	Chapter 5: Open Channel Hydraulics	
YN	normal depth in a trapezoidal channel	5.8
YC	critical depth in a trapezoidal channel	5.9
STDSTEP	standard step method water surface profile	5.18, 5.21
OPNCHNET	open channel networks	5.22
MOC	method of characteristics, unsteady flow, trapezoidal channel	5.28
MACK	MacCormack scheme, unsteady flow, trapezoidal channel	5.28
MACKS	MacCormack scheme, unsteady flow, channels in series	5.29
	Chapter 6: Flood Routing	
RESROUTE	hydrologic reservoir routing	6.1, 6.2
KINRT	kinematic routing of overland flow	6.3
HYRT	hydraulic routing through a trapezoidal channel	6.4
HYRT2	hydraulic routing through trapezoidal channels in series	6.4
HYRT3	hydraulic routing through natural channels in series	6.4
HYRTIMP	hydraulic routing with implicit Preissman scheme	6.5
HYRTI	routing with implicit scheme and a pointer matrix	6.5
HYRTEDB	explicit hydraulic routing of dam break flood wave in a natural channel	6.6
HYRTEDBS	HYRTEDB with submerged weir	6.6
HYRTIDB	implicit hydraulic routing of a dam break flood wave in a natural channel	6.6
HYRTIDBS	HYRTIDB with submerged weir	6.6
HYRTWD1	program one of two for watershed routing	6.7
HYRTWD2	hydraulic routing in a branching channel watershed system	6.7
	Chapter 9: Groundwater Engineering	
WELLU	table of values for the well function *u*	9.10
CJAQTEST	Cooper–Jacob method of pump test analysis	9.11
SEEPAG2D	two-dimensional steady seepage analysis	9.15
THOM1DU	one-dimensional unsteady unconfined aquifer with Thomas Algorithm	9.16
ADI2DC	two-dimensional unsteady confined with ADI method	9.17
ADI2DU	two-dimensional unsteady unconfined with ADI method	9.17
GS2DU	two-dimensional unsteady unconfined with Gauss–Seidel method	9.17
PLUME2D	two-dimensional unsteady relative concentration from a continuous point source	9.18
	Chapter 10: Urban Stormwater Management	
DETENT	detention basin analysis with the unit hydrograph methodology	10.10
WATERSHD	regional detention basin analyses with the unit hydrograph methodology	10.11
HYWEIR	hydraulic routing in a trapezoidal channel with a side channel weir	10.12
CHANNEL	data for open channel design curves	10.18

Water resources engineering models are often components of decision support systems. The models are even more often applied in other planning, design, and resource management situations that do not exhibit all the characteristics attributed above to decision support systems. The relationships between analysis strategies, design and other decision-making processes, and modeling systems vary, depending on the particular water resources engineering application.

1.4.5 Hydraulic Modeling Package

The *Hydraulic Modeling Package (HMP)* is a set of Fortran programs listed in Table 1.8, developed in conjunction with Chapters 4–6, 9, and 10 of this textbook. The programs implement the computational methods outlined in these chapters and are introduced through application in the example problems listed in Table 1.8. Source code, executable programs, and example data sets are available from the Prentice Hall companion web site for this book found at http://www.prenhall.com.

These Fortran programs can be extremely useful in mastering the techniques presented in the book, particularly those methods involving numerical solutions of hydraulic equations. These relatively small programs focus on specific computational methods. Students/engineers can work directly with the code to develop a step-by-step understanding of the algorithms. You can experiment with solution schemes by changing and expanding the programs. The programs are provided primarily as a supplemental option for use in learning computational hydraulics. However, they also provide a starting point for developing your own models for research studies or practical applications.

1.5 UNITS OF MEASURE

The metric system of units is the standard adopted by most countries. The metric system has evolved from its roots in France in the 1790's to its current version known as SI units (Système Internationale d'Unites). The United States is the only major country in the world that still uses the English (U.S. customary or British Gravitational, BG) system. Although the English system of units began in England in the 1200's, England now uses the metric system. The U.S. Congress passed the Metric Conversion Act in 1975, which called for a voluntary change to the metric system. This Act was amended in 1988, making the metric system the preferred system of weights and measures for U.S. trade and commerce and requiring federal agencies to switch to the metric system for business-related activities. The federal water agencies are shifting toward the metric system. However, the English system still dominates throughout the U.S. water management community. Literature and data compiled over many decades, as well as current practice, are based on English units. Water resources engineers working in the U.S. must be familiar with both systems.

TABLE 1.9 METRIC AND ENGLISH UNITS

Dimension	SI units	English (BG) units	Conversion
time (t)	second (s)	second (s)	
length (L)	meter (m)	inch (in), foot (ft), mile (mi)	1 ft = 0.3048 m
mass (M)	kilogram (kg)	slug = lb · s^2/ft	1 slug = 14.59 kg
force (F)	Newton (N) = kg · m/s^2	pound (lb), ton	1 lb = 4.448 N
temperature (T)	Celsius (°C)	Fahrenheit (°F)	°F = 1.8°C + 32

TABLE 1.10 METRIC PREFACES

Prefix	Symbol	Multiplier	Definition
giga	G	1,000,000,000	one billion
mega	M	1,000,000	one million
kilo	k	1,000	one thousand
hecto	h	100	one hundred
deka	da	10	ten
deci	d	0.1	one-tenth
centi	c	0.01	one-hundredth
milli	m	0.001	one-thousandth

This book emphasizes metric units, but also presents English units. Commonly used metric and English units for basic dimensions are noted in Table 1.9. Prefaces showing multiples of a base metric unit are listed in Table 1.10. Dimensions, units, and conversion factors are discussed in pertinent chapters, including Section 2.6, Section 3.1, and other sections, as particular quantities are introduced.

REFERENCES

CHATURVEDI, M. C., "Water for Food and Rural Development: Developing Countries," *Water International, Journal of the International Water Resources Association,* Vol. 25, No. 1, 2000.

GLEICK, P. H., *The World's Water 2000–2001, The Biennial Report on Freshwater Resources,* Island Press, Washington, D.C., 2000.

NEW, L., and G. FIPPS, *Center Pivot Irrigation*, Bulletin 6096, Texas Agricultural Extension Service, Texas A&M University System, College Station, TX, 2001.

WURBS, R. A., *Water Management Models: A Guide to Software,* Prentice Hall, Upper Saddle River, NJ, 1995.

WURBS, R. A., "Dissemination of Generalized Water Resources Models in the United States," *Water International, Journal of the International Water Resources Association,* Vol. 23, No. 3, 1998.

2

Hydrology

Hydrology is the science that treats the waters of the earth, their occurrence, circulation, and distribution, their chemical and physical properties, and their reaction with their environment, including their relation to living things. Hydrology is concerned with water on, under, and above the land surface; ocean waters are the domain of oceanography and the marine sciences. Scientific and engineering hydrology cover a broad spectrum of interdisciplinary subjects that may be approached from various perspectives, including those of the meteorologist, geologist, chemist, biologist, geographer, agricultural scientist or engineer, environmental engineer, and civil engineer, as well as hydrologist. Chapter 2 is an introductory overview of hydrology. Later chapters cover water resources engineering applications of particular aspects of hydrology.

This chapter covers a broad spectrum of subjects briefly that are explored in depth by many other publications. Hydrology textbooks covering hydrology in general with a hydrologic engineering perspective include Linsley, Kohler, and Paulhus (1982), Chow, Maidment, and Mays (1988), Ponce (1989), Bras (1990), Bedient and Huber (1992), Singh (1992), Ward and Elliot (1995), Viessman and Lewis (1996), Wanielista, Kersten, and Eaglin (1997), McCuen (1998), and Dingman (2002). Handbooks providing broad treatments of hydrology include Maidment (1993), Mays (1996), the American Society of Civil Engineers (1996), and Herschy and Fairbridge (1998). Numerous other publications focus on specific aspects of the subjects introduced in this chapter.

2.1 WATER

Water is a marvelous substance. It is the most common substance on earth. Water covers more than 70 percent of the earth's surface. The human body is two-thirds water. A tomato is about 95 percent water. All forms of life need water to survive. Water is the only substance on earth that is naturally present in all three forms of liquid, solid (ice), and gas (water vapor). The same amount of water is present on earth today as when the dinosaurs inhabited the planet millions of years ago. Water is reused over and over again. Every glass of water you drink consists of billions of H_2O molecules that have been used countless times before.

The water molecule, H_2O in chemistry notation, is composed of two hydrogen atoms sharing electrons with an oxygen atom. The asymmetry of the distribution of electrons causes a positive charge on one side of the molecule and a negative charge on the other side. This is called a polar covalent bond. The polarity results in an electrostatic attraction between molecules. Molecules of a substance are always moving, with their speed being related to temperature. Temperature is a key factor governing whether water is a liquid, solid (ice), or gas (water vapor). In the solid state, ice molecules bond together in hexagonal crystals. Individual molecules are vibrating, but not fast enough to break out of the cluster. In the liquid state, increases in temperature cause the molecules to move too fast to be locked into crystals, but still slowly enough to be attached to each other. In the vapor state, the temperature has been increased such that faster moving molecules are not attached. The phase changes for water are evaporation (liquid to vapor), condensation (vapor to liquid), freezing (liquid to solid), melting (solid to liquid), and sublimation (vapor to solid or solid to vapor).

The molecular structure of H_2O results in unique properties. Temperature changes occur very slowly in water. The weather is affected by water bodies warming and cooling slower than land masses as the seasons change. Unlike many other substances, water freezes to form a less dense solid. Lakes freeze from the top down, which has significant implications for aquatic ecosystems. Compared with most common liquids, water has high surface tension. This produces capillary rise of water in soils and causes rain to form into spherical droplets. Water acts as a solvent into which many other substances dissolve. It has been called the (almost) universal solvent. Dissolved water quality constituents greatly affect the use of water.

Water resources have shaped history and will be a major determinant of mankind's future. Since ancient times, civilizations have risen and prospered where water supplies were plentiful and have fallen when these supplies failed in quantity and/or quality. People have killed each other in fights between neighbors and wars between nations over access to water. Floods and droughts have devastated human populations throughout history. Dramatic population growth during the 20th century has made effective water management even more crucial for human survival and prosperity and environmental vitality in the 21st century.

2.1.1 The Hydrosphere

The term *hydrosphere* refers to all of the waters of the earth. Shiklomanov (1993, 2000) and Gleick (1993, 2000a) have compiled data from various sources regarding the distribution of water in the hydrosphere, including the data summarized in Tables 2.1 and 2.2. These types of data are necessarily approximate. Planet Earth has an estimated 1.4 billion cubic kilometers of water. About 97 percent of the water is in the oceans. Over half of the remaining 3 percent is contained in glaciers and permanent ice and snow cover. Freshwater in streams, lakes, and aquifers account for less than 1 percent of the global water, and even much of this water is not accessible for human use.

Seawater in the oceans and inland seas is too salty for most uses. Although rivers, lakes, and aquifers are mostly freshwater, they have salinity levels ranging from fresh to greater than seawater, due primarily to dissolution of salts and minerals from the earth. The total dissolved solids (TDS) concentration of seawater averages about 35,000 milligrams/liter (mg/l), of which about 30,000 mg/l is sodium chloride. The limit between saline and freshwater is typically defined to be in the range of 500–1,000 mg/l. The term *brackish* is often applied to saline water with TDS concentrations falling between that of freshwater and seawater. Generally accepted quality standards for drinking water include maximum TDS limits ranging from 500 to 1,000 mg/l. Tolerable TDS concentrations for irrigation vary greatly, depending on the crop and mix of precipitation and supplemental irrigation, but are generally much lower than 10,000 mg/l.

Water balances for each of the six continents are presented in Table 2.2. The last column is a water balance for the combined total land area of all six continents. An

TABLE 2.1 THE EARTH'S WATER

	Volume 10^3 km^3	Percentage of total	Percentage of freshwater
Salt water			
Oceans	1,338,000	96.54	—
Saline groundwater	12,870	0.93	—
Saline lakes	85	0.006	—
Freshwater			
Antarctic permanent ice	21,600	1.56	61.66
Other permanent ice	2,764	0.20	7.89
Groundwater	10,530	0.76	30.06
Lakes	91	0.007	0.26
Soil moisture	17	0.001	0.05
Atmospheric water	13	0.001	0.04
Wetlands	11	0.0008	0.03
River flows	2	0.0002	0.006
Biota	1	0.0001	0.003
Subtotal–freshwater	35,029	2.53	100.0
Total—salt and freshwater	1,385,984	100.0	—

Source: Shiklomanov (1993) and Gleick (2000a).

TABLE 2.2 MEAN ANNUAL WATER BALANCE OF THE CONTINENTS

	South America	Australia	Europe	Asia	Africa	North America	Total land
Land area (10^6 km^2)	17.8	8.7	9.8	45.0	30.3	20.7	132.3
Precipitation (km^3/yr)	29,355	6,405	7,165	32,690	20,780	13,910	110,305
Precipitation (mm/yr)	1,648	736	734	726	686	670	834
Precipitation (in./yr)	65	29	29	29	27	26	33
Water balance as a percentage of precipitation							
Evapotranspiration	64.6%	69.3%	56.6%	59.7%	79.7%	57.2%	64.8%
Surface-water flow	22.6%	23.4%	28.5%	29.9%	13.3%	30.3%	24.4%
Groundwater flow	12.8%	7.3%	14.9%	10.4%	7.0%	12.5%	10.8%

Source: Gleick (1993, 2000a).

Notes: Data for Australia includes New Zealand, New Guinea, and Tasmania.
Data for Europe includes Iceland.
North America includes Central America and excludes the Canadian archipelago.
Total land excludes Antarctica, Greenland, and the Canadian archipelago.

estimated average annual precipitation of 110,000 km^3/yr falling on 132 million km^2 of land may be expressed as a mean equivalent depth of 833 mm or 33 inches. The long-term mean annual precipitation falling on the land is accounted for ultimately as evapotranspiration and flow to the oceans as surface runoff/streamflow and subsurface flow. Evapotranspiration, surface-water flow, and groundwater flow account for 65 percent, 24 percent, and 11 percent, respectively, of the precipitation falling on the land. The total of water supply withdrawals worldwide for human use is less than 5 percent of the total amount of precipitation falling on the continents. Much of the water withdrawn by people from streams and aquifers is returned to streams after use and, after further cycles of reuse, eventually reaches the oceans. Most withdrawals for irrigation and other consumptive use are accounted for as evapotranspiration.

In terms of total quantity, enough precipitation falls on land surfaces to meet all human needs for water. However, crucial water management problems are driven by drastic geographic and temporal variations in precipitation. Contaminants in surface and groundwater resources are also a major constraint to human use. The global perspective is extremely important, but most water resources engineering problems are solved locally. Suitable quality water must be reliably supplied for use in individual homes, farms, and businesses. Flooding and other problems of excessive water must also be addressed at scales covering the full spectrum from local to international. Water resources engineering deals with the superimposing of human activity on the natural hydrologic cycle.

2.1.2 The Hydrologic Cycle

The hydrologic cycle is the central concept of the science of hydrology, the study of the Earth's water. Solomon, King of Israel, in writing Ecclesiastes 3,000 years ago, provides a concise description of the hydrologic cycle: *"All rivers flow into the*

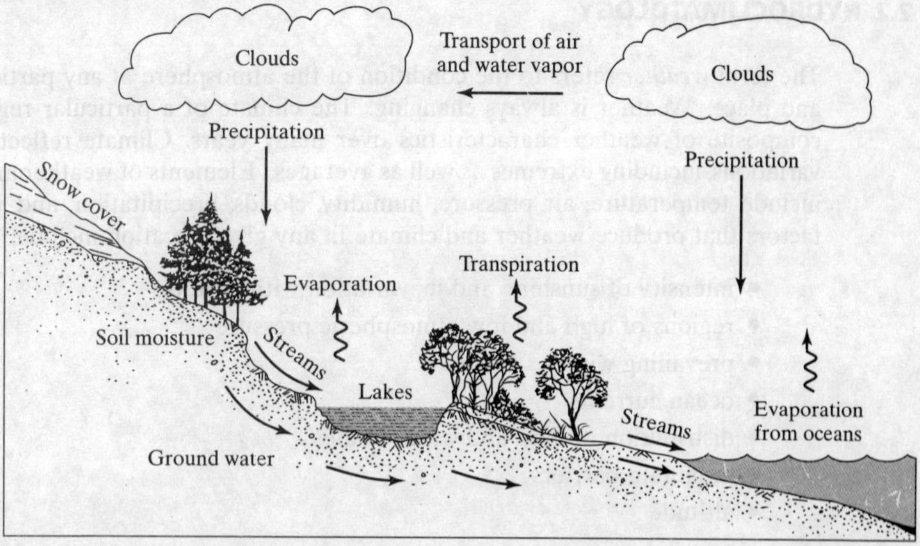

Figure 2.1 The hydrologic cycle.

sea, yet the sea is never full. To the place the streams come from, there they return again."

The hydrologic cycle is illustrated by Fig. 2.1 and provides an organizational framework for this chapter. Energy from the sun results in evaporation of water from ocean and land surfaces and also causes differential heating and resultant movement of air masses. Water vapor is transported with the air masses and under the right conditions becomes precipitation. Evaporation from the oceans is the primary source of atmospheric vapor for precipitation, but evaporation from soil, streams, and lakes and transpiration from vegetation also contribute. Precipitation runoff from the land becomes streamflow. Soil moisture replenishment, groundwater storage, and subsurface flow occurs as a result of water infiltrating into the ground. Stream and groundwater flow convey water back to the oceans.

Although the complete hydrologic cycle is global in nature, many subcycles exist. For example, water vapor resulting from evapotranspiration from land surfaces may be transformed back to precipitation again before returning to the sea. Water quality also changes within the cycle. For example, seawater is converted to freshwater through evaporation. Salts are dissolved from the land by surface and subsurface water. A myriad of other dissolved and suspended constituents also enter and leave the water during various phases of the cycle.

The locations at which water may be stored include the oceans and seas, glaciers and other ice, the atmosphere, lakes and reservoirs, streams and rivers, wetlands, soils, geologic formations, plants and animals, and man-made structures. Residence times may range from seconds to thousands of years. Hydrologic processes by which water moves through the hydrologic cycle includes atmospheric movement of air masses, precipitation, evaporation, transpiration, infiltration, percolation, groundwater flow, surface runoff, and streamflow.

2.2 HYDROCLIMATOLOGY

The term *weather* refers to the condition of the atmosphere at any particular time and place. Weather is always changing. The climate of a particular region is the composite of weather characteristics over many years. Climate reflects weather variations, including extremes as well as averages. Elements of weather and climate include temperature, air pressure, humidity, clouds, precipitation, and wind. The factors that produce weather and climate in any given location include:

- intensity of sunshine and its variation with latitude
- regions of high and low atmospheric pressure
- prevailing winds
- ocean currents
- distribution of land and water bodies
- mountain barriers
- altitude

2.2.1 Types of Climate

No two places experience exactly the same climate. However, climates may be generally categorized as outlined below (Hidore and Oliver, 1993; Ahrens, 2000):

- Tropical Moist Climates (no real winters)
  tropical wet (rain forest)
  tropical monsoon
  tropical wet and dry (savanna)
- Dry Climates
  arid (desert)
  semi-arid (steppe)
- Moist Subtropical Mid-Latitude Climates (moist with mild winters)
  humid subtropical
  marine
  dry-summer subtropical or Mediterranean
- Moist Continental Climates (moist with cold winters)
  humid continental
  subpolar
- Polar
  polar tundra
  polar ice caps
- Highland Climates

Mean annual precipitation, a key parameter describing climate, is shown in Fig. 2.2 for the world's land areas.

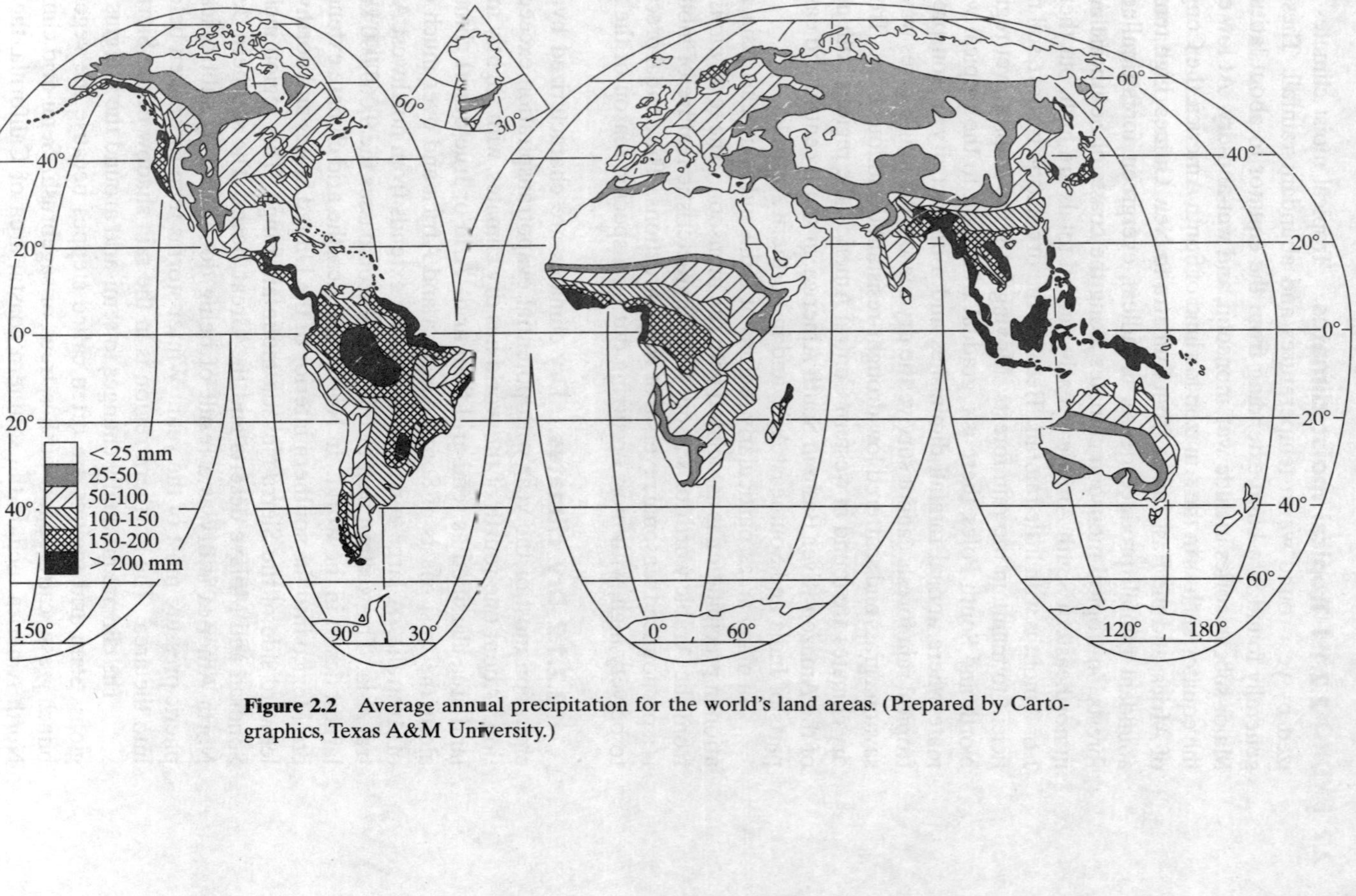

Figure 2.2 Average annual precipitation for the world's land areas. (Prepared by Cartographics, Texas A&M University.)

2.2.1.1 Tropical moist climates. Tropical moist climates are characterized by year-round warm temperatures and abundant rainfall. These climates are generally found in a band extending from the equator to about latitude 15° to 25°. Major subcategories include wet, monsoon, and wet-and-dry. At low elevations near the equator, such as in the Amazon lowland of South America, the Congo River Basin of Africa, and the East Indies from Sumatra to New Guinea, high temperatures and abundant rainfall produce dense, broadleaf, evergreen forests called tropical rain forests. In tropical monsoon climates along the coasts of Southeast Asia, India, and in northeastern South America, yearly rainfall totals are high, but there are perhaps 1 or 2 months with little rainfall. Because the dry season is short, soil moisture is sufficient to maintain the rain forests. Moving from the tropical wet region toward the North and South Poles, there is a gradual transition to the tropical wet-and-dry climate where annual rainfall diminishes and a distinct dry season prevails. Because tropical rain forests cannot survive the dry season, the jungle gives way to tall, coarse savanna grass and scattered low drought-resistant deciduous trees. Tropical wet-and-dry climates are found in western Central America, the regions both north and south of the Amazon River Basin in South America, in south-central and eastern Africa, in parts of India and Southeast Asia, and in northern Australia.

In many areas, particularly regions of India and Southeast Asia, a marked variation in precipitation is associated with the monsoon. The term *monsoon* is derived from the Arabic word for season. The characteristic feature of monsoon climates is a pronounced seasonal reversal of wind directions. The annual precipitation tends to be very high, with most occurring during a specific season of the year.

2.2.1.2 Dry climates. Dry climates are characterized by deficient precipitation most of the year with potential evapotranspiration exceeding precipitation. About one-fourth of the world has dry climates, with slightly more of the dry land area classified as semi-arid than arid. Arid or true desert climates are found along the west coasts of South America and Africa and over much of the interior of Australia. A large zone of arid climate extends from northwest Africa into central Asia. The western part of this zone encompasses the 9,065,000 km^2 Sahara, the largest desert in the world. In North America, the arid climate extends from northern Mexico into the southern interior of the United States and northward along the leeward side of the Sierra Nevada mountain range. This region includes both the Sonoran and Mojave deserts and the Great Basin. The southern desert region of North America is dry as a result of being dominated by subtropical high atmospheric pressure most of the year. Winter storms tend to weaken before they move into the area. The northern region is in the rain shadow of the Sierra Nevada.

The climate gradually changes to semi-arid around the margins of the arid regions. Semi-arid regions are often called steppes because the vegetation is short bunch grass, scattered low bushes, trees, or sagebrush. Semi-arid climate regions in North America include the southern coastal area of California, the northern valleys of the Great Basin, and most of the Great Plains. As average rainfall increases, the climate gradually transitions to humid climate regions.

2.2.1.3 Moist subtropical and continental climates. Moist subtropical climates occur in regions of most continents, from about 25° to 40° latitude, including the southeastern United States, eastern China, southern Japan, southeastern South America, and along the southeastern coasts of Africa and Australia. Characteristics of humid subtropical climates include distinct summer and winter seasons, hot muggy summers and mild winters, and fairly abundant precipitation distributed throughout the year.

Marine climates are strongly influenced by prevailing winds from the ocean that moderate the climate, keeping winters milder. Marine climates exhibit low clouds, fog, and drizzle. Adequate precipitation occurs in all months, with much of it falling as light-to-moderate rain associated with maritime polar air masses. Where mountains parallel the coastline, such as along the west coasts of North and South America, the marine influence is restricted to narrow belts. Western Europe is unobstructed by high mountains, and prevailing westerly winds carrying ocean air provide this region with a marine climate.

Moving toward the equator away from marine climates, the climate transitions to dry-summer subtropical as the influence of the subtropical high atmospheric pressure becomes greater. This climate is also called Mediterranean, because it is found in the coastal areas around the Mediterranean Sea. Along the West Coast of the United States, the transition between marine climate to the north and dry-summer subtropical to the south occurs in Oregon. Extremely dry summers are caused by the sinking air of the subtropical high pressure zones. The dry season in California may last for 5 months. Summers are not as dry along the Mediterranean Sea as in California. During the winter, when the subtropical highs move toward the equator, mid-latitude storms from the ocean bring rainfall. Thus, Mediterranean climates are marked by mild, wet winters and mild-to-hot, dry summers.

Moist continental climates are found only in the Northern Hemisphere, extending across North America, Europe, and Asia between latitude 40°N to 70°N. This includes the north-central and northeast U.S., Alaska, and much of Canada. The climate is controlled by large continental land masses. Characteristics include warm-to-cool summers and cold winters with severe snowstorms. The moist continental climate classification is subdivided into humid continental and subpolar. Humid continental climates are characterized by annual precipitation in the range of 50–100 cm (20–40 in.), well distributed throughout the year. Native vegetation typically includes forests of spruce, fir, pine, and oak. A subpolar climate is characterized by severe winters and short, cool summers. Mean annual precipitation is typically less than 50 cm (20 in.). Subpolar climates occur in a broad zone across Alaska and Canada and in Eurasia extending from Norway over much of Siberia.

2.2.1.4 Polar and highland climates. Polar climates are found in northern areas of North America and Eurasia and in Greenland and Antarctica. In the polar tundra, the ground is permanently frozen to great depths, but the temperature rises above freezing a short time each year allowing the surface to briefly thaw. Tundra vegetation consists of mosses, lichens, and dwarf trees only a few centimeters tall when fully grown. The polar ice cap climate has temperatures essentially always below freezing, perpetual snow and ice cover, and no vegetation.

Polar climates are also experienced at extremely high altitudes on mountains. Because temperature decreases with altitude, climatic changes observed when climbing 1 km in elevation are about equivalent in high latitudes to horizontal changes experienced when traveling 1,000 km northward. Thus, a variety of climates may be found at different heights on the same high mountain.

2.2.2 Climate Variability and Change

Water resources engineers routinely deal with daily, seasonal, and year-to-year fluctuations in precipitation and other aspects of weather and the resulting impacts on water management. Hydrologic extremes of floods and droughts, discussed in Sections 2.2.3 and 2.2.4, are particularly important in water resources engineering. River control structures, water supply systems, stormwater management facilities, and various other water management systems must function for the full spectrum of hydrologic conditions, including periods of relatively normal precipitation and streamflow, as well as extremes. However, extreme flood and drought conditions set the standards to which hydrosystems must be designed and for which the water management community must be prepared. Major floods and droughts in the past have greatly influenced political processes and the evolution of water management policies and practices.

Water resources management may also be impacted by both multiple-year cycles of fluctuating weather patterns and gradual changes in climate over spans of multiple decades. The El Nino–Southern Oscillation (ENSO) phenomenon is an example of fluctuations in global weather patterns that occur at somewhat cyclic multiple-year intervals. Human-induced impacts on global warming associated with greenhouse gases is a long-term climate change issue receiving much attention in the scientific research community (Lettenmaier, McCabe, and Stakhiv, 1996; Gleick, 2000b; National Assessment Synthesis Team, 2000).

2.2.2.1 El Nino–Southern Oscillation (ENSO). Atmospheric and oceanographic scientists study cyclic changes and irregularities in global climate. ENSO is perhaps the best known anomaly resulting from connections between oceanic and weather systems. ENSO is a quasicyclic phenomenon that has occurred every 3 to 7 years during at least the last 450 years (Dingman, 2002). El Nino refers to abnormally high sea surface temperatures off the coast of Peru. The term *El Nino,* referring to the Christ child, was adopted by Peruvian fishermen because the sea warming usually becomes pronounced around Christmas. The land near the coast of Peru is a desert, and the nearby ocean is an extremely productive fishery. However, El Nino years are marked by a disruption in the usually abundant supply of fish, and flooding rains may fall on the desert.

The southern oscillation refers to the accompanying low atmospheric pressure over the eastern Pacific and the high atmospheric pressure in the western Pacific. ENSO is the product of large-scale, long-period waves in the surface of the tropical Pacific Ocean. Resulting major temporary changes in winds and atmospheric circulation cause unusual weather patterns in low- and mid-latitude regions around the world. Particular regions may experience unusually warm or cold winters. Droughts may occur in normally productive agricultural areas. Torrential rains may fall in normally arid regions.

2.2.2.2 Greenhouse gases and global warming. Global climate has undergone slow but continuous change throughout earth's history. Warming and cooling cycles have normally spanned thousands of years. Scientists have long studied the climate changes that have occurred over geologic time. One of the most hotly debated scientific topics in recent years has been the effect of human activity on climate change. One key issue is related to potential warming due to humans causing increased concentrations of carbon dioxide and other trace gases in the atmosphere, known colloquially as the *greenhouse effect* (Mimikou, 1995).

The natural greenhouse effect is a major determinant of climate and crucial to existence of life on earth. As the sun's energy reaches the earth, some of it is absorbed, and some is reflected back. Clouds, water vapor, carbon dioxide, and other gases absorb some of the outgoing infrared energy, and then radiate the energy in all directions including back to earth. This natural greenhouse effect is important in maintaining the temperature of the earth, which would otherwise be unbearably cold. The problem is that humans are changing the amount of greenhouse gases contained in the atmosphere. We are increasing the concentration of carbon dioxide (CO_2) by burning carbon-based fuels and by deforesting large areas. Nitrous oxide (N_2O) also comes from burning fossil fuels. Methane (CH_4) is added to the atmosphere by agriculture and municipal landfills. Chlorofluorocarbons are released through various means, including air conditioner coolants, plastic foams, and aerosol propellants. These gases may all contribute to the greenhouse effect. Scientists are investigating their potential impacts on global warming and climate. Hydrologists and water resources engineers are concerned regarding the extent to which the characteristics of precipitation, water availability, droughts, and floods may change during the 21st century.

2.2.3 Floods

All streams and rivers naturally overflow their channels periodically. Floodplains are a natural component of a stream system. However, problems result because people tend to live and work in floodplains. The rich alluvial soils deposited by centuries of periodic flooding result in prime agricultural lands being located in floodplains. Cities are developed near rivers to facilitate water supply, wastewater disposal, electric energy production, and transportation. People like to build their homes adjacent to the scenic beauty of tree-lined streams. Human activities imposed on natural hydrologic processes keep water resources engineers busy.

In any given year, flood damages are incurred at locations throughout the nation and world. In any local region, flooding may range from the inconvenience of streets and yards being inundated fairly frequently to rare extreme flood events with recurrence intervals of many years that cause loss of life and devastating damage. Although extreme floods do not occur often at any one location, with countless streams throughout the United States, multiple extreme floods occur at different locations nationwide every year.

Several major historical floods in the U.S. are described to illustrate the broad range of characteristics reflected in different types of floods. The 1927 and 1993 floods in the Mississippi River Basin are examples of flooding conditions over a large geographical area occurring over several weeks. The 1976 flood on the Big Thompson River in Colorado is illustrative of devastating flash floods of limited spatial extent that occur quickly without warning throughout the U.S. and the world. The 1900 Galveston Island disaster demonstrates the dangers of hurricane flooding in coastal areas.

2.2.3.1 Large-scale, slow-rising floods. The history of numerous floods along the Mississippi River and its tributaries is illustrative of large river basin floods characterized by slowly rising and falling streamflows that inundate large floodplain areas. The Mississippi River Basin is shown in Fig. 2.3 and described later in Section 2.7.1. A hydrograph provided later as Fig. 2.14 illustrates periodic flooding on the river. Communities in this basin, like many other regions throughout the nation and world, have experienced many major floods in the past and will continue to deal with floods in the future. Flooding during April 2001 is shown in Figs. 1.5 and 1.6.

In addition to being an extreme hydrologic event, the Great Mississippi River Flood of 1927 is historically significant because of its influence on flood control policies and practices. The 1927 flood was the most severe of a series of floods that motivated the U.S. Congress to enact the Flood Control Acts of 1928 and 1936 establishing U.S. Army Corps of Engineers flood control programs. Precipitation for the entire basin was above normal from the fall 1926 throughout the winter and following spring. By late March 1927, the Mississippi and tributaries—including the Ohio, Tennessee, and Cumberland Rivers—were in flood stage. Melting snow was feeding the rivers from northern watersheds. Additional heavy rains fell during the first 3 weeks of April in southern Missouri and most of Arkansas. More than 67,000 km^2 (26,000 mi^2) of land flooded in seven states, forcing 600,000 people from their homes. The flood killed 246 people. At its peak, the Mississippi River was 130 km (80 mi) wide in Louisiana and Mississippi.

The Midwest Flood of 1993 lasted 3 months and resulted in 500 counties in nine states being declared major disaster areas (Changnon, 1996; Perry and Combs, 1998). Major floods occurred simultaneously on the upper Mississippi and Missouri Rivers and their tributaries. The 1993 flood was historic in its geographical extent

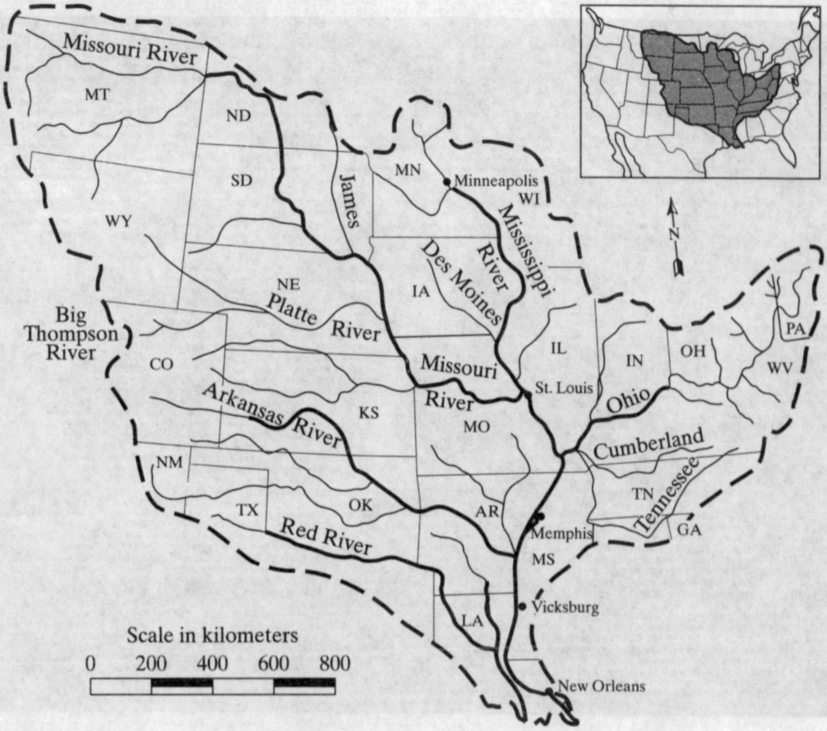

Figure 2.3 Mississippi River Basin.

and duration. It was also unusual because it occurred in summer rather than spring like most floods in the Midwest. However, although the worst flooding occurred from June into August 1993, flooding conditions actually began in the fall 1992 with heavier rains than normal in the north-central U.S. that saturated the ground prior to the summer rains that caused the rivers to overflow.

During the summer 1993, flood waters covered 69,000 km^2 (27,000 mi^2) in the Missouri and upper Mississippi River valleys. More than 22,000 homes were damaged, and 85,000 residents were forced to evacuate. Flood waters completely covered more than 75 towns and partially inundated many other communities. The death toll of 48 people is small compared with the floods discussed next because flood waters rose slowly enough for effective warning and evacuation. The damage has been estimated to be about \$20 million. The Corps of Engineers estimates indicate that the damages would have been twice as much without flood control reservoirs, levees, and other structures. Construction of extensive flood control improvements had been motivated particularly by the extensive flooding in 1927, as well as by other past floods. For example, an 18-km-long, 16-m high flood wall protects the city of St. Louis, Missouri. Many agricultural levees overtopped or otherwise failed, but most levees held.

Figure 2.4 These bridges crossing the Mississippi River at Quincy, Illinois, were among the many inundated during the 1993 flood. (Courtesy of Federal Emergency Management Agency.)

The Mississippi River below St. Louis, including flows from the Missouri River, peaked on August 1, 1993 at 30,600 m^3/s (1,080,000 ft^3/s). Historic streamflow records show that this discharge is the highest at St. Louis since 1861. The highest flow on record is an estimated 37,000 m^3/s in 1844. A 1903 flood peaked at 28,800 m^3/s at St. Louis. However, many flood control reservoirs have been constructed since these earlier floods. In addition to the Midwest Flood of 1993, several other regions of the U.S. suffered notable flood damages during 1992–1993, ranging from river flooding in several different states from California to the southeast to the effects of Hurricane Andrew in Florida.

2.2.3.2 Flash floods. Floods like the 1976 Big Thompson Flood described next are called flash floods because streams rise quickly, peaking in a matter of hours after the rain begins. Conditions associated with the Big Thompson Flood are typical of flash floods throughout the western states. Floods may occur on sunny days after long dry periods, with intense rainfall running off steep, rocky watersheds. In the eastern U.S., where interception and infiltration are typically higher, flash floods are more likely after long wet periods have saturated the soil. Hurricanes and tropical storms bringing large amounts of precipitation inland from the ocean may also cause flash floods. Floods kill on average over 100 people each year in the U.S. (Fig. 1.7). Over half of the fatalities are associated with vehicles being washed away by rapidly rising flood waters.

The Big Thompson River rises in the Colorado Rocky Mountains on the western edge of the Mississippi River Basin, shown in Fig. 2.3, and flows 125 km (78 mi) to the South Platte River. The mountain stream descends over 1,500 m in its first 34 km. Below the scenic tourist town of Estes Park, the stream enters the 40-km- long Big Thompson Canyon where it drops another 760 m. The city of Loveland, 80 km northwest of Denver, is located just downstream of the mouth of the canyon, where the mountains transition to flat plains. Big Thompson Canyon contains restaurants, motels, campgrounds, and vacation cabins, but few permanent residents. Two-lane U.S. Highway 34 runs along the scenic river. The canyon mouth is called the Narrows because the canyon narrows to a width accommodating only the highway and river, with high, vertical walls on either side.

Saturday, July 31, 1976 was a sunny day, perfect for recreation in the mountains. By late afternoon an estimated 3,000 people were in the canyon. Motels and campgrounds were filled to capacity. In late afternoon, a thunderstorm developed with little prior warning. Most thunderstorms that form over the Colorado Rockies drift eastward over the plains. This storm stalled above the top of Big Thompson Canyon pouring out a total of about 28 cm (11 in) of rain, with 20 cm falling in less than 2 hours. At Drake halfway down the canyon, the flow of the river increased from about 4 m^3/s at 6:00 p.m. just before the rain began to 880 m^3/s at 9 p.m. Runoff from the steep rocky slopes of the canyon was rapid. A 5.8-m high wall of water rushed down the canyon, destroying everything in its path and killing 139 people. Much of Highway 34 and many of the buildings in the canyon were washed away. Victims had little warning. Many of the 139 deaths were the result of people staying in their cars. Warning signs in Big Thompson and other Colorado canyons now read: *Climb to safety in case of flash flood!*

2.2.3.3 Hurricane floods. Hurricanes are common along the coasts of the U.S., as well as in many other regions of the world. The longest period on record without a hurricane hitting the U.S. was the 3 year period between Hurricane Allen hitting Brownsville, Texas, in August 1980 and Hurricane Alicia hitting Galveston in August 1983. The 10 deadliest hurricanes during the 20th century in the U.S. were: Galveston, 1900, 7,200 deaths; South Florida, 1928, 1,836 deaths; Florida Keys and Corpus Christi, 1919, 850 deaths; New England, 1938, 600 deaths; Florida Keys, 1935, 408 deaths; Louisiana and Texas, 1957, 390 deaths; Virginia to Massachusetts, 1944, 390 deaths; Louisiana, 1909, 350 deaths; Galveston, 1915, 275 deaths; and New Orleans, 1915, 275 deaths (Williams, 1997).

In addition to high winds and tide surges, hurricanes bring large amounts of rainfall that cause flooding significant distances inland. Rainstorms associated with hurricanes are often extremely intense. For example, the record greatest 24-hour precipitation in the U.S. is 109 cm (43 in.) measured at Alvin, Texas, during Hurricane Claudette in July 1979.

In terms of loss of life, the greatest natural disaster in U.S. history was the hurricane that killed over 7,000 people in Galveston, Texas, on September 8, 1900 (Williams, 1997; Bixel and Turner, 2000). The city of Galveston is on Galveston Island, a low barrier island along the Texas coast between Galveston Bay and the Gulf of Mexico. In 1900, the population of the wealthy port city was 38,000. On September 6, 1900, the U.S. Weather Bureau warned its Galveston office about a tropical storm disturbance moving northward over Cuba. This was not of major concern to the citizens of Galveston because they were accustomed to weathering storms. Winds and tides became severe by the morning of September 8. About 4 p.m. that afternoon, tides surging from both the ocean and bay met, flooding part of the island and cutting off bridges to the mainland. At about 7 p.m., a wave of more than 1 m surged through the city, followed by a 6-m wave. During the remainder of that evening, storm waves on top of the rising tide, along with winds estimated to exceed 190 km/hr, destroyed the city. More than 3,600 houses were ruined. Although the exact number will never be known, the death toll is estimated at 7,200. The storm dropped 25 cm of rain. The tide reached a record 4.6 m above normal.

In 1900, the highest point on Galveston Island was only 2.7 m above sea level. A 5-km-long, 5-m-high seawall along the Gulf of Mexico has since been constructed. Fill material was used to raise the city by as much as 5 m in some places. Galveston is now much better prepared to deal with hurricanes. However, decades of overpumping of groundwater has caused land subsidence in the Houston–Galveston area. Highways used to evacuate Galveston Island during hurricanes are now as much as 2-m lower and thus even more susceptible to inundation than earlier in the century.

2.2.4 Droughts

A drought is an extended period of time during which water availability is significantly below normal. The beginning and end of a drought and its geographic coverage are much more difficult to delineate than for a flood. Droughts may last from several months to several years. A drought may be limited to a local region or be national or international in extent. Drought is defined from various perspectives. Meteorological drought refers to a period of below normal precipitation and above normal temperature. Hydrologic drought refers to below normal streamflow and reservoir storage. Agricultural drought refers to conditions affecting agriculture, such as soil moisture and crop yield. A drought may not affect all aspects of the hydrologic cycle to the same extent or at the same time. For example, soil moisture may be abnormally low, while streamflow and reservoir storage are near normal.

Drought is a normal, recurring feature of climate, occurring in essentially all types of climate. Drought is a temporary aberration, in contrast to aridity, which is a permanent characteristic of climate in regions of low precipitation. Unlike floods and other natural disasters, the beginning and end of droughts are not necessarily obvious. A few sunny days with no rain are pleasant. However, too long with below normal precipitation may eventually evolve into drought conditions. The severity of a drought depends on the demands of human activities and vegetation on the water available in the region, as well as precipitation, temperature, and other weather variables.

Major droughts can have severe impacts. For example, 40 million people reportedly suffered due to droughts of the early to mid-1980s in sub-Sahara, Africa. The 1991–1992 drought in southern Africa affected 20 million people. In the U.S., the 1988 drought resulted in estimated costs of $40 billion, primarily agricultural losses, making this single-year drought the most economically costly disaster in American history (Wilhite, 2000).

Droughts have played a major role in American history. The Homestead Act of 1862, providing settlers free title to homesteads farmed for 5 years, set the stage for the socially disruptive drought in the Great Plains during 1887–1896. Over a million settlers moved west to Kansas, Nebraska, and the surrounding region in the 1860's–1870's, not realizing that they were experiencing an unusually wet period for the region soon to be followed by drought. More than half of the settlers lost their farms. The effects of the next extreme drought occurring in 1934–1941 were felt to various extents by over half of the nation, but the most severe conditions were in the *"Dust Bowl"* of Nebraska, Kansas, Oklahoma, and Texas. Large areas of land on the plains dried out, turned to dust, and at times blew all the way to the East Coast. Thousands of farmers were displaced. The most hydrologically severe drought on record for much of Texas and Oklahoma occurred during 1950–1957, ending in April 1957 with one of the largest floods on record in parts of the region. The series of major droughts in various regions of the country during 1986–1992, 1994, 1996, and 1998–2000 continued to demonstrate the nation's vulnerability to drought (Wilhite, 2000).

2.3 ATMOSPHERIC PROCESSES

The sun provides the energy that drives the weather. The earth revolves around the sun in an elliptical path in slightly longer than 365 days, causing annual seasons. The earth also spins on its own axis, completing one spin in 24 hours, causing daily cycles of sunlight. The sun's electromagnetic energy, including light and ultraviolet and infrared energy, flows into the atmosphere to warm the air, oceans, and land. Portions of the solar radiation reaching the earth's atmosphere are reflected back to space, absorbed by the atmosphere, and absorbed by the earth. Seasons are regulated by the amount of solar energy received at the earth's surface. This amount is determined primarily by the angle at which the sunlight strikes the surface and by the length of time the sun shines each day on any latitude. The equator receives the most energy. The North and South Poles receive the least energy. Land masses heat and cool much faster than water bodies.

2.3.1 Atmospheric Circulation

Differential heating results in pressure differences that move air masses and drive the weather. Water vapor moves with air masses. Winds are the movement of air masses. Molecules of air, like any gas, are in constant motion, zipping around at incredible speeds and bouncing off each other and anything else in their path. These impacts of billions of molecules cause pressure. The speed at which air molecules move and resulting pressure increase as the air is warmed and decrease as it cools.

Pressure gradients work in combination with friction at the earth's surface and the Coriolis effect to generate the winds and govern the global circulation patterns of air masses. The Coriolis effect refers to the curving motion of the winds caused by the rotation of the earth. The equator receives more solar radiation than the higher latitudes. Equatorial air tends to rise due to being warmer and thus less dense. As the air rises, it is replaced by cooler air from higher latitudes. Likewise, the air from the higher latitudes is replaced from above by the poleward flow of air rising from the equator. The thermal-induced pressure gradient driven circulation is modified by the earth's rotation under the air masses and also by the effects of land and sea distribution and land forms.

2.3.2 Atmospheric Moisture

Dry air with little water vapor is comprised of about 78 percent nitrogen (N_2) and 21 percent oxygen (O_2) by volume with small amounts of other gases. The atmosphere becomes less dense with altitude, eventually merging with empty space without a definite upper boundary. Almost 99 percent of the mass of the atmosphere lies within 30 km of the earth's surface. More than half of the water vapor is within 2 km of the earth's surface.

Water vapor (H_2O) is an invisible gas that makes up from a fraction of a percent to up to 4 percent of the atmosphere at various places and times. Water vapor is an extremely important atmospheric gas in determining weather. It forms into both liquid and frozen cloud particles that grow in size and fall to earth as precipitation. Large amounts of latent heat are released in the transformation between phases that are an important source of energy for storms, such as thunderstorms and hurricanes.

Moisture is always present in the atmosphere, even on cloudless days. The quantity of atmospheric water vapor varies with location and time as a function of temperature and source of supply considerations. The greatest moisture concentrations are found near the ocean surface in the tropics. Concentrations generally decrease with latitude, altitude, and distance inland from the sea.

2.3.2.1 Humidity. Humidity refers to any one of a number of ways of specifying the amount of water vapor in the air. For purposes of visualization, consider a small parcel of air contained in an imaginary thin elastic container.

Absolute humidity is defined as

$$\text{absolute humidity} = \frac{\text{mass of water vapor}}{\text{volume of air}} \tag{2.1}$$

Absolute humidity represents the water vapor density and is commonly expressed in units of grams of water vapor in a cubic meter of air. However, a rising or descending parcel of air will experience a change in its volume because of changes in surrounding air pressure. Consequently, when the volume of air fluctuates, the absolute humidity changes even though the amount of water vapor in the parcel has not changed. Thus, the following measures of humidity are normally adopted because they are not influenced by changes in air volume.

Specific humidity is defined as

$$\text{specific humidity} = \frac{\text{mass of water vapor}}{\text{total mass of air}} \tag{2.2}$$

Specific humidity compares the mass of the water vapor with the total mass of air in the parcel, including the water vapor. Alternatively, the mixing ratio compares the mass of water vapor in the parcel to the mass of the remaining dry air. Thus, the *mixing ratio* is defined as

$$\text{mixing ratio} = \frac{\text{mass of water vapor}}{\text{mass of dry air}} \tag{2.3}$$

Specific humidity and mixing ratio are both expressed as grams of water vapor per kilogram of air.

The moisture content of the air may also be described in terms of the pressure exerted by the water vapor in the air. Pressure is caused by the collisions of the myriad of moving molecules. The total pressure inside the parcel of air is equal to the sum of the pressures of the individual gases. The individual gases exert pressures in proportion to their volume. For example, if water vapor accounts for 2 percent of the volume of the air, the actual vapor pressure or partial pressure of the water vapor is also 2 percent of the air pressure.

Saturation vapor pressure describes the amount of water vapor required to saturate the air at a given temperature. It is the pressure that water vapor molecules would exert if the air was saturated with water vapor at a given temperature. Water vapor is supplied to the air by evaporation from water surfaces, such as oceans, lakes, rivers, soil water, plants, and rain drops (or by sublimation from ice). However, molecules of water vapor also return back to the liquid form. This is called condensation. Saturation represents a condition of equilibrium between evaporation and condensation. If additional water vapor were added, it would condense back to liquid water. The number of water vapor molecules required to reach saturation increases with air temperature. Saturation vapor pressure depends primarily on temperature.

Relative humidity is the most commonly used expression of the moisture content of the air. It is defined as

$$\text{relative humidity} = \left(\frac{\text{actual vapor pressure}}{\text{saturation vapor pressure}}\right) 100 \text{ percent} \tag{2.4}$$

and may also be expressed as

$$\text{relative humidity} = \left(\frac{\text{actual mixing ratio}}{\text{saturation mixing ratio}}\right) 100 \text{ percent} \qquad (2.5)$$

The amount of water vapor actually in the air is expressed as a percentage of the maximum amount of water vapor required for saturation at that particular temperature and air pressure. Air with a 60 percent relative humidity contains 60 percent of the number of H_2O molecules required for saturation. Air with a 100 percent relative humidity is filled to capacity with water vapor. A change in relative humidity may result from either a change in water vapor content or a change in temperature.

The dew point provides another way to express the concept that the saturation vapor pressure is a function of temperature. The *dew-point temperature* is defined as the temperature to which air would have to be cooled, with no change in air pressure or moisture content, for saturation to occur. The dew-point temperature is a measurement used to predict the formation of dew or fog.

2.4 PRECIPITATION

Precipitation includes all liquid and frozen water that falls from the atmosphere to the earth's surface. Rain consists of liquid water drops mostly larger than about 0.5 mm in diameter. Drizzle or mist consists of smaller liquid water droplets. Frozen precipitation includes snow, hail, sleet, and freezing rain. Fog and dew are normally not considered precipitation. Fog consists of water droplets so small that their fall velocities are negligible. Dew is the condensation of water vapor at the ground surface due to cooling of the ground at night.

2.4.1 Precipitation Processes

Atmospheric water vapor is the source of precipitation. An ample amount of water vapor is a necessary but not sufficient condition for precipitation to occur. The amount of precipitation in a region is not necessarily proportional to the amount of water vapor in the overlying air. For example, the amount of water vapor over dry areas of the southwestern U.S. at times exceeds that over considerably wetter northeastern areas of the country. For precipitation to occur, some mechanism is required to cool the air sufficiently to bring it to near saturation. The large-scale cooling needed for significant precipitation is caused by lifting the air. Assuming the air is close to saturation, condensation or freezing nuclei are also required on which liquid droplets or ice crystals form. These nuclei are very small particles of various substances that might include salt from the oceans, oxides of nitrogen, clay minerals, dust, or products of combustion.

A liquid droplet or ice crystal grows to visible size through diffusion of water vapor to a particle. However, this initial diffusion leads only to cloud or fog elements that are much too small to fall. Only a slight upward motion of the air is required to support clouds. Further growth of drops large enough to fall under the force of gravity as rain or snow is required. Some growth of cloud elements through diffusion occurs due to differences in vapor pressure associated with differences in size and temperature. However, collision and coalescence of cloud and precipitation elements are important processes leading to significant precipitation. Collisions occur primarily as a result of differences in falling velocities. For precipitation to occur, the drops must increase in size until their fall speed exceeds the rising rate of the air. Rain drops must also be large enough to descend through the unsaturated air below the cloud base without completely evaporating before reaching the ground.

Cooling of the air by lifting is the primary cause of condensation. Air cools as it rises. Precipitation may be classified as convective, orographic, and cyclonic based on the mechanisms that generate vertical air motion. These different lifting mechanisms are interrelated and may work together to cause a particular precipitation event.

Convective lifting results from solar heating of the air near the ground. Warmer and thus less dense air rises in colder, denser surroundings. As the heated air expands, increasing amounts of water vapor are taken up, the warm moist air becomes unstable, and pronounced vertical currents are developed. Dynamic cooling occurs, causing condensation and precipitation. Convective precipitation may range from light showers to storms of extremely high intensity, including thunderstorms. Many areas of the U.S. experience severe convective storms, which are often called thunderstorms because they are accompanied by lightning. Convective precipitation is typical of the tropics.

Orographic precipitation results from the mechanical lifting of moist air currents over mountain barriers. This type of precipitation is common on the west coast of the U.S. where moisture-laden winds from the Pacific Ocean are intercepted by coastal mountain ranges.

Cyclonic precipitation is associated with the movement of air masses from high-pressure regions to low-pressure regions created by unequal heating of the earth's surface. The term *cyclone* refers to a large-scale atmospheric pressure and wind system characterized by low pressure at its center and by circular wind motion, counterclockwise in the Northern Hemisphere and clockwise in the Southern Hemisphere. Cyclonic precipitation may be categorized as either frontal or nonfrontal. Frontal precipitation results from the lifting of a warmer air mass over the frontal surface of a colder, denser air mass. A low-pressure region can also result in nonfrontal precipitation as air is lifted as a result of horizontal convergence of the inflow into the low-pressure area.

2.4.2 Variations in Precipitation

Precipitation varies greatly both geographically and temporally. Temporal variations in precipitation include variations in rain intensity during a storm, seasonal patterns, random variations within the year and from year to year, and the extremes of floods and droughts. Variations in precipitation are a governing concern in water resources engineering. For example, construction and operation of the massive Central Valley and State Water Projects in California (Section 12.7.8), consisting of extensive systems of large dams and conveyance facilities, are governed by the following characteristics of precipitation. Most of California's precipitation occurs in the northern third of the state, but most of the people live in the southern and central regions. The Central Valley of California, among the most productive agricultural regions of the world, receives essentially no precipitation during the April through August growing season. The state is also subject to extended periods of below-normal annual precipitation and associated drought conditions. Conversely, extreme storm events periodically cause massive flood damages.

Globally, as shown in Fig. 2.2, average annual precipitation is generally highest in the tropical regions on either side of the equator and decreases toward the North and South Poles, indicative of the diminishing capacity of air to hold moisture with decreasing temperature. However, there are significant deviations from this general trend. Latitudes near 30° have relatively little precipitation because of the climatic propensity for air to rise near the equator and then descend at these latitudes. This sinking of air is counterproductive to formation of precipitation. Some of the world's great deserts are in this zone, including the Sahara of Africa, southwestern Africa, the Middle East, the Australian Interior, and the American Southwest.

In addition to the cellular structure of poleward moving air, other dominant forces shaping regional precipitation are the general circulation of both the oceans and the atmosphere, and their relationship to shape and position of the continents. In general, flow in the major oceans is cyclonic, which means tropical water and atmospheric moisture are transported poleward along the east sides of continents. Therefore, the eastern portions of North America, Eurasia, and Australia have more precipitation than their western coasts, especially during summers. Exceptions include the abundant precipitation on the Pacific coast of the northern United States and Canada resulting from a persistently strong jet stream and semi-permanent low-pressure area located in the vicinity of the Aleutian Islands, enhanced by orographic lifting by the coastal mountain ranges. Polar regions have very little precipitation because of sinking air and little atmospheric capacity for water vapor.

Geographical variations in precipitation are also influenced by distance from the sea and orographic effects. Continental interiors typically have less precipitation because of less water vapor. Also, ocean-derived salt particles are better nucleating materials than dust and other materials from land. The orographic effects of mountains depend on several factors, including wind direction relative to topography, atmospheric moisture, elevation rise, and slope. Mountains can cause dramatic differences in precipitation patterns over short distances.

The geographical variation in mean annual precipitation in the U.S. is shown in Fig. 2.5. The variation in annual precipitation nationwide in the United States

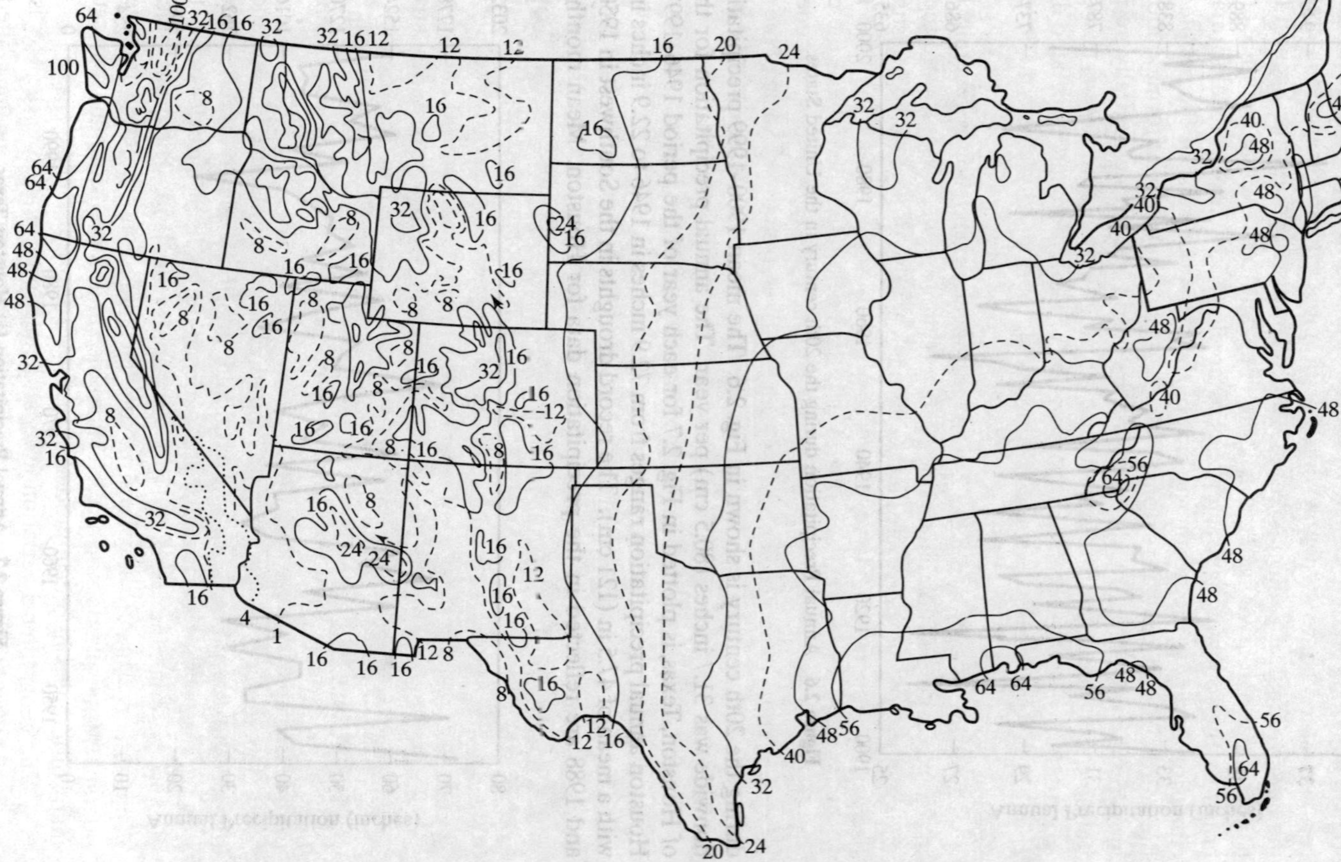

Figure 2.5 Mean Annual Precipitation in the United States in inches.

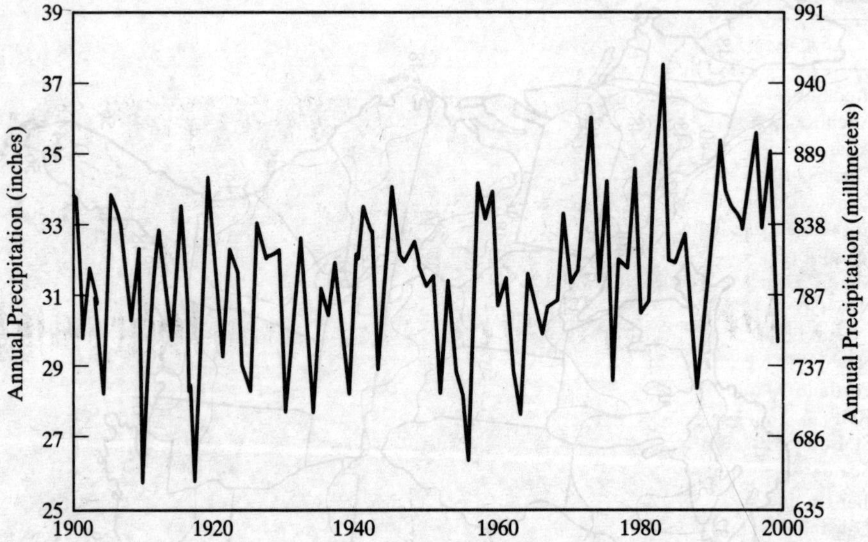

Figure 2.6 Annual Precipitation during the 20th century in the United States.

during the 20th century is shown in Fig. 2.6. The mean 1900–1999 precipitation nationwide was 31.7 inches (80.5 cm) per year. The annual precipitation for the City of Houston, Texas, is plotted in Fig. 2.7 for each year of the period 1946–1997. The Houston annual precipitation ranges from 72.9 inches in 1946 to 22.9 inches in 1988, with a mean of 47.5 in (121 cm). The record droughts in the Southwest in 1950–1956 and 1988 are reflected in the precipitation data for Houston. Mean monthly and

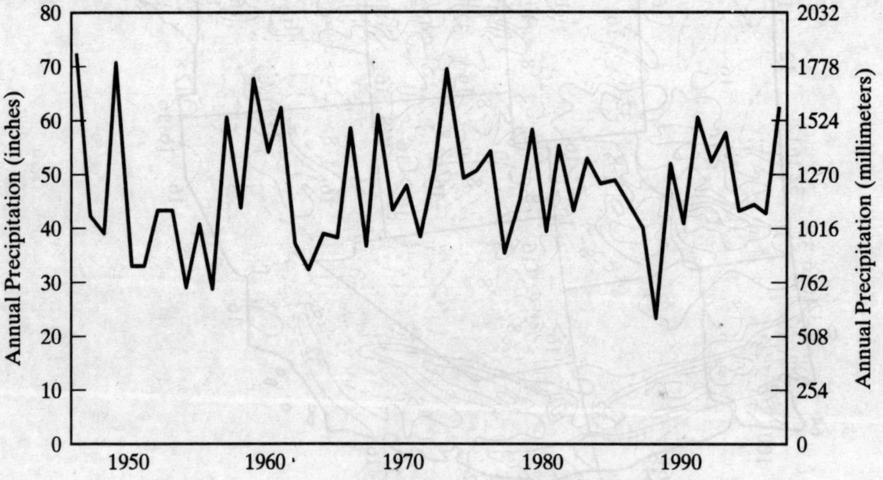

Figure 2.7 Annual Precipitation for Houston, Texas.

TABLE 2.3 MEAN MONTHLY AND ANNUAL PRECIPITATION FOR SELECTED CITIES

City	Jan	Feb	Mar	Apr	May	Jun	Jul	Aug	Sep	Oct	Nov	Dec	Annual	Annual
United States						(in.)							(in.)	(cm)
Anchorage, Alaska	0.8	0.7	0.5	0.4	0.5	1.0	1.9	2.6	2.5	1.9	1.0	0.9	14.7	37
Atlanta, Georgia	4.5	4.5	5.4	4.5	3.2	3.8	4.7	3.6	3.3	2.4	3.0	4.4	47.1	120
Chicago, Illinois	1.9	1.6	2.7	3.0	3.7	4.1	3.4	3.2	2.7	2.8	2.2	1.9	33.2	84
Denver, Colorado	0.6	0.7	1.2	2.1	2.7	1.4	1.5	1.3	1.1	1.0	0.7	0.5	14.8	38
Los Angeles, California	2.7	2.9	1.8	1.0	0.1	0.1	0.0	0.0	0.2	0.4	1.1	2.4	12.6	32
Miami, Florida	2.0	1.9	2.3	3.9	6.4	7.4	6.8	7.0	9.5	8.2	2.8	1.7	59.8	152
New Orleans, Louisiana	4.8	4.2	6.6	5.4	5.4	5.6	7.1	6.4	5.8	3.7	4.0	4.6	63.5	161
New York, New York	3.2	2.9	4.2	3.5	3.7	3.4	4.0	5.0	4.2	3.2	3.5	3.2	43.9	112
Portland, Maine	4.4	3.8	4.3	3.7	3.4	3.2	2.9	2.4	3.5	3.2	4.2	3.8	42.8	109
Reno, Nevada	1.0	1.0	0.7	0.5	0.5	0.04	0.2	0.2	0.2	0.6	0.6	0.9	7.0	18
St. Louis, Missouri	2.0	2.0	3.1	3.7	3.7	4.3	3.3	3.0	2.8	2.9	2.6	2.0	35.3	90
Seattle, Washington	5.7	4.2	3.8	2.4	1.7	1.6	0.8	1.0	2.0	4.0	5.4	6.3	38.9	99
Other Countries														
Calgary, Canada	0.5	0.5	0.8	1.0	2.3	3.1	2.5	2.3	1.5	0.7	0.7	0.6	16.7	42
Montreal, Canada	3.8	3.0	3.5	2.6	3.1	3.4	3.7	3.5	3.7	3.1	3.5	3.6	40.8	104
Mexico City, Mexico	0.2	0.3	0.5	0.7	1.9	4.1	4.5	4.3	4.1	1.6	0.5	0.3	23.0	58
Lima, Peru	0.1	0.0	0.0	0.0	0.2	0.2	0.3	0.3	0.3	0.1	0.1	0.0	1.6	4.1
Cherrapunji, India	0.7	2.1	7.3	26.3	50.4	106.1	96.3	70.1	43.3	19.4	2.7	0.5	425.0	1,080
Moscow, Russia	1.5	1.4	1.1	1.9	2.2	2.9	3.0	2.9	1.9	2.7	1.7	1.6	24.8	63
Paris, France	2.2	1.8	1.4	1.7	2.2	2.1	2.3	2.5	2.2	2.0	2.0	2.0	24.4	62
Tokyo, Japan	1.9	2.9	4.2	5.3	5.8	6.5	5.6	6.0	9.2	8.2	3.8	2.2	61.6	156
Cape Town, South Africa	0.6	0.3	0.7	1.9	3.1	3.3	3.5	2.6	1.7	1.2	0.7	0.4	20.0	51

annual precipitation for several cities in the U.S. and abroad are tabulated in Table 2.3. The table illustrates seasonal variations in precipitation for different locations.

World-record high precipitation observations for various durations of time are listed in Table 2.4. Many of the places that experience extreme precipitation amounts are located on the windward side of mountains. For example, Mount Waialeale on the island of Kauai, Hawaii, holds the world record of 1,168 cm of mean

TABLE 2.4 WORLD-RECORD EXTREME HIGH PRECIPITATION MEASUREMENTS

Duration	Depth (cm)	Depth (in.)	Location and date
Greatest mean annual	1,168	460	Mount Waialeale, Hawaii
Greatest 1 year	2,647	1,042	Cherrapunji, India; Aug 1860–July 1861
Greatest 1 month	930	366	Cherrapunji, India; July 1861
Greatest 15 days	480	189	Cherrapunji, India; June–July 1931
Greatest 24 hours	187	74	Reunion Island, Indian Ocean; March 1952
Greatest 12 hours	135	53	Reunion Island, Indian Ocean; February 1964
Greatest 42 minutes	30	12	Holt, Missouri; June 1947

annual precipitation. Cherrapunji on the southern slopes of the Khasi Hills in northeastern India has an average annual rainfall of 1,080 cm, which is a little less than Mt. Waialeale, but holds the record for periods between 15 days and 2 years. The greatest recorded 12-month rainfall in the world is 2,647 cm measured at Cherrapunji during the period from August 1860 through July 1861. Locations like Cherrapunji with extremely high rainfall accumulations over periods of more than a few days are often associated with the Asian monsoon. Extreme intensities over several hours are frequently associated with cyclones and hurricanes. For example, a tropical cyclone in February 1964 dropped 135 cm in 12 hours on the island of La Reunion in the Indian Ocean. The 24-hour precipitation record for the continental United States is 109 cm (43 in) of rain that fell in 24 hours at Alvin, Texas, near Houston, during Hurricane Claudette in July 1979.

Linsley, Kohler, and Paulhus (1982) compiled world-record extreme precipitation depths for 29 durations, ranging from 1 minute to 2 years, including the six records shown in Table 2.4. The data were found to plot as essentially a straight line on log–log paper. The envelope curve describing the upper bound is given by

$$R = 425D^{0.47} \tag{2.6}$$

where R is the maximum recorded precipitation depth in millimeters and D is the duration in hours over which the precipitation depth occurred.

The driest regions of the world are located on the leeward side of mountains, in the belt of subtropic high pressure between latitude 15° and 30°, and in the polar region. The record lowest annual precipitation is 0.08 cm (0.03 in.) for Arica in northern Chile. Death Valley, California, holds the record for the U.S. with a mean annual precipitation of 4.5 cm (1.78 in.). The longest recorded period in the U.S. with no rain was 767 days (October 3, 1912–November 8, 1914) at Bagdad, California. The longest record of zero rain in the world is 14 years (October 1903–January 1918) held by Arica, Chile.

2.4.3 Measurement of Precipitation

Rain gages can be categorized as either nonrecording or recording. A nonrecording gage measures the rainfall depth since the gage was last read and emptied. Any open container with vertical sides makes a convenient nonrecording rain gage. However, depth measurements in different receptacles may not be comparable due to wind and splash effects unless the receptacles are of the same shape and size and similarly exposed. The standard nonrecording gage used by the National Weather Service (NWS) in the U.S. consists of an 8-inch (20.3 cm) diameter rain receptor that funnels into a measuring tube with a cross-sectional area of one-tenth the top area of the receptor. Thus, the rainfall depth is amplified 10 times as the water passes from the receptor to the measuring tube, increasing the accuracy of the measurement. The 24-inch (61 cm) high metal gage also includes an outer overflow can from which water can be poured back into the measuring tube as necessary. An observer reads and empties the gage periodically, typically daily.

Measurements of snow include both freshly fallen snow and accumulated snowpack. Snow measurements are expressed in terms of water equivalent, the depth of water obtained after melting the snow. The outer overflow can of NWS standard nonrecording gages is often used as snow gages. The snow is melted and poured into the measuring tube. Standpipe gages, also known as storage gages, are one of several other types of gages commonly used to measure snowfall. The standpipe gage consists of one or more 5-ft (1.6 m) long sections of 12-inch (30.5 cm) diameter galvanized thin-wall tube erected high enough to extend above the anticipated accumulation of snow.

A recording gage automatically records precipitation depth accumulations and corresponding times. The slope of the accumulated depth versus time relationship is a measure of precipitation intensity. Data may be recorded by paper charts or digital or electronic recorders at the gaging station or transmitted to a central data collection site. Recording gages use either a tipping bucket, weighing mechanism, or float device. A tipping bucket gage is based on funneling the collected rain to a small bucket that tilts and empties each time it fills. An electronic pulse is generated with each tilt. The number of bucket tilts per time interval provides a basis for determining the precipitation depth over time relationship. A weighing-type gage is based on continuously recording the weight of the accumulated precipitation. Float-type recording gages operate by catching rainfall in a tube containing a float, the rise of which is recorded over time.

Radar measurements are also used to estimate precipitation intensities. Use of radar in combination with rain gage data provides useful estimates for areas between the gages. Radar is less accurate than gage measurements. However, radar provides the advantage of covering large areas with high spatial and temporal resolution. Electromagnetic waves from a radar antenna are reflected back by raindrops and other matter in the atmosphere. The strength of reflected radar pulses is a function of the number and size of raindrops. The distance from the radar site to the precipitation area is measured by the time between emission of the radar pulse and receipt of its echo. A new network of 161 Doppler radar systems covering the entire U.S., installed by the NWS during the 1990's, is routinely used to provide weather information. Weather radar systems operated by the Federal Aviation Administration at airports and by various other entities provide detailed weather information for local areas for various purposes.

2.4.4 Spatially Averaged Precipitation

Precipitation depths at multiple gaging stations are commonly averaged to obtain a mean depth over a particular area, for a specific storm, month, season, or year or for monthly, seasonal, or annual means. Areal mean rainfall depths and intensities are often computed for watersheds. Average depths are also of interest for various other land areas such as cities, states, regions, and nations, and/or water bodies such

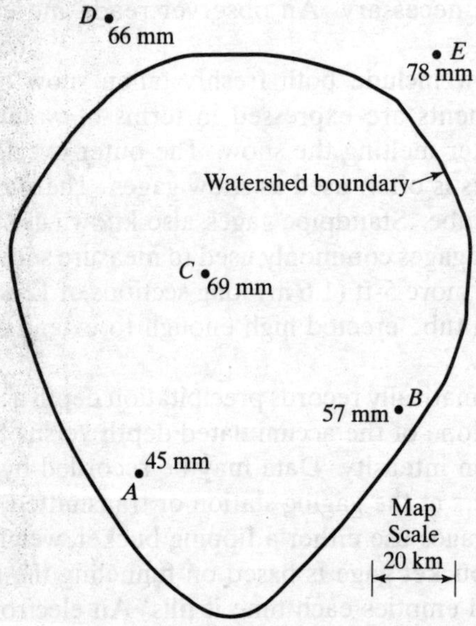

Figure 2.8 Map of watershed and five precipitation gage stations for Example 2.1.

as lake surfaces. The simple arithmetic average of depths for all gages located within an area of interest is the simplest method for obtaining an average depth over the area. For example, the locations of five precipitation gages are shown in Fig. 2.8. Three gages are located in the watershed; two are outside. The area average precipitation may be estimated as the simple arithmetic mean of the observations at gages *A*, *B*, and *C*. However, various other weighted-average methods address the nonuniform spatial distribution of the gages. The Thiessen method is the most commonly used.

The Thiessen method is based on weighting the precipitation at each gage in proportion to the land area within the basin that is closer to that gage than any other gage. The areal mean precipitation depth $\overline{P}$ is computed as

$$\overline{P} = \sum_{i=1}^{n} a_i P_i \tag{2.7}$$

where P_i is the observed precipitation at the *i*th gage, a_i is the fraction of the basin area that is closer to that gage than any other gage, and n is the number of gages. Thiessen polygons are constructed to delineate the portion of the basin assigned to each gage. From geometry, the perpendicular bisector of a straight line connecting points *A* and *B* divides a plane such that all points that are closer to *A* than to *B* are on one side of the perpendicular bisector. As illustrated by Fig. 2.9, straight lines are drawn to connect adjacent gages. Perpendicular bisectors of the connecting lines form polygons around each gage. Any point within a polygon is closer to that gage than any other gage. The areas enclosed by each polygon are determined and expressed as a fraction a_i of the total basin area.

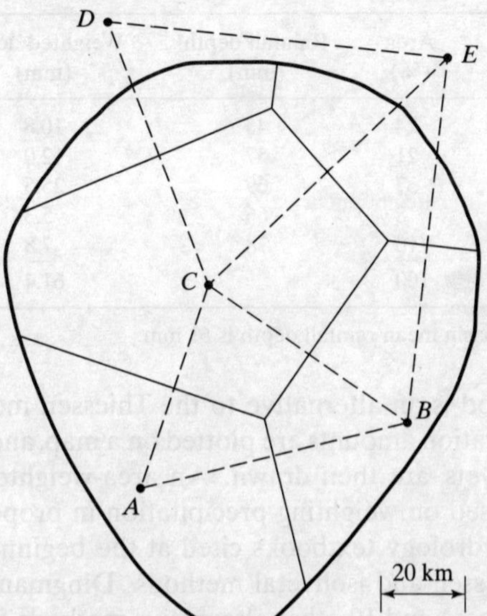

Figure 2.9 Thiessen polygon network for Example 2.1.

Example 2.1 Average Precipitation

The location of precipitation gages A, B, C, D, and E are shown in the map of Fig. 2.8. Gages A, B, and C are located within the watershed boundary shown on the map. Gages D and E are located just outside the 10,700 km^2 watershed. Total precipitation depths during the month of June 2001 of 45, 57, 69, 66, and 78 mm were observed at gages A, B, C, D, and E, respectively. Estimate the mean basin precipitation for June 2001 using (a) a simple arithmetic average and (b) the Thiessen method.

Solution

(a) If only the three gages located within the basin are included in the arithmetic average:

$$\overline{P} = \frac{45 + 57 + 69}{3} = 57 \text{ mm}$$

An arithmetic average for all five gages is also computed for comparison:

$$\overline{P} = \frac{45 + 57 + 69 + 66 + 78}{5} = 63 \text{ mm}$$

(b) The Thiessen polygons of Fig. 2.9 are constructed using perpendicular bisectors of the straight lines connecting the gages. The watershed area encompassed within each polygon is determined from the map using a planimeter or computer digitizer or by simply counting squares on graph paper. The watershed has a total area of 10,700 km^2. However, since the individual polygon areas are expressed as a fraction of their summation, map measurements in cm^2 can be used directly without necessarily scaling to ground measurements in km^2. The area assigned to each gage is expressed below as a percentage of the total.

Gage	Area (%)	Rainfall depth (mm)	Weighted depth (mm)
A	24	45	10.8
B	21	57	12.0
C	37	69	25.5
D	8	66	5.3
E	10	78	7.8
Total	100		61.4

The basin mean rainfall depth is 61 mm.

The isohyetal method is an alternative to the Thiessen method. Gaging station locations and precipitation amounts are plotted on a map, and contours of equal precipitation, called isohyets, are then drawn. An area-weighted mean precipitation is then computed based on weighting precipitation in proportion to the areas between isohyets. The hydrology textbooks cited at the beginning of this chapter all cover at least the Thiessen and isohyetal methods. Dingman (2002) provides a detailed presentation of these and 10 other alternative methods for estimating areal precipitation from point measurements.

2.5 EVAPORATION AND TRANSPIRATION

As previously indicated in Table 2.2, well over half of the precipitation that falls on the land surfaces of the earth is returned to the atmosphere through evapotranspiration. *Evaporation* is the transformation of water from its liquid to its vapor phase. Evaporation occurs from water bodies, soils, and various wet surfaces. *Transpiration* is evaporation from plant leaves. Plants remove moisture from the soil and release it to the air as water vapor. *Evapotranspiration* is the combined total of evaporation and transpiration. The two phenomena are difficult to separate for typical land areas that include transpiration from trees and plants along with evaporation from soil, water, and other surfaces. The physics of water vapor loss from open water surfaces, soils, and plant leaves are essentially the same.

2.5.1 Evaporation Processes

Evaporation occurs by exchange of H_2O molecules between air and a water surface, which could be a rain drop, soil moisture, wetted surface, water within a plant leaf, or the surface of a stream, lake, or ocean. The molecules in liquid water are held close together by attractive intermolecular forces. In water vapor, the molecules are much further apart and the intermolecular forces are much less. During evaporation, work is done against the attractive forces to greatly separate the molecules, and energy is absorbed. The energy required is called the latent heat of vaporization.

Water molecules are continuously exchanged both to and from the atmosphere. The term *evaporation* refers to the net exchange of molecules. Evaporation is the difference between two rates, a vaporization rate determined by temperature and a condensation rate determined by vapor pressure. If molecules can diffuse away from the surface, vapor pressure remains low, vaporization exceeds condensation, and evaporation continues. On the other hand, if the air above the water is thermally insulated and enclosed, the vapor pressure increases until the rates of vaporization and condensation are the same, and there is no more evaporation. The air is saturated.

Key meteorological factors affecting evaporation rates include temperature, atmospheric pressure, vapor-pressure gradient, and wind. Solar radiation provides the required energy. The movement of water molecules increases with temperature. Evaporation increases if water is heated. Difference in vapor pressure at the evaporating water surface and in the surrounding air provides the driving force. The relationship between vapor-pressure gradient and evaporation is affected by the temperature of the air and water. Vapor pressure and humidity are interdependent. As the air becomes saturated, evaporation will slow and stop. Thus, wind plays a key role in removing water vapor from the zone near the water surface. Increases in wind speed increase evaporation rates. Evaporation is greater at lower atmospheric pressures. Both atmospheric pressure and temperature decrease with altitude, having somewhat offsetting effects on the relationship between altitude and evaporation, However, evaporation is generally higher at higher altitudes.

Evaporation rates are also affected by the nature of the evaporating surface and by water quality. Shallow water bodies are more rapidly heated than deeper bodies and thus incur greater evaporation rates. The rate of evaporation from a saturated soil surface is about the same as from an adjacent lake surface at the same temperature. As the soil dries, evaporation decreases and its temperature rises to maintain the heat balance. Eventually, evaporation essentially ceases, since no effective mechanism exists for transporting water from appreciable depths in the soil. Thus, evaporation from soil surfaces is limited by water availability. The vapor pressure of water is reduced with dissolved solids. Thus, evaporation rates for seawater are slightly less than for freshwater.

2.5.2 Transpiration Processes

Transpiration is the process by which water contained in plant tissues is vaporized and discharged to the atmosphere through plant stomata. Stomata are intercellular openings in the lower side of leaves, that for most plants open during the day to take in carbon dioxide for photosynthesis and respiration. Water coming from the roots is used in the process. Most of the water taken up by plants is vaporized from leaf surfaces. The rate of transpiration is typically much greater than the rate of consumption of water in the formation of vegetative matter. The transpiration rate is a function of the type, stage, and growth of the plants, soil type, and moisture, as well as meteorological conditions. Temperature, pressure, and wind affect transpiration similarly as evaporation from water bodies.

The *potential transpiration rate* is the rate that occurs when the water supply to the leaves exceeds the evaporative capacity of the air. Otherwise, actual transpiration is controlled by the plant and soil system. Water supply to the leaves and resulting transpiration are enhanced by capillary potential caused by intercellular openings in the leaves and osmotic pressure caused by the difference in moisture between the sap at the roots and the surrounding soil. Working against transpiration are gravity and capillary tension in the soil, which increases with soil dryness.

2.5.3 Determination of Lake Evaporation Rates

Mean annual lake evaporation rates in inches are shown in Fig. 2.10 for the contiguous United States based on information available from the NWS. Mean annual lake evaporation varies from 20 inches (51 cm) in the northeast and northwest to 80 inches (200 cm) in the southwest. Evaporation may vary significantly between years, being much higher during drought conditions than during wet years. Rates also vary greatly during the year. In regions of extreme temperature differences between hot summers and freezing winters, essentially all of the annual evaporation may occur during the summer months. The impacts of evaporation losses vary greatly among regions and among reservoir projects. For many water supply reservoirs, the annual evaporation volume in a typical year exceeds the volume of water withdrawn for municipal, industrial, and agricultural uses.

An array of techniques for estimating evaporation from water surfaces are described by the hydrology books cited at the beginning of this chapter. The alternative methods involve water budgets, energy budgets, mass transfer equations, aerodynamics equations, and combinations thereof. Data requirements are significant for these evaporation prediction approaches.

Evaporation can be measured with atmometers, which are calibrated instruments used to measure the evaporation potential of the air. The different types of atmometers generally utilize a porous bulb that draws water from a container as evaporation occurs from the bulb surface. Coefficients must be established for correlating evaporation with water loss in the container.

The most widely applied technique for estimating lake evaporation involves combining pan evaporation measurements and pan coefficients. Water is maintained in a pan with the water surface exposed to the atmosphere, and the changes in depth due to evaporation and rainfall are periodically measured. A variety of different types of pans have been used. The most common is the standard Class A pan developed by the NWS (formerly the U.S. Weather Bureau) in 1916. The Class A pan is a circular container 47.5 inches (1.21 m) in diameter and with vertical walls 10 inches (25.4 cm) deep. The water depth is maintained at 8 inches, which is 2 inches below the rim of the pan. The unpainted galvanized iron or monel sheet metal pan is placed on a wooden platform just above ground level.

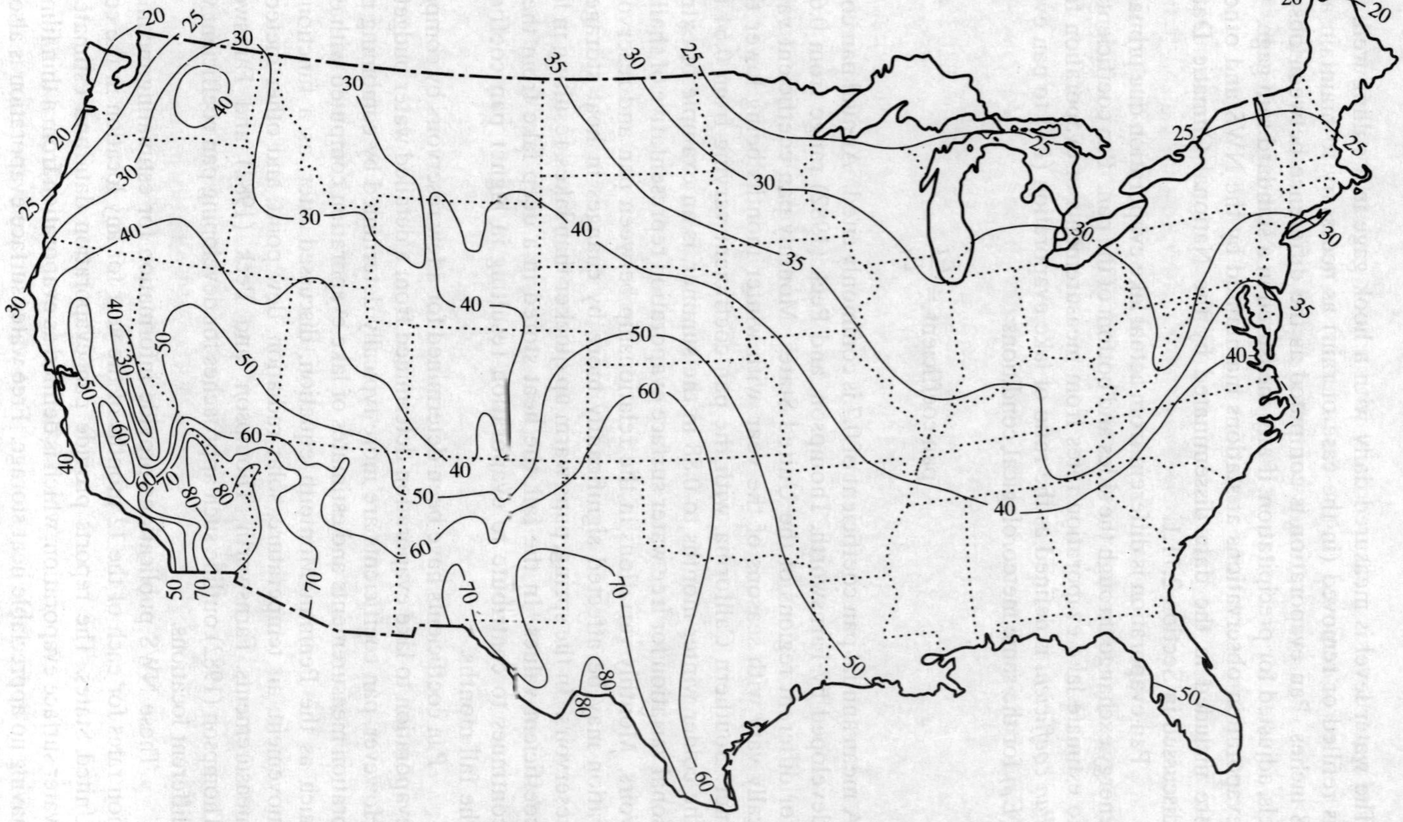

Figure 2.10 Mean annual lake evaporation in inches.

The water level is measured daily with a hook gage in a stilling well, and water is refilled or removed (in the case of rain) as necessary to maintain a depth of 8 inches. Pan evaporation is computed as the difference between observed levels, adjusted for precipitation, if any, measured in a standard rain gage. Daily pan evaporation observations at stations maintained by the NWS and other entities are included in the data disseminated by the National Climatic Data Center discussed in Section 2.11.1.

Pan evaporation is different from actual lake evaporation due primarily to the energy exchange through the sides and bottom of the pan. Pan coefficients are used to estimate lake evaporation rates from measurements of evaporation from pans. *Pan coefficient* is defined as the ratio of lake evaporation (E_L) to pan evaporation (E_P) for the same meteorological conditions:

$$\text{pan coefficient} = \frac{E_L}{E_P} \tag{2.8}$$

A mean annual pan coefficient of 0.7 is commonly cited. Annual pan coefficients developed by Farnsworth, Thompson, and Peck (1982) range from 0.64 to 0.88 for different regions of the United States. Monthly pan coefficient values typically vary with seasons of the year, with winter months being lower than summer. Southern California, with the pan coefficient varying from 0.64 to 0.68 in the colder winter months to 0.88 in the summer, is an example of extreme seasonal variation for free water surface evaporation representative of shallow reservoirs. Monthly variations in the relationship between pan and reservoir evaporation may be affected significantly more by changes in heat storage in deep reservoirs. In the spring, pans warm up quicker than lakes, resulting in lower pan coefficient values. In the fall, the heat stored in a deep lake from the summer continues to contribute to evaporation resulting in higher pan coefficients for the fall months.

Pan coefficients have been determined for a few reservoirs by comparing pan evaporation to lake evaporation determined from a detailed water budget analysis. However, pan coefficients are more typically determined by combining pan evaporation measurements and estimates of lake evaporation computed with equations, such as the Penman–Monteith equation, discussed later, as a function of wind movement, air temperature, solar radiation, dew point, and other meteorological measurements. Farnsworth, Thompson, and Peck (1982) and Farnsworth and Thompson (1982) outline such approaches for developing pan coefficient values for different locations.

These NWS publications provide information for estimating mean evaporation rates for each of the 12 months of the year for any location in the contiguous United States. The reports provide pan evaporation data and estimates of free water surface evaporation, which is defined as evaporation from a thin film of water having no appreciable heat storage. Free water surface evaporation is also referred to as shallow lake evaporation and potential evapotranspiration. The mean annual free water surface evaporation is also representative of deep lakes, but heat storage effects can be expected to cause variations in individual months.

The evaporation atlas (Farnsworth, Thompson, and Peck, 1982) includes maps of (1) evaporation observed from Class A pans from May through October, (2) estimates of shallow reservoir evaporation from May through October, (3) estimates of shallow reservoir evaporation for the entire year, and (4) pan coefficients used to convert pan evaporation to shallow reservoir evaporation. Additional information provided by Farnsworth and Thompson (1982) includes tabulations of mean monthly pan evaporation for each of the 12 months of the year for several hundred stations. A procedure is outlined for combining the monthly tabulations and atlas maps to develop monthly reservoir evaporation rates for any ungaged location in the contiguous U.S.

Mean monthly pan evaporation rates for selected stations are reproduced in Table 2.5. Approximate values of shallow reservoir evaporation at these locations can be obtained by multiplying the pan evaporation rates by an average pan coefficient of roughly 0.7. More accurate estimates are obtained using pan coefficients provided in the atlas. The NWS publications also outline computational methods for developing more accurate pan coefficients based on measurements of temperature, wind movement, and other meteorological variables.

Methods for estimating areal means from point measurements are discussed in Section 2.2.4 from the perspective of precipitation depths. The Thiessen polygon method and other methods are also applicable to determining spatially averaged evaporation depths from measurements at multiple gaging stations.

TABLE 2.5 PAN EVAPORATION AT SELECTED LOCATIONS (FARNSWORTH AND THOMPSON, 1982)

Month	Folsom Dam, California	Davis Dam, Arizona	Ft. Peck Dam, Montana	Elephant Butte Dam, New Mexico	Whitney Dam, Texas	Wolf Creek Dam, Kentucky	Kerr Dam, Virginia
				(inches)			
Jan	0.90	5	—	3.28	2.95	2	—
Feb	1.62	6	—	4.85	3.88	2	—
Mar	3.46	9	—	8.53	6.05	4	—
Apr	5.38	11	—	11.75	7.02	4.68	5.27
May	8.09	14	7.49	14.45	8.46	5.47	6.22
Jun	10.13	16.68	8.68	16.17	10.65	6.35	6.81
Jul	11.46	14.43	10.67	13.64	12.39	6.57	7.20
Aug	10.18	14.62	9.86	11.63	11.38	5.88	6.12
Sep	7.66	11.8	5.88	9.72	8.33	4.58	4.87
Oct	4.96	8.93	3.56	7.70	6.24	3.24	3.37
Nov	2.03	7.45	—	4.75	4.02	2	—
Dec	0.94	5.73	—	3.21	3.12	2	—
				Annual pan evaporation			
inches	66.81	124.00	46.14	109.68	84.67	49.00	39.86
cm	170	315	117	279	215	124	101

2.5.4 Determination of Evapotranspiration Rates

Evaporation from land areas such as an agricultural field or a large region of pastures and forests includes transpiration from plants and evaporation from soil, ponds, and various wetted surfaces. The term evapotranspiration encompasses the total evaporation from all sources. Either potential or actual evapotranspiration may be of concern. *Potential evapotranspiration* is the rate that would occur under some specified standard condition of vegetation with an unlimited supply of water in the soil and without advection or heat storage effects. With dry soil, the actual evapotranspiration may be much less than the potential. A key factor determining the impact of vegetation on evapotranspiration is shading of the soil or the fraction of the solar radiation reaching the soil surface. Also, actual evapotranspiration will vary between different plants in various stages of growth.

Free water surface evaporation was defined in Section 2.5.3 as evaporation from a thin film of water having no appreciable heat storage. Free water surface evaporation is also referred to as shallow lake evaporation and potential evapotranspiration. Evaporation from saturated soil that is bare or covered with relatively short vegetation is similar to evaporation from a shallow lake. Thus, the use of pan evaporation data discussed in the previous section is pertinent to determining potential evapotranspiration, as well as lake evaporation. Various methods for computing potential evapotranspiration using climatic data are also reported in the literature. The precise operational definition of potential evapotranspiration varies depending on the estimation technique.

The determination of actual evaporation may also be based on field measurements, empirical relationships, mathematical models representing mass and/or energy budgets, or combinations thereof. Tanks and lysimeters are designed to directly measure evapotranspiration. The alternative approach of using mathematical models to calculate actual evapotranspiration for agricultural lands usually involves two steps. First, an estimate is made for either potential evapotranspiration or evapotranspiration for a reference crop with standard canopy characteristics. The two most common reference crops are grass and alfalfa. Then the evapotranspiration for the actual vegetation is obtained by multiplying by an empirically determined coefficient. Variations of the Penman–Monteith equation are the most commonly used of various mathematical models based on weather data to determine the potential or reference crop evaporation (Jensen, Burman, and Allen, 1990; Allen, Pereira, Raes, and Smith, 1998).

2.5.4.1 Lysimeter measurements. A lysimeter is an artificially enclosed volume of soil for which the inflows and outflows of liquid water can be measured, and changes in storage can be monitored by weighing. A lysimeter is essentially a tank filled with soil in which vegetation is grown. The lysimeter is placed in or near the field of interest and is designed to simulate actual field conditions as closely as possible. Carefully obtained lysimeter measurements are usually considered to give the best determination of actual evapotranspiration during a time period and adopted as standards against which other methods are compared. However, lysimeters are expensive to install and monitor and are impractical for many types of water resources planning and management studies.

2.5.4.2 Mathematical modeling of evapotranspiration. Consequently, various methods for estimating evapotranspiration from weather data have been developed. Variations of the Penman–Monteith equation are commonly used to compute either potential or reference crop evapotranspiration, which is then combined with empirical factors representing the characteristics of particular types of vegetation to determine actual evapotranspiration.

Penman (1948) combined mass and energy balance concepts to derive an equation that was further expanded and refined by Monteith (1965) and other researchers. The American Society of Civil Engineers (Jensen, Burman, and Allen, 1990) and the Food and Agricultural Organization (FAO) of the United Nations (Allen, Pereira, Raes, and Smith, 1998) outline detailed procedures for applying the Penman–Monteith equation and empirical factors for particular crops in the estimation of evapotranspiration requirements for use in irrigation planning and management. The hydrology books cited at the beginning of this chapter include brief discussions of the Penman–Monteith equation and other alternative methods for estimating evapotranspiration.

The FAO version (Allen, Pereira, Raes, and Smith, 1998) of the basic equation is reproduced below simply to illustrate the various terms involved in the calculations. The FAO report provides detailed instructions for compiling data and applying the equation and also provides factors for relating the results to particular types of actual crops.

$$ET_O = \frac{0.408\Delta\,(R_n - G) + \lambda \dfrac{900}{T + 273}\, u_2\,(e_s - e_a)}{\Delta + \gamma\,(1 + 0.34u_2)} \tag{2.9}$$

The terms in the FAO Penman–Monteith equation are defined as follows:

ET_O	reference evaporation (mm/day)
R_n	net radiation at the crop surface (MJ m^2 day^{-1})
G	soil heat flux density (MJ m^{-2} day^{-1})
T	mean daily air temperature at 2 m height (°C)
u_2	wind speed at 2 m height (m s^{-1})
e_s	saturation vapor pressure (kPa)
e_a	actual vapor pressure (kPa)
$e_s - e_a$	saturation vapor pressure deficit (kPa)
Δ	slope of vapor pressure curve (kPa °C^{-1})
λ	latent heat of vaporization (MJ/kg)
γ	psychrometric constant (kPa °C^{-1})

2.6 UNITS OF MEASURE FOR DEPTH, AREA, VOLUME, AND VOLUMETRIC RATES

Hydrology deals with quantities of water and quantities of other substances dissolved or suspended in the water. Common units of measure for depths, volumes, intensities, and volumetric rates are listed in Table 2.6. Metric and English units are listed with conversion factors in Table 2.7. Units of measure for dissolved or suspended constituents are addressed in a later section on water quality.

Precipitation, evapotranspiration, infiltration, and runoff volumes are often expressed in terms of a volume equivalent to covering a watershed area to a specified depth. Volumes per unit area may be viewed in units of mm $\cdot$ km^2/km^2 expressed simply as mm. Likewise, the unit of inch $\cdot$ acre/acre is reduced to inch for convenience. For example, assume the volume of runoff from a watershed resulting from a particular rainfall event is expressed as 25 mm. This is a volume equivalent to uniformly covering the watershed to a depth of 25 mm. It may be viewed as a volume of 25 mm $\cdot$ km^2/km^2 of watershed area. A 1-mm depth over 1 km^2 is 1,000 m^3. A 1-inch depth over 1 mi^2 equals 53.33 ac-ft or 26.89 sfd.

The acre-foot (ac-ft) and second-foot-day (sfd) are among the common volume units used traditionally in the U.S. An ac-ft is the volume of water required to cover 1 acre of land to a depth of 1 foot. Since there are 43,560 ft^2 in 1 acre. an ac-ft equals 43,560 ft^3. The ac-ft is commonly used to express volumes of irrigation water requirements, reservoir storage, reservoir releases, and streamflow.

Streamflow volume may also be expressed in units of sfd. An sfd or (ft^3/s) $\cdot$ day is a volume equivalent to 1 ft^3/s flowing for 1 day. The use of the sfd is motivated by the practice of recording streamflow gage data as the mean daily flow each day. For example, a mean daily flow of 48.5 ft^3/s multiplied by 1 day results in a volume of 48.5 sfd flowing by the gaging station that day. Since there are 86,400 seconds in 1 day, an sfd equals 86,400 ft^3. An sfd equals 1.9835 ac-ft (86,400 ft^3/sfd divided by 43,560 ft^3/ac-ft).

TABLE 2.6 EXAMPLES OF COMMON UNITS OF MEASURE

Quantity	Metric units	English units
watershed area	hectares (ha), square kilometers (km^2)	acres (ac), square miles (mi^2)
lake surface area	hectares (ha), square kilometers (km^2)	acres (ac), square miles (mi^2)
precipitation depth	millimeters (mm), centimeters (cm)	inches (in)
precipitation rate	millimeters/hour (mm/hr), mm/month	inches/hour (in/hour), in/month
evaporation rate	millimeters/hour (mm/hr), mm/month	inches/hour (in/hour), in/month
infiltration rate	millimeters/hour (mm/hr), mm/month	inches/hour (in/hour), in/month
volume of water	cubic meters (m^3)	cubic feet (ft^3), acre-feet (ac-ft)
streamflow rate	cubic meters/second (m^3/s or cms)	cubic feet/second (ft^3/s or cfs), ac-ft/year
human water use	liters per day (l/day)	gallons per day (gal/day)
flow velocity	m/s	ft/s

TABLE 2.7 CONVERSION FACTORS FOR LENGTH, AREA, VOLUME, AND DISCHARGE UNITS

Multiply English units	By	To obtain metric units
Length		
inches (in)	2.540	centimeters (cm)
feet (ft)	0.3048	meters (m)
miles (mi)	1.6093	kilometers (km)
foot = 12 inches		m = 100 cm
mile = 5,280 feet		km = 1,000 m
Area		
acres (ac)	0.4047	hectares (ha)
acres (ac)	4047	square meters (m^2)
square miles (mi^2)	2.590	square kilometers (km^2)
mi^2 = 640 acres		km^2 = 100 ha
acre = 43,560 ft^2		ha = 10,000 m^2
Volume		
cubic feet (ft^3)	0.02832	cubic meters (m^3)
acre-feet (ac-ft)	1,234	cubic meters (m^3)
second-foot-day (sfd)	621.0	cubic meters (m^3)
gallons (gal)	3.785	liters (l)
ft^3 = 7.481 gal		m^3 = 1,000 liters
sfd = (ft^3/s) day = 86,400 ft^3		
ac-ft = 43,560 ft^3		
Flow Rate		
ft^3/s or cfs	0.02832	m^3/s
million gallons/day (mgd)	0.04381	m^3/s
ac-ft/year	3.91×10^{-5}	m^3/s

A flow rate or discharge is the volume per unit of time:

$$\text{Discharge} = \frac{\text{Volume}}{\text{Time}} \tag{2.10}$$

Streamflow rates are typically measured in units of m^3/s or ft^3/s. Streamflow rates, reservoir releases, and irrigation water use rates may also be expressed in units of ac-ft/yr, ac-ft/day, or ac-ft per some other time period. Municipal water use in the U.S. is often measured in million gallons per day (mgd).

Examples 2.2 through 2.5 illustrate unit conversions and manipulation of water quantities. Hydrologic engineering computations are often based on various water budget equations representing the fundamental concept that, for a given container (continent, country, watershed, stream reach, lake, aquifer, etc.) and a given time period (multiple-years, year, month, day, hour, second, etc.), the sum of inflows less the sum of outflows equals the change in storage. For example, in Example 2.2, the container is a reservoir on a river, and the time interval is 30 days.

Example 2.2 Water Budget

During a particular 30-day period, the streamflow flowing into a reservoir averaged 5.0 m^3/s. Water supply withdrawals from the reservoir averaged 136 mgd. The only other outflow from the reservoir was 9.4 cm of evaporation from the reservoir water surface. The average water surface area of the reservoir during the 30 days was 3.75 km^2. The reservoir had 12,560 ac-ft of water in storage at the beginning of the 30-day period. Determine the storage content at the end of the 30 days alternatively in units of m^3 and ac-ft.

Solution End-of-period storage S_2 is determined as a function of beginning-of-period storage S_1, inflow I, water supply withdrawal W, and evaporation E based on the following water budget equation with all terms expressed in consistent units.

$$S_2 = S_1 + I - W - E$$

$$S_1 = (12{,}560 \text{ ac-ft})\left(43{,}560\,\frac{\text{ft}^3}{\text{ac-ft}}\right)\left(0.02832\,\frac{\text{m}^3}{\text{ft}^3}\right) = 15{,}494{,}300 \text{ m}^3$$

$$I = \left(5.00\,\frac{\text{m}^3}{\text{s}}\right)(30 \text{ days})\left(\frac{86{,}400\text{s}}{\text{day}}\right) = 12{,}960{,}000 \text{ m}^3$$

$$W = \left(136{,}000{,}000\,\frac{\text{gal}}{\text{day}}\right)\left(0.003785\,\frac{\text{m}^3}{\text{gal}}\right)(30 \text{ days}) = 15{,}442{,}800 \text{ m}^3$$

$$E = (3.75 \text{ km}^2)(9.4 \text{ cm})\left(1{,}000{,}000\,\frac{\text{m}^2}{\text{km}^2}\right)\left(0.01\,\frac{\text{m}}{\text{cm}}\right) = 352{,}500 \text{ m}^3$$

$$S_2 = 15{,}494{,}300 + 12{,}960{,}000 - 15{,}442{,}800 - 352{,}500 \text{ m}^3 = 12{,}659{,}000 \text{ m}^3$$

$$S_2 = (12{,}659{,}000 \text{ m}^3)\left(\frac{\text{ac-ft}}{1{,}234 \text{ m}^3}\right) = 10{,}260 \text{ ac-ft}$$

Example 2.3 Reservoir Evaporation

Table 2.5 indicates that the average June pan evaporation at Folsom Reservoir on the American River in California is 10.13 inches. Folsom Reservoir has a storage capacity of 1,140 million m^3. At capacity, the water surface has an area of 4,610 ha.

(a) Estimate the average June evaporation volume that would occur with a full reservoir. Use a pan coefficient of 0.7. Express the evaporation volume alternatively in units of m^3 and ac-ft.

(b) Assuming a per capita domestic water use rate of 80 gal/day, the June evaporation loss from Folsom Reservoir is equivalent to supplying domestic water for how many people?

(c) Assuming a June irrigation demand of 30 cm of water, the evaporation loss is equivalent to irrigating how many square miles of crop land?

Solution

$$\text{Evaporation volume} = (0.7)(10.13\text{ in})(4{,}610\text{ ha})\left(10{,}000\frac{\text{m}^2}{\text{ha}}\right)\left(0.0254\frac{\text{m}}{\text{in}}\right) = 8{,}303{,}000\text{ m}^3$$

$$\text{Evaporation volume} = (8{,}303{,}000\text{ m}^3)\left(0.0008104\frac{\text{ac-ft}}{\text{m}^3}\right) = 6{,}730\text{ ac-ft}$$

$$\text{Number of people} = \frac{8{,}303{,}000\text{ m}^3}{\left(\dfrac{80\dfrac{\text{gal}}{\text{day}}}{\text{person}}\right)(30\text{ days})\left(0.003785\dfrac{\text{m}^3}{\text{gal}}\right)} = 914{,}000\text{ people}$$

$$\text{Area of irrigated crop land} = \left(\frac{6{,}730\text{ ac-ft}}{(30\text{ cm})\left(0.03281\dfrac{\text{ft}}{\text{cm}}\right)}\right)\left(\frac{\text{mi}^2}{640\text{ ac}}\right) = 10.7\text{ mi}^2$$

Example 2.4 Watershed Precipitation-runoff Units

The watershed above a particular site on a river has a drainage area of 77,700 hectares. The mean annual precipitation for this area is 812 mm. About 25 percent of the precipitation reaches the basin outlet as streamflow. Estimate the mean flow rate Q at that particular site on the river alternatively in units of m^3/yr, m^3/s, ft^3/s, and ac-ft/yr.

Solution

$$\text{Precipitation} = (77{,}700\text{ ha})\left(10{,}000\frac{\text{m}^2}{\text{ha}}\right)\left(812\frac{\text{mm}}{\text{yr}}\right)\left(0.001\frac{\text{m}}{\text{mm}}\right) = 6.31\times 10^8\text{ m}^3/\text{yr}$$

$$Q = \left(6.31\times 10^8\frac{\text{m}^3}{\text{yr}}\right)(0.25) = 1.58\times 10^8\frac{\text{m}^3}{\text{yr}}$$

$$Q = \left(1.58\times 10^8\frac{\text{m}^3}{\text{yr}}\right)\left(\frac{\text{year}}{365\text{ days}}\right)\left(\frac{\text{day}}{86{,}400\text{ s}}\right) = 5.00\text{ m}^3/\text{s}$$

$$Q = \left(5.00\frac{\text{m}^3}{\text{s}}\right)\left(\frac{\text{ft}^3}{0.02832\text{ m}^3}\right) = 176.6\text{ ft}^3/\text{s}$$

$$Q = \left(176.6\frac{\text{ft}^3}{\text{s}}\right)\left(\frac{\text{ac-ft}}{43{,}560\text{ ft}^3}\right)\left(86{,}400\frac{\text{s}}{\text{day}}\right)\left(365\frac{\text{days}}{\text{yr}}\right) = 127{,}900\text{ ac-ft/yr}$$

alternatively

$$Q = (77{,}700\text{ ha})\left(812\frac{\text{mm}}{\text{yr}}\right)\left(\frac{\text{acre}}{0.4047\text{ ha}}\right)\left(\frac{\text{ft}}{304.8\text{ mm}}\right)(0.25) = 127{,}900\text{ ac-ft/yr}$$

Example 2.5 Stream Flow Units

The mean daily discharge at a stream gaging station for each day of a 7-day period (Sunday through Saturday) is recorded as follows: 4,630 ft^3/s, 7,620 ft^3/s, 7,290 ft^3/s, 5,640 ft^3/s, 4,110 ft^3/s, 2,580 ft^3/s, and 4,080 m^3/s. Determine the total volume of water that passed the gaging station that week in units of sfd and ac-ft. What was the mean discharge for the week in ft^3/s and m^3/s?

Solution

$$\text{Volume} = (4{,}630 + 7{,}620 + 7{,}290 + 5{,}640 + 4{,}110 + 2{,}580 + 4{,}080\ \text{ft}^3/\text{s})(1\ \text{day}) = 35{,}950\ \text{sfd}$$

$$\text{Volume} = \left(35{,}950\ \frac{\text{ft}^3}{\text{s}}\ \text{day}\right)\left(\frac{\text{acre}}{43{,}560\ \text{ft}^2}\right)\left(86{,}400\ \frac{\text{s}}{\text{day}}\right) = 71{,}306\ \text{ac-ft}$$

$$\text{mean discharge} = \left(\frac{4{,}630 + 7{,}620 + 7{,}290 + 5{,}640 + 4{,}110 + 2{,}580 + 4{,}080\ \text{ft}^3/\text{s}}{7}\right) = 5{,}136\frac{\text{ft}^3}{\text{s}}$$

alternatively

$$\text{mean discharge} = \frac{\left(35{,}950\ \frac{\text{ft}^3}{\text{s}}\ \text{day}\right)}{7\ \text{days}} = 5{,}136\ \frac{\text{ft}^3}{\text{s}}$$

$$\text{mean discharge} = \left(5{,}136\ \frac{\text{ft}^3}{\text{s}}\right)\left(0.02832\ \frac{\text{m}^3}{\text{ft}^3}\right) = 145\ \frac{\text{m}^3}{\text{s}}$$

2.7 WATERSHED HYDROLOGY AND STREAMFLOW

Surface runoff and streamflow along with groundwater are the components of the hydrologic cycle that are typically of most interest to water resources engineers. Chapter 8 focuses specifically on modeling the watershed precipitation-runoff processes that result in streamflow. Chapters 5 and 6 cover the hydraulics of flow in natural and man-altered streams and constructed channels. Chapter 10 deals with stormwater management in urban watersheds. Chapter 12 explores river basin management. Groundwater is the focus of Chapter 9.

2.7.1 Watersheds and Streams

A *stream* is a flowing body of water. Larger streams are usually called rivers. Smaller streams are often called brooks, creeks, or bayous, depending on regional customs. Natural stream systems are composed of channels, floodplains, and watersheds. Channels usually have well-defined banks, with much flatter, wider floodplains on either side of the channel that are not as clearly defined. Floodplains convey channel overflows. For most natural rivers, flows exceed the bankful channel capacity with flood frequencies on the order of once every 1 to 10 years.

During peak flow conditions of infrequent extreme flood events, the amount of flow in the floodplain may be many times the flow within the channel banks.

A *watershed* is the land that contributes runoff to a given site. The terms *watershed, basin, catchment,* and *drainage area* are used synonymously. Larger watersheds are typically called *river basins*. For any point of interest on the land (the drainage outlet), the land area draining to that point (the watershed) is delineated primarily by the slope of the topography, although man-made storm sewers and other water control structures may divert flows across the drainage divides defined by topography. Watersheds vary greatly in size and other characteristics. For example, the watershed above a particular urban storm sewer inlet consists of a 0.1 hectare portion of an asphalt parking lot. The Mississippi River Basin shown in Fig. 2.3 consists of 41% of the contiguous United States and a small part of Canada. The drainage area of the watershed of the Mississippi River above its mouth at the Gulf of Mexico near New Orleans is 323,000,000 hectares (3,230,000 km^2 or 1,250,000 mi^2).

Watersheds are composed of smaller subwatersheds. Each subwatershed is itself a watershed that can be subdivided into smaller subwatersheds that can each be further subdivided into more subwatersheds. The four largest subwatersheds of the Mississippi River watershed are the Missouri River Basin (1,370,000 km^2), the Ohio River Basin (520,000 km^2), the Arkansas River Basin (416,000 km^2), and Red River Basin (93,200 km^2). The Big Thompson River, discussed in Section 2.2.3.3, in northern Colorado, is a tributary of the South Platte River that joins with the North Platte in Nebraska to form the Platte River. The Platte River, shown in Fig. 2.3, is a tributary of the Missouri River. Likewise, the Big Thompson River is fed by many tributary streams not shown in Fig. 2.3, each of which has its own tributaries.

The authors live in the Burton Creek and Bee Creek watersheds that are subwatersheds of the 130 km^2 Carter Creek watershed shown in Fig. 2.11. Most of the

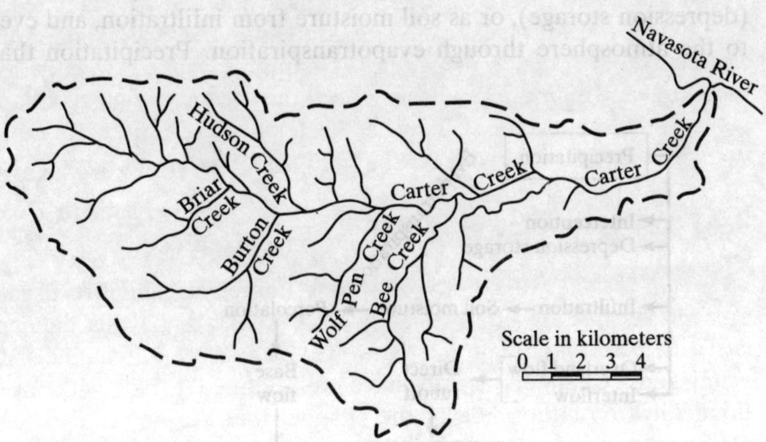

Figure 2.11 Carter Creek Watershed in Bryan and College Station, Texas.

Carter Creek Watershed lies within the corporate limits of the adjoining cities of Bryan and College Station, Texas. Carter Creek flows into the Navasota River, which flows into the Brazos River, which flows into the Gulf of Mexico. The 23 km^2 watershed of Bee Creek above its confluence with Carter Creek is a small portion of the 5,640 km^2 Navasota River Basin and the 118,000 km^2 Brazos River Basin, but can be further subdivided into numerous much smaller watersheds, each having its own outlet.

Most streams are components of stream networks that eventually discharge into oceans or other water bodies connected to oceans. However, exceptions include closed basins, such as watersheds of the numerous small playa lakes in the High Plains of the southwestern U.S., the Great Salt Lake near Salt Lake City, Utah, and the Dead Sea in Jordan and Israel into which the Jordan River flows.

2.7.2 Hydrologic Processes

The precipitation falling within the boundary of a watershed drains to its outlet. The hydrologic processes shown in Fig. 2.12 result in precipitation being divided into direct runoff and abstractions. Direct runoff is the rainfall that, upon reaching the ground, flows to the watershed outlet without being stored for long periods of time. Direct runoff may include interflow as well as surface flow. Interflow is flow through relatively pervious soil that reaches the outlet about as quickly as surface flow as a direct response to the rainfall event. Surface runoff typically begins as overland sheet flow and then concentrates into small rivulets of flow that grow larger in route to man-made drainage improvements and natural streams. Abstractions are losses of precipitation due to interception, depression storage, and infiltration. The abstracted precipitation may be stored temporarily on the leaves of trees and other objects (interception), in puddles or pools on the ground surface (depression storage), or as soil moisture from infiltration, and eventually returned to the atmosphere through evapotranspiration. Precipitation that infiltrates into

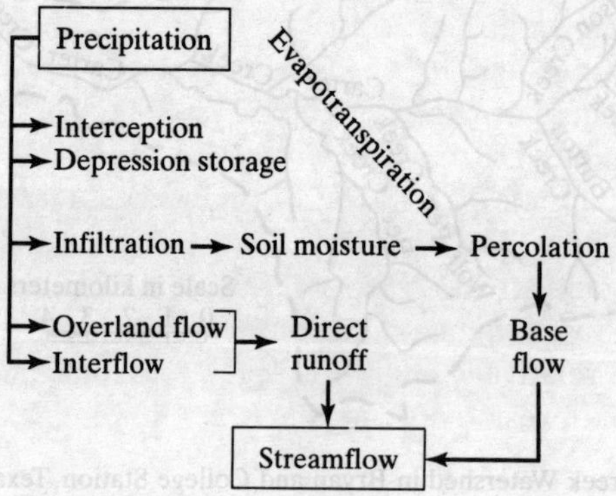

Figure 2.12 Watershed precipitation-runoff processes.

the ground may also percolate downward, replenish groundwater, and eventually contribute to stream baseflow.

Interception is the part of precipitation that is caught by vegetation and other objects above the ground. The precipitation adheres to these surfaces until it evaporates. A forest canopy may intercept most of the precipitation falling during a storm. The amount of water intercepted is a function of storm characteristics, density and types of vegetation and land cover, and season of the year. Precipitation that reaches the ground may runoff, infiltrate, or be trapped in small depressions until it infiltrates or evaporates. The nature of depression storage depends on natural topography, vegetation, land-use practices, and man-made drainage systems. Infiltration is the entry of water into the ground. The quantity and rate of infiltration are functions of soil type, soil moisture, soil permeability, ground cover, drainage conditions, depth-to-water table, and volume and intensity of precipitation.

Baseflow is the portion of streamflow derived from sources other than recent precipitation. Baseflow is typically from groundwater. Effluent from municipal wastewater treatment plants is another source of baseflow. Cities pump water from rivers, reservoirs, and groundwater aquifers, and then return much of the water to a stream after it is used. Melting of the winter snowpack provides flows in some streams throughout the summer.

Streams may be classified as ephemeral, perennial, or intermittent. An ephemeral stream flows only after rainfall. There is no baseflow. A perennial stream hardly ever dries up. Baseflow is continuous. The streambed of perennial streams is usually below the groundwater table. Mountain streams may be fed from springs. An intermittent stream during dry seasons may have some reaches with flowing water interspersed with other reaches in which the water flows below dry streambeds.

2.7.3 Variations in Streamflow

Streamflow is a function of precipitation, drainage area, and other watershed characteristics. Streamflow, like precipitation, varies drastically both geographically and temporally. The largest rivers, in terms of 1931–1970 mean flow rates, in the contiguous United States are shown in Fig. 2.13, along with a graphics comparison of their mean flows (Iseri and Langbein, 1974). The rivers shown are those with mean flows at their mouths that equal or exceed 17,000 ft^3/s (480 m^3/s). The outflow of all streams from the contiguous U.S. into the oceans or across national borders is about 2,000,000 ft^3/s (56,600 m^3/s). The rivers shown in Fig. 2.13 account for more than 75 percent of this total flow. The remaining flows to the oceans are discharged by numerous smaller rivers and streams. Several rivers in the humid East (such as the Hudson, Delaware, and Susquehanna) and Southeast (such as the Appalachicola and Alabama Rivers) have relatively small drainage areas, but much larger flows than rivers in the arid Southwest (such as the Rio Grande) with much larger watershed areas.

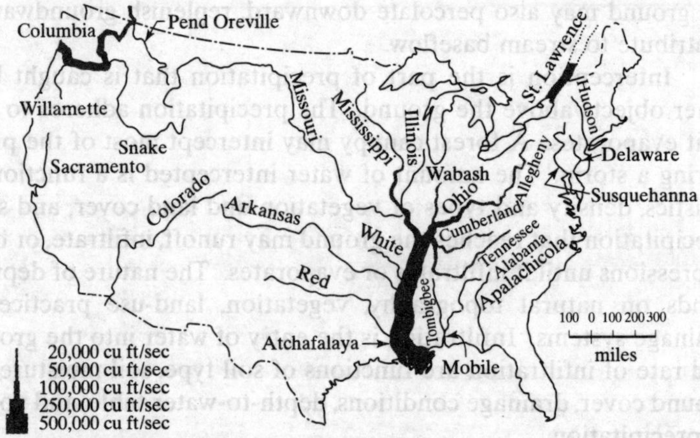

Figure 2.13 Mean flows of the largest rivers in the conterminous United States (Iseri and Langbein, 1974). Rivers shown are those whose average flow at the mouth is 17,000 ft^3/s or more. The average flow of the Yukon River in Alaska (not shown) is 240,000 ft^3/s.

A hydrograph for a location on a stream shows discharge as a function of time. Discharge values at instants in time or averaged over short time intervals of less than 1 hour are desirable for hydrographs of particular floods to capture rapid changes in flow. However, as discussed in Section 2.11.2, long-term discharge measurements at gaging stations are often recorded as daily flows. The hydrographs plotted in Figs. 2.14 and 2.15 are mean daily flows in ft^3/s at two U.S. Geological Survey gaging stations. The gage on the Mississippi River at St. Louis, shown in Fig. 2.3, has a watershed drainage area of 1,810,000 km^2 (697,000 mi^2). The headwaters of the Big Thompson River, also shown in Fig. 2.3, flow from the continental divide at the western edge of the Mississippi River Basin. The gaging station near Drake has a drainage area of 790 km^2 (305 mi^2). The 1976 flood on the Big

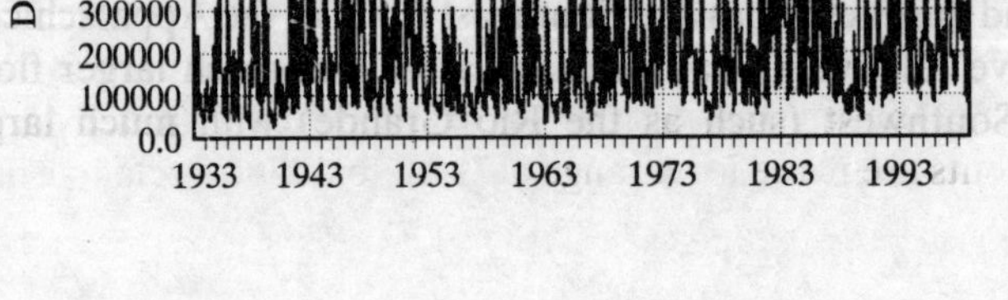

Figure 2.14 Mean daily flow of the Mississippi River at St. Louis, Missouri. (USGS Gaging Station 07010000.)

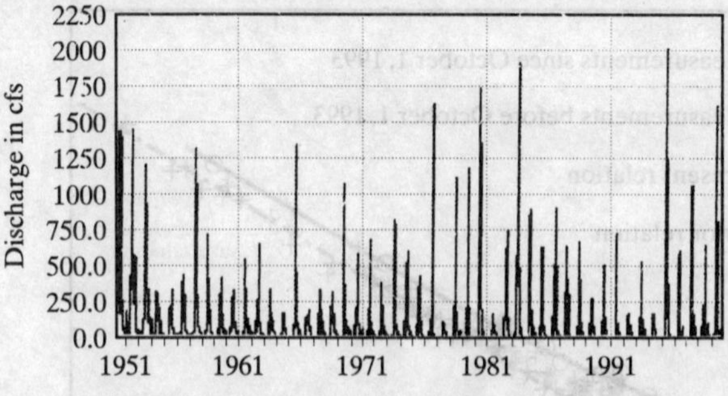

Figure 2.15 Mean daily flow of the Big Thompson River at the Mouth of Canyon near Drake, Colorado. (USGS Gaging Station 06738000.)

Thompson River and 1927 and 1933 floods in the Mississippi River Basin are described in Section 2.2.3.

The two watersheds reflected in the streamflow hydrographs of Figs. 2.14 and 2.15 are very different. One is a relatively small, steep mountain watershed. The other is very large encompassing diverse geography and land uses. The annual cycles of winter snowpack accumulation and summer snowmelt are evident in the Big Thompson hydrograph. The mean daily flows do not portray the rapid flow changes during flash floods on the Big Thompson River. For example, the previously discussed 1976 flood had an instantaneous peak of 31,200 ft^3/s (884 m^3/s) at this gage, but most of the flood wave passed in a couple of hours, with negligible flow most of the day. The mean daily peak shown in Fig. 2.15 is 1,650 ft^3/s (47 m^3/s). Instantaneous flood peaks vary relatively little from the peak daily mean flows plotted in the hydrograph for the much slower rising and falling Mississippi River. The drastic difference in the flow of the Mississippi River comparing the 1988 drought with the 1993 flood are evident from Fig. 2.14. The great seasonal and random variability of streamflows exhibited by these two hydrographs are characteristic of essentially all streams and is a governing concern in water resources engineering.

2.7.4 Streamflow Measurements

Streamflow discharge measurement normally involves:

(1) establishing the relationship (rating curve) between water surface elevation or height above a reference datum (stage) and discharge (flow rate) at a gaging station

(2) continuously or periodically measuring stage at the gaging station

(3) transforming the record of stage into a record of discharge by applying the rating curve.

Measuring stage is much easier than measuring discharge. Therefore, a limited number of discharge measurements are made for a range of stage to define a relationship

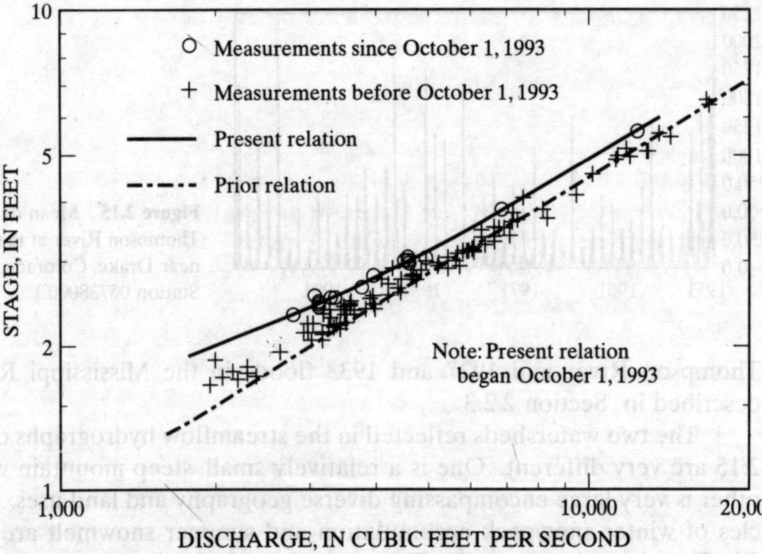

Figure 2.16 Sections of stage-discharge relations for the Colorado River at the Colorado-Utah State Line. (Wahl, Thomas, and Hirsch, 1995).

between stage and discharge at a gaging station. The stage-discharge relation, often called a rating curve, is then combined with continuing periodic stage measurements to record discharge as well as stage.

Stage-discharge relations may have to be periodically updated due to changes in the hydraulic characteristics of a stream reach over time caused by erosion and sedimentation, bank vegetation, and other changes. This problem is illustrated by the two rating curves shown in Figure 2.16. Several discharge and stage measurements made at various times after October 1, 1993 result in a different rating curve than earlier measurements, indicating the hydraulics of the channel has changed (Wahl, Thomas, and Hirsch, 1995).

2.7.4.1 Measurement of stage. A staff gage is the simplest device for measuring river stage or water surface elevation. A staff gage is a graduated strip of wood or metal marked in feet or meters and fractions thereof. Water levels are read by an observer. Simple crest-stage gages have been devised to mark the highest level reached since the gage was last read, for use typically in measuring peak stage during floods.

Self-recording gages that maintain a continuous record of stage are based on various types of sensors. The three most commonly used types of sensors are float-driven, pressure, and ultrasonic. In a typical installation of a float-driven, water-level sensor, the vertical movement of a float in a stilling well, resulting from fluctuations in water level, are translated by a mechanical movement or an electronic signal. Pressure sensors are based on the relationship of water depth to pressure. Ultrasonic sensors use acoustic pulses to sense water levels either by contact or noncontact methods.

A stream gaging station usually includes a stilling well located on the bank of a stream or attached to a bridge pier, with a shelter on top holding the recording equipment and instruments. The stilling well is connected to the stream by intakes such that the water level is the same in the well as the stream. The stilling well damps out momentary fluctuations in water levels due to waves.

2.7.4.2 Measurement of Discharge. A relationship between stage and discharge is required to convert stage measurements to flow rates. Discharge measurement in small channels is sometimes accomplished by installing a weir or flume for which a head–discharge relationship has been determined in a laboratory. Measurements of head are converted to flow rates. Hydraulic structures such as a spillway for a dam may also act as a weir for measuring flow. However, constructed weirs and flumes are not practical for most locations on rivers.

The velocity-area method is the most common approach for measuring the discharge of a stream (Wahl, Thomas, and Hirsch, 1995). Mechanical, electromagnetic, ultrasonic, and thermal-pulse sensors may be used to measure flow velocity. The Price current meter has been widely used for several decades. The meter assembly consists of a wheel of cups rotating around a vertical axis, tail vanes to keep the meter headed into the current, and a weight to keep the meter cable as vertical as possible. The flowing water rotates the cup wheel. Electrical contacts are used to count its revolutions in an interval of time. Flow velocity is related to the rotational speed of the cup wheel. The meter is calibrated by towing through a tank of still water at known towing speeds. In shallow streams, the current meter may be attached to a rod for wading measurements. In deep rivers, the meter is suspended from a bridge or boat.

Discharge Q (m^3/s, ft^3/s) is related to mean velocity V (m/s, ft/s) and cross-sectional area A (m^2, ft^2) as follows:

$$Q = VA \tag{2.11}$$

V is a flow area weighted-average velocity. Point velocity varies throughout a stream cross-section. Flow velocities at point locations vary both vertically and horizontally across the channel. In a vertical, the velocity approaches zero at the streambed and is greatest near the surface. Horizontally, the velocity approaches zero at the banks and is greatest toward the middle or deepest part of the channel. In a typical stream, the mean velocity in a vertical is closely approximated by averaging the velocities at 20 and 80 percent of the depth. A single measurement at 60 percent of the depth from the surface provides a slightly less accurate estimate of the mean velocity in a vertical.

A stream cross-section is subdivided into multiple widths for purposes of determining the discharge for a given stage. Current meter measurements of point velocities at 0.2 and 0.8 of the depth (or alternatively 0.6 of the depth) for multiple verticals are combined with Eq. 2.12 to estimate total discharge. The total Q is the summation of Q_i in each of n subsections of the stream cross-section:

$$Q = \sum_{i=1}^{n} Q_i = \sum_{i=1}^{n} V_i A_i \tag{2.12}$$

The procedure for estimating discharge from current meter measurements of velocities at points in the flow cross-section is illustrated by Example 2.6.

Example 2.6 Stream Discharge

The discharge at a stream location for a given stage is computed based on the current meter measurements provided below. The horizontal distance across the stream is measured from the edge of the water at one bank. Depths are measured from the water surface.

Distance, m	1.0	2.0	3.0	4.0	5.0	6.0	7.0	8.0	9.0	10.0
Depth, m	0.5	2.2	3.7	4.3	3.7	2.8	2.4	1.8	1.1	0.7
Velocity, m/s										
at 0.2 depth	0.040	0.054	0.074	0.082	0.076	0.072	0.065	0.060	0.051	0.042
at 0.8 depth	0.020	0.034	0.050	0.058	0.054	0.048	0.043	0.038	0.031	0.028

Solution The cross-section is shown in Fig. 2.17 with the location of each of the current meter measurements of point velocity. Applying Eq. 2.7, the Q_i for each of the ten 1.0-m wide flow subareas are estimated and summed to obtain the total flow. The cross-section area A_i for each subarea is estimated as depth multiplied by 1-m width. The mean velocity in each subarea is estimated as the average of the flows at 0.2 and 0.8 depth.

A_i, m^2	0.5	2.2	3.7	4.3	3.7	2.8	2.4	1.8	1.1	0.7
V_i, m/s	0.030	0.044	0.062	0.070	0.065	0.060	0.054	0.049	0.041	0.035
$Q_i = V_iA_i$, m^3/s	0.0150	0.0968	0.2294	0.3010	0.2405	0.1680	0.1296	0.0882	0.0451	0.0245

$$Q = 0.0150 + 0.0968 + 0.2294 + 0.3010 + 0.2405 + 0.1680 + 0.1296 + 0.0882 + 0.0451 + 0.0245$$

$$Q = 1.34 \text{ m}^3\text{/s}$$

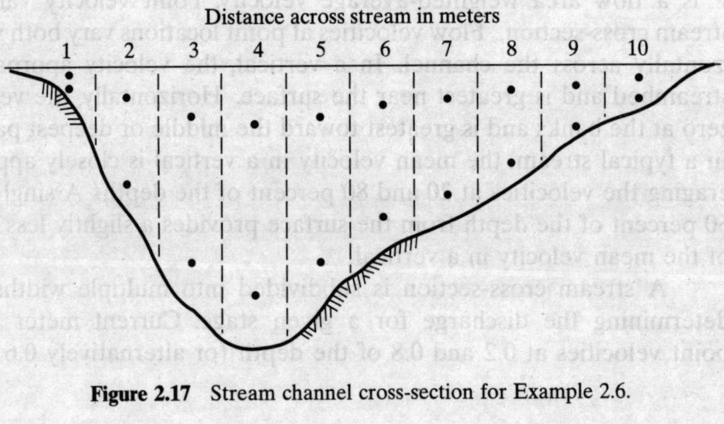

Figure 2.17 Stream channel cross-section for Example 2.6.

2.8 SUBSURFACE WATER

As indicated in Table 2.1, groundwater accounts for most of the water on earth, excluding the oceans and permanent ice. Much of the groundwater is saline, but groundwater also accounts for most of the freshwater on earth. Although much of this water is deep below the surface of the ground, important hydrologic processes near the ground surface involve both surface and subsurface water. Much of the flow of perennial streams is derived from subsurface water. Flow in ephemeral streams may be lost through seepage into the ground beneath the stream bed. Flow may be intermixed between flow above the streambed and through alluvium material below the streambed. Precipitation is supplied to the soil through infiltration. Soil moisture taken through the roots of plants is returned to the atmosphere through evapotranspiration. Water is also evaporated directly from the soil near the surface. Groundwater aquifers are a major source of water supply. Return flows from irrigation or municipal wastewater treatment plants may be discharged into surface streams.

The schematic of Fig. 2.18 shows the hydrologic cycle from a subsurface perspective. The water below the ground surface is divided into zones of aeration and saturation, separated by the water table. Below the water table, pore spaces or interstices in soil and rock, are filled with water. This zone of saturation is also called the phreatic zone or groundwater. Above the water table, in the unsaturated zone

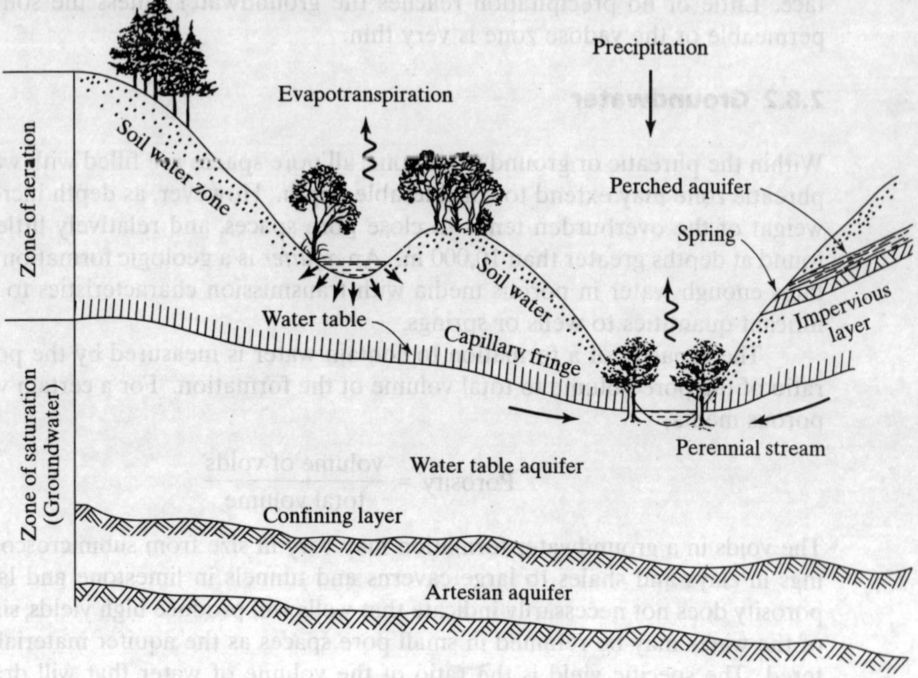

Figure 2.18 Hydrologic cycle from a subsurface water perspective.

of aeration, soil pores may contain air and/or water. The unsaturated region between the water table and ground surface is also called the vadose zone.

2.8.1 Moisture in the Vadose Zone

Three moisture regions occur in the zone of aeration. The soil water zone is penetrated by the roots of vegetation that may extend as much as 10 m below the ground surface, depending on the types of trees and other vegetation present. In the capillary fringe above the water table, water is raised by capillarity, which varies inversely with pore size and surface tension. The intermediate zone below the soil moisture and above the capillary fringe may be nonexistent for high water table conditions or range in thickness to over 100 m in arid regions. Water percolates downward toward the water table under the force of gravity and is held back by hygroscopic and capillary forces.

Infiltration is the movement of water through the soil surface into the soil. Percolation is the subsequent movement of the water through the soil. When water is applied to the soil surface, it moves through larger soil openings by gravity while smaller pores are filled by capillarity. The downward moving gravity water is also taken in by capillary pores. In homogeneous soils, infiltration decreases gradually with saturation. Soil is typically stratified with the infiltration rate being eventually limited by the rate of percolation through the least pervious subsoil layer. Precipitation infiltrating into the soil is usually distributed relatively near the ground surface. Little or no precipitation reaches the groundwater unless the soil is highly permeable or the vadose zone is very thin.

2.8.2 Groundwater

Within the phreatic or groundwater zone, all pore spaces are filled with water. The phreatic zone may extend to considerable depth. However, as depth increases, the weight of the overburden tends to close pore spaces, and relatively little water is found at depths greater than 10,000 m. An *aquifer* is a geologic formation that contains enough water in porous media with transmission characteristics to yield significant quantities to wells or springs.

The capacity of a formation to contain water is measured by the porosity, or ratio of the pore volume to total volume of the formation. For a certain volume of porous media,

$$\text{Porosity} = \frac{\text{volume of voids}}{\text{total volume}} \tag{2.13}$$

The voids in a groundwater formation may vary in size from submicroscopic openings in clays and shales to large caverns and tunnels in limestone and lava. High porosity does not necessarily indicate that wells will produce high yields, since much of the water may be retained in small pore spaces as the aquifer material is dewatered. The specific yield is the ratio of the volume of water that will drain freely from the material to the total volume of the formation. Specific yield is always less than porosity, since some water is retained by molecular or capillary forces. Fine-grained clay may have a porosity around 45 percent and a specific yield of 3 percent. In contrast, typical values of porosity and specific yield for sand are 34 and 25 percent, respectively. Clay contains much water but yields little of it to a pumping well. Most of the more productive wells tend to be in deposits of sand and gravel.

Aquifers generally extend over large areas and may be underlain or overlain by a confining layer of relatively impermeable material such as clay or rock. Aquifers may be either confined or unconfined. Confined or artesian aquifers are formed under an impermeable layer with recharge occurring at some distant location where the aquifer intersects the ground surface. The aquifer is under pressure such that water will rise in a well above the overlying confining layer. An unconfined or water table aquifer is a permeable formation having a water surface, called the water table, at atmospheric pressure. Flow depends on the slope of the water table, in an unconfined aquifer, or piezometric surface, in a confined aquifer. The piezometric surface is the water level in a group of wells and is the artesian equivalent of the water table. If the piezometric surface is above the ground surface, water will flow from a well without pumping. Groundwater engineering is covered in Chapter 9.

2.9 EROSION AND SEDIMENTATION

Land erosion, stream bank erosion, sediment transport, and sediment deposition accompany the precipitation-runoff-streamflow processes that occur in a watershed over temporal scales ranging from a storm event to geologic time. Water greatly influences geomorphology. Erosion processes, which involve forces associated with water, wind, ice, and gravity, are constantly at work reshaping the earth's surface. Water is recognized as the dominant agent of the erosion process and is responsible for the bulk of the sediment transported from the land to the sea. Precipitation, along with other natural forces, break down rocks over geologic time to form soils. Displays of the great influence of these processes on the earth's landscape include the picturesque Niagara Falls, Grand Canyon, other river falls and canyons, and the meanders of the mighty Mississippi and other rivers. Precipitation over short time spans, particularly during extreme floods, may transport large amounts of soil and streambed materials. Loss of topsoil, gully formation, streambed movement, and meandering of alluvial rivers result in severe problems for humans using the land and water resources.

Land erosion involves detachment, transport, and subsequent deposition of solid particles typically called sediments. Soil is detached both by raindrop impact and the shearing force of flowing water. Sediment is transported down-slope by the flowing water. The sediment may be composed of various materials, including individual primary soil particles, aggregates, organic matter, and associated chemicals. Sediment is a major water quality constituent itself and also serves as carrier and storage agent for a full spectrum of other pollutants.

Soil is protected from erosion by vegetation. Human activities that disturb vegetation, such as logging, mining, agriculture, and construction, may greatly increase land erosion. For example, when land is disturbed by construction projects, soil erosion may increase from 2 to 40,000 times the normal or geological erosion rate. Mine dumps and spoil areas may continue to be eroded by rainfall discharging sediment loads into streams many years after mining operations have ceased.

Channel improvements and hydraulic structures may significantly affect stream bank erosion. Channel straightening, which increases slope and flow velocity, may also cause channel erosion. If the bed of a main stream is lowered, the bed slopes on tributaries are increased and erosion processes may be affected in the upstream watershed. Construction of dams results in sediment deposition upstream and increased stream erosion downstream.

Streams may be described as straight, meandering, or braided. A meandering stream flows in large loops or bends. Bank erosion tends to occur in the concave curves with sediment deposition at the subsequent point bar. A typical meandering channel may be 1.5 to 2 times as long as its relatively straight nonmeandering floodplain valley. A braided stream consists of many intertwined channels separated by islands. Braided channels are usually found in relatively steep reaches with easily eroded sandy bank material with little vegetative cover. Meandering tends to occur on flatter slopes with more cohesive banks.

2.10 WATER QUALITY

Water quality refers to the quantity of suspended and dissolved solids, dissolved gases, heat, and microorganisms in a given quantity of water. In nature, a myriad of impurities enter and leave water throughout the hydrologic cycle. Human activities add and remove pollutants superimposed with naturally occurring water quality constituents.

Solvent action, dissociation, and transparency are unique properties of water that affect its interactions with other substances in the hydrologic cycle. Of all natural occurring liquids, water is the most effective solvent. Sodium chloride and salts of potassium are readily dissolved in water. Water is effective in forming bonds with both cations and anions, creating a number of dissolved compounds that are vital for organisms in an aquatic environment. Water is not only a solvent for other substances, but also is itself capable of dissociating into two charged ions. Thus, water acts as both an acid and a base. The transparency of water allows both absorption of heat from solar radiation and transmission of light critical to stream and lake environments.

2.10.1 Impurities in Water

Water quality is typically defined relative to water use. Most human and ecosystem uses of water have desirable and/or necessary ranges for certain water quality constituent concentrations. Throughout history, the most important characteristic of water has been the concentration of dissolved solids, commonly referred to as salinity. Salinity constrains the use of the vast sea, as well as large quantities of ground and surface water for irrigated agriculture and human consumption, as well as governing ecosystem development. As human populations grew, health-related characteristics such as disease-causing (pathogenic) microorganisms became important. Human development also resulted in organic waste loads associated with sewage and runoff from agricultural lands that exceeded natural oxidation capabilities of streams. With industrial development, characteristics such as specific ion content and temperature became significant. Most recently, manufactured chemicals that can have an impact on human health when present in even trace amounts have become a major concern in pollution control.

A number of water impurities originating from various sources are listed in Table 2.8. The water quality constituents are categorized as (1) suspended solids, (2) dissolved ions, and (3) gases. Suspended solids are substances that are not dissolved, including material that will settle in still water and smaller colloidal

TABLE 2.8 PRINCIPAL WATER QUALITY CONSTITUENTS

Source of impurity	Suspended solids	Dissolved ions		Gases
		Positive ions	Negative ions	
atmosphere	salt from sea, dust, pollen, and volcanic ash from land	hydrogen, H^+	bicarbonate, HCO_3^- chloride, Cl^- sulfate, SO_4^{2-}	carbon dioxide, CO_2 nitrogen, N_2 oxygen, O_2 sulfur dioxide, SO_2
soils and rocks	clay, silt, sand, and other inorganic soils silica, SiO_2 ferric oxide, Fe_2O_3 aluminum oxide, Al_2O_3 magnesium dioxide, MnO_2	calcium, Ca^{2+} iron, Fe^{2+} magnesium, Mg^{2+} manganese, Mn^{2+} potassium, K^+ sodium, Na^+ zinc, Zn^{2+}	bicarbonate, HCO_3^- borates, $H_2BO_3^-$ carbonate, CO_3^{2-} chloride, Cl^- fluoride, F^- hydroxides, OH^- nitrate, NO_3^- phosphate, PO_4^{3-} silicates, $H_3SiO_4^-$ sulfate, SO_4^{2-}	carbon dioxide, CO_2
decomposition of organic matter	organic soil and wastes coloring matter	ammonia, NH_4^+ hydrogen, H^+ sodium, Na^+	chloride, Cl^- bicarbonate, HCO_3^- hydroxide, OH^- nitrite, NO_2^- nitrate, NO_3^- sulfide, HS^- organic radicals	ammonia, NH_3 carbon dioxide, CO_2 hydrogen sulfide, H_2S hydrogen, H_2 methane, CH_4 nitrogen, N_2 oxygen, O_2
human activities	inorganic solids, natural & synthetic organic compounds, viruses, bacteria	inorganic ions, including a variety of heavy metals	inorganic ions, organic molecules	chlorine, Cl_2 sulfur dioxide, SO_2

particles that can be filtered out. Dissolved solids are ions that have separated and chemically associated with water molecules. An ion is an atom or group of atoms that has lost or gained one or more electrons. Ions are either positively charged cations or negatively charged anions. All natural waters contain dissolved ionic constituents. All water exposed to the atmosphere has dissolved gases. Gases are also commonly derived from the decomposition of organic matter.

2.10.2 Physical, Chemical, and Biological Characteristics and Parameters

Water quality characteristics, and the parameters used to measure them, are commonly categorized as physical, chemical, and biological. Tchobanoglous and Schroeber (1985), McCutcheon, Martin, and Barnwell (1993), Malina (1996), and Tebbutt (1998) explore water quality characteristics and parameters in detail. Water quality measurements involve field sampling and analysis, laboratory analysis, and *in situ* monitoring. Temperature, conductivity, dissolved oxygen, pH, turbidity, and other parameters with very short sample holding times are measured in the field. Many of the methods of measuring inorganic, organic, and radioactive substances in water utilize analytical procedures that can be performed only in a laboratory. Analytical procedures used to analyze water samples for most of the physical, chemical, and biological parameters used to assess water quality are presented in detail by the American Public Health Association, the American Water Works Association, and the Water Environment Federation (2001).

Concentrations of impurities in water are expressed in a variety of units. The concentrations of dissolved gases, dissolved solids, and suspended solids are expressed as the mass concentration of constituent per either unit volume or unit mass of solution or suspension. The most common unit for concentration is milligrams per liter (mg/l), which is equal to grams per cubic meter (g/m^3). Concentrations are also often expressed as parts per million (ppm), which is related to mg/l as follows

$$\text{ppm} = \frac{\text{mg/l}}{\text{specific gravity of liquid}} \tag{2.14}$$

where the specific gravity is the ratio of density of the liquid (water) to the density of pure water at standard conditions of temperature and pressure. In chemical analyses, concentrations are commonly expressed as molality or molarity

$$\text{molality, mol/kg} = \frac{\text{moles of solute, mol}}{\text{1.0 kg of solvent}} \tag{2.15}$$

$$\text{molarity, mol/kg} = \frac{\text{moles of solute, mol}}{\text{1.0 liter of solvent}} \tag{2.16}$$

where 1 mol of a compound is equal to the molecular mass expressed in grams. A mole is 6.02×10^{23} molecules of a substance. The concentration of heat may be expressed as temperature. Various other related parameters and units of measure have been adopted as well to measure water quality.

Pollutants can be classified as either conservative or nonconservative substances. Conservative substances do not lose mass to chemical reactions or biochemical degradation. Examples include chlorides, other dissolved salts, and certain metals. Nonconservative substances lose mass through chemical reactions, biochemical degradation, radioactive decay, and settling of particulates out of the water column. Examples include oxidizable organic matter, nutrients, volatile chemicals, and bacteria.

2.10.2.1 Physical characteristics. Parameters used to measure the physical quality of water include total suspended and dissolved solids, turbidity, color, taste, odor, and temperature. The total solids content of water may be measured by evaporating a water sample and weighing the dry residue. Suspended solids are the portion of the total solids removed by a standard filter. Dissolved solids are the difference between total and suspended solids. Turbidity is the presence of small particles suspended in water that scatter and absorb light, giving the water a murky or turbid appearance. Color is caused by dissolved or suspended colloidal particles, usually from decaying leaves or plants. Odors also often stem from decay of organic matter. Temperature affects various other physical properties of water such as density, specific weight, viscosity, surface tension, vapor pressure, conductance, salinity, and solubility of dissolved gases.

2.10.2.2 Chemical characteristics. Natural waters are actually solutions of many substances including those listed in Table 2.8. Of the numerous chemical water quality parameters, the following are particularly important: dissolved oxygen content, biochemical oxygen demand, chemical oxygen demand, acidity, and hardness.

The higher the concentration of dissolved oxygen, the better the water quality. However, since oxygen is only slightly soluble in water, the saturation is low and varies significantly with temperature, dropping substantially as water warms. At 0°C (32°F), the saturation concentration of oxygen in water is 14.6 mg/l. At 30°C (80°F), the saturation concentration is 7.6 mg/l.

The biochemical oxygen demand (BOD) is a measure of the amount of oxygen used by the indigenous microbial population in water in response to the introduction of degradable organic material. The chemical oxygen demand (COD) test of natural water yields the oxygen equivalent of the organic matter that can be oxidized by a strong chemical oxidizing agent in an acidic medium. A COD test can be performed much quicker than a standard BOD test. BOD and COD are among the most widely used parameters in pollution control and water quality management.

Acidity refers to the capacity of water to react with hydroxyl ions. Acidity is the capacity of water to neutralize alkalis and may be expressed in mg/l of equivalent calcium carbonate ($CaCO_3$). Conversely, alkalinity is the capacity of water to neutralize acids and also is measured in mg/l of equivalent $CaCO_3$. A related parameter, the pH is defined as the logarithm of the reciprocal of the hydrogen ion concentration. Pure water at 24°C is balanced with 10^{-7} mol/l of both H^+ and OH^- ions. Thus, the pH is 7. A pH less than 7 is acidic and greater than 7 is basic. Most microorganisms can survive only within a narrow range of pH. These parameters also represent conditions controlling the solubility of most metals.

The term *hardness* refers to the properties of certain highly mineralized waters. Hardness is measured in mg/l of equivalent $CaCO_3$ Medium-hard water has hardness equivalent to a concentration of 60–120 mg/l of $CaCO_3$. Calcium and magnesium ions usually account for the greatest portion of hardness in natural waters, but other dissolved ions also contribute. Hardness causes scale deposits in pipes and equipment and difficulty in lathering soap. Groundwater is generally harder than surface water because it remains in contact with mineral deposits for longer periods of time.

2.10.2.3 Biological characteristics. The diversity of fish and insects, as well as other living organisms, provides an indication of whether streams and lakes are polluted. A large number of different species is usually an indication of good water quality. A smaller diversity with the absence of certain species and excess of others is an indicator of pollution.

Microorganisms are present in all surface waters. Concentrations typically are much lower in groundwater, due to the filtering action of aquifer materials and long retention times. A myriad of bacteria, fungi, algae, protozoa, worms, rotifers, crustaceans, and viruses live in streams, lakes, and other water bodies. Pathogenic (disease-causing) bacteria, viruses, protozoa, and parasitic worms are particularly important. In developing countries, 80 percent of all illness is water-related (Tebbutt, 1998). Human diseases caused by ingesting contaminated water include typhoid fever, cholera, dysentery, hepatitis, gastroenteritis, giardiasis, diarrhea, meningitis, leptospirosis, and salmonellosis.

Testing water for individual pathogens is difficult. A more practical and reliable approach is to test for a single species that would indicate the possible presence of contamination. A species serving this purpose is known as an indicator organism. The most common biological indicator of pollution from untreated sewage is the coliform bacteria species *Escherichia coli* or *E. coli*. Coliform bacteria are present in the intestines of warm-blooded animals, including humans. These bacteria are not pathogens. However, their presence in water samples indicate the presence of fecal matter in a stream or lake and thus the likelihood of other microorganisms that are pathogenic.

Algae are an important component of aquatic ecosystems. Algae are microscopic plants that survive by converting inorganic matter into organic matter using energy from the sun. In the photosynthesis process, they take in carbon dioxide from the air and give off oxygen. Even though most species of algae are microscopic, they can be easily observed when their numbers proliferate in the water to become an algal bloom.

2.10.3 The Hydrologic Cycle from a Water Quality Perspective

Water vapor is basically the only *pure* water in the natural hydrologic cycle. Evapotranspiration results in essentially pure water vapor. However, since the impurities are left behind, the concentrations of the impurities in soil or a water body are increased as water is loss to evapotranspiration. Thus, the oceans and seas have high salt concentrations. Irrigation increases the salinity of soils, decreasing the long-term agricultural productivity of land.

2.10.3.1 Precipitation. Impurities begin to accumulate as soon as precipitation droplets start to form through condensation. Every raindrop or snowflake is formed around a solid nucleus of sea salt, windblown dust, smoke particles, or various pollutants in the air. Subsequently, the chemical composition of the droplets change as gases are dissolved from the air at rates proportional to the solubility and concentration of each gas. The most important gases naturally present in precipitation are oxygen (O_2) and carbon dioxide (CO_2). The increased use of combustion engines and fossil fuels has resulted in the generation of sulfur dioxide (SO_2) and oxides of nitrogen (NO_x). When these substances dissolve in rain drops, mineral acids are produced and the hydrogen ion concentration of the water may increase by 1 to 3 orders of magnitude. The gases can travel long distances before precipitation occurs. Acid rain is a significant problem in Europe, the northeastern United States, and Canada.

2.10.3.2 Surface water. As precipitation strikes the ground and flows over the land surface, various materials are dissolved or suspended and transported with the runoff. These materials may include sediment, mineral salts, organics, microbial organisms, nutrients, heavy metals, and pollutants associated with agricultural, industrial, or urban activities. Pollutants in runoff may be in solution or suspension, or attached to particles of sediment. The water quality characteristics of surface runoff depend on both the accumulation of pollutants on the land surface and the transport mechanisms through which materials interact with the water. Water quality of runoff is highly dependent on land use. Agricultural activities, urban areas, industries, and mining operations produce various mixes of sediments, nutrients, pesticides, petrochemicals, debris, acids, heavy metals, and salts in runoff water.

The water quality characteristics of streams are determined by the constituents in the watershed runoff entering the stream; interactions between the streamflow and channel rocks, soils, and vegetation; interactions at the air–water interface; and hydraulic mixing characteristics of the flow. Turbulence mixes substances longitudinally and across the channel section. The capacity for transporting sediments and other suspended material is a function of flow velocity. Most biota are found primarily near the banks and bottom. The air above the surface serves as a source and sink for oxygen and carbon dioxide, which are important to microorganisms, fish, and ecosystems.

Water quality in lakes and reservoirs is subject to the effects of eutrophication. This is the process by which lakes gradually become shallower and more biologically productive through the introduction and cycling of nutrients. Eutrophication is a natural process that may be significantly accelerated by human activities, resulting in addition of sediments and nutrients.

Thermal stratification is another phenomenon that may affect lake water quality. During the summer, solar radiation causes water near the surface to be warmer and less dense than the water deeper in the lake. A stratification results in which dissolved gases and nutrients near the surface are not readily distributed to lower levels. An aerobic ecosystem is maintained in the upper zone. Nutrients settle to the lower zones, but the depletion of oxygen limits biological activity. In the autumn, as the surface water cools and becomes less dense, nutrient-rich, oxygen-depleted water from the bottom may circulate to the surface causing water quality problems.

2.10.3.3 Subsurface water. Water quality constituents are also transported and transformed as precipitation infiltrates into the ground and percolates through the soil. The interactions between the water and soil depend on soil type and texture, temperature, and the concentration and characteristics of the water quality constituents. In general, passage through the soil profile tends to result in purification of water through filtering, adsorption, decomposition or degradation, volatilization, nitrification, denitrification, and plant uptake.

The adsorption process entails the removal of constituents from solution and retention on the surface of soil particles by chemical or physical bonding. Adsorption keeps the chemicals in the soil until processes such as decomposition and plant uptake occur. Organic materials decompose to form carbon dioxide, water, and inorganic elements such as nitrogen and chloride. Nitrification is a process by which ammonia is oxidized to nitrite and then to nitrate. This is an important reaction in the soil-water system because ammonia, a largely immobile form of nitrogen, is converted to highly mobile nitrate that may be absorbed by plants or lost by leaching or denitrification. In biological denitrification, bacteria convert nitrate to gaseous nitrogen species. In soils with thick vegetal cover, uptake by plants is the primary mechanism for removal of inorganic nitrogen and phosphorous. Flow of water toward roots in response to transpiration results in the transport of nonadsorbed nutrients with high solubilities such as nitrate. Diffusion is the primary mechanism for transporting adsorbed species such as phosphorous, potassium, and iron to plant roots.

Water in aquifers may have fallen as precipitation as recently as several hours ago to as long as many centuries ago. The percolation of water through the soils in recharge areas generally results in significant purification. Where the water table is shallow and the soil porous, water quality constituents from the ground surface may readily reach the groundwater system. In areas where the water table is relatively deep and the soil less porous, the purification process is more thorough, and the aquifer recharge may be void of organic materials and suspended sediments. Water stored in and flowing through an aquifer contacts soluble rocks, resulting in dramatic increases in dissolved minerals, such as calcium bicarbonate, magnesium bicarbonate, calcium sulfate, and magnesium sulfate. Salt water intrusion from the oceans may affect aquifers in coastal areas.

2.11 CLIMATIC, HYDROLOGIC, AND WATER QUALITY DATA

Many federal, state, and local entities in the United States collect water-related data measurements for various purposes. However, the federal agencies noted in the following sections play particularly key roles in collecting hydrologic, water quality, and climatic data, and making these data available to the water management community. The data collection programs of these federal agencies are accomplished in cooperation with other federal, state, and local agencies and the public.

2.11.1 Climatic Data

The National Climatic Data Center (NCDC) of the National Oceanic and Atmospheric Administration of the U.S. Department of Commerce is located in Asheville, North Carolina. NCDC is the world's largest archive of weather data. A wealth of information is available through the NCDC web site: http://www.ncdc.noaa.gov. NCDC archives, manages, and distributes weather data collected by the NWS, Military Services, Coast Guard, Federal Aviation Administration, and cooperative observers. The majority of the daily precipitation and temperature observations at ground stations throughout the U.S. are reported by volunteer local observers in cooperation with the NWS.

NCDC also maintains the World Data Center for Meteorology, Asheville, which is a partner with similar centers in other countries. NCDC fosters global exchange of data through numerous international agreements with individual nations and groups like the World Meteorological Organization. Data centers in other countries can also be located through the NCDC web site.

The numerous climatological databases maintained by the NCDC include precipitation, temperature, solar radiation, dewpoint, relative humidity, pressure, wind speed and direction, evaporation, Palmer drought index, and various other types of meteorological data in a variety of formats for daily and other time intervals. Precipitation data include (1) rainfall measurements at ground stations; (2) rainfall estimates obtained remotely by radar, satellites, or other sensors; (3) depth and water equivalent of snow accumulation as determined from ground measurements or remote sensors; and (4) snow-covered area as determined from aerial or ground reconnaissance, or remote sensors from satellites or aircraft. Data are available in electronic format, as well as in published reports. Numerous weather and climate-related reports and serial publications are available from the NCDC. The printed report series *Climatological Data* is distributed to subscribers monthly, along with an annual summary issue. This series dates back to the 1880's; is published for each state or (in a few cases) combinations of states; and presents basic daily and monthly climatological data.

An NCDC online database accessible through the internet includes 15-minute, hourly, daily, and monthly precipitation data and daily and monthly data for other climatic variables measured at ground stations. The monthly and daily climatic data cover the period since typically 1948 at more than 19,000 stations maintained historically, with over 8,000 of the stations being active in 2001. Hourly precipitation data cover the period since typically 1948 at more than 6,000 stations historically, with over 2,800 active in 2001. Fifteen-minute precipitation data cover the period since typically 1971 at more than 3,400 stations historically, with over 2,400 active in 2001. These stations are primarily in the U.S. The NCDC also has online databases for many other stations located outside the U.S.

Various other entities collect and distribute climatic data to meet specific regional or local needs. For example, in addition to snow data available through the NCDC, the Natural Resource Conservation Service (NRCS) of the U.S. Department of Agriculture administers a cooperative federal–state–private system for conducting snow surveys in the western states. From January through May of each year, the NRCS publishes a monthly report entitled *Water Supply Outlook* that provides snowpack and streamflow forecast data for each state and region. The NRCS also publishes snowpack and water content data in the monthly *Basin Outlook Reports.* Snow data are also collected by various other entities. The California Department of Water Resources maintains a snow survey system. In the eastern United States, several federal, state, and private agencies perform snow surveys.

2.11.2 Water Quantity and Quality

The Water Resources Division of the U.S. Geological Survey (USGS) has primary federal responsibility for collection and dissemination of measurements of stream discharge and stage, reservoir and lake stage and storage content, groundwater levels, well and spring discharge, and the quality of surface and groundwater. The USGS works closely with other federal, state, and local agencies in its cooperative data collection programs. Hydrologic data collected by the USGS at stations nationwide are available to other agencies and the public in computer-readable form or as tables or graphs, statistical analyses, and digital plots. The data are published by water year for each state in a publication series entitled *Water-Data Reports.* Much of the data are available online and on CD-ROM.

The USGS began operating its first stream-gaging station in 1889. The majority of the more than 7,000 currently active stream-gaging stations operated by the USGS are funded through cooperative federal-state programs, with participants that include more than 600 federal, state, and local agencies (Wahl et al., 1995). The National Water Information System, maintained by the USGS, provides internet access to historical records collected by the USGS and others of daily streamflow and peak flows for almost 20,000 stations. Many stations were operated during various periods over the past century but are not currently active. The hydrographs reproduced as Figs. 2.14 and 2.15 were downloaded through the internet from the National Water Information System (http://www.usgs.gov).

The various water quality data collection programs of the USGS include the Hydrologic Benchmark Network and National Stream Quality Accounting Network (Alexander, Slack, Ludtke, Fitzgerald, and Schertz, 1998). Measurements from these two nationwide water quality sampling station networks include data for 122 physical, chemical, and biological properties of water collected at 680 locations beginning in 1962. The data are distributed via CD-ROM and the internet.

2.11.3 Automated Data Acquisition Systems

Historically, hydrologic data collection systems have relied largely on manual observations. Even with more advanced systems, manual observations serve as backup and are necessary for observing data that cannot feasibly be collected by automated means. However, advances in electronics and computer technology have enabled development of automated data observation, transmission, and storage systems. Automated data collection systems include precipitation, streamflow, and various other types of measurements. Many remote gaging stations may be linked to a central office through radio or telephone. Real-time hydrologic data from self-reporting gages are useful in a variety of applications such as flood forecasting and operating reservoir systems.

Self-reporting gaging stations include a data collection platform with equipment for transmitting the data to a central station for storage and processing. A variety of technologies are available for transmitting data, including: ground-based VHF radio; environmental or general purpose communication satellites; meteor-burst-based communication systems; land line equipment utilizing hard wire or switched commercial telephone circuits; and microwave communication systems (U.S. Army Corps of Engineers, 1987; Latkovich and Leavesley, 1993). The choice of one technology over another depends on several factors, including the minimum allowable time between data collection and reporting, the spatial scale of the application, and the available funds.

The USGS collects real-time streamflow data available online for 3,000 stations throughout the U.S. The Corps of Engineers, with the support of the USGS and NWS, uses automated data acquisition systems in operating its multipurpose reservoir systems. The USGS and Corps of Engineers are the largest users of Geostationary Operational Environmental Satellites (GOES), which have been widely used to transmit streamflow, precipitation, and other data in support of real-time reservoir system operations and other water management activities. GOES satellites are operated by the National Earth Satellite Data and Information Service of the National Oceanic and Atmospheric Administration. The major components of a GOES-based data collection system are the field gaging sites, the satellite, and direct readout groundstations located at central receiving sites for data retrieval. The equipment located at the gaging station site necessary for the use of a GOES-based data collection system is referred to as a data collection platform. Different types of equipment (1) send data at any time upon request, (2) periodically report at prespecified times, or (3) report whenever specified thresholds are exceeded.

PROBLEMS

2.1. Lake Mead impounded by Hoover Dam is the largest reservoir in the U.S., with a water supply storage capacity of 21.4 billion m^3. Ignoring evaporation and other losses, how many years would be required for a city of 1.0 million people to use 21.4 billion m^3 of water at a per capita water use rate of 100 gal/day?

2.2. Assume that 1.0 m of water is applied to crop land during an irrigation season. How much land would the 21.4 billion m^3 of Problem 2.1 cover, in units of km^2, hectares, mi^2, and acres?

2.3. At the top of the conservation pool, Lake Mead has a water surface area of 65,800 hectares. In a certain month, the reservoir is full, and the pan evaporation rate is 25 cm. Using a pan coefficient of 0.7, estimate the evaporation volume for the month in m^3, ft^3, and ac-ft.

2.4. The mean annual rainfall depth over a 280 km^2 watershed is 725 mm. What is the mean annual volume of rain falling on the watershed in m^3, ft^3, gallons, and ac-ft?

2.5. A streamflow volume of 8,250 second-foot-day (sfd) passed a stream gage during a period of 30 days. Convert this volume to ac-ft. What is the mean flow rate in ft^3/s and m^3/s?

2.6. A city with a population of 950,000 people has a mean per capita total municipal water use requirement of 175 gal/day averaged over the year. How much water is used in a year, in units of gallons, ac-ft, second-foot-day, liters, and m^3?

2.7. A certain crop has a consumptive use requirement of 1.2 ac-ft of water per acre of crop land. The crop is planted on 780 acres of land. Precipitation falling during the growing season is 3.2 inches. The remainder of the 1.2 ac-ft/ac requirement is met by irrigation. Determine the volume of irrigation water needed in ac-ft, ft^3, and m^3.

2.8. The mean annual precipitation for a certain 132 mi^2 watershed is 25 inches. Assume that 20 percent of the annual precipitation reaches the watershed outlet as streamflow. Determine the mean streamflow rate in ac-ft/yr, sfd/yr, ft^3/s, and m^3/s.

2.9. Use the internet to find and print the historical daily flow hydrograph for USGS gaging station 09380000 on the Colorado River at Lees Ferry, Arizona. What is the drainage area of the watershed above this gage?

2.10. Use the internet to find and print the historical daily flow hydrographs for USGS gaging station 01636500 on the Shenandoah River at Millville, West Virginia, and station 11446500 on the American River at Fair Oaks, California. What are the watershed areas above these gages?

2.11. Use the internet to find and print the historical daily flow hydrograph for a USGS gaging station on a river in your state.

2.12. Compute the discharge in the stream based on the current meter measurements provided below for different points in the cross-section.

Distance, ft	0	3	6	9	12	15	18	21	24	26
Depth, ft	0.0	1.2	3.7	6.5	8.6	7.6	6.4	4.8	2.3	0.0
Velocity, ft/s										
at 0.2 depth	—	—	1.11	1.43	1.63	1.36	1.24	1.18	—	—
at 0.6 depth	—	0.72	—	—	—	—	—	—	0.84	—
at 0.8 depth	—	—	0.91	1.16	1.24	1.10	1.03	0.87	—	—

2.13. Compute the discharge in the stream based on the current meter measurements provided below for different points in the cross-section.

Distance, m	0	2	4	6	8	10	12	14	16	18
Depth, m	0.0	0.95	1.98	2.64	3.14	3.55	2.68	1.53	0.49	0.0
Velocity, m/s										
at 0.2 depth	—	0.36	0.49	0.51	0.56	0.67	0.59	0.46	0.32	—
at 0.8 depth	—	0.33	0.36	0.43	0.45	0.51	0.48	0.38	0.28	—

2.14. A large river basin, with a watershed area of 125,000 km^2 is divided into six subbasins with drainage areas of (a) 5,200 km^2, (b) 18,600 km^2, (c) 32,400 km^2, (d) 20,800 km^2, (e) 37,200 km^2, and (f) 10,800 km^2. The average annual precipitation for each subbasin is (a) 981 mm, (b) 752 mm, (c) 678 mm, (d) 495 mm, (e) 520 mm, and (f) 504 mm. What is the average annual precipitation for the overall 125,000 km^2 river basin?

2.15. A storm resulted in the following rainfall measurements for the precipitation gages of Example 2.1: gage A, 11 mm; gage B, 14 mm; gage C, 26 mm; gage D, 42 mm; and gage E, 36 mm. Estimate the average rainfall depth for the watershed for this storm.

2.16. The map following Problem 2.17 shows the location of 8 rain gages and the boundary of a watershed. The rainfall depths for a certain storm are as follows: gage A, 5.6 inches; gage B, 4.2 inches; gage C, 3.9 inches; gage D, 2.5 inches; gage E, 1.8 inches; gage F, 0.9 inches; gage G, 2.4 inches; and gage H, 0.3 inches. Use the Thiessen polygon method to determine the mean rainfall depth over the watershed for this storm event.

2.17. The following map shows the location of 8 rain gages and the boundary of a watershed. The rainfall depths for a certain storm are as follows: gage A, 25 mm; gage B, 18 mm; gage C, 92 mm; gage D, 95 mm; gage E, 192 mm; gage F, 175 mm; gage G, 152 mm; and gage H, 168 mm. Use the Thiessen polygon method to determine the mean rainfall depth over the watershed for this storm event.

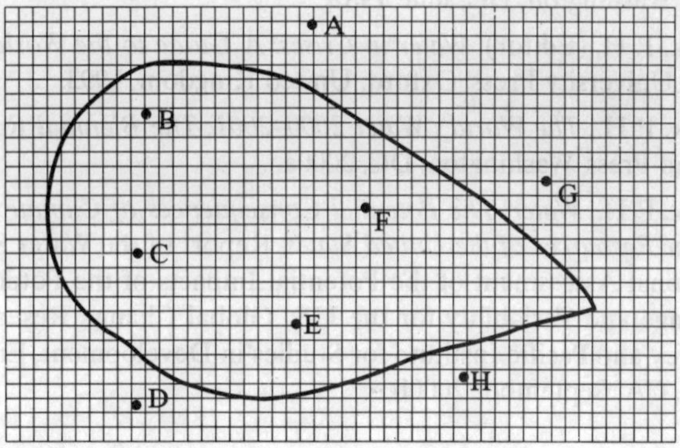

REFERENCES

AHRENS, C. D., *Meteorology Today: An Introduction to Weather, Climate, and the Environment*, 6th Ed., Brooks and Cole Publishers, Pacific Grove, CA, 2000.

ALEXANDER, R. B., J. R. SLACK, A. S. LUDTKE, K. K. FITZGERALD, and T. L. SCHERTZ, "Data from Selected U.S. Geological Survey National Stream Water Quality Monitoring Networks," *Journal of Water Resources Research,* American Geophysical Union, Vol. 34, No. 9, September 1998.

ALLEN, R. G., L. S. PEREIRA, D. RAES, and M. SMITH, *Crop Evapotranspiration, Guidelines for Computing Crop Water Requirements,* FAO Irrigation and Drainage Paper No. 56, Food and Agricultural Organization of the United Nations, Rome, Italy, 1998.

American Public Health Association, American Water Works Association, and Water Environment Federation, *Standard Methods for Examination of Water and Wastewater,* 20th Ed., American Public Health Association, New York, NY, 2001.

American Society of Civil Engineers, *Hydrology Handbook,* 2nd Ed., ASCE Manuals and Reports on Engineering Practice, No. 28, New York, NY, 1996.

BEDIENT, P. B., and W. C. HUBER, *Hydrology and Floodplain Analysis,* 2nd Ed., Addison.-Wesley, Reading, MA, 1992.

BIXEL, P. A., and E. H. TURNER, *Galveston and the 1900 Storm: Catastrophe and Catalyst,* University of Texas Press, Austin, TX, 2000.

BRAS, R. L., *Hydrology, An Introduction to Hydrologic Science,* Addison-Wesley, Reading, MA, 1990.

CHANGNON, S. A. (Editor), *The Great Flood of 1993: Causes, Impacts, and Responses,* Westview Press, Boulder, CO, 1996.

CHOW, V. T., D. R. MAIDMENT, and L. W. MAYS, *Applied Hydrology,* McGraw-Hill, New York, NY, 1988.

DINGMAN, S. L., *Physical Hydrology*, 2nd Ed., Prentice Hall, Upper Saddle River, NJ, 2002.

FARNSWORTH, R. K., and E. S. THOMPSON, *Mean Monthly, Seasonal, and Annual Pan Evaporation for the United States,* Technical Report NWS 34, National Oceanic and Atmospheric Administration, Washington, DC, December 1982.

FARNSWORTH, R. K., E. S. THOMPSON, and E. L. PECK, *Evaporation Atlas for the Contiguous United States*, Technical Report NWS 33, National Oceanic and Atmospheric Administration, Washington, DC, June 1982.

GLEICK, P. H. (Editor), *Water in Crises: A Guide to the World's Fresh Water Resources,* Oxford University Press, Oxford, United Kingdom, 1993.

GLEICK, P. H., *The World's Water 2000–2001*, *The Biennial Report on Freshwater Resources,* Island Press, Washington, DC, 2000a.

GLEICK, P. H., *Water: The Potential Consequences of Climate Variability and Change for the Water Resources of the United States,* Report of the Water Sector Assessment Team of the National Assessment of the Potential Impacts of the Potential Consequences of Climate Change and Variability for the U.S. Global Change Research Program, U.S. Geological Survey and Pacific Institute for Studies in Development, Environment, and Security, Oakland, CA, September 2000b.

HERSCHY, R. W., and R. W. FAIRBRIDGE (Editors), *Encyclopedia of Hydrology and Water Resources*, Kluwer Academic Publishers, Boston, MA, 1998.

HIDORE, J. J., and J. E. OLIVER, *Climatology: An Atmospheric Science,* Macmillan, New York, NY, 1993.

ISERI, K. T., and W. B. LANGBEIN, *Large Rivers in the United States,* Circular 686, U.S. Geological Survey, Reston, VA, 1974.

JENSEN, M. E., R. D. BURMAN, and R. G. ALLEN, *Evapotranspiration and Irrigation Water Requirements,* Manuals and Reports on Engineering Practice No. 70, American Society of Civil Engineers, New York, NY, 1990.

LATKOVICH, V. J., and G. H. LEAVESLEY, "Chapter 25 Automatic Data Acquisition and Transmission," *Handbook of Hydrology* (D. R. Maidment, Ed.), McGraw-Hill, New York, NY, 1993.

LETTENMAIER, D. P., G. MCCABE, and E. Z. STAKHIV, "Chapter 29 Global Climate Change: Effect on Hydrologic Cycle," *Water Resources Handbook* (L. W. Mays, Ed.), McGraw-Hill, New York, NY, 1996.

LINSLEY, R. K., Jr., M. A. KOHLER, and J. L. H. PAULHUS, *Hydrology for Engineers*, McGraw-Hill, New York, NY, 1982.

MAIDMENT, D. R. (Editor), *Handbook of Hydrology,* McGraw-Hill, New York, NY, 1993.

MALINA, J. F., "Chapter 8 Water Quality," *Water Resources Handbook* (L. W. Mays, Ed.), McGraw-Hill, New York, NY, 1996.

MAYS, L. W. (Editor), *Water Resources Handbook,* McGraw-Hill, New York, NY, 1996.

MCCUEN, R. H., *Hydrologic Analysis and Design,* 2nd Ed., Prentice Hall, Upper Saddle River, NJ, 1998.

MCCUTCHEON, S. C., J. L. MARTIN, and T. O. BARNWELL, "Chapter 11 Water Quality," *Handbook of Hydrology* (D. R. Maidment, Ed.), McGraw-Hill, New York, NY, 1993.

MIMIKOU, M. A., "Climatic Change," *Environmental Hydrology* (V. P. Singh, Ed.), Kluwer Academic Publishers, Boston, MA, 1995.

MONTEITH, J. L., "Evaporation and the Environment," *The State and Movement of Water in Living Organisms, XIX Symposium of Society for Experimental Biology,* Cambridge University Press, Cambridge, United Kingdom, 1965.

National Assessment Synthesis Team, U.S. Global Change Research Program, *Climate Change Impacts on the United States,* Cambridge University Press, Cambridge, United Kingdom, 2000.

PENMAN, H. L., "Natural Evaporation from Open Water, Bare Soil, and Grass," *Proceedings of Royal Society of London,* London, Great Britain, 1948.

PERRY, C. A., and L. J. COMBS (Editors), *Summary of Floods in the United States, January 1992 Through September 1993,* U.S. Geological Survey, Water Supply Paper 2499, Denver, CO, 1998.

PONCE, V. M., *Engineering Hydrology, Principles and Practices,* Prentice Hall, Inc., Upper Saddle River, NJ, 1989.

SHIKLOMANOV, I. A., "World Freshwater Resources," *Water in Crises: A Guide to the World's Fresh Water Resources* (P. H. Gleick, Ed.), Oxford University Press, Oxford, United Kingdom, 1993.

SHIKLOMANOV, I. A., "Appraisal and Assessment of World Water Resources," *Water International, Journal of the International Water Resources Association,* Vol. 25, No. 1, 2000.

SINGH, V. P., *Elementary Hydrology,* Prentice Hall, Inc., Upper Saddle River, NJ, 1992.

TCHOBANOGLOUS, G., and E. D. SCHROEDER, *Water Quality,* Addison-Wesley, Reading, MA, 1985.

TEBBUTT, T. H. Y., *Principles of Water Quality Control,* 5th Ed., Butterworth-Heinemann, Oxford, England, 1998.

U.S. Army Corps of Engineers, *Management of Water Control Systems,* Engineering Manual 1110-2-3600, Washington, DC, November 1987.

VIESSMAN, W., and G. L. LEWIS, *Introduction to Hydrology,* 4th Ed., Harper-Collins, New York, NY, 1996.

WAHL, K. L., W. O. THOMAS, and R. M. HIRSCH, Stream-Gaging Program of the U.S. Geological Survey, Circular 1123, U.S. Geological Survey, Reston, VA, 1995.

WANIELISTA, M., R. KERSTEN, and R. EAGLIN, *Hydrology: Water Quantity and Quality Control,* 2nd Ed., John Wiley & Sons, New York, NY, 1997.

WARD, A. D., and W. J. ELLIOT, *Environmental Hydrology,* Lewis Publishers, New York, NY, 1995.

WILHITE, D. A., "Drought as a Natural Hazard," *Drought: A Global Assessment* (D. A. Wilhite, Ed.), Routledge, New York, NY, 2000.

WILLIAMS, J., *The Weather Book,* 2nd Ed., Random House, New York, NY, 1997.

3 Fluid Mechanics

The design or evaluation of hydraulic systems requires the application of fluid mechanics. Pressure conduit, open channel, and groundwater hydraulics, as well as various areas of hydrology are built on the principles of fluid mechanics. This chapter provides a brief review of several basic aspects of fluid mechanics prior to covering hydraulic engineering applications in subsequent chapters. Several of the many textbooks covering fluid mechanics in detail are referenced in the bibliography.

3.1 UNITS

Both the British Gravitational (BG) System and the International System (SI) of units are used in this text. In the SI system, the unit of mass (M) is the kilogram (kg), the unit of length (L) is the meter (m), the unit of time (t) is the second (s), and the unit of temperature (T) is the Kelvin (K). Based on Newton's second law of motion, the unit of force (F) is the Newton (N) and is defined as

$$F = M \cdot a \tag{3.1}$$

$$1\ \text{N} = 1\ \text{kg} \cdot 1\ \text{m/s}^2$$

where a 1 N force acting on a 1 kg mass will accelerate the mass at 1 m/s^2. The standard gravity (g) in SI units on earth is 9.81 m/s^2, and a 1 kg mass weighs 9.81 N.

In the BG (English) system, the unit of force (F) is the pound (lb), the unit of length (L) is the foot (ft), the unit of time (t) is the second (s), and the unit of temperature (T) is the Rankine (R). The unit of mass (M) is the slug and is defined as

$$1 \text{ lb} = 1 \text{ slug} \cdot 1 \text{ ft/s}^2$$

where a 1 lb force acting on a mass of 1 slug will give the mass an acceleration of 1 ft/s^2. With the standard gravitational acceleration (g) of 32.2 ft/s^2, a mass of 1 slug weighs 32.2 lbs.

The basic system of dimensions is MLtT in the SI system and FLtT in the BG system. The Celsius (°C) scale is commonly used for the temperature scale in the SI system, while the Fahrenheit (°F) scale is commonly used for the temperature scale in the BG system.

Work (W) is the application of a force through a distance (L). In SI units, joule (J) is the work unit and is defined as

$$W = F \cdot L \tag{3.2}$$

$$1 \text{ J} = 1 \text{ N} \cdot 1 \text{ m}$$

where 1 J of work is done when a 1 N force is displaced 1 m. The watt (w) is the unit of power and is defined as 1 J/s.

In the BG system of units, the ft-lb is the work unit, and horsepower (hp) is the power unit, where 1 hp = 550 ft-lbs/s. Conversions between SI units and BG units are shown in Table 3.1.

TABLE 3.1 UNIT CONVERSIONS

Length (L)	1 m = 3.281 ft	1 ft = 0.3048 m
Mass (M)	1 kg = 0.0685 slugs	1 slug = 14.59 kg
Force (F)	1 N = 0.2248 lbs	1 lb = 4.448 N
Work (W)	1 J = 0.7376 ft-lbs	1 ft-lb = 1.356 J
Power (P)	1 kw = 1.341 hp	1 hp = 0.7457 kw

3.2 PROPERTIES OF WATER

Fluids include all liquids and gases. We are primarily concerned with water in this book. Physical properties of water at atmospheric pressure are tabulated in Tables 3.2A and 3.2B as a function of temperature.

3.2.1 Phases

Water can exist in three phases or forms: solid, liquid or gas. Snow and ice are solid forms, rain and streamflow are liquid forms, and water vapor in the air is a

TABLE 3.2A PHYSICAL PROPERTIES OF WATER AT ATMOSPHERIC PRESSURE IN SI UNITS

Temperature °C	Vapor pressure p_v kN/m², abs	density ρ kg/m³	Specific weight γ kN/m³	Viscosity $\mu \times 10^3$ N · s/m²	Kinematic viscosity $\nu \times 10^6$ m²/s	Surface tension* σ N/m	Bulk modulus of elasticity $E_b \times 10^{-6}$ kN/m²
0	0.61	999.8	9.805	1.781	1.785	0.0756	2.02
5	0.87	1000.0	9.807	1.518	1.519	0.0749	2.06
10	1.23	999.7	9.804	1.307	1.306	0.0742	2.10
15	1.70	999.1	9.798	1.139	1.139	0.0735	2.14
20	2.34	998.2	9.789	1.002	1.003	0.0728	2.18
25	3.17	997.0	9.777	0.890	0.893	0.0720	2.22
30	4.24	995.7	9.764	0.798	0.800	0.0712	2.25
40	7.38	992.2	9.730	0.653	0.658	0.0696	2.28
50	12.33	988.0	9.689	0.547	0.553	0.0679	2.29
60	19.92	983.2	9.642	0.466	0.474	0.0662	2.28
70	31.16	977.8	9.589	0.404	0.413	0.0644	2.25
80	47.34	971.8	9.530	0.354	0.364	0.0626	2.20
90	70.10	965.3	9.466	0.315	0.326	0.0608	2.14
100	101.33	958.4	9.399	0.282	0.294	0.0589	2.07

TABLE 3.2B PHYSICAL PROPERTIES OF WATER AT ATMOSPHERIC PRESSURE IN BG UNITS

Temperature °F	Vapor pressure p_v psia	Density ρ slugs/ft³	Specific weight γ lb/ft³	Viscosity $\mu \times 10^5$ lb-s/ft²	Kinematic viscosity $\nu \times 10^5$ ft²/s	Surface tension* $\sigma \times 10^3$ lb/ft	Bulk modulus of elasticity $E_b \times 10^{-3}$ psi
32	0.09	1.940	62.42	3.746	1.931	5.18	293
40	0.12	1.940	62.43	3.229	1.664	5.13	297
50	0.18	1.940	62.41	2.735	1.410	5.09	305
60	0.26	1.938	62.37	2.359	1.217	5.03	311
70	0.36	1.936	62.30	2.050	1.059	4.97	320
80	0.51	1.934	62.22	1.799	0.930	4.91	322
90	0.70	1.931	62.11	1.595	0.826	4.86	323
100	0.95	1.927	62.00	1.424	0.739	4.79	327
120	1.69	1.918	61.71	1.168	0.609	4.67	333
140	2.89	1.908	61.38	0.981	0.514	4.53	330
160	4.74	1.896	61.00	0.838	0.442	4.40	326
180	7.51	1.883	60.58	0.726	0.385	4.26	318
200	11.52	1.868	60.12	0.637	0.341	4.12	308
212	14.70	1.860	59.83	0.593	0.319	4.04	300

*Surface tension of water in contact with air.

gaseous form. The partial pressure exerted by water vapor in the atmosphere is called the vapor pressure (P_v) and is a function of temperature (Table 3.2).

Cavitation occurs in a hydraulic system when the pressure drops below the vapor pressure and the water vaporizes. In a pipeline, cavitation can cause column separation of the water and damage to the pipeline. At the crest of a spillway, the sharp curvature of the streamlines can cause cavitation and damage to the spillway. Low pressure at the intake of a pump can result in cavitation and damage to the impeller.

3.2.2 Mass and Weight

The density (ρ) of water is defined as the mass per unit volume and has units of kg/m^3. Water has maximum density at 4°C and becomes less dense at higher or lower temperature (Table 3.2). This causes a turnover of water in lakes and reservoirs when the surface water (wind mixed layer above the thermocline) cools in the fall and becomes more dense than the subsurface water.

The specific weight (γ) is the weight per unit volume (N/m^3 or lb/ft^3). Specific weight and density are related as follows, where g is the gravitational acceleration constant.

$$\gamma = \rho g \tag{3.3}$$

The specific weight of water ranges from 9.81 kN/m^3 at 4°C to 9.40 kN/m^3 at 100°C. The specific weight of fresh water in BG units is commonly used as 62.4 lbs/ft^3.

3.2.3 Viscosity

Water is a Newtonian fluid and there is a linear relation between shear stress (τ) and velocity gradient (du/dy) or

$$\tau = \mu \frac{du}{dy} \tag{3.4}$$

where the constant of proportionally (μ) is the viscosity. Viscosity has units of N · s/m^2 and is dependent on temperature (Tables 3.2A, B). Water at 20°C has a viscosity of 1.00×10^{-3} N · s/m^2 (2.09×10^{-5} lb-s/ft^2).

The term μ/ρ often occurs in hydraulic engineering computations. This ratio is called the kinematic viscosity (ν) or

$$\nu = \frac{\mu}{\rho} \tag{3.5}$$

The kinematic viscosity of water at 20°C is 1.00×10^{-6} m^2/s (1.08×10^{-5} ft^2/s).

3.2.4 Surface Tension

The surface tension force is generally small compared with the other forces in a hydraulic system. However, in some groundwater studies, the capillary fringe above the water table must be considered. The height of the capillary fringe depends on the pore size of the aquifer.

If a small glass tube is inserted into water, water will rise in the tube. The height of the capillary rise (h) is determined by the balance of the force caused by surface tension and the weight of the water in the tube above the surface or

$$h = \frac{4\,\sigma\cos\theta}{\gamma D} \tag{3.6}$$

where σ is the surface tension, θ is the angle that the water film meets the solid surface, and D is the diameter of the tube. The surface tension for water at 20°C is 7.28×10^{-2} N/m (4.99×10^{-3} lb/ft). The height of the capillary rise is inversely proportional to the pore size and the capillary fringe is more pronounced for fine-grain aquifers.

3.2.5 Elasticity

In most hydraulic engineering applications, water is considered incompressible. However, for groundwater and unsteady flow in pipelines, the compressibility of water must be considered. The bulk modulus of elasticity (E_b) is defined as

$$E_b = -\frac{dP}{d\forall/\forall} \tag{3.7}$$

where $\forall$ is the volume and $d\forall$ is the change in volume caused by a change in pressure dP.

Since the mass ($M = \rho\forall$) is constant, the bulk modulus of elasticity can also be expressed as

$$E_b = \frac{dP}{d\rho/\rho} \tag{3.8}$$

The modulus of elasticity of water at 20°C is 2.18×10^6 kN/m² (3.18×10^5 lb/in²), approximately 100 times more compressible than steel.

3.3 STATICS

The free surface of water in a reservoir is subject to atmospheric pressure. Pressure measured relative to atmospheric pressure is called gage pressure. Pressure is a force per unit area, and 1 N/m² is called a Pascal (Pa). In the BG system of units, pressure is commonly measured in lb/in² or psi. Absolute pressure is gage pressure plus atmospheric pressure.

3.3.1 Pressure Head

With uniform atmospheric pressure and no wind stress, the surface of still water will be horizontal. The pressure at all points on a horizontal plane below the surface will be the same. The pressure at any point is independent of direction. As shown in Fig. 3.1, the pressure difference (ΔP) between any two points is the

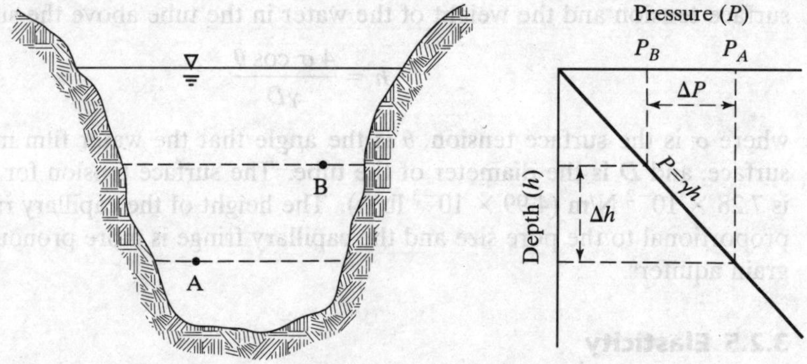

Figure 3.1 Hydrostatic pressure distribution

specific weight of water times the difference in water depth (Δh) between the two points.

$$P_A - P_B = \gamma \Delta h \tag{3.9}$$

or

$$\frac{dP}{dz} = -\gamma \tag{3.10}$$

where z represents the elevation and is positive upward.

Pressure (P) may be expressed in the height of a column of fluid (h) by the relation

$$h = \frac{P}{\gamma} \tag{3.11}$$

If P is expressed in kN/m^2 (lbs/ft^2) and γ is expressed in kN/m^3 (lbs/ft^3), then h is in m (ft). h is commonly referred to as the pressure head.

3.3.2 Hydrostatic Forces

Pressure forces are developed on submerged surfaces of hydraulic structures. If the water is at rest, the force is perpendicular to the surface.

Figure 3.2 represents the hydrostatic force on a submerged plane surface inclined at an angle θ. The hydrostatic force (F) is the product of the area of the plane surface (A) and the pressure at the centroid of the area ($h_c\gamma$)

$$F = \gamma h_c A \tag{3.12}$$

The hydrostatic force acts at the center of pressure, which is located below the centroid a distance Δy.

$$\Delta y = \frac{I_o}{A y_c} \tag{3.13}$$

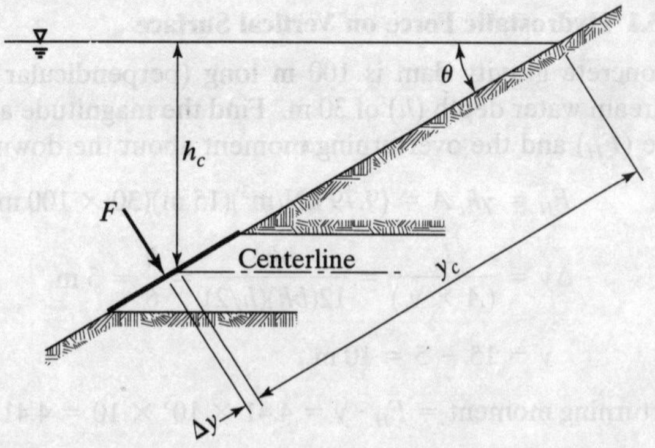

Figure 3.2 Hydrostatic force on an inclined plane surface

where I_o is the moment of inertia about the horizontal axis through the centroid of the area. For a rectangular gate with a height (h) and width (b), $I_o = bh^3/12$.

When finding the hydrostatic force on a plane or curved surface, it is sometimes easier to compute the horizontal (F_H) and vertical (F_v) components of the force. In Fig. 3.3, A–B represents a gate having a curved surface with a unit length normal to the plane of the figure. The horizontal component (F_H) of the hydrostatic force is equal to the hydrostatic force on the vertical projection ($A'B$) of the curved surface. The magnitude of F_H can be computed from Eq. 3.12, and the location of F_H can be determined from Eq. 3.13.

The vertical component (F_v) of the hydrostatic force is equal to the weight of the water above the gate or

$$F_v = W_1 + W_2 \tag{3.14}$$

where W_1 is the weight of the water above $A'A$ (Fig. 3.3), and W_2 is the weight of the water below $A'A$. F_V is located at the center of gravity of the water column.

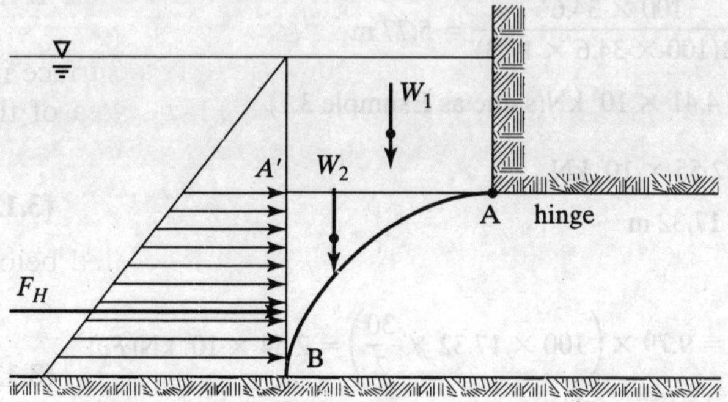

Figure 3.3 Hydrostatic forces on a curved surface

Example 3.1 Hydrostatic Force on Vertical Surface

A concrete gravity dam is 100 m long (perpendicular to plane of sketch) with an upstream water depth (h) of 30 m. Find the magnitude and location of the hydrostatic force (F_H) and the overturning moment about the downstream toe of the dam (0):

$$F_H = \gamma h_c A = (9.79\ \text{kN/m}^3)(15\ \text{m})(30 \times 100\ \text{m}^2) = 4.41 \times 10^5\ \text{kN}$$

$$\Delta y = \frac{I_o}{(A \times y_c)} = \frac{bh^3}{12(bh)(h/2)} = \frac{h}{6} = 5\ \text{m}$$

$$y = 15 - 5 = 10\ \text{m}$$

$$\text{overturning moment} = F_H \cdot y = 4.41 \times 10^5 \times 10 = 4.41 \times 10^6\ \text{kN} - \text{m}$$

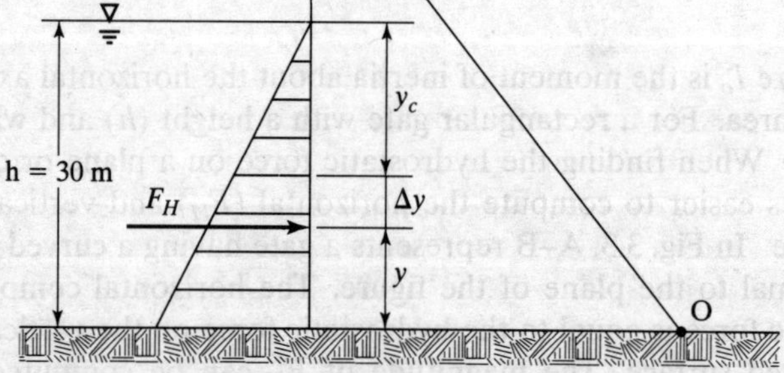

Example 3.2 Hydrostatic Force on Sloping Surface

The front face of the dam in Example 3.1 has been modified and is at an angle of 60° with the horizontal. Find the location and magnitude of the hydrostatic force (F) and the horizontal (F_H) and vertical (F_V) components.

$$F = \gamma h_c A = (9.79\ \text{kN/m}^3) \times (15\ \text{m}) \times \left(\frac{30\ \text{m}}{\sin 60°} \times 100\ \text{m}\right) = 5.09 \times 10^5\ \text{kN}$$

$$y_c = \frac{30}{(2 \times \sin 60°)} = 17.32\ \text{m}$$

$$\Delta y = \frac{I_o}{(A \times y_c)} = \frac{100 \times 34.6^3}{12(100 \times 34.6 \times 17.3)} = 5.77\ \text{m}$$

$$F_H = F \times \cos 30° = 4.41 \times 10^5\ \text{kN(same as Example 3.1)}$$

$$F_V = F \times \sin 30° = 2.55 \times 10^5\ \text{kN}$$

$$X = h \times \tan 30° = 17.32\ \text{m}$$

Weight of water ABC

$$= \gamma \times \text{volume} = 9.79 \times \left(100 \times 17.32 \times \frac{30}{2}\right) = 2.54 \times 10^5\ \text{kN}(F_V)$$

If the uplift pressure in the foundation of the dam varies linearly from γh at C to zero the downstream toe (0), compute the overturning moment caused by the horizontal component of the hydrostatic force on the front face of the dam and the uplift pressure.

The uplift force (F_u) acts at the centroid of the pressure triangle and is equal to

$$F_u = \gamma \frac{bh}{2} L = 9.79 \times 90 \times 30 \times \frac{100}{2} = 1.32 \times 10^6\,\text{kN}$$

overturning moment

$$= F_H \times \frac{h}{3} + F_u \times b \times \frac{2}{3} = 4.41 \times 10^5 \times \frac{30}{3} + 1.32 \times 10^6 \times 90 \times \frac{2}{3}$$

$$= 4.41 \times 10^6 + 7.92 \times 10^7 = 8.36 \times 10^7\ \text{kN} \cdot \text{m}$$

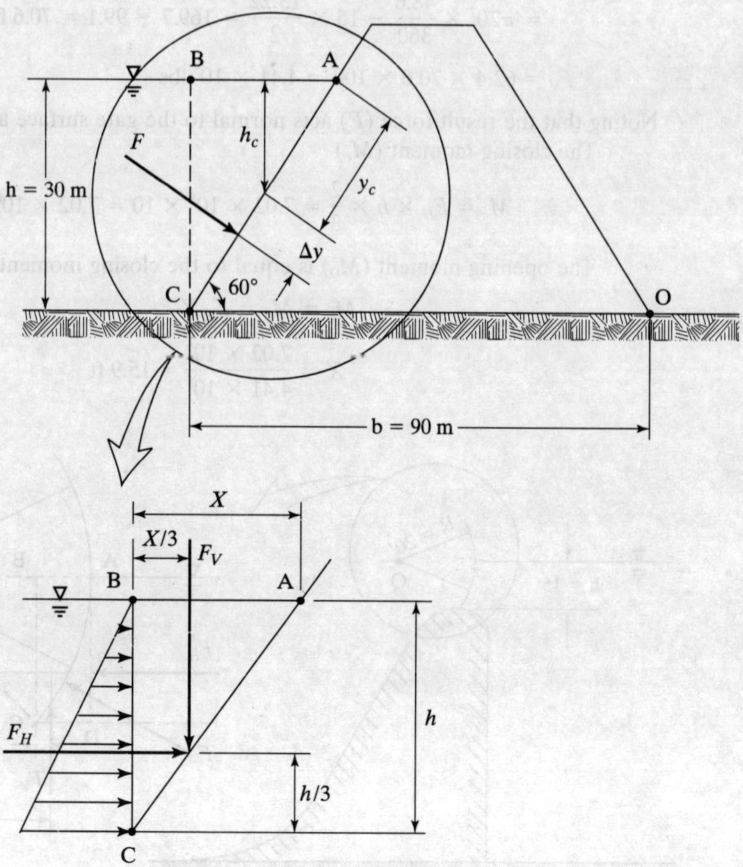

Example 3.3 Hydrostatic Force on Curved Surface

The spillway discharge for a dam is controlled with a 50-ft-long radial gate. The radial gate pivots about point "0" and consists of a segment of a cylinder with a radius of 20 ft and a length of 50 ft. Determine the closing moment on the gate caused by the horizontal hydrostatic force (F_H) and the location (X) of the vertical force (F_V).

$$F_H = \gamma h_c A = 62.4 \times \left(\frac{15}{2}\right) \times (15 \times 100) = 7.02 \times 10^5 \text{ lbs}$$

$$F_V = \text{weight of volume water displaced} = \gamma(\text{Area ABC}) \times 100$$

Area of ABC

$$\measuredangle BOC = \sin^{-1}\left(\frac{15}{20}\right) = 48.6°$$

$$BO = 20 \cos 48.6° = 13.22 \text{ ft}$$

Area ABC = Area AOC − Area BOC

$$= \pi 20^2 \times \frac{48.6}{360} - 15 \times \frac{13.22}{2} = 169.7 - 99.1 = 70.6 \text{ ft}^2$$

$$F_V = 62.4 \times 70.6 \times 100 = 4.41 \times 10^5 \text{ lbs}$$

Noting that the result force (F) acts normal to the gate surface and passes through 0. The closing moment (M_c)

$$M_c = F_H \times h \times \frac{2}{3} = 7.02 \times 10^5 \times 10 = 7.02 \times 10^6 \text{ ft-lbs}$$

The opening moment (M_0) is equal to the closing moment

$$M_0 = M_c = F_V X$$

$$X = \frac{7.02 \times 10^6}{4.41 \times 10^5} = 15.9 \text{ ft}$$

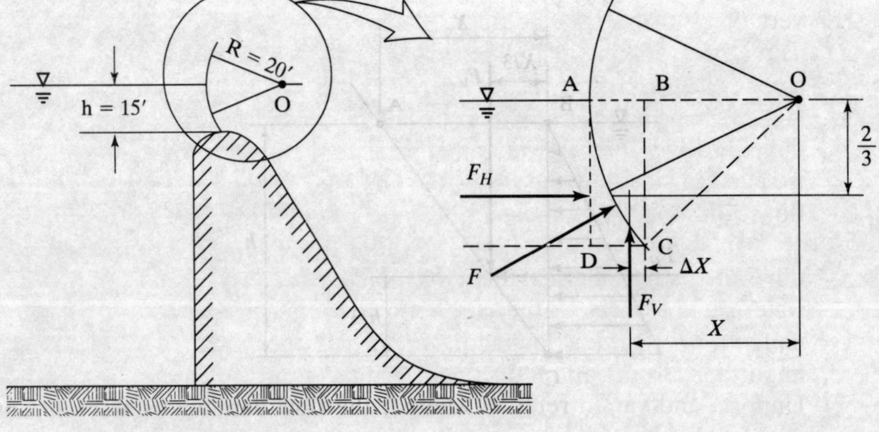

3.4 CLASSIFICATION OF FLUID FLOW

In fluid dynamics, fluids are referred to as *real* or *ideal* (*viscid* or *inviscid*). *Ideal* is an assumption of a hypothetical frictionless condition with zero viscosity that simplifies certain analyses. Real fluid flow implies frictional (viscous) effects. Hydraulic engineering is concerned with real fluid flow.

Incompressible means constant density. Gases are compressible, with their densities being a function of pressure and temperature. Liquids are relatively incompressible and are generally treated as being wholly incompressible. As noted in Section 3.2.5, although liquid water is slightly compressible, it is assumed to be incompressible in most, though not all, applications.

Flow may be either *laminar* or *turbulent*. In laminar flow, the fluid appears to move by the sliding of laminations of infinitesimal thickness over adjacent layers, with relative motion of fluid particles occurring at a molecular scale. The particles move in definite paths or streamlines. Viscosity plays a significant role. Conversely, turbulent flow is characterized by random instabilities in the flow field. The particles move in erratic, irregular motions. Sheet flow across a smooth pavement may be laminar. A laminar boundary layer occurs in pipe flow. However, hydraulic engineering usually deals with turbulent flow in pipes and open channels. Groundwater flow is nearly always laminar. Laminar and turbulent flows are characterized by relatively low and high, respectively, Reynolds numbers (Section 3.6).

Flow is actually three-dimensional. However, one- or two-dimensional analyses are sufficient for most applications. One-dimensional analyses are based on assuming the flow occurs along a central pathway. The average velocity ($V = Q/A$) through a cross-section is considered to be representative of the aggregate flow. In two-dimensional flow, all streamlines are identical in a series of parallel planes. For example, for many applications, flow in a river is considered one-dimensional in the longitudinal direction of the channel. However, two-dimensional (horizontally across the river as well as longitudinally downstream) flow effects may be important as the flow contracts and expands through a culvert or as rising flood waters flow from the channel into adjacent floodplains. Three-dimensional, including vertical, effects may be important near the outlet of a conduit discharging into the river.

Steady and *unsteady* refer to whether changes occur over time. In steady flow, discharge and other flow parameters at a location are constant over time. The flow is called unsteady if the discharge changes with time. Both steady and unsteady flow conditions are important in water resources engineering.

In pressure conduit flow, also called pipe flow, the flow occurs under pressures other than atmospheric (Chapter 4). In gravity or open channel flow, there is a free water (liquid) surface at atmospheric pressure (Chapters 5 and 6). *Uniform* versus *nonuniform (varied)* flow and *subcritical, critical,* and *supercritical* conditions are important classifications in open channel flow and are discussed in Chapter 5. Uniform and varied refer to whether the flow parameters are constant along the channel. Subcritical (Fr < 1), critical (Fr = 1), and supercritical (Fr > 1) flows are differentiated by their Froude numbers (Section 3.6).

3.5 REYNOLDS TRANSPORT THEOREM

The Reynolds transport theorem for a fixed, nondeforming control volume is used in later chapters to develop the basic equations for pipe, open channel, and groundwater flow. An example of a control volume for a flood wave in an open channel is shown in Fig. 3.4. A control volume in a pipe is shown in Fig. 3.5. This theorem can be expressed as

$$\frac{DB_{sys}}{Dt} = \frac{\partial}{\partial t}\int_{cv} \rho b d\forall + \int_{cs} \rho b \mathbf{V} \cdot \hat{n} dA \tag{3.15}$$

where B represents a parameter of the system such as mass, momentum, or energy; b represents the amount of B per unit mass; ρ is the density of the fluid; $\forall$ is the volume; cv represents the control volume; cs represents the control surface; $\mathbf{V}$ is the velocity vector; $\hat{n}$ is the unit vector of area with outward positive; and A is the area. The first term on the right of the equal sign in Eq. 3.15 represents the time rate of change of the parameter within the control volume and is equal to zero for steady-

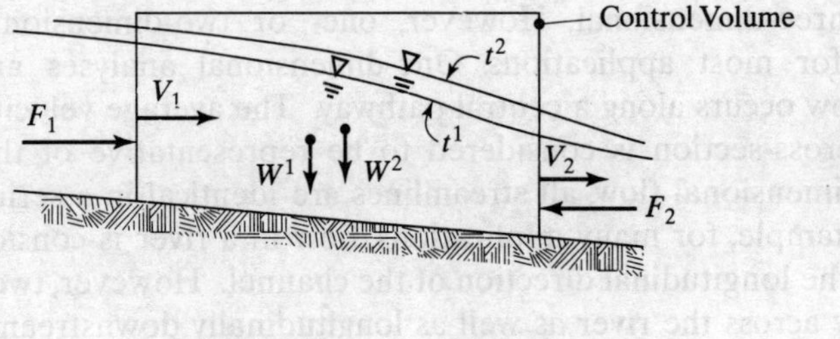

Figure 3.4 Control volume for a flood wave in an open channel (subscripts denote location, and superscripts denote time.)

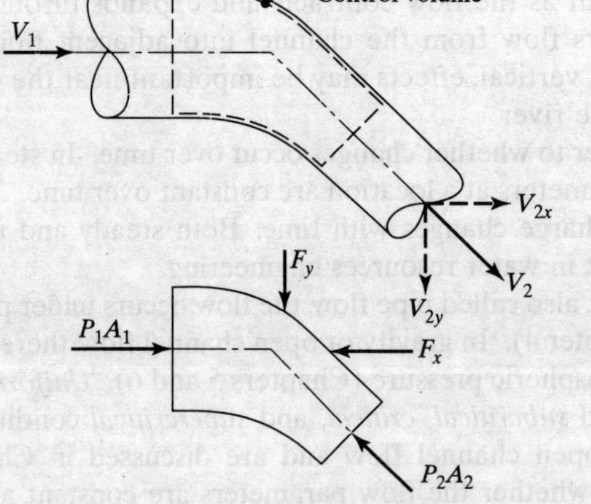

Figure 3.5 Control volume for a reducing pipe bend in a horizontal plane.

state conditions. The second term on the right of the equal sign in Eq. 3.15 represents the net rate of flow of the parameter through the control surface. The control volume is generally selected to simplify the evaluation of the two integrals. The sides of the control volume are typically selected to be either parallel or normal to the velocity vectors in the system.

If B represents the mass, then $b = 1$ and the control volume expression for conservation of mass is

$$\frac{\partial}{\partial t}\int_{cv} \rho d\forall + \int_{cs} \rho \mathbf{V} \cdot \hat{n} dA = 0 \tag{3.16}$$

If B represents the momentum, then $b = \mathbf{V}$ and the control volume expression for linear momentum is

$$\frac{\partial}{\partial t}\int_{cv} \mathbf{V}\rho d\forall + \int_{cs} \rho \mathbf{V}\mathbf{V} \cdot \hat{n} dA = \Sigma F \tag{3.17}$$

where the term on the right of the equal sign is the sum of the external forces acting on the control volume.

If B represents energy, then $b = V^2/2 + gz + u$ (sum of kinetic, potential, and internal energy per unit mass), and the control volume expression for conservation of energy is

$$\frac{\partial}{\partial t}\int_{cv} \rho b d\forall + \int_{cs} \rho b \mathbf{V} \cdot \hat{n} dA = \dot{Q} + \dot{W} \tag{3.18}$$

where $\dot{Q}$ represents the heat transfer rate and W is the work transfer rate, including work by pressure forces, pumps, and turbines. Heat and work transfer rates into the control volume are considered positive.

3.6 DIMENSIONAL ANALYSIS

The basic principle of dimensional analysis is that equations expressing a physical relationship between variables must be dimensionally homogeneous. For example power (P) of a pump or turbine is dependent on the flow rate (Q), specific weight of the fluid (γ), and energy (E) added or subtracted per unit of fluid or

$$P = f(Q, \gamma, E)$$

The general form of the equation is

$$P = CQ^a\gamma^bE^c \tag{3.19}$$

where C is the dimensionless constant. The units on each side of the equal sign in Eq. 3.19 must be equal. Writing the equations dimensionally based on the force

(F), length (L), and time (T) dimensions gives

$$\frac{FL}{T} = \left(\frac{L^3}{T}\right)^a \left(\frac{F}{L^3}\right)^b (L)^c$$

Comparing the following units

$$F: \quad 1 = b, \qquad b = 1$$
$$T: \quad -1 = -a, \qquad a = 1$$
$$L: \quad 1 = 3 - 3 + c, \qquad c = 1$$

Substituting into Eq. 3.19 gives

$$P = CQ\gamma E \tag{3.20}$$

Since there are only three dimensions (F, L, and T), this approach is limited to four variables (one dependent and three independent).

The Buckingham π-theorem allows the expansion of dimensional analysis to n variables by forming $n - k$ dimensionless π grouping of variables (where k is number of dimensions). The first pi term (π_1) includes the dependent variable and three repeating independent variables. The second pi term includes one nonrepeating independent variable and the three repeating independent variables. The procedure is continued until all n variables are grouped into $n - k$ dimensionless pi terms. The variable exponents are determined by making each pi term dimensionless. The general form of the equation relating pi terms is

$$\pi_1 = \phi(\pi_2, \pi_3, \ldots, \pi_{n-k}) \tag{3.21}$$

For example, the pressure drop (ΔP) in a pipe of diameter (D) and length (ℓ) can be written in functional form as

$$\Delta P = f(V, D, \rho, \mu, \ell, \varepsilon) \tag{3.22}$$

where V is the average velocity, ρ is the density of the fluid, μ is the viscosity, and ε is the roughness of the pipe. Since there are seven variables ($n = 7$) that can be written in terms of three dimensions ($k = 3$), Eq. 3.22 can be written in terms of $n - k = 4$ dimensionless groups.

Taking V, D, and ρ as repeating independent variables gives

$$\pi_1 = \Delta P V^a D^b \rho^c$$
$$F^0 L^0 T^0 = (FL^{-2})(LT^{-1})^a (L)^b (FL^{-4}T^2)^c$$

$$F: \quad 0 = 1 + c \qquad c = -1$$

$$T: \quad 0 = -a - 2 \qquad a = -2$$

$$L: \quad 0 = -2 - 2 + b + 4 \qquad b = 0$$

$$\pi_1 = \frac{\Delta P}{\rho V^2} \tag{3.23}$$

Replacing the dependent variable with the nonrepeating independent variable μ gives

$$\pi_2 = \mu V^a D^b \rho^c$$

$$F^0L^0T^0 = FL^{-2}T(LT^{-1})^a(L)^b(FL^{-4}T^2)^c$$

$$F: \quad 0 = 1 + c \qquad c = -1$$

$$T: \quad 0 = 1 - a - 2 \qquad a = -1$$

$$L: \quad 0 = -2 - 1 + b + 4 \qquad b = -1$$

$$\pi_2 = \frac{\mu}{VD\rho} = \frac{1}{\text{Re}} \tag{3.24}$$

where Re is the Reynolds number. Selecting ℓ as the next nonrepeating variable gives

$$\pi_3 = \ell V^a D^b \rho^c$$

$$F^0L^0T^0 = L(LT^{-1})^a L^b (FL^{-4}T^2)^c$$

$$F: \quad 0 = c \qquad c = 0$$

$$T: \quad 0 = -a \qquad a = 0$$

$$L: \quad 0 = 1 + b \qquad b = -1$$

$$\pi_3 = \frac{L}{D} \tag{3.25}$$

The remaining nonrepeating variable is ε, so that

$$\pi_4 = \varepsilon V^a D^b \rho^c$$

solving for exponents gives

$$\pi_4 = \frac{\varepsilon}{D} \tag{3.26}$$

where π_4 is the relative roughness of the pipe.

The equation for the pressure drop in a pipe (Eq. 3.22) can be written in terms of

$$\frac{\Delta P}{\rho V^2} = \phi\left(\frac{1}{\text{Re}}, \frac{L}{D}, \frac{\varepsilon}{D}\right) \tag{3.27}$$

Equation 3.27 forms the basis for the Darcy–Weisbach equation for headloss ($h_l = \Delta P/\gamma$) in a pipe

$$h_L = f\frac{L}{D}\frac{V^2}{2g} \tag{3.28}$$

where the friction factor (f) is a function of Re and ε/D and is given in the Moody chart presented in Chapter 4.

Free surface flow occurs in channels, flow around piers, and control structures. In this type of flow, viscous, gravitational, and inertial forces are important. The pressure drop between two points in open channel flow is expected to be a function of velocity (V), length (L), density (ρ), viscosity (μ), and gravity (g). The functional relationship is given as

$$\Delta P = f(V, L, \rho, \mu, g)$$

Using V, L, and ρ as repeating variables, dimensional analysis yields

$$\pi_1 = \frac{\Delta P}{\rho V^2}$$

$$\pi_2 = \frac{\mu}{VL\rho} = \frac{1}{R_e}$$

and

$$\pi_3 = \frac{gL}{V^2} = \frac{1}{F_r^2}$$

or

$$\frac{\Delta P}{\rho V^2} = \phi(\mathrm{Re}, \mathrm{Fr}) \quad \textbf{(3.29)}$$

where the pressure drop is proportional to the velocity squared and is a function of Reynolds number (Re) and Froude number (Fr).

The Re and Fr numbers are dimensionless groups that are particularly relevant in hydraulic engineering applications.

$$\mathrm{Re} = \frac{VL\rho}{\mu} = \frac{VL}{\nu} \quad \textbf{(3.30)}$$

$$\mathrm{Fr} = \frac{V}{\sqrt{gL}} \quad \textbf{(3.31)}$$

Re is a function of velocity (V), viscosity (μ), and density (ρ) [or alternatively kinematic viscosity ($\nu = \mu/\rho$)], and a characteristic length (L). The Re number physically represents the ratio of inertial force to viscous force and is especially useful in pressure conduit and groundwater hydraulics. In flow through circular pipes, the pipe diameter is adopted for L. The Fr number reflects the velocity (V), gravitational acceleration constant (g), and a characteristic length L. Fr represents the ratio of inertial force to gravitational force and is particularly pertinent in open channel flow in which gravitational forces tend to dominate. The hydraulic depth is normally adopted for L.

3.7 CONSERVATION OF MASS, ENERGY, AND MOMENTUM

The conservation laws apply to all control volumes. Conservation of mass is also known as the continuity equation. Conservation of energy is also known as the first law of thermodynamics. Conservation of momentum is also known as Newton's second law of motion. The conservation laws are expressed in various formats in fluid mechanics depending on premises regarding whether the fluid flow is real or ideal; compressible or incompressible; steady or unsteady; uniform or varied; and one-, two-, or three-dimensional. Common forms of equations expressing the conservation laws are presented in this section and applied in subsequent chapters. Equations representing more complex flow conditions are developed in subsequent chapters.

3.7.1 One-Dimensional, Steady-Flow Continuity Equation

For an incompressible (constant ρ) fluid under steady (time invariant) flow conditions, conservation of mass reduces to conservation of volume, as expressed by the following form of the continuity equation

$$Q = V_1A_1 = V_2A_2 \tag{3.32}$$

where A is the cross-sectional flow area (m^2 or ft^2), V is the mean velocity (m/s or ft/s) through A ($V = Q/A$), and Q is the constant discharge (m^3/s or ft^3/s). Subscripts 1 and 2 refer to upstream and downstream cross-section locations.

3.7.2 One-Dimensional, Steady-Flow Energy Equation

The first law of thermodynamics, also called the law of conservation of energy, states that for any closed system, the heat added to the system minus the work done by the system is equal to the change in internal energy within the system. In hydraulics, for one-dimensional, steady flow, conservation of energy between two cross-sections of flow is expressed as follows

$$\alpha_1 \frac{V_1^2}{2g} + \frac{P_1}{\gamma} + Z_1 + E_p - E_t = \alpha_2 \frac{V_2^2}{2g} + \frac{P_2}{\gamma} + Z_2 + H_L \tag{3.33}$$

Head is energy per unit weight of water, typically expressed in units of meters ($N \cdot m/N = m$) or feet ($ft \cdot lb/lb = ft$). Subscripts 1 and 2 denote upstream and downstream locations. The energy head at the downstream location equals the head upstream plus head added by pumps (E_p), minus head subtracted by turbines (E_t), less the head loss H_L occurring between the two locations. H_L is the heat loss generated by fluid friction per unit weight of water.

From physics, a body of mass m when moving at a velocity V possesses a kinetic energy (KE)

$$KE = \frac{1}{2} m V^2 \tag{3.34}$$

If a fluid was flowing with all particles moving at the same velocity, the kinetic energy per unit weight term in Eq. 3.33 would be

$$\frac{KE}{\text{weight}} = \frac{\frac{1}{2} mV^2}{\gamma \forall} = \frac{\frac{1}{2}(\gamma \forall) mV^2}{\gamma g \forall} = \frac{V^2}{2g} \tag{3.35}$$

where $\forall$ represents the volume of the fluid mass, and the unit weight $\gamma = \rho g$. However, in most situations, the velocities of different fluid particles are not the same, so integration of the kinetic energies at all points in the flow cross-section is required to obtain the true total kinetic energy. In practice, a kinetic energy correction factor α is adopted to express the relation between Eq. 3.35 and the true kinetic energy.

$$\frac{\text{True } KE}{\text{Weight}} = \alpha \frac{V^2}{2g} \tag{3.36}$$

An expression for α may be developed by considering the velocity (u) at a point in the cross-section of flow. The mass flow through an elementary area dA is $\rho\, dQ$ or $\rho u\, dA$. Thus, the true kinetic energy per unit of time across area dA is

$$\frac{1}{2}(\rho u\, dA)u^2 = \frac{1}{2}\rho u^3\, dA \tag{3.37}$$

The weight rate of flow through dA is $\gamma dQ = \rho g u\, dA$. For the entire section of flow

$$\frac{\text{True } KE/\text{time}}{\text{Weight/time}} = \frac{\text{True } KE}{\text{Weight}} = \frac{\frac{1}{2}\rho \int u^3\, dA}{\rho g \int u\, dA} = \frac{\int u^3\, dA}{2g \int u\, dA} \tag{3.38}$$

An expression for the kinetic energy correction factor is obtained by comparing Eqs. 3.36 and 3.38.

$$\alpha = \frac{1}{V^2}\frac{\int u^3\, dA}{\int u\, dA} = \frac{1}{AV^3}\int u^3\, dA \tag{3.39}$$

For turbulent flow in pipes, α ranges from 1.01 to 1.15. Estimating α for open channels is covered in Chapter 5.

The potential energy term in Eq. 3.33 consists of the elevation head plus pressure head.

$$\text{Potential energy} = \frac{P}{\gamma} + Z \tag{3.40}$$

A fluid particle of weight W situated a distance Z above an elevation datum possesses a potential energy $W \cdot Z$. The potential energy per unit weight is $W \cdot Z/W = Z$. Pressure head (Eq. 3.11) is discussed in Section 3.3.

3.7.3 Linear Momentum Equation

Newton's second law of motion may be applied to a control volume to develop various forms of the momentum equation. For the control volume shown in Fig. 3.5, the time rate of change in linear momentum is equal to the sum of external forces acting on the control volume. For steady flow, the linear momentum equation is

$$\int_{cs} \mathbf{V}\rho\mathbf{V} \cdot \hat{n}dA = \Sigma\mathbf{F} \quad \textbf{(3.41)}$$

where cs is the control volume surface, $\mathbf{V}$ is the velocity vector, $\hat{n}$ is the area unit vector outward positive, and $\mathbf{F}$ is the force vector. When the control volume has only one inflow and one outflow boundary with uniform velocity distribution at each boundary, Eq. 3.41 can be written in x, y, and z components

$$\Sigma F_x = \rho Q(V_{2x} - V_{1x}) \quad \textbf{(3.42a)}$$

$$\Sigma F_y = \rho Q(V_{2y} - V_{1y}) \quad \textbf{(3.42b)}$$

$$\Sigma F_z = \rho Q(V_{2z} - V_{1z}) \quad \textbf{(3.42c)}$$

where the subscripts 1 and 2 refer to the inflow and outflow boundaries, respectively.

3.8 NEWTON–RAPHSON FORMULA

Many equations of fluid mechanics and hydraulics require iterative solutions. The Newton–Raphson formula, developed in numerical methods textbooks, is one of the most widely used of the various iterative methods for finding roots of an equation. Given an initial estimate or previously computed value x_i, an improved value x_{i+1} is determined as follows.

$$x_{i+1} = x_i - \frac{f(x_i)}{f'(x_i)} \quad \textbf{(3.43)}$$

where $f'(x)$ is the derivative of the function $f(x)$. Following an initial estimate (guess) for x_i, Eq. 3.43 is repeated, iteratively updating x_i to x_{i+1}, until x_{i+1} essentially equals x_i. The Newton–Raphson formula (Eq. 3.43) is applied in several example problems in the following chapters.

PROBLEMS

3.1. A rectangular gate is 2.0 m (6.6 ft) wide and 6.3 m (20.7 ft) long and is used on the front face of an earth dam to control the discharge through the outlet works. As in the sketch below, the slope of the front face of the dam is 3H:1V and the water depth is 5.0 m (16.4 ft) above the centroid of the gate. Determine the total hydrostatic force on the gate and the distance between the center of pressure and the centroid of the gate.

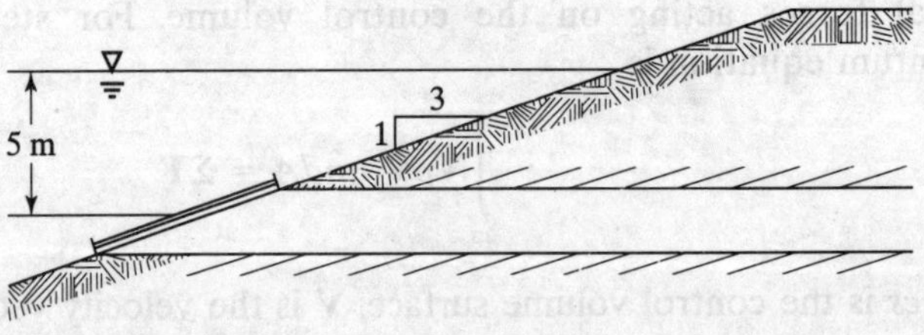

3.2. A vertical lift gate is square 3.0 m (9.8 ft) on a side and weighs 1,000 kg (2,205 lbs). The centroid of the gate is 61.0 m (200.1 ft) below the water surface. Determine the vertical force required to lift the gate if the coefficient of friction between the gate and guides is 0.05.

3.3. The sliding gate in Problem 3.1 has a coefficient of friction against the guides of 0.10 and weighs 3,000 kg (6,615 lbs). Compute the force required to move the gate parallel with the face of the dam.

3.4. A circular gate is used to control the discharge into a 3.05 m (10.0 ft) diameter pipe. The gate pivots about a horizontal axis through its centroid. If the water depth is 10.0 m (32.8 ft) above the centroid of the gate, determine the horizontal force (F) necessary to prevent the gate from moving.

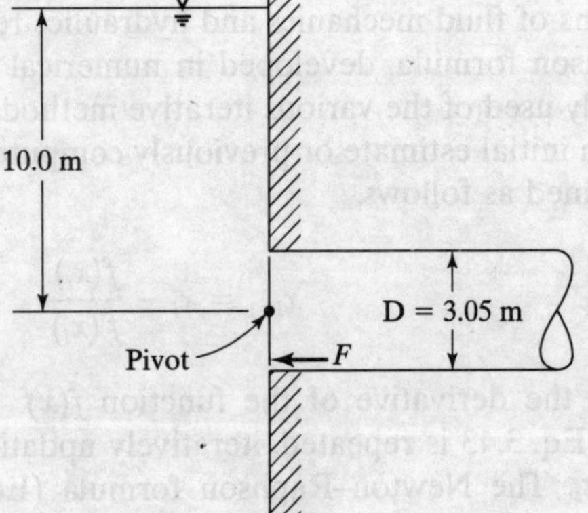

3.5. The hinge gate is used to control the flow of water over the spillway. Neglect the weight of the gate. Compute the depth h that will cause the gate to open for $R = 2.0$ m (6.6 ft).

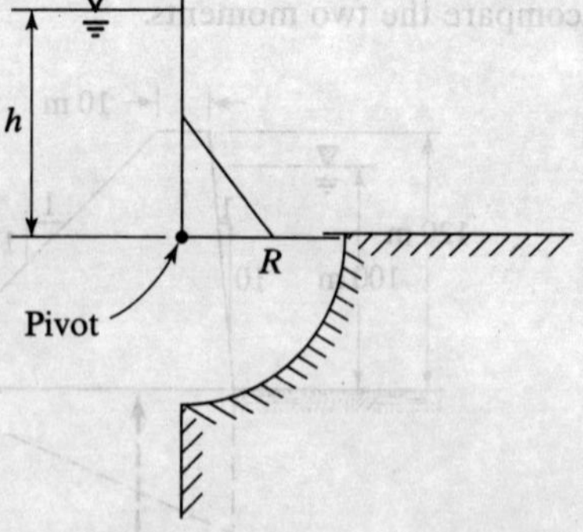

3.6. The hinged gate in Problem 3.5 was modified as shown below. Neglect the weight of the gate. Compute the water depth h that will cause the gate to open for $R = 2.0$ m (6.6 ft).

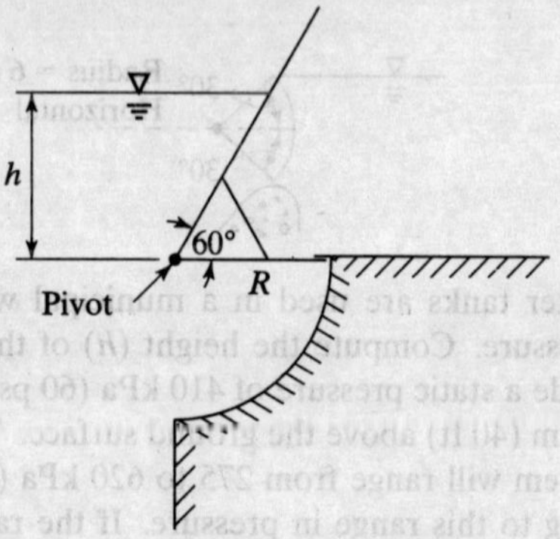

3.7. A wicket dam is designed to collapse when the upstream water depth exceeds a depth h. The strut holding the gate in position can pivot at each end and forms a 90° angle with the face of the dam. Compute the water depth h that will cause the dam to collapse for $b = 3.0$ m (9.8 ft).

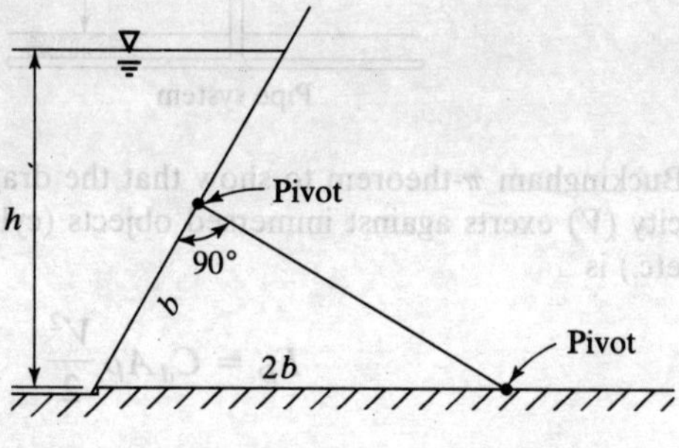

3.8. A concrete gravity dam is 120 m (394 ft) high with an upstream slope of 1H:10V and a downstream slope of 1H:1V. The uplift varies linearly from zero at the downstream toe to full hydrostatic pressure at the upstream toe. Compute the overturning moment if the water depth in the reservoir is 100 m (328 ft). The top width of the dam is 10.0 m (32.8 ft), and specific gravity of the concrete is 2.40. Compute the righting moment for the dam and compare the two moments.

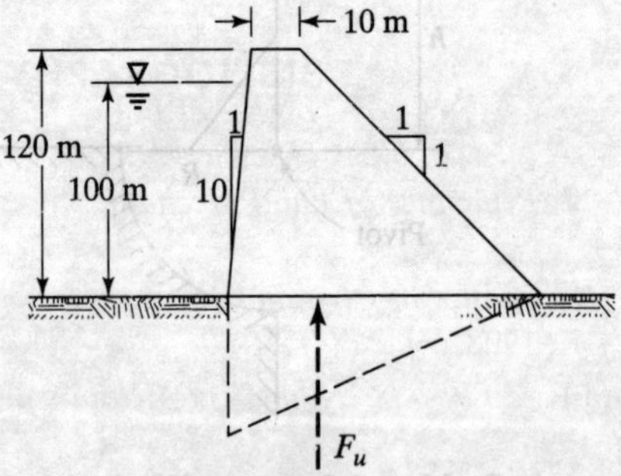

3.9. Compute the horizontal and vertical forces per unit length on the radial gate shown below.

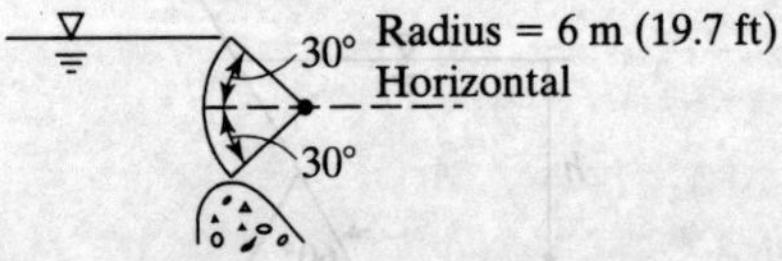

3.10. Elevated water tanks are used in a municipal water distribution system to provide adequate pressure. Compute the height (h) of the water tank above the ground surface to provide a static pressure of 410 kPa (60 psi). Compute the pressure in a building that is 12 m (40 ft) above the ground surface. Typically, the pressure in a water distribution system will range from 275 to 620 kPa (40–90 psi). Compute the range in h corresponding to this range in pressure. If the range in pressure is greater than this, multiple pressure zones may be required for the water distribution system.

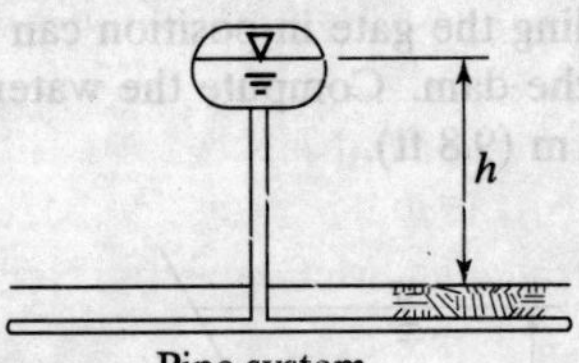

3.11. Use the Buckingham π-theorem to show that the drag force (F_D) that water flowing at a velocity (V) exerts against immersed objects (cylinders, people, piers, flat plates, vehicles, etc.) is

$$F_D = C_d A \rho \frac{V^2}{2}$$

where C_d is the drag coefficient and A is the cross-sectional area of the object perpendicular to the flow. The functional relationship is

$$F_D = f(V, A, \rho, \mu, g)$$

Use V, A, and ρ as repeating variables.

BIBLIOGRAPHY

ALEXANDROU, A. N., *Principles of Fluid Mechanics*, Prentice Hall, Upper Saddle River, NJ, 2001.

CHADWICK, A. J., *Hydraulics in Civil and Environmental Engineering*, 3rd Ed., Chapman & Hall, London, UK, 1998.

CHAPRA, S. C., and R. P. CANALE, *Numerical Methods for Engineers*, 2nd Ed., McGraw-Hill, New York, NY, 1998.

FOX, R. W., and A. T. MCDONALD, *Introduction to Fluid Mechanics*, John Wiley & Sons, New York, NY, 1985.

FRANZINI, J. B., and E. J. FINNEMORE, *Fluid Mechanics with Engineering Applications*, 9th Ed., McGraw-Hill, New York, NY, 1997.

LEVI, E., *The Science of Water: The Foundation of Modern Hydraulics*, American Society of Civil Engineers Press, Reston, VA, 1995.

LIGGETT, J. A., *Fluid Mechanics*, McGraw-Hill, New York, NY, 1994.

MOTT, R. L., *Applied Fluid Mechanics*, Prentice Hall, Upper Saddle River, NJ, 2000.

MUNSON, B. R., D. F. YOUNG, and T. H. OKIISHI, *Fundamentals of Fluid Mechanics*, John Wiley & Sons, New York, NY, 1999.

ROBERSON, J. A., and C. T. CROWE, *Engineering Fluid Mechanics*, Houghton Mifflin Co., Boston, MA, 1993.

ROUSE, H., and S. INCE, *History of Hydraulics*, Iowa Institute of Hydraulic Research, Iowa City, Iowa, 1957.

SHAMES, I. H., *Mechanics of Fluids*, McGraw-Hill, New York, NY, 1992.

SIMON, A. L., and S. F. KOROM, *Hydraulics*, 4th Ed., Prentice Hall, Upper Saddle River, NJ, 1997.

STREET, R. L, S. G. WATTERS, and J. K. VENNARD, *Elementary Fluid Mechanics*, John Wiley & Sons, New York, NY, 1996.

WHITE, F. M., *Fluid Mechanics*, McGraw-Hill, New York, NY, 1994.

4

Hydraulics of Pipelines and Pipe Networks

Hydraulics presented in this chapter is limited to turbulent flow of water in closed conduits (pipes) flowing full. Closed conduits flowing partially full, such as stormwater collection systems, are analyzed as open channel flow. Pipes are connected together in various configurations (called networks) to transport water from the supply to the user. When all pipes are connected in series, the system is called a pipeline. A pipe system can include many pipes of various lengths and diameters, along with valves to control the flow rate and pumps and turbines to convert between hydraulic energy and thermal, mechanical, and electrical energy.

Flow in a pipe can either be steady or unsteady. For unsteady flow, the velocity is a function of time and will change within a few seconds. For steady flow, the velocity is not considered a function of time, although it may change gradually. Steady flow equations can be used to analyze a water distribution system where the demands on the system change hourly, pumps turn on and off, and storage tanks fill and drain during the numerical simulation.

4.1 BASIC EQUATIONS FOR STEADY FLOW

The hydraulics of steady flow in pipe systems is described by the continuity and energy equations. Headloss is a key term in the energy equation.

4.1.1 Continuity Equation

For steady flow in pipelines and pipe networks, water is considered incompressible, and the conservation of mass equation (continuity equation) reduces to the volumetric flow rate (Q)

$$Q = AV \tag{4.1}$$

where A is the cross-section area of the pipe, and V is the average velocity. The flow rate (Q) is measured in cu m per sec (cms) or cu ft per sec (cfs). Discharge rate (Q) may also be specified in liters per second (lps), gallons per minute (gpm), or million gallons per day (mgd). The continuity equation between cross-sections 1 and 2 of a pipe is

$$A_1V_1 = A_2V_2 \tag{4.2}$$

Junction nodes are located where two or more pipes join together. A three-pipe junction node with a constant demand (C) is shown in Fig. 4.1. The continuity equation for the junction node is

$$Q_1 - Q_2 - Q_3 - C = 0 \tag{4.3}$$

In modeling pipe networks, all demands on the system are located at junction nodes, and the flow in pipes connecting nodes is assumed to be uniform. If a major demand is located between nodes, then an additional junction node is established at the location of the demand. Equation 4.2 serves as the continuity equation for a two-pipe junction node without a demand, where the subscripts refer to the pipe number.

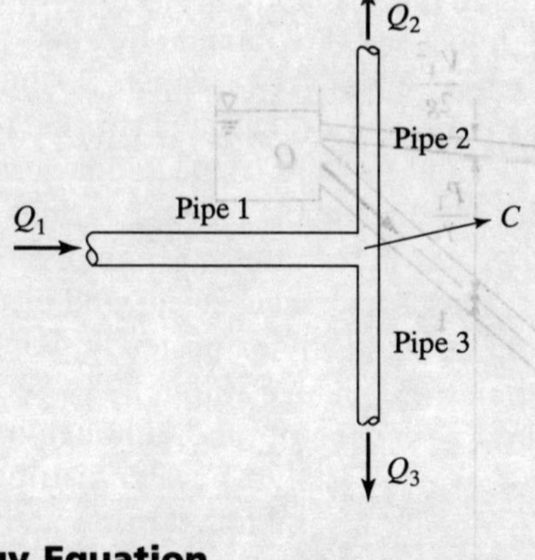

Figure 4.1 Three-pipe junction node with a constant demand.

4.1.2 Energy Equation

Figure 4.2 represents a pumped-storage hydroelectric plant where water is pumped to the upper reservoir during the off-peak power period and used to generate electricity during the peak power period. In Fig. 4.2a, water is being pumped from the lower supply reservoir through a pipeline to an upper storage reservoir. Water discharges from the upper reservoir in Fig. 4.2b through a pipeline and turbine into the lower reservoir.

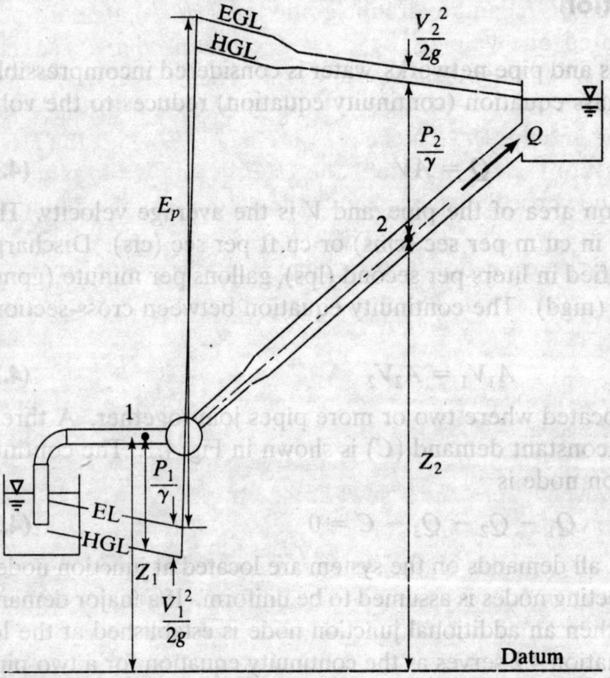

Figure 4.2a Pump system.

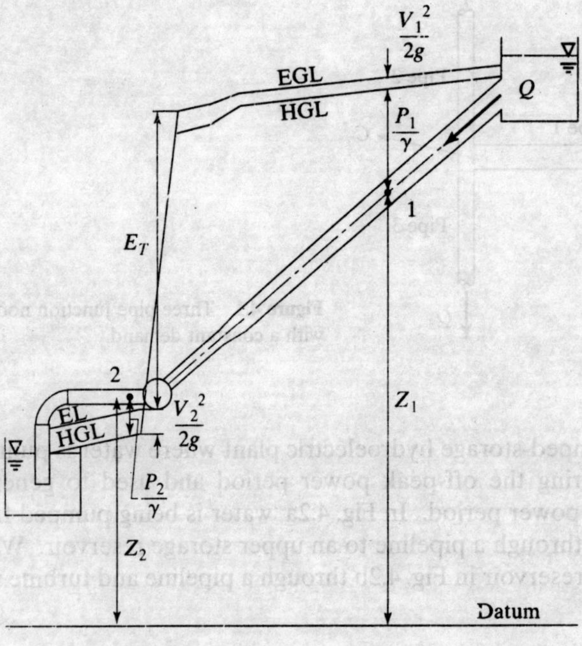

Figure 4.2b Turbine system.

The energy grade line (EGL) and hydraulic grade line (HGL) are shown in Fig. 4.2. The HGL is located one velocity head ($V^2/2g$) below the EGL. The EGL and HGL are parallel when the pipe size is uniform. The EGL slopes downward in the direction of flow because of energy loss. The vertical distance between the center of the pipe and the HGL is the pressure head (P/γ). If the HGL is above the pipe, the pressure is positive, and if the HGL is below the pipe, the pressure is negative. Z is the vertical distance above the datum (usually mean sea level—msl).

The energy equation written for flow in a pipeline (Fig. 4.3) is

$$\frac{V_1^2}{2g} + \frac{P_1}{\gamma} + z_1 + E_P = \frac{V_2^2}{2g} + \frac{P_2}{\gamma} + z_2 + H_L \tag{4.4}$$

where z is the elevation of the pipe, P/γ is the pressure head, $V^2/2g$ is the velocity head, E_P is the energy head added by the pump, and H_L is the total headloss between points 1 and 2. Each term in the energy equation has units of length and represents energy per unit weight of fluid (Newton-meters per Newton of fluid flowing or ft-lbs per lb)

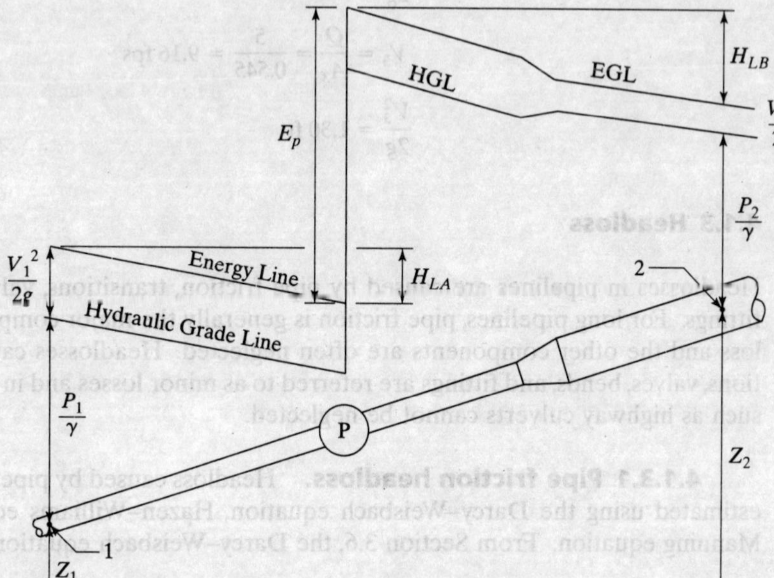

Figure 4.3 Energy equation for pipeline flow.

Example 4.1 Continuity Equation

If 5 cfs of water is flowing in the pipeline from the upper reservoir to the lower reservoir, determine the velocity in each line. Sketch the energy grade line (EGL) and hydraulic grade line (HGL) on the figure.

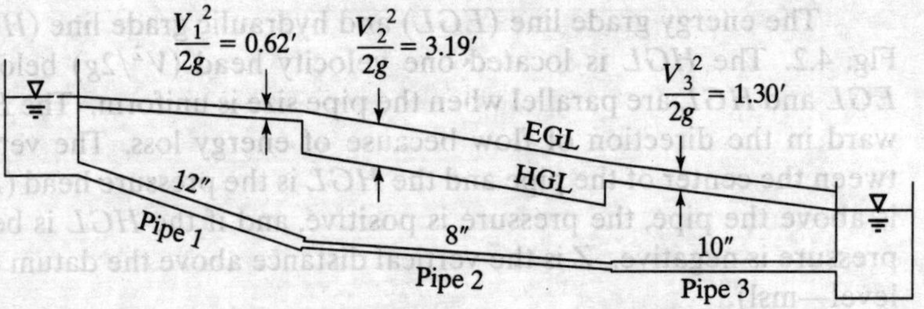

$$V_1 = \frac{Q}{A_1} = \frac{5}{0.785} = 6.34 \text{ fps}$$

$$\frac{V_1^2}{2g} = 0.62 \text{ ft}$$

$$V_2 = \frac{Q}{A_2} = \frac{5}{0.349} = 14.3 \text{ fps}$$

$$\frac{V_2^2}{2g} = 3.19 \text{ ft}$$

$$V_3 = \frac{Q}{A_3} = \frac{5}{0.545} = 9.16 \text{ fps}$$

$$\frac{V_3^2}{2g} = 1.30 \text{ ft}$$

4.1.3 Headloss

Headlosses in pipelines are caused by pipe friction, transitions, valves, bends, and fittings. For long pipelines, pipe friction is generally the major component of headloss and the other components are often neglected. Headlosses caused by transitions, valves, bends, and fittings are referred to as minor losses and in short pipelines such as highway culverts cannot be neglected.

4.1.3.1 Pipe friction headloss. Headloss caused by pipe friction can be estimated using the Darcy–Weisbach equation, Hazen–Williams equation, or the Manning equation. From Section 3.6, the Darcy–Weisbach equation is

$$H_L = f\frac{L}{D}\frac{V^2}{2g} \tag{4.5}$$

where f is the friction factor, L is the length of the pipe, and D is the diameter of the pipe. The friction factor can be estimated from the Moody diagram in Fig. 4.4 and is a function of the Reynolds number ($\text{Re} = DV/\nu$) of the flow and the relative roughness of the pipe (ε/D). In most pipelines, the flow will be fully turbulent.

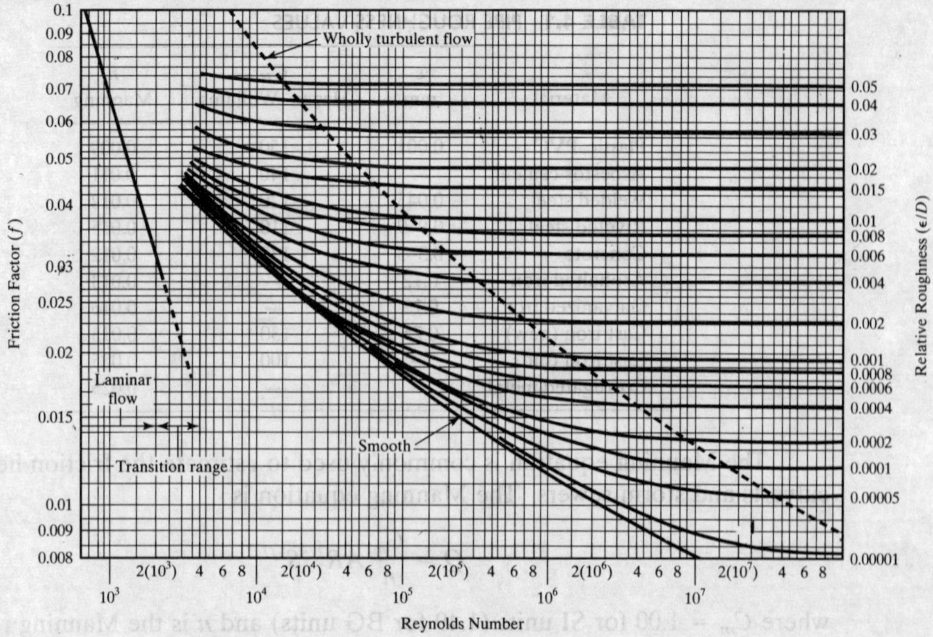

Figure 4.4 Moody diagram of Darcy-Weisbach friction factors.

The hydraulic radius (R) is defined as the area (A) of the flow cross-section divided by the wetted perimeter (P). For a circular pipe flowing full

$$R = \frac{A}{P} = \frac{\frac{1}{4}\pi D^2}{\pi D} = \frac{1}{4}D \tag{4.6}$$

For noncircular pipes, the headloss can be estimated using the equations for circular pipe with $4R$ substituted for D. Pipe roughness values are listed in Table 4.1 for common pipe materials.

After a pipe has been in service for some time, the diameter and roughness of the pipe may change, and it may be difficult to estimate the roughness of the pipe. The Hazen–Williams equation is often used in pipe network analysis. Tables are available relating the Hazen–Williams coefficient (C_H) to the age of the pipe.

The Hazen–Williams equation is

$$Q = C_w C_H A R^{0.63} S^{0.54} \tag{4.7}$$

where $C_w = 0.85$ for International System (SI) units [1.318 for British Gravitational (BG) units] and S is the slope of energy line. Writing the Hazen–Williams equation for headloss gives

$$H_L = S \times L = L\left(\frac{4}{D}\right)^{1.17}\left(\frac{V}{C_w C_H}\right)^{1.85} \tag{4.8}$$

TABLE 4.1 PIPE ROUGHNESS VALUES

Material	ε mm	C_H Hazen–Williams	n Manning
Plastic, PVC	0.001	150	0.009
Asbestos cement	—	140	0.011
Welded steel	0.045	120	0.012
Riveted steel	0.9–9	110	0.015
Concrete	0.3–3	130	0.012
Asphalted iron	0.12	—	0.013
Galvanized iron	0.15	—	0.016
Cast iron (new)	0.25	130	0.013
Cast iron (old)	—	100	0.025
Corrugated metal	—	—	0.025

The Manning equation is commonly used to estimate the friction headloss in culverts and storm sewers. The Manning equation is

$$Q = \frac{C_m}{n} AR^{2/3}S^{1/2} \tag{4.9}$$

where $C_m = 1.00$ for SI units (1.49 for BG units) and n is the Manning roughness coefficient. Writing the Manning equation for headloss gives

$$H_L = S \times L = \frac{n^2V^2L}{C_m^2R^{4/3}} \tag{4.10}$$

Example 4.2 Pipe Friction

Two reservoirs are connected with a 10-inch diameter pipeline 4500 ft long. If the pipe roughness is 0.005 inches, determine the discharge rate in the pipeline. Neglect minor losses.

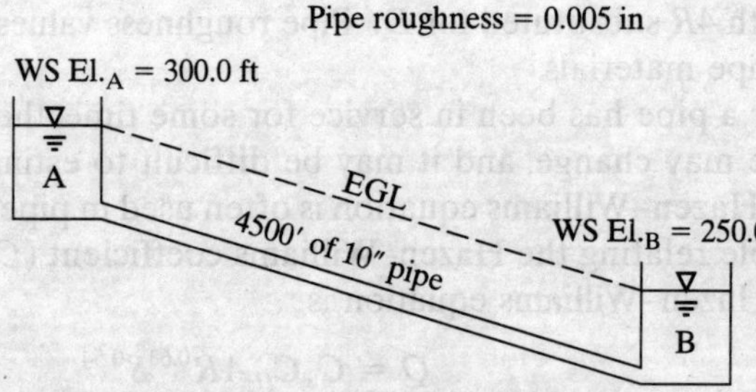

$$\text{Relative roughness } \frac{\varepsilon}{D} = 0.0005$$

from the Moody diagram $f = 0.017$ for $R_e > 10^6$

$$\text{headloss } H_L = \frac{fL}{D}\frac{V^2}{2g}$$

$$300.0 - 250.0 = \frac{0.017 \times 4500}{10/12}\frac{V^2}{2g} = 91.8\frac{V^2}{2g}$$

$$\frac{V^2}{2g} = 0.545 \text{ ft}$$

$$V = (2g \times 0.545)^{1/2} = 5.90 \text{ fps}$$

$$R_e = \frac{DV}{\nu} = \frac{0.833 \times 5.9}{10^{-5}} = 4.9 \times 10^5$$

From the Moody diagram $f = 0.018$

$$H_L = 97.2\frac{V^2}{2g}$$

$$\frac{V^2}{2g} = 0.514 \text{ ft}$$

$$V = 5.75 \text{ fps}$$

$$R_e = 4.8 \times 10^5$$

$$Q = AV = 0.545 \times 5.75 = 3.14 \text{ cfs}$$

4.1.3.2 Minor losses. Minor losses are caused by excessive turbulence generated by a change in flow geometry. They represent the headloss that is in excess of the normal pipe friction at transitions, bends, valves, and other fittings. The coefficient (K) is used to give the minor headloss (H_M) as a function of the velocity head

$$H_M = K\frac{V^2}{2g} \tag{4.11}$$

At transitions V is the velocity in the smaller pipe. Minor loss coefficients are listed in Table 4.2.

TABLE 4.2 MINOR LOSS COEFFICIENTS (K)

Transitions		
Diameter ratio	Expansion	Contraction
0	1.0	0.5
0.2	0.92	0.45
0.4	0.70	0.38
0.6	0.40	0.29
0.8	0.12	0.12
1.0	0.0	0.0
Entrance		
Pipe projection	0.8	
Square edge	0.5	
Rounded	0.1	
Exit	1.0	

TABLE 4.2 (*Continued*)

Bends		
Radius/diameter	90°	45°
1	0.5	0.37
2	0.3	0.22
4	0.25	0.19
6	0.15	0.11
Valves		
Globe (open)	10	
Swing check	2.0	
Gates (open)	0.2	
Gate (1/2 open)	5.6	
Butterfly (open)	1.2	
Ball (open)	0.05	

Example 4.3 Short Pipe Problem

The two reservoirs are connected with 450 ft of 10-in. diameter pipe ($f = 0.020$). The entrance loss coefficient is 0.5 at the upper reservoir, and the exit loss coefficient is 1.0 at the lower reservoir. Determine the discharge rate in the pipe. Draw the *EGL* and *HGL* on the sketch

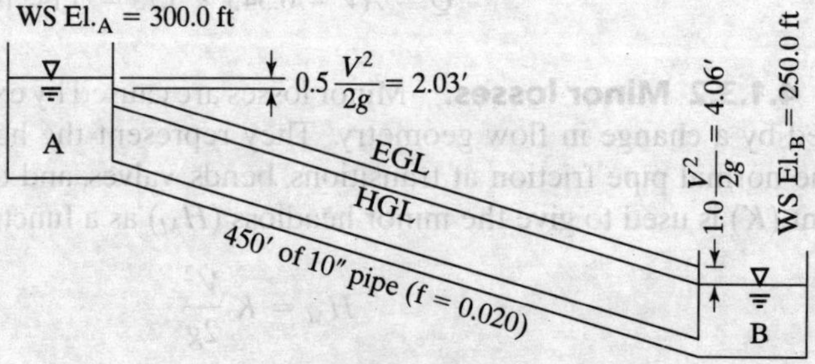

$$H_L = 0.5\frac{V^2}{2g} + \frac{fL}{D}\frac{V^2}{2g} + \frac{V^2}{2g}$$

$$300.0 - 250.0 = \left(1.5 + \frac{0.02 \times 450}{0.833}\right)\frac{V^2}{2g}$$

$$50.0 = 12.3\frac{V^2}{2g}$$

$$\frac{V^2}{2g} = 4.06 \text{ ft}$$

$$V = 16.2 \text{ fps}$$

$$Q = AV = 0.545 \times 16.2 = 8.83 \text{ cfs}$$

Example 4.4 Minor Losses

A pipeline consisting of three pipes in series ($f = 0.02$) extends from an upper reservoir (Elevation 200.0 m) to a lower reservoir (Elevation 180.0 m). Compute the discharge rate in the pipeline using minor loss coefficients of 0.5 for entrance, 0.15 for contraction at junction 1, 0.40 for expansion at junction 2, and 1.0 for exit. The minor loss coefficients at the two pipe junctions are based on the velocity in the smaller pipe (D = 152 mm). The headloss between A and B is the difference in water surface elevation ($WSEL$).

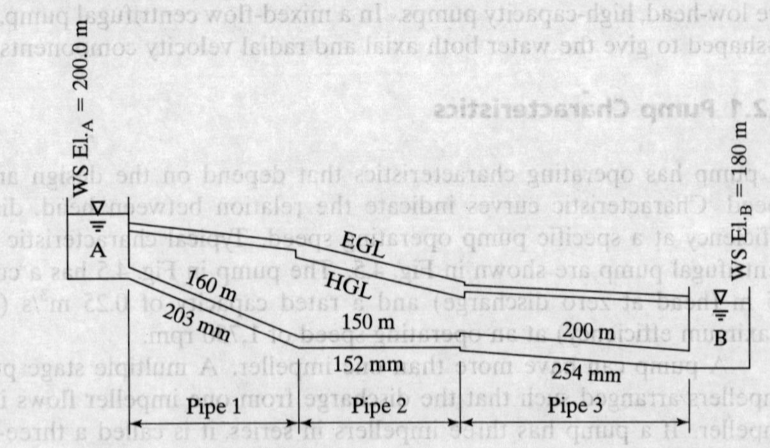

$$H_L = WSEL_A - WSEL_B$$

$$= 0.5\frac{V_1^2}{2g} + \frac{fL_1}{D_1}\frac{V_1^2}{2g} + 0.15\frac{V_2^2}{2g} + \frac{fL_2}{D_2}\frac{V_2^2}{2g} + 0.40\frac{V_2^2}{2g} + \frac{fL_3}{D_3}\frac{V_3^2}{2g} + 1.0\frac{V_3^2}{2g}$$

$$= (0.5 + 15.8)\frac{V_1^2}{2g} + (0.15 + 19.74 + 0.40)\frac{V_2^2}{2g} + (15.7 + 1.0)\frac{V_3^2}{2g}$$

$$= 16.3\frac{V_1^2}{2g} + 20.3\frac{V_2^2}{2g} + 16.7\frac{V_3^2}{2g}$$

Based on the continuity equation

$$V_1 = \frac{A_2}{A_1}V_2 = \left(\frac{D_2}{D_1}\right)^2 V_2 = 0.56V_2$$

$$V_3 = \left(\frac{D_2}{D_3}\right)^2 V_2 = 0.36V_2$$

$$H_L = 20\text{ m} = \left[16.3(0.56)^2 + 20.3 + 16.7(0.36)^2\right]\frac{V_2^2}{2g} = 27.6\frac{V_2^2}{2g}$$

$$\frac{V_2^2}{2g} = \frac{20}{27.6} = 0.72\text{ m}$$

$$V_2 = (2g \times 0.72)^{1/2} = (2.0 \times 9.81 \times 0.72)^{1/2} = 3.76\text{ mps}$$

$$Q = A_2V_2 = 0.0181 \times 3.76 = 0.0682\text{ cms}$$

4.2 PUMPS IN PIPELINES

Centrifugal pumps are used in pipelines to add energy to the water. The rotating element that transfers the energy from the motor to the water is called the impeller. In a radial flow centrifugal pump, the impeller is shaped to force water outwaru in a direction at a right angle to its axis. Radial flow pumps are high-head, low-capacity pumps. In an axial flow centrifugal pump, the impeller is shaped like a propeller, forcing the water in an axial direction. Axial flow pumps are low-head, high-capacity pumps. In a mixed-flow centrifugal pump, the impeller is shaped to give the water both axial and radial velocity components.

4.2.1 Pump Characteristics

A pump has operating characteristics that depend on the design and operating speed. Characteristic curves indicate the relation between head, discharge, and efficiency at a specific pump operating speed. Typical characteristic curves for a centrifugal pump are shown in Fig. 4.5. The pump in Fig. 4.5 has a cutoff head of 45 m (head at zero discharge) and a rated capacity of 0.25 m^3/s (discharge at maximum efficiency) at an operating speed of 1,750 rpm.

A pump can have more than one impeller. A multiple stage pump has the impellers arranged such that the discharge from one impeller flows into the next impeller. If a pump has three impellers in series, it is called a three-stage pump. The total energy transferred to the water by the pump (E_p) is equal to the head per stage (h_s) times the number of stages.

The water power (P) transferred from the impeller to the water in *kw* is

$$P_{\text{water}} = T\omega = \gamma Q E_p \tag{4.12}$$

where T is the torque exerted by the impeller on the water and ω is the rotation speed in radians/second (rad/s). The shaft power in *kw* is

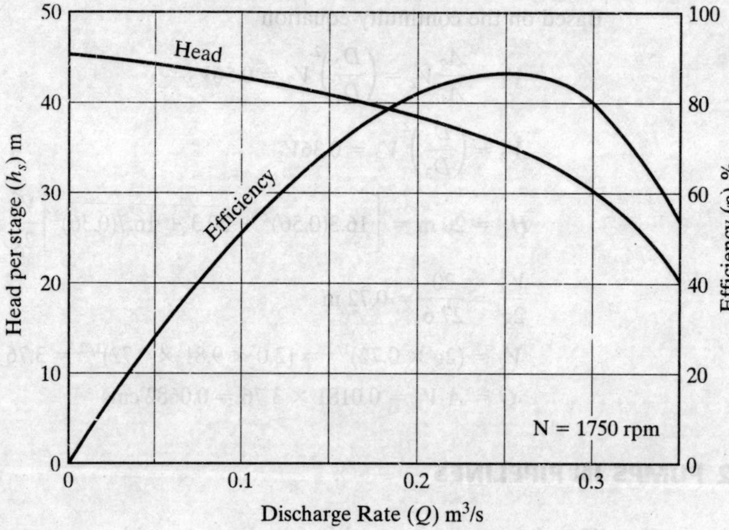

Figure 4.5 Pump characteristic curves.

$$P_{\text{shaft}} = \frac{\gamma Q E_p}{\eta} \tag{4.13}$$

where η is the efficiency of the pump.

Changing the pump speed or impeller diameter will change the pump characteristic curve. The head added per stage (h_s) is a function of impeller diameter (D), rotational speed (ω), discharge rate (Q), fluid density (ρ), and viscosity (μ). Written in functional form

$$gh_s = f(D, \omega, Q, \rho, \mu)$$

where gh_s is the head rise in terms of energy per unit mass. The pi one term (Section 3.6) is

$$\pi_1 = gh_s \omega^a D^b$$

where gh_s is the dependent variable and ω and D are repeating independent variables.

$$\begin{aligned} F^0L^0T^0 &= L^2T^{-2}(T^{-1})^a(L)^b \\ L: \quad 0 &= 2 + b \quad b = -2 \\ T: \quad 0 &= -2 - a \quad a = -2 \\ \pi_1 = C_H &= \frac{gh_s}{\omega^2 D^2} \end{aligned} \tag{4.14}$$

π_1 is called the head rise coefficient (C_H). Replacing the dependent variable with the nonrepeating variable Q gives

$$\begin{aligned} \pi_2 &= Q\omega^a D^b \\ F^0L^0T^0 &= L^3T^{-1}(T^{-1})^a(L)^b \\ L: \quad 0 &= 3 + b \quad b = -3 \\ T: \quad 0 &= -1 - a \quad a = -1 \\ \pi_2 = C_Q &= \frac{Q}{\omega D^3} \end{aligned} \tag{4.15}$$

π_2 is called the flow coefficient (C_Q).

For the last pi term, there are four variables

$$\begin{aligned} \pi_3 &= \rho\mu^a\omega^b D^c \\ F^0L^0T^0 &= FT^2L^{-4}(FTL^{-2})^a(T^{-1})^b(L)^c \\ F: \quad 0 &= 1 + a \quad a = -1 \\ T: \quad 0 &= 2 - 1 - b \quad b = 1 \\ L: \quad 0 &= -4 + 2 + c \quad c = 2 \\ \pi_3 &= \rho\frac{\omega D^2}{\mu} = \frac{\omega D^2}{\nu} \end{aligned} \tag{4.16}$$

The π_3 term represents the Reynolds number. For high Reynolds number flow, the effects of Reynolds number can be neglected. For geometrically similar pumps

$$\frac{gh_s}{\omega^2 D^2} = \phi\left(\frac{Q}{\omega D^3}\right) \tag{4.17}$$

For two geometrically similar pumps

$$\left(\frac{gh_s}{\omega^2 D^2}\right)_1 = \left(\frac{gh_s}{\omega^2 D^2}\right)_2$$

and

$$\left(\frac{Q}{\omega D^3}\right)_1 = \left(\frac{Q}{\omega D^3}\right)_2$$

If the speed of a pump changes from N_o to N, then

$$Q = Q_o\left(\frac{N}{N_o}\right) \tag{4.18}$$

$$E_p = E_{po}\left(\frac{N}{N_o}\right)^2 \tag{4.19}$$

and

$$P = P_o\left(\frac{N}{N_o}\right)^3 \tag{4.20}$$

where subscript o indicates a point on the pump characteristic curve.

The specific speed (N_s) is a dimensionless parameter used to characterize the type of pump. N_s is obtained by eliminating the diameter (D) in the ratio of flow coefficient and head rise coefficient. Raising the flow coefficient to an exponent of 1/2 and dividing by the head coefficient raised to an exponent of 3/4 gives

$$N_s = \frac{\omega Q^{1/2}}{(gh_s)^{3/4}} \tag{4.21}$$

where ω is the pump rotation speed in rad/s, Q is the discharge rate in m^3/s, and h_s is the head per stage in m. The specific speed is sometimes called the shape number as it reflects the shape of the impeller. A specific speed between 0.2 and 1.3 indicates a radial flow pump, between 1.3 and 2.8 indicates a mixed flow pump, and between 2.8 and 4.4 indicates an axial flow pump. The specific speed of a pump is determined by the operating characteristics at the point of maximum efficiency.

Example 4.5 Booster Pump

A booster pump is installed in the pipeline between the two reservoirs shown below. If the energy added by the pump (E_p) is 50.0 ft, determine the flow rate in the pipeline. Neglect minor losses.

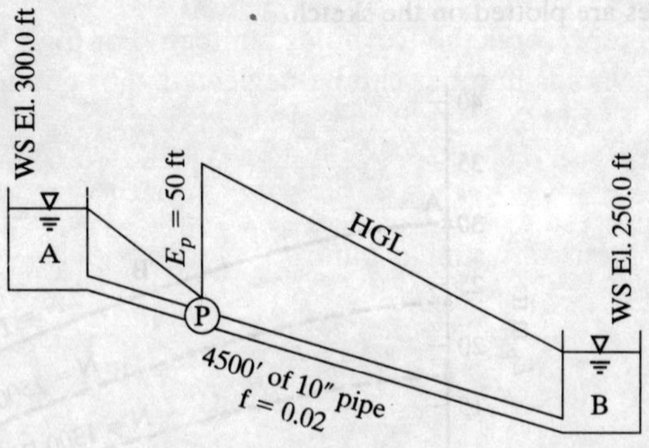

Write the energy equation from A to B in terms of water surface elevation ($WSEL$).

$$E_p + WSEL_A = H_L + WSEL_B$$

$$H_L = E_p + WSEL_A - WSEL_B$$

$$\frac{fL}{D}\frac{V^2}{2g} = 50.0 + 300.0 - 250.0$$

$$\frac{0.02 \times 4500}{0.833}\frac{V^2}{2g} = 100.0 \text{ ft}$$

$$\frac{V^2}{2g} = \frac{100.0}{108.0} = 0.926 \text{ ft}$$

$$V = 7.72 \text{ fps}$$

$$Q = AV = 0.545 \times 7.72 = 4.2 \text{ cfs}$$

Example 4.6 Pump Characteristic Curve

The pump characteristic curve shown below is for a pump speed of 1,700 rpm. Determine the pump characteristic curve for a pump speed of 1,500 rpm and 1,300 rpm, using Eqs. 4.18 and 4.19.

$$Q_2 = Q_1\left(\frac{N_2}{N_1}\right)$$

$$E_{p2} = E_{p1}\left(\frac{N_2}{N_1}\right)^2$$

Select 3 points (A, B, and C) on the pump characteristic curve.

	$N = 1700$		$N = 1500$		$N = 1300$	
Point	Q cfs	E_p ft	Q cfs	E_p ft	Q cfs	E_p ft
A	0	30	0	23.4	0	17.5
B	1.0	24.5	0.88	19.1	0.76	14.3
C	2.0	14.5	1.76	11.3	1.53	8.5

Curves are plotted on the sketch.

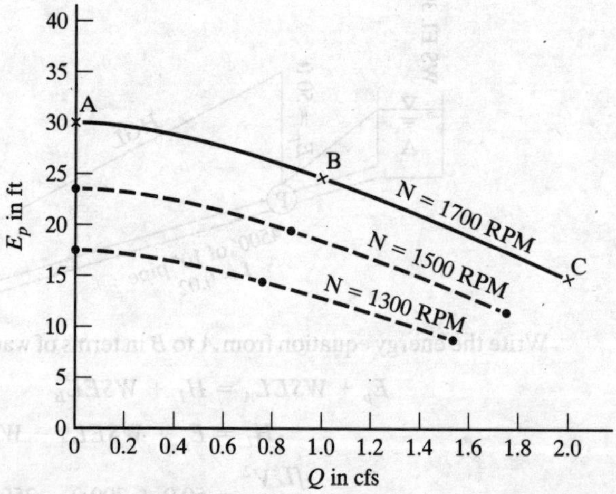

4.2.2 Pipeline with Pump

The pump and pipeline in Fig. 4.6 are used to transport water from the lower reservoir to the upper reservoir. Writing the energy equation between reservoirs A and B gives

$$El_A + E_p = El_B + H_L \tag{4.22}$$

or

$$E_p = El_B - El_A + K_{eq}Q^n \tag{4.23}$$

where El denotes elevation, $K_{eq}Q^n$ represents the headloss in the pipeline, and

$$K_{eq} = K_1 + K_2 + K_3$$

The energy added by the pump is equal to the difference in elevation between the two reservoirs plus the headloss in the pipeline. The discharge rate in the pipeline for a given pump can be determined graphically or numerically.

Graphic solution requires that the system curve (Eq. 4.23) be plotted on the same plot as the pump characteristic curve in Fig. 4.6. The pump operating point is at the intersection of the system curve and the pump characteristic curve. The discharge rate at the pump operating point can be read from the graph.

The pump characteristic curve shown in Fig. 4.6 can be approximated by the polynomial

$$E_p = AQ^2 + BQ + H_c \tag{4.24}$$

where H_c is the pump cutoff head (head when $Q = 0$). Coefficients A and B are evaluated from two points on the pump characteristic curve selected to represent the operating range of the pump. To ensure that the slope of pump characteristic

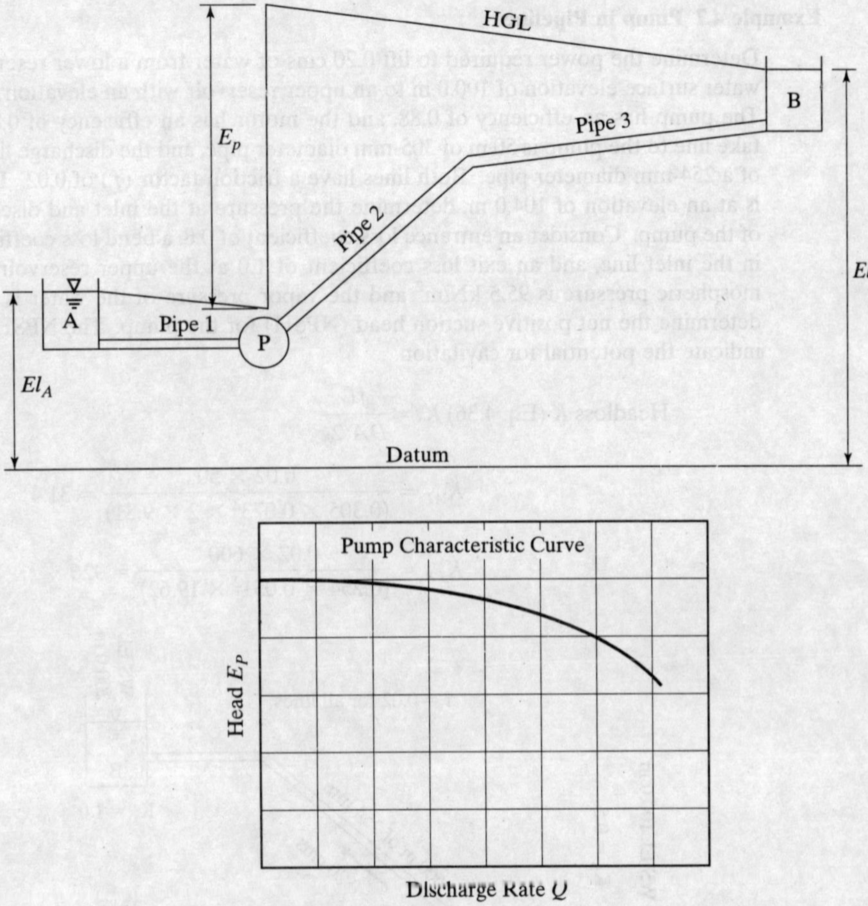

Figure 4.6 Pump in pipeline.

curve is always negative, both A and B should be negative. The energy equation becomes

$$AQ^2 + BQ + H_c = El_B - El_A + K_{eq}Q^n \tag{4.25}$$

Based on the Darcy–Weisbach headloss ($n = 2$), Eq. 4.25 can be solved using the quadratic equation

$$Q = \frac{-B - \sqrt{B^2 - 4ac}}{2a} \tag{4.26}$$

where

$$a = A - K_{eq}, \quad \text{and}$$

$$c = H_c + El_A - El_B$$

Example 4.7 Pump in Pipeline

Determine the power required to lift 0.20 cms of water from a lower reservoir with a water surface elevation of 100.0 m to an upper reservoir with an elevation of 200.0 m. The pump has an efficiency of 0.88, and the motor has an efficiency of 0.83. The intake line to the pump is 50 m of 305-mm diameter pipe, and the discharge line is 600 m of a 254-mm diameter pipe. Both lines have a friction factor (f) of 0.02. If the pump is at an elevation of 104.0 m, determine the pressure at the inlet and discharge sides of the pump. Consider an entrance loss coefficient of 0.8, a bend loss coefficient of 0.5 in the inlet line, and an exit loss coefficient of 1.0 at the upper reservoir. If the atmospheric pressure is 95.5 kN/m^2 and the vapor pressure of the water is 2.3 kN/m^2, determine the net positive suction head (NPSH) for the pump. The NPSH is used to indicate the potential for cavitation.

$$\text{Headloss } K \text{ (Eq. 4.36) } K = \frac{fL}{DA^2 2g}$$

$$K_{AP} = \frac{0.02 \times 50}{(0.305 \times 0.073^2 \times 2 \times 9.81)} = 31.4$$

$$K_{PB} = \frac{0.02 \times 600}{(0.254 \times 0.051^2 \times 19.62)} = 926$$

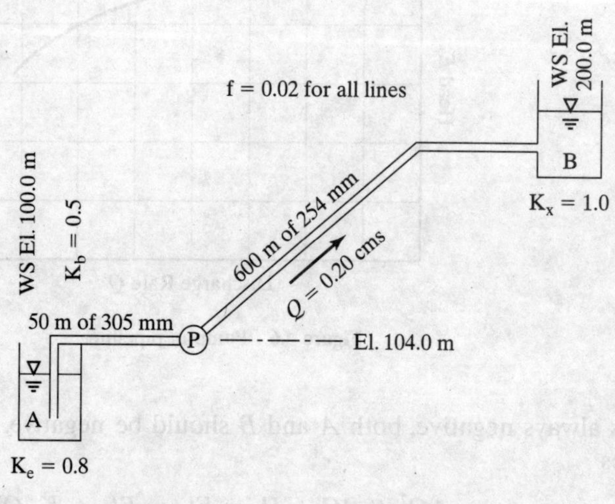

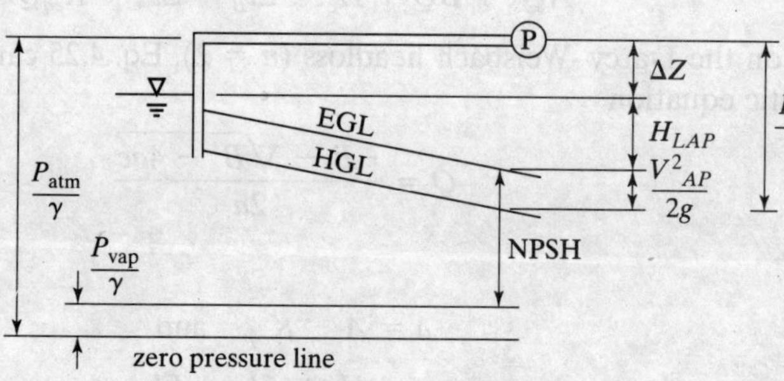

$$\text{Pumping head } E_p \text{ (Eq. 4.23)} = El_B - El_A + \text{Headloss} = 100.0 + \text{Headloss}$$

$$\text{Pipe friction losses} = K_{AP}Q^2 + K_{PB}Q^2 = 957(0.2)^2 = 38.3 \text{ m}$$

$$\text{Minor losses} = 0.8\frac{V_{AP}^2}{2g} + 0.5\frac{V_{AP}^2}{2g} + 1.0\frac{V_{PB}^2}{2g} = 0.8 \times \frac{2.74^2}{19.62} + 0.5 \times \frac{2.74^2}{19.62} + 1.0 \times \frac{3.92^2}{19.62}$$

$$= 0.31 + 0.19 + 0.78 = 1.3 \text{ m}$$

$$\text{Pumping head } E_p = 100.0 + 38.3 + 1.3 = 139.6 \text{ m}$$

$$\text{Water power} = Q\gamma\frac{E_p}{1000} = 0.20 \times 9.79 \times 139.6 = 273 \text{ kw}$$

$$\text{Pump input power} = \frac{\text{water power}}{\text{efficiency pump}} = \frac{273}{0.88} = 310 \text{ kw}$$

$$\text{Motor input power} = \frac{\text{pump power}}{\text{efficiency motor}} = \frac{310}{0.83} = 373 \text{ kw}$$

Pressure at Pump Intake (P_s)

Energy equation between A and pump

$$Z_A = \frac{V_{AP}^2}{2g} + \frac{P_s}{\gamma} + Z_p + H_{LAP}$$

$$\frac{P_s}{\gamma} = (Z_A - Z_P) - \frac{V_{AP}^2}{2g} - H_{LAP}$$

$$H_{LAP} = K_{AP}Q^2 + (0.8 + 0.5)\frac{V_{AP}^2}{2g} = 31.4(0.2)^2 + 1.3 \times \frac{2.74^2}{19.62} = 1.25 + 0.50 = 1.75 \text{ m}$$

$$\frac{P_s}{\gamma} = (100.0 - 104.0) - 0.4 - 1.7 = -6.1 \text{ m}$$

$$P_s = -6.1 \times 9.79 = -59.7 \text{ kN/m}^2$$

Pressure at Pump Discharge (P_d)

Energy equation between pump and B

$$\frac{V_{PB}^2}{2g} + \frac{P_d}{\gamma} + Z_p = Z_B + H_{LPB}$$

$$\frac{P_d}{\gamma} = (Z_B - Z_p) - \frac{V_{PB}^2}{2g} + H_{LPB}$$

$$H_{LPB} = K_{PB}Q^2 + 1.0\frac{V_{PB}^2}{2g} = 926(0.2)^2 + 1.0\frac{3.92^2}{19.62} = 37.0 + 0.8 = 37.8 \text{ m}$$

$$\frac{P_d}{\gamma} = (200.0 - 104.0) - 0.8 + 37.8 = 133.0 \text{ m}$$

$$P_d = 133.0 \times 9.79 = 1{,}302 \text{ kN/m}^2$$

An enlargement of the intake side of the pump is included in the figure. By definition the NPSH is

$$\text{NPSH} = \frac{(P_{\text{atm}} - P_{\text{vap}})}{\gamma} - \Delta Z - H_{LAP} = \frac{(95.5 - 2.3)}{9.79} - 4.0 - 1.8 = 3.7 \text{ m}$$

The cavitation parameter (σ) is defined as

$$\sigma = \frac{\text{NPSH}}{h_s}$$

where h_s is the head per stage. For a 4 stage pump

$$\sigma = \frac{3.7(4)}{139.6} = 0.11$$

To prevent cavitation of the pump, the value of σ must be greater than a critical value which is normally provided by the pump manufacturer.

Example 4.8 Pump Characteristic Curve

(a) Determine the discharge rate for the pump and pipeline system shown below. The vertical turbine pump lifts water from the Columbia River (elevation 200 ft) and discharges through a 10,000-ft, 30-inch diameter pipeline ($f = 0.02$) to an upper reservoir at elevation 800 ft. The pump characteristic curve is also shown below. Determine the discharge rate using Eq. 4.26 and check the result using the graphic method.

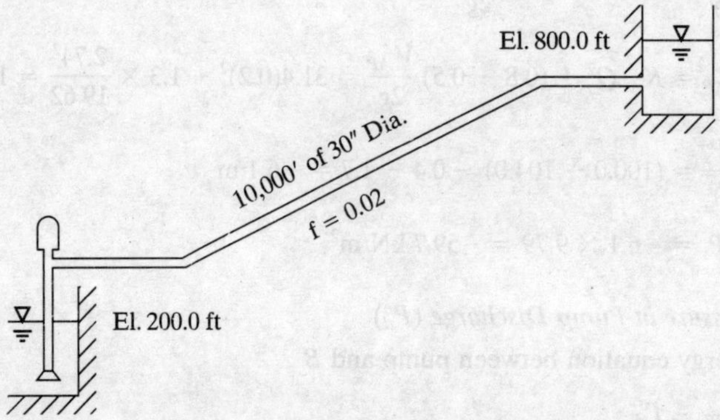

Compute the headloss K value

$$K = \frac{fL}{DA^2 2g} = \frac{0.02 \times 10{,}000}{(2.5) \times 4.91^2 \times 64.4} = 0.0515$$

Pump curve

Three points

$$E_p = 800, \quad Q = 0$$

$$E_p = 777, \quad Q = 20$$

$$E_p = 664, \quad Q = 50$$

Pump curve

$$E_p = -0.0522Q^2 - 0.11Q + 800$$

Discharge rate, Eq. 4.26

$$Q = \frac{-B - \sqrt{B^2 - 4ac}}{2a}$$

where

$$B = -0.11$$

$$a = A - K = -0.0515 - 0.0522 = -0.1037$$

$$c = H_c - El_B + El_A = 800 - 800 + 200 = 200$$

$$Q = \frac{+0.11 - \sqrt{(0.11)^2 + 4 \times 0.1037 \times 200}}{-2 \times 0.1037} = 43.4 \text{ cfs}$$

Graphic solution is shown on characteristic curve plot for pump head Eq. 4.23

$$E_p = El_B - El_A + KQ^n = 600 + 0.0515Q^2$$

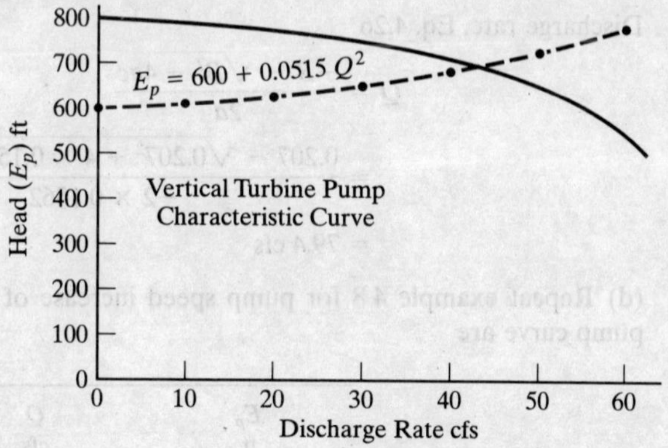

(b) Repeat example 4.8 for two pumps in parallel. For pumps in parallel, the points on the combined characteristic curve are

E_p ft	Q cfs
800	0
777	40
664	100

Pump curve

$$E_p = -0.01308Q^2 - 0.052Q + 800$$

Discharge rate, Eq. 4.26

$$Q = \frac{-B - \sqrt{B^2 - 4ac}}{2a}$$

$$= \frac{0.052 - \sqrt{0.052^2 + 4 \times 0.0646 \times 200}}{-2 \times 0.0646}$$

$$= 55.2 \text{ cfs}$$

(c) Repeat example 4.8 for two pumps in series: The points on the combined characteristic curve are

E_p ft	Q cfs
1,600	0
1,554	20
1,328	50

Pump curve

$$E_p = -0.1047Q^2 - 0.207Q + 1{,}600$$

Discharge rate, Eq. 4.26

$$Q = \frac{-B - \sqrt{B^2 - 4ac}}{2a}$$

$$= \frac{0.207 - \sqrt{0.207^2 + 4 \times 0.1562 \times 1{,}000}}{-2 \times 0.1562}$$

$$= 79.4 \text{ cfs}$$

(d) Repeat example 4.8 for pump speed increase of 10 percent. The points on the pump curve are

E_p ft	Q cfs
$800 \times 1.1^2 = 968$	$0 = 0$
$777 \times 1.1^2 = 940$	$20 \times 1.1 = 22$
$664 \times 1.1^2 = 803$	$50 \times 1.1 = 55$

Pump curve

$$E_p = -0.0523Q^2 - 0.121Q + 968$$

Discharge rate, Eq. 4.26

$$Q = \frac{-B - \sqrt{B^2 - 4ac}}{2a}$$

$$= \frac{0.12 - \sqrt{0.12^2 + 4 \times 0.1038 \times 368}}{-2 \times 0.1038}$$

$$= 59.5 \text{ cfs}$$

4.3 CULVERTS

A culvert is a cross-drainage structure used to convey water from one side of a highway, railroad, or canal to the other side. If the culvert exit and entrance are submerged, the hydraulics is the same as a pipe connecting two reservoirs. If the entrance is not submerged, then the structure is analyzed as open channel flow. The entrance is generally considered submerged if $H > 1.2D$, where H is the headwater depth and D is the pipe diameter or the height of a box culvert (Fig. 4.7). The energy equation for the culvert in Fig. 4.7 is

$$Z_{HW} - Z_{TW} = H_L \tag{4.27}$$

where Z_{HW} is the headwater surface elevation, Z_{TW} is the tailwater elevation, and H_L is the headloss. When the tailwater is below the crown (top) of the pipe at the outlet, the elevation of the crown of the pipe is used for Z_{TW} in Eq. 4.27 to compute the headloss. If there is no downstream channel and the outlet of the culvert discharges into the atmosphere, the elevation of the centerline of the culvert at the outlet is used for Z_{TW} in Eq. 4.27 to compute the headloss.

The headloss in the culvert includes both pipe friction and minor losses. Traditionally, the Mannings equation has been used in culvert hydraulics to compute friction losses. The headloss equation is

$$H_L = K_e \frac{V^2}{2g} + L \frac{n^2 V^2}{R^{4/3} C_m^2} + K_x \frac{V^2}{2g} \tag{4.28}$$

or

$$H_L = \frac{V^2}{2g}\left[K_e + K_f + K_x\right] \tag{4.29}$$

where K_e is the entrance loss coefficient (Fig. 4.8), K_f is the friction loss coefficient, K_x is the exit loss coefficient, and R is the hydraulic radius. Assuming the tailwater

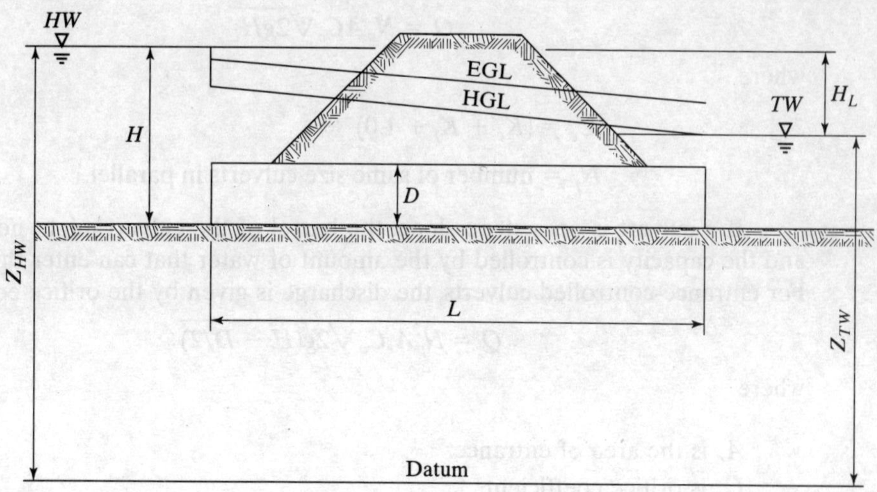

Figure 4.7 Culvert hydraulics.

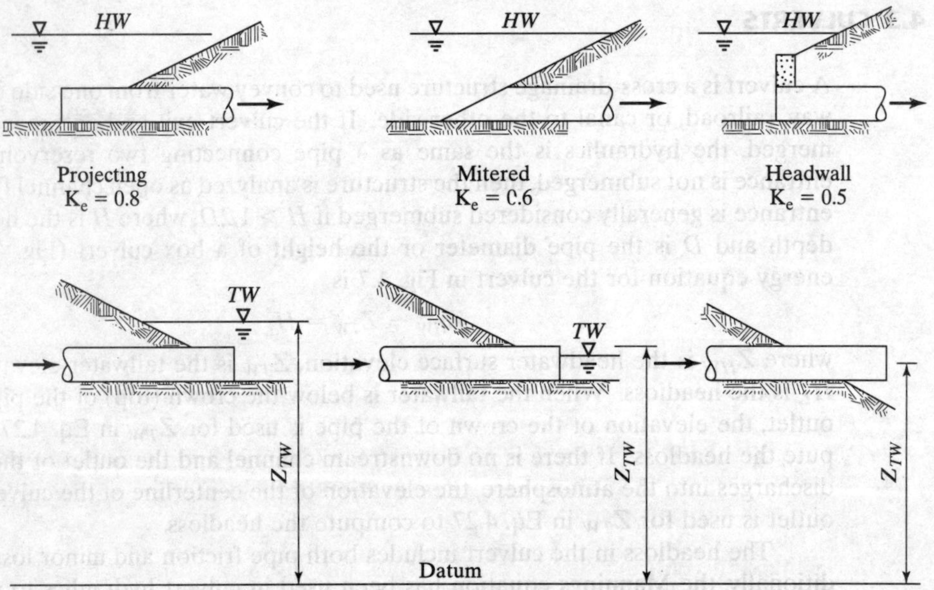

Figure 4.8 Culvert entrance and exit conditions.

velocity is small relative to the velocity in the culvert, K_x is approximately 1.0. The friction loss coefficient is

$$K_f = \frac{Ln^2 2g}{R^{4/3} C_m^2} \tag{4.30}$$

By substituting $V = Q/A$ in Eq. 4.29 and solving for Q gives the discharge rate for culverts flowing full (called outlet control)

$$Q = N_p A C_c \sqrt{2gH_L} \tag{4.31}$$

where

$$C_c = (K_e + K_f + 1.0)^{-1/2} \tag{4.32}$$

N_p = number of same size culverts in parallel.

If a culvert is on a steep slope, the barrel of the culvert might not flow full, and the capacity is controlled by the amount of water that can enter the entrance. For entrance-controlled culverts, the discharge is given by the orifice equation

$$Q = N_p A_e C_o \sqrt{2g(H - D/2)} \tag{4.33}$$

where

A_e is the area of entrance,
C_o is orifice coefficient,

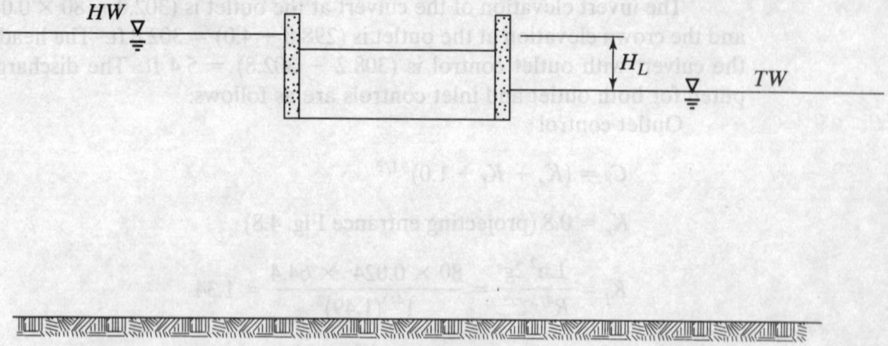

Figure 4.9 Submerged bridge.

H is the headwater depth (headwater elevation minus pipe invert elevation at inlet), and $H - D/2$ is the pressure head at the centroid of the area of the entrance.

Unless the entrance is flared, the area of the entrance is the same as the area of the pipe. The orifice coefficient (C_o) ranges from 0.62 for a sharp-edged entrance to 1.0 for a rounded entrance. The capacity of a culvert is the smaller of the two discharge rates computed from Eq. 4.31 for outlet control or Eq. 4.33 for entrance control.

The orifice equation is also used to compute the discharge rate through a submerged bridge (Fig. 4.9).

$$Q = A_b C_c \sqrt{2gH_L} \tag{4.34}$$

where A_b is the net area of the bridge opening (gross area less area of piers), and C_c is defined by Eq. 4.32. Experimental values of C_c by the Bureau of Public Roads for typical bridges range from 0.7 to 0.9. Bridges are normally not designed to operate under submerged conditions.

Example 4.9 Culvert Capacity

Determine the discharge rate through three parallel corrugated metal pipe culverts shown in the sketch below. The 48-inch diameter culverts have a Manning "n" value of 0.024. The culverts are 80 ft long, laid on a slope of 4.0 percent, and have a projecting inlet and outlet. The culvert inlet invert (bottom) elevation is 302.0 ft with a headwater elevation of 308.2 ft and a tailwater elevation of 302.0 ft.

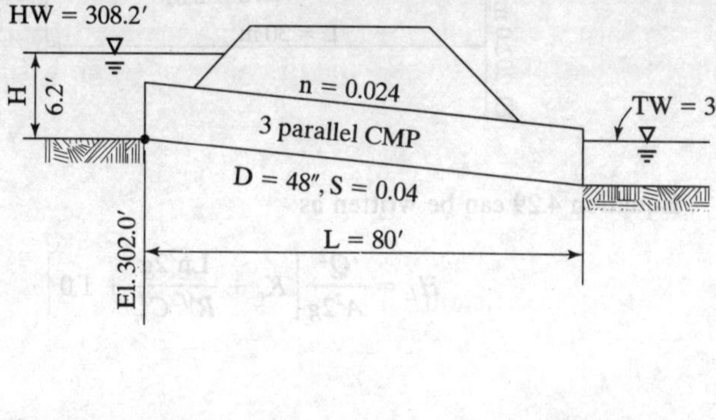

The invert elevation of the culvert at the outlet is (302.0 − 80 × 0.04) = 298.8 ft and the crown elevation at the outlet is (298.8 + 4.0) = 302.8 ft. The headloss through the culvert with outlet control is (308.2 − 302.8) = 5.4 ft. The discharge rate computed for both outlet and inlet controls are as follows:

Outlet control

$$C_c = (K_e + K_f + 1.0)^{-1/2}$$

$K_e = 0.8$ (projecting entrance Fig. 4.8)

$$K_f = \frac{Ln^2 2g}{R^{4/3} C_m^2} = \frac{80 \times 0.024^2 \times 64.4}{1^{4/3}(1.49)^2} = 1.34$$

$$C_c = (0.8 + 1.34 + 1.0)^{-1/2} = 0.56$$

$$Q = N_p A C_c \sqrt{2gH_L} = 3 \times \frac{\pi 4^2}{4} \times 0.56\sqrt{64.4 \times 5.4} = 394 \text{ cfs}$$

Inlet control

$$Q = N_p A C_o \sqrt{2g\left(H - \frac{D}{2}\right)} = 3 \times \frac{\pi 4^2}{4} \times 0.62\sqrt{2g(6.2 - 2.0)} = 384 \text{ cfs}$$

Because the discharge with inlet control is less than the discharge with outlet control, the discharge through the culverts is 384 cfs.

Example 4.10 Culvert Size

Determine the size of two parallel reinforced concrete pipe culverts ($n = 0.012$) to carry a discharge of 50 cms with a headloss of 3.0 m for outlet control. As shown in the sketch below, the culverts are 30 m long with headwalls ($K_e = 0.5$) on a slope of 1.0 percent, and both exit and entrance submerged. The inlet invert elevation is 202.0 m with a headwater elevation of 207.4 m and a tailwater elevation of 204.4 m.

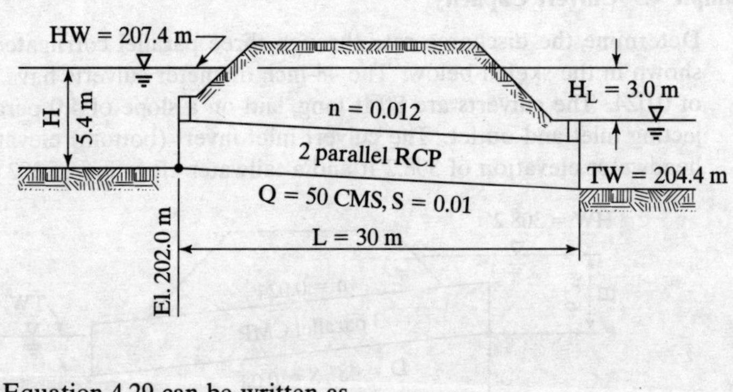

Equation 4.29 can be written as

$$H_L = \frac{Q^2}{A^2 2g}\left[K_e + \frac{Ln^2 2g}{R^{4/3} C_m^2} + 1.0\right]$$

where Q is the discharge per pipe, A is the area ($\pi D^2/4$), and R is the hydraulic radius ($D/4$) of the pipe. Substituting the known values into the headloss equation gives

$$3.0 = \frac{25^2}{\left(\frac{\pi D^2}{4}\right)^2 2 \times 9.81}\left[0.5 + \frac{30 \times 0.012^2 \times 19.62}{\left(\frac{D}{4}\right)^{4/3}(1.0)^2} + 1.0\right]$$

$$3.0 = \frac{51.64}{D^4}\left[1.5 + \frac{0.54}{D^{4/3}}\right]$$

or

$$D^4 - 25.82 - 9.30D^{-1.33} = 0$$

The above equation can be solved for the diameter by either trial and error or by a numerical procedure. For demonstration purposes, the Newton–Raphson numerical procedure will be used to solve for the pipe diameter.

$$F(D) = D^4 - 9.30D^{-1.33} - 25.82$$

$$F'(D) = 4D^3 + 12.37D^{-2.33}$$

The new estimate of diameter (D^+) is

$$D^+ = D - \frac{F(D)}{F'(D)}$$

where D is the old estimate of diameter. Using an initial estimate of the diameter of 2.0 m

$$F(D) = 16 - 3.7 - 25.8 = -13.5$$

$$F'(D) = 32 + 2.5 = 34.5$$

$$D^+ = 2.0 + \frac{13.5}{34.5} = 2.4 \text{ m}$$

Based on a diameter of 2.4 m

$$F(D) = 33.2 - 2.9 - 25.8 = 4.5$$

$$F'(D) = 55.3 + 1.6 = 56.9$$

$$D^+ = 2.4 - \frac{4.5}{56.9} = 2.32 \text{ m}$$

and finally based on $D = 2.32$ m

$$F(D) = 28.97 - 3.03 - 25.82 = 0.12$$

$$F'(D) = 49.9 + 1.7 = 51.6$$

$$D^+ = 2.32 - \frac{0.12}{51.6} = 2.32 \text{ m}$$

Culverts are available in sizes 7.5 ft (2.28 m) or 8 ft (2.44 m) diameter. The next culvert size is 2.44 m in diameter.

4.4 PIPELINES CONNECTING RESERVOIRS

Reservoirs in a pipeline or pipe network system are considered fixed-grade nodes, indicating a node with a constant EGL. A pipe is considered hydraulically long when pipe friction is the dominant headloss term and minor losses can be neglected. A pipe is considered hydraulically short when minor losses account for a significant part of the total headloss. A culvert (Section 4.3) is an example of short pipe flow.

4.4.1 Pipes in Series

The pipeline shown in Fig. 4.10 is designed to transport water from the upper reservoir to the lower reservoir. The pipeline consists of three pipes in series, two junction nodes and two fixed-grade nodes (A and B). For long pipe problems where the friction losses are much greater than the minor losses, minor losses are neglected and the friction headloss equation is written as

$$H_L = KQ^n \tag{4.35}$$

For the Darcy–Weisbach equation

$$K = \frac{fL}{DA^2 2g} \quad \text{and} \quad n = 2. \tag{4.36}$$

For the Hazen–Williams equation

$$K = \frac{L}{(C_w C_H A R^{0.63})^{1.85}} \quad \text{and} \quad n = 1.85. \tag{4.37}$$

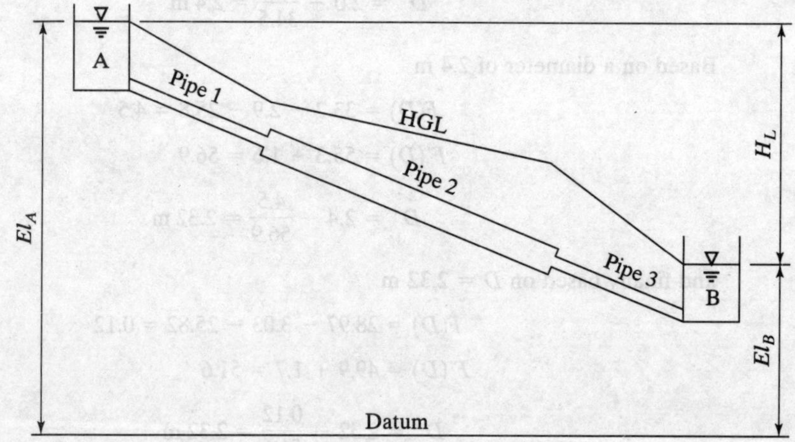

Figure 4.10 Pipes in series.

For the pipeline in Fig. 4.10, the flow rate in each pipe is unknown. Writing the continuity equation for the two junction nodes yields

$$Q_1 - Q_2 = 0 \tag{4.38}$$

$$Q_2 - Q_3 = 0 \tag{4.39}$$

and writing the energy equation between the two fixed-grade nodes yields

$$K_1Q_1^n + K_2Q_2^n + K_3Q_3^n = El_A - El_B \tag{4.40}$$

Combining Eqs. 4.38–4.40

$$H_L = El_A - El_B = K_{eq}Q^n \tag{4.41}$$

where

$$K_{eq} = K_1 + K_2 + K_3 \tag{4.42}$$

Example 4.11 Pipes in Series

As shown in the sketch below, a pipeline ($f = 0.02$) consisting of three pipes in series (6,000 ft of 12-in. diameter pipe, 10,000 ft of 18-in. diameter pipe, and 4,000 ft of 10-in. diameter pipe) runs from an upper reservoir to a lower reservoir with water surface elevations of 300 ft and 250 ft, respectively. Determine discharge through the pipeline and the pressure at each of the junction nodes.

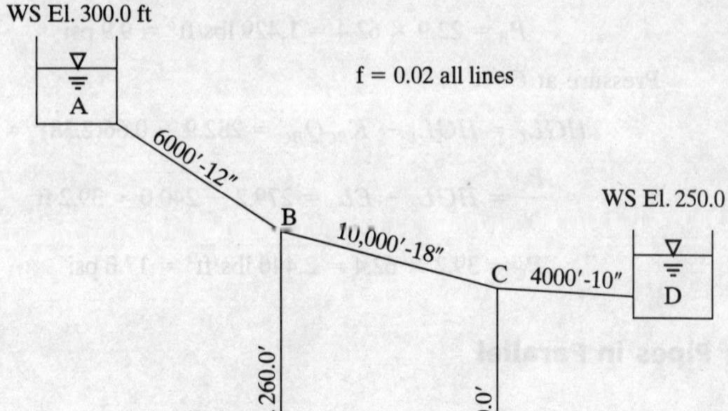

The headloss K value is computed for each line from Eq. 4.36

$$K = \frac{fL}{DA^2 2g}$$

Line A–B

$$K_{AB} = \frac{0.02 \times 6000}{(1.0 \times 0.78^2 \times 64.4)} = 3.02$$

Line B–C

$$K_{BC} = \frac{0.02 \times 10{,}000}{(1.5 \times 1.77^2 \times 64.4)} = 0.66$$

Line C–D

$$K_{CD} = \frac{0.02 \times 4000}{(0.83 \times 0.54^2 \times 64.4)} = 5.13$$

From Eq. 4.42, the equivalent K value

$$K_{eq} = K_{AB} + K_{BC} + K_{CD} = 3.02 + 0.66 + 5.13 = 8.81$$

The discharge rate computed from Eq. 4.41

$$Q = \left[\frac{El_A - El_D}{K_{eq}}\right]^{1/n} = \left[\frac{300.0 - 250.0}{8.81}\right]^{1/2} = 2.38 \text{ cfs}$$

Pressure at B

$$El\ HGL \text{ at } B$$

$$HGL_B = El_A - K_{AB}Q_{AB}^2 = 300.0 - 3.02(2.38)^2 = 282.9 \text{ ft}$$

$$\frac{P_B}{\gamma} = HGL_B - El_B = 282.9 - 260 = 22.9 \text{ ft}$$

$$P_B = 22.9 \times 62.4 = 1{,}429 \text{ lbs/ft}^2 = 9.9 \text{ psi}$$

Pressure at C

$$HGL_C = HGL_B - K_{BC}Q_{BC} = 282.9 - 0.66(2.38)^2 = 279.2 \text{ ft}$$

$$\frac{P_C}{\gamma} = HGL_c - EL_c = 279.2 - 240.0 = 39.2 \text{ ft}$$

$$P_C = 39.2 \times 62.4 = 2{,}446 \text{ lbs/ft}^2 = 17.0 \text{ psi}$$

4.4.2 Pipes in Parallel

The pipeline connecting the two reservoirs in Fig. 4.11 consists of pipes in series and pipes in parallel. Pipes 2 and 4 extend from the same junction nodes and are considered to be in parallel with each other. The flow rates in the four pipes shown in Fig. 4.11 are unknown. For the two pipes in parallel, the headloss in pipe 2 is equal to the headloss in pipe 4. Pipes 2 and 4 can be considered to form a closed pipe loop, and the summation of headloss (clockwise plus) around any closed pipe loop is equal to zero or

$$K_2Q_2^n - K_4Q_4^n = 0 \qquad \textbf{(4.43)}$$

Writing the continuity equation for the two junction nodes

$$Q_1 - Q_2 - Q_4 = 0 \qquad \textbf{(4.44)}$$

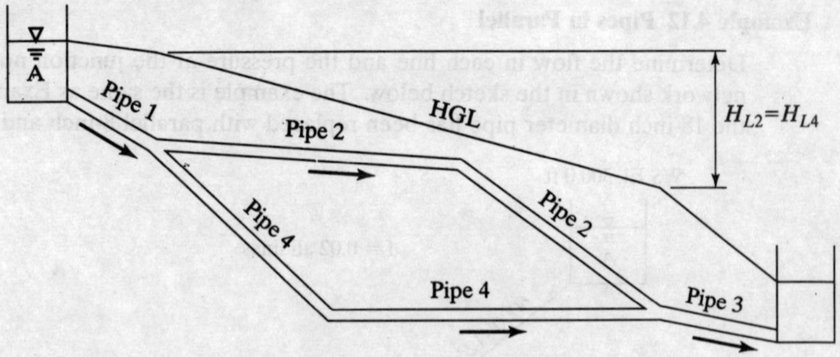

Figure 4.11 Pipes in parallel.

and

$$Q_2 + Q_4 - Q_3 = 0 \tag{4.45}$$

and writing the energy equation between the two fixed-grade nodes

$$K_1Q_1^n + K_2Q_2^n + K_3Q_3^n = El_A - El_B \tag{4.46}$$

or

$$K_1Q_1^n + K_4Q_4^n + K_3Q_3^n = El_A - El_B \tag{4.47}$$

yields four equations to solve for the four unknown flow rates. (Either Eq. 4.46 or Eq. 4.47 can be used in the computations.)

A different procedure is used to solve the equations manually than is used in computer analysis. For manual computation, the two parallel pipes are replaced with a single equivalent (imaginary) pipe that gives the same headloss based on total flow (Q) as pipe 2 or pipe 4 based on partial flow or

$$H_{Leq} = H_{L2} = H_{L4} = H_L \tag{4.48}$$

From Eqs. 4.44 and 4.45, the total flow is

$$Q = Q_1 = Q_3 \tag{4.49}$$

Solving the headloss Eq. 4.35 for Q and substituting into Eq. 4.44 yields

$$\left[\frac{H_L}{K_{eq}}\right]^{1/n} = \left[\frac{H_L}{K_2}\right]^{1/n} + \left(\frac{H_L}{K_4}\right)^{1/n} \tag{4.50}$$

or

$$K_{eq} = \frac{1}{\left[\dfrac{1}{K_2^{1/n}} + \dfrac{1}{K_4^{1/n}}\right]^n} \tag{4.51}$$

The equivalent K value (Eq. 4.51) is used as K_2 in Eq. 4.42 to solve for Q in Eq. 4.41. Equations 4.35 and 4.48 are then used to solve for Q_2 and Q_4.

Example 4.12 Pipes in Parallel

Determine the flow in each line and the pressure at the junction nodes for the pipe network shown in the sketch below. The example is the same as Example 4.11, except the 18-inch diameter pipe has been replaced with parallel 8-inch and 10-inch pipes.

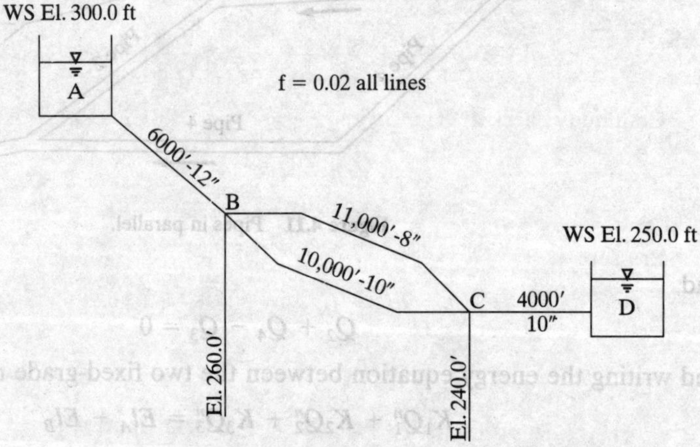

Compute the headloss K value

$$K = \frac{fL}{DA^2\,2g}$$

$$K_{BC-8} = \frac{0.02 \times 11{,}000}{(0.67 \times 0.35^2 \times 64.4)} = 41.6$$

$$K_{BC-10} = \frac{0.02 \times 10{,}000}{(0.83 \times 0.54^2 \times 64.4)} = 12.8$$

The equivalent K value for parallel pipes (Eq. 4.51)

$$K_{eqBC} = \frac{1}{\left[\dfrac{1}{K_8^{1/n}} + \dfrac{1}{K_{10}^{1/n}}\right]^n} = \frac{1}{\left[\dfrac{1}{(41.6)^{1/2}} + \dfrac{1}{(12.8)^{1/2}}\right]^2} = 5.30$$

The equivalent K value for pipes in series (Eq. 4.42)

$$K_{eq} = K_{AB} + K_{eqBC} + K_{CD} = 3.02 + 5.30 + 5.13 = 13.45$$

The discharge rate from Eq. 4.41

$$Q = \left[\frac{El_A - El_D}{K_{eq}}\right]^{1/n} = \left[\frac{300.0 - 250.0}{13.45}\right]^{1/2} = 1.92 \text{ cfs}$$

Discharge rates in parallel pipes

Compute headloss for parallel pipes

$$H_{LBC} = K_{eqBC}Q^n = 5.30(1.92)^2 = 19.5 \text{ ft}$$

Discharge 8-inch line (Eq. 4.35)

$$Q_{8''} = \left[\frac{H_{LBC}}{K_{BC-8}}\right]^{1/n} = \left[\frac{19.5}{41.6}\right]^{1/2} = 0.68 \text{ cfs}$$

Discharge 10-inch line (Eq. 4.35)

$$Q_{10} = \left[\frac{H_{LBC}}{K_{BC-10}}\right]^{1/n} = \left[\frac{19.5}{12.8}\right]^{1/2} = 1.23 \text{ cfs}$$

Continuity check at B

$$Q_{AB} = Q_{BC-8} + Q_{BC-10}$$

$$1.92 \approx 0.68 + 1.23 = 1.91 \text{ cfs}$$

Pressure at B

$$HGL_B = El_A - H_{LAB} = 300.0 - 3.02 \times 1.92^2 = 288.9 \text{ ft}$$

$$\frac{P_B}{\gamma} = HGL_B - El_B = 28.9 \text{ ft}$$

$$P_B = 28.9 \times 62.4 = 1{,}803 \text{ lbs/ft}^2 = 12.5 \text{ psi}$$

Pressure at C

$$HGL_C = HGL_B - H_{LBC} = 288.9 - 19.5 = 269.4 \text{ ft}$$

$$\frac{P_C}{\gamma} = HGL_C - El_C = 269.4 - 240.0 = 29.4 \text{ ft}$$

$$P_C = 29.4 \times 62.4 = 1{,}835 \text{ lbs/ft}^2 = 12.7 \text{ psi}$$

4.4.3 Three-Reservoir System

Figure 4.12 shows a pipe network system connecting three reservoirs. The problem is included as an introduction to more complex pipe network systems. Since there are three unknowns in this problem (the discharge rate in the three pipes), three equations are required. The continuity equation can be written at the junction node, and two energy equations can be written between the reservoirs. If there are N

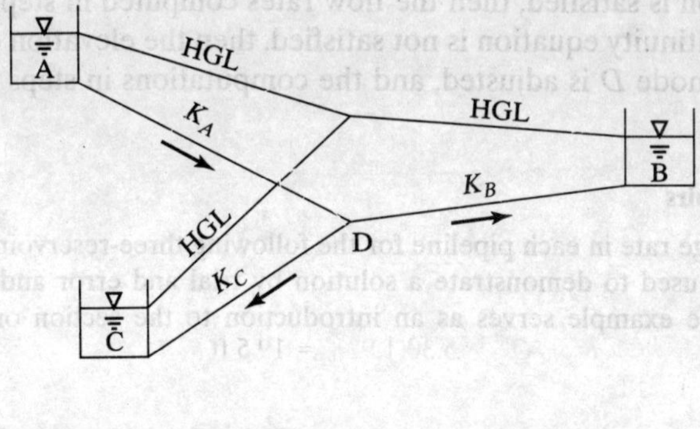

Figure 4.12 Pipe network connecting three reservoirs.

FGN (fixed-grade nodes or reservoirs) in a system, then there are $N - 1$ independent energy equations written between the reservoirs.

The continuity equation written at junction node D is

$$Q_A - Q_B - Q_C = 0 \quad (4.52)$$

and the two energy equations

$$El_A - K_A Q_A^n - K_C Q_C^n = El_C$$

or

$$K_A Q_A^n + K_C Q_C^n = El_A - El_C \quad (4.53)$$

and

$$El_C + K_C Q_C^n - K_B Q_B^n = El_B$$

or

$$K_C Q_C^n - K_B Q_B^n = El_B - El_C \quad (4.54)$$

For a computer solution, Eqs. 4.52–4.54 are solved simultaneously. For a manual solution, the problem is normally solved as follows:

1. The elevation of the HGL at junction node D is assumed HGL_D.
2. Based on the assumed HGL_D, the discharge rate in each line is computed

$$Q_A = \left[\frac{El_A - HGL_D}{K_A}\right]^{1/n} \quad (4.55)$$

$$Q_B = \left[\frac{HGL_D - El_B}{K_B}\right]^{1/n} \quad (4.56)$$

$$Q_C = \left[\frac{HGL_D - El_C}{K_C}\right]^{1/n} \quad (4.57)$$

3. The continuity equation (Eq. 4.52) at the junction node is checked. If the continuity equation is satisfied, then the flow rates computed in step 2 are correct. If the continuity equation is not satisfied, then the elevation of the HGL at junction node D is adjusted, and the computations in steps 2 and 3 are repeated.

Example 4.13 Three Reservoirs

Determine the discharge rate in each pipeline for the following three-reservoir problems. This example is used to demonstrate a solution by trial and error and a numerical procedure. The example serves as an introduction to the section on pipe networks.

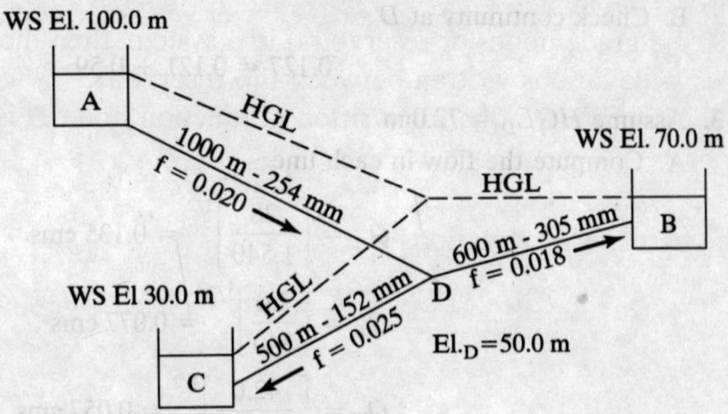

Compute the headloss K values

$$K = \frac{fL}{DA^2\,2g}$$

$$K_A = 1{,}540$$

$$K_B = 340$$

$$K_C = 12{,}900$$

The assumed direction of flow is shown on the sketch.
Solve by trial and error

1. Assume $HGL_D = 70.0$ m

A. Compute the flow in each line from Eq. 4.35

$$Q_A = \left[\frac{30.0}{1{,}540}\right]^{1/2} = 0.14 \text{ cms}$$

$$Q_B = 0$$

$$Q_C = \left[\frac{40}{12{,}900}\right]^{1/2} = 0.056 \text{ cms}$$

B. Check continuity at D

$$0.14 \neq 0 + 0.056$$

2. Assume $HGL_D = 75.0$ m

A. Compute the flow in each line

$$Q_A = \left[\frac{25}{1{,}540}\right]^{1/2} = 0.127 \text{ cms}$$

$$Q_B = \left[\frac{5}{340}\right]^{1/2} = 0.121 \text{ cms}$$

$$Q_C = \left[\frac{45}{12{,}900}\right]^{1/2} = 0.059 \text{ cms}$$

B. Check continuity at D

$$0.127 \neq 0.121 + 0.59$$

3. Assume $HGL_D = 72.0$ m

A. Compute the flow in each line

$$Q_A = \left[\frac{28}{1{,}540}\right]^{1/2} = 0.135 \text{ cms}$$

$$Q_B = \left[\frac{2.0}{340}\right]^{1/2} = 0.077 \text{ cms}$$

$$Q_C = \left[\frac{42.0}{12{,}900}\right]^{1/2} = 0.057 \text{ cms}$$

B. Check continuity at D

$$0.135 = 0.077 + 0.057$$

flows balance.

Trial and error solution

$$Q_A = 0.135 \text{ cms}, \quad Q_B = 0.077 \text{ cms}, \quad Q_C = 0.057 \text{ cms}$$

Simultaneous Equations Solution

The three equations include the linear continuity or node equation (Eq. 4.52) and the two nonlinear energy or loop equations. The nonlinear terms in the energy equations are linearized by using the first two terms in Taylor's series.

$$KQ^2 = Kq^2 + 2Kq(Q - q) = -Kq^2 + 2Kq(Q)$$

where Q is the new estimate of flow and q is the previous estimate of flow.

The three linear equations are

Continuity equation

$$(1) \quad Q_A - Q_B - Q_C = 0$$

Energy equations

$$(2) \quad (2K_Aq_A)Q_A + (2K_Cq_C)Q_C = K_Aq_A^2 + K_Cq_C^2 + El_A - El_C$$

$$(3) \quad -(2K_Bq_B)Q_B + (2K_Cq_C)Q_C = -K_Bq_B^2 + K_Cq_C^2 + El_B - E_C$$

Solution to the above equations is an iterative procedure. The initial estimate of flow is based on a velocity of 1.0 mps in each line giving

$$q_A = 0.05 \text{ cms}, \quad q_B = 0.07 \text{ cms}, \quad q_C = 0.02 \text{ cms}$$

Based on the estimated initial flow rates, the equations are

$$(1) \quad Q_A - Q_B - Q_C = 0$$

$$(2) \quad 154Q_A + 516Q_C = 79.0$$

$$(3) \quad -47.6Q_B + 516Q_C = 43.5$$

Solution yields

$$Q_A = 0.198, \quad Q_B = 0.104, \quad Q_C = 0.094$$

The linear equations based on the new flow rates are

$$(1)\quad Q_A - Q_B - Q_C = 0$$

$$(2)\quad 610Q_A + 2{,}425Q_C = 244$$

$$(3)\quad -70.7Q_B + 2{,}425Q_C = 150$$

Solution yields

$$Q_A = 0.145, \quad Q_B = 0.081, \quad Q_C = 0.064$$

The new linear equations are

$$Q_A - Q_B - Q_C = 0$$

$$447Q_A + 1{,}651Q_C = 155$$

$$-55.1Q_B + 1{,}651Q_C = 90.6$$

Solution yields

$$Q_A = 0.135, \quad Q_B = 0.077, \quad Q_C = 0.058$$

The iterative procedure continues until the maximum change in flow is within some limit, for example, $|Q - q|_{max} \leq 0.002$. Based on the previously computed flow rates, the new linear equations are

$$Q_A - Q_B - Q_C = 0$$

$$416Q_A + 1{,}496Q_C = 141.5$$

$$-52.4Q_B + 1{,}496Q_C = 81.4$$

Final iteration yields

$$Q_A = 0.135 \text{ cms}$$

$$Q_B = 0.078 \text{ cms}$$

$$Q_C = 0.057 \text{ cms}$$

where $|Q - q|_{max} < 0.002$.

A set of computer programs developed in conjunction with this textbook is described by Section 1.4.5. A program for solving linear equations has been included under file SIMEQ.FOR. The input file for this example is SIMEQ.DAT. The FORTRAN program can be modified to compute K values and coefficients for the linearized energy equations.

4.5 PIPE NETWORK SYSTEMS

A municipal water distribution system is used to deliver water to the consumer. Although water is withdrawn from along the pipes in a pipe network system, for computational purposes all demands on the system are assumed to occur at the junction nodes. Pressure is the main concern in a water distribution system. At no time should the water pressure in the system be so low that contaminated groundwater could enter the system at points of leakage even with water hammer pressure waves.

The total water consumption for the water distribution service area is usually available from past records. The total water demand on the system is allocated to the junction nodes based on the estimated residential, industrial, and commercial water demands for each node.

A simple water distribution system is shown in Fig. 4.13. The system is connected to fixed-grade nodes (elevated water tanks labeled A and B in Fig. 4.13) to permit the computation of pressures throughout the system after the flow rates for all pipes have been determined. Pipes 1–9 and junction nodes 1–6 have been numbered on the sketch in Fig. 4.13. The demands (C) at the junction nodes are also shown on the sketch. The number of equations required for solution of the system is equal to the number of unknowns (number of pipes = 9) in the network. The number of independent equations (N_{eq}) available for solution is equal to the number of junction nodes (N_j) plus the number of loops (N_ℓ) plus the number of FGN (N_f) minus 1 or

$$N_{eq} = N_j + N_\ell + N_f - 1 \tag{4.58}$$

or for the example

$$N_{eq} = 6 + 2 + 2 - 1 = 9$$

The estimated direction of flow is also shown on the sketch. If the assumed direction of flow is wrong, the computed flow rate will be negative. Writing the six continuity equations for six junction nodes based on the assumed direction of flow gives

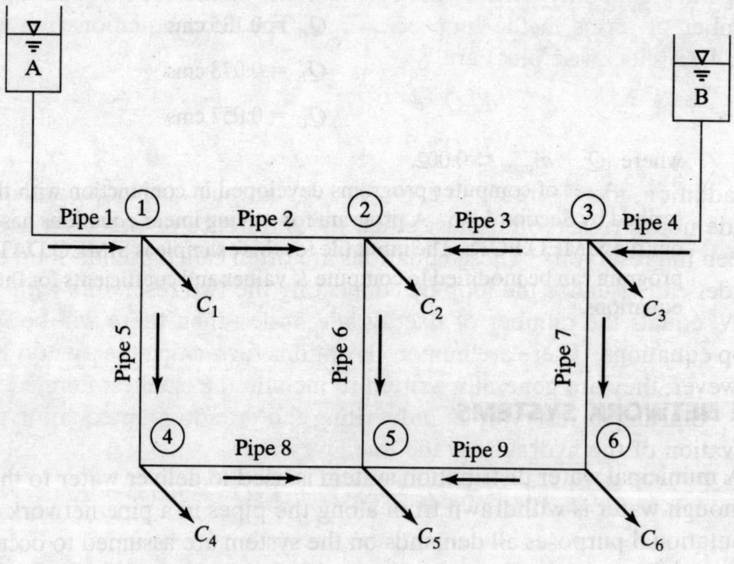

Figure 4.13 Pipe network system.

$$Q_1 - Q_2 - Q_5 - C_1 = 0 \tag{4.59.1}$$

$$Q_2 + Q_3 - Q_6 - C_2 = 0 \tag{4.59.2}$$

$$Q_4 - Q_3 - Q_7 - C_3 = 0 \tag{4.59.3}$$

$$Q_5 - Q_8 - C_4 = 0 \tag{4.59.4}$$

$$Q_6 + Q_8 + Q_9 - C_5 = 0 \tag{4.59.5}$$

$$Q_7 - Q_9 - C_6 = 0 \tag{4.59.6}$$

or in matrix notation

$$\begin{array}{ll} & \text{Pipe} \quad 1 \quad 2 \quad 3 \quad 4 \quad 5 \quad 6 \quad 7 \quad 8 \quad 9 \\ \text{Equation number} & \begin{array}{c} 1 \\ 2 \\ 3 \\ 4 \\ 5 \\ 6 \end{array} \begin{bmatrix} 1 & -1 & 0 & 0 & -1 & 0 & 0 & 0 & 0 \\ 0 & 1 & 1 & 0 & 0 & -1 & 0 & 0 & 0 \\ 0 & 0 & -1 & 1 & 0 & 0 & -1 & 0 & 0 \\ 0 & 0 & 0 & 0 & 1 & 0 & 0 & -1 & 0 \\ 0 & 0 & 0 & 0 & 0 & 1 & 0 & 1 & 1 \\ 0 & 0 & 0 & 0 & 0 & 0 & 1 & 0 & -1 \end{bmatrix} \begin{bmatrix} Q_1 \\ Q_2 \\ Q_3 \\ Q_4 \\ Q_5 \\ Q_6 \\ Q_7 \\ Q_8 \\ Q_9 \end{bmatrix} = \begin{bmatrix} C_1 \\ C_2 \\ C_3 \\ C_4 \\ C_5 \\ C_6 \end{bmatrix} \end{array} \tag{4.59.7}$$

The loop equations are written for the condition that the algebraic sum of headloss around any closed loop in the system is equal to zero. Nonoverlapping loops are generally used in the formulation of the loop equations to reduce the number of terms in the loop equations. The loop equations for the network in Fig. 4.13 (clockwise plus) are

$$K_2 Q_2^n + K_6 Q_6^n - K_8 Q_8^n - K_5 Q_5^n = 0 \tag{4.60.1}$$

$$-K_3 Q_3^n + K_7 Q_7^n + K_9 Q_9^n - K_6 Q_6^n = 0 \tag{4.60.2}$$

In addition, the algebraic sum of headloss around any loop containing two fixed-grade nodes (reservoirs) must equal the difference in water surface elevation between the reservoirs. These equations are often called pseudo-loop equations. The reader can visualize the loop by connecting the two reservoirs with a no flow pipe. If N_f equals the number of fixed-grade nodes, then there will be $N_f - 1$ pseudo-loop equations. There are numerous options for writing the pseudo-loop equations; however, they are generally written to include the smallest number of terms.

Starting at reservoir B and writing the pseudo-loop equation in terms of the elevation of the hydraulic grade line gives

$$El_B - K_4 Q_4^n - K_3 Q_3^n + K_2 Q_2^n + K_1 Q_1^n = El_A \tag{4.61}$$

or

$$K_4 Q_4^n + K_3 Q_3^n - K_2 Q_2^n - K_1 Q_1^n = El_B - El_A \tag{4.62}$$

To solve for the flow rates in each of the pipes in Fig. 4.13, the nine equations (six linear and three nonlinear) must be solved.

Many variations of methods are available for solving the system of equations representing conservation of mass and energy in a pipe network. The Hardy Cross and linear methods are presented here. The Hardy Cross method traditionally included in most texts since the 1940's is amiable to manual computations as well as computer programs. The alternative more computationally efficient linear method is incorporated in the widely applied computer models discussed in Section 4.7. The linear method is a very stable numerical procedure that is very effective in computer modeling of pipe systems, including complex networks with thousands of pipes. Both the linear and Hardy Cross methods compute the discharge in each pipe of a network.

After the flows have been computed for all pipes in the network, the elevation of the hydraulic grade line and the pressure are computed for each junction node. The hydraulic grade line elevation for any junction node is equal to the elevation of a fixed-grade node minus the algebraic sum of headlosses in the pipes connecting the fixed-grade node and the junction node. The headloss in a pipe is considered positive if the flow in the pipe is away from the fixed-grade node and toward the junction node. The pressure (P) at the junction node (at ground elevation) is computed as

$$P = (El_{HGL} - El_{GD})\gamma_{\omega} \tag{4.63}$$

where El_{HGL} is the elevation of the HGL at the junction node, El_{GD} is the ground elevation at the junction node, and γ_{ω} is the unit weight of water.

4.5.1 Hardy Cross Method

The Hardy Cross method of pipe network analyses requires an initial estimate of flow in each pipe so that the continuity equation for the junction nodes are satisfied. The loop (both closed and pseudo) equations are solved iteratively one at a time until the correction for each loop is within an acceptable magnitude. When the loop equations are solved simultaneously, it is generally referred to as the Newton–Raphson method of pipe network analysis.

If Q_i is the correct flow rate for pipe i and q_i is the assumed flow rate (or the flow rate from the previous iteration), then

$$Q_i = q_i + \Delta \tag{4.64}$$

where Δ is a correction term to be applied to all (N) pipes in the loop. The closed loop equation becomes

$$\sum_{i=1}^{N} K_i(q_i + \Delta)^n = 0 \tag{4.65}$$

Expanding

$$\sum_{i=1}^{N} K_i q_i^n + \sum_{i=1}^{N} nK_i\Delta q_i^{n-1} + \frac{n-1}{2}\sum_{i=1}^{N} nK_i\Delta^2 q_i^{n-2} + \cdots = 0 \tag{4.66}$$

Using only the first two terms in the binomial expansion and solving for Δ yields

$$\Delta = -\frac{\sum_{i=1}^{N} K_i q_i^n}{\sum_{i=1}^{N} |nK_i q_i^{n-1}|} \tag{4.67}$$

for the closed loops and

$$\Delta = -\frac{\sum_{i=1}^{N} K_i q_i^n - (El_B - El_A)}{\sum_{i=1}^{N} |nK_i q_i^{n-1}|} \tag{4.68}$$

for the pseudo-loop.

The flow (q) and headloss term ($H_L = Kq^n$) for each pipe is considered positive if the flow is in the clockwise direction around the loop. Each term in the denominator can be considered as $n \times H_L/q$ and is always positive. The same correction term (Δ) is applied to all pipes in a loop. A positive correction term is added to the flow in all pipes that have flow in a clockwise direction around the loop and subtracted from the flow in all pipes that have flow in the counter clockwise direction around the loop. The continuity equations remain in balance after the flow in the pipes for a loop have been corrected.

Because only the first two terms were used in the binomial expansion of the headloss equation and because pipes that are in more than one loop have multiple corrections, the process is iterative. After applying one iterative correction to all loops, the process is repeated until convergence is achieved.

Example 4.14 Hardy Cross Pipe Network Problem

Determine the flow rate in each line and the pressure at each junction node for the pipe network in the sketch below using the Hardy Cross method of analysis. The pipe and junction data are listed below. The pipe area and headloss K values have also been computed and are listed in the table below.

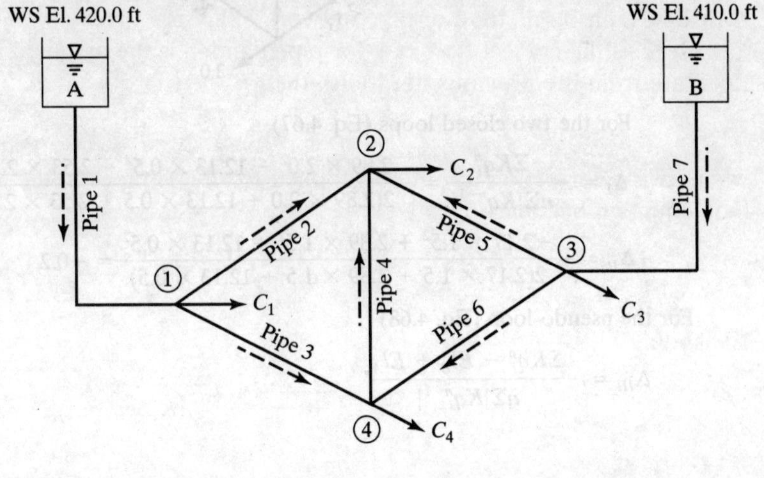

Line	Nodes	Length ft	Diameter in.	f	A ft^2	K
1	A–1	1,000	12	0.015	0.78	0.38
2	1–2	800	8	0.019	0.35	2.89
3	1–4	700	8	0.019	0.35	2.53
4	4–2	750	6	0.020	0.196	12.13
5	3–2	600	8	0.019	0.35	2.17
6	3–4	800	8	0.019	0.35	2.89
7	B–3	900	10	0.017	0.55	0.94

Junction	Elevation ft	Demand cfs
1	320	2.0
2	330	4.0
3	310	1.0
4	300	3.0
Total Demand		10.0

The Hardy Cross method requires an initial estimate of flow in each pipe such that the continuity equation is satisfied for each junction node. The estimated flows are shown below along with the K values for each pipe and the three loops.

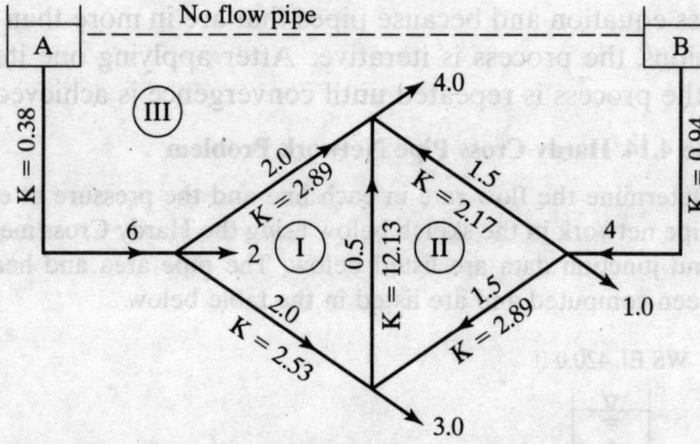

For the two closed loops (Eq. 4.67)

$$\Delta_I = -\frac{\Sigma K q_i^n}{n\Sigma|Kq^{n-1}|} = -\frac{2.89 \times 2.0^2 - 12.13 \times 0.5^2 - 2.53 \times 2.0^2}{2(2.89 \times 2.0 + 12.13 \times 0.5 + 2.53 \times 2.0)} = +0.05$$

$$\Delta_{II} = -\frac{-2.17 \times 1.5^2 + 2.89 \times 1.5^2 + 12.13 \times 0.5^2}{2(2.17 \times 1.5 + 2.89 \times 1.5 + 12.13 \times 0.5)} = -0.2$$

For the pseudo-loop (Eq. 4.68)

$$\Delta_{III} = -\frac{\Sigma K q_i^n - El_B + El_A}{n\Sigma|Kq_i^{n-1}|}$$

$$= -\frac{0.94 \times 4^2 + 2.17 \times 1.5^2 - 2.89 \times 2.0^2 - 0.38 \times 6.0^2 + 10.0}{2(0.94 \times 4.0 + 2.17 \times 1.5 + 2.89 \times 2.0 + 0.38 \times 6.0)} = -0.15$$

The adjusted flows for each line are shown below. Lines 2, 4, and 5 are included in two loops and are adjusted twice. After the flows are adjusted, the continuity equation remains satisfied at each junction node.

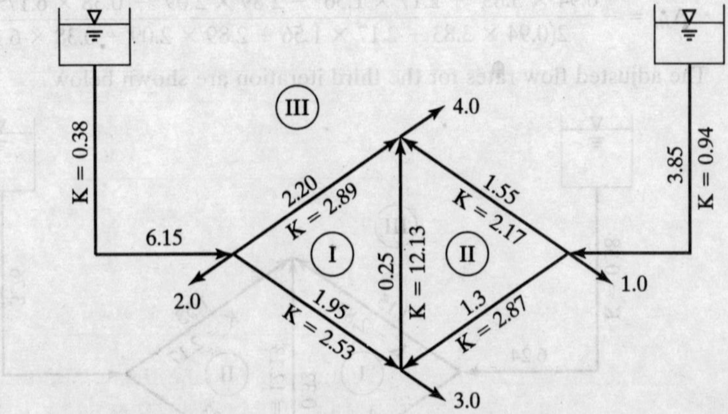

The correction terms for the second iteration are

$$\Delta_I = -\frac{2.89 \times 2.2^2 - 12.13 \times 0.25^2 - 2.53 \times 1.95^2}{2(2.89 \times 2.2 + 12.13 \times 0.25 + 2.53 \times 1.95)} = -0.13$$

$$\Delta_{II} = -\frac{-2.17 \times 1.55^2 + 2.89 \times 1.3^2 + 12.13 \times 0.25^2}{2(2.17 \times 1.55 + 2.89 \times 1.3 + 12.13 \times 0.25)} = -0.03$$

$$\Delta_{III} = -\frac{0.94 \times 3.85^2 + 2.17 \times 1.55^2 - 2.89 \times 2.20^2 - 0.38 \times 6.15^2 + 10}{2(0.94 \times 3.85 + 2.17 \times 1.55 + 2.89 \times 2.20 + 0.38 \times 6.15)} = -0.02$$

The adjusted flow rates for the second iteration are shown below.

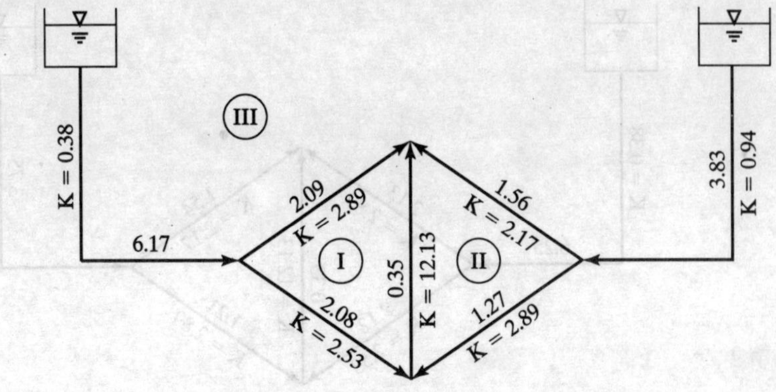

The correction terms for the third iteration are

$$\Delta_I = -\frac{2.89 \times 2.09^2 - 12.13 \times 0.35^2 - 2.53 \times 2.08^2}{2(2.89 \times 2.09 + 12.13 \times 0.35 + 2.53 \times 2.08)} = -0.01$$

$$\Delta_{II} = -\frac{-2.17 \times 1.56^2 + 2.89 \times 1.27^2 + 12.13 \times 0.35^2}{2(2.17 \times 1.56 + 2.89 \times 1.27 + 12.13 \times 0.35)} = -0.04$$

$$\Delta_{III} = -\frac{0.94 \times 3.83^2 + 2.17 \times 1.56^2 - 2.89 \times 2.09^2 - 0.38 \times 6.17^2 + 10}{2(0.94 \times 3.83 + 2.17 \times 1.56 + 2.89 \times 2.09 + 0.38 \times 6.17)} = -0.07$$

The adjusted flow rates for the third iteration are shown below

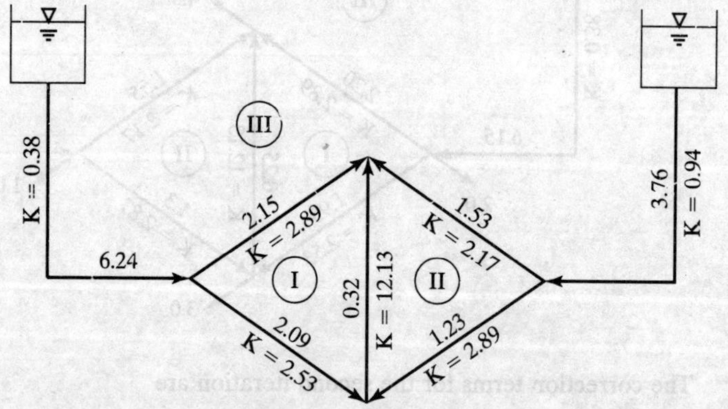

The correction terms for the fourth iteration are

$$\Delta_I = -\frac{2.89 \times 2.15^2 - 12.13 \times 0.32^2 - 2.53 \times 2.09^2}{2(2.89 \times 2.15 + 12.13 \times 0.32 + 2.53 \times 2.09)} = -0.03$$

$$\Delta_{II} = -\frac{-2.17 \times 1.53^2 + 2.89 \times 1.23^2 + 12.13 \times 0.32^2}{2(2.17 \times 1.53 + 2.89 \times 1.23 + 12.13 \times 0.32)} = -0.02$$

$$\Delta_{III} = -\frac{0.94 \times 3.76^2 + 2.17 \times 1.53^2 - 2.89 \times 2.15^2 - 0.38 \times 6.24^2 + 10}{2(0.94 \times 3.76 + 2.17 \times 1.53 + 2.89 \times 2.15 + 0.38 \times 6.24)} = -0.01$$

The adjusted flow rates for the fourth iteration are shown below.

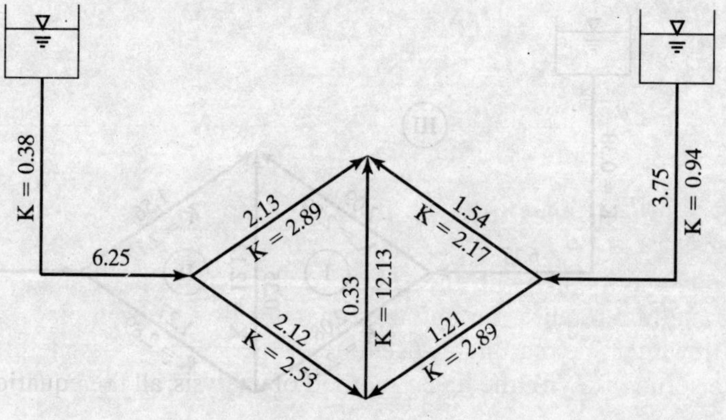

The correction terms for the fifth iteration are

$$\Delta_I = -\frac{2.89 \times 2.13^2 - 12.13 \times 0.33^2 - 2.53 \times 2.12^2}{2(2.89 \times 2.13 + 12.13 \times 0.33 + 2.53 \times 2.12)} = -0.01$$

$$\Delta_{II} = -\frac{-2.17 \times 1.54^2 + 2.89 \times 1.21^2 + 12.13 \times 0.33^2}{2(2.17 \times 1.54 + 2.89 \times 1.21 + 12.13 \times 0.33)} = -0.02$$

$$\Delta_{III} = -\frac{0.94 \times 3.75^2 + 2.17 \times 1.54^2 - 2.89 \times 2.13^2 - 0.38 \times 6.25^2 + 10}{2(0.94 \times 3.75 + 2.17 \times 1.54 + 2.89 \times 2.13 + 0.38 \times 6.25)} = -0.01$$

The final flow rates are listed below along with the headloss for each line.

Line	Nodes	K	Q cfs	H_L ft
1	A–1	0.38	6.26	14.9
2	1–2	2.89	2.13	13.1
3	1–4	2.53	2.13	11.5
4	4–2	12.13	0.32	1.2
5	3–2	2.17	1.55	5.2
6	3–4	2.89	1.19	4.1
7	B–3	0.94	3.74	13.1

The pressure at each node is computed in the table below.

Node	Elevation ft	Line	H_L ft	HGL ft	P/γ ft	P psi
A	420			420.0		
		A–1	−14.9			
1	320			405.1	85.1	37
		1–2	−13.1			
2	330			392.0	62.0	27
		2–3	+5.2			
3	310			397.2	87.2	38
		3–4	−4.1			
4	300			393.1	93.1	40
		3–B	+13.1			
B	410			410.3		

4.5.2 Linear Method

A major difficulty with the Hardy Cross method of pipe network analysis is that the method requires an initial estimate of flow in each pipe. The initial estimates of flow must be reasonably accurate or the Hardy Cross method will not converge to a solution. With the linear method of analysis, all the equations (both continuity and loop equations) are solved simultaneously. This is a very stable numerical procedure that does not require the user to provide initial estimates of flow and will nearly always converge to a solution. The nonlinear loop equations are linearized, and all equations are solved iteratively. The user must specify the direction of flow in each line. If the computed flow rate is negative, the direction of flow in that line is reversed. For pipes with pumps, check valves, or pressure regulating valves, the flow direction cannot be reversed, and if the computed flow rate is negative the flow rate must be set equal to zero.

Each nonlinear headloss term $[f(Q)]$ in the loop equations is linearized using the first two terms in the Taylor series

$$f(Q) = f(q) + \frac{\partial f}{\partial q}(Q - q) \tag{4.69}$$

where q is the flow rate from the previous iteration and Q is the unknown flow rate. Each pipe friction term (KQ^n) in the loop equations is replaced with the linear term

$$Kq^n + nKq^{n-1}(Q - q) \tag{4.70}$$

or

$$DQ + D' \tag{4.71}$$

where

$$D = nKq^{n-1}$$

$$D' = (1 - n)Kq^n$$

and for loops with pumps the nonlinear pump characteristic equation ($AQ^2 + BQ + H_c$) is replaced with the linear equation

$$Aq^2 + Bq + H_c + (2Aq + B)(Q - q) \tag{4.72}$$

or

$$EQ + E' \tag{4.73}$$

where

$$E = 2Aq + B$$

$$E' = H_c - Aq^2$$

The nonlinear loop equations (Eqs. 4.60 and 4.62) in the example problem become

$$D_2Q_2 + D_6Q_6 - D_8Q_8 - D_5Q_5 = -D_2' - D_6' + D_8' + D_5' \tag{4.74}$$

$$-D_3Q_3 + D_7Q_7 + D_9Q_9 - D_6Q_6 = D'_3 - D'_7 - D'_9 + D'_6 \tag{4.75}$$

and

$$D_4Q_4 + D_3Q_3 - D_2Q_2 - D_1Q_1 = -D'_4 - D'_3 + D'_2 + D'_1 + El_B - El_A \tag{4.76}$$

In matrix notation, the set of linear equations are

$$\begin{array}{cc} & \text{Pipe} \quad 1 \quad 2 \quad 3 \quad 4 \quad 5 \quad 6 \quad 7 \quad 8 \quad 9 \\ \text{Equation number} & \begin{array}{c} 1 \\ 2 \\ 3 \\ 4 \\ 5 \\ 6 \\ 7 \\ 8 \\ 9 \end{array} \begin{bmatrix} 1 & -1 & 0 & 0 & -1 & 0 & 0 & 0 & 0 \\ 0 & 1 & 1 & 0 & 0 & -1 & 0 & 0 & 0 \\ 0 & 0 & -1 & 1 & 0 & 0 & -1 & 0 & 0 \\ 0 & 0 & 0 & 0 & 1 & 0 & 0 & -1 & 0 \\ 0 & 0 & 0 & 0 & 0 & 1 & 0 & 1 & 1 \\ 0 & 0 & 0 & 0 & 0 & 0 & 1 & 0 & -1 \\ 0 & D_2 & 0 & 0 & -D_5 & D_6 & 0 & -D_8 & 0 \\ 0 & 0 & -D_3 & 0 & 0 & -D_6 & D_7 & 0 & D_9 \\ -D_1 & -D_2 & D_3 & D_4 & 0 & 0 & 0 & 0 & 0 \end{bmatrix} \begin{bmatrix} Q_1 \\ Q_2 \\ Q_3 \\ Q_4 \\ Q_5 \\ Q_6 \\ Q_7 \\ Q_8 \\ Q_9 \end{bmatrix} = \begin{bmatrix} C_1 \\ C_2 \\ C_3 \\ C_4 \\ C_5 \\ C_6 \\ C_7 \\ C_8 \\ C_9 \end{bmatrix} \end{array} \tag{4.77}$$

where

$$C_1 - C_6 = \text{Demands at junctions}$$

$$D = nKq^{n-1}$$

$$C_7 = -D'_2 - D'_6 + D'_8 + D'_5$$

$$C_8 = D'_3 - D'_7 - D'_9 + D'_6$$

$$C_9 = -D'_4 - D'_3 + D'_2 + D'_1 + El_B - El_A$$

and

$$D' = (1 - n)Kq^n$$

Equation 4.77 can be written as

$$[A][Q] = [C] \tag{4.78}$$

where $[A]$ is the coefficient matrix and $[C]$ is a column vector of constants.

Solving for the new flow rates

$$[Q] = [A]^{-1}[C] \tag{4.79}$$

where $[A]^{-1}$ is the inverse of coefficient matrix.

For the initial iteration, the value of q for each pipe can be computed based on a velocity of 1 mps (3 fps). The flow rates (Q) computed from Eq. 4.79 became the estimated flow rates (q) for the next iteration. This iterative process is continued until the equations converge to a solution. Convergence is assumed to occur when the maximum change $|Q - q|$ is within a specified limit or the relative accuracy

$$\frac{\sum_{i=1}^{N} |Q_i - q_i|}{\sum_{i=1}^{N} |Q_i|} \tag{4.80}$$

is within a specified value.

Typically the demands specified at the junction nodes represent the average daily demand on the network systems and the model can be rerun using demand peaking factors to represent the peak daily demand on the system or the peak hourly demand on the system. Fire flows can be added at selected junction nodes to determine if the system can provide the fire demand at the required pressure.

Pressure-regulating valves (PRVs) are designed to maintain a specified discharge pressure (P_{RV}), which is lower than the upstream pressures. As shown in Fig. 4.14, the PRV can be modeled as a fixed-grade node located downstream of a junction node. The elevation of the fixed-grade node is equal to the ground elevation of the upstream junction node plus the PRV pressure head setting (P_{RV}/γ_ω). The computed flow rate in the downstream pipe is added to the demand at the upstream junction node. If the computed flow in the downstream pipe is negative, the pipe is closed and the flow through the valve is zero. If the computed pressure at the upstream junction node is less than P_{RV}, the PRV is fully open and the system is analyzed without the PRV.

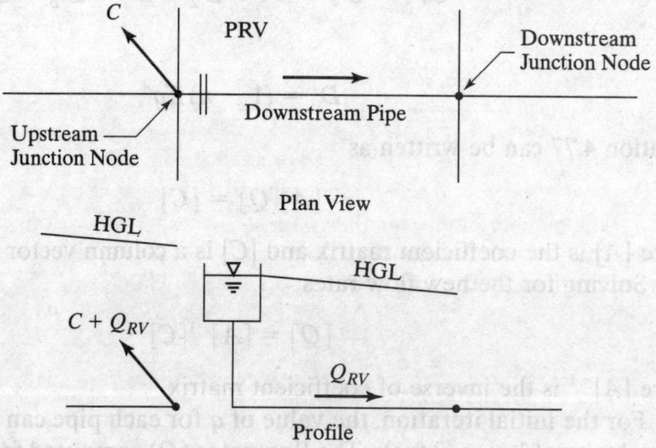

Figure 4.14 Modeling a pressure regulating valve.

Example 4.15 Linear Method of Pipe Network Analysis

For this problem, Example 4.14 was modified by changing the elevated storage tank at A to a ground-level storage tank and adding a pump in line A–1. The pump characteristic curve is shown below. The pipe and junction data for this problem are the same as Example 4.14. Determine the flow rate in each line and the pressure at each junction node using the linear method of pipe network analysis.

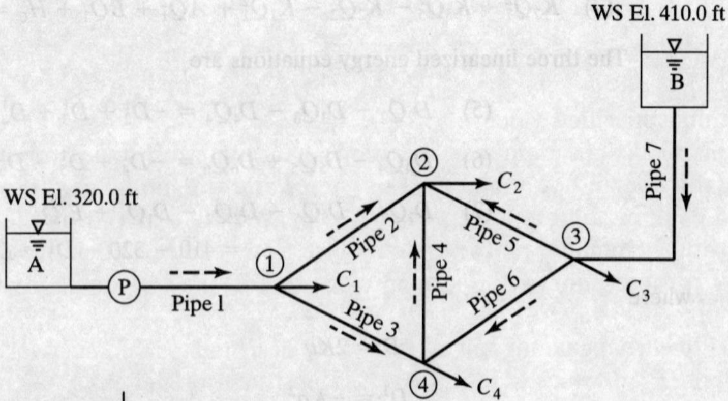

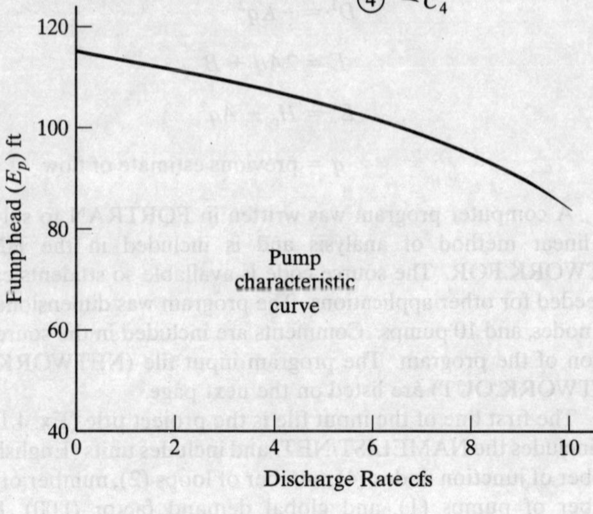

The number of pipes in the system ($N_P = 7$) must equal the number of junction nodes ($N_j = 4$) plus the number of loops ($N_l = 2$) plus the number of fixed-grade nodes ($N_f = 2$) minus 1. The four continuity (node) equations are

$$(1)\quad Q_1 - Q_2 - Q_3 = C_1$$

$$(2)\quad Q_2 + Q_4 + Q_5 = C_2$$

$$(3)\quad Q_7 - Q_5 - Q_6 = C_3$$

$$(4)\quad Q_3 - Q_4 + Q_6 = C_4$$

The two loop equations are

$$(5)\quad K_2Q_2^2 - K_4Q_4^2 - K_3Q_3^2 = 0$$

$$(6)\quad K_4Q_4^2 - K_5Q_5^2 + K_6Q_6^2 = 0$$

The one pseudo-loop equation is

$$(7)\quad K_7Q_7^2 + K_5Q_5^2 - K_2Q_2^2 - K_1Q_1^2 + AQ_1^2 + BQ_1 + H_C = 410 - 320$$

The three linearized energy equations are

$$(5)\quad D_2Q_2 - D_3Q_3 - D_4Q_4 = -D_2^1 + D_3^1 + D_4^1$$

$$(6)\quad D_4Q_4 - D_5Q_5 + D_6Q_6 = -D_4^1 + D_5^1 - D_6^1$$

$$(7)\quad D_7Q_7 + D_5Q_5 - D_2Q_2 - D_1Q_1 + E_1Q_1 = 410 - 320 - D_7^1 - D_5^1 + D_2^1 + D_1^1 - E_1^1$$

where

$$D = 2Kq$$

$$D^1 = -Kq^2$$

$$E = 2Aq + B$$

$$E^1 = H_c - Aq^2$$

$$q = \text{previous estimate of flow}$$

A computer program was written in FORTRAN to solve pipe networks using the linear method of analysis and is included in the HMP (Section 1.4.5) as NETWORK.FOR. The source code is available so students can modify the program as needed for other applications. The program was dimensioned for 50 pipes, 25 junction nodes, and 10 pumps. Comments are included in the source code explaining each section of the program. The program input file (NETWORK.DAT) and output file (NETWORK.OUT) are listed on the next page.

The first line of the input file is the project title (Ex. 4.15 pipe network). Line two includes the NAMELIST/NET/ and includes units (English), number of pipes (7), number of junction nodes (4), number of loops (2), number of fixed-grade nodes (2), number of pumps (1), and global demand factor (1.00). Pipe data follows the NAMELIST and is entered in the same order the pipes are numbered. Pipe data include pipe number upstream and downstream node numbers, pipe length (ft), pipe diameter (in.), and the Darcy–Weisbach friction factor and minor loss coefficient. If either node number is zero, indicating a fixed-grade node, the elevation of the fixed-grade node is entered on the following line. The junction data follow the pipe data and include the junction node number, the ground elevation (ft), and the demand (gpm). Pump data follow the junction data and include pipe number, cutoff head (ft), followed by the discharge (gpm) and head (ft) for two additional points on the pump characteristic curve. Loop data follow the pump data and include number of pipes in loop followed by pipe numbers in loop (clockwise plus). Pseudo-loops are listed last with pipe numbers starting with first pipe clockwise past the imaginary no-flow pipe.

```
EXAMPLE 4.15 PIPE NETWORK
&NET  IUN = 2,     NP = 7,     NJ = 4,    NL = 2,    NF = 2,    NPU = 1,    GDF = 1.0/
 1     0     1     1000.0      12.0       .015       .5
  320.0
 2     1     2      800.0       8.0       .019       .0
 3     4     1      700.0       8.0       .019       .0
 4     4     2      750.0       6.0       .020       .0
 5     3     2      600.0       8.0       .019       .0
 6     3     4      800.0       8.0       .019       .0
 7     3     0      900.0      10.0       .017       .5
  410.0
 1       320.0      896.0
 2       330.0     1792.0
 3       310.0      448.0
 4       300.0     1344.0
 1       115.0     2240.0      105.0     4480.0      85.0
 3
 2    -4     3
 3
-5     6     4
 4
-7     5    -2    -1
```

RESULTS

PIPE	NODES		LENGTH	DIAMETER	DISCHARGE	VELOCITY	HEADLOSS
1	0	1	1000.00	12.00	2835.22	8.06	15.63
2	1	2	800.00	8.00	961.66	6.15	13.39
3	4	1	700.00	8.00	−977.56	−6.25	−12.11
4	4	2	750.00	6.00	146.00	1.66	1.28
5	3	2	600.00	8.00	684.33	4.38	5.08
6	3	4	800.00	8.00	512.45	3.28	3.80
7	3	0	900.00	10.00	−1644.78	−6.73	−13.27

JUNCTION DATA

NODE	DEMAND	GROUND	HGL	PRESSURE
1	896.00	320.00	405.03	36.85
2	1792.00	330.00	391.65	26.71
3	448.00	310.00	396.73	37.58
4	1344.00	300.00	392.93	40.27

Example 4.16 Branching Pipeline

Compute the discharge in each line for the branching pipeline. There are 5 pipes, no loop, 2 pumps, and 4 fixed-grade nodes in the system. Use a friction factor of 0.02 for all lines.

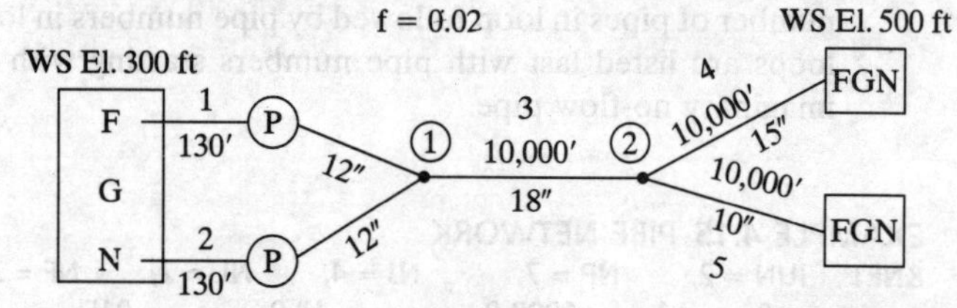

PLAN

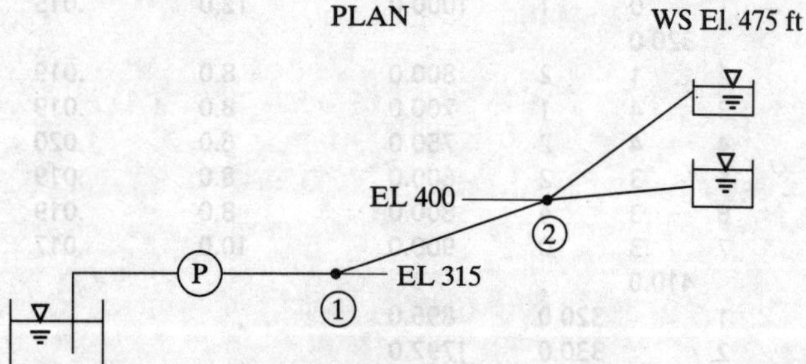

PROFILE

Pump 1	
E_p ft	Q cfs
400	0
350	6
250	12

Pump 2	
E_p ft	Q cfs
400	0
350	6
250	12

The input file and a summary of the output file are listed below.

```
EXAMPLE 4.16  PIPELINE BRANCHING NETWORK
&NET  IUN = 2,    NP = 5,     NJ = 2,    NL = 0,    NF = 4,    NPU = 2,    GDF = 1.0/
   1    0     1   130.0       12.0      .020        1.0
   300.0
   2    0     1   130.0       12.0      .020        1.0
   300.0
   3    1     2   10000.0     18.0      .020        .0
   4    2     0   10000.0     15.0      .020        1.0
   500.0
   5    2     0   10000.0     10.0      .020        1.0
   475.0
   1        315.0   0.0
   2        400.0   0.0
   1        400.0   2688.0    350.0     5376.0      250.0
   2        400.0   2688.0    350.0     5376.0      250.0
   2
   1   -2
   3
  -4   -3   -1
   2
  -5    4
```

RESULTS

PIPE	NODES		LENGTH	DIAMETER	DISCHARGE	VELOCITY	HEADLOSS
1	0	1	130.00	12.00	2301.20	6.54	2.39
2	0	1	130.00	12.00	2301.20	6.54	2.39
3	1	2	10000.00	18.00	4602.39	5.81	69.97
4	2	0	10000.00	15.00	3260.21	5.93	87.91
5	2	0	10000.00	10.00	1342.19	5.49	112.91

JUNCTION DATA

NODE	DEMAND	GROUND	HGL	PRESSURE
1	.00	315.00	657.88	148.58
2	.00	400.00	587.91	81.43

Example 4.17(a) Irrigation System

A 500-ac farm is to be irrigated with a sprinkler system. The crop requires 12.0 inches of water in July of which 2.0 inches is provided by rain and 10.0 inches is to be provided by the sprinkler system. Determine the water supply flow rate if the sprinkler system is 80 percent efficient in delivering water to the crop and the operation requires a 20 percent peaking factor.

$$\text{Discharge rate} = \text{Volume/time}$$

$$\text{Volume} = 500 \text{ ac} \times 43{,}560 \frac{\text{ft}^2}{\text{ac}} \times \frac{10}{12} \text{ft} \times \frac{1}{0.80} \times 1.20$$

$$= 27.2 \times 10^6 \text{ ft}^3$$

$$Q = \frac{27.2 \times 10^6}{31 \times 86{,}400} = 10.2 \text{ cfs}$$

Example 4.17(b) Golf Course Sprinkler System

A golf course is to be watered by pumping from a lake (water surface elevation 300 ft) into a branching pipe network system shown below. The ground elevation varies from

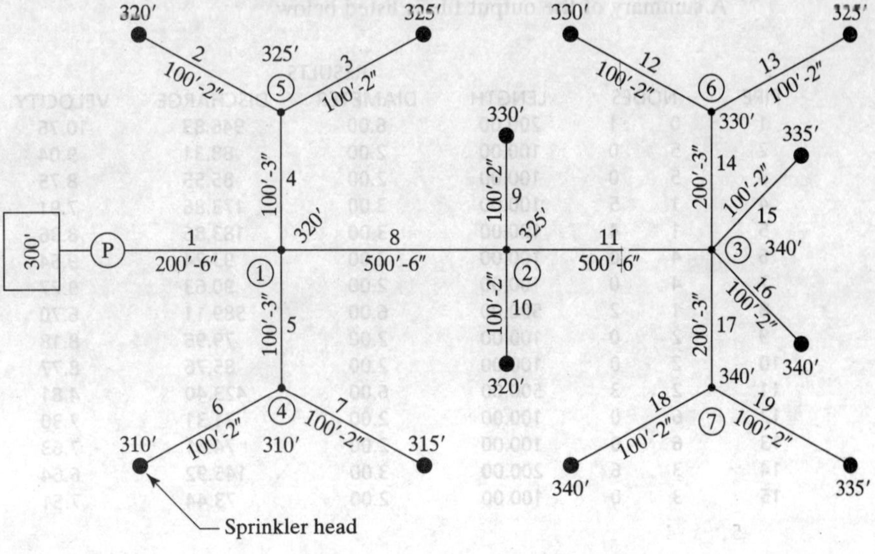

310 to 340 ft. In this 19-pipe system, there are no demands at the junctions and all the flow is through the sprinkler heads. The sprinkler heads are modeled as a fixed-grade node at ground level plus a minor loss. For the example problem, the sprinkler heads were selected for a discharge of 75 gpm at a head of 50 ft. All lines connected to sprinkler heads are 2 inches in diameter. At the design discharge of 75 gpm, the velocity in the 2-inch pipe is 7.68 fps. The minor loss coefficient (M) for a head of 50 ft is

$$50 = M\frac{V^2}{2g} = M\frac{7.68^2}{64.4}$$

$$M = 55$$

This minor loss coefficient is included in all lines connected to a sprinkler head. The pump has the following characteristics

Head ft	Discharge cfs
200	0
175	1
125	2

Compute the flow at each sprinkler head. In this problem, there are 19 pipes, 7 junction nodes, 0 loops, and 13 fixed-grade nodes.

$$N_p = N_j + N_\ell + N_f - 1$$

$$19 = 7 + 0 + 13 - 1$$

A summary of the output file is listed below.

RESULTS

PIPE	NODES		LENGTH	DIAMETER	DISCHARGE	VELOCITY	HEADLOSS
1	0	1	200.00	6.00	946.83	10.76	10.79
2	5	0	100.00	2.00	88.31	9.04	81.13
3	5	0	100.00	2.00	85.55	8.75	76.13
4	1	5	100.00	3.00	173.86	7.91	5.82
5	1	4	100.00	3.00	183.86	8.36	6.51
6	4	0	100.00	2.00	93.24	9.54	90.44
7	4	0	100.00	2.00	90.63	9.27	85.44
8	1	2	500.00	6.00	589.11	6.70	10.45
9	2	0	100.00	2.00	79.96	8.18	66.51
10	2	0	100.00	2.00	85.76	8.77	76.51
11	2	3	500.00	6.00	423.40	4.81	5.40
12	6	0	100.00	2.00	71.31	7.30	52.91
13	6	0	100.00	2.00	74.61	7.63	57.91
14	3	6	200.00	3.00	145.92	6.64	8.20
15	3	0	100.00	2.00	73.44	7.51	56.11

16	3	0	100.00	2.00	70.09	7.17	51.11
17	3	7	200.00	3.00	133.95	6.09	6.91
18	7	0	100.00	2.00	65.18	6.67	44.20
19	7	0	100.00	2.00	68.77	7.04	49.20

JUNCTION DATA

NODE	DEMAND	GROUND	HGL	PRESSURE
1	.00	320.00	406.95	37.68
2	.00	325.00	396.51	30.99
3	.00	340.00	391.11	22.15
4	.00	310.00	400.44	39.19
5	.00	325.00	401.13	32.99
6	.00	330.00	382.91	22.93
7	.00	340.00	384.20	19.15

Example 4.17(c) Building Sprinkler System

The 15-pipe branching system represents a 2-story building sprinkler system. Water supply is from a main pipeline with a total head of 438.5 ft. Sprinkler heads are all located at the end of a 1.0-inch diameter pipe and have a minor loss coefficient (M) equal to 6.0. Compute the discharge through each sprinkler head.

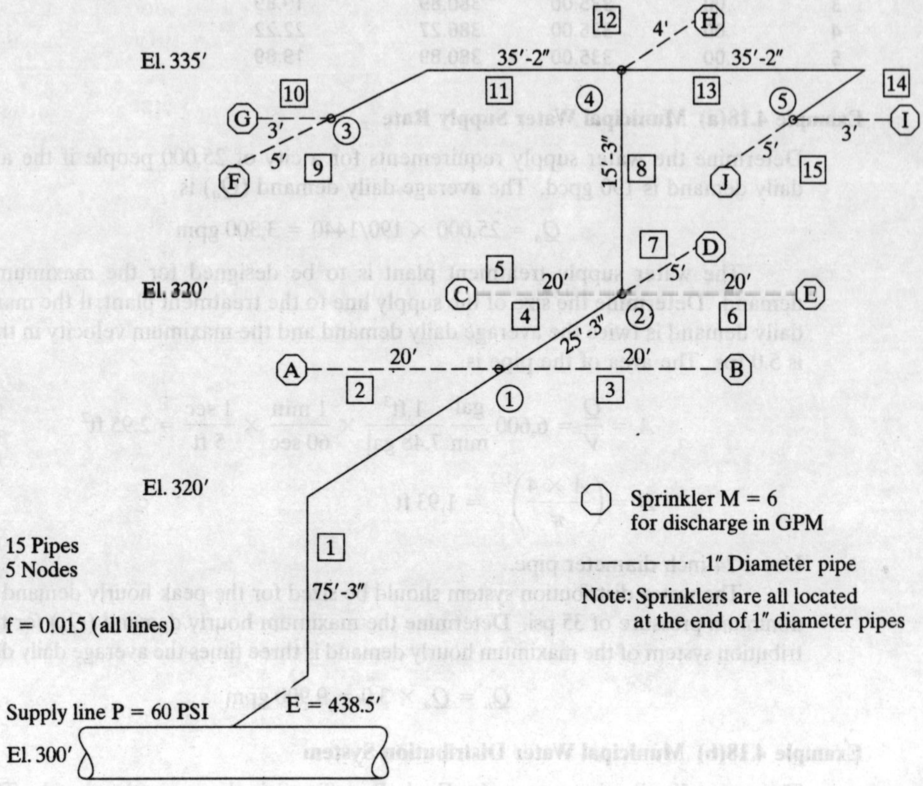

RESULTS

PIPE	NODES		LENGTH	DIAMETER	DISCHARGE	VELOCITY	HEADLOSS
1	0	1	75.00	3.00	536.40	24.39	41.57
2	1	0	20.00	1.00	55.51	22.72	76.93
3	1	0	20.00	1.00	55.51	22.72	76.93
4	1	2	25.00	3.00	425.38	19.34	8.71
5	2	0	20.00	1.00	52.27	21.39	68.21
6	2	0	20.00	1.00	52.27	21.39	68.21
7	2	0	5.00	1.00	61.65	25.23	68.21
8	2	4	15.00	3.00	259.19	11.79	1.94
9	3	0	5.00	1.00	50.57	20.70	45.89
10	3	0	3.00	1.00	51.94	21.26	45.89
11	4	3	35.00	2.00	102.51	10.49	5.38
12	4	0	4.00	1.00	54.16	22.17	51.27
13	4	5	35.00	2.00	102.51	10.49	5.38
14	5	0	3.00	1.00	51.94	21.26	45.89
15	5	0	5.00	1.00	50.57	20.70	45.89

JUNCTION DATA

NODE	DEMAND	GROUND	HGL	PRESSURE
1	.00	320.00	396.93	33.34
2	.00	320.00	388.21	29.56
3	.00	335.00	380.89	19.89
4	.00	335.00	386.27	22.22
5	.00	335.00	380.89	19.89

Example 4.18(a) Municipal Water Supply Rate

Determine the water supply requirements for a city of 25,000 people if the average daily demand is 190 gpcd. The average daily demand (Q_a) is

$$Q_a = 25{,}000 \times 190/1440 = 3{,}300 \text{ gpm}$$

The water supply treatment plant is to be designed for the maximum daily demand. Determine the size of the supply line to the treatment plant, if the maximum daily demand is twice the average daily demand and the maximum velocity in the pipe is 5.0 fps. The area of the pipe is

$$A = \frac{Q}{V} = 6{,}600 \frac{\text{gal}}{\text{min}} \frac{1 \text{ ft}^3}{7.48 \text{ gal}} \times \frac{1 \text{ min}}{60 \text{ sec}} \times \frac{1 \text{ sec}}{5 \text{ ft}} = 2.95 \text{ ft}^2$$

$$D = \left(\frac{A \times 4}{\pi}\right)^{1/2} = 1.93 \text{ ft}$$

Use a 24-inch diameter pipe.

The water distribution system should be sized for the peak hourly demand with a minimum pressure of 35 psi. Determine the maximum hourly demand (Q_h) for the distribution system of the maximum hourly demand is three times the average daily demand

$$Q_h = Q_a \times 3.0 = 9{,}900 \text{ gpm}$$

Example 4.18(b) Municipal Water Distribution System

The water distribution system for Eagle Pass, Texas, is shown on the sketch. The low pressure zone has ground elevations ranging from 700 to 775 ft msl while the high pressure zone has ground elevations ranging from 800 to 825 ft msl. Contours showing the ground elevations are included in the sketch. Pump 1 pumps water from the ground-level sto.age into the lower pressure zone, whereas booster pumps 2 and 3 pump water from the low pressure zone to the high pressure zone. There are four elevated storage tanks (2 in each zone) in the system. Compute pressures in the system for

average daily demand (GDF = 1.0),
peak day demand (GDF = 2.0), and
peak hour demand (GDF = 3.0).

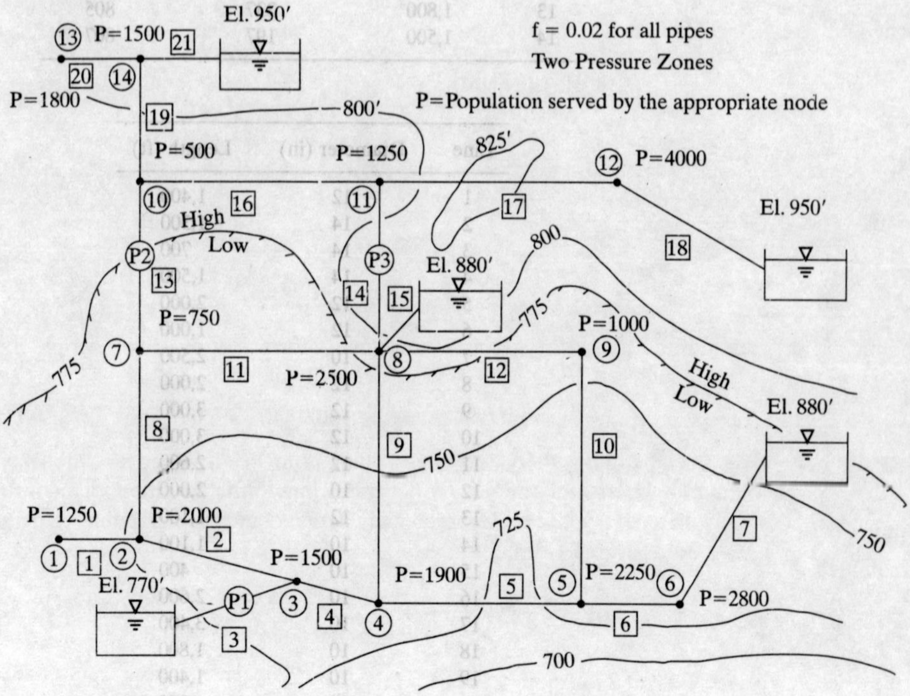

Pump Data

Pump 1 E_p ft	Pump 1 Q cfs	Pump 2 E_p ft	Pump 2 Q cfs	Pump 3 E_p ft	Pump 3 Q cfs
200	0.0	150	0.0	160	0.0
130	20.0	90	1.1	100	3.3
80	25.0	60	1.6	60	4.5

Node	Population	Demand (gpm)	Elevation
1	1,250	166	760
2	2,000	264	749
3	1,500	197	735
4	1,900	250	730
5	2,250	296	726
6	2,800	367	726
7	750	98	765
8	2,500	332	787
9	1,000	130	760
10	500	67	785
11	1,250	166	790
12	4,000	529	815
13	1,800	237	805
14	1,500	197	807

Line	Diameter (in)	Length (ft)
1	12	1,400
2	14	1,300
3	14	700
4	14	1,500
5	12	2,000
6	12	1,000
7	10	2,500
8	12	2,000
9	12	3,000
10	12	3,000
11	12	2,600
12	10	2,000
13	12	1,100
14	10	1,100
15	10	400
16	10	2,600
17	10	3,400
18	10	1,800
19	10	1,400
20	10	900
21	10	550

Summaries of the output files are listed below. During the average demand, all elevated storage tanks are filling, and during the peak hourly demand, all elevated storage tanks are emptying. The minimum pressure during the peak hourly demand is 38.7 psi.

RESULTS (AVERAGE DAY)

PIPE	NODES		LENGTH	DIAMETER	DISCHARGE	VELOCITY	HEADLOSS
1	2	1	1400.00	12.00	166.00	.47	.10
2	3	2	1300.00	14.00	2347.65	4.90	8.32
3	0	3	700.00	14.00	5897.95	12.32	28.26
4	3	4	1500.00	14.00	3353.29	7.00	19.58
5	4	5	2000.00	12.00	1633.35	4.64	13.38
6	5	6	1000.00	12.00	882.39	2.51	1.95
7	6	0	2500.00	10.00	515.39	2.11	4.15
8	2	7	2000.00	12.00	1917.65	5.45	18.45
9	4	8	3000.00	12.00	1469.94	4.18	16.26
10	5	9	3000.00	12.00	454.96	1.29	1.56
11	7	8	2600.00	12.00	1179.33	3.35	9.07
12	8	9	2000.00	10.00	–324.96	–1.33	–1.32
13	7	10	1100.00	12.00	640.32	1.82	1.13
14	8	11	1100.00	10.00	1506.21	6.16	15.58
15	8	0	400.00	10.00	1136.03	4.65	3.22
16	10	11	2600.00	10.00	–506.62	–2.07	–4.17
17	11	12	3400.00	10.00	833.59	3.41	14.75
18	12	0	1800.00	10.00	304.59	1.25	1.04
19	10	14	1400.00	10.00	1079.94	4.42	10.19
20	14	13	900.00	10.00	237.00	.97	.32
21	14	0	550.00	10.00	645.94	2.64	1.43

JUNCTION DATA

NODE	DEMAND	GROUND	HGL	PRESSURE
1	166.00	760.00	910.65	65.28
2	264.00	749.00	910.74	70.09
3	197.00	735.00	919.06	79.76
4	250.00	730.00	899.48	73.44
5	296.00	726.00	886.10	69.38
6	367.00	726.00	884.14	68.53
7	98.00	765.00	892.29	55.16
8	332.00	787.00	883.22	41.70
9	130.00	760.00	884.54	53.97
10	67.00	785.00	961.62	76.54
11	166.00	790.00	965.79	76.18
12	529.00	815.00	951.04	58.95
13	237.00	805.00	951.12	63.32
14	197.00	807.00	951.43	62.59

RESULTS (PEAK DAY)

PIPE	NODES		LENGTH	DIAMETER	DISCHARGE	VELOCITY	HEADLOSS
1	2	1	1400.00	12.00	332.00	.94	.39
2	3	2	1300.00	14.00	2598.03	5.42	10.18
3	0	3	700.00	14.00	6272.02	13.10	31.96
4	3	4	1500.00	14.00	3279.99	6.85	18.73
5	4	5	2000.00	12.00	1509.48	4.29	11.43
6	5	6	1000.00	12.00	614.06	1.75	.95
7	6	0	2500.00	10.00	–119.94	–.49	–.22
8	2	7	2000.00	12.00	1738.03	4.94	15.15
9	4	8	3000.00	12.00	1270.51	3.61	12.15
10	5	9	3000.00	12.00	303.41	.86	.69
11	7	8	2600.00	12.00	921.46	2.62	5.54
12	8	9	2000.00	10.00	–43.41	–.18	–.02

13	7	10	1100.00	12.00	620.57	1.76	1.06
14	8	11	1100.00	10.00	1526.20	6.25	15.99
15	8	0	400.00	10.00	45.19	.18	.01
16	10	11	2600.00	10.00	−445.42	−1.82	−3.22
17	11	12	3400.00	10.00	748.78	3.06	11.90
18	12	0	1800.00	10.00	−309.22	−1.27	−1.07
19	10	14	1400.00	10.00	931.99	3.81	7.59
20	14	13	900.00	10.00	474.00	1.94	1.26
21	14	0	550.00	10.00	63.99	.26	.01

JUNCTION DATA

NODE	DEMAND	GROUND	HGL	PRESSURE
1	332.00	760.00	900.31	60.80
2	528.00	749.00	900.70	65.74
3	394.00	735.00	910.88	76.22
4	500.00	730.00	892.15	70.27
5	592.00	726.00	880.72	67.05
6	734.00	726.00	879.78	66.64
7	196.00	765.00	885.54	52.24
8	664.00	787.00	880.01	40.30
9	260.00	760.00	880.03	52.01
10	134.00	785.00	957.60	74.80
11	332.00	790.00	960.82	74.02
12	1058.00	815.00	948.93	58.03
13	474.00	805.00	948.75	62.29
14	394.00	807.00	950.01	61.97

RESULTS (PEAK HOUR)

PIPE	NODES		LENGTH	DIAMETER	DISCHARGE	VELOCITY	HEADLOSS
1	2	1	1400.00	12.00	498.00	1.42	.87
2	3	2	1300.00	14.00	2820.42	5.89	12.00
3	0	3	700.00	14.00	6624.81	13.83	35.65
4	3	4	1500.00	14.00	3213.40	6.71	17.98
5	4	5	2000.00	12.00	1408.38	4.00	9.95
6	5	6	1000.00	12.00	486.55	1.38	.59
7	6	0	2500.00	10.00	−614.45	−2.51	−5.89
8	2	7	2000.00	12.00	1530.42	4.35	11.75
9	4	8	3000.00	12.00	1055.02	3.00	8.38
10	5	9	3000.00	12.00	33.83	.10	.01
11	7	8	2600.00	12.00	631.38	1.79	2.60
12	8	9	2000.00	10.00	356.17	1.46	1.58
13	7	10	1100.00	12.00	605.04	1.72	1.01
14	8	11	1100.00	10.00	1555.27	6.37	16.61
15	8	0	400.00	10.00	−1221.05	−5.00	−3.72
16	10	11	2600.00	10.00	−298.32	−1.22	−1.44
17	11	12	3400.00	10.00	758.95	3.11	12.22
18	12	0	1800.00	10.00	−828.05	−3.39	−7.70
19	10	14	1400.00	10.00	702.35	2.87	4.31
20	14	13	900.00	10.00	711.00	2.91	2.84
21	14	0	550.00	10.00	−599.65	−2.45	−1.23

JUNCTION DATA

NODE	DEMAND	GROUND	HGL	PRESSURE
1	498.00	760.00	889.76	56.23
2	792.00	749.00	890.63	61.37
3	591.00	735.00	902.63	72.64

4	750.00	730.00	884.65	67.02
5	888.00	726.00	874.70	64.44
6	1101.00	726.00	874.11	64.18
7	294.00	765.00	878.88	49.35
8	996.00	787.00	876.28	38.69
9	390.00	760.00	874.69	49.70
10	201.00	785.00	953.08	72.83
11	498.00	790.00	954.52	71.29
12	1587.00	815.00	942.30	55.16
13	711.00	805.00	945.93	61.07
14	591.00	807.00	948.77	61.43

4.5.3 Design of Municipal Water Distribution Systems

As shown in the sketch of a water distribution system in Fig. 4.15, the major system components are the treatment unit, ground level storage, pumps, elevated storage, and pipe networks. The type of water treatment depends on the source of the water. Groundwater typically requires little treatment, whereas surface water

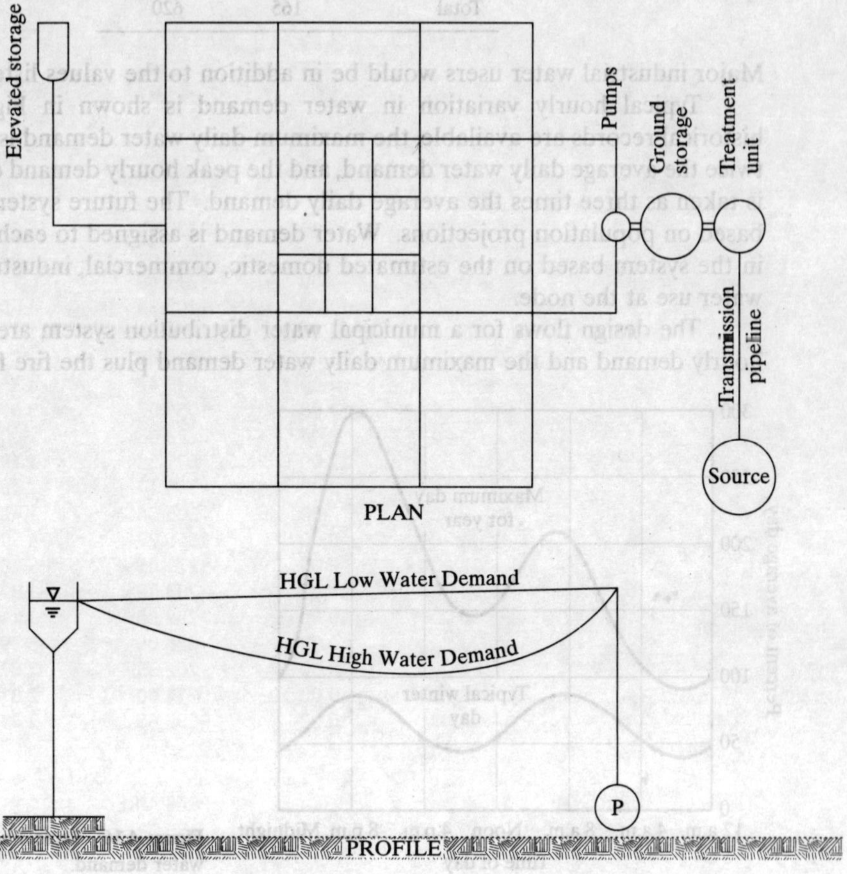

Figure 4.15 Sketch of water distribution system.

supply may require removal of suspended solids. Water is pumped from the ground level storage into the pipe network system. The elevated storage fills with water when the pumping rate is greater than the demand and empties when the demand is greater than the pumping rate. The height of the elevated storage is selected to provide the required system pressure.

Factors affecting the water demands for a system include water pressure, conservation programs, rainfall, water rates, and the economic status of residents. Average water requirements for a typical water distribution system are listed below.

Use	Average use per person Gal/day	Liters/day
Domestic	70	260
Commercial	25	90
Industrial	45	170
Public	15	60
Loss	10	40
Total	165	620

Major industrial water users would be in addition to the values listed above.

Typical hourly variation in water demand is shown in Fig. 4.16. Unless historical records are available, the maximum daily water demand is often taken as twice the average daily water demand, and the peak hourly demand during the year is taken as three times the average daily demand. The future system demands are based on population projections. Water demand is assigned to each junction node in the system based on the estimated domestic, commercial, industrial, and public water use at the node.

The design flows for a municipal water distribution system are the maximum hourly demand and the maximum daily water demand plus the fire flow. Fire flows

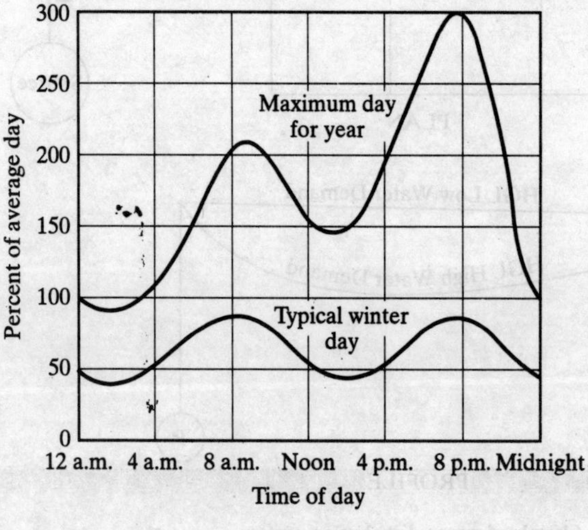

Figure 4.16 Typical hourly variation in water demand.

range from 500 gpm (1,900 lpm) for a residential fire to 12,000 gpm (45,000 lpm) for a fire in a major central business district. The duration of the fire ranges from 2 hr for a fire flow of less than 2,500 gpm (10,000 lpm) to 10 hr for a fire flow of 12,000 gpm (45,000 lpm). Fire hydrants are located throughout the service area generally at street intersections. Typical fire flows might be 500 gpm (1,900 lpm) for a residential fire, 2,000 gpm (7,500 lpm) for a small commercial or industrial area, and 3,000 gpm (11,400 lpm) in the main commercial area. For a large system, simultaneous fires may require consideration. The required maximum fire flow depends on the size of the city.

Elevated and ground level storage for a water distribution system is required to provide (a) equalization storage so the pumping does not have to match the demand, (b) fire demand, and (c) emergency storage. A schematic of system storage allocation is shown in Fig. 4.17. Emergency storage is typically a 1- to 2-day supply at the average daily demand. Emergencies that typically occur are line breaks on critical lines and system-wide power outages. Elevated equalization storage provides flexibility in the pumping schedule. With adequate elevated equalization storage, much of the pumping can be done at night during off-peak power demand. Typically, the amount of elevated storage may range from less than 10 percent for a large water distribution system to more than 50 percent of the total storage for a small water distribution system.

The height and location of the elevated storage in a water distribution system are selected to provide adequate pressure throughout the service area under various demand scenarios. During the peak hourly demand, the pressure should be greater than 35 psi (2.4×10^5 N/m^2) and greater than 20 psi (1.4×10^5 N/m^2) for the peak daily demand plus the fire flow. To reduce leakage and water use, the pressure in the water distribution system should be less than 90 psi (6.5×10^5 N/m^2).

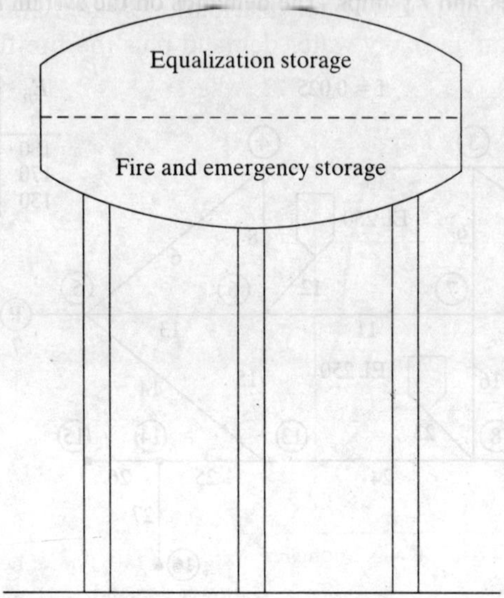

Figure 4.17 Schematic of system storage allocation.

Steady-state water distribution system modeling can be used to evaluate water velocities, pipe headlosses, and system pressure. Generally, water velocities should be less than 5 fps (1.5 mps) and headlosses (expressed as S_f) should be less than 0.01 for small pipes and less than 0.003 for large pipes in the system. Usually the minimum pipe diameter is 6 inches (150 mm) in residential areas and 8 inches (203 mm) in mercantile areas. Time-dependent computer simulations of the water distribution system can be used to evaluate pumping schedules and the operation of elevated storage tanks.

Air-relief valves should be installed at high points in the water distribution system to prevent the accumulation of air, and drain valves should be provided at low points. Gate valves are commonly installed throughout the service area and are used to isolate segments of the system during repairs. Pressure-regulating valves may be used to divide the service area into pressure zones.

Drinking water standards set maximum limits on the level of contaminants that may be hazardous to the health of consumers. Treatment processes commonly used for public water supply do not remove all undesirable contaminants from the raw water. Many of the contaminants originate from industrial, agricultural, and domestic pollution. Safe drinking water standards apply to drinking water systems (publicly or privately owned) that serve at least 25 people (or 15 service connections) for at least 60 days per year. The Safe Drinking Water Act was passed in 1974 and amended in 1986 and 1996, and gives the Environmental Protection Agency the authority to set drinking water standards. Information on the drinking water standards can be obtained from http://www.epa.gov/water/.

Example 4.19(a) Municipal Water Distribution System Simulation

Conduct a 24-hour simulation of the operation of the water distribution system shown below using a 1-hour computational time step. The system includes 27 pipes, 7 closed loops, 5 fixed-grade nodes, and 2 pumps. The demands on the system are adjusted for

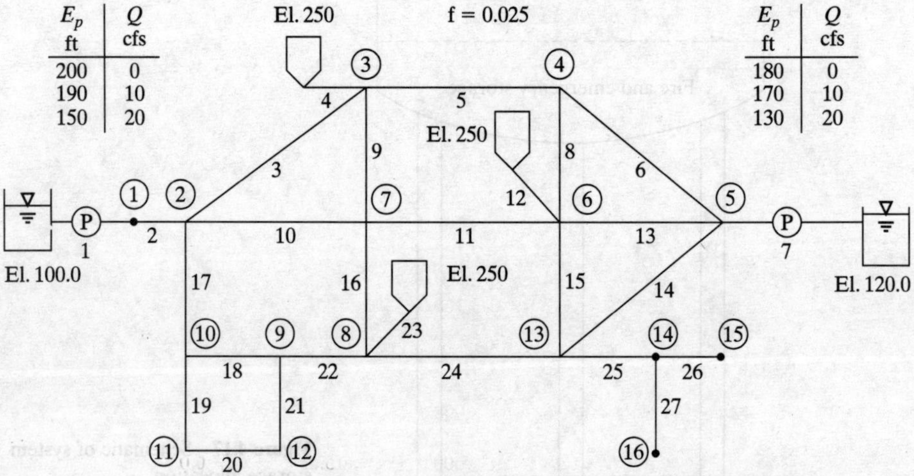

Example 4.19a Municipal water distribution system.

each time step using a global demand factor from Fig. 4.16 for the maximum day of the year. In addition, a fire demand is added at nodes 7 and 8 of 3.0 cfs each between the hours of 10 and 12.

The elevated storage tanks in the system are variable-level tanks, and the water level in the tank is computed after each time step. The elevation/storage information for each tank is listed below.

Elevation ft	Storage ft^3
220.0	0
230.0	40,000
240.0	120,000
250.0	200,000
260.0	250,000

The pumps in the system are variable-speed pumps, and the speed of the pump is adjusted after each time step depending on the elevation of the hydraulic grade line (HGL) at a specified junction node. The junction node for changing pump 1 is node 3, and the junction node for changing pump in line 7 is node 6 according to the following.

HGL Elevation	Speed ratio
<240	1.5
250	1.0
>260	0.5

The pump characteristic curve is adjusted using Eqs. 4.18 and 4.19.

The pipe network program (NETWORK.FOR) was modified for variable demands, variable-level elevated storage tanks and variable-speed pumps and is called NETOPSIM.FOR. The input data and summary of the output file are listed below. Comment lines in the program explain the input requirements.

Pipe no.	Node US	Node DS	Length (ft)	Diameter (in)	Minor loss coefficient	Fixed grade (ft)
1	0	1	2,000.0	24.0	0.5	100.0
2	1	2	800.0	24.0	0.0	
3	2	3	5,000.0	18.0	0.0	
4	3	0	700.0	18.0	0.5	250.0
5	3	4	3,700.0	12.0	0.0	
6	5	4	3,900.0	15.0	0.0	
7	0	5	2,100.0	24.0	0.5	120.0
8	6	4	2,500.0	10.0	0.0	
9	3	7	3,100.0	12.0	0.0	
10	2	7	5,500.0	18.0	0.0	
11	6	7	3,700.0	15.0	0.0	
12	0	6	900.0	18.0	0.5	250.0
13	5	6	2,900.0	15.0	0.0	
14	5	13	4,500.0	15.0	0.0	
15	6	13	2,500.0	15.0	0.0	
16	7	8	2,700.0	15.0	0.0	

Pipe no.	Node US	Node DS	Length (ft)	Diameter (in)	Minor loss coefficient	Fixed grade (ft)
17	2	10	3,100.0	18.0	0.0	
18	10	9	1,900.0	15.0	0.0	
19	10	11	1,600.0	8.0	0.0	
20	11	12	1,500.0	6.0	0.0	
21	9	12	1,650.0	8.0	0.0	
22	8	9	2,900.0	15.0	0.0	
23	0	8	1,900.0	18.0	7.5	250.0
24	13	8	3,100.0	15.0	0.0	
25	13	14	1,600.0	8.0	0.0	
26	14	15	1,750.0	6.0	0.0	
27	14	16	1,500.0	6.0	0.0	

Junction no.	Elevation (ft)	Demand (gpm)
1	90.00	0
2	110.00	694
3	95.00	694
4	105.00	2,083
5	100.00	694
6	103.00	2,428
7	97.00	2,083
8	103.00	1,044
9	107.00	0
10	112.00	0
11	115.00	350
12	112.00	350
13	110.00	0
14	120.00	0
15	135.00	175
16	130.00	175

RESULTS AT TIME = 18.000000 HOURS

PIPE	NODES		DISCHARGE	VELOCITY	HEADLOSS
1	0	1	9811.37	6.97	19.24
2	1	2	9811.37	6.97	7.55
3	2	3	2490.56	3.15	12.81
4	3	0	−1025.06	−1.29	−.32
5	3	4	958.79	2.72	10.67
6	5	4	3708.57	6.75	55.11
7	0	5	12657.70	8.99	33.60
8	6	4	540.14	2.21	5.69
9	3	7	821.83	2.34	6.57
10	2	7	2920.58	3.69	19.37
11	6	7	646.91	1.18	1.59
12	0	6	1969.76	2.49	1.49

13	5	6	4072.61	7.41	49.42
14	5	13	3141.52	5.71	45.63
15	6	13	–1214.69	–2.21	–3.79
16	7	8	–818.17	–1.49	–1.86
17	2	10	2665.23	3.37	9.09
18	10	9	1742.27	3.17	5.93
19	10	11	922.95	5.90	32.45
20	11	12	47.95	.55	.35
21	9	12	827.05	5.29	26.87
22	8	9	–915.23	–1.66	–2.50
23	0	8	1461.11	1.85	1.75
24	13	8	1051.83	1.91	3.52
25	13	14	875.00	5.60	29.17
26	14	15	437.50	4.97	33.61
27	14	16	437.50	4.97	28.81

JUNCTION DATA

NODE	DEMAND	GROUND	HGL	PRESSURE
1	.00	90.00	267.18	76.78
2	1735.00	110.00	259.63	64.84
3	1735.00	95.00	246.82	65.79
4	5207.50	105.00	236.16	56.84
5	1735.00	100.00	291.27	82.88
6	6070.00	103.00	241.85	60.17
7	5207.50	97.00	240.26	62.08
8	2610.00	103.00	242.11	60.28
9	.00	107.00	244.61	59.63
10	.00	112.00	250.54	60.03
11	875.00	115.00	218.08	44.67
12	875.00	112.00	217.74	45.82
13	.00	110.00	245.64	58.78
14	.00	120.00	216.47	41.80
15	437.50	135.00	182.86	20.74
16	437.50	130.00	187.66	24.99

TOTAL DEMANDS ON SYSTEM = 60.1

VARIABLE-LEVEL TANK DATA

TANK	ELEVATION	STORAGE
4	246.11	168880.80
12	241.36	130881.80
23	242.40	139163.80

Example 4.19(b) Water Quality

Assume that the water being pumped into the network at pipe 1 is from an unconfined aquifer. The area adjacent to the well field was an industrial storage site, and the area is contaminated with toxic chemicals. Residents living on the left side of the water distribution system have experienced a high rate of cancers, neurological problems, birth defects, and other health problems. Testing of the groundwater indicates high levels of arsenic and phenols. Determine the fraction of water at each junction node that is from this source.

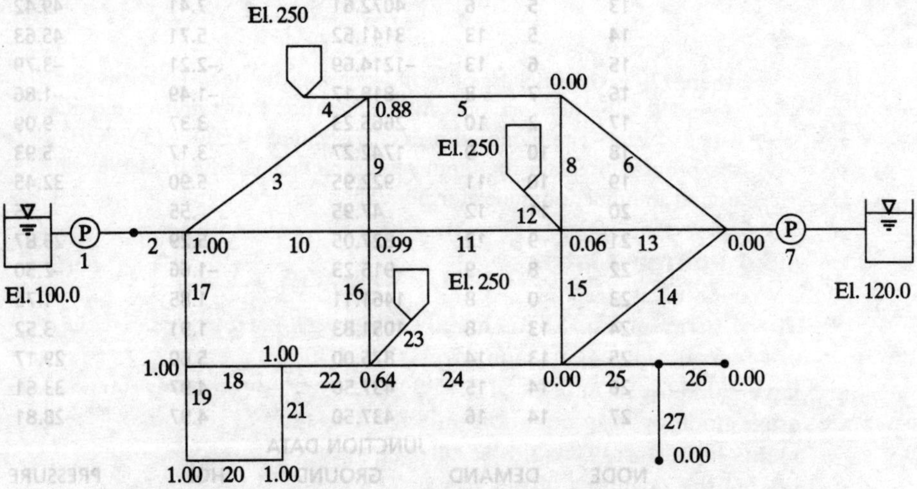

Example 4.19b Municipal water distribution system concentrations at hour 96.

The variation in concentration along a pipe can be estimated using the one-dimensional water quality model.

$$\frac{\partial C}{\partial t} = -V\frac{\partial C}{\partial x} + Dx\frac{\partial^2 C}{\partial x^2} - \text{losses} \tag{4.19.1}$$

where C is the concentration of the contaminate, V is the average velocity in the pipe, and Dx is the longitudinal dispersion coefficient. The first term on the left is the convective term, the second term on the left is the dispersion term, and the last term is the reaction/decay rate. For the purpose of this study, the losses will be represented using first-order reaction/decay (KC). Equation 4.19.1 can be written

$$C_i^+ = \frac{C_{i+1} + C_{i-1}}{2} - \left[\frac{V(C_{i+1} - C_{i-1})}{2dx} - \frac{Dx(C_{i+1} - 2C_i + C_{i-1})}{dx^2} + KC_i\right]dt \tag{4.19.2}$$

for pipe nodes 2 through $N - 1$ and

$$C_N^+ = C_N - \left[\frac{V(C_N - C_{N-1})}{dx} + KC_N\right]dt \tag{4.19.3}$$

for pipe node N.

The concentration at a junction node (C_J) is the blended concentration of the flow into the node or

$$C_J^+ = \frac{\sum^{\text{in}} C_{N,K}^+ Q_K}{\sum^{\text{in}} Q_K} \tag{4.19.4}$$

where C_J^+ is the concentration of the flow leaving the junction. Concentration of the contaminate in elevated storage tanks (C_t) depends on the concentration of the water flowing into the tank ($C_{N,K}$) and the volume of water in the storage tank ($\forall_t$) or

$$C_t^+ = \frac{C_t \forall_t + C_{N,K}^+ Q_K dt}{\forall_t + Q_K dt} \tag{4.19.5}$$

where Q_K is the discharge rate flowing into the tank and dt is the computation time step. The pipe network operation simulation program was modified to include Eqs. 4.19.2–4.19.5. The modified program is called NETOPCON.FOR. Comment lines in the program explain the input requirements. The figure 4.19b shows the concentration at the junction nodes for hour 96.

4.5.4 Pointer Matrix

The A array in Eq. 4.77 is mainly filled with zero values. Of the 81 values in the 9 × 9 array, only 28 are nonzero. A large municipal water distribution system may have 1,000 pipes, and the A array would be dimensioned 1,000 × 1,000. If the average node or loop equation only has five terms, more than 99 percent of the elements in the array would be zeros. A pointer matrix is a matrix of integers containing the positions of the nonzero elements in the A array.

An example of a pointer matrix is shown in Fig. 4.18 for the pipe network shown in Fig. 4.13. The square A array has been replaced with an A array

$$[A]_{old} = \begin{array}{c} \\ 1 \\ 2 \\ 3 \\ 4 \\ 5 \\ 6 \\ 7 \\ 8 \\ 9 \end{array} \begin{array}{c} \begin{array}{ccccccccc} 1 & 2 & 3 & 4 & 5 & 6 & 7 & 8 & 9 \end{array} \\ \begin{bmatrix} X & X & 0 & 0 & X & 0 & 0 & 0 & 0 \\ 0 & X & X & 0 & 0 & X & 0 & 0 & 0 \\ 0 & 0 & X & X & 0 & 0 & X & 0 & 0 \\ 0 & 0 & 0 & 0 & X & 0 & 0 & X & 0 \\ 0 & 0 & 0 & 0 & 0 & X & 0 & X & X \\ 0 & 0 & 0 & 0 & 0 & 0 & X & 0 & X \\ 0 & X & 0 & 0 & X & X & 0 & X & 0 \\ 0 & 0 & X & 0 & 0 & X & X & 0 & X \\ X & X & X & X & 0 & 0 & 0 & 0 & 0 \end{bmatrix} \end{array}$$

$$[A]_{new} = \begin{array}{c} \\ 1 \\ 2 \\ 3 \\ 4 \\ 5 \\ 6 \\ 7 \\ 8 \\ 9 \end{array} \begin{array}{c} \begin{array}{ccccc} 1 & 2 & 3 & 4 & 5 \end{array} \\ \begin{bmatrix} X & X & X & 0 & C_1 \\ X & X & X & 0 & C_2 \\ X & X & X & 0 & C_3 \\ X & X & 0 & 0 & C_4 \\ X & X & X & 0 & C_5 \\ X & X & 0 & 0 & C_6 \\ X & X & X & X & C_7 \\ X & X & X & X & C_8 \\ X & X & X & X & C_9 \end{bmatrix} \end{array} \qquad [P] = \begin{array}{c} \\ 1 \\ 2 \\ 3 \\ 4 \\ 5 \\ 6 \\ 7 \\ 8 \\ 9 \end{array} \begin{array}{c} \begin{array}{ccccc} 1 & 2 & 3 & 4 & 5 \end{array} \\ \begin{bmatrix} 1 & 2 & 5 & 0 & 3 \\ 2 & 3 & 6 & 0 & 3 \\ 3 & 4 & 7 & 0 & 3 \\ 5 & 8 & 0 & 0 & 2 \\ 6 & 8 & 9 & 0 & 3 \\ 7 & 9 & 0 & 0 & 2 \\ 2 & 5 & 6 & 8 & 4 \\ 3 & 6 & 7 & 9 & 4 \\ 1 & 2 & 3 & 4 & 4 \end{bmatrix} \end{array}$$

Figure 4.18 Example of pointer matrix.

dimensioned 5×9 that contains the nonzero values in the old A array and a 5×9 P array that includes the column position index for the nonzero values. An extra column was included in the new A array to store the C vector of constants and an extra column was added to the P array to store the number of nonzero values in each row.

For a municipal water distribution system, the A array and P array (integer) might each be dimensioned $25 \times 1{,}000$. This would handle a 1,000 pipe system with up to 24 pipes in one loop. This would represent a reduction in memory requirements of about 96 percent. The use of the pointer matrix also eliminates the need to multiply most zero elements, and the procedure saves computation time.

4.6 UNSTEADY FLOW

Unsteady flow in pipe systems can result in extremely high or low pressures. High pressures can cause rupture of the pipe or damage to pipeline appurtenances. Excessively low pressure can cause collapse of the pipe or vaporization of the water. When the pressure in the pipe drops below the vapor pressure of water, a vapor cavity is formed causing water column separation. The high pressure generated during vapor cavity closure can also cause pipe failure.

The most common cause of unsteady flow in a pipe system is valve movement. Either opening or closing a valve will cause pressure waves to travel through the system. Severe unsteady flow conditions can occur in a pump system both during pump run-down after a power failure or during pump start-up. Water hammer pressure can also occur during air removal from the pipe system while filling empty lines. Extreme unsteady flow pressures can be prevented by controlling valve movement or installing surge relief valves, surge tanks, air chambers, or air-vacuum valves in a pipe system.

4.6.1 Basic Equations for Unsteady Flow

Unsteady flow occurs in a pipe system during pump start-up, pump shutdown, valve changes, and air removal. Figure 4.19 shows the pressure wave propagation resulting from a sudden valve closure in a single pipeline of length (L). Figure 4.19(a) represents steady-state conditions with a uniform velocity (V) in the pipeline. Because the velocity head is small compared with the water hammer pressure head, only the HGL is shown in Fig. 4.19. At time = 0, the valve is closed and a compression wave is generated at the valve. The wave travels upstream toward the reservoir at a velocity $\mathbf{a}$. The water velocity at the valve decreases to zero and the pressure head at the valve increases by ΔH. The increase pressure causes an increase in water density and an expansion of the pipe. At time $L/2\mathbf{a}$, the pressure wave is half-way to the reservoir. Water upstream of the wave front is moving toward the valve at the initial velocity. Velocity reversal occurs when the wave front reaches the reservoir, and the pressure head decreases to that in

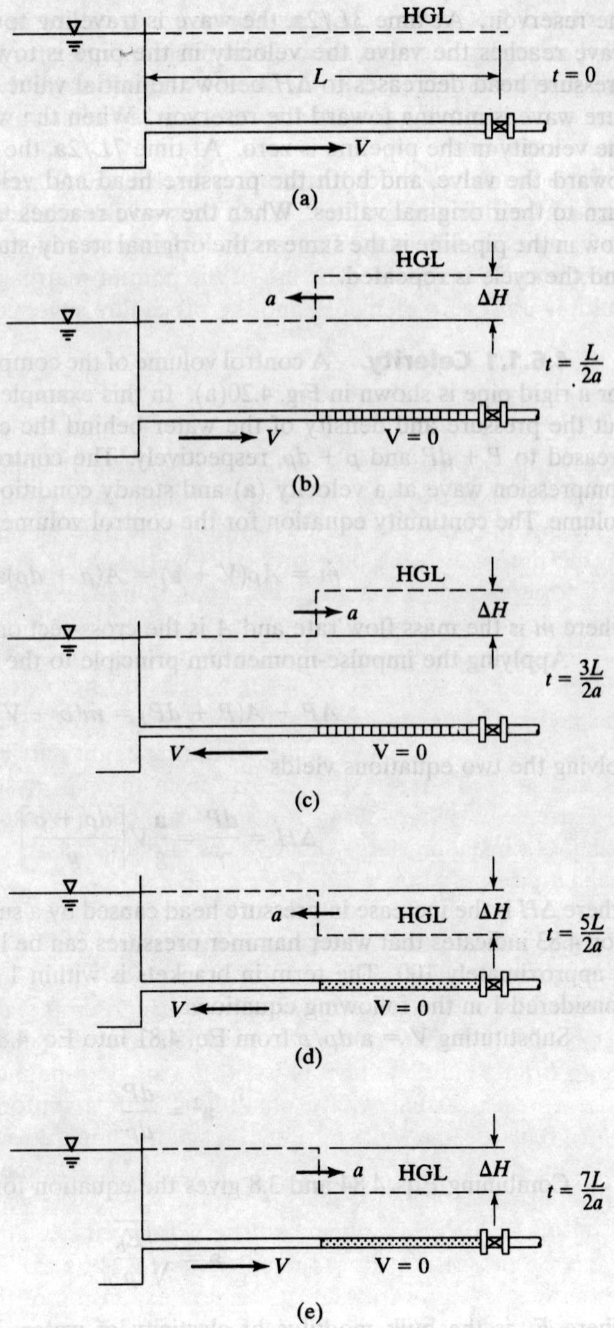

Figure 4.19 Water hammer in a simple pipe system.

the reservoir. At time $3L/2\mathbf{a}$, the wave is traveling toward the valve. When the wave reaches the valve, the velocity in the pipe is toward the reservoir and the pressure head decreases to ΔH below the initial value. At time $5L/2\mathbf{a}$, the pressure wave is moving toward the reservoir. When the wave reaches the reservoir, the velocity in the pipeline is zero. At time $7L/2\mathbf{a}$, the pressure wave is traveling toward the valve, and both the pressure head and velocity behind the wave return to their original values. When the wave reaches the valve at time $4L/\mathbf{a}$, the flow in the pipeline is the same as the original steady-state conditions [Fig. 4.19(a)] and the cycle is repeated.

4.6.1.1 Celerity. A control volume of the compression wave in Fig. 4.19(b) for a rigid pipe is shown in Fig. 4.20(a). In this example, the pipe does not expand but the pressure and density of the water behind the compression wave have increased to $P + dP$ and $\rho + d\rho$, respectively. The control volume moves with the compression wave at a velocity ($\mathbf{a}$) and steady conditions exist within the control volume. The continuity equation for the control volume is

$$\dot{m} = A\rho(V + \mathbf{a}) = A(\rho + d\rho)\mathbf{a} \tag{4.81}$$

where $\dot{m}$ is the mass flow rate and A is the cross-section area of the pipe.

Applying the impulse-momentum principle to the control volume yields

$$AP - A(P + dP) = \dot{m}(o - V) \tag{4.82}$$

solving the two equations yields

$$\Delta H = \frac{dP}{\gamma} = \frac{\mathbf{a}}{g}V\left[\frac{d\rho + \rho}{\rho}\right] \tag{4.83}$$

where ΔH is the increase in pressure head caused by a sudden valve closure. Equation 4.83 indicates that water hammer pressures can be large since the value of $\mathbf{a}/g$ is approximately 100. The term in brackets is within 1 percent of 1.0 and will be considered 1 in the following equations.

Substituting $V = \mathbf{a}\, d\rho/\rho$ from Eq. 4.81 into Eq. 4.83 gives

$$\mathbf{a}^2 = \frac{dP}{d\rho} \tag{4.84}$$

Combining Eqs. 4.84 and 3.8 gives the equation for celerity in a rigid pipe.

$$\mathbf{a} = \sqrt{\frac{E_b}{\rho}} \tag{4.85}$$

where E_b is the bulk modulus of elasticity of water. Under normal conditions, $\mathbf{a}$ = 1,440 m/s. Because of the elasticity of the pipe, the compression wave velocity created by water hammer in a pipe is less than that given by Eq. 4.85.

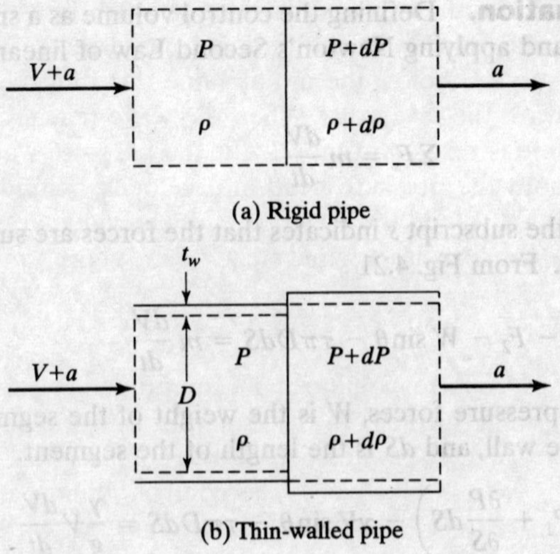

Figure 4.20 Control volume for compression wave.

Figure 4.20(b) represents the control volume for a thin-walled pipe ($D/t_w > 40$) with a diameter D and a wall thickness t_w. The kinetic energy of the moving fluid is converted into work done in compressing the water and expanding the pipe. In this example, only the circumferential expansion of the pipe is considered. Applying the first law of thermodynamics to the control volume

$$\dot{m}\frac{V^2}{2} = \frac{dP^2 A\mathbf{a}}{2E_b} + \frac{dP^2 AD\mathbf{a}}{2t_w E_p} \tag{4.86}$$

where E_p is the modulus of elasticity of the pipe wall. Solving Eqs. 4.81, 4.83, and 4.86 gives the equation for celerity in a thin-walled pipe.

$$\mathbf{a} = [E_b/\rho]^{1/2}\left[1 + \frac{DE_b}{t_w E_p}\right]^{-1/2} \tag{4.87}$$

Example 4.20 Celerity

Determine the wave speed or celerity in a 12-inch diameter steel pipe ($E_p = 30 \times 10^6$ psi) with a wall thickness (t_w) of 0.20 inches and in a 12-inch diameter PVC pipe ($E_p = 4 \times 10^5$ psi) with a wall thickness of 0.35 inches. Use a bulk modulus of elasticity for water of 300,000 psi.

$$a = \left(\frac{E_b}{\rho}\right)^{1/2}\left[1 + \frac{D\,E_b}{t_w\,E_p}\right]^{-1/2} = \left[\frac{300{,}000 \times 144}{1.93}\right]^{1/2}\left[1 + \frac{D\,E_b}{t_w\,E_p}\right]^{-1/2} = 4{,}731\left[1 + \frac{D\,E_b}{t_w\,E_p}\right]^{-1/2}$$

$$a_{st} = 4{,}731\left[1 + \frac{12}{0.20}\,\frac{30 \times 10^4}{30 \times 10^6}\right]^{-1/2} = 3{,}740 \text{ fps}$$

$$a_{pvc} = 4{,}731\left[1 + \frac{12}{0.35}\,\frac{30 \times 10^4}{40 \times 10^4}\right]^{-1/2} = 915 \text{ fps}$$

4.6.1.2 Euler equation. Defining the control volume as a small segment of the water in the pipe and applying Newton's Second Law of linear momentum gives

$$\Sigma F_s = m\frac{dV}{dt} \tag{4.88}$$

where m is the mass and the subscript s indicates that the forces are summed along the centerline of the pipe. From Fig. 4.21

$$F_1 - F_2 - W\sin\theta - \tau\pi D dS = m\frac{dV}{dt}$$

where F_1 and F_2 are the pressure forces, W is the weight of the segment, τ is the shear stress along the pipe wall, and dS is the length of the segment.

$$AP_1 - A\left(P_1 + \frac{\partial P}{\partial S}dS\right) - \gamma \forall \sin\theta - \tau\pi D dS = \frac{\gamma}{g}\forall\frac{dV}{dt}$$

where $\forall$ is the volume of the segment and γ is the specific weight of water. Simplifying the equation gives the one-dimensional Euler equation.

$$-\frac{1}{\gamma}\frac{\partial P}{\partial S} - \frac{dZ}{dS} - \frac{4\tau}{\gamma D} = \frac{1}{g}\frac{dV}{dt}$$

Relating the wall shear stress (τ) to the Darcy–Weisbach friction factor (f) by

$$\tau = \frac{1}{8}f\rho V|V| \tag{4.89}$$

gives

$$-\frac{1}{\gamma}\frac{\partial P}{\partial S} - \frac{dZ}{dS} - \frac{fV|V|}{D2g} = \frac{1}{g}\frac{dV}{dt}$$

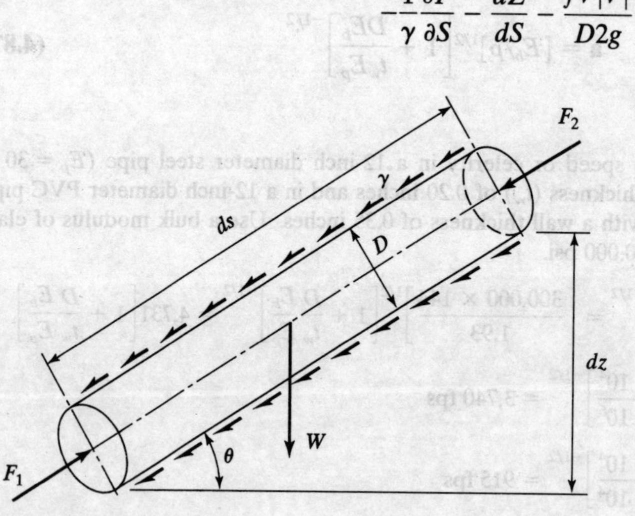

Figure 4.21 Control volume for Euler equation.

or

$$\frac{dV}{dt} + \frac{1}{\rho}\frac{\partial P}{\partial S} + g\frac{dZ}{dS} + \frac{fV|V|}{2D} = 0 \tag{4.90}$$

This one-dimension dynamic equation for unsteady flow in a pipe has two dependent variables, V and P. A second equation (continuity) is required to solve for V and P along the pipe.

Example 4.21 Valve Closure

The open valve in the 10,000-ft-long, 12-inch diameter pipeline closes so that the velocity in the pipe decreases linearly to zero in 10 seconds. Determine the maximum pressure upstream of the valve.

Initial conditions

$$H_L = \frac{fL}{D}\frac{V^2}{2g}$$

$$100 = \frac{0.02 \times 10{,}000}{1}\frac{V^2}{2g}$$

$$V = (2g \times 0.5)^{1/2} = 5.67 \text{ fps}$$

Based on Eq. 4.90

$$\frac{dV}{dt} + \frac{g}{\gamma}\frac{\partial P}{\partial S} + \frac{gdz}{ds} + \frac{fV|V|}{2D} = 0$$

$$\frac{1}{g}\int_0^L \frac{dV}{dt}ds + \int_A^B \frac{dP}{\gamma} + \int_A^B dz + \frac{fV^2}{D2g}\int_0^L dS = 0$$

$$\frac{P_A}{\gamma} + Z_A - \left(\frac{P_B}{\gamma} + Z_B\right) - \frac{fL}{D}\frac{V^2}{2g} = \frac{L}{g}\frac{dV}{dt}$$

$$\frac{P_B}{\gamma} = \left(\frac{P_A}{\gamma} + Z_A - Z_B\right) - \frac{fL}{D}\frac{V^2}{2g} - \frac{L}{g}\frac{dV}{dt}$$

$$\frac{dV}{dt} = -\frac{5.67}{10} = -0.567 \text{ ft/sec}^2$$

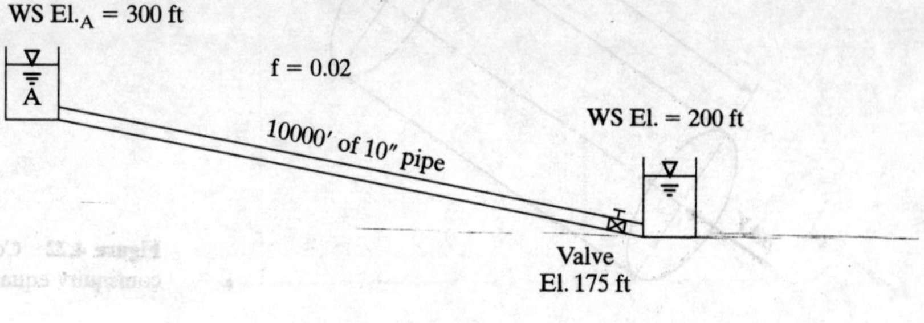

for maximum pressure evaluate at $V = 0$

$$\frac{P_B}{\gamma} = 300 - 175 - \frac{10{,}000}{32.2}(-0.567) = 125 + 176 = 301 \text{ ft}$$

$$P_B = \frac{301 \times 62.4}{144} = 130 \text{ psi}$$

4.6.6.3 Continuity equation. The control volume for the continuity equation is shown in Fig. 4.22. The diameter of the control volume will expand and contract, depending on the pressure in the control volume. In this formulation, the length of the control volume (dS) remains constant. Conservation of mass gives

$$\rho AV - \left[\rho AV + \frac{\partial}{\partial S}(\rho AV)\,dS\right] = \frac{d}{dt}(\rho A dS)$$

or

$$-\frac{\partial}{\partial S}(\rho AV)\,dS = \frac{d}{dt}(\rho A dS) \qquad \textbf{(4.91)}$$

Taking the derivative of the product of the three terms and collecting terms yields

$$\frac{1}{\rho}\frac{d\rho}{dt} + \frac{1}{A}\frac{dA}{dt} + \frac{\partial V}{\partial S} = 0 \qquad \textbf{(4.92)}$$

Noting that

$$\frac{1}{\rho}\frac{d\rho}{dt} = \frac{1}{E_b}\frac{dP}{dt}$$

and

$$\frac{1}{A}\frac{dA}{dt} = \frac{D}{t_w E_p}\frac{dP}{dt}$$

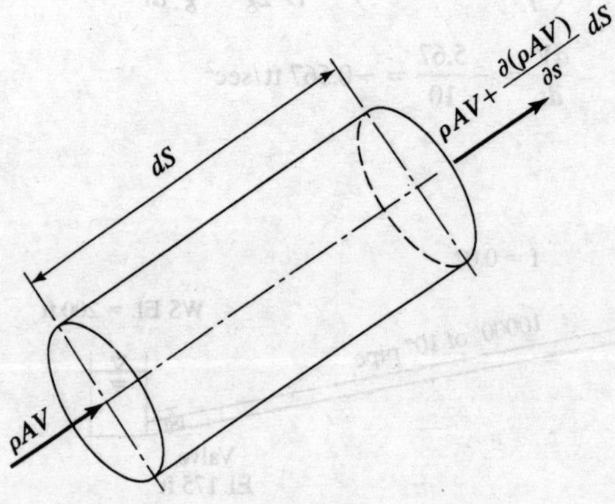

Figure 4.22 Control volume for continuity equation.

yields

$$\frac{dP}{dt}\left[\frac{1}{E_b} + \frac{D}{t_w E_p}\right] + \frac{\partial V}{\partial S} = 0 \tag{4.93}$$

Combining Eqs. 4.87 and 4.93 gives the continuity equation for unsteady flow in a pipe.

$$a^2\frac{\partial V}{\partial S} + \frac{1}{\rho}\frac{dP}{dt} = 0 \tag{4.94}$$

We now have two equations (Euler—Eq. 4.90 and Continuity—Eq. 4.94) to solve for V and P along the pipeline. The method of characteristics is compatible with numerical solutions by digital computer and will be used to solve these equations. The method is not limited to single pipelines, but can include branching systems and networks with pumps, air chambers, surge tanks, and valves.

4.6.2 Method of Characteristics

The solution to unsteady flow in pipelines and pipe networks requires that pressure (P) and velocity (V) be determined as a function of time for any point in the system. The two independent, one-dimensional, partial differential equations used to solve for V and P are the Euler equation (Eq. 4.90) and the continuity equation (Eq. 4.94). These equations can be solved by either implicit finite-difference or explicit finite-difference methods.

In the method of characteristics, the partial differential equations are converted to ordinary differential equations that are then solved by explicit finite-difference techniques. The method of characteristics is very compatible with numerical solution of complex pipe systems and will be the only method of solution presented. The method requires that small time steps be used to satisfy the Courant condition for stability. This is generally not a problem, because unsteady flow conditions in a pipe system occur during a short time period and the duration of computer simulation needed to analyze unsteady flow conditions is generally less than 1 minute.

Multiplying the Euler equation by a linear scaling factor (λ) and adding the result to the continuity equation gives

$$\lambda\left[\frac{dV}{dt} + \frac{1}{\rho}\frac{\partial P}{\partial S} + g\frac{dZ}{dS} + \frac{f}{2D}V|V|\right] + a^2\frac{\partial V}{\partial S} + \frac{1}{\rho}\frac{dP}{dt} = 0 \tag{4.95}$$

Replacing dV/dt and dP/dt with their partial components yields

$$\lambda\frac{\partial V}{\partial t} + \lambda V\frac{\partial V}{\partial S} + \frac{\lambda}{\rho}\frac{\partial P}{\partial S} + \lambda g\frac{dZ}{dS} + \frac{\lambda f}{2D}V|V| + a^2\frac{\partial V}{\partial S} + \frac{1}{\rho}\frac{\partial P}{\partial t} + \frac{V}{\rho}\frac{\partial P}{\partial S} = 0 \tag{4.96}$$

Combining terms

$$\left[\lambda\frac{\partial V}{\partial t} + (\lambda V + a^2)\frac{\partial V}{\partial S}\right] + \left[\frac{1}{\rho}\frac{\partial P}{\partial t} + \left(\frac{\lambda}{\rho} + \frac{V}{\rho}\right)\frac{\partial P}{\partial S}\right] + \lambda g\frac{dZ}{dS} + \frac{\lambda f}{2D}V|V| = 0 \tag{4.97}$$

Noting that

$$\lambda\frac{\partial V}{\partial t} + (\lambda V + a^2)\frac{\partial V}{\partial S} = \lambda\frac{dV}{dt}$$

only if

$$\lambda\frac{dS}{dt} = \lambda V + a^2$$

and

$$\frac{1}{\rho}\frac{\partial P}{\partial t} + \left(\frac{\lambda}{\rho} + \frac{V}{\rho}\right)\frac{\partial P}{\partial S} = \frac{1}{\rho}\frac{dP}{dt}$$

only if

$$\frac{1}{\rho}\frac{dS}{dt} = \frac{\lambda}{\rho} + \frac{V}{\rho}$$

The restriction on the equations are

$$\frac{dS}{dt} = V + \frac{a^2}{\lambda} \quad \text{and} \quad \frac{dS}{dt} = \lambda + V$$

Solving for the scaling factor gives

$$\lambda = \pm a.$$

The plus characteristic equation (C^+) is obtained when

$$\frac{dS}{dt} = V + a$$

and the negative characteristic equation (C^-) is obtained when

$$\frac{dS}{dt} = V - a$$

Substituting $(H - Z)\gamma$ for P gives ordinary differential equations

$$C^+: \quad \frac{dV}{dt} + \frac{g}{a}\frac{dH}{dt} - \frac{g}{a}V\frac{dZ}{dS} + \frac{f}{2D}V|V| = 0 \quad \text{for} \quad \frac{dS}{dt} = V + a \tag{4.98}$$

and

$$C^-: \quad \frac{dV}{dt} - \frac{g}{a}\frac{dH}{dt} + \frac{g}{a}V\frac{dZ}{dS} + \frac{f}{2D}V|V| = 0 \quad \text{for} \quad \frac{dS}{dt} = V - a \tag{4.99}$$

where H is the elevation of the hydraulic grade line.

4.6.3 Solution

Figure 4.23 shows an s-t plot of a pipe where the origin is placed at the upstream end and the pipe extends downstream in the positive direction for a length L. Numerical solution of the C^+ and C^- differential equations requires that the pipe be divided into N segments, with each segment $\Delta S = L/N$ in length. Values of H and V at each of the computational nodes for $t = 0$, represent initial conditions. Boundary conditions are applied at $S = 0$ and $S = L$. For example, if the pipe extends from a reservoir, H is constant at $S = 0$ for all values of t. Similarly, if a valve is located at the downstream end of the pipe, V must be specified as a function of time at $S = L$.

Once H and V have been computed for all nodes along a time line, then the characteristic equations can be used to determine H and V for all nodes at the next time line. Two characteristic equations can be used to solve for H and V at all interior nodes. One characteristic equation plus the boundary condition are used to solve for H and V at the boundaries.

Figure 4.24 shows the characteristic lines extending from old time line (J) to the new time line ($J + 1$), where

$$C^+: \quad \frac{\Delta X}{\Delta t} = a + V$$

and

$$C^-: \quad \frac{\Delta X}{\Delta t} = V - a$$

Because V may change along the length of the pipe and from one time line to the next, the characteristic lines are curved. However, because a is much greater than V, the characteristic lines can be approximated by straight lines within each grid element.

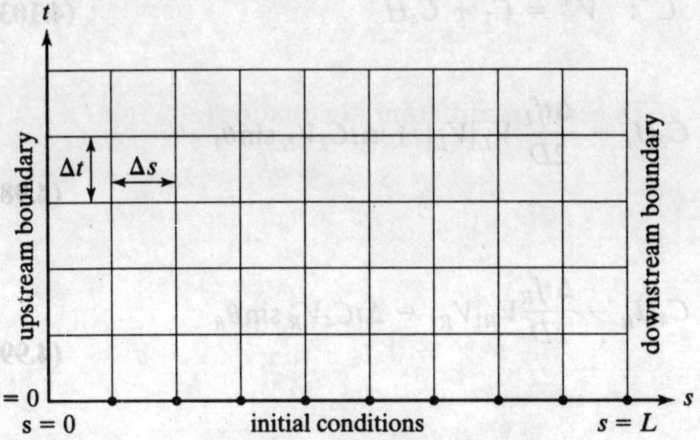

Figure 4.23 s-t plot for simple pipe.

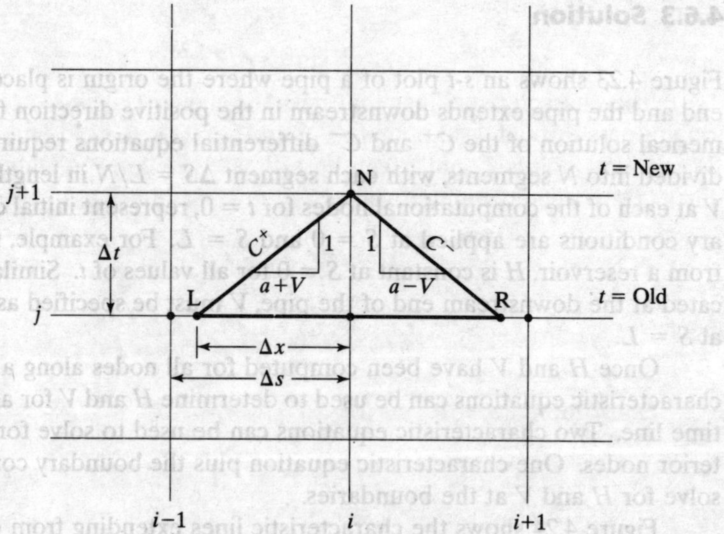

Figure 4.24 Characteristic lines.

Node N on the new time line represents any interior node where H^+ and V^+ are to be computed using the two characteristic equations. The finite-difference approximation to the characteristic equations are

$$C^+:\quad \frac{V^+ - V_L}{\Delta t} + \frac{g}{a}\frac{H^+ - H_L}{\Delta t} - \frac{g}{a}V_L\frac{dZ}{dS} + \frac{f}{2D}V_L|V_L| = 0 \tag{4.100}$$

$$C^-:\quad \frac{V^+ - V_R}{\Delta t} - \frac{g}{a}\frac{H^+ - H_R}{\Delta t} + \frac{g}{a}V_R\frac{dZ}{dS} + \frac{f}{2D}V_R|V_R| = 0 \tag{4.101}$$

Because the only two unknowns in the equations are V^+ and H^+, they can be written as

$$C^+:\quad V^+ = C_1 - C_2H^+ \tag{4.102}$$

$$C^-:\quad V^+ = C_3 + C_4H^+ \tag{4.103}$$

where

$$C_1 = V_L + C_2H_L - \frac{\Delta t f_L}{2D}V_L|V_L| + \Delta t C_2 V_L \sin\theta_L$$

$$C_2 = \left(\frac{g}{a}\right)_L$$

$$C_3 = V_R - C_4H_R - \frac{\Delta t f_R}{2D}V_R|V_R| - \Delta t C_4 V_R \sin\theta_R$$

$$C_4 = \left(\frac{g}{a}\right)_R$$

and

$$\sin\theta = dZ/dS$$

The characteristic lines that pass through N (Fig. 4.24) on the new time line do not pass through grid points $i-1$ and $i+1$ on the old time line. The values of V and H at points L and R can be estimated by interpolating along the old time lines (j) between computed values at $i-1$, i, and $i+1$.

$$V_L = V_i + (V_{i-1} - V_i)\frac{\Delta X}{\Delta S} \tag{4.104}$$

$$H_L = H_i + (H_{i-1} - H_i)\frac{\Delta X}{\Delta S} \tag{4.105}$$

Substituting $(a + V_L)\Delta t$ for ΔX gives

$$V_L = \frac{V_i + a\dfrac{\Delta t}{\Delta S}(V_{i-1} - V_i)}{1 - \dfrac{\Delta t}{\Delta S}(V_{i-1} - V_i)} \tag{4.106}$$

and

$$H_L = H_i + \frac{\Delta t}{\Delta S}(H_{i-1} - H_i)(a + V_L) \tag{4.107}$$

Because $\Delta t/\Delta S(V_{i-1} - V_i)$ is much smaller than 1 in the denominator of Eq. 4.106, the equation can be rewritten as

$$V_L = V_i + a\frac{\Delta t}{\Delta S}(V_{i-1} - V_i) \tag{4.108}$$

Similarly,

$$V_R = V_i + a\frac{\Delta t}{\Delta S}(V_{i+1} - V_i) \tag{4.109}$$

and

$$H_R = H_i + \frac{\Delta t}{\Delta S}(H_{i+1} - H_i)(a - V_R) \tag{4.110}$$

The stability of the scheme requires the Courant number (C_n) to be less than 1, where

$$C_n = \frac{\max|a + V|}{\Delta S/\Delta t}$$

The value of Δt is selected so that

$$\Delta t \leq \frac{\Delta S}{\max|a + V|} \tag{4.111}$$

where max$|a + V|$ is the maximum absolute value of the sum of wave speed and velocity. If the value of Δt is selected to satisfy Eq. 4.111, the points R and L in Fig. 4.24 will always fall between points $i - 1$ and $i + 1$.

The same Δt value is used for all pipes in the system. The minimum value of Δt is computed based on dividing all pipes into a minimum of N segments

$$\Delta t_K = \frac{L_K}{N(V_K + a_K)} \tag{4.112}$$

After $\Delta t_{\min}$ has been determined, then each pipe is divided into N_K parts.

$$N_K = \frac{L_K}{\Delta t_{\min}(V_K + a_K)} \tag{4.113}$$

4.6.4 Junction Nodes

4.6.4.1 Two pipes. A two pipe junction (pipes in series) is shown in Fig. 4.25. When writing the characteristic equations for the junction node, the values of C_1 and C_2 are determined for pipe 1 and values of C_3 and C_4 are computed for pipe 2 or

$$C^+:\ V_1^+ = C_{1,1} - C_{1,2}H_1^+ \tag{4.114}$$

$$C^-:\ V_2^+ = C_{2,3} + C_{2,4}H_2^+ \tag{4.115}$$

The continuity equation for the junction node is

$$V_1^+A_1 = V_2^+A_2 \tag{4.116}$$

where A is the area of the pipe.

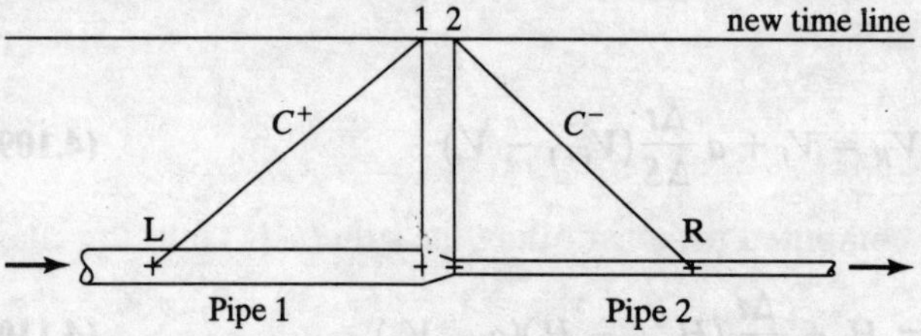

Figure 4.25 Two-pipe junction node.

Writing the energy equation across the junction and neglecting minor losses

$$H_1^+ = H_2^+ \tag{4.117}$$

Combining equations and solving for the elevation of the hydraulic grade line gives

$$H_1^+ = H_2^+ = \frac{A_1C_{1,1} - A_2C_{2,3}}{A_1C_{1,2} + A_2C_{2,4}} \tag{4.118}$$

The pipe velocities can then be computed from the characteristic equations (4.114 and 4.115).

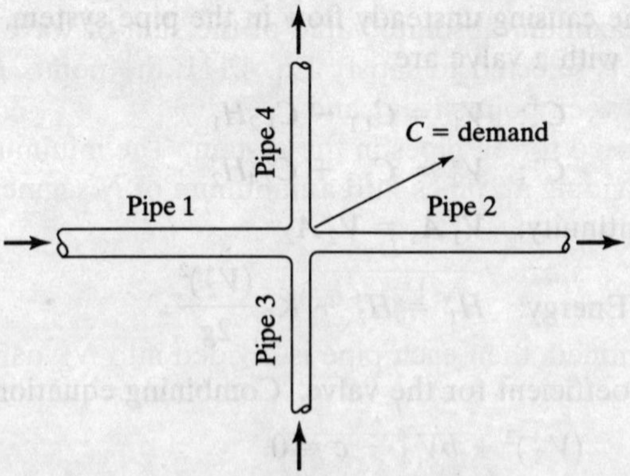

Figure 4.26 Four-pipe junction node with demand.

4.6.4.2 Four pipes. A four-pipe junction with an external demand is shown in Fig. 4.26. For the two pipes flowing into the junction, the C^+ characteristic equations are written whereas the C^- characteristic equations are written for the other two pipes flowing out of the junction. The equations for the four-pipe junction are

$$\text{Pipe 1} \quad C^+: \quad V_1^+ = C_{1,1} - C_{1,2}H_1^+ \tag{4.119}$$

$$\text{Pipe 2} \quad C^-: \quad V_2^+ = C_{2,3} + C_{2,4}H_2^+ \tag{4.120}$$

$$\text{Pipe 3} \quad C^+: \quad V_3^+ = C_{3,1} - C_{3,2}H_3^+ \tag{4.121}$$

$$\text{Pipe 4} \quad C^-: \quad V_4^+ = C_{4,3} + C_{4,4}H_4^+ \tag{4.122}$$

$$\text{Continuity:} \quad V_1^+A_1 + V_3^+A_3 - V_2^+A_2 - V_4^+A_4 - C = 0 \tag{4.123}$$

$$\text{Energy:} \quad H_1^+ = H_2^+ = H_3^+ = H_4^+ = H^+ \tag{4.124}$$

Combining equations gives

$$H^+ = \frac{C_{1,1}A_1 - C_{2,3}A_2 + C_{3,1}A_3 - C_{4,3}A_4 - C}{C_{1,2}A_1 + C_{2,4}A_2 + C_{3,2}A_3 + C_{4,4}A_4} \tag{4.125}$$

The velocity in each pipe is computed by substituting the head (H^+) into the characteristic equations.

4.6.4.3 Valve. Figure 4.27 is a two-pipe junction node with a valve. The valve setting may be constant representing a minor loss or the valve setting may

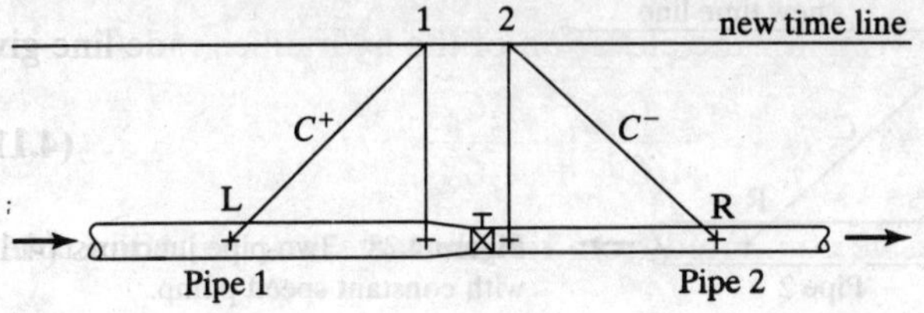

Figure 4.27 Two-pipe junction node with a valve.

change as a function of time causing unsteady flow in the pipe system. The equations for the junction node with a valve are

$$C^+: \quad V_1^+ = C_{1,1} - C_{1,2}H_1^+ \tag{4.126}$$

$$C^-: \quad V_2^+ = C_{2,3} + C_{2,4}H_2^+ \tag{4.127}$$

$$\text{Continuity:} \quad V_1^+A_1 = V_2^+A_2 \tag{4.128}$$

$$\text{Energy:} \quad H_1^+ = H_2^+ + K_L\frac{(V_2^+)^2}{2g} \tag{4.129}$$

where K_L is the headloss coefficient for the valve. Combining equations gives

$$(V_2^+)^2 + bV_2^+ - c = 0 \tag{4.130}$$

where

$$b = \frac{2g}{K_L}\left(\frac{A_2}{A_1}\frac{1}{C_{1,2}} + \frac{1}{C_{2,4}}\right)$$

$$c = \frac{2g}{K_L}\left(\frac{C_{1,1}}{C_{1,2}} + \frac{C_{2,3}}{C_{2,4}}\right)$$

b is always positive because both $C_{1,2}$ and $C_{2,4}$ are always positive. Solving for V_2^+ gives

$$V_2^+ = \frac{b}{2}\left[-1 + \sqrt{1 + \frac{4c}{b^2}}\right] \tag{4.131}$$

If flow is reversed through the valve, the equations for the junction node will need to be reformulated. Generally, the headloss coefficient (K_L) will be different for forward flow through the valve than it is for reverse flow through the valve.

4.6.4.4 Constant speed pump. Writing the equations for a two-pipe junction node with a constant speed pump (Fig. 4.28) gives

$$C^+: \quad V_1^+ = C_{1,1} - C_{1,2}H_1^+ \tag{4.132}$$

$$C^-: \quad V_2^+ = C_{2,3} + C_{2,4}H_2^+ \tag{4.133}$$

$$\text{Energy:} \quad H_1^+ + E_p = H_2^+ \tag{4.134}$$

$$\text{Continuity:} \quad V_1^+A_1 = V_2^+A_2 \tag{4.135}$$

$$\text{Pump:} \quad E_p = AQ^2 + BQ + Hc \tag{4.136}$$

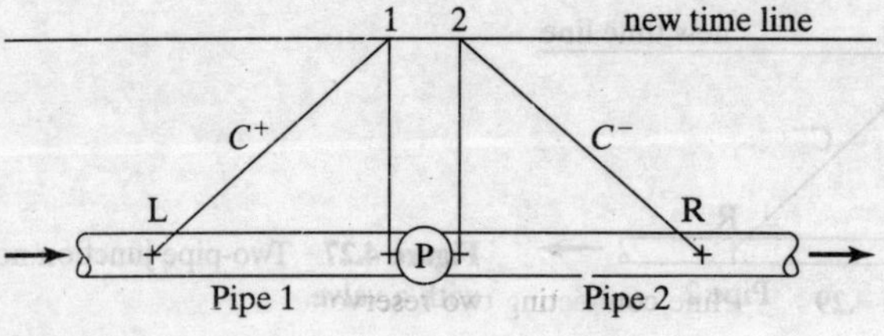

Figure 4.28 Two-pipe junction node with constant speed pump.

The pump characteristic equation can be rewritten in terms of V_2^+

$$E_p = A^1(V_2^+)^2 + B^1V_2^+ + Hc \tag{4.137}$$

where

$$A^1 = AA_2^2 \quad \text{and}$$
$$B^1 = BA_2$$

Combining equations gives

$$(V_2^+)^2 + bV_2^+ + c = 0 \tag{4.138}$$

where

$$b = \frac{1}{A^1}\left(B^1 - \frac{A_2}{A_1C_{1,2}} - \frac{1}{C_{2,4}}\right)$$
$$c = \frac{1}{A^1}\left(\frac{C_{1,1}}{C_{1,2}} + \frac{C_{2,3}}{C_{2,4}} + Hc\right)$$

Solving for V_2^+

$$V_2^+ = \frac{b}{2}\left(-1 + \sqrt{1 - \frac{4c}{b^2}}\right) \tag{4.139}$$

A check valve would normally be installed to prevent backflow through the pump. If V_2^+ is negative, then V_2^+ is set equal to zero.

4.6.5 Boundary Nodes

4.6.5.1 Reservoir. A pipeline connecting two reservoirs is shown in Fig. 4.29. The upstream reservoir represents a constant head boundary node. The two equations for the upstream boundary are

$$C^-: \quad V_1^+ = C_{1,3} + C_{1,4}H_1^+ \tag{4.140}$$
$$\text{Energy:} \quad H_1^+ = H_0 \tag{4.141}$$

where H_0 is the elevation of the upstream reservoir.

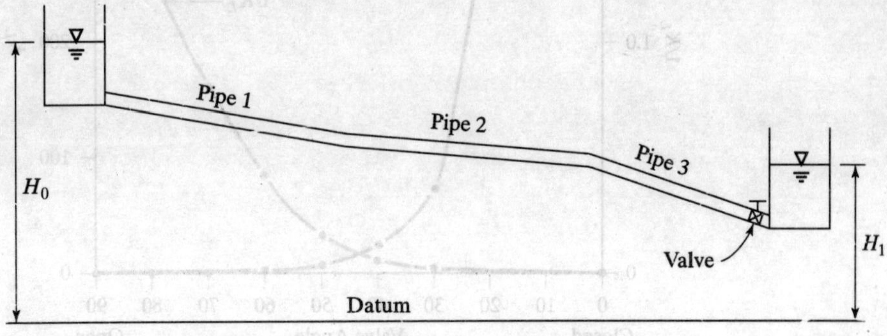

Figure 4.29 Pipeline connecting two reservoirs.

4.6.5.2 Velocity. The downstream boundary in Fig. 4.29 is a valve and constant head reservoir. Changing the valve setting causes unsteady flow in the pipeline. If the valve was to close in such a manner as to cause the velocity in the pipe to decrease linearly from V_0 to zero in T_c seconds, the valve would be considered a velocity boundary condition and the equations would be

$$C^+: \quad V_3^+ = C_{3,1} - C_{3,2}H_3^+ \tag{4.142}$$

$$\text{Valve:} \quad V_3^+ = V_0\left(1 - \frac{t}{T_c}\right), \quad 0 \le t \le T_c \tag{4.143}$$

$$V_3^+ = 0, \qquad t > T_c \tag{4.144}$$

4.6.5.3 Valve. Generally, the relation between the valve setting and pipe velocity is not known, making it difficult to represent a valve as a velocity boundary condition. However, valves can be tested in the laboratory for headloss at various valve settings and valve manufacturers will often provide this information. Figure 4.30 shows the headloss coefficients for a butterfly valve. If the valve is closed at a specified rate, the headloss coefficient can be determined as a function of time. The most critical time during valve closure is just before the valve closes (0–20 degrees in Fig. 4.30) and the value of K_L is usually interpolated from the $1/K_L$ curve.

The equations for a valve at the downstream end of pipe 3 next to a reservoir (Fig. 4.29) are

$$C^+: \quad V_3^+ = C_{3,1} - C_{3,2}H_3^+ \tag{4.145}$$

$$\text{Energy:} \quad H_3^+ = H_1 + K_L\frac{(V_3^+)^2}{2g} \tag{4.146}$$

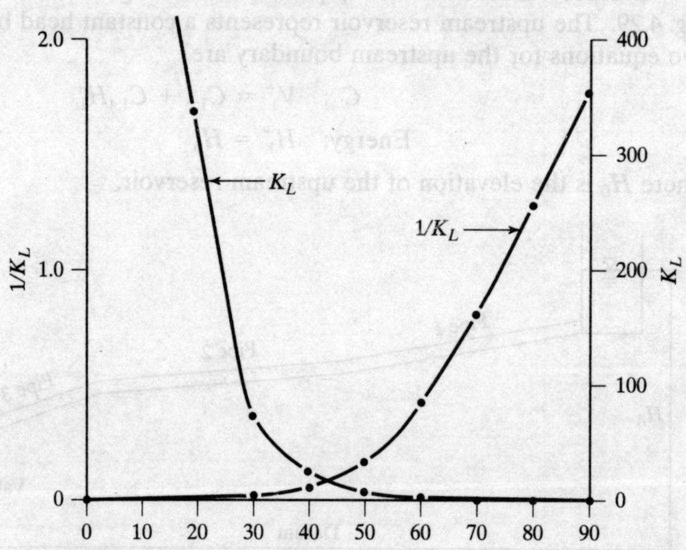

Figure 4.30 Butterfly valve headloss coefficients for forward flow.

Combining equation

$$(V_3^+)^2 + bV_3^+ + c = 0 \tag{4.147}$$

$$b = \frac{2g}{K_L C_{3,2}}$$

$$c = \frac{2g}{K_L}\left(H_1 - \frac{C_{3,1}}{C_{3,2}}\right)$$

Solving for V_3^+

$$V_3^+ = \frac{b}{2}\left[-1 + \sqrt{1 - \frac{4c}{b^2}}\right] \tag{4.148}$$

for forward flow and

$$V_3^+ = \frac{b}{2}\left[1 - \sqrt{1 + \frac{4c}{b^2}}\right] \tag{4.149}$$

for reverse flow. The headloss coefficient for the valve may be different for reverse flow through the valve than it is for forward flow.

4.6.5.4 Variable speed pump. The pump characteristic curve in Fig. 4.31 is for a speed (N_o). The discharge (Q), head (E_p), and power (P) at any other speed N can be determined from the following equations from Section 4.2.

$$Q = Q_o \frac{N}{N_o} \tag{4.150}$$

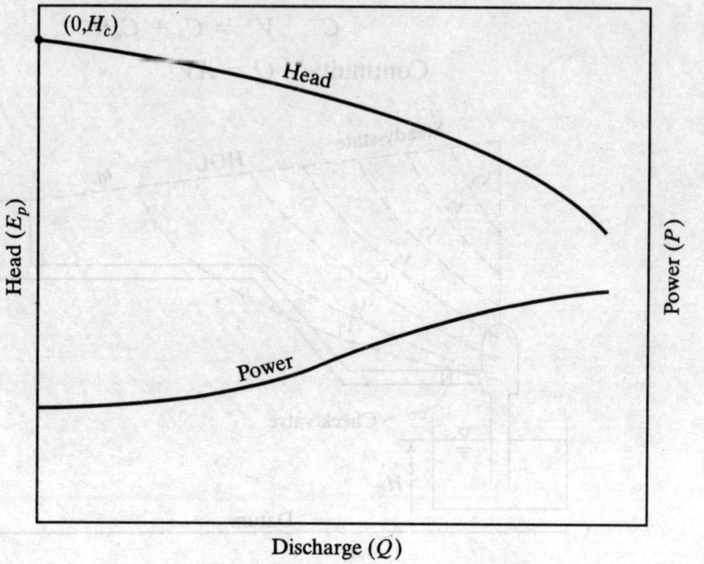

Figure 4.31 Pump characteristic curves for speed N_o.

$$E_p = E_{p_o}\left(\frac{N}{N_o}\right)^2 \tag{4.151}$$

$$P = P_o\left(\frac{N}{N_o}\right)^3 \tag{4.152}$$

Power is proportional to torque times the angular velocity (ω) or $T = P/\omega$. The torque that the motor exerts against the resisting torque of the water is

$$T = T_o\left(\frac{N}{N_o}\right)^2 \tag{4.153}$$

A sudden loss of energy to the motor (power failure) will cause the pump speed to decrease (pump rundown). During power failure, the driving torque is zero, but the resisting torque continues and causes the pump speed to decelerate according to

$$T = \frac{2\pi}{60} I \frac{dN}{dt} \tag{4.154}$$

where I is the rotational inertia of the pump shaft, impeller, water, and motor armature and shaft.

The new pump speed (N^+) can be estimated from the pump speed (N) and torque (T) at the old time step using

$$N^+ = N - \frac{60T\Delta t}{2\pi I} \tag{4.155}$$

The pump characteristic curve for the new pump speed can now be determined ($E_p^+ = A''(V^+)^2 + B''V^+ + C''$).

The equations for the left boundary in Fig. 4.32 are

$$C^-: \quad V^+ = C_3 + C_4H^+ \tag{4.156}$$

$$\text{Continuity:} \quad Q = AV^+ \tag{4.157}$$

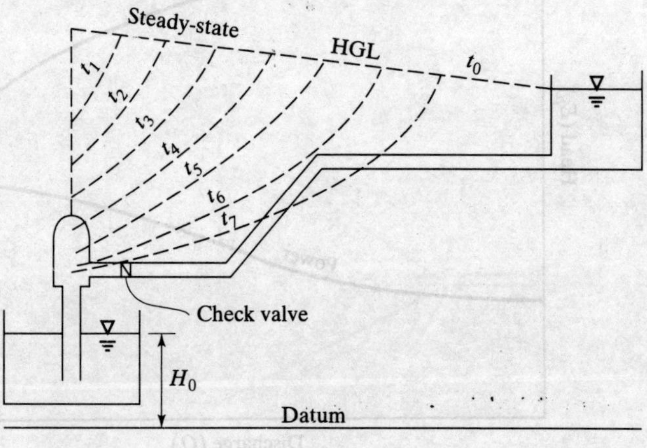

Figure 4.32 Pump boundary with power failure.

$$\text{Energy:} \quad H^+ = H_o + E_p^+ \tag{4.158}$$

$$\text{Pump:} \quad E_p = f(Q,N) \tag{4.159}$$

Combining equations yields

$$V^+ = \frac{b}{2}\left(-1 + \sqrt{1 - \frac{4c}{b^2}}\right) \tag{4.160}$$

where

$$b = \frac{1}{A''}\left[B'' - \frac{1}{C_4}\right] \quad \text{and}$$

$$c = \frac{1}{A''}\left[\frac{C_3}{C_4} + H_o + C''\right]$$

where the double prime indicates coefficients for the equation of the pump characteristic curve at the new pump speed.

The pump characteristic curve in Fig. 4.31 is only for forward rotation of the impeller and forward flow through the pump. A check valve upstream of the pump will prevent back flow through the pump and backward rotation of the impeller. The *HGL* for steady flow and various times after power failure are shown in Fig. 4.32. At time t_7, the *HGL* is below the pipe, and column separation will occur if the pressure drops below the vapor pressure of the water.

Example 4.22 Unsteady Flow in a Pipe Network Caused by Valve Closure

Conduct an unsteady flow analysis for the pipe network shown in the sketch below. The valve in line 7 closes from 100% at time = 0 to 5% open at time = 5 seconds and 0% open at time = 10 seconds. The valve has the following headloss K values ($H_L = KV^2/2g$).

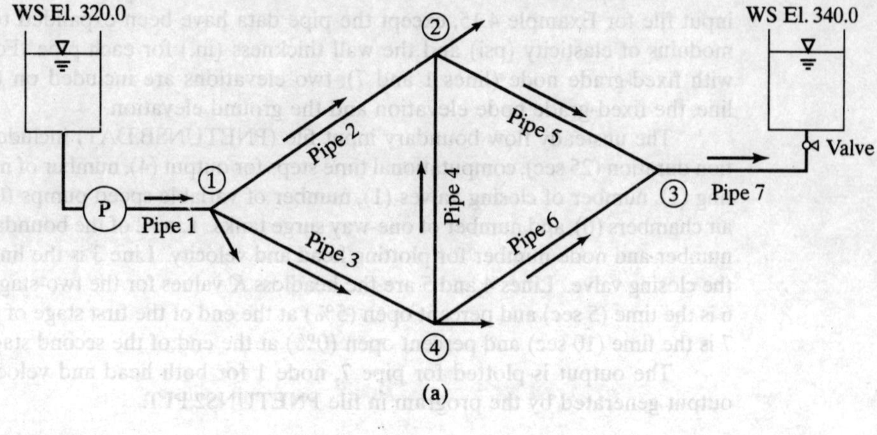

(a)

% Open	K
10	550
20	340
30	75
40	25
50	6.0
60	3.0
70	1.0
80	0.05
90	0.02
100	0.01

The pipes in the sketch have the same diameters, lengths and friction factors as Example 4.14. The modulus of elasticity for all pipes is 30×10^6 psi, and the pipe wall thicknesses are 0.20 inches for pipes 1 and 7; 0.18 inches for pipes 2, 3, 5, and 6; and 0.15 inches for pipe 4. The pump on line 1 is a constant speed pump that has the same pump characteristic curve as the pump in Example 4.15. The elevations of the junction nodes are the same as Example 4.14, but the demands have been changed to 1.0 cfs at each junction node.

Run the simulation for 25 seconds with output every 4th computational time step. Using EXCEL, plot the head and velocity for node 1 in pipe 7.

Two FORTRAN programs were written to solve the problem. The source codes are included as PNETUNS1.FOR and PNETUNS2.FOR in the HMP described in Section 1.4.5. The source codes are for student use and can be modified as necessary for other applications. The program includes comments so the student can follow the analysis and equations presented in the text. The first program (PNETUNS1.FOR) is a modification of the pipe network program (NETWORK.FOR from Example. 4.15) and is used to establish initial conditions for the unsteady flow program (PNETUNS2.FOR). The unsteady flow program reads the output file (PNETUNS2.DAT) generated by the first program along with input file PNETUNSB.DAT and conducts the unsteady flow analysis. The second program was coded to handle valves located at reservoirs and two-pipe junctions.

The input file for PNETUNS1.FOR is listed. The input file is the same as the input file for Example 4.15, except the pipe data have been expanded to include the modulus of elasticity (psi) and the wall thickness (in.) for each pipe. For those lines with fixed-grade node (lines 1 and 7), two elevations are included on the following line, the fixed-grade node elevation and the ground elevation.

The unsteady flow boundary input file (PNETUNSB.DAT) includes the simulation duration (25 sec), computational time steps for output (4), number of nodes for plotting (1), number of closing valves (1), number of variable speed pumps (0), number of air chambers (0), and number of one-way surge tanks. Line 2 of the boundary file is pipe number and node number for plotting head and velocity. Line 3 is the line number for the closing valve. Lines 4 and 5 are the headloss K values for the two-stage valve. Line 6 is the time (5 sec) and percent open (5%) at the end of the first stage of closure. Line 7 is the time (10 sec) and percent open (0%) at the end of the second stage of closure.

The output is plotted for pipe 7, node 1 for both head and velocity using the output generated by the program in file PNETUNS2.PLT.

```
EXAMPLE 4.22  UNSTEADY FLOW IN A PIPE NETWORK CAUSED BY VALVE INITIAL CONDITIONS
 &NET    IUN = 1,         NP = 7,    NJ = 4,    NL = 2,    NF = 2,     NPU = 1/
     1      0        1     1000.0      12.0      .015      30.E + 06      .20
        320.0       310.0
     2      1        2      800.0       8.0      .019      30.E + 06      .18
     3      1        4      700.0       8.0      .019      30.E + 06      .18
     4      4        2      750.0       6.0      .020      30.E + 06      .15
     5      2        3      600.0       8.0      .019      30.E + 06      .18
     6      4        3      800.0       8.0      .019      30.E + 06      .18
     7      3        0      900.0      10.0      .017      30.E + 06      .20
        340.0       330.0
     1          320.0        1.0
     2          330.0        1.0
     3          310.0        1.0
     4          300.0        1.0
     1          115.0        5.0     105.0      10.0       85.0
     3
     2     -4       -3
     3
     5     -6        4
     4              -1
    -7     -5       -2
EXAMPLE 4.22  PNETUNSB.DAT INPUT FILES FOR UNSTEADY FLOW ANALYSIS
 &SPECS   DURT = 25.0,  IOUT = 4,  NPLT = 1,  NVAL = 1,  NPUP = 0,  NAIR = 0,  NSTK = 0/
    7           1
    7
              550.        340.       75.0       25.0       6.0
              3.0         1.0        0.05       0.02       0.01
              5.0         5.0
              10.0        0.0
```

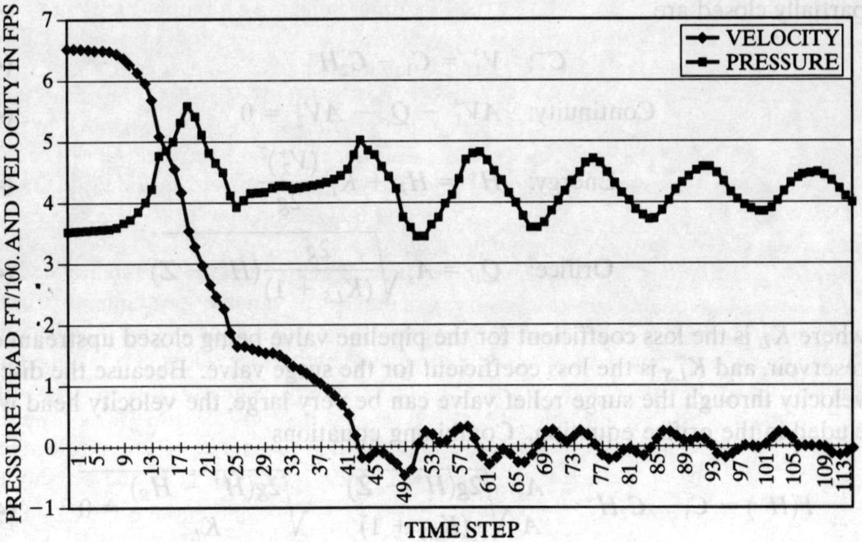

4.6.6 Control Devices

Control devices are installed to reduce water hammer pressures to acceptable levels. This section will show how control devices can be added to a pipe system and evaluated using numerical simulation. The size of the control device is generally determined by changing size and/or location and rerunning the numerical simulation of the system until the water hammer pressures are within acceptable levels. Surge relief valves, surge tanks, air chambers, and other techniques will be discussed in this section.

4.6.6.1 Surge relief valve. A surge relief valve is designed to open when the pressure in the pipeline at the valve exceeds a specified value. The valve opens quickly when the specified pressure is exceeded and discharges into the atmosphere according to the orifice equation

$$Q_o = C_o A_s \sqrt{2g(H^+ - Z)} \tag{4.161}$$

where C_o is the orifice coefficient, A_s is the cross-sectional area of the valve when open, and $(H^+ - Z)$ is pressure head in the pipeline. The valve then closes in a manner that minimizes loss of water from the pipeline and avoids generating water hammer pressures during closure. For example, the surge relief valve may be designed to open when the pressure in the pipeline exceeds P_x pressure and then closes linearly in T_c seconds. The loss coefficients (K_{LS}) for the surge valve must be provided.

A surge relief valve is shown in Fig. 4.33 just upstream of a valve at the downstream end of a pipeline connecting two reservoirs. The equations for the boundary node at the downstream reservoir with the surge valve open and pipeline valve partially closed are

$$C^+: \quad V_1^+ = C_1 - C_2 H^+ \tag{4.162}$$

$$\text{Continuity:} \quad AV_1^+ - Q_o - AV_2^+ = 0 \tag{4.163}$$

$$\text{Energy:} \quad H^+ = H_R + K_L \frac{(V_2^+)^2}{2g} \tag{4.164}$$

$$\text{Orifice:} \quad Q_o = A_s \sqrt{\frac{2g}{(K_{LS} + 1)}(H^+ - Z)} \tag{4.165}$$

where K_L is the loss coefficient for the pipeline valve being closed upstream of the reservoir, and K_{LS} is the loss coefficient for the surge valve. Because the discharge velocity through the surge relief valve can be very large, the velocity head was included in the orifice equation. Combining equations

$$F(H^+) = C_1 - C_2 H^+ - \frac{A_s}{A} \sqrt{\frac{2g(H^+ - Z)}{(K_{LS} + 1)}} - \sqrt{\frac{2g(H^+ - H_R)}{K_L}} = 0 \tag{4.166}$$

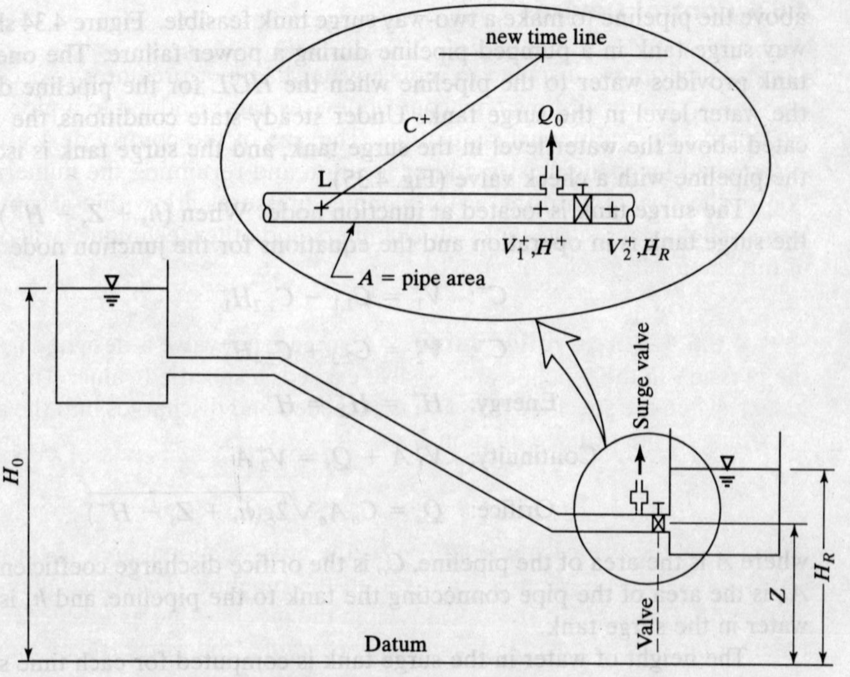

Figure 4.33 Surge valve in pipeline.

This equation can be solved for H^+ iteratively using the Newton–Raphson method

$$H^{j+1} = H^j - F(H^j)/F'(H^j) \tag{4.167}$$

where H^j is the estimate of H from the previous iteration, H^{j+1} is the new estimate of H, and $F'(H)$ is dF/dH or

$$F'(H^j) = -C_2 - \frac{A_s}{A}\left(\frac{2g}{K_{LS}+1}(H^j - Z)\right)^{-1/2}\frac{g}{K_{LS}+1} - \left(\frac{2g}{K_L}(H^j - H_R)\right)^{-1/2}\frac{g}{K_L} \tag{4.168}$$

The first estimate of H^j is usually taken from the previous time step.

The iterative process is continued until $F(H) = 0$ is satisfied within acceptable limits. Four conditions can occur at the downstream boundary node (1) pipeline valve partially closed and surge valve closed, (2) pipeline valve partially closed and surge valve open or partially open, (3) pipeline valve closed and surge valve partially open, and (4) pipeline valve closed and surge valve closed.

4.6.6.2 Surge tank. Surge tanks often used in hydroelectric installations to prevent excessively high or low pressures in the penstock when there is a rapid change in demand for energy. In pumped pipelines, one-way surge tanks are often used to prevent low pressures in the pipeline because the *HGL* is located too far above the pipeline to make a two-way surge tank feasible. Figure 4.34 shows a one-way surge tank in a pumped pipeline during a power failure. The one-way surge tank provides water to the pipeline when the *HGL* for the pipeline drops below the water level in the surge tank. Under steady-state conditions, the *HGL* is located above the water level in the surge tank, and the surge tank is isolated from the pipeline with a check valve (Fig. 4.35).

The surge tank is located at junction node. When $(h_s + Z_s - H^+)$ is positive, the surge tank is in operation and the equations for the junction node are

$$C^+: \quad V_1^+ = C_{1,1} - C_{1,2}H_1^+ \tag{4.169}$$

$$C^-: \quad V_2^+ = C_{2,3} + C_{2,4}H_2^+ \tag{4.170}$$

$$\text{Energy:} \quad H_1^+ = H_2^+ = H^+ \tag{4.171}$$

$$\text{Continuity:} \quad V_1^+A + Q_s = V_2^+A \tag{4.172}$$

$$\text{Orifice:} \quad Q_s = C_oA_n\sqrt{2g(h_s + Z_s - H^+)} \tag{4.173}$$

where A is the area of the pipeline, C_o is the orifice discharge coefficient (0.6–0.9), A_n is the area of the pipe connecting the tank to the pipeline, and h_s is the height water in the surge tank.

The height of water in the surge tank is computed for each time step from

$$h_s^+ = h_s - \frac{\Delta t}{A_s}Q_s \tag{4.174}$$

where A_s is the cross-sectional area of the surge tank.

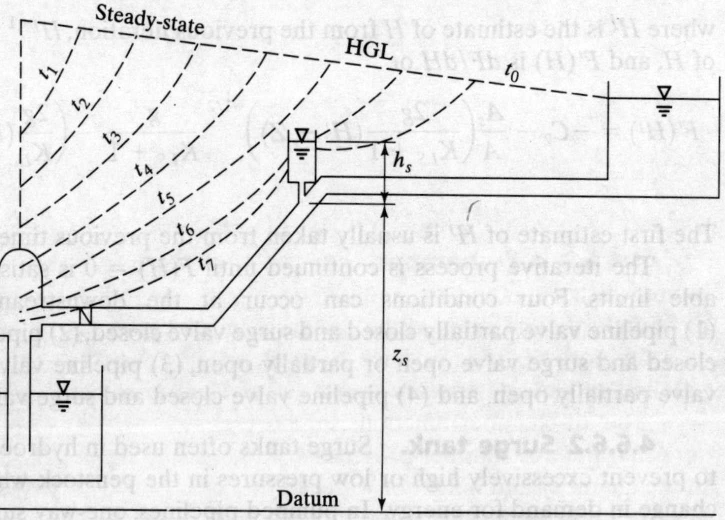

Figure 4.34 One-way surge tank in pipeline.

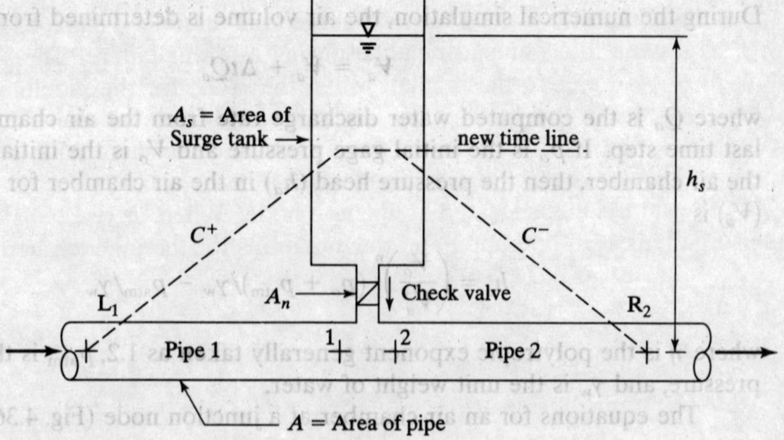

Figure 4.35 One-way surge tank.

Let $H_s = h_s^+ + Z_s$

$$C_5 = (C_{2,3} - C_{1,1})/(C_{1,2} + C_{2,4})$$

$$C_6 = C_o A_n/(A(C_{1,2} + C_{2,4}))$$

Combining equations

$$(H^+)^2 + (2C_5 + 2gC_6^2)H^+ + C_5^2 - 2gC_6^2 H_s = 0 \tag{4.175}$$

or

$$H^+ = \frac{b}{2}\left(-1 + \sqrt{1 - \frac{4c}{b^2}}\right) \tag{4.176}$$

where

$$b = 2C_5 + 2gC_6^2$$
$$c = C_5^2 - 2gC_6^2 H_s$$

4.6.6.3 Air chamber. An air chamber is a hydropneumatic tank (containing both water and air) and is connected to the pipeline at either a junction node or boundary node, often at the discharge side of pumps. The air chamber acts both as a temporary storage device to prevent high pressures from developing in the pipeline and as a temporary water supply to prevent low pressures in the pipeline.

The air chamber functions similarly to a two-way surge tank except the pressure that the air chamber provides at the pipeline is a function of both air volume in the chamber and height of water in the tank. As water flows from the air chamber, the air volume increases and the pressure decreases. Similarly, when water flows into the air chamber, the air volume decreases and the pressure increases. During the numerical simulation, the air volume is determined from

$$\forall_a^+ = \forall_a + \Delta t Q_a \tag{4.177}$$

where Q_a is the computed water discharge rate from the air chamber during the last time step. If p_o is the initial gage pressure and $\forall_o$ is the initial air volume in the air chamber, then the pressure head (h_a) in the air chamber for any air volume ($\forall_a$) is

$$h_a = \left(\frac{\forall_o}{\forall_a^+}\right)^{\eta} (p_o + p_{\text{atm}})/\gamma_w - p_{\text{atm}}/\gamma_w \tag{4.178}$$

where η is the polytropic exponent generally taken as 1.2, p_{atm} is the atmospheric pressure, and γ_w is the unit weight of water.

The equations for an air chamber at a junction node (Fig. 4.36) are

$$C^+: \quad V_1^+ = C_{1,1} - C_{1,2}H_1^+ \tag{4.179}$$

$$C^-: \quad V_2^+ = C_{2,3} + C_{2,4}H_2^+ \tag{4.180}$$

$$\text{Energy:} \quad H^+ = H_1^+ = H_2^+ \tag{4.181}$$

$$\text{Continuity:} \quad AV_1^+ + Q_a = AV_2^+ \tag{4.182}$$

$$\text{Orifice:} \quad Q_a = C_o A_n \sqrt{2g(h_a + Z_a - H^+)} \tag{4.183}$$

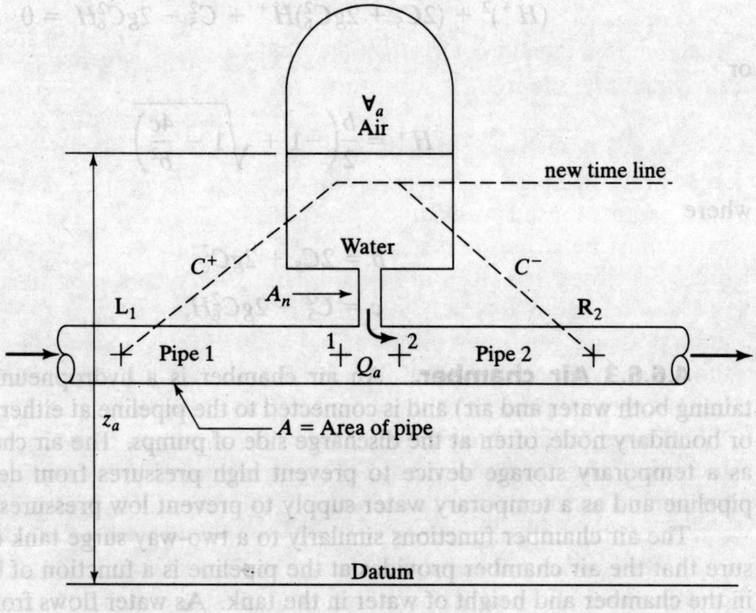

Figure 4.36 Air chamber.

for flow out of the air chamber, or

$$Q_a = -C_i A_n \sqrt{2g(H^+ - h_a - Z_a)} \tag{4.184}$$

for flow into the air chamber. Where C_o is the orifice coefficient for flow out of the air chamber, C_i is the orifice coefficient for flow into air chamber, A_n is the cross-section area of pipe connecting the air chamber to the pipeline, and A is the cross-section area of the pipeline.

For flow into the air chamber, let

$$H_a = h_a + Z_a$$

$$C_5 = (C_{2,3} - C_{1,1})/(C_{1,2} + C_{2,4})$$

$$C_6 = C_i A_n/(A(C_{1,2} + C_{2,4}))$$

Combining equations

$$(H^+)^2 + (2C_5 - 2gC_6^2)H^+ + C_5^2 + 2gC_6^2 H_a = 0 \tag{4.185}$$

Solving

$$H^+ = \frac{b}{2}\left(-1 + \sqrt{1 - \frac{4c}{b^2}}\right) \tag{4.186}$$

where

$$b = 2C_5 - 2gC_6^2$$

$$c = C_5^2 + 2gC_6^2 H_a$$

The equation developed for the one-way surge tank can be used for flow out of the air chamber, except H_a is substituted for H_s.

4.6.6.4 Other control techniques. The most common cause of unsteady flow in a pipe system is valve movement. The equations presented in Section 4.6.5.3 can be used to evaluate various valve closure schemes. For example, if a valve must be closed in 5 seconds, the valve could be (1) closed linearly in 5 seconds, (2) closed 100–10% in 1 second and 10–0% in 4 seconds, (3) closed 100–5% in 1 second and 5–0% in 4 seconds, or (4) closed using other schemes. The maximum pressure generated using various valve closure schemes can be computed and the results compared to see if the pressures are within acceptable levels. Generally, the last 5 to 10 percent of the valve closure is the most critical.

Power failure in a pumped system is a common cause of column separation. Increasing the rotational inertia of the pump might prevent the vapor cavity from forming. The inertia can be increased by adding a flywheel to the drive shaft and can be evaluated using the equations presented in Section 4.6.5.4.

The air-vacuum valve will open when the pressure drops below atmospheric pressure and allow air to be pulled into the pipeline. The valve does not prevent column separation, but does reduce the high pressure generated during cavity closure.

Example 4.23 Water Hammer in a Pipe

A 12-inch diameter steel pipeline with a friction factor of 0.015 and a wall thickness of 0.20 inches extends from an upper reservoir ($WSEL$ = 100 ft) 7,480 ft to a lower reservoir ($WSEL$ = 30 ft). A valve in the pipeline at the lower reservoir has an elevation of 10 ft and closes in 1.0 second. Compute the steady state velocity (V_0) in the pipeline with the valve open neglecting minor losses.

Steady-state velocity (V_0)

$$H_L = \frac{fL}{D}\frac{V_0^2}{2g}$$

$$70 = \frac{0.015 \times 7{,}480}{1.0}\frac{V_0^2}{2g}$$

$$V_0 = 6.34 \text{ fps}$$

Compute the velocity of the pressure wave (a) and the travel time (t) for the pressure wave to travel from the valve to the reservoir and back to the valve.

From Example 4.20, celerity (a) = 3,740 fps

$$\text{travel time } (t) = \frac{2L}{a} = \frac{2 \times 7{,}480}{3{,}740} = 4.0 \text{ seconds}$$

Closing the valve in less than 4.0 seconds is considered an instantaneous closure, and the water hammer pressure can be estimated as follows.

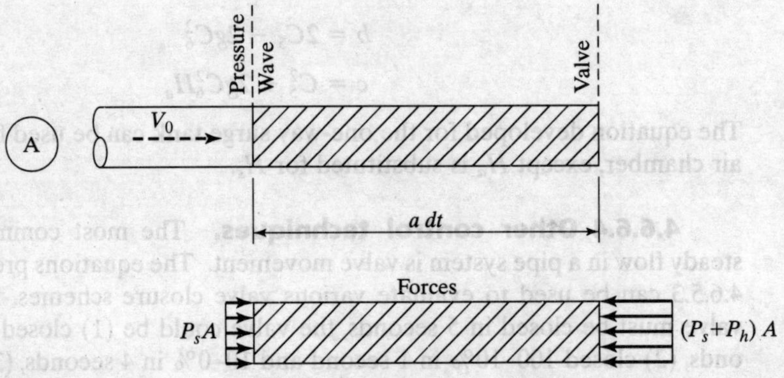

where

A = area of pipe

P_s = static water pressure

P_h = water hammer pressure

a = celerity

Neglecting pipe friction

$$F = M\frac{dV}{dt}$$

$$-P_h A = \left(\frac{\gamma}{g} Aadt\right)\frac{dV}{dt}$$

$$H_h = \frac{P_h}{\gamma} = \frac{a}{g} dV$$

If the velocity decreases to zero, then $dV = -V_0$ and the water hammer pressure head is

$$H_h = \frac{a}{g} V_0 = \frac{3{,}740}{32.2} 6.34 = 736 \text{ ft}$$

Total pressure head at the valve is equal to the water hammer pressure head plus the static pressure head

$$H_t = H_h + H_s = 736 + 90 = 826 \text{ ft}$$

Use the unsteady state model to compute the velocity in the pipeline at the upper reservoir and the pressure head in the pipeline upstream of the valve for a period of 3 minutes after valve closure.

A plot of the pressure head at the valve and the velocity in the pipeline at the upper reservoir is shown below. The maximum pressure head computed at the valve was 810 ft at a time of 4.0 seconds. Each cycle of the pressure wave takes 8.0 seconds to complete or 22.5 cycles during the 180 seconds of simulation as shown in the plot. A negative pressure of less than 34 ft is not possible, because column separation will occur and the pipeline will probably fail.

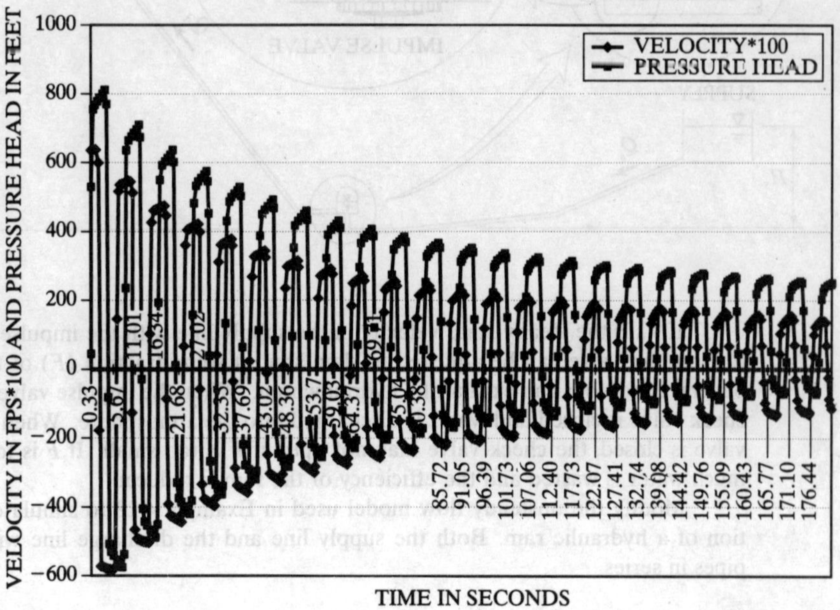

Example 4.24 Numerical Simulation of a Hydraulic Ram

Assume that you are to provide a water supply for a remote village (without power) using a hydraulic ram shown in the sketch below. The components of the ram installation are the supply pipeline, impulse valve, check valve, air chamber, and discharge line. Water flows into the ram from the supply source through the supply pipeline. Some of the water escapes through the impulse valve and some of the water is pumped through the discharge line to the storage reservoir. When the impulse valve is open, water flows through the impulse valve. The drag force of the water on the disk of the impulse valve causes the valve to slam shut, creating water hammer pressure in the supply pipe. The pressure increase is large enough to open the check valve and cause water to flow into the air chamber and discharge line to the storage reservoir. The impulse valve remains closed until the pressure in the supply line at the valve drops to nearly zero. The impulse valve opens and remains open until the drag force of the water causes the valve to slam shut and the cycle is repeated.

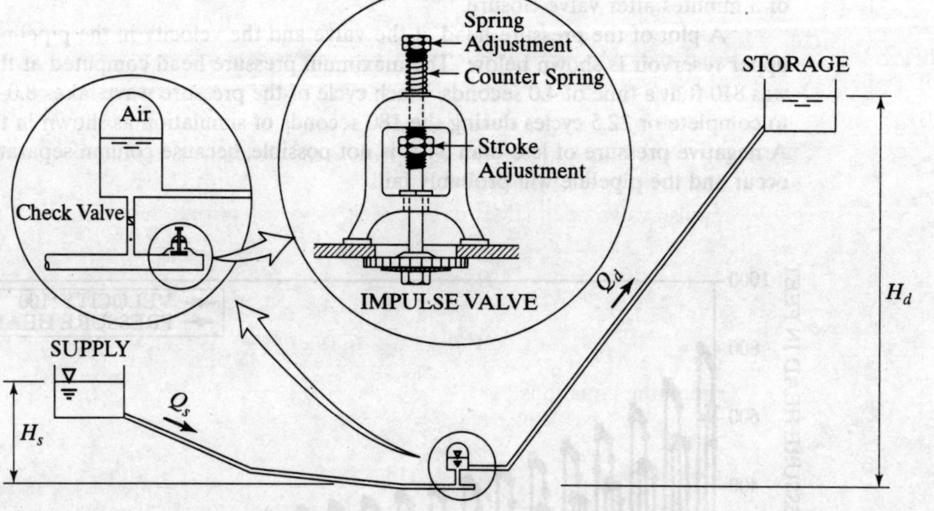

If V_0 is the steady-state velocity in the supply line with the impulse valve open, the impulse valve can be adjusted to slam shut at some fraction (F) of this velocity. Several conditions can exist at the hydraulic ram. When the impulse valve is open, the check valve is closed and water is discharged into the atmosphere. When the impulse valve is closed, the check valve may not open if F is too small. If F is too large, too much water is wasted and the efficiency of the ram is reduced.

Modify the unsteady flow model used in Example 4.22 to simulate the operation of a hydraulic ram. Both the supply line and the discharge line will consist of pipes in series.

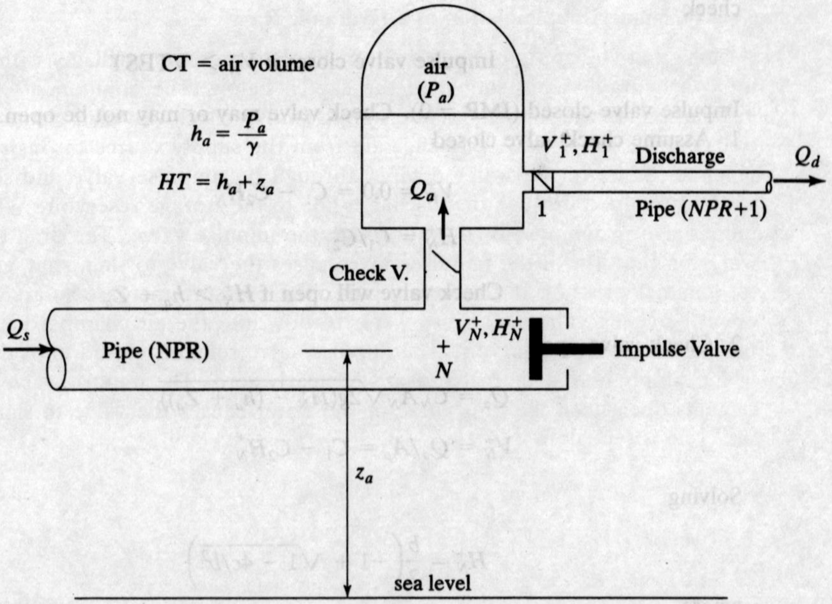

Referring to the sketch above, the equations at the junction representing the hydraulic ram are implemented as follows:

Impulse valve open (IMP = 1) (check valve will always be closed)

$$H_N^+ = \frac{K(V_N^+)^2}{2g} + Z_a \tag{4.146}$$

$$V_N^+ = C_1 - C_2 H_N^+ \tag{4.145}$$

Combining equations

$$H_N^+ = \frac{K(C_1 - C_2 H_N^+)^2}{2g} + Z_a$$

$$H_N^+ = \frac{b}{2}\left(-1 + \sqrt{1 - 4c/b^2}\right) \tag{4.24.1}$$

where

$$b = -(2C_1/C_2 + 2g/(KC_2^2))$$

$$c = (C_1/C_2)^2 + \frac{2gZ_a}{KC_2^2}$$

and

$$V_N^+ = C_1 - C_2 H_N^+ \tag{4.145}$$

check

Impulse valve closes if $V_N^+ >$ VTEST

Impulse valve closed (IMP = 0). Check valve may or may not be open.
1. Assume check valve closed

$$V_N^+ = 0.0 = C_1 - C_2 H_N^+ \tag{4.24.2}$$

$$H_N^+ = C_1/C_2 \tag{4.24.3}$$

Check valve will open if $H_N^+ > h_a + Z_a$

2. Check valve open

$$Q_a = C_N A_N \sqrt{2g(H_N^+ - (h_a + Z_a))} \tag{4.184}$$

$$V_N^+ = Q_a/A_p = C_1 - C_2 H_N^+$$

Solving

$$H_N^+ = \frac{b}{2}\left(-1 + \sqrt{1 - 4c/b^2}\right) \tag{4.24.4}$$

where

$$b = -\left(2C_1/C_2 + 2g\left(\frac{A_N C_N}{A_p C_2}\right)^2\right)$$

$$c = (C_1/C_2)^2 + 2g(h_a + Z_a)\left(\frac{A_N C_N}{A_p C_2}\right)^2$$

Check if $H_N^+ < Z_a$ impulse valve will open
Discharge line (NPR + 1) at air chamber

$$V_1 = C_3 + C_4 H_1^+ \tag{4.140}$$

$$H_1^+ = h_a + Z_a - K_m (V_1^+)^2/2g \qquad \text{(minor loss)} \tag{4.24.5}$$

Solving

$$H_1^+ = \frac{b}{2}\left[-1 + \sqrt{1 - 4c/b^2}\right] \tag{4.24.6}$$

where

$$b = \frac{2C_3}{C_4} + \frac{2g}{K_m C_4^2}$$

$$c = \left(\frac{C_3}{C_4}\right)^2 - \frac{2g(h_a + Z_z)}{K_m C_4^2}$$

and

$$V_1^+ = C_3 + C_4 H_1^+ \tag{4.140}$$

$$Q_d = V_1^+ (A_p)_{NPR+1} \tag{4.24.7}$$

$$Q_s = V_N^+ (A_p)_{NPR} \tag{4.24.8}$$

$$\forall_a^+ = \forall_a + (Q_d - Q_a)\Delta t \tag{4.177}$$

$$h_a = \left((h_a)_0 + \frac{P_{atm}}{\gamma}\right)\left(\frac{\forall_0}{\forall_a^+}\right)^{1.2} - \frac{P_{atm}}{\gamma} \tag{4.178}$$

Model the following example where the supply line is a 200-ft, 6-inch diameter steel pipe with $H_s = 30$ ft. The discharge line is 400 ft of 2-inch diameter PVC pipe with $H_d = 130$ ft. The air chamber has an initial volume of 6 ft^3 under a head of 130 ft. The air in the air chamber is enclosed in a rubber bladder so that the air does not dissolve into the water. The air chamber is connected to the supply line with a 2-inch diameter pipe with a check valve.

Compute the efficiency of the ram as

$$E_f = \frac{Q_d H_d}{Q_s H_s} \times 100$$

Compare the computed ram efficiency with that of a conventional electrical turbine generator and pump-motor system of approximately 50 percent.

The pressure head and velocity in the supply pipe and discharge line at the ram are plotted in the sketch below for a simulation time of 22 seconds. During this time, the impulse valve opened and closed 10 times. During the operation of the hydraulic ram, the discharge in the supply line was 0.725 cfs, and the discharge in the discharge line was 0.085 cfs. The efficiency of the ram is

$$E_f = \frac{0.085 \times 130}{0.725 \times 30} \times 100 = 51\%$$

```
EXAMPLE 4.24 HYDRAULIC RAM WITH PIPES IN SERIES BETWEEN TWO RESERVOIRS
&RAM IUN = 2, NP = 4, DURT = 100, HSUPP = 1000, HSTOR = 1100, ZPEND = 1090,
        NPR = 2, HATM = 32, CVML = 1.0, CKML = 0.5, FCV = 0.52, HOV = 1.0/
&AIRC CTZERO = 6.00, CIN = 0.6, DNOZ = 2.00, EXPON = 1.2/
 1    100.0    6.0    0.020    30.E + 06    0.25    990.00
 2    100.0    6.0    0.020    30.E + 06    0.25    970.00
 3    200.0    2.0    0.015    4.0E + 05    0.25    970.00
 4    200.0    2.0    0.015    4.0E + 05    0.25   1000.00
```

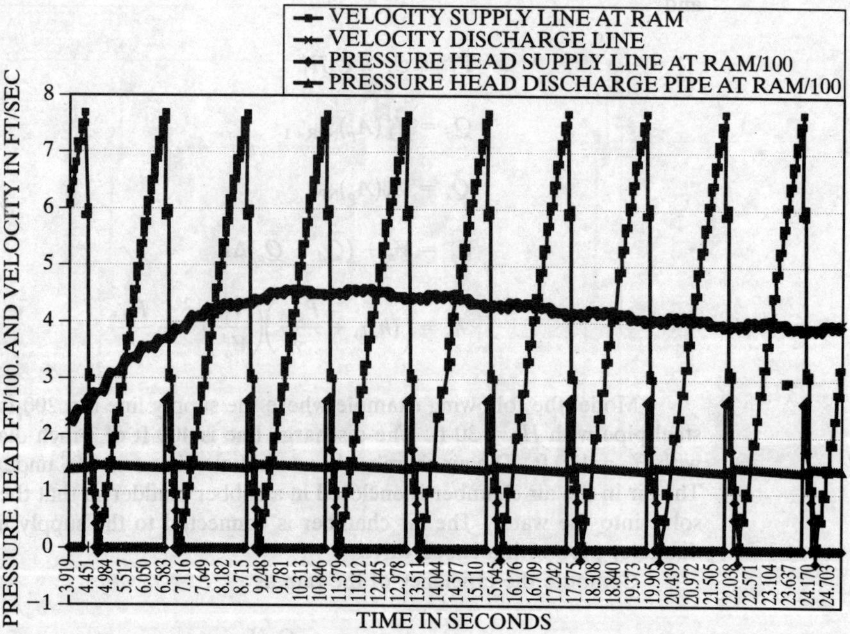

Example 4.25 Hydraulic Ram Water Supply Simulation

A hydraulic ram is to be used to pump water from a lower reservoir (elevation 1,050 ft) to an upper reservoir at an elevation of 1,200 ft. The supply line is 1,000 ft of 10-inch diameter steel pipe while the discharge line is 6,000 ft of 6-inch diameter pipe. The hydraulic ram is located at an elevation of 1,000 ft. Use a friction factor of 0.015 for the supply line and 0.02 for the discharge line. Compute the discharge and efficiency for the hydraulic ram.

The input file was developed and the model run for 10 minutes. The flow rate in the supply line was 2.625 cfs, and the flow rate in the discharge line was 0.398 cfs. The efficiency was

$$E_f = \frac{Q_d H_d}{Q_s H_s} \times 100 = \frac{0.398 \times 200}{2.625 \times 50} \times 100 = 60.6\%$$

The input file is listed below along with a plot of pressure head and velocity in the supply line at the ram.

```
EXAMPLE 4.25 HYDRAULIC RAM WATER SUPPLY SIMULATION
&RAM IUN = 2, NP = 5, DURT = 600, HSUPP = 1050, HSTOR = 1200, ZPEND = 1175,
        NPR = 2, HATM = 32, CVML = 1.0, CKML = 0.5, FCV = 0.75, HOV = 1.0/
&AIRC CTZERO = 250.0, CIN = 0.6, DNOZ = 6.00, EXPON = 1.2/
1      500.0     10.0     0.015     30.E + 06     0.35     1040.00
2      500.0     10.0     0.015     30.E + 06     0.35     1000.00
3     2000.0      6.0     0.020     30.E + 06     0.25     1000.00
4     2000.0      6.0     0.020     30.E + 06     0.25     1100.00
5     2000.0      6.0     0.020     30.E + 06     0.25     1150.00
```

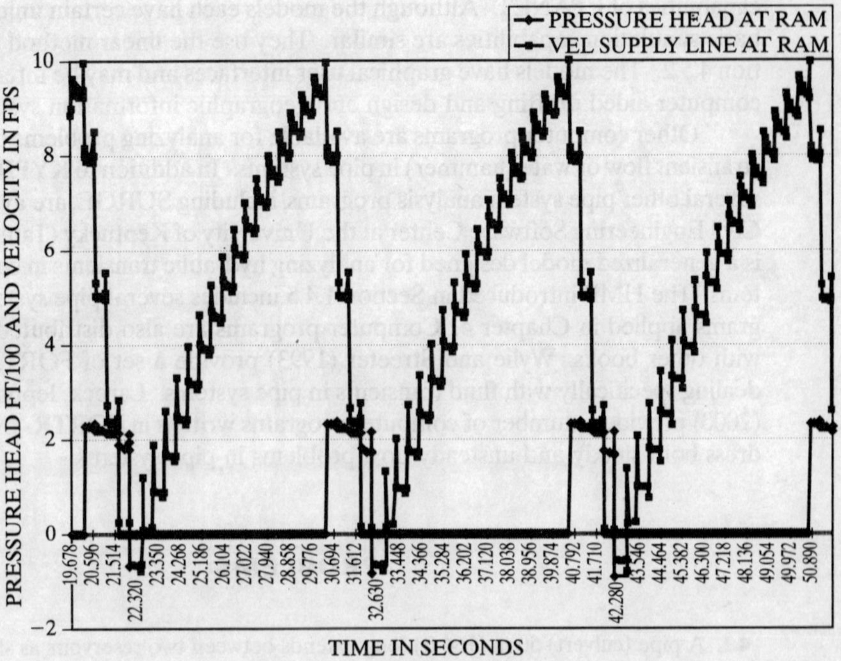

4.7 GENERALIZED PIPE SYSTEM SIMULATION MODELS

Water distribution system analyses are routinely performed in professional practice, using generalized software packages reflecting the concepts and methods outlined in this chapter. Widely used comprehensive steady-flow pipe network simulation models include KYPIPE, WaterCAD, WADISO (Water Distribution Simulation and Optimization), and EPANET. The original KYPIPE developed at the University of Kentucky during the 1970's has since evolved through various versions into a modeling system with a graphical user interface and various auxiliary programs. The Civil Engineering Software Center at the University of Kentucky (Table 1.7) continues to update, expand, and distribute KYPIPE. The WaterCAD modeling system is distributed by Haestad Methods (Table 1.7). KYPIPE and WaterCAD are proprietary software products sold by these entities. The WADISO model originally developed by the USACE Waterways Experiment Station, later renamed the Engineering Research and Development Center (Table 1.7), is public domain. A refined proprietary version of WADISO is marketed by GSL Engineering Software of South Africa. EPANET is a public domain model developed and distributed by the Environmental Protection Agency (Table 1.7).

EPANET, KYPIPE, WaterCAD, and WADISO are comprehensive modeling systems that compute steady-state flows and pressures throughout complex pipe networks. They also have extended period simulation capabilities. EPANET simulates both hydraulics and water quality. EPANET is a stand-alone model, but the three other models have features for interconnecting with the water quality simulation capabilities of EPANET. Although the models each have certain unique features, the basic simulation capabilities are similar. They use the linear method outlined in Section 4.5.2. The models have graphical user interfaces and may be interconnected with computer-aided drafting and design and geographic information system software.

Other computer programs are available for analyzing problems of unsteady flow (transient flow or water hammer) in pipe systems. In addition to KYPIPE noted above, several other pipe system analysis programs, including SURGE, are available from the Civil Engineering Software Center at the University of Kentucky (Table 1.7). SURGE is a generalized model designed for analyzing hydraulic transients in complex pipe systems. The HMP introduced in Section 1.4.5 includes several pipe system analysis programs applied in Chapter 4. Computer programs are also distributed in conjunction with other books. Wylie and Streeter (1993) provide a set of FORTRAN programs dealing specifically with fluid transients in pipe systems. Larock, Jeppson, and Watters (2000) provide a number of computer programs written in FORTRAN and C that address both steady and unsteady flow problems in pipe systems.

PROBLEMS

4.1. A pipe (culvert) 50 m (164 ft) long extends between two reservoirs as shown below. The difference in water surface elevation (Δh) between the two reservoirs is 2.0 m (6.6 ft). The pipe is 2.44 m (8.0 ft) in diameter and has a friction factor of 0.02. Compute the discharge in the pipe if the entrance loss coefficient is 0.5 and the exit loss coefficient is 1.0.

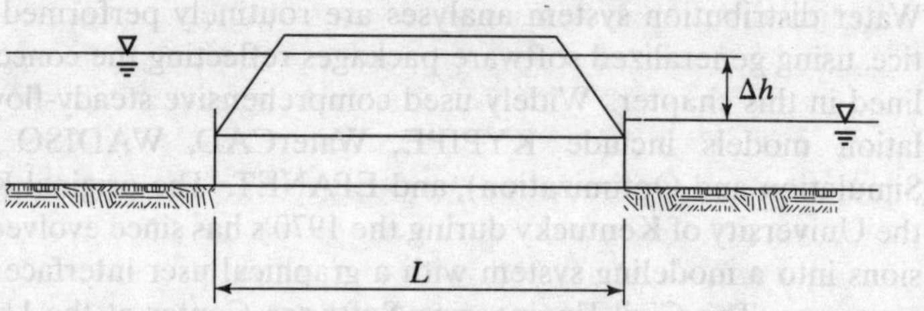

4.2. Compute the diameter of the pipe in Problem 4.1 for a discharge rate of 10.0 cms (353 cfs). Assume all other factors remain the same.

4.3. The culvert shown below has a projecting entrance and a tailwater below the pipe crown at the outlet. Compute the discharge rate through the culvert for $L = 50$ m (164 ft), $D = 2.44$ m (8.0 ft), $\Delta h = 2.0$ m (6.6 ft), and $K_{en} = 0.7$.

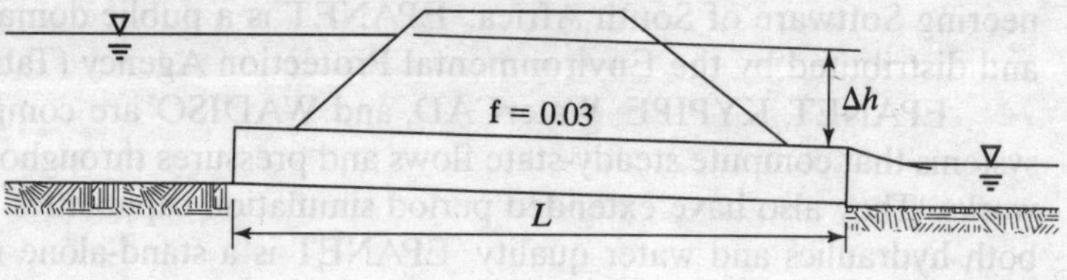

4.4. The culvert shown below, has a length of 100 m (328 ft) and a diameter of 2.13 m (7.0 ft). The culvert has a mitered inlet and a free outlet. Compute the discharge in the culvert for $K_{ent} = 0.8$, $f = 0.02$, and $\Delta h = 3.0$ m (9.8 ft).

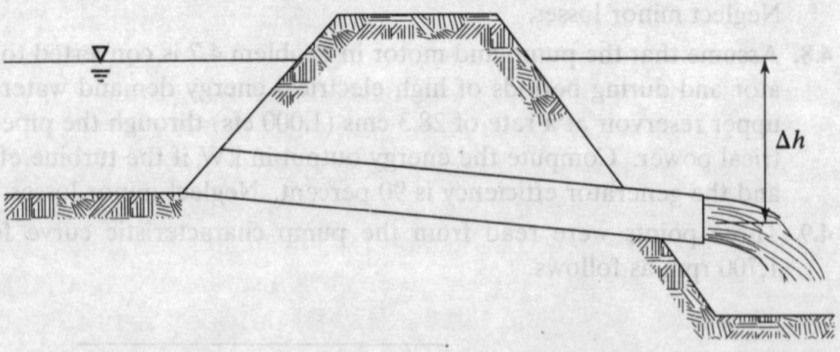

4.5. A pipeline extends between two reservoirs as shown below with $h_1 = 80$ m (262 ft) and $h_2 = 40$ m (131 ft). The pipeline is horizontal and consists of three pipes in series each 500 m (1,640 ft) long. Pipes 1 and 3 are 305 mm (12.0 in.) in diameter and pipe 2 is 203 mm (8.0 in.) in diameter. The entrance and exit loss coefficients are 0.5 and 1.0, respectively. Compute the discharge rate and the minimum pressure in the pipeline, neglecting minor losses at the junctions.

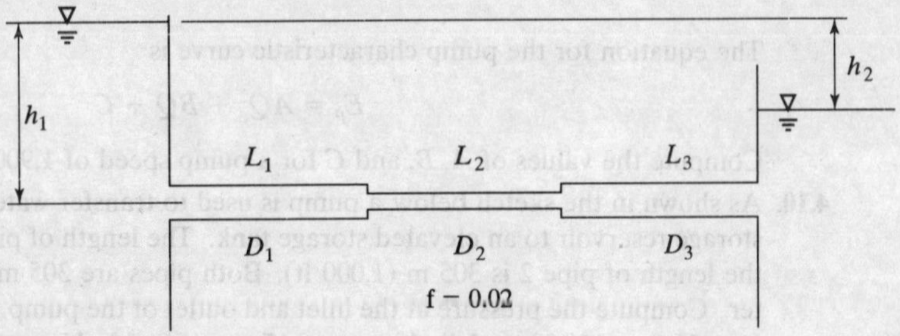

4.6. A pipeline consisting of two pipes in series extends from a reservoir and discharges into the atmosphere. Each pipe is 1,000 m (3,280 ft) in length. Pipe 1 is 305 mm (12.0 in.) in diameter and pipe 2 is 152 mm (6.0 in.) in diameter. Compute the flow rate in the pipeline for $h = 25$ m (82 ft). Neglect minor loss at the junction node. If a 76-mm (3.0 in.) diameter nozzle is added to the end of the pipeline, compute the velocity in the nozzle using a minor loss coefficient in the nozzle of 0.05.

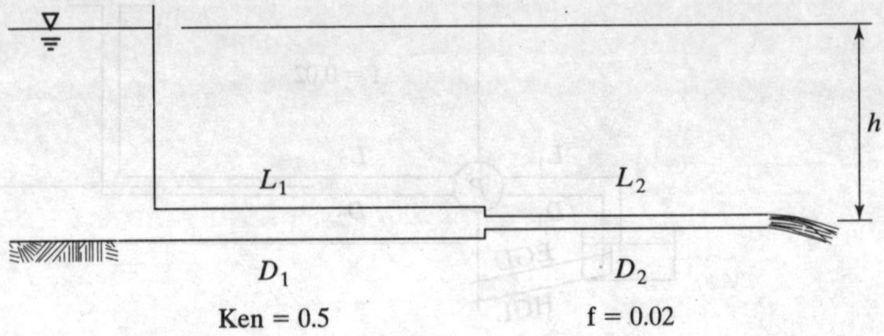

4.7. Compute the motor size required to pump 28.3 cms (1,000 cfs) through a 3,050 m (10,000 ft) long, 3.66-m (12.0-ft) diameter pipeline to an upper reservoir 610 m (2,000 ft) above the lower reservoir. Use a friction factor of 0.02 and a pump efficiency of 80 percent. Compute the energy input to the motor, if the motor efficiency is 90 percent. Neglect minor losses.

4.8. Assume that the pump and motor in Problem 4.7 is converted to a turbine and generator and during periods of high electrical energy demand water is released from the upper reservoir at a rate of 28.3 cms (1,000 cfs) through the pipeline to generate electrical power. Compute the energy output in kW, if the turbine efficiency is 80 percent and the generator efficiency is 90 percent. Neglect minor losses.

4.9. Three points were read from the pump characteristic curve for a pump speed of 1,700 rpm as follows

E_p		Q	
m	ft	cms	cfs
51.8	170	0	0
48.8	160	0.14	5.0
36.6	120	0.28	10.0

The equation for the pump characteristic curve is

$$E_p = AQ^2 + BQ + C$$

Compute the values of A, B, and C for a pump speed of 1,900 rpm.

4.10. As shown in the sketch below, a pump is used to transfer water from the ground-level storage reservoir to an elevated storage tank. The length of pipe 1 is 30 m (100 ft) and the length of pipe 2 is 305 m (1,000 ft). Both pipes are 305 mm (12 inches) in diameter. Compute the pressure at the inlet and outlet of the pump for h_1 = 3.05 m (10.0 ft), h_2 = 36.6 m (120 ft), and discharge = 0.17 cms (6.0 cfs). Use an entrance loss coefficient of 0.9 and an exit loss coefficient of 1.0. Neglect bend minor losses.

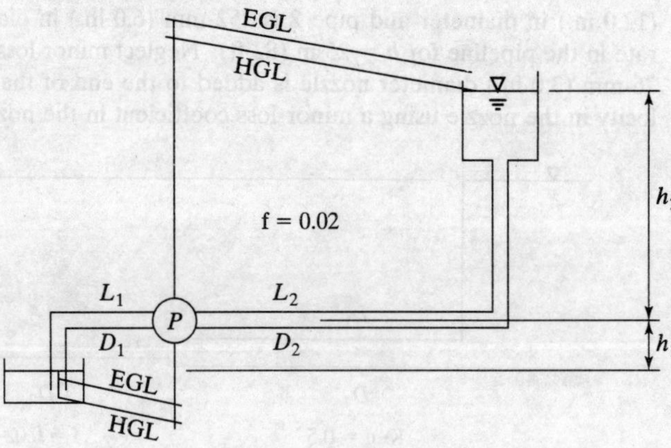

4.11. A 152-mm (6-in.) diameter pipeline is 305 m (1,000 ft) long and ends with a 76-mm (3-in.) diameter nozzle. Determine the head on the pump for h_1 = 3.05 m (10.0 ft), h_2 = 36.6 m (120 ft), and a discharge of 31.5 lps (500 gpm). Neglect entrance, bend, and nozzle losses.

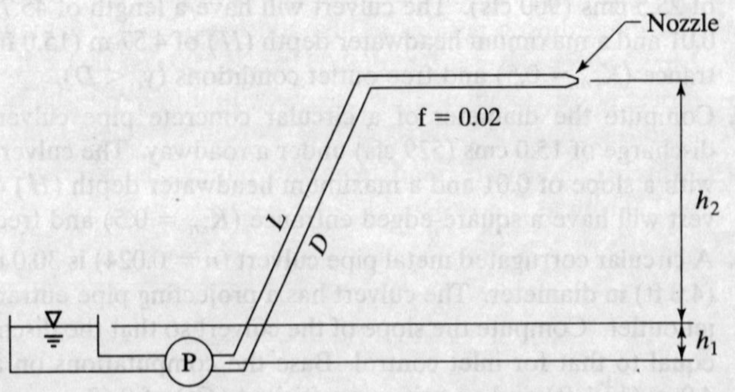

4.12. What resisting force is required to anchor a 1.22-m (4.0-ft) diameter pipeline at a horizontal 90° bend if the line is carrying 3.0 cms (106 cfs) of water under a pressure head of 40 m (131 ft)? Neglect headloss in the bend.

4.13. A 1.22-m (4.0-ft) diameter pipeline is carrying a discharge of 4.0 cms (141.0 cfs) under a head of 30.5 m (100 ft). A transition in the line reduces the diameter to 1.07 m (3.5 ft). What force does the water exert on the transition?

4.14. Determine the capacity of a circular culvert under a roadway (n = 0.024) as shown below for the following conditions:

a. H = 3.0 m (9.8 ft), D = 1.52 m (5.0 ft), L = 30.5 m (100.0 ft), y_t = 1.22 m (4.0 ft), S = 0.005

b. H = 3.0 m (9.8 ft), D = 1.52 m (5.0 ft), L = 30.5 m (100.0 ft), y_t – 1.22 m (4.0 ft), S = 0.03

c. H = 5.0 m (16.4 ft), D = 1.52 m (5.0 ft), L = 100 m (328 ft), y_t = 1.22 m (4.0 ft), S = 0.03

d. H = 5.0 m (16.4 ft), D = 1.52 m (5.0 ft), L = 100 m (328 ft), y_t = 1.83 m (6.0 ft), S = 0.01.

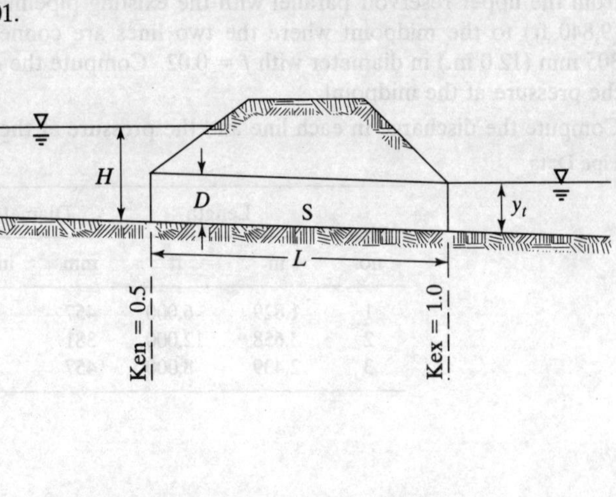

4.15. Compute the diameter of a circular concrete pipe culvert ($n = 0.015$) to carry a discharge of 6.0 cms (212 cfs) under a roadway. The culvert will be 50 m (164 ft) long on a slope of 0.01. The upstream headwater depth (H) is 4.0 m (13.1 ft) with free-outlet conditions ($y_t < D$). The entrance loss coefficient is 0.6 and the exit loss coefficient is 1.0.

4.16. Compute the size of a square concrete box culvert ($n = 0.015$) to handle a discharge of 25.5 cms (900 cfs). The culvert will have a length of 45.7 m (150 ft) with a slope of 0.01 and a maximum headwater depth (H) of 4.57 m (15.0 ft). Use a square-edged entrance ($K_{en} = 0.5$) and free outlet conditions ($y_t < D$).

4.17. Compute the diameter of a circular concrete pipe culvert ($n = 0.015$) to carry a discharge of 15.0 cms (529 cfs) under a roadway. The culvert will be 40 m (131 ft) long with a slope of 0.01 and a maximum headwater depth (H) of 4.0 m (11.2 ft). The culvert will have a square-edged entrance ($K_{en} = 0.5$) and free-jet outlet.

4.18. A circular corrugated metal pipe culvert ($n = 0.024$) is 30.0 m (98.4 ft) long and 1.22 m (4.0 ft) in diameter. The culvert has a projecting pipe entrance ($K_{en} = 0.8$) and a free-jet outlet. Compute the slope of the culvert so that the discharge for exit control is just equal to that for inlet control. Base the computations on a headwater depth (H) of 4.0 m (13.1 ft) and an orifice coefficient (C_o) of 0.62.

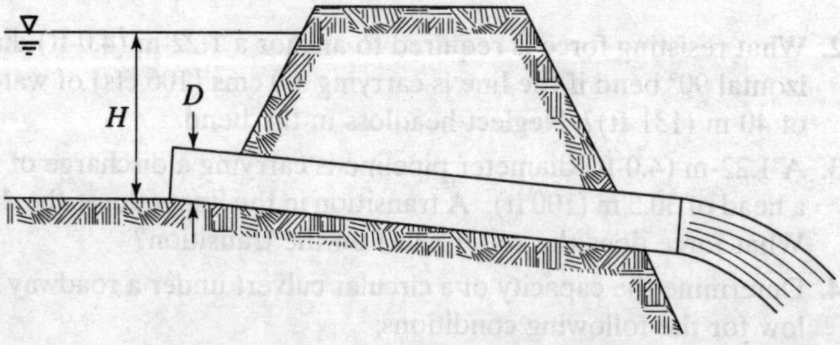

4.19. Two reservoirs are connected with a 762-mm (30-in.) diameter steel pipeline ($f = 0.02$) with a length of 6,000 m (19,680 ft). Compute the discharge in the pipeline and the pressure at midpoint in the pipeline. The water surface elevation of the upper reservoir is 300 m (984 ft), lower reservoir is 270 m (886 ft), and the elevation of the pipeline at midpoint is 280 m (918 ft).

4.20. To increase the capacity of the pipeline in Problem 4.19, a second pipeline was installed from the upper reservoir parallel with the existing pipeline for a distance of 3,000 m (9,840 ft) to the midpoint where the two lines are connected. The new pipeline is 305 mm (12.0 in.) in diameter with $f = 0.02$. Compute the discharge in each pipe and the pressure at the midpoint.

4.21. Compute the discharge in each line and the pressure at the two junction nodes.

Pipe Data

Pipe	Length		Diameter	
no.	m	ft	mm	inches
1	1,829	6,000	457	18
2	3,658	12,000	381	15
3	2,439	8,000	457	18

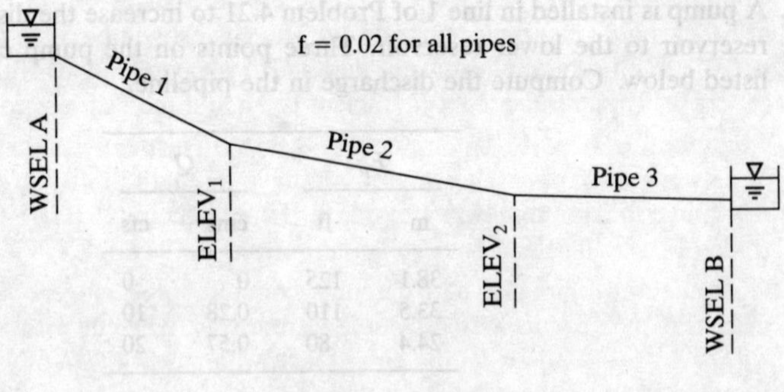

Node Data

Node	Elevation m	Elevation ft
WSEL A	304.9	1,000
ELEV1	289.6	950
ELEV2	282.0	925
WSEL B	274.4	900

4.22. A second identical pipe was installed parallel to pipe 2 in Problem 4.21. Compute the discharge in each line and the pressure at the two junction nodes.

4.23. Water is removed from the pipeline in Problem 4.21 at the two junction nodes. If the demand at each junction node is 31.5 ℓps (500 gpm), compute the discharge in each line and the pressure at the two junction nodes.

4.24. A pump is installed in line 3 of Problem 4.21 to lift 0.34 cms (12.0 cfs) of water from the lower reservoir to the upper reservoir. Compute the power output of the pump.

4.25. A pump is installed in line 3 of Problem 4.21 to lift water from the lower reservoir to the upper reservoir. Three points on the pump characteristic curve are listed below. Compute the discharge in the pipeline.

E_p (m)	E_p (ft)	Q (cms)	Q (cfs)
76.2	250	0	0
68.6	225	0.28	10
45.7	150	0.57	20

4.26. Two pumps are installed in line 3 of Problem 4.25 to lift water from the lower reservoir to the upper reservoir. Both pumps have the same characteristics as the pump in Problem 4.25. Compute the discharge rate in the pipeline for (a) pumps installed in parallel and (b) pumps installed in series.

4.27. A pump is installed in line 1 of Problem 4.21 to increase the discharge from the upper reservoir to the lower reservoir. Three points on the pump characteristic curve are listed below. Compute the discharge in the pipeline.

E_p		Q	
m	ft	cms	cfs
38.1	125	0	0
33.5	110	0.28	10
24.4	80	0.57	20

4.28. Determine the flow in each line and the pressure at junction node *D*.

		Elevation			Length		Diameter	
Problem	Reservoir	m	ft	Pipe	m	ft	m	ft
(a)	A	200	656	AD	6,000	19,680	1.0	3.28
	B	180	590	BD	4,000	13,120	0.762	2.5
	C	170	558	CD	9,000	29,520	0.762	2.5
(b)	A	200	656	AD	6,000	19,680	1.0	3.28
	B	175	574	BD	4,000	13,120	1.0	3.28
	C	150	492	CD	9,000	29,520	0.762	2.5
(c)	A	200	656	AB	6,000	19,680	1.0	3.28
	B	180	590	BD	6,000	19,680	1.0	3.28
	C	150	492	CD	6,000	19,680	1.0	3.28

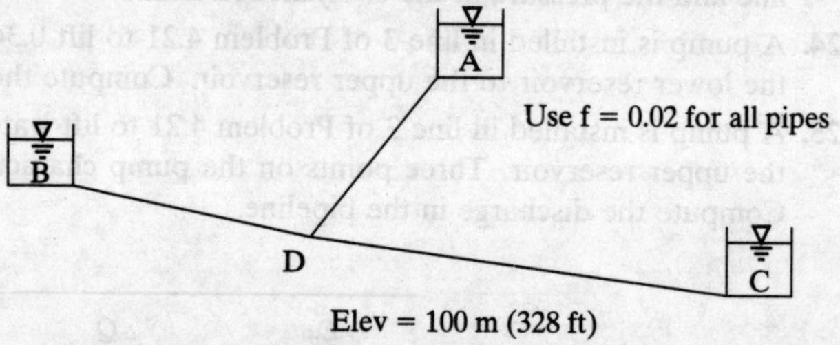

4.29. A pump is installed in line CD of Problem 4.28(c) to lift water from reservoir C into reservoirs A and B. Three points on the pump characteristic curve are listed below. Compute the discharge in each line and the pressure at junction node *D*.

E_p		Q	
m	ft	cms	cfs
100	328	0.0	0.0
85	279	0.5	17.6
60	197	1.0	35.3

4.30. Compute the discharge in each pipe and the pressure at each junction node for the 8-pipe system shown below. The water surface elevation in the elevated storage tank is 96.0 m (315.0 ft).

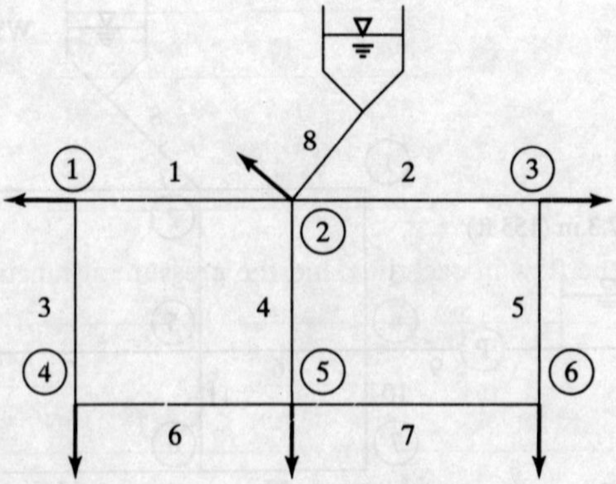

Pipe Data

Pipe no.	Length m	Length ft	Diameter mm	Diameter in.	Friction factor
1	1,220	4,000	254	10	0.024
2	1,829	6,000	254	10	0.024
3	1,829	6,000	305	12	0.022
4	1,982	6,500	610	24	0.018
5	2,134	7,000	254	10	0.024
6	915	3,000	457	18	0.020
7	1,524	5,000	254	10	0.024
8	91	300	305	12	0.022

Junction Data

Junction node	Ground elevation m	Ground elevation ft	Demand ℓps	Demand gpm
1	51.8	170	31.5	500
2	54.9	180	31.5	500
3	50.3	165	31.5	500
4	47.3	155	94.6	1,500
5	45.7	150	63.1	1,000
6	44.2	145	94.6	1,500

4.31. The pipe network system in Problem 4.30 was expanded to a 14-pipe system and includes a ground-level storage tank located near junction node 4. A pump is installed in a 457-mm (18-in.) diameter pipe extending 152 m (500 ft) from the ground-level storage tank [WSEL = 47.3 m (155 ft)] to junction node 4. Three points on the pump characteristic curve are listed below. Compute the flow in each line and the pressure at each junction node.

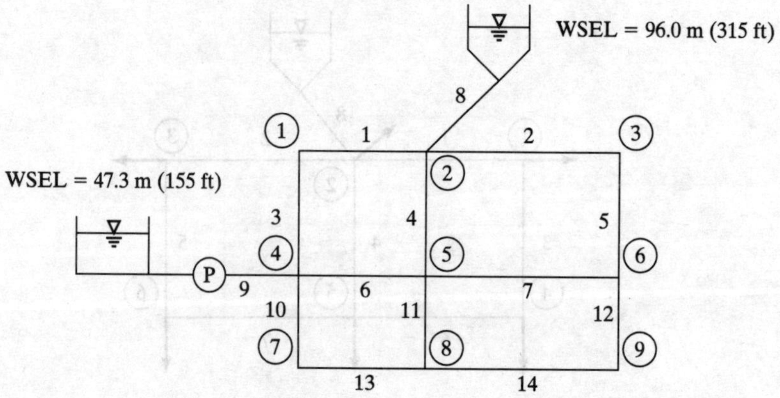

Pump Data

E_P		Q	
m	ft	cms	cfs
61.0	200	0	0
53.4	175	0.28	10.0
33.5	110	0.57	20.0

Pipe Data

Pipe no.	Length m	Length ft	Diameter mm	Diameter in.	Friction factor
9	152	500	457	18	0.020
10	1,220	4,000	254	10	0.024
11	1,220	4,000	610	24	0.018
12	1,220	4,000	305	12	0.022
13	915	3,000	203	8	0.026
14	1,524	5,000	305	12	0.022

Junction Data

Junction no.	Ground elevation m	Ground elevation ft	Demand ℓps	Demand gpm
7	50.3	165	31.5	500
8	51.8	170	63.1	1,000
9	50.3	165	31.5	500

4.32. The pipe network system in Problem 4.31 was expanded to a 24-pipe system shown below. The water levels in the two elevated storage tanks are 96.0 m (315 ft) and the water levels in the two ground-level storage tanks are the same 47.3 m (155.0 ft). The pump characteristics for the new pump in line 17 are the same as the existing pump in line 9. Compute the flow in each line and the pressure at each junction node.

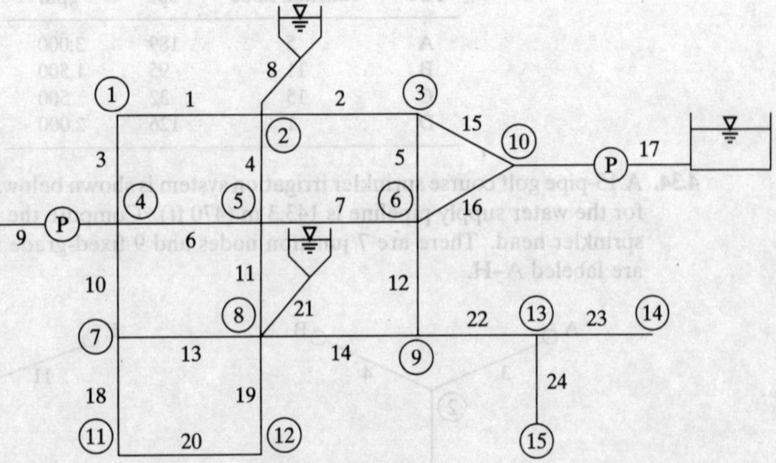

Pipe Data

Pipe no.	Length m	Length ft	Diameter mm	Diameter in.	Friction factor
15	1,829	6,000	457	18	0.020
16	1,829	6,000	457	18	0.020
17	152	500	610	24	0.018
18	1,524	5,000	305	12	0.022
19	1,524	5,000	305	12	0.022
20	915	3,000	254	10	0.024
21	92	300	457	18	0.020
22	915	3,000	305	12	0.022
23	915	3,000	305	12	0.022
24	305	1,000	152	6	0.028

Junction Data

Junction no.	Ground elevation m	Ground elevation ft	Demand ℓps	Demand gpm
10	53.4	175	63.1	1,000
11	53.4	175	31.5	500
12	54.9	180	31.5	500
13	48.8	160	0	0
14	45.7	150	94.6	1,500
15	48.8	160	6.3	100

4.33. Can the pipe network in Problem 4.32 provide the following fire demand without the pressure at any junction node dropping below 140 kPa (20 psi)? The fire demand is in addition to the existing demand at the junction. Consider only one fire at a time.

Fire	Junction node	Fire ℓps	Demand gpm
A	5	189	3,000
B	11	95	1,500
C	15	32	500
D	1	126	2,000

4.34. A 15-pipe golf course sprinkler irrigation system is shown below. Elevation of the *HGL* for the water supply pipeline is 143.3 m (470 ft). Compute the discharge through each sprinkler head. There are 7 junction nodes and 9 fixed-grade nodes. Sprinkler heads are labeled A–H.

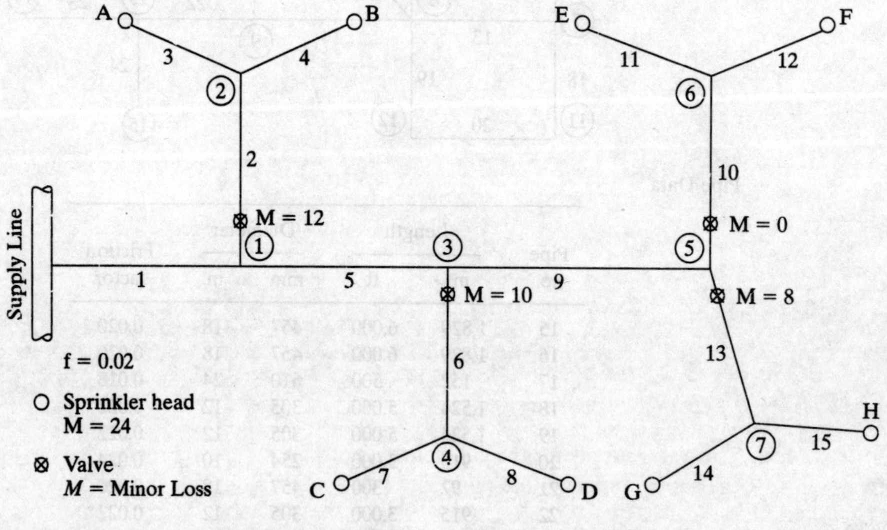

Pipe Data

Line no.	Length m	Length ft	Diameter mm	Diameter in.	Minor loss
1	183	600	102	4	
2	152	500	51	2	12
3	91	300	51	2	
4	91	300	51	2	
5	183	600	102	4	
6	152	500	76	3	10
7	76	250	51	2	
8	91	300	51	2	
9	183	600	102	4	

Line no.	Length m	Length ft	Diameter mm	Diameter in.	Minor loss
10	152	500	76	3	0
11	91	300	51	2	
12	76	250	51	2	
13	122	400	76	3	8
14	76	250	51	2	
15	76	250	51	2	

Junction Data

Junction no.	Elevation m	Elevation ft
1	97.6	320
2	99.1	325
3	97.6	320
4	93.0	305
5	99.1	325
6	100.6	330
7	91.5	300
A	100.6	330
B	97.6	320
C	91.5	300
D	89.9	295
E	103.7	340
F	102.1	335
G	88.4	290
H	91.5	300

4.35. The water supply for the golf course sprinkler system in Problem 4.34 was changed from the supply pipeline to a pond and pump. The pump has the following characteristics:

E_p m	E_p ft	Q cms	Q cfs
67.1	220	0	0
61.0	200	0.014	0.5
48.8	160	0.028	1.0

The water surface elevation of the pond is 88.4 m (290 ft). The diameter and length of pipe 1 were changed to 152 mm (6.0 in.) and 305 m (1,000 ft). Compute the discharge at each sprinkler head.

4.36. The City of Fredericksburg water supply is provided by five groundwater wells located approximately 6 km (4 mi) from town, each with a capacity of 56.8 ℓps (900 gpm).

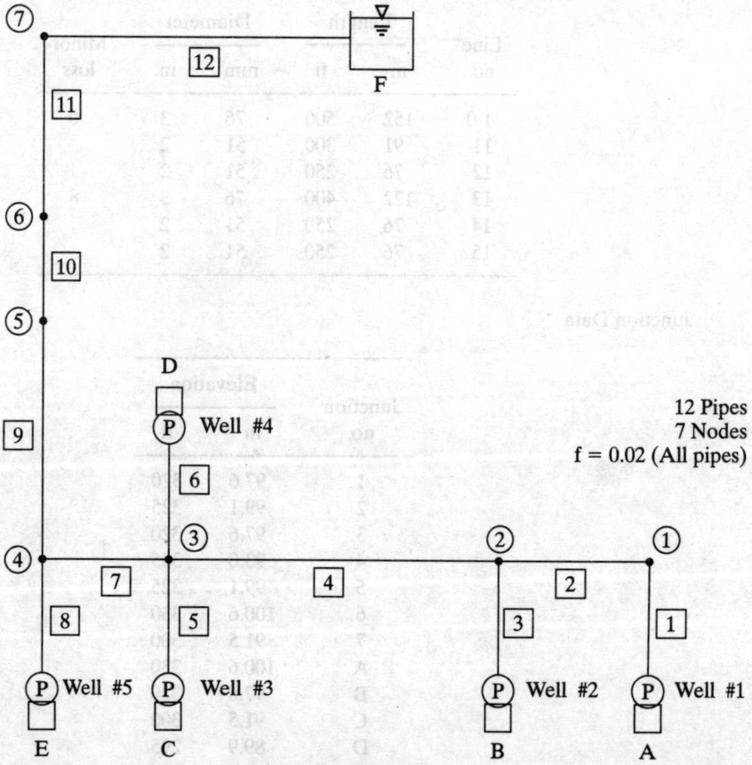

The well pumps are used to lift the water from the aquifer into a pipeline that discharges into a ground-level storage tank in town. In the past, additional wells have been added to meet increased water demands. However, as the flow increases in the pipeline, the head on the well pumps also increases and discharge from each well decreases. The objective of the problem is to size a booster pump in the 406-mm (16 in.) pipeline that will lower the head on the well pumps and increase the discharge rate from the well field.

First run the model for existing conditions. Then run the model with a booster pump in line 10. You have a choice of three pumps listed below. Select the pump that will give a discharge of about 252 lps (4,000 gpm).

Pump 1				Pump 2				Pump 3			
E_p		Q		E_p		Q		E_p		Q	
m	ft	ℓps	gpm	m	ft	ℓps	gpm	m	ft	ℓps	gpm
610	2,000	0	0	122.0	400	0	0	152.0	500	0	0
564	1,850	252	4,000	76.2	250	252	4,000	106.7	350	252	4,000
549	1,800	315	5,000	61.0	200	315	5,000	91.5	300	315	5,000

Pipe Data

Line	Length m	Length ft	Diameter mm	Diameter in.
1	42.7	140	152	6
2	640.2	2,100	152	6
3	42.7	140	203	8
4	1,341.5	4,400	305	12
5	42.7	140	203	8
6	42.7	140	203	8
7	146.3	480	406	16
8	42.7	140	203	8
9	1,432.9	4,700	406	16
10	12.2	40	406	16
11	4,908.5	16,100	406	16
12	61.0	200	305	12

Junction Data

Node	Elevation m	Elevation ft
1	480.2	1,575
2	477.1	1,565
3	483.2	1,585
4	482.9	1,584
5	487.8	1,600
6	487.8	1,600
7	510.7	1,675

Fixed-Grade Nodes

Node	Elevation m	Elevation ft
A	446.6	1,465
B	454.3	1,490
C	454.3	1,490
D	454.3	1,490
E	451.2	1,480
F	518.0	1,699

Well Pumps

E_p m	E_p ft	Q ℓps	Q gpm
94.5	310	0	0
76.2	250	25.2	400
45.7	150	63.1	1,000

4.37. Conduct a 24-hour simulation of the operation of the water distribution system in Problem 4.32 using a 1-hour computational time step. The demands are to be adjusted for each time step using global demand factors from Fig. 4.16 for the maximum day of the year. The elevated storage tanks are variable-level tanks with the elevation and storage information listed below.

Elevation		Storage	
m	ft	m^3	ft^3
88.4	290.0	0	0
91.5	300.0	1,134	40,000
94.5	310.0	3,401	120,000
97.6	320.0	5,669	200,000
100.0	328.0	7,086	250,000

The two pumps in the system are variable speed pumps, and the speed is adjusted according to the elevation of the *HGL* at node 8 for pumps in line 9 and line 17 according to the following:

Elevation		Pump speed ratio
HGL m	Node 8 ft	
<94.5	310	1.5
96.0	315	1.0
>97.6	320	0.5

Can the system meet the demands with pressures greater than 240 kPa (35 psi)? If the system cannot, what changes in the pipe network system are needed?

4.38. A steel pipeline 152-mm (6.0-in.) diameter with a wall thickness of 6 mm (0.24 in.) extends 1,000 m (3,280 ft) between two reservoirs as shown below. Compute the steady-state flow rate in the pipeline for $h_1 = 10.0$ m (32.8 ft) and $h_2 = 50$ m (164 ft), neglecting minor losses. If a valve in the pipeline at the lower reservoir is suddenly closed, compute the time that it will take the pressure wave to travel from the valve to the upper reservoir and back to the valve.

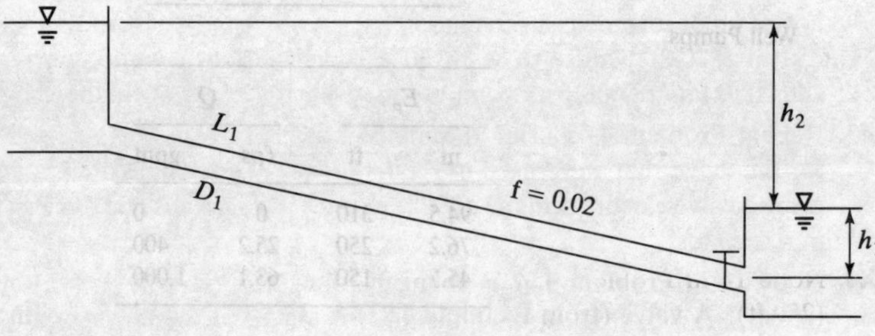

4.39. If the valve in Problem 4.38 is closed such that $dV/dt = -0.61\ \text{m/s}^2$ ($-2.0\ \text{ft/s}^2$), compute the maximum pressure at the valve.

4.40. Compute the maximum safe pressure (P) in a 457-mm (18-in.) diameter steel pipe if the allowable stress in the steel is taken as 124,000 kPa (18,000 psi) and the wall thickness (t) of the pipe is 5.1 mm (0.20 in.). Compute the celerity in the pipe.

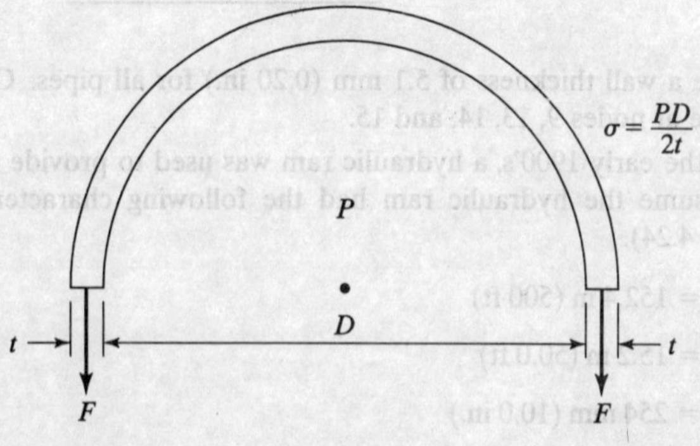

4.41. The pipe in Problem 4.40 ($f = 0.02$) is 3,049 m (10,000 ft) long and extends from an upper reservoir [WSEL = 304.9 m (1,000 ft)] to a lower reservoir [WSEL = 259.1 m (850 ft)]. What water-hammer pressure would develop if a valve in the line near the lower reservoir [Elevation = 253.0 m (830 ft)] was closed based on the following approximate equation

$$H_h = \frac{a}{g}\Delta V\left(\frac{T}{T_c}\right)$$

where H_h is the water-hammer pressure head, ΔV is the change in velocity in the pipeline caused by the valve change, T_c is the time for the valve change, and T is $2L/a$ with the limitation that (T/T_c) cannot exceed 1.0. Compute the minimum time for valve closure, if the allowable stress in the steel pipe is 240,000 kPa (18,000 psi). Note that the total pressure head is H_h plus the static pressure head of 51.9 m (170 ft).

4.42. A 2.44-m (8-ft) diameter steel ($f = 0.015$) penstock 2,439 m (8,000 ft) long is used to deliver water from a reservoir [WSEL = 304.9 m (1,000 ft)] to a turbine with a tailwater elevation 246.9 m (810.0 ft) under a head of 42.7 m (140 ft) [Elevation = *HGL* upstream of turbine is 289.6 m (950 ft)]. A valve in the line near the turbine [Elevation 250 m (820 ft)] must close in 30 seconds. Using a working stress for the steel as 240,000 kPa (18,000 psi), compute the wall thickness of the penstock.

4.43. Repeat Problem 4.41 using the method of characteristics. Use the headloss K values for the valve that were used in Example 4.22. Determine the maximun pressure head using a valve closure scheme of 90% closure in 3 seconds and the remaining 10% closure in 4 seconds.

4.44. Node 14 in Problem 4.32 is changed to a fixed-grade node with a WSEL of 76.2 m (250 ft). A valve (from Example 4.22) is placed at the downstream end of line 23 and is closed using the following scheme

Closure (percent)	Time (seconds)
90	5
100	10

Use a wall thickness of 5.1 mm (0.20 in.) for all pipes. Compute the maximum pressure at nodes 9, 13, 14, and 15.

4.45. In the early 1900's, a hydraulic ram was used to provide water for the city of Seattle. Assume the hydraulic ram had the following characteristics (see figure in Example 4.24).

L_s = 152.4 m (500 ft)

H_s = 15.2 m (50.0 ft)

D_s = 254 mm (10.0 in.)

t_s = 8.6 mm (0.34 in.)

f_s = 0.015

L_d = 1,524 m (5,000 ft)

D_d = 102 mm (4.0 in.)

t_d = 6.4 mm (0.25 in.)

f_d = 0.015

The water surface elevation of the supply reservoir is 152.4 m (500 ft) and 176.8 m (580 ft) in the storage reservoir. Use an upstream pipe elevation of 149.4 m (490 ft), a downstream pipe elevation of 175.3 m (575 ft), a ram elevation of 137.2 m (450 ft), a nozzle diameter of 51 mm (2.0 in.), and an air chamber volume of 0.28 m^3 (10 ft^3). Determine the maximum efficiency for the operation of the ram. All lines are steel pipes.

BIBLIOGRAPHY

CHAPRA, S. C., and R. P. CANALE, *Numerical Methods for Engineers,* 2nd Ed., McGraw-Hill, New York, NY, 1998.

CHEREMISINOFF, P., and P. N. CHEREMISINOFF, *Pumps and Pumping Operations,* Prentice Hall, Upper Saddle River, NJ, 1992.

HWANG, N. H. C., and C. E. HITA, *Fundamentals of Hydraulic Engineering Systems,* Prentice Hall, Upper Saddle River, NJ, 1987.

JEPPSON, R. W. *Analysis of Flow in Pipe Networks*, Butterworth Publishers, Boston, MA, 1976.

LAROCK, B. E., R. W. JEPPSON, and G. Z. WATTERS, *Hydraulics of Pipeline Systems,* CRC Press, Boca Raton, FL, 2000.

MAYS, L. W. (Editor), *Hydraulic Design Handbook,* McGraw-Hill, New York, NY, 1999.

MAYS, L. W. (Editor), *Water Distribution Systems Handbook,* McGraw-Hill, New York, NY, 2000.

ROBERSON, J. A., J. J. CASSIDY, and M. H. CHAUDHRY, *Hydraulic Engineering,* 2nd Ed., John Wiley & Sons, New York, NY, 1998.

TULLIS, J. P., *Hydraulics of Pipelines: Pumps, Valves, Cavitation, Transients,* John Wiley & Sons, New York, NY, 1989.

WALSKI, T. M., *Analysis of Water Distribution Systems,* Van Nostrand Reinhold, New York, NY, 1984.

WATTERS, G. Z., *Analysis and Control of Unsteady Flow in Pipelines,* Butterworths Publishers, Boston, MA, 1984.

WYLIE, E. B., and V. L. STREETER, *Fluid Transients in Systems,* Prentice Hall, Upper Saddle River, NJ, 1993.

5

Open Channel Hydraulics

Open channel flow occurs in canals, ditches, and natural streams. It also occurs in closed conducts flowing partially full with a free water surface at atmospheric pressure. Except for pressure lines, stormwater and wastewater collection systems are designed as open channels. Gravity is the primary driving force for open channel flow.

5.1 OPEN CHANNELS

Streams and rivers are the most common form of open channels. They vary in size from small drainage channels that flow only part of the time to large rivers that flow continuously. Natural channels are generally irregular in shape and alignment. They are often modified to change location, reduce bank erosion, increase flow capacity, or confine the flow during flood events. The capacity of a channel can be increased by controlling vegetation and smoothing the banks to reduce roughness, enlarging the cross-section, and straightening the alignment. Levees can be used to confine the flow during flood events. Concrete channels can be used to increase capacity, confine the flow, and prevent erosion. Dikes, revetments, and other stabilization structures can be used to control erosion.

Floodplain management along streams and rivers is a major concern. Under the impetus of the National Flood Insurance Program, in most communities in the U.S., construction of homes and businesses in floodplains is restricted to reduce susceptibility to flood damages. Floodplain management is typically based on delineating the limits of the 100-year return period flood (Chapter 7). Flood heights for other recurrence intervals are also determined for use in setting flood insurance rates. Because of the irregular alignment and cross-section of natural channels, flow is nonuniform, and hydraulic analyses are performed using computer models.

Canals, flumes, inverted siphons, and tunnels can be used to transport large volumes of water from the supply to the user. Water usually flows in these channels for an extended period of time, depending on the use (municipal, industrial, or irrigation). Surface drainage water is generally not allowed to enter a water supply canal. Cross-drainage culverts are used to carry small drainage channels under the canal, while flumes may be used to carry the canal water over large drainage channels, or inverted siphons may be used to carry the canal water under large drainage channels.

The longitudinal slope of a canal is a design parameter and is usually selected to conserve energy and prevent channel erosion. Earth canals are generally of trapezoidal shape with side slopes determined by the stability of the bank material. Canal lining may be required to reduce seepage from the canal. The rate of seepage depends on the permeability of the soil, the shape of the channel, and the groundwater level. Channel seepage is discussed in the chapter on groundwater.

Flumes, inverted siphons, and tunnels are generally used to shorten the canal length between the supply and demand. A tunnel may be used to carry water through a hill or ridge, while an inverted siphon is used to carry water under a stream valley. Flumes may be either elevated, supported by piers, or ground level. Flumes can be many shapes and constructed of several materials, but are usually rectangular, concrete structures.

Stormwater drainage channels are similar to water supply canals except water generally flows in the channel for only short periods of time after precipitation-runoff events. They usually have a trapezoidal cross section and are located in topographic low areas. Drain inlet structures are placed along the channel to allow surface runoff to enter the channel. The longitudinal slope of the channel is usually dependent on the slope of the terrain, and channel erosion can be a major problem. Concrete channels are often used to increase capacity and prevent erosion. Drainage channels are typically designed for a selected return period flood (for example, a 25-year flood event).

Circular and box culverts flowing partially full, stormwater collection systems, and subsurface perforated pipe drains are other examples of open channel flow. A box culvert flowing partially full is a rectangular channel, while gutter flow along a street is a triangular channel.

A chute is an open channel on a steep slope. Because of the high velocities, they are generally constructed of concrete. Chutes are often used as the inflow channel to the stilling basin in a drop structure or spillway. The depth of flow for most open channel flow problems is taken as the vertical distance between the channel bottom and the water surface. Because of the steep slope, the depth of flow in a chute is the perpendicular distance between the channel bottom and the water surface.

5.2 FLOW CLASSIFICATION

This chapter explores turbulent, one-dimensional, open channel flow. The chapter is organized based on the common practice of classifying flow as uniform versus nonuniform (gradually or rapidly varied), steady versus unsteady, and subcritical/critical/supercritical.

Steady versus unsteady refers to whether flow characteristics change over time. Uniform versus nonuniform (varied) refers to whether flow characteristics change spatially along the length of a channel. Flow characteristics include discharge, velocity, depth, and other related quantities. Since flow is seldom simultaneously both unsteady and uniform in nature, uniform generally also implies steady flow. Steady, nonuniform flow is common. Unsteady, nonuniform flow is also common. Nonuniform flow is classified as gradually varied or rapidly varied, depending on whether the flow switches between subcritical and supercritical.

Critical flow is an interesting phenomenon with practical significance covered in this chapter. As discussed in Section 5.4, flow is classified as subcritical ($F_r < 1$), critical ($F_r = 1$), or supercritical ($F_r > 1$) based on energy considerations. Velocities for supercritical and subcritical flow, respectively, are greater and less than critical velocity.

5.3 UNIFORM FLOW

Uniform flow occurs in an open channel when the depth of flow does not vary along the length of the channel. The slope of the channel bottom (S_o), the slope of the water surface, and the slope of the energy line (S_f) are all equal. Figure 5.1 is the control volume for steady, uniform flow in a segment of open channel of length L.

The momentum equation (Eq. 3.41) applied to the control volume in Fig. 5.1 gives

$$F_1 - F_2 - \tau_w PL + W \sin\theta = \rho Q(V_2 - V_1) \tag{5.1}$$

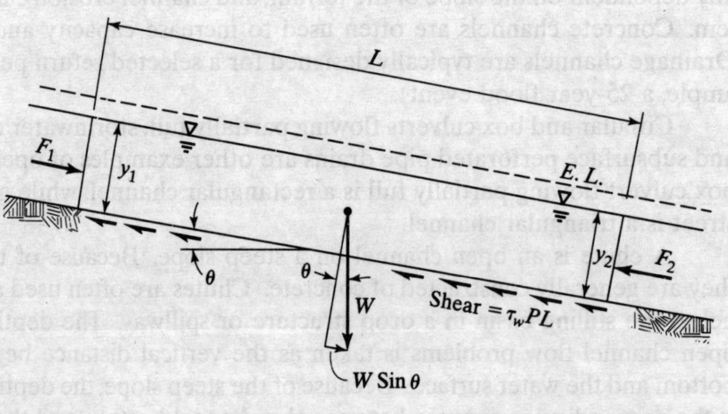

Figure 5.1 Control volume for steady, uniform flow in an open channel.

where F_1 and F_2 are the hydrostatic forces, τ_w is the shear stress between the water and the channel perimeter, P is the perimeter of the channel, W is the weight of the water in the channel segment (γLA), and θ is the slope angle of the channel. Since the flow in the channel is uniform, $y_1 = y_2$, $F_1 = F_2$, and $V_2 = V_1$. For small slopes usually encountered in open channel flow, $\sin\theta$ may be replaced with the channel slope (S_o) and Eq. 5.1 reduces to

$$\tau_w = \gamma R S_o \tag{5.2}$$

where R is the hydraulic radius ($R = A/P$).

Turbulent flow occurs in most open channel flow problems, and the shear stress (τ_w) may be estimated as

$$\tau_w = f\rho\frac{V^2}{8} \tag{5.3}$$

where f is a friction factor dependent on the surface roughness of the channel perimeter and the Reynolds number. Substituting Eq. 5.3 into Eq. 5.2 and solving for V gives the Chezy equation

$$V = C\sqrt{RS_o} \tag{5.4}$$

where C is the Chezy coefficient and is equal to $\sqrt{8g/f}$.

Based on experimental data, Manning developed the following empirical relation

$$C = \frac{R^{1/6}}{n} \tag{5.5}$$

where n is the Manning roughness coefficient. Substituting Eq. 5.5 into Eq. 5.4 gives the Manning equation for open channel flow

$$V = \frac{C_m}{n}R^{2/3}S_o^{1/2} \tag{5.6}$$

or since $Q = VA$

$$Q = \frac{C_m}{n}AR^{2/3}S_o^{1/2} \tag{5.7}$$

where

C_m = 1.0 for International System (SI) units and 1.49 for British Gravitational (BG) units
A = cross sectional area
R = hydraulic radius = A/P
P = wetted perimeter
S_o = longitudinal slope of channel
n = Manning's roughness coefficient (Table 5.1)

TABLE 5.1 MANNING ROUGHNESS VALUES FOR OPEN CHANNELS

	n
Natural channels	
Clean, straight	0.025–0.033
Clean, irregular	0.033–0.045
Weedy, irregular	0.045–0.080
Brush, irregular	0.07–0.16
Floodplains	
Pasture, no brush	0.030–0.050
Brush, scattered	0.035–0.070
Brush, dense	0.070–0.15
Timber and brush	0.10–0.20
Excavated uniform earth channels	
Straight with short grass	0.02–0.03
Winding with short grass	0.025–0.035
Cobble, stony	0.03–0.05
Dense vegetation	0.05–0.12
Lined channels	
Concrete, finished	0.012–0.015
Gravel	0.02–0.03
Asphalt	0.015–0.02
Closed conduits (partly full)	
Steel, welded	0.010–0.015
Cast iron	0.011–0.016
Concrete	0.010–0.015
Corrugated metal	0.020–0.030

The Manning equation is widely used for open channel flow, and considerable information is available on the selection of n values. Typical Manning n values are listed in Table 5.1. Manning n values are dimensionless and the same values are used for SI and BG units. Uniform flow occurs essentially only in prismatic channels of constant cross-section and longitudinal bottom slope. Typical cross-sections for prismatic channels are shown in Fig. 5.2.

The capacity of a prismatic channel to carry uniform flow is called normal discharge and is computed using the Manning equation. The equations for the area and wetted perimeter for typical prismatic channels are given in Fig. 5.2. The graph in Fig. 5.3 is often used for manual computation of flow in a circular channel flowing partially full. The ratio of water depth (y_n) to pipe diameter (D) is used to determine ratio of discharge rate partially full to discharge rate flowing full.

Because the side slope (SS) of a street is nearly flat, the wetted perimeter for gutter flow is often taken as the top width or

$$P \sim SS \times y_n \tag{5.8}$$

The Manning equation for gutter flow reduces to

$$Q = \frac{C_m}{n}\frac{SS}{3.2} y_n^{8/3} S_o^{1/2} \tag{5.9}$$

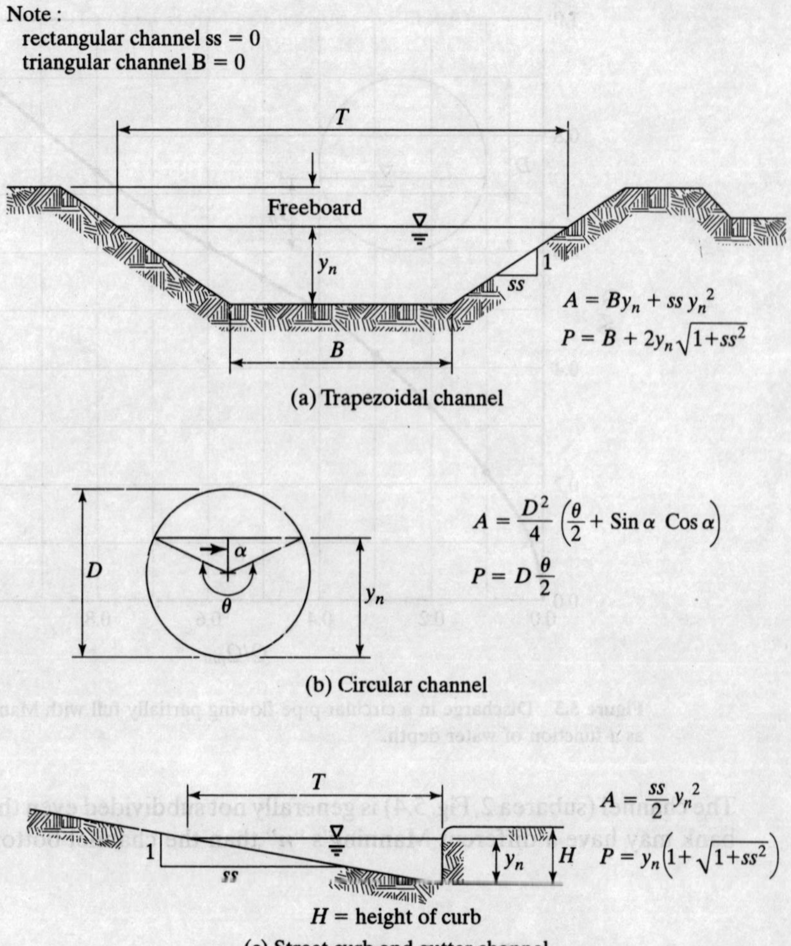

Figure 5.2 Prismatic channels.

A natural channel with over bank flooding (Fig. 5.4) is generally a compound channel. The floodplain is relatively flat, often with dense vegetation and a high resistance to flow. The Manning equation is used to compute the discharge rate by dividing the cross-section into subareas and computing the discharge rate for each subarea. The total discharge for the cross section is

$$Q = S_o^{1/2}\Sigma k_i \tag{5.10}$$

where k_i = conveyance for subarea i

$$= C_m A_i R_i^{2/3}/n_i$$

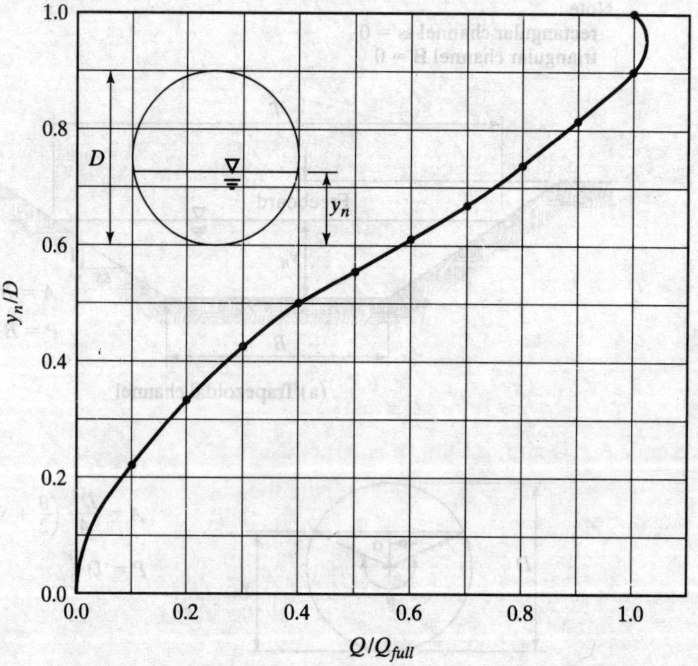

Figure 5.3 Discharge in a circular pipe flowing partially full with Manning "n" as a function of water depth.

The channel (subarea 2, Fig. 5.4) is generally not subdivided even though the channel bank may have a different Manning's "n" than the channel bottom.

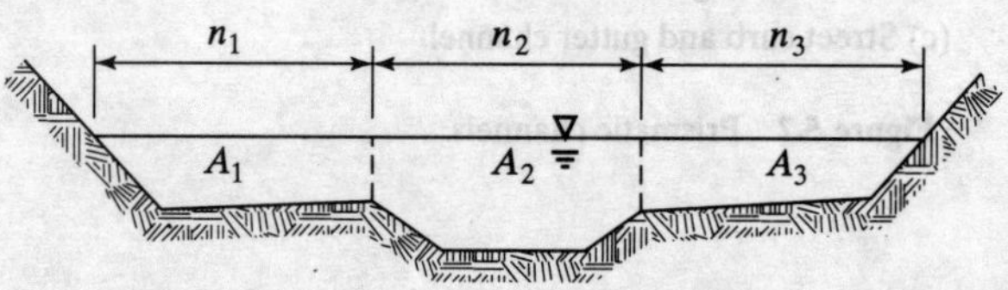

Figure 5.4 Compound channel.

Example 5.1 Discharge Rate in an Open Channel

Water is flowing in a trapezoidal earth channel ($n = 0.030$) at a depth (y) of 10.0 ft. The channel has a bottom width (B) of 50 ft and has 4H:1V (4:1) side slopes (SS). If the channel is on a slope (S) of 0.0005, determine the discharge rate (Q).

$$Q = \frac{1.49}{n} AR^{2/3}S^{1/2}$$

$$A = (B + SS \times y)y = (50 + 4 \times 10)10 = 900 \text{ sq ft}$$

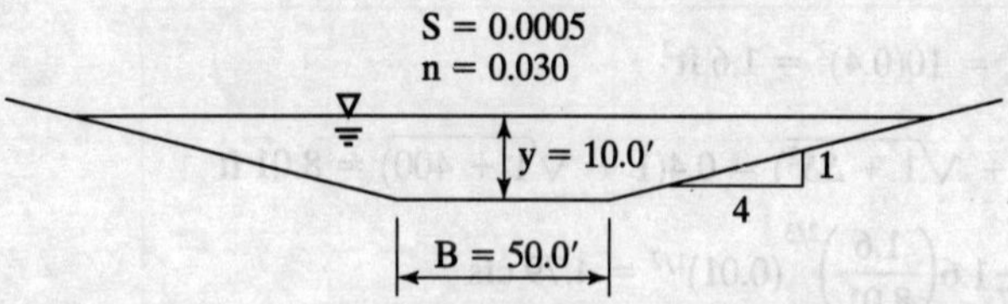

Example 5.1 Discharge in open channel.

$$P = B + 2 \times y \times \sqrt{SS^2 + 1} = 50 + 2 \times 10 \times \sqrt{17} = 132.5 \text{ ft}$$

$$R = \frac{A}{P} = \frac{900}{132.5} = 6.79 \text{ ft}$$

$$Q = \frac{1.49}{0.030} 900(6.79)^{2/3}(0.0005)^{1/2} = 3{,}585 \text{ cfs}$$

Example 5.2 Depth of Flow in an Open Channel

If the discharge rate is 2,000 cfs in the channel in Example 5.1, determine the depth of flow.

$$Q = \frac{1.49}{n} AR^{2/3}S^{1/2}$$

$$AR^{2/3} = \frac{Qn}{1.49S^{1/2}}$$

$$\frac{A^{5/3}}{P^{2/3}} = \frac{2000 \times 0.030}{1.49 \times (0.0005)^{1/2}}$$

$$\frac{(50y + 4y^2)^{5/3}}{(50 + 2 \times y \times 4.123)^{2/3}} = 1801$$

Solve for y by trial and error

y (ft)	$AR^{2/3}$	
6	1232	
8	2083	
7.5	1840	
7.4	1795	
7.41	1801	OK

Normal depth of flow is 7.41 ft.

Example 5.3 Gutter Flow

Water is flowing in a gutter at a depth of 0.40 ft. The gutter has a Manning n value of 0.017, a side slope of 20:1, and a longitudinal slope of 0.01. Determine the discharge rate in the gutter using both Eqs. 5.9 and 5.7 (see Fig. 5.2c). Estimate the error in using Eq. 5.9.

$$Q = \frac{1.49}{n} AR^{2/3}S_o^{1/2}$$

$$A = \frac{SS}{2}y_n^2 = 10(0.4)^2 = 1.6 \text{ ft}^2$$

$$P = y_n(1 + \sqrt{1 + SS^2}) = 0.4(1 + \sqrt{1 + 400}) = 8.01 \text{ ft}$$

$$Q = \frac{1.49}{0.017}1.6\left(\frac{1.6}{8.01}\right)^{2/3}(0.01)^{1/2} = 4.79 \text{ cfs}$$

$$Q = \frac{C_m}{n}\frac{SS}{3.2}y_n^{8/3}S_o^{1/2} = \frac{1.49}{0.017}\frac{20}{3.2}(0.4)^{8/3}(0.01)^{1/2} = 4.75 \text{ cfs}$$

$$\text{Error} = -\frac{0.04}{4.79} \times 100 = -0.8\%$$

Example 5.4 Normal Discharge

Determine the normal discharge for the channel shown below. The water in the left floodplain is 42 m wide, and the water in the right floodplain is 26 m wide. (Traditionally, stream cross-sections are defined looking downstream.) The Manning roughness value for the floodplain is 0.10, while the Manning roughness value is 0.05 for the channel. The longitudinal slope of the channel is 0.0005.

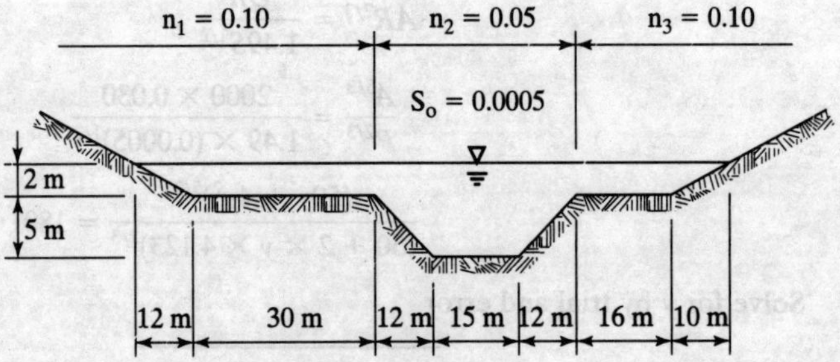

Left overbank	Channel	Right overbank
Area (A)		
$A_1 = 36 \times 2$	$A_2 = 27 \times 5 + 39 \times 2$	$A_3 = 21 \times 2$
$= 72 \text{ m}^2$	$= 213 \text{ m}^2$	$= 42 \text{ m}^2$
Wetted Perimeter (P)		
$P_1 = 30 + \sqrt{12^2 + 2^2}$	$P_2 = 15 + 2\sqrt{12^2 + 5^2}$	$P_3 = 16 + \sqrt{10^2 + 2^2}$
$= 42.2$ m	$= 41$ m	$= 26.2$ m
Hydraulic Radius $\left(R = \frac{A}{P}\right)$		
$R_1 = \frac{72}{42.2} = 1.71$ m	$R_2 = \frac{213}{41} = 5.20$ m	$R_3 = \frac{42}{26.2} = 1.60$ m
Conveyance $\left(k = \frac{C_m A R^{2/3}}{n}\right)$		

Left overbank	Channel	Right overbank
$k_1 = \dfrac{1.0 \times 72 \times 1.71^{2/3}}{0.10}$ = 1,030 cms	$k_2 = \dfrac{1.0 \times 213 \times 5.20^{2/3}}{0.05}$ = 12,790 cms	$k_3 = \dfrac{1.0 \times 42 \times 1.60^{2/3}}{0.10}$ = 575 cms

Discharge $Q = S_o^{1/2}\Sigma k_i = (0.0005)^{1/2}(1{,}030 + 12{,}790 + 575) = 322$ cms

Example 5.5 Bank Full Capacity

Determine the bankful capacity of the channel in Example 5.4 with and without subdividing the channel.

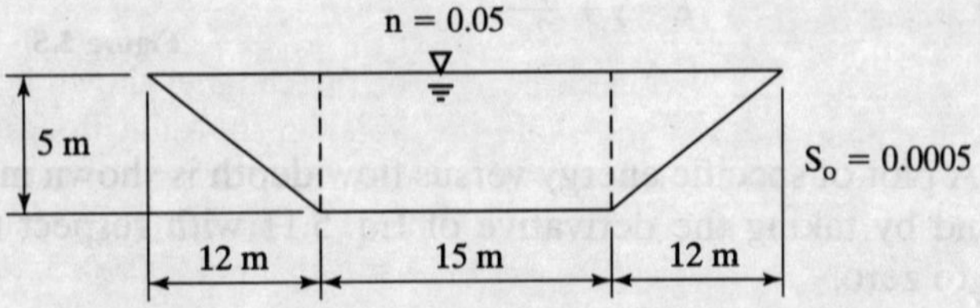

Without subdividing channel

$$Q = \frac{1.0}{n} S_o^{1/2} \frac{A^{5/3}}{P^{2/3}} = \frac{1.0}{0.05}(0.0005)^{1/2} \frac{(5 \times 27)^{5/3}}{(15 + 26)^{2/3}} = 0.447 \times 299 = 133.8 \text{ cms}$$

With subdividing channel

$$Q = \frac{1.0}{n} S_o^{1/2}\left(\frac{A_1^{5/3}}{P_1^{2/3}} + \frac{A_2^{5/3}}{P_2^{2/3}} + \frac{A_3^{5/3}}{P_3^{2/3}}\right) = \frac{1.0 \times 0.0005^{1/2}}{0.05}\left(\frac{30^{5/3}}{13^{2/3}} + \frac{75^{5/3}}{15^{2/3}} + \frac{30^{5/3}}{13^{2/3}}\right)$$
$$= 0.447(52.4 + 219.4 + 52.4) = 144.9 \text{ cms}$$

obviously

$$\frac{(\Sigma A_i)^{5/3}}{(\Sigma P_i)^{2/3}} \neq \Sigma\left(\frac{A_i^{5/3}}{P_i^{2/3}}\right)$$

The channel is normally not subdivided for flow computations.

5.4 CRITICAL FLOW

Critical flow occurs when the specific energy is minimum for a given discharge. The specific energy (E) is equal to the depth of flow (y) plus the velocity head ($V^2/2g$).

$$E = y + \frac{V^2}{2g} = y + \frac{Q^2}{A^2 2g} \tag{5.11}$$

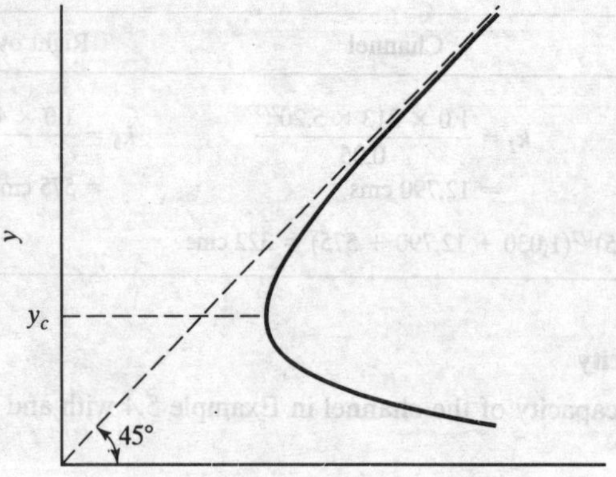

Figure 5.5 Specific energy diagram.

A plot of specific energy versus flow depth is shown in Fig. 5.5. Critical depth is found by taking the derivative of Eq. 5.11 with respect to y and setting dE/dy equal to zero.

$$\frac{dE}{dy} = 1 - \frac{Q^2}{gA^3}\frac{dA}{dy} = 0 \tag{5.12}$$

As shown in Fig. 5.6, $dA = Tdy$, where T is the top width of the channel at the water surface. Equation 5.12 becomes

$$\frac{Q^2}{g} = \frac{A_c^3}{T_c} \tag{5.13}$$

or

$$\frac{V_c^2}{2g} = \frac{A_c}{2T_c} \tag{5.14}$$

where the subscript c refers to critical flow. In open channel flow, cross-section area (A) divided by the top width (T) is called the hydraulic depth (D). Equation 5.14 becomes

$$\frac{V_c^2}{2g} = \frac{D_c}{2} \tag{5.15}$$

Rewriting Eq. 5.15 in terms of the Froude Number (F_r)

$$F_r = \frac{V_c}{\sqrt{gD_c}} = 1 \tag{5.16}$$

Critical flow occurs when the Froude Number is equal to 1.

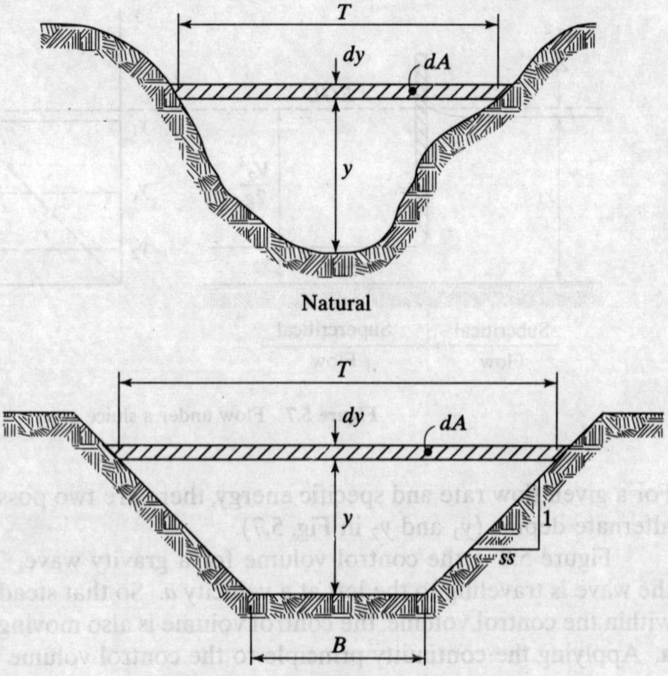

Figure 5.6 Open channel geometry.

If the channel is of rectangular cross-section, the hydraulic depth (D) is equal to the water depth (y) and Eq. 5.16 becomes

$$\frac{V_c^2}{2g} = \frac{y_c}{2} \tag{5.17}$$

or

$$y_c = \tfrac{2}{3} E_c \tag{5.18}$$

For critical flow in a rectangular channel, the velocity head is one-half the flow depth. Critical depth in a rectangular channel is

$$y_c = \left[\frac{q^2}{g}\right]^{1/3} \tag{5.19}$$

where q is the discharge per unit width of the channel.

If the depth of flow is greater than critical depth, the flow is called subcritical flow, and if the depth of flow is less than critical depth, the flow is called supercritical flow. Figure 5.7 is an example of nonuniform flow where subcritical flow occurs to the left of the sluice gate and supercritical flow occurs to the right of the sluice gate.

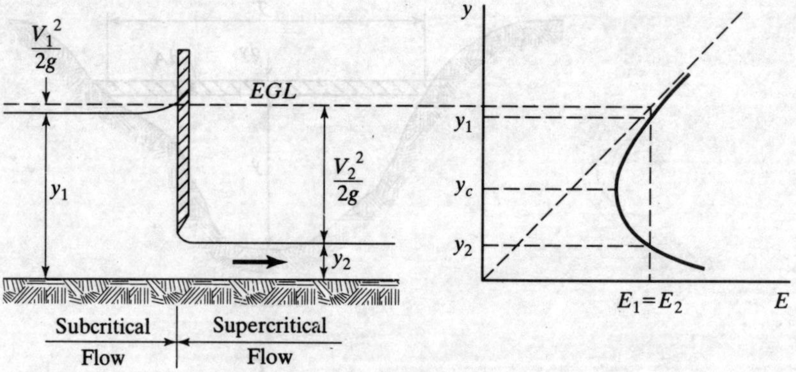

Figure 5.7 Flow under a sluice gate.

For a given flow rate and specific energy, there are two possible flow depths, called alternate depths (y_1 and y_2 in Fig. 5.7).

Figure 5.8 is the control volume for a gravity wave. The fluid is at rest and the wave is traveling to the left at a velocity *a*. So that steady-state conditions exist within the control volume, the control volume is also moving to the left at a velocity **a**. Applying the continuity principle to the control volume gives

$$\mathbf{a}y = V(y + dy) \tag{5.20}$$

where *V* is the velocity at the center of the wave. Applying the conservation of linear momentum principle to the control volume gives

$$\frac{\gamma y^2}{2} - \frac{\gamma(y + dy)^2}{2} = \mathbf{a}y\frac{\gamma}{g}(V - \mathbf{a}) \tag{5.21}$$

Substituting the value for *V* from Eq. 5.20 into Eq. 5.21 and dropping the *dy* squared term gives

$$\mathbf{a}^2 = g(y + dy) \tag{5.22}$$

For a small gravity wave, *dy* approaches zero and

$$\mathbf{a} = \sqrt{gy} \tag{5.23}$$

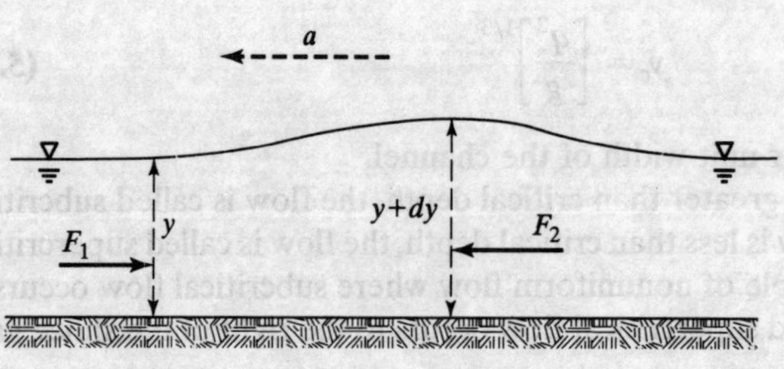

Figure 5.8 Control volume for gravity wave.

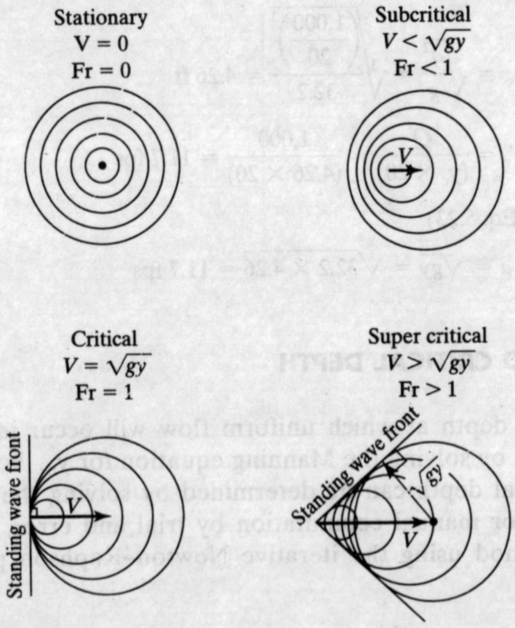

Figure 5.9 Surface wave patterns in free surface flow.

At critical flow in an open channel, the velocity in the channel is equal to the velocity of a gravity wave. Wave patterns generated in an open channel are illustrated in Fig. 5.9 for stationary water, subcritical flow, critical flow, and supercritical flow. To compute the velocity of a gravity wave in an open channel, y is replaced with the hydraulic depth (D) in Eq. 5.23.

Example 5.6 Critical Depth

If the discharge is 4,000 cfs in the earth channel in Example 5.1, determine the critical depth (y_c).

$$\frac{Q^2}{g} = \frac{A_c^3}{T_c}$$

$$\frac{4{,}000^2}{32.2} = \frac{(50 \times y_c + 4 \times y_c^2)^3}{50 + 2 \times 4 \times y_c}$$

$$496{,}900 = \frac{(50y_c + 4y_c^2)^3}{50 + 8y_c}$$

Solving by trial and error,

$$y_c = 5.06 \text{ ft}$$

Example 5.7 Critical Velocity

A rectangular channel is 20 ft wide and has a discharge rate of 1,000 cfs. Determine critical depth and critical velocity for the flow. Compare critical velocity with the velocity of gravity wave (Eq. 5.23).

$$y_c = \sqrt[3]{\frac{q^2}{g}} = \sqrt[3]{\frac{\left(\frac{1{,}000}{20}\right)^2}{32.2}} = 4.26 \text{ ft}$$

$$V_c = \frac{Q}{(y_c \times 20)} = \frac{1{,}000}{(4.26 \times 20)} = 11.7 \text{ fps}$$

Gravity wave velocity (Eq. 5.23)

$$a = \sqrt{gy} = \sqrt{32.2 \times 4.26} = 11.7 \text{ fps}$$

5.5 COMPUTING NORMAL AND CRITICAL DEPTH

Normal depth (y_n) is the depth at which uniform flow will occur in an open channel and is determined by solving the Manning equation for y_n. For a trapezoidal cross-section, normal depth can be determined by solving the Manning equation numerically or for manual computation by trial and error, tables, or graphs. A numerical method using the iterative Newton–Raphson procedure (Section 3.8) is

$$F(y_n) = Q - \frac{C_m}{n} A^{5/3} P^{-2/3} S_o^{1/2} = 0$$

$$F(y_n) = A^{5/3} P^{-2/3} - \frac{Qn}{C_m S_o^{1/2}} = 0 \qquad (5.24)$$

$$F'(y_n) = \tfrac{5}{3} A^{2/3} P^{-2/3} \frac{dA}{dy} - \tfrac{2}{3} A^{5/3} P^{-5/3} \frac{dP}{dy}$$
$$= \tfrac{5}{3} R^{2/3} T - \tfrac{2}{3} R^{5/3} (2\sqrt{1 + SS^2}) \qquad (5.25)$$

where $dA/dy = T$ = top width and

$$\frac{dP}{dy} = 2\sqrt{1 + SS^2}$$

The new value of depth (y^+) is

$$y_n^+ = y_n - \frac{F(y_n)}{F'(y_n)} \qquad (5.26)$$

The initial value of depth may be estimated or taken as one unit.

Critical depth (y_c) occurs in an open channel when the specific energy is minimum. The equation for critical depth from (Eq. 5.13) is

$$\frac{Q^2}{g} = \frac{A_c^3}{T_c} \qquad (5.13)$$

A numerical procedure to solve for critical depth in a trapezoidal channel using the iterative Newton–Raphson method is

$$y_c^+ = y_c - \frac{F(y_c)}{F'(y_c)} \tag{5.27}$$

where y_c^+ is the new estimate of critical depth, and y_c is the old estimate of critical depth.

$$F(y_c) = \frac{Q^2}{g} - \frac{A^3}{T} \tag{5.28}$$

$$F'(y_c) = \frac{2A^3SS}{T^2} - 3A^2 \tag{5.29}$$

Example 5.8 Normal Depth

Determine normal depth of flow for the earth channel ($n = 0.03$) shown below using the Newton–Raphson procedure. The channel carries 500 cms on a slope of 0.0004. The channel has a bottom width of 25 m and 3:1 side slopes.

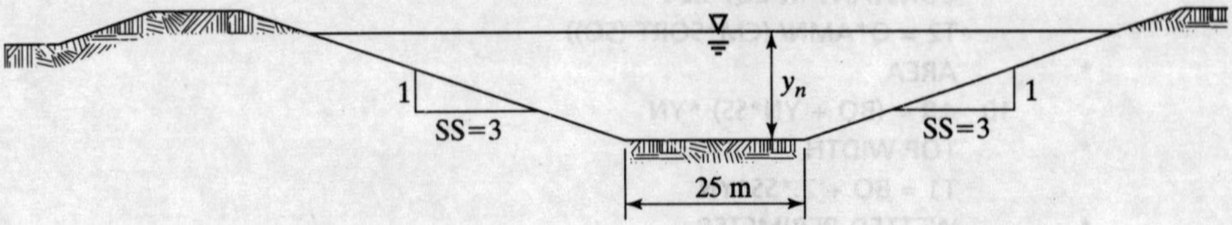

Assume a depth of 6 m

$$A = By_n + SSy_n^2 = 25 \times 6 + 3 \times 6^2 = 150 + 108 = 258 \text{ m}^2$$

$$P = B + 2 \times y_n\sqrt{SS^2 + 1} = 25 + 12\sqrt{10} = 62.9 \text{ m}$$

$$R = \frac{A}{P} = \frac{258}{62.9} = 4.10 \text{ m}$$

$$F(y_n) = \frac{A^{5/3}}{P^{2/3}} - \frac{Qn}{C_m S_o^{1/2}} = \frac{(258)^{5/3}}{(62.9)^{2/3}} - \frac{500 \times 0.03}{1.0(0.0004)^{1/2}} = 661 - 750 = -89$$

$$F'(y_n) = \tfrac{5}{3}R^{2/3}\,T - \tfrac{2}{3}R^{5/3}(2\sqrt{1 + SS^2})$$

$$= \tfrac{5}{3}(4.10)^{2/3}(61) - \tfrac{2}{3}(4.10)^{5/3}(2\sqrt{10}) = 261.1 - 44.5 = 217$$

$$y_n^+ = y_n - \frac{F(y_n)}{F'(y_n)} = 6.0 + \frac{89}{217} = 6.4 \text{ m}$$

Iteration 2 $y_n = 6.4$ m

$$A = 283 \text{ m}^2$$

$$P = 65.5 \text{ m}$$

$$R = 4.32 \text{ m}$$

$$T = 63.4 \text{ m}$$

$$F(y_n) = 1.1$$

$$F'(y_n) = 232$$

$$y_n^+ = y_n - \frac{F(y_n)}{F'(y_n)} = 6.4 - \frac{1.1}{232} = 6.40 \text{ m}$$

A FORTRAN function to solve for normal depth in a trapezoidal channel is listed as follows.

```
      REAL FUNCTION YN (Q, AMN, SO, BO, SS, CM)
*     THIS FUNCTION WILL SOLVE FOR NORMAL DEPTH IN TRAPEZOIDAL CHANNEL
*     Q = DISCHARGE RATE, AMN = MANNINGS N VALUE, SO = LONGITUDINAL SLOPE,
*     BO = BOTTOM WIDTH, SS = CHANNEL SIDE SLOPE, CM = 1.0 FOR SI AND 1.49 BG
*     INITIAL ESTIMATE OF NORMAL DEPTH
      YN = BO/4.
      IF (YN.LT.1.0) YN = 1.0
*     CONSTANT IN EQ. 5.24
      T2 = Q*AMN/ (CM*SQRT (SO))
*     AREA
10    AR = (BO + YN*SS) *YN
*     TOP WIDTH
      T1 = BO + 2.*SS*YN
*     WETTED PERIMETER
      PR = BO + 2.*YN*SQRT (1. + SS*SS)
*     HYDRAULIC RADIUS
      HR = AR/PR
*     EQUATION 5.24
      FY = AR**1.67/PR**.667 – T2
*     EQUATION 5.25
      FPY = 1.67*T1*HR**.667 – 1.333*HR**1.67*SQRT (1. + SS*SS)
*     EQUATION 5.26
      DY = FY/FPY
      YN = YN – DY
      IF (ABS (DY).GT..001) GO TO 10
      RETURN
      END
```

Example 5.9 Critical Depth

Compute critical depth for the channel in Example 5.8. Write a FORTRAN function to solve for critical depth in a trapezoidal channel.

Solution

$$F(y_c) = \frac{Q^2}{g} - \frac{A_c^3}{T_c}$$

$$F'(y_c) = 2A_c^3 \frac{SS}{T_c^2} - 3A_c^2$$

$$y_c^+ = y_c - \frac{F(y_c)}{F'(y_c)}$$

Initial estimate

$$y_c = \frac{B}{6} = \frac{25}{6} = 4.2 \text{ m}$$

$$A_c = y_c(B + SSy_c) = 4.2(25 + 3 \times 4.2) = 158 \text{ m}^2$$

$$T_c = B + 2 \times SS \times y_c = 25 + 2 \times 3 \times 4.2 = 50.2 \text{ m}$$

$$F(y_c) = \frac{500^2}{9.81} - \frac{158^3}{50.2} = -53{,}000$$

$$F'(y_c) = \frac{2 \times 158^3 \times 3}{50.2^2} - 3 \times 158^2 = -66{,}000$$

$$y_c^+ = 4.2 - \frac{53{,}000}{66{,}000} = 4.2 - 0.8 = 3.4 \text{ m}$$

Iteration 2 $y_c = 3.4$ m

$$A_c = 120 \text{ m}^2$$

$$T_c = 45 \text{ m}$$

$$F(y_c) = -13{,}000$$

$$F'(y_c) = -38{,}000$$

$$y_c^+ = 3.4 - \frac{13{,}000}{38{,}000} = 3.4 - 0.3 = 3.1$$

Iteration 3 $y_c = 3.1$ m

$$A_c = 106 \text{ m}^2$$

$$T_c = 43.6 \text{ m}$$

$$F(y_c) = -1{,}830$$

$$F'(y_c) = -30{,}000$$

$$y_c^+ = 3.1 - \frac{1{,}830}{30{,}000} = 3.1 - 0.06 = 3.04 \text{ m}$$

A FORTRAN function to solve for critical depth in a trapezoidal channel is listed as follows.

```
      REAL FUNCTION YC (Q, BO, SS, CM)
*     THIS FUNCTION WILL SOLVE FOR CRITICAL DEPTH IN TRAPEZOIDAL CHANNEL
*     Q = DISCHARGE RATE, BO = CHANNEL BOTTOM WIDTH, SS = CHANNEL SIDE SLOPE
*     CM = 1.0 FOR SI UNITS AND 1.49 FOR BG UNITS
*     INITIAL ESTIMATE OF CRITICAL DEPTH
      YC = BO/6.
      IF (YC.LT.1.0) YC = 1.0
      T2 = Q*Q/9.81
      IF (CM .GT. 1.0) T2 = Q*Q/32.2
*     AREA
   10 AR = (BO + YC*SS) *YC
*     TOP WIDTH
      T1 = BO + 2. *SS*YC
*     EQ. 5.28
      FY = T2 - AR**3/T1
*     EQ. 5.29
      FPY = 2. *AR**3*SS/ (T1*T1) - 3.*AR*AR
*     EQ. 5.27
      DY = FY/FPY
      YC = YC - DY
      IF (ABS (DY) .GT..001) GO TO 10
      RETURN
      END
```

5.6 CHANNEL DESIGN

Channels are usually designed for subcritical, uniform flow. Chutes and some concrete channels are designed for supercritical flow. Channels are generally designed to avoid critical depth. As the depth of flow approaches critical depth, large standing waves are formed, and a larger channel is required to contain the flow. To avoid standing waves, the Froude number should normally be less than 0.6 for subcritical flow or greater than 1.4 for supercritical flow.

Freeboard (Fig. 5.2a) is the vertical distance between the design water surface and the top of the channel bank. Freeboard is provided to account for uncertainty in design, construction, and operation of the channel. The U.S. Bureau of Reclamation recommends a minimum freeboard of 0.3 m (1.0 ft) for small channels and the following formula for estimating the freeboard (FB) for larger channels

$$\text{FB} = Cy^{1/2}, \tag{5.30}$$

where y is the water depth in m (ft) and C is a coefficient that varies from 0.7 (1.2) for small channels with a capacity of 0.6 cms (20 cfs) to 0.9 (1.6) for large channels with a capacity of 85 cms (3,000 cfs) or greater.

Because of the centrifugal force, additional freeboard should be provided along the outside of bends. As shown in Fig. 5.10, the water surface along the outside of the bend will be higher than the water surface along the inside of this bend

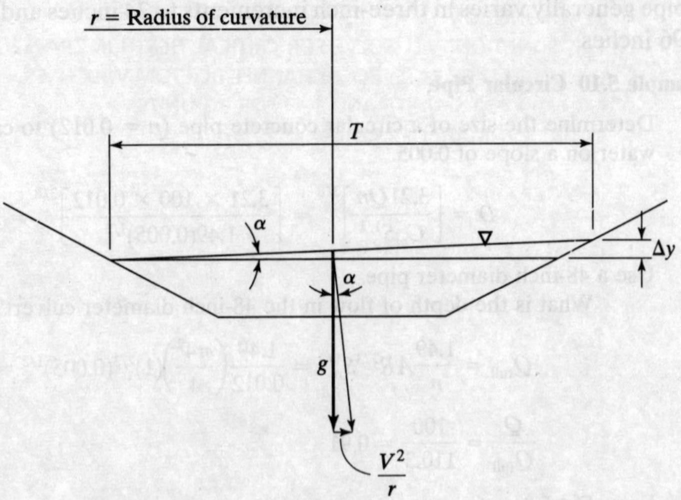

Figure 5.10 Open channel flow around bend.

by an amount (Δy)

$$\Delta y = \frac{V^2 T}{gr} \tag{5.31}$$

where T = top width, r = radius of curvature, and V = average velocity. The height of the outside bank should be increased by approximately two-thirds of Δy.

The most efficient hydraulic cross-section occurs when the wetted perimeter is minimum for a given area. The most efficient channel is a half-circle, the most efficient rectangular channel is a half-square, and the most efficient trapezoidal channel is a half-hexagon.

If the discharge contains sediments, the channel should be designed for a velocity greater than 0.7 mps (2 fps) to prevent sedimentation in the channel.

5.6.1 Circular Channels

Stormwater collection systems are typically designed for a specific return period flood event. If the design discharge of the storm sewer is exceeded, some street flooding can be expected. A stormwater collection system might be designed for a 10-year return period flood event flowing full. Once the discharge rate (Q) and the pipe slope (S_o) have been determined, the pipe diameter (D) can be computed from the Manning equation for

$$D = \left[\frac{3.21 Qn}{C_m S_o^{1/2}}\right]^{3/8} \tag{5.32}$$

where the hydraulic radius of a circular pipe flowing full is $D/4$ (Eq. 4.6). The minimum pipe size considered for stormwater drainage is typically 12 inches. The size of pipe generally varies in three-inch increments to 24 inches and six-inch increments to 96 inches.

Example 5.10 Circular Pipe

Determine the size of a circular concrete pipe ($n = 0.012$) to carry 100 cfs of stormwater on a slope of 0.005.

$$D = \left[\frac{3.21Qn}{C_m S_o^{1/2}}\right]^{3/8} = \left[\frac{3.21 \times 100 \times 0.012}{1.49(0.005)^{1/2}}\right]^{3/8} = 3.86 \text{ ft}$$

Use a 48-inch diameter pipe.

What is the depth of flow in the 48-inch diameter culvert?

$$Q_{\text{full}} = \frac{1.49}{n} AR^{2/3} S_o^{1/2} = \frac{1.49}{0.012}\left(\frac{\pi 4^2}{4}\right)(1)^{2/3}(0.005)^{1/2} = 110.3 \text{ cfs}$$

$$\frac{Q}{Q_{\text{full}}} = \frac{100}{110.3} = 0.91$$

From Fig. 5.3

$$\frac{y_n}{D} = 0.82$$

$$y_n = 0.82 \times 4.0 = 3.3 \text{ ft}$$

5.6.2 Rectangular Channels

Rectangular channels are usually designed for the most efficient cross-section that occurs when the depth is equal to half the bottom width. Rectangular channels are normally constructed of concrete and are used for both subcritical and supercritical flow.

5.6.3 Concrete Trapezoidal Channels

Concrete lining in a canal is generally used to reduce seepage. Concrete lining in a drainage channel is generally used to increase capacity and prevent erosion. The lining is often installed with a slip-form paving machine. The side slopes typically vary from 1:1 for small channels to 1.5(H):1(V) for large channels. Typically, the value of B/y_n varies from 1 for small channels to 3 for large channels.

Drainage channels are generally located where the groundwater table is high. Seep holes are often placed through the concrete lining to allow groundwater to enter the channel. Without the seep holes, the lining may float when the water table is high and the channel flow is low.

Example 5.11 Trapezoidal Channel Design

Design a water supply concrete-lined canal ($n = 0.015$) to serve a population of 2.0 million people. The estimated per capita water consumption is 140 gallons on an average day with a water conservation program. The peak daily water consumption is estimated at 280 gpcd. The canal is to have a B/y_n of 2.0, 2H:1V side slopes and a velocity of 5.0 fps. Assume 15 percent evaporation and seepage loss in the canal.

Design discharge rate (Q)

$$Q = \frac{2{,}000{,}000 \times 280 \times 1.15}{86{,}400 \times 7.48} = 1{,}000 \text{ cfs}$$

Required area

$$A = \frac{Q}{V} = \frac{1{,}000}{5} = 200 \text{ ft}^2$$

Required depth

$$A = (B + SS \times y_n)y_n$$

where $A = 200$ and $B = 2y_n$

or $200 = 2y_n^2 + 2y_n^2$

$$y_n = \sqrt{\frac{200}{4}} = 7.1 \text{ ft}$$

Freeboard (FB)

$$FB = Cy_n^{1/2}$$

Assume $C = 1.3$

$$FB = 1.3(7.1)^{1/2} = 3.5 \text{ ft}$$

Canal slope (S_o)

$$V = \frac{C_m}{n} R^{2/3} S_o^{1/2}$$

$$S_o = \left[\frac{Vn}{C_m R^{2/3}}\right]^2$$

$$P = (14.2 + 2 \times 7.1 \times \sqrt{5}) = 46.0 \text{ ft}$$

$$R = \frac{A}{P} = \frac{200}{46.0} = 4.35 \text{ ft}$$

$$S_o = \left[\frac{5.0 \times 0.015}{1.49 \times 4.35^{2/3}}\right]^2 = 0.00036$$

Check Froude Number ($F_r < 0.6$)

$$F_r = \frac{V}{\sqrt{gD}}$$

$$\text{where } D = \frac{A}{T} = \frac{200}{14.2 + 2 \times 7.1 \times 2} = 4.69 \text{ ft}$$

$$F_r = \frac{5.0}{\sqrt{32.2 \times 4.69}} = 0.41 \qquad \text{Design is OK.}$$

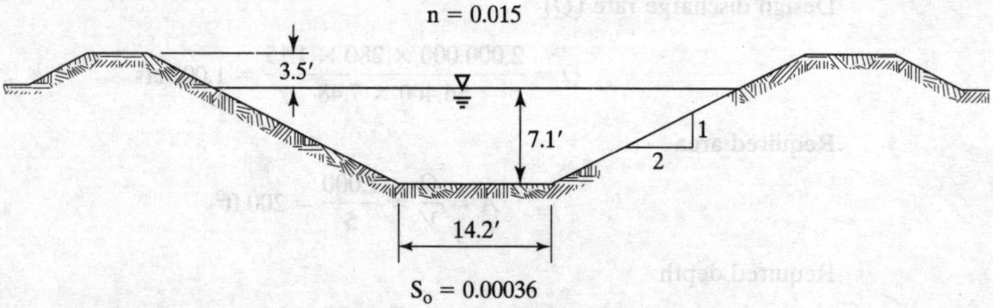

Example 5.11 Trapezoidal channel design.

5.6.4 Earth Channels

Earth channels are designed for subcritical flow. Because erosion is a major concern in the design of earth channels, they tend to be wider and shallower than concrete channels. The side slopes of an earth canal are determined by the stability of the bank material (Table 5.2) and maintenance. Generally, the maintenance cost of a channel decreases as the side slope of the channel becomes flatter. The side slope of a drainage channel may be as flat as 4:1 to allow mowing the vegetation on the banks. The value of B/y_n will generally range from 3 to 6 (small to large canal) for canals and 3 to 10 for drainage channels. Increasing the value of b/y_n increases the resistance to flow and reduces the velocity in the channel.

Because water flows in a surface water drainage channel for only short periods of time, they are often grass-lined. The longitudinal slope of a drainage channel is generally not a design parameter but is determined from the topography. The velocity in the channel can be adjusted by the channel roughness and the b/y_n value. They are often designed based on the permissible velocity (Table 5.3). Some channel erosion can normally be tolerated during major flood events.

The permissible tractive force method is often used in the design of canals where water may flow in the channel for an extended period of time. Tractive force is the force exerted by the flowing water on the channel and is equal to the

TABLE 5.2 TYPICAL SIDE SLOPES FOR CHANNELS

Material	Side slope H:V
Rock	$\frac{1}{4}$:1
Stiff clay	1:1
Stone lining	1:1
Firm clay	$1\frac{1}{2}$:1
Gravelly loam	$1\frac{1}{2}$:1
Sandy loam	3:1

TABLE 5.3 PERMISSIBLE VELOCITIES FOR CHANNELS CARRYING WATER WITH SEDIMENTS

Material	Velocity	
	mps	fps
Sandy loam	0.8	2.5
Coarse sand	1.2	4.0
Fine gravel	1.5	5.0
Coarse gravel	1.8	6.0
Silt loam	0.9	3.0
Alluvial silt	1.1	3.5
Silty clay	1.1	3.5
Clay	1.5	5.0
Shale	1.8	6.0
Grass-lined with slopes less than 5 percent		
Bermuda grass		
Sandy silt	1.8	6.0
Silty clay	2.4	8.0
Kentucky bluegrass		
Sandy silt	1.5	5.0
Silty clay	2.1	7.0
Grass mixture		
Sandy silt	1.2	4.0
Silty clay	1.5	5.0

component of the weight of water acting in the direction of flow. For the channel bottom, the tractive force per unit area (τ) is

$$\tau = \gamma y_n S_o \tag{5.33}$$

where γ is the unit weight of water. Permissible tractive force is given in Figs. 5.11 and 5.12 for noncohesive and cohesive channel material, respectively.

Example 5.12 Grass-Lined Channel Design

A grass-lined drainage channel is to carry a discharge of 2,000 cfs at a maximum velocity of 4.0 fps. The side slopes of the channel will be 4:1, and the longitudinal slope of the channel will be 0.001. Design the channel for a Manning n value of 0.030 and 0.035.

Required area (A)

$$A = \frac{Q}{V} = \frac{2{,}000}{4.0} = 500 \text{ ft}^2$$

Required hydraulic radius (R) $\quad n = 0.30 \quad\quad n = 0.35$

$$V = \frac{1.49}{n} R^{2/3} S_o^{1/2}$$

$$R = \left[\frac{Vn}{1.49 S_o^{1/2}}\right]^{3/2} = \left[\frac{4.0 \times 0.030}{1.49 \times (0.001)^{1/2}}\right]^{3/2} = 4.06 \text{ ft} \quad\quad R = 5.12 \text{ ft}$$

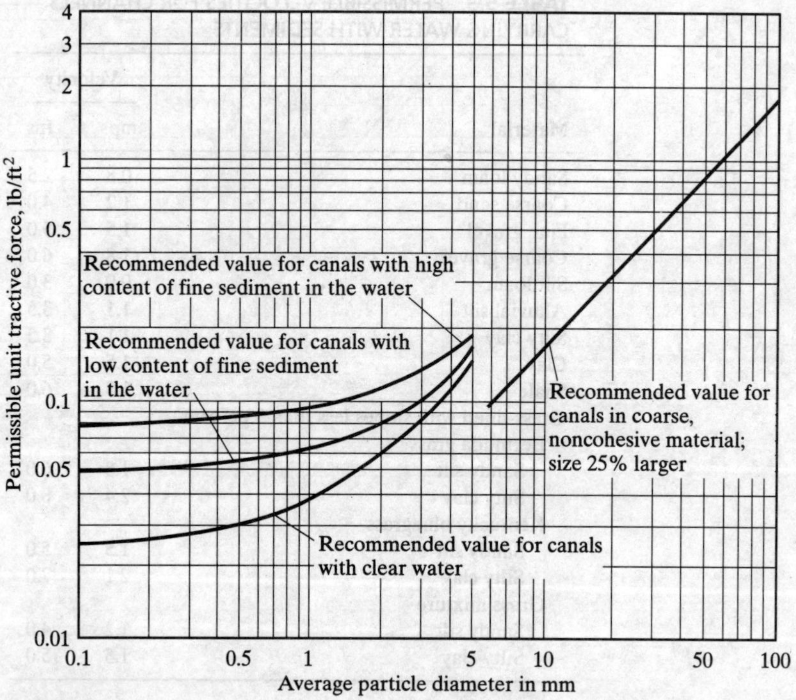

Figure 5.11 Permissible tractive force for noncohesive materials (U. S. Bureau of Reclamation).

Wetted perimeter (P)

$$R = \frac{A}{P}$$

$$P = \frac{A}{R} = \frac{500}{4.06} = 123.2 \text{ ft} \qquad P = 97.7 \text{ ft}$$

Solve for depth (y_n) and bottom width (B)

$$A = By_n + SSy_n^2$$

$$P = B + 2 \times y_n \times \sqrt{SS^2 + 1}$$

$$500 = By_n + 4y_n^2$$

$$123.2 = B + 8.25y_n$$

or

$$500 = (123.2 - 8.25y_n)y_n + 4y_n^2$$

$$y_n^2 - 29.0y_n + 117.6 = 0$$

$$y_n = \frac{+29.0 \pm \sqrt{29.0^2 - 4 \times 1 \times 117.6}}{2} = \frac{(29.0 - 19.2)}{2} = 4.9 \text{ ft} \qquad y_n = 7.7 \text{ ft}$$

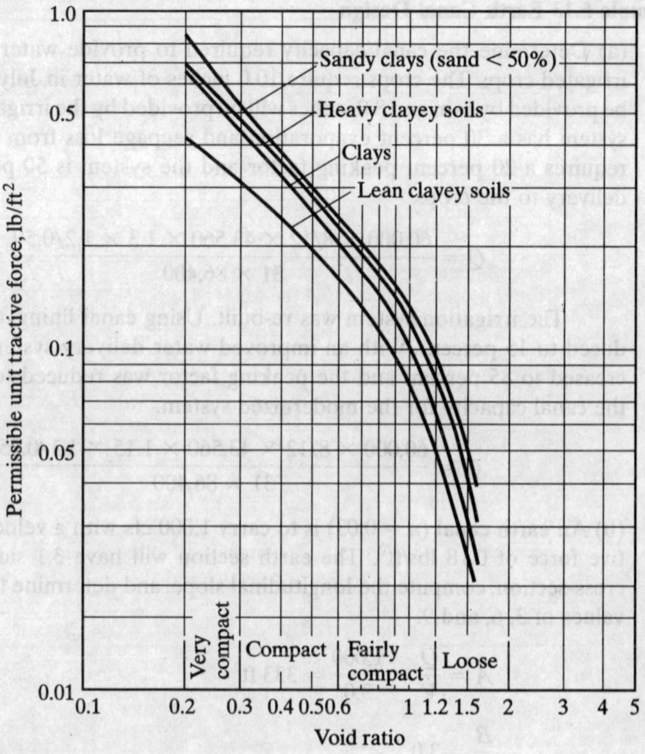

Figure 5.12 Permissible tractive force for cohesive materials (after Chow).

$$B = 123.2 - 8.25 \times 4.9 = 83 \text{ ft} \qquad B = 34.2 \text{ ft}$$

Freeboard (FB) assuming $C = 1.4$

$$FB = Cy_n^{1/2} = 1.4(4.9)^{1/2} = 3.1 \text{ ft} \qquad FB = 3.9 \text{ ft}$$

Use a Manning n value of 0.035 for the design. Maintain the height of grass between 6 and 12 inches.

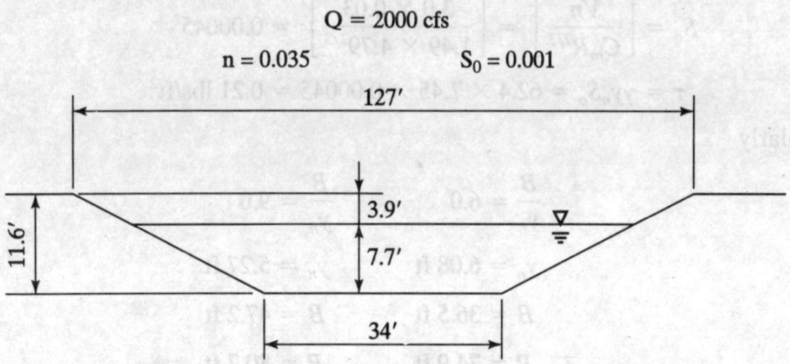

Example 5.12 Grass-lined channel design.

Example 5.13 Earth Canal Design

(a) Determine the canal capacity required to provide water supply for 60,000 ac of irrigated crops. The crops require 10.0 inches of water in July of which 2.0 inches will be provided by rain and 8.0 inches will be provided by the irrigation system. The existing system has a 30 percent evaporation and seepage loss from the canal, the operation requires a 20 percent peaking factor and the system is 50 percent efficient in water delivery to the crops.

$$Q = \frac{60{,}000 \times 8/12 \times 43{,}560 \times 1.3 \times 1.2/0.50}{31 \times 86{,}400} = 2{,}030 \text{ cfs}$$

The irrigation system was re-built. Using canal lining, the canal losses were reduced to 15 percent. With an improved water delivery system, the efficiency was increased to 85 percent and the peaking factor was reduced to 10 percent. Determine the canal capacity for the modernized system.

$$Q = \frac{60{,}000 \times 8/12 \times 43{,}560 \times 1.15 \times 1.1/0.85}{31 \times 86{,}400} = 970 \text{ cfs}$$

(b) An earth canal ($n = 0.03$) is to carry 1,000 cfs with a velocity of 3.0 fps and a tractive force of 0.18 lbs/ft^2. The earth section will have 3:1 side slopes. Size the canal cross-section, compute the longitudinal slope, and determine the tractive force for B/y values of 3, 6, and 9.

$$A = \frac{Q}{V} = \frac{1{,}000}{3.0} = 333 \text{ ft}^2$$

$$\frac{B}{y} = 3.0$$

$$A = (B + SS\, y_n)y_n$$

$$y_n = 7.45 \text{ ft}$$

$$B = 3.0 \times y_n = 22.4 \text{ ft}$$

$$P = B + 2y_n\sqrt{SS^2 + 1} = 22.4 + 2 \times 7.45 \times 3.16 = 69.5 \text{ ft}$$

$$R = \frac{A}{P} = \frac{333}{69.5} = 4.79 \text{ ft}$$

$$S_o = \left[\frac{Vn}{C_m R^{2/3}}\right]^2 = \left[\frac{3.0 \times 0.03}{1.49 \times 4.79^{2/3}}\right]^2 = 0.00045$$

$$\tau = \gamma y_n S_o = 62.4 \times 7.45 \times 0.00045 = 0.21 \text{ lbs/ft}^2$$

Similarly

$\frac{B}{y_n} = 6.0$	$\frac{B}{y_n} = 9.0$
$y_n = 6.08$ ft	$y_n = 5.27$ ft
$B = 36.5$ ft	$B = 47.2$ ft
$P = 74.9$ ft	$P = 80.7$ ft

$R = 4.44$ ft $\qquad R = 4.13$ ft

$S_o = 0.00050 \qquad S_o = 0.00055$

$\tau = 0.19 \qquad \tau = 0.18$

(c) Design an earth canal ($n = 0.03$) to carry 1,000 cfs with a tractive force of 0.18 lbs/ft^2, a longitudinal slope of 0.0004 and 2H:1V side slopes.

$$y_n = \frac{\tau}{\gamma S_o} = \frac{0.18}{62.4 \times 0.0004} = 7.21 \text{ ft}$$

$$AR^{2/3} = \frac{Qn}{1.49\, S_o^{1/2}} = \frac{1{,}000 \times 0.03}{1.49 \times (0.0004)^{1/2}} = 1{,}007$$

$$1{,}007 = \frac{(By_n + 2y_n^2)^{5/3}}{(B + 4.47y_n)^{2/3}} = \frac{(7.21B + 104.0)^{5/3}}{(B + 32.24)^{2/3}}$$

$$B = 32.0 \text{ ft}, A = 336.2 \text{ ft}^2, V = 3.0 \text{ fps}$$

$$\text{FB} = 1.3 \times 7.21^{1/2} = 3.5 \text{ ft}$$

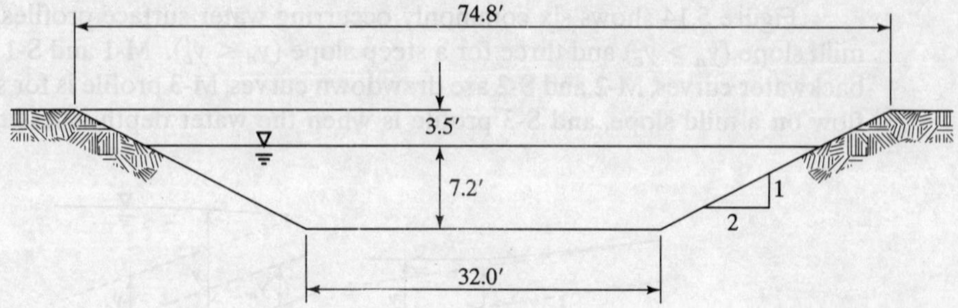

5.7 GRADUALLY VARIED STEADY FLOW

For steady, nonuniform flow in an open channel, the flow depth and velocity do not change with time but vary along the length of the channel. By computing the water surface profile for given discharge rate in the channel, the depth and velocity of water are determined. There are three water depths that must be considered in water surface profile computations. The normal depth (y_n) is computed using the Manning equation for uniform flow, the critical depth (y_c) is computed using the minimum specific energy equation, and the flow depth (y) is computed from the energy equation.

The one-dimensional energy equation for open channel flow is

$$\frac{V_1^2}{2g} + y_1 + z_1 = \frac{V_2^2}{2g} + y_2 + z_2 + H_L \tag{5.34}$$

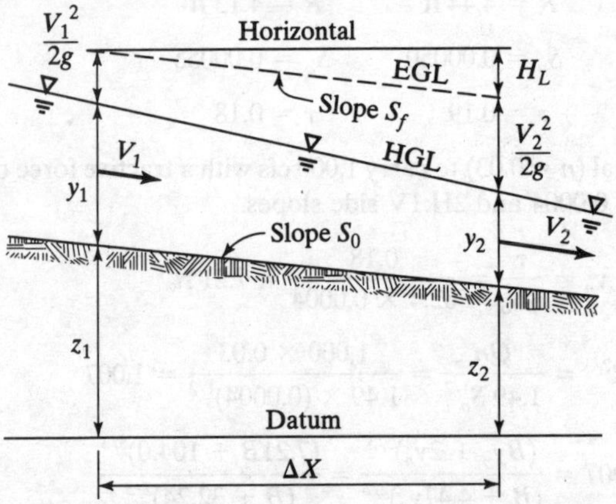

Figure 5.13 Energy equation for nonuniform flow.

where the subscripts 1 and 2 refer to the upstream and downstream sections, respectively (Fig. 5.13). If the flow in the channel is subcritical, the water surface computations begin downstream and proceed upstream. For supercritical flow, computations begin upstream and continue downstream.

Figure 5.14 shows six commonly occurring water surface profiles, three for a mild slope ($y_n > y_c$) and three for a steep slope ($y_n < y_c$). M-1 and S-1 profiles are backwater curves, M-2 and S-2 are drawdown curves, M-3 profile is for supercritical flow on a mild slope, and S-3 profile is when the water depth is less than normal

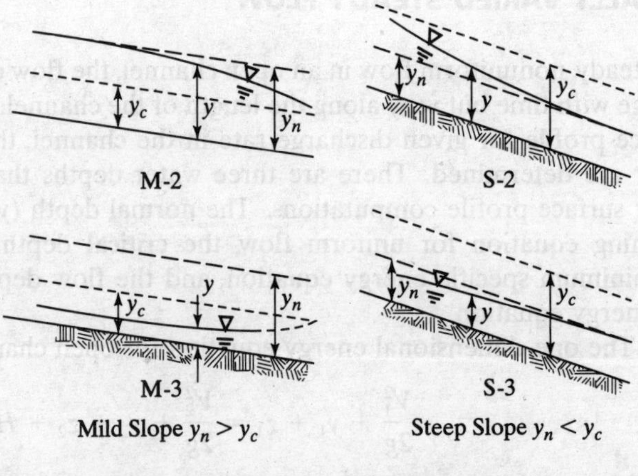

Figure 5.14 Typical water surface profiles.

depth on a steep slope. Other water surface profiles will occur for critical, horizontal, and adverse bottom slopes.

The direct step method of water surface profile computation is used for channels with uniform cross-section, such as canals and drainage channels. In the direct step method of water surface profile computation for subcritical flow, the downstream water depth is known, the upstream water depth is assumed, and the distance upstream where that depth occurs is computed from the energy equation. Substituting ΔXS_o for $z_1 - z_2$ and $\Delta X\overline{S}_f$ for H_L in Eq. 5.34, the direct step form of the energy equation is

$$\Delta X = \frac{E_1 - E_2}{\overline{S}_f - S_o} \tag{5.35}$$

where $E = V^2/2g + y$ and $\overline{S}_f = (S_{f_1} + S_{f_2})/2$. The friction slope ($S_f$) is computed from the Manning equation

$$S_f = \frac{n^2V^2}{C_m^2 R^{4/3}} \tag{5.36}$$

The standard step method of water surface profile computation (Section 5.7.3) is generally used for natural channels. The channel geometry is defined by a series of cross-sections either surveyed or from detailed topographic maps. Equation 5.34 is solved for the correct water depth at section 1 (y_1) either by trial and error or by a numerical procedure such as the Newton–Raphson Method.

A longitudinal profile of nonuniform flow in an open channel is shown in Fig. 5.15. Critical depth occurs at the break in slope with M-2 water surface profile on the mild slope upstream and an S-2 water surface profile on the steep slope downstream. Supercritical flow on a mild slope (M-3 profile) occurs just upstream of the hydraulic jump.

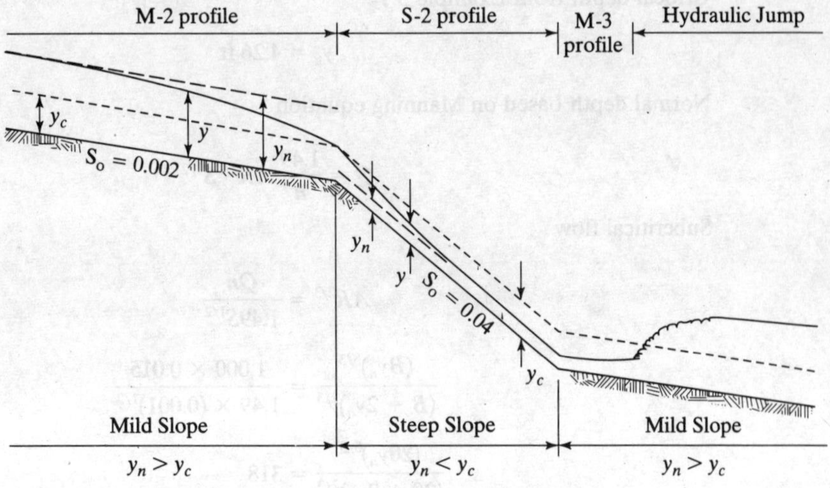

Figure 5.15 Nonuniform flow in an open channel.

Example 5.14 Direct Step Water Surface Profile Computation

A concrete rectangular channel ($n = 0.015$) is 20 ft wide and carries a discharge of 1,000 cfs. The slope of the channel changes from $S = 0.001$ (mild) to $S = 0.01$ (steep). Critical depth occurs where the channel slope changes from mild to steep. Compute the water surface profile both upstream (subcritical flow) and downstream (supercritical flow) of the break in channel slope.

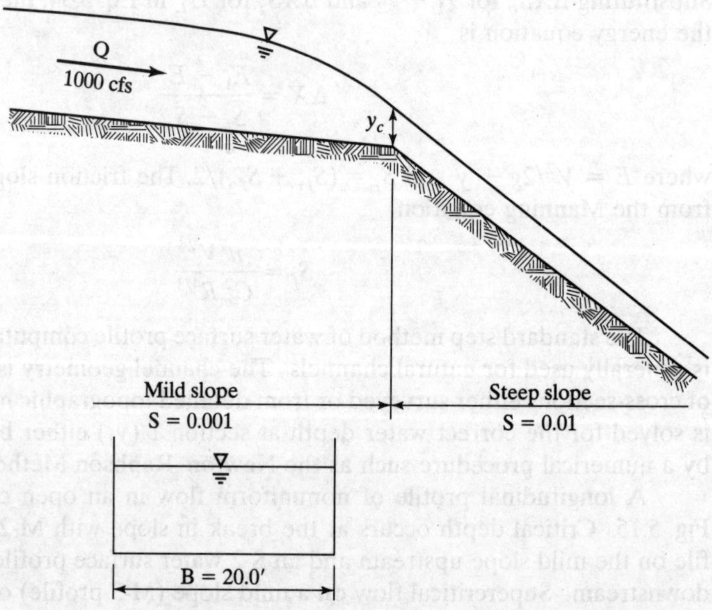

Critical depth from Example 5.7

$$y_c = 4.26 \text{ ft}$$

Normal depth based on Manning equation

$$Q = \frac{1.49}{n} AR^{2/3}S^{1/2}$$

Subcritical flow

$$AR^{2/3} = \frac{Qn}{1.49S^{1/2}}$$

$$\frac{(By_n)^{5/3}}{(B + 2y_n)^{2/3}} = \frac{1{,}000 \times 0.015}{1.49 \times (0.001)^{1/2}}$$

$$\frac{(20y_n)^{5/3}}{(20 + 2y_n)^{2/3}} = 318$$

Solve for y_n by trial and error

$$y_n = 6.34 \text{ ft}$$

Supercritical flow

$$AR^{2/3} = \frac{1{,}000 \times 0.015}{1.49 \times (0.01)^{1/2}}$$

$$\frac{(20y_n)^{5/3}}{(20 + 2y_n)^{2/3}} = 100.7$$

Solve for y_n by trial and error

$$y_n = 2.90 \text{ ft}$$

Develop two spreadsheets to solve Eq. 5.35 for Δx upstream and downstream of the change in channel slope. The depth of flow for the first spreadsheet will range from 4.26 ft to 6.34 ft in increments of 0.1 ft. The depth of flow for the second spreadsheet will range from 4.26 to 2.90 in increments of 0.1 ft. Sum the Δx's and plot distance versus water depth.

Depth	Area	P	R	V	E	$E_1 - E_2$	S_f	$\overline{S}_f$	$\overline{S}_f - S_o$	ΔX	$\Sigma\Delta X$
4.26	85.2	28.52	2.99	11.74	6.40		0.000964				
4.36	87.2	28.72	3.04	11.47	6.40	−0.003	0.000944	0.000954	−4.6E-05	65	65
4.46	89.2	28.92	3.08	11.21	6.41	−0.00945	0.000927	0.000936	−6.4E-05	147	212
4.56	91.2	29.12	3.13	10.96	6.43	−0.01534	0.000913	0.00092	−8E-05	192	404
4.66	93.2	29.32	3.18	10.73	6.45	−0.02073	0.000901	0.000907	−9.3E-05	223	627
4.76	95.2	29.52	3.22	10.50	6.47	−0.02568	0.000891	0.000896	−0.0001	246	873
4.86	97.2	29.72	3.27	10.29	6.50	−0.03022	0.000882	0.000886	−0.00011	266	1,139
4.96	99.2	29.92	3.32	10.08	6.54	−0.0344	0.000876	0.000879	−0.00012	284	1,423
5.06	101.2	30.12	3.36	9.88	6.58	−0.03825	0.00087	0.000873	−0.00013	301	1,724
5.16	103.2	30.32	3.40	9.69	6.62	−0.0418	0.000866	0.000868	−0.00013	317	2,041
5.26	105.2	30.52	3.45	9.50	6.66	−0.04509	0.000863	0.000865	−0.00014	334	2,375
5.36	107.2	30.72	3.49	9.33	6.71	−0.04813	0.000862	0.000863	−0.00014	350	2,725
5.46	109.2	30.92	3.53	9.16	6.76	−0.05096	0.000861	0.000861	−0.00014	367	3,092
5.56	111.2	31.12	3.57	8.99	6.82	−0.05358	0.000861	0.000861	−0.00014	386	3,478
5.66	113.2	31.32	3.61	8.83	6.87	−0.05602	0.000862	0.000862	−0.00014	405	3,883
5.76	115.2	31.52	3.65	8.68	6.93	−0.05829	0.000864	0.000863	−0.00014	425	4,308
5.86	117.2	31.72	3.69	8.53	6.99	−0.06041	0.000866	0.000865	−0.00013	448	4,756
5.96	119.2	31.92	3.73	8.39	7.05	−0.06238	0.000869	0.000868	−0.00013	472	5,228
6.06	121.2	32.12	3.77	8.25	7.12	−0.06423	0.000873	0.000871	−0.00013	499	5,726
6.16	123.2	32.32	3.81	8.12	7.18	−0.06596	0.000877	0.000875	−0.00012	529	6,255
6.26	125.2	32.52	3.85	7.99	7.25	−0.06758	0.000882	0.00088	−0.00012	562	6,817
6.34	126.8	32.68	3.88	7.89	7.31	−0.05516	0.000886	0.000884	−0.00012	477	7,294

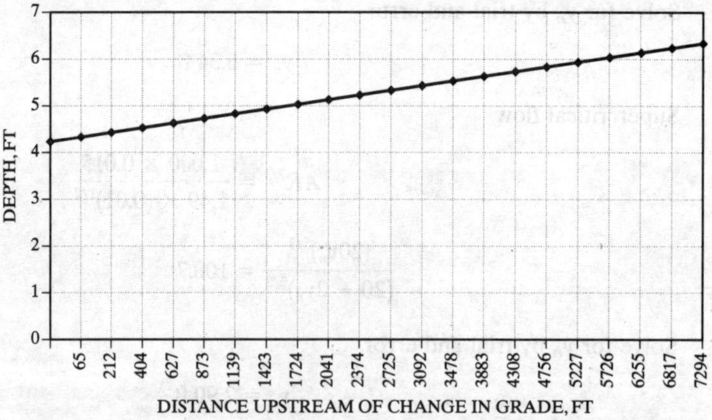

Depth	Area	P	R	V	E	$E_1 - E_2$	S_f	$\bar{S}_f$	$\bar{S}_f - S_o$	ΔX	$\Sigma\Delta X$
4.26	85.2	28.52	2.99	11.74	6.40		0.000964				
4.16	83.2	28.32	2.94	12.02	6.40	−0.00408	0.000987	0.000975	−0.00902	0.45	0.45
4.06	81.2	28.12	2.89	12.32	6.42	−0.01186	0.001014	0.001	−0.009	1.32	1.77
3.96	79.2	27.92	2.84	12.63	6.44	−0.02044	0.001045	0.001029	−0.00897	2.28	4.05
3.86	77.2	27.72	2.78	12.95	6.47	−0.02993	0.00108	0.001063	−0.00894	3.35	7.40
3.76	75.2	27.52	2.73	13.30	6.51	−0.04043	0.001122	0.001101	−0.0089	4.54	11.94
3.66	73.2	27.32	2.68	13.66	6.56	−0.0521	0.00117	0.001146	−0.00885	5.88	17.82
3.56	71.2	27.12	2.63	14.04	6.62	−0.06509	0.001227	0.001199	−0.0088	7.40	25.22
3.46	69.2	26.92	2.57	14.45	6.70	−0.07961	0.001292	0.001259	−0.00874	9.11	34.33
3.36	67.2	26.72	2.51	14.88	6.80	−0.09589	0.001369	0.00133	−0.00867	11.06	45.39
3.26	65.2	26.52	2.46	15.34	6.91	−0.11419	0.001459	0.001414	−0.00859	13.30	58.69
3.16	63.2	26.32	2.40	15.82	7.05	−0.13484	0.001565	0.001512	−0.00849	15.89	74.57
3.06	61.2	26.12	2.34	16.34	7.21	−0.15824	0.00169	0.001627	−0.00837	18.90	93.47
2.96	59.2	25.92	2.28	16.89	7.39	−0.18486	0.001839	0.001765	−0.00824	22.45	115.92
2.9	58	25.8	2.25	17.24	7.52	−0.12524	0.001943	0.001891	−0.00811	15.44	131.37

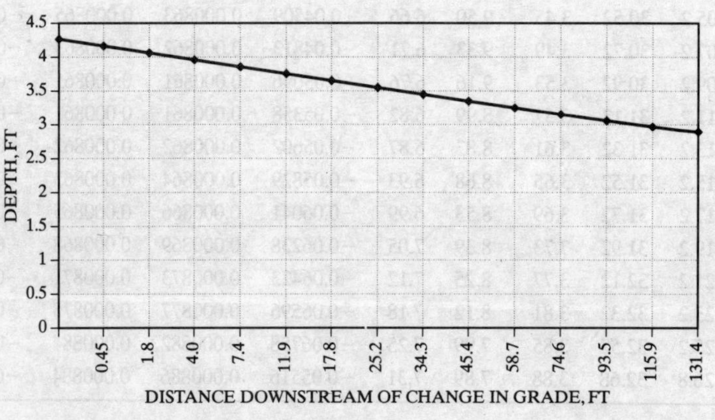

5.7.1 Transitions

Transitions are used to connect channels with different characteristics. A properly designed transition will provide smooth flow through the transition with little headloss. If the cross-sectional area is reduced, the transition is called a contraction, and if the area is increased, it is called an expansion. Straight-line transitions are generally used to reduce construction costs. A gradual tapering of the channel in the transition of 1:1 for a contraction and 4:1 for an expansion is generally adequate for subcritical flow. Transitions can also be used at the inlet and outlet of culverts, siphons, flumes, and tunnels. A straight-line expansion transition is shown in Fig. 5.16. The energy equation (Eq. 5.34) can be written for the transition and the headloss is computed

$$H_L = \frac{S_{f1} + S_{f2}}{2}\Delta x + \frac{K_m|V_1^2 - V_2^2|}{2g} \tag{5.37}$$

where K_m is the minor loss coefficient and Δx is the length of the transition.

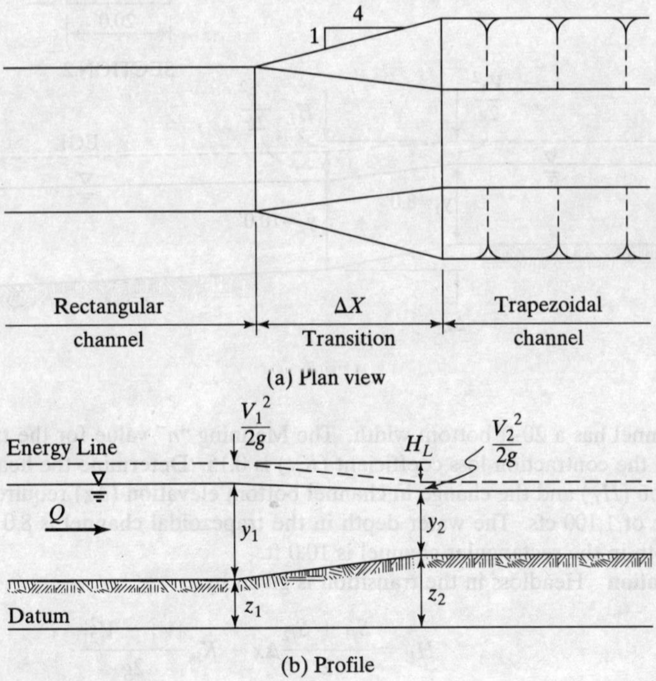

Figure 5.16 Straight line expansion transition.

Example 5.15 Contraction Loss

The transition from a trapezoidal to a rectangular channel is 25 ft long (Δx). The trapezoidal channel has a 30-ft bottom width with 2:1 side slopes, while the rectangular

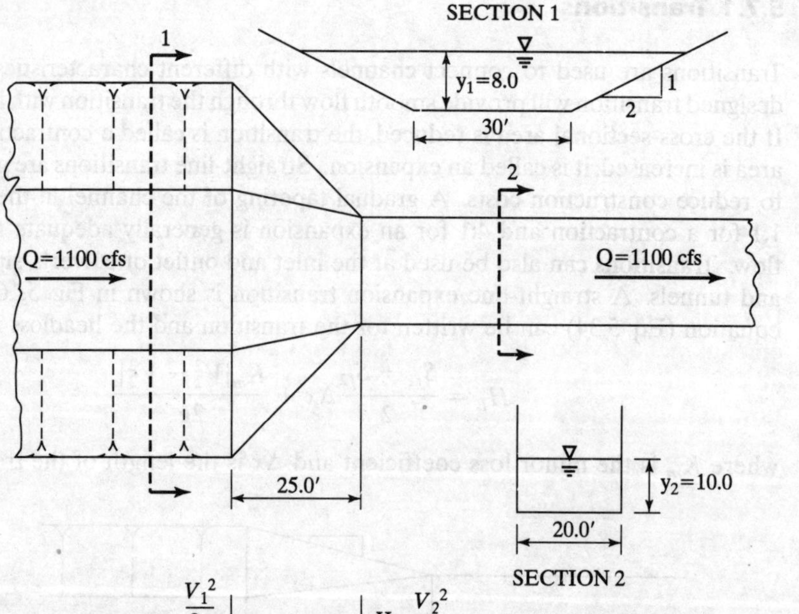

channel has a 20-ft bottom width. The Manning "n" value for the transition is 0.015, and the contraction loss coefficient (K_m) is 0.15. Determine the headloss in the transition (H_L) and the change in channel bottom elevation (Δz) required for a discharge rate of 1,100 cfs. The water depth in the trapezoidal channel is 8.0 ft, and the water depth in the rectangular channel is 10.0 ft.

Solution Headloss in the transition is given by

$$H_L = \frac{S_{f1} + S_{f2}}{2}\Delta x + K_m \frac{|V_1^2 - V_2^2|}{2g}$$

where

$$S_f = \frac{n^2 V^2}{C_m^2 R^{4/3}}$$

$V_1 = Q/A = 1{,}100/(240 + 128) = 2.99$ fps $\qquad V_2 = Q/A = 1{,}100/200 = 5.5$ fps

$R_1 = A/P = 368/65.8 = 5.59$ ft $\qquad R_2 = A/P = 200/40 = 5.0$ ft

$$S_{f1} = 0.000092 \qquad S_{f2} = 0.000361$$

$$H_L = \frac{0.000092 + 0.000361}{2} \times 25 + (0.15)\frac{|2.99^2 - 5.5^2|}{64.4} = 0.006 + 0.050 = 0.056 \text{ ft}$$

$$\Delta z = y_2 + \frac{V_2^2}{2g} + H_L - y_1 - \frac{V_1^2}{2g} = 10.0 + 0.47 + 0.06 - 8.0 - 0.14 = 2.39 \text{ ft}$$

5.7.2 Encroachments

Encroachments into a waterway commonly occur at bridges where the approach embankments to the bridge extend into the channel. Plan and profile views for flow at a bridge are shown in Fig. 5.17. As the encroachment distance (Δ) increases, the flow area decreases and the velocity increases through the bridge opening. If the encroachment is large enough to cause critical flow in the bridge opening, the encroachment is called a choke and the upstream water depth is independent of the downstream water depth.

Backwater from a bridge is the difference in water surface elevation upstream of the bridge (Section 4) with and without the bridge. The water surface elevation with the bridge is computed using the energy equation (Eq. 5.34) in three

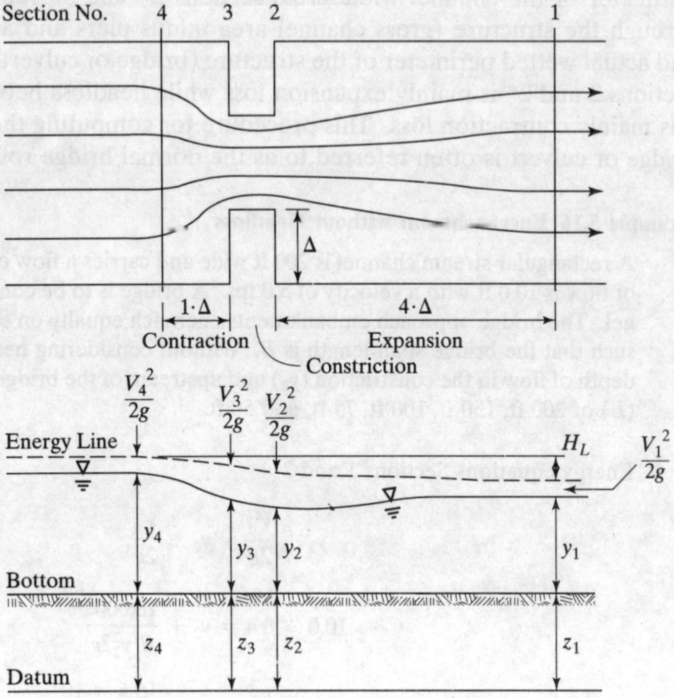

Figure 5.17 Encroachment.

TABLE 5.4 MINOR LOSS COEFFICIENTS (K_m)

Transition	Contraction	Expansion
Gradual	0.15	0.3
Bridge	0.3	0.5
Abrupt	0.6	0.8

segments—the expansion segment from Section 1 to Section 2, the constriction segment from Section 2 to Section 3, and the contraction segment from Section 3 to Section 4. Equation 5.37 can be used to compute the headloss in each segment. Minor loss coefficients for contractions and expansions are listed in Table 5.4. The length of the expansion section has considerable affect on the computed backwater at the bridge. The length of the expansion section will normally range from 2 times the encroachment distance for a densely vegetative channel to 4 times the encroachment distance for a sparsely vegetative channel.

If the bridge is submerged or includes piers such that the resistance to flow is increased through the structure, two additional cross-sections are generally used. Cross-section 2* is located just upstream of 2 and cross-section 3* is located just downstream of 3. Cross-sections 2 and 3 represent the gross area and wetted perimeter of the channel while cross-sections 2* and 3* represent net flow area through the structure (gross channel area minus piers and area above low cord) and actual wetted perimeter of the structure (bridge or culvert). Headloss between sections 2 and 2* is mainly expansion loss while headloss between sections 3* and 3 is mainly contraction loss. This procedure for computing the headloss through a bridge or culvert is often referred to as the normal bridge routine.

Example 5.16 Encroachment without Headloss

A rectangular stream channel is 200 ft wide and carries a flow of 10,000 cfs. The depth of flow is 10.0 ft with a velocity of 5.0 fps. A bridge is to be constructed over the channel. The bridge approach embankments encroach equally on each side of the channel such that the bridge span length is B. Without considering headlosses, determine the depth of flow in the constriction (y_2) and upstream of the bridge (y_3) for bridge lengths (B) of 200 ft, 150 ft, 100 ft, 75 ft, and 50 ft.

Energy equations Sections 1 and 2

$$y_1 + \frac{V_1^2}{2g} = y_2 + \frac{V_2^2}{2g}$$

$$10.0 + 0.4 = y_2 + \frac{10{,}000^2}{B^2 y_2^2 2g}$$

$$10.4 = y_2 + \frac{15.5 \times 10^5}{B^2 y_2^2}$$

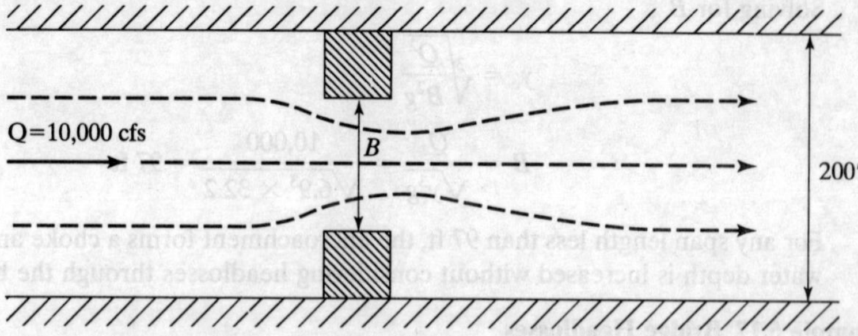

PLAN

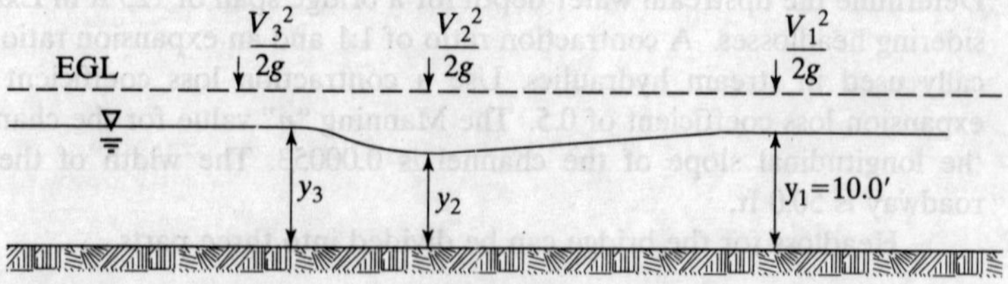

PROFILE

Minimum depth in construction is critical depth

$$y_c = \sqrt[3]{\frac{Q^2}{B^2 g}}$$

Energy equation between Sections 2 and 3

$$y_2 + \frac{V_2^2}{2g} = y_3 + \frac{V_3^2}{2g} = y_3 + \frac{38.8}{y_3^2}$$

Table of depths

Span	Downstream depth	Constriction		Upstream depth
		Depth	Critical depth	
B ft	y_1 ft	y_2 ft	y_c ft	y_3 ft
200	10.0	10.0	4.3	10.0
150	10.0	9.6	5.3	10.0
100	10.0	7.8	6.7	10.0
75	10.0	8.2	8.2	12.3
50	10.0	10.7	10.7	16.0

The minimum span length that causes critical depth to occur in the constriction is determined as follows:

$$y_c = \tfrac{2}{3}E_c = \tfrac{2}{3}(10.4) = 6.9 \text{ ft}$$

Solving for B

$$y_c = \sqrt[3]{\frac{Q^2}{B^2 g}}$$

$$B = \frac{Q}{\sqrt{y_c^3 g}} = \frac{10{,}000}{\sqrt{6.9^3 \times 32.2}} = 97 \text{ ft}$$

For any span length less than 97 ft, the encroachment forms a choke and the upstream water depth is increased without considering headlosses through the bridge.

Example 5.17 Bridge Headlosses

Determine the upstream water depth for a bridge span of 125 ft in Example 5.16 considering headlosses. A contraction ratio of 1:1 and an expansion ratio of 4:1 are typically used in stream hydraulics. Use a contraction loss coefficient of 0.3 and an expansion loss coefficient of 0.5. The Manning "n" value for the channel is 0.03, and the longitudinal slope of the channel is 0.00053. The width of the encroachment roadway is 50.0 ft.

Headloss for the bridge can be divided into three parts.

Expansion Sections 1–2

$$H_{L1-2} = \frac{S_{f1} + S_{f2}}{2} \times 150 + 0.5\frac{(V_2^2 - V_1^2)}{2g}$$

Constriction Sections 2–3

$$H_{L2-3} = \frac{S_{f2} + S_{f3}}{2} 50.0$$

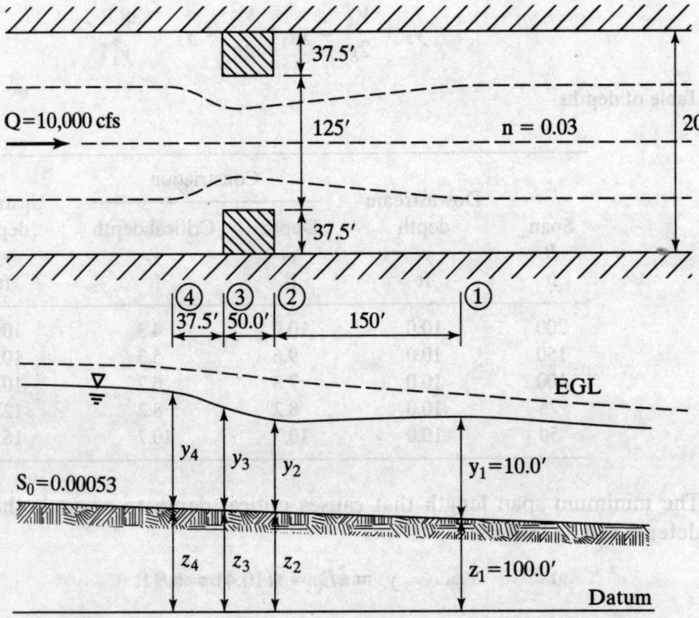

Contraction Sections 3–4

$$H_{L3-4} = \frac{S_{f3} + S_{f4}}{2} 37.5 + 0.3 \frac{(V_3^2 - V_4^2)}{2g}$$

Writing the energy equation between 1 and 2

$$\frac{V_2^2}{2g} + y_2 + z_2 = \frac{V_1^2}{2g} + y_1 + z_1 + H_{L1-2}$$

$$\frac{V_2^2}{2g} + y_2 + 100.08 = 0.39 + 10.0 + 100.00 + \frac{S_{f1} + S_{f2}}{2} 150 + 0.5 \frac{V_2^2}{2g} - 0.5 \frac{V_1^2}{2g}$$

$$0.5 \frac{V_2^2}{2g} + y_2 - 75S_{f2} = 10.39 - 0.08 - 0.19 + 75S_{f1}$$

where

$$S_f = \frac{n^2 V^2}{C_m^2 R^{4/3}} = \frac{29n^2}{R^{4/3}} \frac{V^2}{2g}$$

and

$$S_{f1} = \frac{29 \times 0.03^2}{9.09^{4/3}} \frac{5^2}{64.4} = 0.00054$$

$$S_{f2} = \frac{0.0261}{R_2^{4/3}} \frac{V_2^2}{2g}$$

The energy equation reduces to

$$F(y_2) = 10.16 - y_2 - \left(0.5 - \frac{1.96}{R_2^{4/3}}\right) \frac{Q^2}{A_2^2 2g} = 0$$

Solving for y_2 yields

$$y_2 = 9.76 \text{ ft}$$

$$V_2 = 8.20 \text{ fps}$$

$$S_{f2} = 0.0016$$

Writing the energy equation between Sections 2 and 3 gives

$$\frac{V_3^2}{2g} + y_3 + z_3 = \frac{V_2^2}{2g} + y_2 + z_2 + H_{L2-3}$$

$$\frac{V_3^2}{2g} + y_3 + 100.11 = 1.04 + 9.76 + 100.08 + \frac{50}{2}(S_{f2} + S_{f3})$$

$$F(y_3) = 10.81 - y_3 - \left(1.0 - \frac{0.65}{R_3^{4/3}}\right) \frac{Q^2}{A_3^2 2g} = 0$$

Solving for y_3 yields

$$y_3 = 9.82 \text{ ft}$$
$$V_3 = 8.15 \text{ fps}$$
$$S_{f3} = 0.0016$$

Finally, writing the energy equation between Sections 3 and 4

$$\frac{V_4^2}{2g} + y_4 + z_4 = \frac{V_3^2}{2g} + y_3 + z_3 + H_{L3-4}$$

$$\frac{V_4^2}{2g} + y_4 + 100.13 = 1.03 + 9.82 + 100.11 + \frac{37.5}{2}(S_{f3} + S_{f4}) + 0.3\frac{V_3^2}{2g} - 0.3\frac{V_4^2}{2g}$$

$$F(y_4) = 11.17 - y_4 - \left[1.3 - \frac{0.49}{R_4^{4/3}}\right]\frac{Q^2}{A_4^2 2g}$$

Solving for y_4 yields

$$y_4 = 10.74 \text{ ft}$$

If the normal depth at Section 4 is 10.00 ft without the bridge, then the backwater effect of the bridge is 0.74 ft.

5.7.3 Standard Step Method of Water Surface Profile Computations

The analysis of gradually varied flow is accomplished by computing the water surface profile using the energy equation. When the locations of the cross-sections are fixed such as a transition, bridge, or natural channel, the standard step method is generally used to solve the energy equation. As shown in Fig. 5.18, the energy equation (Eq. 5.34) is written in terms of the water surface elevation ($WS = y + z$). For subcritical flow in a natural channel the energy equation becomes

$$\alpha_2\frac{V_2^2}{2g} + WS_2 = \alpha_1\frac{V_1^2}{2g} + WS_1 + H_L \qquad (5.38)$$

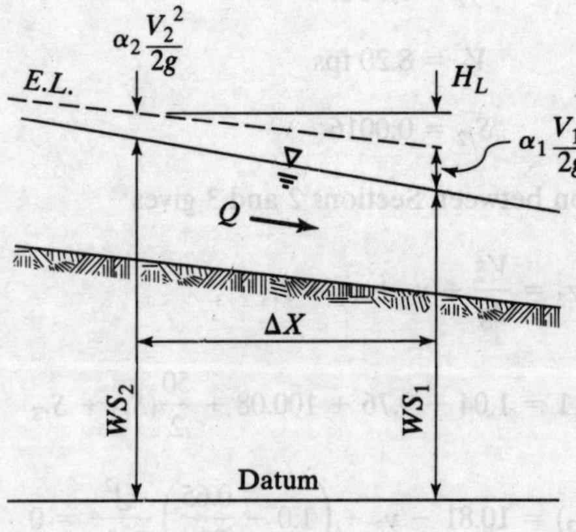

Figure 5.18 Standard step method of water surface profile computation.

where subscript 1 refers to the downstream section, subscript 2 refers to the upstream section, and α is the velocity head coefficient (Section 3.7.2).

$$\alpha = \frac{\int v^2 dQ}{V^2 Q} \tag{5.39}$$

where v is the velocity for the incremental subarea, V is the average velocity for the cross-section, and Q is the total discharge for the cross-section. For a compound section in a natural channel, α is estimated as

$$\alpha = \frac{\Sigma\left[\left(\frac{Q_i}{A_i}\right)^3 A_i\right]}{\left(\frac{\Sigma Q_i}{\Sigma A_i}\right)^3 \Sigma A_i} = \frac{\left[\Sigma\left(\frac{Q_i^3}{A_i^2}\right)\right](\Sigma A_i)^2}{(\Sigma Q_i)^3} \tag{5.40}$$

The flow (Q_i) for each subarea is written in terms of conveyance (k_i)

$$Q_i = k_i S_f^{1/2} \tag{5.41}$$

where S_f is the slope of the energy line for the cross-section. The value of the velocity head coefficient can be computed for each cross-section as

$$\alpha = \frac{\Sigma\left(\frac{k_i^3}{A_i^2}\right) A^2}{K^3} \tag{5.42}$$

where $A = \Sigma A_i$ and $K = \Sigma k_i$.

The value of α for a natural channel generally ranges from about 1.1 to 2.0. The headloss term in the energy equation is computed from Eq. 5.37. Substituting Eq. 5.37 for the headloss term, Eq. 5.38 becomes

$$F(y_2) = \alpha_2(1 + K_m)\frac{V_2^2}{2g} - \left(\frac{Q_2}{K_2}\right)^2 \frac{\Delta x}{2} + WS_2 - \alpha_1(1 + K_m)\frac{V_1^2}{2g} - \left(\frac{Q_1}{K_1}\right)^2 \frac{\Delta x}{2} - WS_1 = 0 \tag{5.43}$$

where subscripts 1 and 2 refer to downstream and upstream cross-sections, respectively, K_m is the minor loss coefficient and is equal to K_m for contraction and equal to $-K_m$ for expansion, K_2 is the total conveyance at Section 2, and K_1 is the total conveyance at Section 1.

Equation 5.43 can be solved using the iterative Newton–Raphson method or

$$WS_2^+ = WS_2 - \frac{F(WS_2)}{F'(WS_2)} \tag{5.44}$$

where WS_2^+ is the new estimate of water surface elevation at section 2, WS_2 is the old estimate, $F(SW_2)$ is the value of Eq. 5.43 based on the old water surface elevation, and $F'(W)$ is the derivative of Eq. 5.43 with respect to the water surface elevation. Generally, the starting water surface elevation for Section 2 is based on normal depth, and Eq. 5.43 will converge to a solution after a few iterations. For natural stream channels, $F'(WS_2)$ is typically evaluated by solving Eq. 5.43 twice for two different water surface elevations. For a prismatic channel, $F'(WS_2)$ can be determined by taking the derivative of Eq. 5.43 and solving for $F'(WS_2)$ directly.

Although the flow in a channel during a flood event is unsteady, the maximum flood level is usually computed as gradually varied flow. It is generally assumed that the peak discharge rate occurs at the same time for the entire length of the channel and that the discharge rate only changes along the channel at major tributaries.

The standard step method of water surface profile computation is commonly used to delineate flood prone areas along natural channels. The computed profiles are used to establish floor elevations for buildings along the channel, and an adequate number of cross-sections is required for accuracy. Cross-sections should be located at changes in channel geometry and slope, above and below major tributaries, and at structures such as bridges, submerged roads, and transitions. The maximum distance between cross-sections depends on the size and slope of the channel. A small channel on a relatively steep slope will require more cross-sections than a large channel on a relatively mild slope. Cross-sectional information can be obtained from topographic maps or field surveys. Mapping standards usually require that the elevation of 90 percent of the points be within a half of a contour interval of the correct elevation. Cross-sections are selected to be representative of the conveyance for a reach of the channel. They are established perpendicular to the streamlines and should not include no flow or dead storage areas. For subcritical flow, computations begin downstream of the area of interest starting with normal depth, critical depth, or a known water surface elevation.

Example 5.18 Standard Step Method of Water Surface Profile

Rock Creek is shown below with cross-sections 1 through 10 on the main channel and 11 through 14 on Branch A. A road crosses the main channel below the confluence with the tributary. The road approach embankment to the bridge encroaches approximately 100 ft into the left and right overbank areas of the channel. Using a 4:1 expansion ratio and a 1:1 contraction ratio, cross-section 2 was established 400 ft below the road and cross-section 5 was established 100 ft above the road. Cross-sections 3 and 4 are located in the bridge constriction. Surveyed eight-point cross-sections are located at sections 1, 3, 5, 6, 7, and 11. Cross-sections not surveyed are a modification of the downstream cross-section. A plot of an eight point cross-section is shown below along with a listing of the cross-sections.

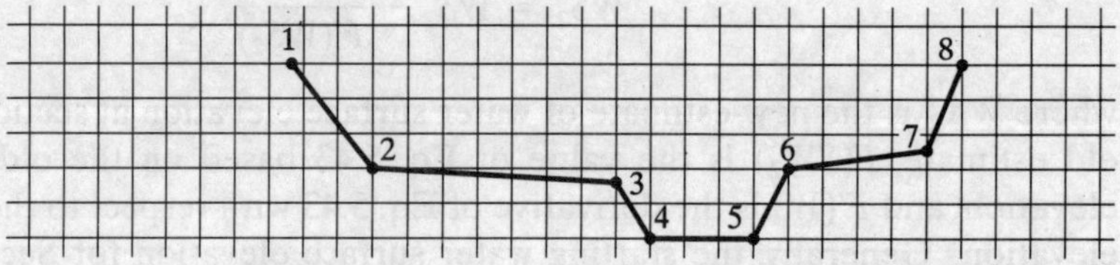

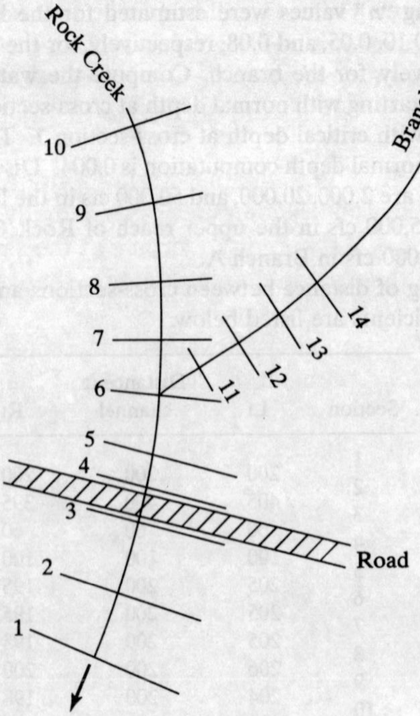

Table of cross-sections for Example 5.18

Section		Point 1	2	3	4	5	6	7	8
1	X	0	50	200	210	260	270	420	450
	Z	300	280	275	265	265	275	280	300
2	$\Delta Z = 0.8$ and X factor $= 0.95$								
3	X	100	150	200	210	260	270	320	350
	Z	301	277	276.5	266.5	266	277	278	302
4	$\Delta Z = 0.2$ and X factor $= 1.00$								
5	X	0	40	180	200	270	280	400	430
	Z	301	281	277	267	268	278	282	302
6	X	0	20	220	240	300	320	400	430
	Z	302	282	278	268	269	278	283	303
7	X	0	20	120	130	180	200	300	350
	Z	303	283	278	270	269	278	283	303
8	$\Delta Z = 0.8$ and X factor $= 0.95$								
9	$\Delta Z = 0.7$ and X factor $= 1.00$								
10	$\Delta Z = 0.7$ and X factor $= 1.00$								
11	X	0	20	120	130	160	170	270	300
	Z	303	283	278	270	269	278	283	303
12	$\Delta Z = 0.8$ and X factor $= 1.00$								
13	$\Delta Z = 0.7$ and X factor $= 1.00$								
14	$\Delta Z = 0.7$ and X factor $= 1.00$								

Manning "n" values were estimated for the left overbank, channel, and right overbank as 0.10, 0.05, and 0.08, respectively, for the main channel, and 0.08, 0.05, and 0.08, respectively, for the branch. Compute the water surface profiles for three discharge rates starting with normal depth at cross-section 1 and three water surface profiles starting with critical depth at cross-section 1. The slope of the channel at cross-section 1 for normal depth computation is 0.004. Discharge rates to be used for profile computations are 2,000, 20,000, and 50,000 cfs in the lower reach of Rock Creek; 1,500, 15,000, and 35,000 cfs in the upper reach of Rock Creek above Branch A; and 500, 5,000, and 15,000 cfs in Branch A.

A listing of distance between cross-sections and minor loss contraction and expansion coefficients are listed below.

Section	Lt	Distance ft channel	Rt	K_m con	K_m exp
1					
	200	200	200	0.15	0.30
2					
	405	400	395	0.30	0.50
3					
	60	60	60	0.30	0.50
4					
	100	100	100	0.30	0.50
5					
	205	200	195	0.15	0.30
6					
	205	200	195	0.15	0.30
7					
	205	200	195	0.15	0.30
8					
	206	200	200	0.15	0.30
9					
	204	200	198	0.15	0.30
10					
	100*	100*	100*	0.15	0.30
11					
	105	100	95	0.15	0.30
12					
	106	100	100	0.15	0.30
13					
	104	100	98	0.15	0.30
14					

*Distance to cross-section 6.

Solution The set of computer programs listed in Section 1.4.5 includes a program (STDSTEP.FOR) to compute water surface profiles for a natural channel and branches. The program is intended for student use and can be modified as necessary for other applications. The program uses the standard step method of profile computation, eight-point cross-sections with three Manning "n" values for each cross-section; will handle both subcritical and supercritical flow; discharge rates can be specified for multiple profiles; and the user can specify one of the three starting conditions—water surface elevation, normal depth, or critical depth.

Input file (STDSTEP.DAT) for Rock Creek is listed below. Each line of the input file is identified with alphabetic characters in the first two columns as follows:

SC line has starting conditions

QT line has table of flow rates

MN line has Manning "n" values and contraction/expansion coefficients

XS line has cross-section data

SZ line has elevation values for eight points

XX line has distance values for eight points

More detail on the input file is given as comment lines in the program listing.

The program generates two output files. The general output file is called STDSTEP.OUT, and file STDSTEP.PLT is for plotting profiles.

The distance between cross-sections is computed as

$$\Delta X = \frac{L_{Lt}Q_{Lt} + L_{Ch}Q_{Ch} + L_{Rt}Q_{Rt}}{Q_{Lt} + Q_{Ch} + Q_{Rt}}$$

```
TEST PROFILE FOR ROCK CREEK USING STD STEP METHOD OF COMPUTATION
SC     00      0.004    0.004    0.004      -.1       -1.      -1.
QT             2000.    20000    50000     2000.    20000.   50000.
MN             0.10     0.05     0.08      .15       0.3
XS             1.0      8.0      0         0         0
XZ     300.    280.     275.     265.      265.      275.     280.     300.
XX     0.      50.      200.     210.      260.      270.     420.     450.
XS             2.0      0.0      200.      200.      200.     0.8      0.95
MN             0        0        0         .3        0.5
XS             3.0      8.0      405.      400.      395.
XZ     301.    277.     276.5    266.5     266.0     277.     278.     302.
XX     100.    150.     200.     210.0     260.0     270.     320.     350.
XS             4.0      0.       60.       60.       60.      0.2      1.0
XS             5.0      8.       100.      100.      100.
XZ     301.    281.0    277.0    267.      268.      278.     282.     302.
XX     0.      40.      180.0    200.      270.      280.     400.     430.
MN             0        0        0         .15       0.3
XS             6.0      8.0      205.      200.      195.
XZ     302.    282.     278.     268.      269.      278.     283.     303.
XX     0.      20.      220.     240.      300.      320.     400.     430.
QT             1500.    15000.   35000.    1500.     15000.   35000.
XS             7.0      8.       205.      200.      195.
XZ     303.    283.     278.     270.      269.      278.     283.     303.
XX     0.      20.      120.     130.      180.      200.     300.     350.
XS             8.0      0.       205.      200.      195.     0.8      0.95
XS             9.0      0.       206.      200.      200.     0.7      1.00
XS             10.      0.       204.      200.      198.     0.7      1.00
XS             -6.0
QT             500.     5000.    15000     500.      5000.    15000.
MN             0.08     0.05     0.08      .15       0.3
XS             11.      8.       100.      100.      100.
XZ     303.    283.     278.     270.      269.      278.     283.     303.
XX     0.      20.      120.     130.      160.      170.     270.     300.
XS             12.0     0.       105.      100.      95.      0.8      1.00
XS             13.0     0.       106.      100.      100.     0.7      1.00
XS             14.0     0.       104.      100.      98.      0.7      1.00
```

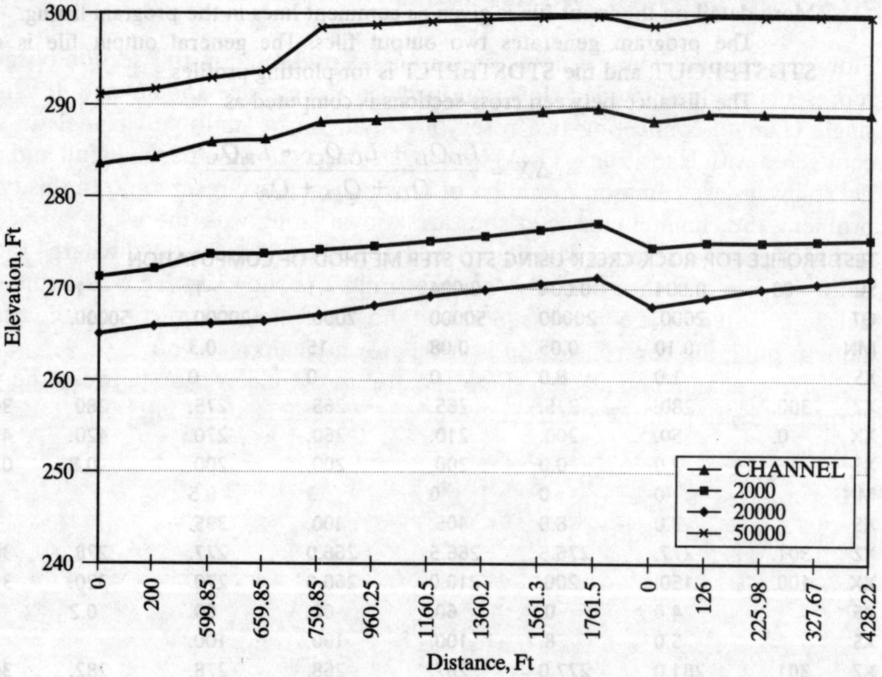
300
290
280
270
260
250
240
Elevation, Ft
CHANNEL
2000
20000
50000
0
200
599.85
659.85
759.85
960.25
1160.5
1360.2
1561.3
1761.5
0
126
225.98
327.67
428.22
Distance, Ft

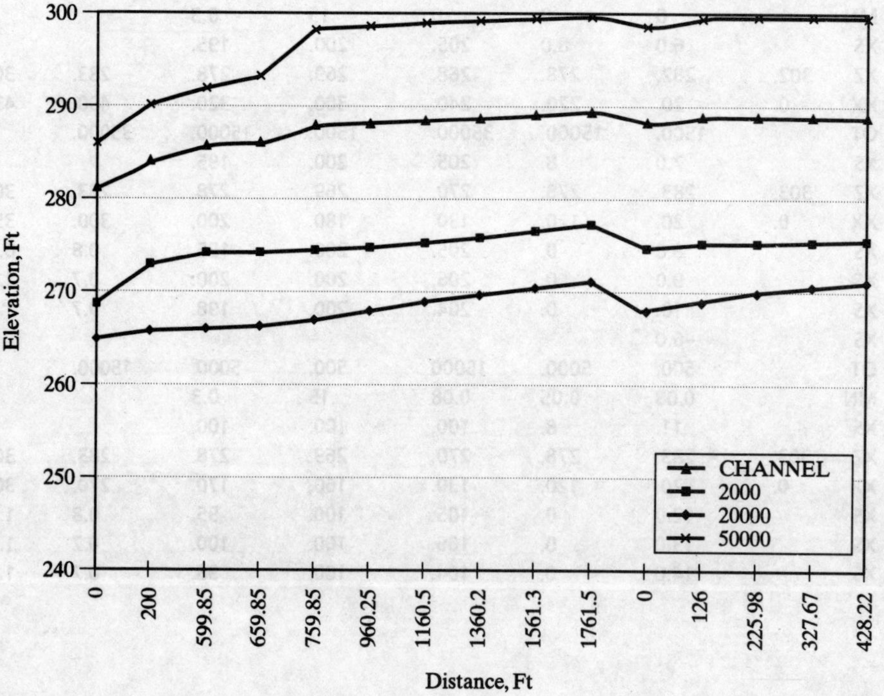
300
290
280
270
260
250
240
Elevation, Ft
CHANNEL
2000
20000
50000
0
200
599.85
659.85
759.85
960.25
1160.5
1360.2
1561.3
1761.5
0
126
225.98
327.67
428.22
Distance, Ft

5.7.4 Flow between Two Reservoirs

Flow in a channel connecting two reservoirs is presented in this section because it represents gradually varied flow that includes an entrance and an exit. It can be a single channel connecting two reservoirs or it can be multiple channels in series connected with transitions. Cross-drainage culverts flowing partially full and channel spillways are common examples of flow between two reservoirs. In this type of problem, the channel characteristics are known along with the water surface elevation of the two reservoirs. The procedure used to solve for the discharge rate depends on whether the channel is steep or mild. As shown in Fig. 5.19, if the channel slope is steep, the control section is at the upstream reservoir, and if the channel slope is mild, the control section is at the downstream reservoir.

For a channel on a steep slope, the difference in elevation between the water surface elevation in the upstream reservoir and the channel inlet invert elevation (H_1) is

$$H_1 = (1 + K_m)\frac{V_c^2}{2g} + y_c \tag{5.45}$$

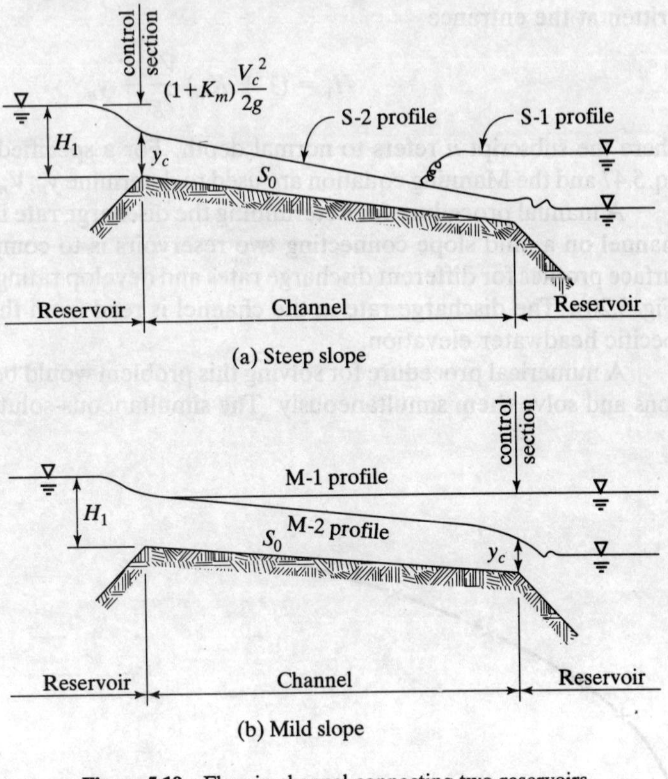

Figure 5.19 Flow in channel connecting two reservoirs.

where K_m is the entrance loss coefficient and subscript c indicates critical velocity and depth. The relationship between the channel geometry and the critical velocity head is given in Eq. 5.14 for any shape channel. For a rectangular channel, Eq. 5.45 reduces to

$$H_1 = (1 + K_m)\frac{y_c}{2} + y_c = (1.5 + 0.5K_m)y_c \tag{5.46}$$

After the value of y_c has been determined, A_c, V_c, and Q can be computed.

A hydraulic jump occurs in the transition from supercritical to subcritical flow (Fig. 5.19a). For a low tailwater elevation (elevation of the downstream reservoir), the hydraulic jump occurs in the reservoir and is a submerged jump. For a high tailwater elevation, the jump occurs in the channel. The higher the tailwater elevation, the further upstream the jump occurs in the channel. As long as the jump occurs downstream of the control section, the tailwater elevation does not affect the discharge rate in the channel.

If the channel is on a mild slope, the control section is at the outlet, and water surface profile(s) will normally have to be computed to determine the discharge rate in the channel. If uniform flow occurs in the upper reach of the channel, it is called a hydraulically long channel and the following energy equation can be written at the entrance

$$H_1 = (1 + K_m)\frac{V_n^2}{2g} + y_n \tag{5.47}$$

where the subscript n refers to normal depth. For a specified headwater elevation, Eq. 5.47 and the Manning equation are used to determine y_n, V_n, and Q in the channel.

A manual procedure for determining the discharge rate in a hydraulically short channel on a mild slope connecting two reservoirs is to compute a series of water surface profiles for different discharge rates and develop rating curve for the channel (Fig. 5.20). The discharge rate in the channel is read from the rating curve for the specific headwater elevation.

A numerical procedure for solving this problem would be to write all the equations and solve them simultaneously. The simultaneous-solution procedure will be

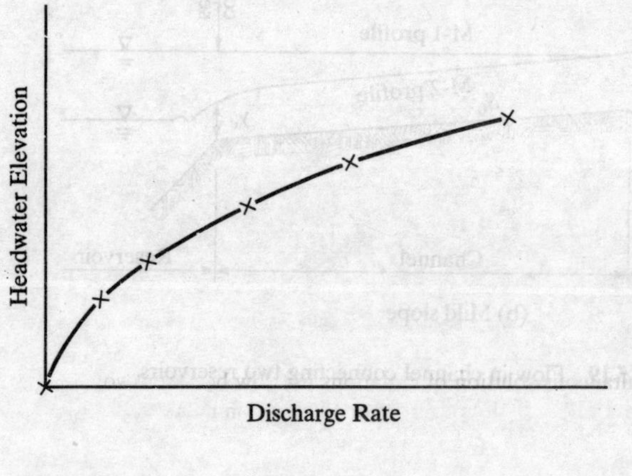

Figure 5.20 Rating curve for a hydraulically short channel on a mild slope.

discussed in this section, because it is also applicable to split flow around an island and flow in channel networks.

Figure 5.21 is a sketch of a hydraulically short channel on a mild slope connecting two reservoirs. Seven cross-sections are used to describe the channel geometry. There are six unknowns—the water surface elevation at cross-sections 2 through 6 and the discharge rate. The water surface elevations at cross-sections 1 and 7 are given as boundary conditions. The six energy equations needed to solve for the six unknowns are obtained by writing Eq. 5.43 for each of the six reaches shown in Fig. 5.21.

$$F_1 = \frac{\alpha_2(1 + K_{m1})}{2g}\left(\frac{Q}{A_2}\right)^2 - \left(\frac{Q}{K_2}\right)^2 \frac{\Delta x_1}{2} + WS_2 - WS_1 = 0 \qquad \textbf{(5.48a)}$$

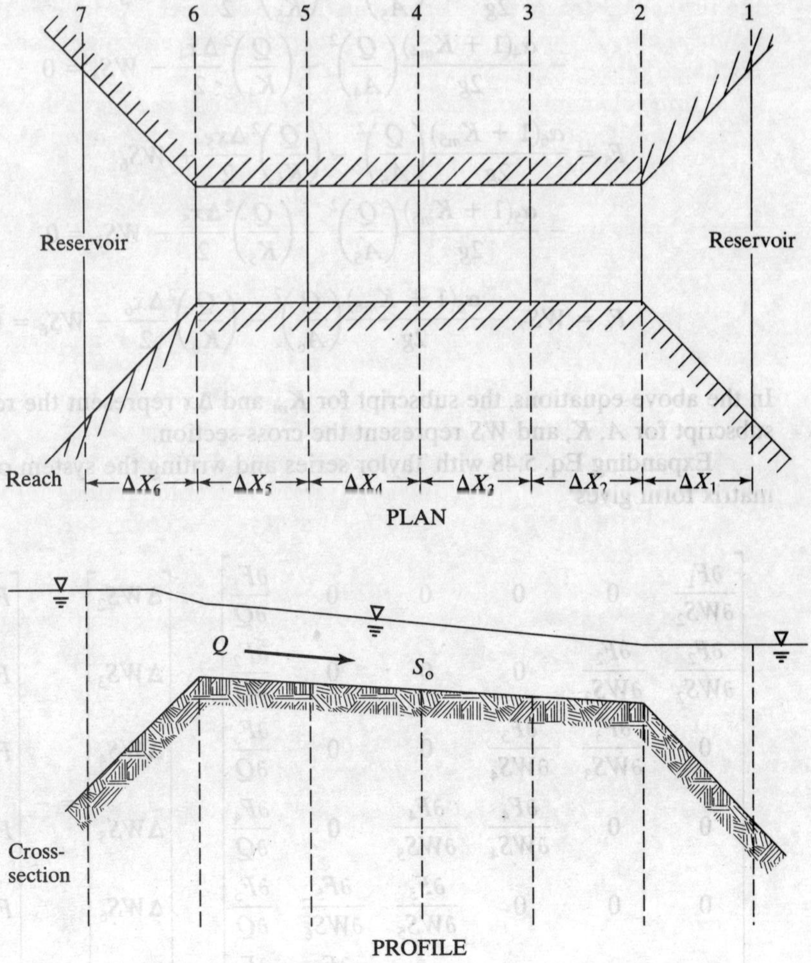

Figure 5.21 Simultaneous-solution of equations for flow between two reservoirs connected with a hydraulically short channel on mild slope.

$$F_2 = \frac{\alpha_3(1 + K_{m2})}{2g}\left(\frac{Q}{A_3}\right)^2 - \left(\frac{Q}{K_3}\right)^2\frac{\Delta x_2}{2} + WS_3$$
$$- \frac{\alpha_2(1 + K_{m2})}{2g}\left(\frac{Q}{A_2}\right)^2 - \left(\frac{Q}{K_2}\right)^2\frac{\Delta x_2}{2} - WS_2 = 0 \qquad \textbf{(5.48b)}$$

$$F_3 = \frac{\alpha_4(1 + K_{m3})}{2g}\left(\frac{Q}{A_4}\right)^2 - \left(\frac{Q}{K_4}\right)^2\frac{\Delta x_3}{2} + WS_4$$
$$- \frac{\alpha_3(1 + K_{m3})}{2g}\left(\frac{Q}{A_3}\right)^2 - \left(\frac{Q}{K_3}\right)^2\frac{\Delta x_3}{2} - WS_3 = 0 \qquad \textbf{(5.48c)}$$

$$F_4 = \frac{\alpha_5(1 + K_{m4})}{2g}\left(\frac{Q}{A_5}\right)^2 - \left(\frac{Q}{K_5}\right)^2\frac{\Delta x_4}{2} + WS_5$$
$$- \frac{\alpha_4(1 + K_{m4})}{2g}\left(\frac{Q}{A_4}\right)^2 - \left(\frac{Q}{K_4}\right)^2\frac{\Delta x_4}{2} - WS_4 = 0 \qquad \textbf{(5.48d)}$$

$$F_5 = \frac{\alpha_6(1 + K_{m5})}{2g}\left(\frac{Q}{A_6}\right)^2 - \left(\frac{Q}{K_6}\right)^2\frac{\Delta x_5}{2} + WS_6$$
$$- \frac{\alpha_5(1 + K_{m5})}{2g}\left(\frac{Q}{A_5}\right)^2 - \left(\frac{Q}{K_5}\right)^2\frac{\Delta x_5}{2} - WS_5 = 0 \qquad \textbf{(5.48e)}$$

$$F_6 = WS_7 - \frac{\alpha_6(1 + K_{m6})}{2g}\left(\frac{Q}{A_6}\right)^2 - \left(\frac{Q}{K_6}\right)^2\frac{\Delta x_6}{2} - WS_6 = 0 \qquad \textbf{(5.48f)}$$

In the above equations, the subscript for K_m and Δx represent the reach while the subscript for A, K, and WS represent the cross-section.

Expanding Eq. 5.48 with Taylor series and writing the system of equations in matrix form gives

$$\begin{bmatrix} \frac{\partial F_1}{\partial WS_2} & 0 & 0 & 0 & 0 & \frac{\partial F_1}{\partial Q} \\ \frac{\partial F_2}{\partial WS_2} & \frac{\partial F_2}{\partial WS_3} & 0 & 0 & 0 & \frac{\partial F_2}{\partial Q} \\ 0 & \frac{\partial F_3}{\partial WS_3} & \frac{\partial F_3}{\partial WS_4} & 0 & 0 & \frac{\partial F_3}{\partial Q} \\ 0 & 0 & \frac{\partial F_4}{\partial WS_4} & \frac{\partial F_4}{\partial WS_5} & 0 & \frac{\partial F_4}{\partial Q} \\ 0 & 0 & 0 & \frac{\partial F_5}{\partial WS_5} & \frac{\partial F_5}{\partial WS_6} & \frac{\partial F_5}{\partial Q} \\ 0 & 0 & 0 & 0 & \frac{\partial F_6}{\partial WS_6} & \frac{\partial F_6}{\partial Q} \end{bmatrix} \begin{bmatrix} \Delta WS_2 \\ \Delta WS_3 \\ \Delta WS_4 \\ \Delta WS_5 \\ \Delta WS_6 \\ \Delta Q \end{bmatrix} = - \begin{bmatrix} F_1 \\ F_2 \\ F_3 \\ F_4 \\ F_5 \\ F_6 \end{bmatrix} \qquad \textbf{(5.49)}$$

The partials in the Jacobian [J] can be evaluated numerically by computing the value of the function (F) for two different values of the dependent variable (Q or WS). The partials can be approximated by

$$\frac{\partial F}{\partial WS} = \frac{\Delta F}{\Delta WS} \tag{5.50}$$

and

$$\frac{\partial F}{\partial Q} = \frac{\Delta F}{\Delta Q} \tag{5.51}$$

The column vector of correction terms [Δ] can be computed

$$[\Delta] = [J]^{-1}[F] \tag{5.52}$$

The new estimate of water surface elevation (WS^+) is

$$WS_i^+ = WS_i + \Delta WS_i \tag{5.53}$$

and discharge rate (Q^+) is

$$Q^+ = Q + \Delta Q \tag{5.54}$$

This iterative procedure is continued until the correction terms are within specified limits.

Example 5.19 Chute Spillway on a Steep Slope

Outflow from a reservoir is through a concrete chute spillway on a steep slope. The channel is trapezoidal in shape and has a bottom width of 10.0 m with 2H:1V side slopes. Determine the discharge rate through the spillway when the head (H) on the spillway is 3.0 m. The minor loss coefficient for the inlet is 0.6.

Solution Critical depth occurs at the entrance to the chute. Neglecting the friction loss in the transition between the reservoir and the chute, the energy equation is

$$H = (1 + K_m)\frac{V_c^2}{2g} + y_c$$

where K_m is the minor loss coefficient.

When critical depth occurs in an open channel, the velocity head is equal to one-half the hydraulic depth (Eq. 5.15) and the energy becomes

$$H = (1 + K_m)\frac{D_c}{2} + y_c$$

where

$$D_c = \frac{A}{T} = (By_c + 2y_c^2)/(B + 4y_c)$$

Substituting for the hydraulic depth and simplifying the energy equation reduces to

$$y_c^2 + 1.07y_c - 5.36 = 0$$

Solving for y_c yields

$$y_c = 1.84 \text{ m}$$

$$A = 25.18 \text{ m}^2$$

$$D_c = 1.45 \text{ m}$$

$$\frac{V_c^2}{2g} = 0.725$$

$$V_c = 3.77 \text{ mps}$$

$$Q = 25.18 \times 3.77 = 95.0 \text{ cms}$$

Example 5.20 Hydraulically Long Channel

A rectangular concrete channel ($n = 0.015$) on a mild slope ($s = 0.001$) extends 3,000 m from a reservoir. The channel is hydraulically long and normal depth occurs downstream of the entrance. The entrance loss coefficient is 0.6, and the channel bottom width is 30 m. Determine the discharge rate for a head (H) of 10 m.

Solution Writing the energy equation between the reservoir and the entrance to the channel gives

$$H = (1 + K_m)\frac{V_n^2}{2g} + y_n$$

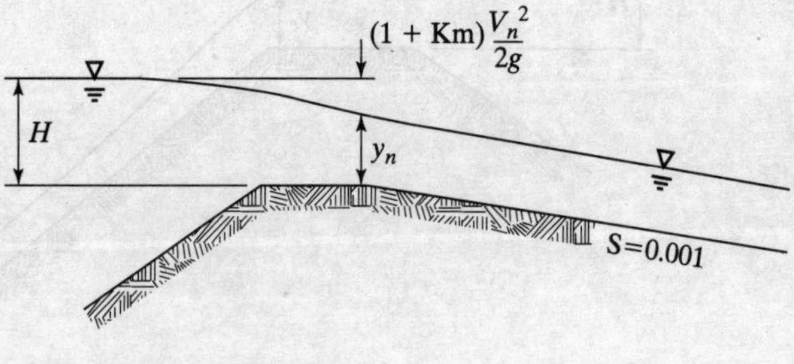

Manning equation for uniform flow is

$$V_n = \frac{1.00}{n} R^{2/3} S_o^{1/2}$$

combining the two equations

$$H = \frac{(1 + K_m)}{2g} \frac{R^{4/3}}{n^2} S_o + y_n$$

where

$$R = \frac{A}{P} = \frac{By_n}{(B + 2y_n)}$$

Substituting for R

$$0 = \frac{(1 + K_m)}{2g\,n^2} \left[\frac{By_n}{B + 2y_n} \right]^{4/3} S_o + y_n - H$$

or

$$0 = 0.362 \left[\frac{30y_n}{30 + 2y_n} \right]^{4/3} + (y_n - 10)$$

Solving by trial and error yields

$$y_n = 7.08 \text{ m}$$

$$A = 212.4 \text{ m}^2$$

$$R = 4.81 \text{ m}$$

$$V_n = 6.01 \text{ mps}$$

$$Q = 1{,}276 \text{ cms}$$

$$y_c = 5.69 \text{ m (Eq. 5.19)}$$

Flow is subcritical.

Example 5.21 Hydraulically Short Channel

A grass-lined spillway channel extends from a reservoir as shown below. Flow in the spillway channel is subcritical for the first 400 ft where the channel becomes steep and critical depth occurs. An eight-point cross-section for the grass spillway is shown below in Section 1. The channel has a slope of 0.002, a Manning "n" value of 0.03 with n = 0.04 in the overbanks, and an entrance loss coefficient of 0.6.

Determine the discharge rate in the spillway for a reservoir water surface elevation of 296.4 ft.

Solution Use the standard step program developed in Ex. 5.18 and compute water surface profiles for discharge rates of 2,000, 4,000, 6,000, 8,000, 10,000, 12,000, and 14,000 cfs. Input file (STDSTEP.DAT) for the program using nine cross-sections spaced at 50-ft intervals is listed below.

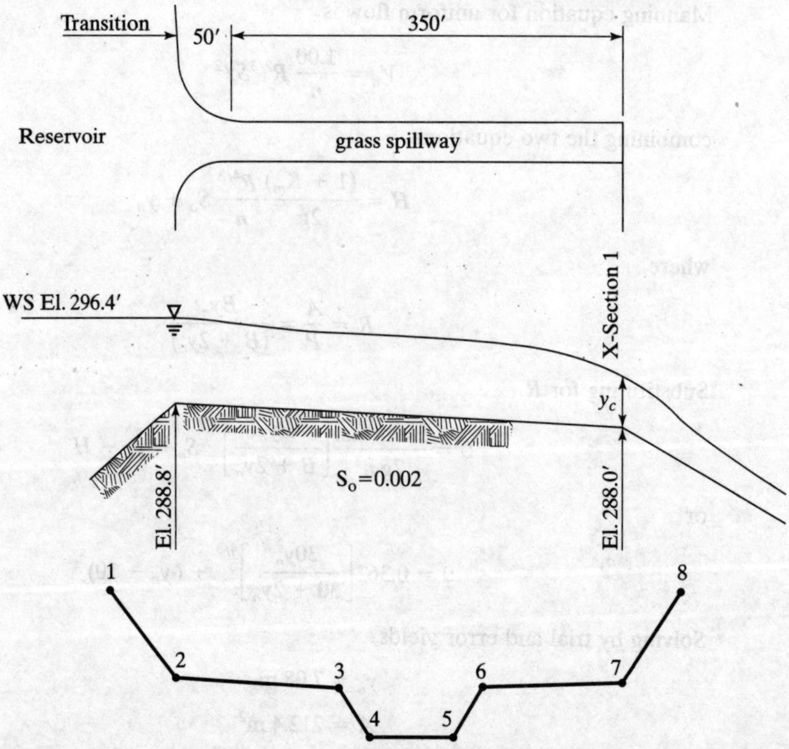

X-Section 1

xx	1	2	3	4	5	6	7	8
z	300	291	290	288	288	290	291	300
x	0	50	150	160	240	250	350	400

RESERVOIR RATING CURVE FOR EXAMPLE PROBLEM 5.21

SC	00	–1.	–1.	–1.	–1.0	–1.	–1.	–1.
QT		2000.	4000.	6000.	8880.8	10000.	12000.	14000.
MN		0.04	0.03	0.04	0.15	0.3		
XS		1.0	8.0	0	0	0		
XZ	300.	291.	290.	288.	288.	290.	291.	300.
XX	0.	50.	150.	160.	240.	250.	350.	400.
XS		1.5	0.0	50.	50.	50.	0.1	1.00
XS		2.0	0.0	50.	50.	50.	0.1	
XS		2.5	0.	50.	50.	50.	0.1	1.00
XS		3.0	0.	50.	50.	50.	0.1	1.00
XS		3.5	0.0	50.	50.	50.	0.1	
XS		4.0	0.	50.	50.	50.	0.1	
XS		4.5	0.	50.	50.	50.	0.1	1.00
MN					0.6	0.8		
XS		5.0	8.	50.	50.	50.		
XZ	310.	251.	250.	249.	249.	250.	251.	310.
XX	0.	50.	150.	160.	240.	250.	350.	400.

The computed water surface elevations at the reservoirs are listed below.

Q cfs	Reservoir $WS\ El$ ft
2,000	292.79
4,000	294.12
6,000	295.15
8,000	296.04
10,000	296.84
12,000	297.57
14,000	298.24

The rating curve was plotted and the discharge rate for a water surface elevation of 296.4 ft was read from the curve as 9,000 cfs.

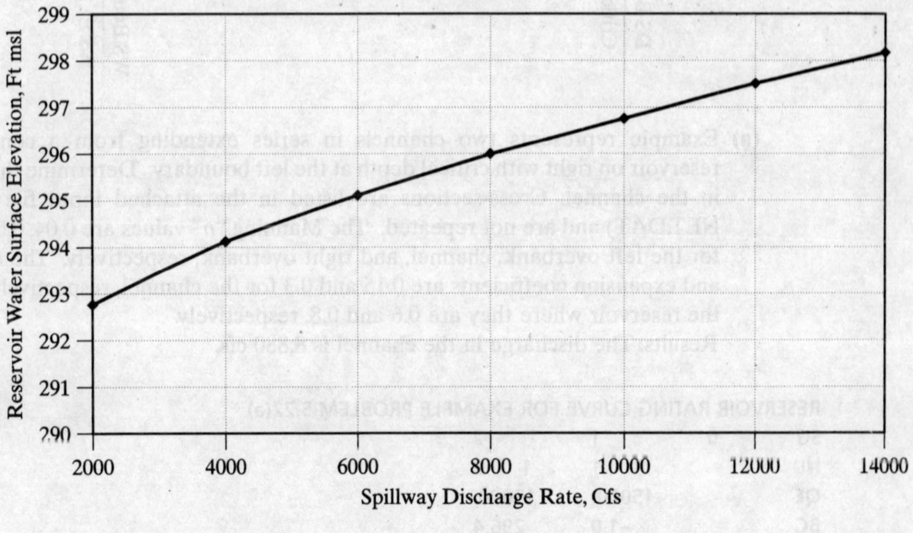

Example 5.22 Open Channel Networks

The standard step model (STDSTEP.FOR from Ex. 5.18) was modified for simultaneous solution of the flow equations by adding Eqs. 5.49 through 5.54 to the model. The new model is called OPNCHNET.FOR and was written to handle 2 and 3 channel junction nodes. Four additional lines of input data were added: the SG line includes a unit code (0) and a listing of the channel segment numbers, the ND line lists the downstream and upstream node numbers for each segment listed on the SG line, the QE line lists the estimated initial discharge rate for each channel segment, and the BC line lists the boundary conditions for each boundary node (critical depth, normal depth, constant discharge, or constant water surface elevation). Boundary nodes have zero node numbers. The XS line was modified to include the channel segment numbers.

To demonstrate the use of the model, two example problems are presented. The model is for subcritical flow and cross-sections for each reach are entered starting downstream and continue upstream. The last cross-section in a reach is labeled 99.

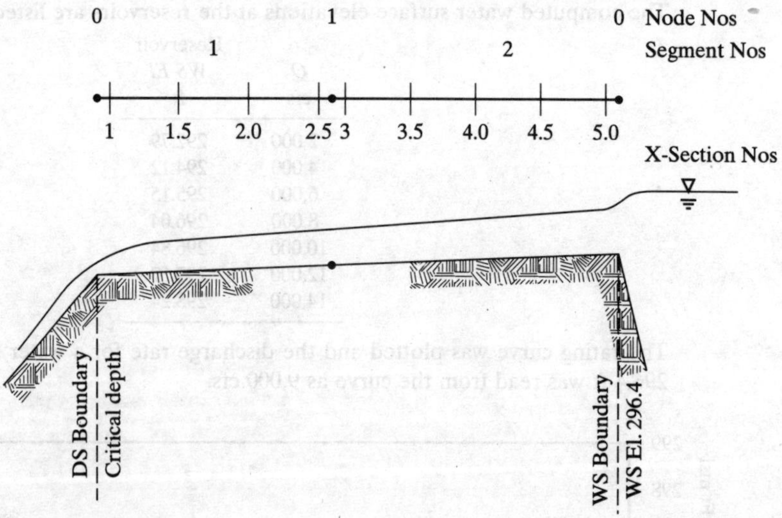

(a) Example represents two channels in series extending from a constant head reservoir on right with critical depth at the left boundary. Determine the discharge in the channel. Cross-sections are listed in the attached input file (OPNCHNET.DAT) and are not repeated. The Manning "*n*" values are 0.04, 0.03, and 0.04 for the left overbank, channel, and right overbank, respectively. The contraction and expansion coefficients are 0.15 and 0.3 for the channel, respectively, except at the reservoir where they are 0.6 and 0.8, respectively.
Results: The discharge in the channel is 8,880 cfs.

RESERVOIR RATING CURVE FOR EXAMPLE PROBLEM 5.22(a)

SG	0	1	2					
ND		0 1	1 0					
QE		15000.	15000.					
BC		−1.0	296.4					
MN		0.04	0.03	0.04	0.15	0.3		
XS	1	1.0	8.0	0	0	0		
XZ	300.	291.	290.	288.	288.	290.	291.	300.
XX	0.	50.	150.	160.	240.	250.	350.	400.
XS	1	1.5	0.0	50.	50.	50.	0.1	1.00
XS	1	2.0	0.0	50.	50.	50.	0.1	
XS	99	2.5	0.	50.	50.	50.	0.1	1.00
XS	2	3.0	0.	50.	50.	50.	0.1	1.00
XS	2	3.5	0.0	50.	50.	50.	0.1	
XS	2	4.0	0.	50.	50.	50.	0.1	
XS	2	4.5	0.	50.	50.	50.	0.1	1.00
MN					0.6	0.8		
XS	99	5.0	8.	50.	50.	50.		
XZ	310.	251.	250.	249.	249.	250.	251.	310.
XX	0.	50.	150.	160.	240.	250.	350.	400.

SUMMARY OUTPUT DATA

CROSS-SECTION ID NO	DISCHARGE RATE	TOTAL LENGTH	CHANNEL ELEVATION	WATER ELEVATION
1.000	8880.753	.000	288.000	293.107
1.500	8880.753	50.000	288.100	294.151
2.000	8880.753	100.000	288.200	294.368
2.500	8880.753	150.000	288.300	294.554
3.000	8880.753	200.000	288.400	294.723
3.500	8880.753	250.000	288.500	294.878
4.000	8880.753	300.000	288.600	295.024
4.500	8880.753	350.000	288.700	295.163
5.000	8880.753	400.000	249.000	296.400

(b) Same as problem (a) except the upstream boundary has been changed to a constant discharge boundary of 12,000 cfs. Compute the water surface elevation in the reservoir. The input file is attached with the boundary at the reşervoir on BC left blank indicating that the discharge listed on the QE line is to remain constant. Results: The water surface elevation in the reservoir is 297.57 ft.

RESERVOIR RATING CURVE FOR EXAMPLE PROBLEM 5.22(b)

SG	0	1	2					
ND		0 1	1 0					
QE		12000.	12000.					
BC		−1.0						
MN		0.04	0.03	0.04	0.15	0.3		
XS	1	1.0	8.0	0	0	0		
XZ	300.	291.	290.	288.	288.	290.	291.	300.
XX	0.	50.	150.	160.	240.	250.	350.	400.
XS	1	1.5	0.0	50.	50.	50.	0.1	1.00
XS	1	2.0	0.0	50.	50.	50.	0.1	
XS	99	2.5	0.	50.	50.	50.	0.1	1.00
XS	2	3.0	0.	50.	50.	50.	0.1	1.00
XS	2	3.5	0.0	50.	50.	50.	0.1	
XS	2	4.0	0.	50.	50.	50.	0.1	
XS	2	4.5	0.	50.	50.	50.	0.1	1.00
MN					0.6	0.8		
XS	99	5.0	8.	50.	50.	50.		
XZ	310.	251.	250.	249.	249.	250.	251.	310.
XX	0.	50.	150.	160.	240.	250.	350.	400.

SUMMARY OUTPUT DATA

CROSS-SECTION ID NO	DISCHARGE RATE	TOTAL LENGTH	CHANNEL ELEVATION	WATER ELEVATION
1.000	12000.000	.000	288.000	293.802
1.500	12000.000	50.000	288.100	294.971
2.000	12000.000	100.000	288.200	295.210
2.500	12000.000	150.000	288.300	295.414
3.000	12000.000	200.000	288.400	295.597
3.500	12000.000	250.000	288.500	295.764
4.000	12000.000	300.000	288.600	295.921
4.500	12000.000	350.000	288.700	296.070
5.000	12000.000	400.000	249.000	297.566

5.7.5 Flow Over a Roadway

Figure 5.22 shows the flow of water over a roadway for a low profile road (b) and a high profile road (c). If the downstream tailwater depth (y_1) and the discharge rate (Q) are known, then the depth of water on the roadway can be determined by writing the energy equation between Section 1 and 2 or

$$\frac{Q^2}{A_1\,2g} + y_1 = \frac{Q^2}{A_2\,2g} + y_2 + h \tag{5.55}$$

for y_2 greater than critical depth. The headloss and the channel slope terms were assumed to be approximately equal and were omitted from Eq. 5.55. h in Eq. 5.55 is the height of the roadway above the channel bottom. If the roadway has a high profile (c) and critical depth occurs on the roadway, then the upstream water depth (y_3) is independent of the tailwater depth (y_1) and is determined

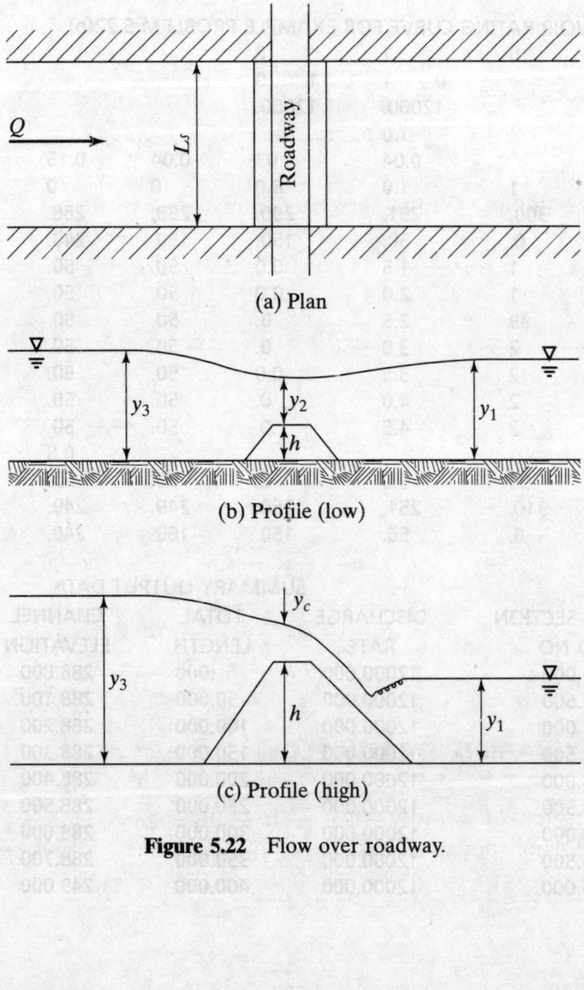

Figure 5.22 Flow over roadway.

from the weir equation

$$Q = K_B L_w \sqrt{2g} H^{3/2} \tag{5.56}$$

where K_B is the weir coefficient, L_w is the length of the weir, and H is $(y_3 - h)$. The flow in Fig. 5.22(b) represents gradually varied flow, while that in Fig. 5.22(c) includes supercritical flow and a hydraulic jump and is termed rapidly varied steady flow.

Example 5.23 Flow Over Roadway

A roadway is to cross a wide open channel where the discharge rate is 50 cfs/ft of width and the water depth is 10.0 ft. The water is to flow over the roadway. Determine the minimum height (h) of the roadway such that critical depth occurs at the crest. Neglect headloss and slope of the channel.

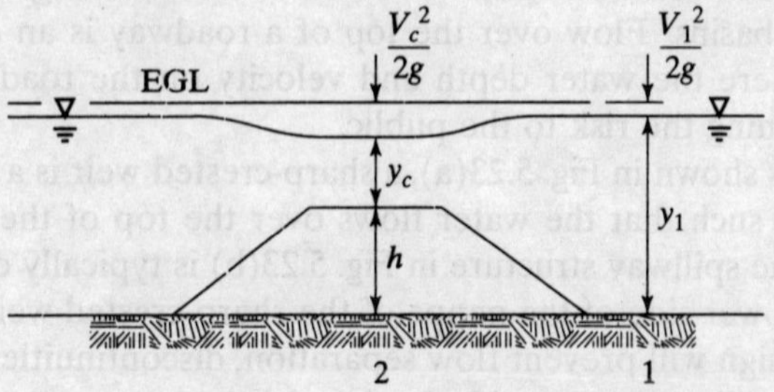

Solution Writing the energy equation between points 1 and 2

$$\frac{V_c^2}{2g} + y_c + h = \frac{V_1^2}{2g} + y_1$$

where

$$V_1 = \frac{q}{y_1} = \frac{50}{10} = 5 \text{ fps}$$

$$y_c = \sqrt[3]{\frac{q^2}{g}} = \left(\frac{50^2}{32.2}\right)^{1/3} = 4.27 \text{ ft}$$

$$\frac{y_c}{2} = \frac{V_c^2}{2g}$$

$$\frac{V_c^2}{2g} = 2.13 \text{ ft}$$

$$h = 10.39 - 4.27 - 2.13 = 4.0 \text{ ft}$$

5.8 RAPIDLY VARIED STEADY FLOW

Examples of rapidly varied flow include weirs, flow control structures, flow measuring devices, hydraulic jump, energy dissipations, and transitions for supercritical flow. Steady, rapidly varied flow occurs over a relatively short distance and losses due to boundary shear are generally small and are often neglected. Because the streamlines for rapidly varied flow are highly curved, hydrostatic pressure distribution cannot be assumed. The change in curvature may be so great that separation zones are formed that distort the velocity distribution making it difficult to define flow boundaries. Because of the complex flow patterns, most of the analyses for rapidly varied flow are based on physical model studies.

5.8.1 Weirs

Weirs can be used to measure the discharge rate in an open channel. They can also be used in water management to control the discharge rate from reservoirs and detention basins. Flow over the top of a roadway is an example of a broad-crested weir where the water depth and velocity on the roadway is of major concern in determining the risk to the public.

As shown in Fig. 5.23(a), a sharp-crested weir is a weir plate placed across the channel such that the water flows over the top of the plate into the downstream pool. The spillway structure in Fig. 5.23(b) is typically designed to fit the curvature of the lower side of the nappe of the sharp-crested weir for a specific design head. This design will prevent flow separation, discontinuities in pressure, and cavitation on the spillway surface. Neglecting the upstream approach velocity, the discharge

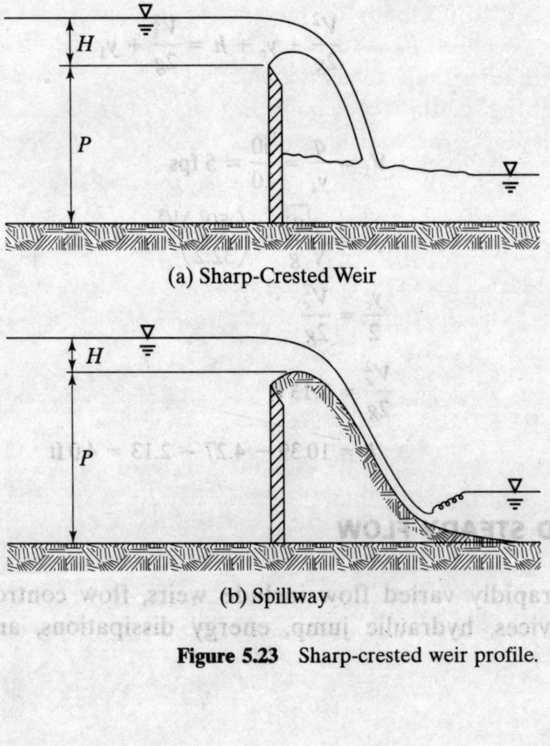

Figure 5.23 Sharp-crested weir profile.

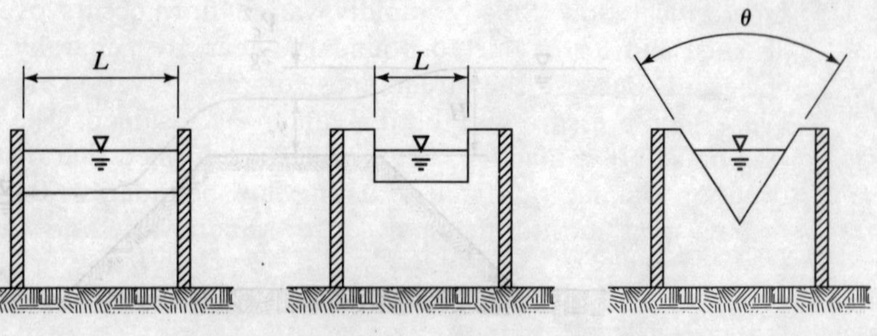

Figure 5.24 Sharp-crested weir geometry.

equation for both structures is

$$Q = K_R \sqrt{2g}\, L_w\, H^{3/2} \tag{5.57}$$

where K_R is the discharge coefficient. For the sharp-crested weir, the discharge coefficient is

$$K_R = 0.40 + 0.05\frac{H}{P} \tag{5.58}$$

Because the head (H) is measured differently (Fig. 5.23), the discharge coefficient for the spillway will normally be larger than that given by Eq. 5.58. Figure 5.24 shows the cross-sections for a rectangular weir that spans the entire channel, a rectangular weir that partially spans the channel and triangular weir. If the weir does not span the entire width of the channel, the effective length of the weir is reduced by end contractions. The weir length to be used in Eq. 5.57 is

$$L_w = L - 0.2H \tag{5.59}$$

for $L/H > 3$.

A triangular weir can be used to measure or control the discharge over a wide range of heads. The discharge for a triangular weir is given as

$$Q = \frac{8}{15} K_T \sqrt{2g} \tan\left(\frac{\theta}{2}\right) H^{5/2} \tag{5.60}$$

The value of K_T ranges from about 0.58 to 0.60 for values of ranging from 20° to 90° and head values greater than 0.3 m.

If the weir is long in the direction of flow (broad-crested), critical flow occurs at the crest (Fig. 5.25). If the upstream velocity head is negligible

$$\frac{V_c^2}{2g} = \frac{H}{3}$$

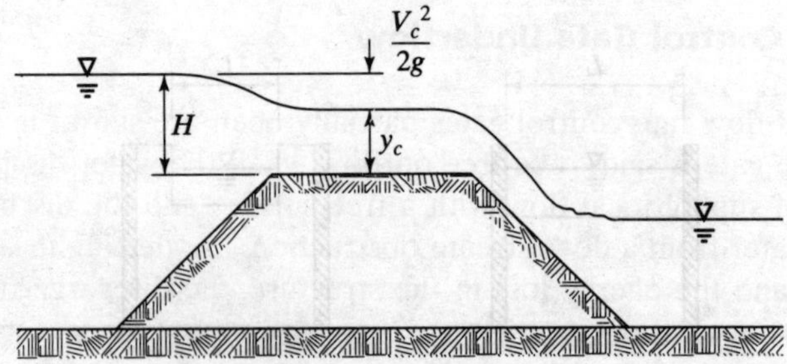

Figure 5.25 Broad-crested weir.

and

$$y_c = \tfrac{2}{3}H$$

the flow rate is

$$Q = K_B L_w \sqrt{2g}\, H^{3/2} \tag{5.61}$$

where K_B has a theoretical value of 0.38. For weir flow over a roadway, the value of K_B will need to be reduced for vegetation, guard rails and other obstructions to the flow.

Example 5.24 Broad-Crested Weir

During major flood events water flows over the top of a roadway. Determine the head (H) on the broad-crested weir for a discharge rate of 10,000 cfs if the overflow section of roadway is horizontal and 600 ft long. Assume a weir coefficient (K_B in Eq. 5.61) of 0.33.

$$Q = K_B L_w \sqrt{2g} H^{3/2}$$

$$H = \left(\frac{Q}{K_B L_w \sqrt{2g}}\right)^{2/3} = \left(\frac{10{,}000}{0.33 \times 600\sqrt{64.4}}\right)^{2/3} = 3.41 \text{ ft}$$

Assume that a vehicle stalled on the roadway will be washed off the roadway into the downstream channel if the water depth exceeds 2.0 ft. Compute the water depth and velocity on the roadway. Will a stalled vehicle on the roadway remain on the road during the flood?

$$y_c = \sqrt[3]{\frac{q^2}{g}} = \sqrt[3]{\frac{\left(\frac{10{,}000}{600}\right)^2}{32.2}} = 2.05 \text{ ft}$$

$$V_c = \frac{q}{y_c'} = \frac{16.7}{2.05} = 8.1 \text{ fps}$$

The stalled vehicle will be washed off the roadway.

5.8.2 Control Gate Underflow

Typical flow rate control gates partially open are shown in Fig. 5.26. The flow under the gate is said to be free outflow when the water discharges from the gate as a jet of supercritical flow with a free surface and the discharge is not reduced by backwater from a downstream obstruction. Neglecting the upstream approach velocity and the energy loss in the structure, the discharge rate (Q) under the gate with a free outflow is

$$Q = C_c A\sqrt{2g(y_1 - y_2)} \tag{5.62}$$

where C_c is a contraction coefficient, A is the area under the gate, y_1 is the upstream depth, and y_2 is the downstream depth. The contraction coefficient is defined as

$$C_c = \frac{y_2}{h_g} \tag{5.63}$$

where h_g is the height of the gate opening.

5.8.3 Hydraulic Jump

A hydraulic jump occurs in the transition from supercritical to subcritical flow. The intense turbulence in the jump causes mixing, air entrainment, and energy dissipation. The hydraulic jump is often used downstream of spillways and drop structures to dissipate energy and prevent erosion in the downstream channel.

An hydraulic jump in a horizontal, rectangular channel is shown in Fig. 5.27. Applying the one-dimensional momentum equation to the hydraulic jump gives

$$\frac{\gamma\, b y_1^2}{2} - \frac{\gamma\, b y_2^2}{2} = \rho V_1 b y_1 (V_2 - V_1) \tag{5.64}$$

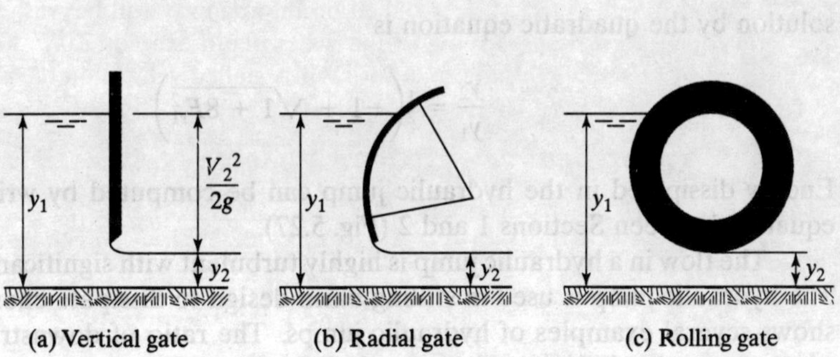

Figure 5.26 Underflow gates.

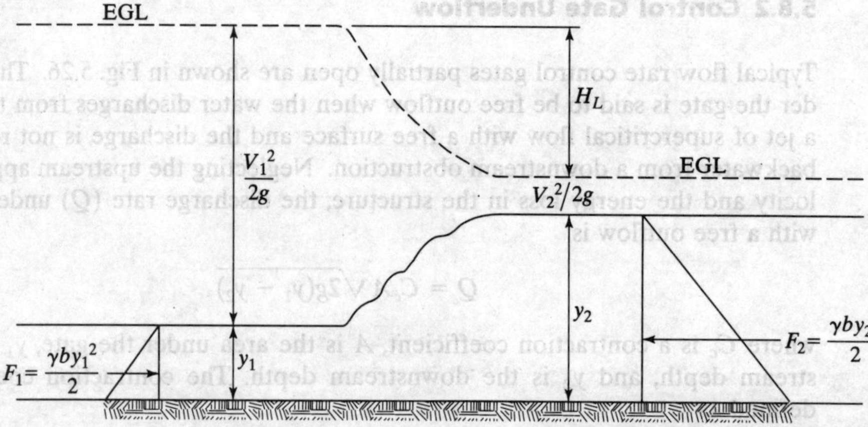

Figure 5.27 Hydraulic jump in a rectangular channel.

where b is the width of the channel. The continuity equation for flow is

$$y_1 b V_1 = y_2 b V_2 = Q \tag{5.65}$$

Combining Eqs. 5.64 and 5.65 gives

$$\frac{y_1^2}{2} - \frac{y_2^2}{2} = \frac{V_1^2 y_1}{g y_2}(y_1 - y_2) \tag{5.66}$$

Substituting, $F_{r1}^2 = V_1^2/g y_1$ on the right, factoring $(y_1 - y_2)$ from each side and rearranging the equation gives

$$\left(\frac{y_2}{y_1}\right)^2 + \frac{y_2}{y_1} - 2F_{r1}^2 = 0 \tag{5.67}$$

solution by the quadratic equation is

$$\frac{y_2}{y_1} = \tfrac{1}{2}\left(-1 + \sqrt{1 + 8F_{r1}^2}\right) \tag{5.68}$$

Energy dissipated in the hydraulic jump can be computed by writing the energy equation between Sections 1 and 2 (Fig. 5.27).

The flow in a hydraulic jump is highly turbulent with significant loss of energy. The hydraulic jump is used in stilling basin design to dissipate energy. Figure 5.28 shows several examples of hydraulic jumps. The ratio of downstream depth (y_2) to upstream depth (y_1) is given by the hydraulic jump equation (Eq. 5.68). The length of the jump can be determined from Fig. 5.29 based on the inflow Froude

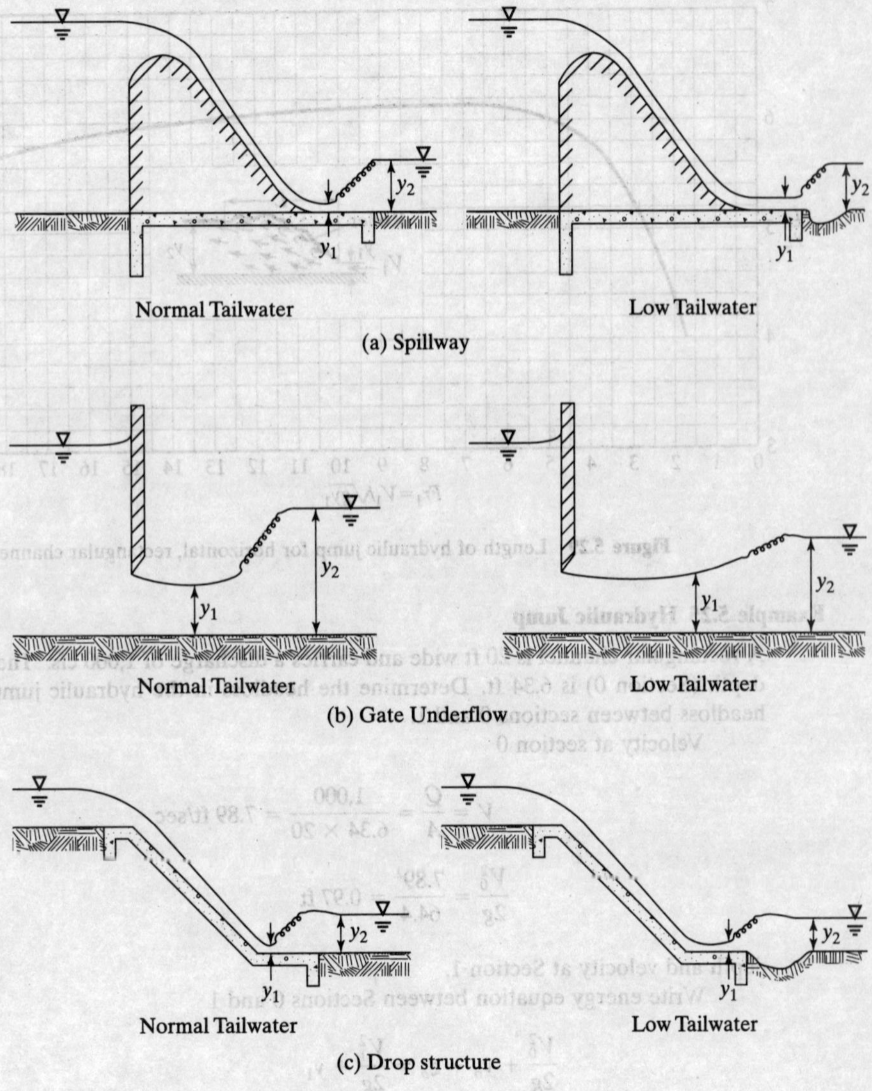

Figure 5.28 Location of hydraulic jump.

Number. The location of the hydraulic jump can be determined by water surface profile computations both upstream (supercritical flow) and downstream (subcritical flow) to identify the location where Eq. 5.68 is satisfied when the length of the jump is considered. As shown in Fig. 5.28, the hydraulic jump moves downstream when the tailwater elevation is low and the jump moves upstream at high tailwater elevations.

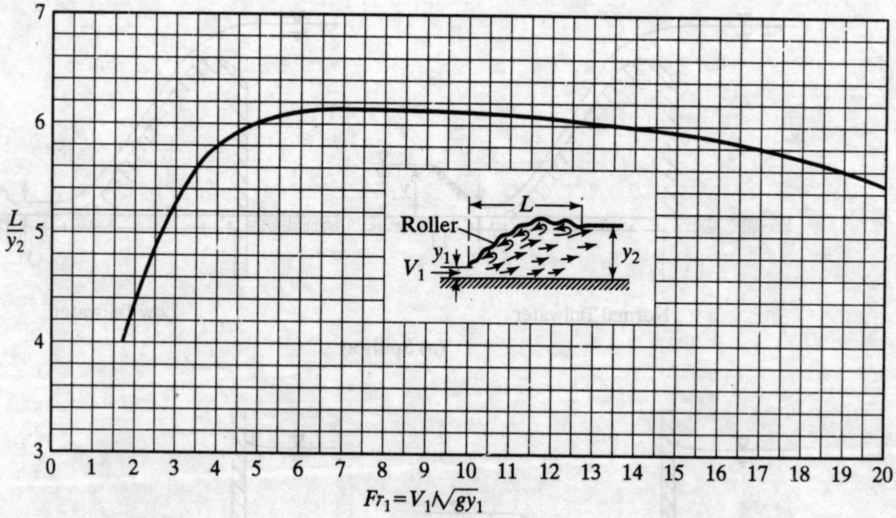

Figure 5.29 Length of hydraulic jump for horizontal, rectangular channel.

Example 5.25 Hydraulic Jump

A rectangular channel is 20 ft wide and carries a discharge of 1,000 cfs. The upstream depth (Section 0) is 6.34 ft. Determine the headloss in the hydraulic jump. Neglect headloss between sections 0 and 1.

Velocity at section 0

$$V = \frac{Q}{A} = \frac{1{,}000}{6.34 \times 20} = 7.89 \text{ ft/sec}$$

$$\frac{V_0^2}{2g} = \frac{7.89^2}{64.4} = 0.97 \text{ ft}$$

depth and velocity at Section 1.

Write energy equation between Sections 0 and 1

$$\frac{V_0^2}{2g} + y_o + z_o = \frac{V_1^2}{2g} + y_1$$

$$0.97 + 6.34 + 20 = \frac{Q^2}{(20y_1)^2\, 64.4} + y_1$$

$$27.31 = \frac{38.82}{y_1^2} + y_1$$

$$y_1 = 1.22 \text{ ft}$$

$$V_1 = \frac{Q}{A} = \frac{1{,}000}{20 \times 1.22} = 41.0 \text{ fps}$$

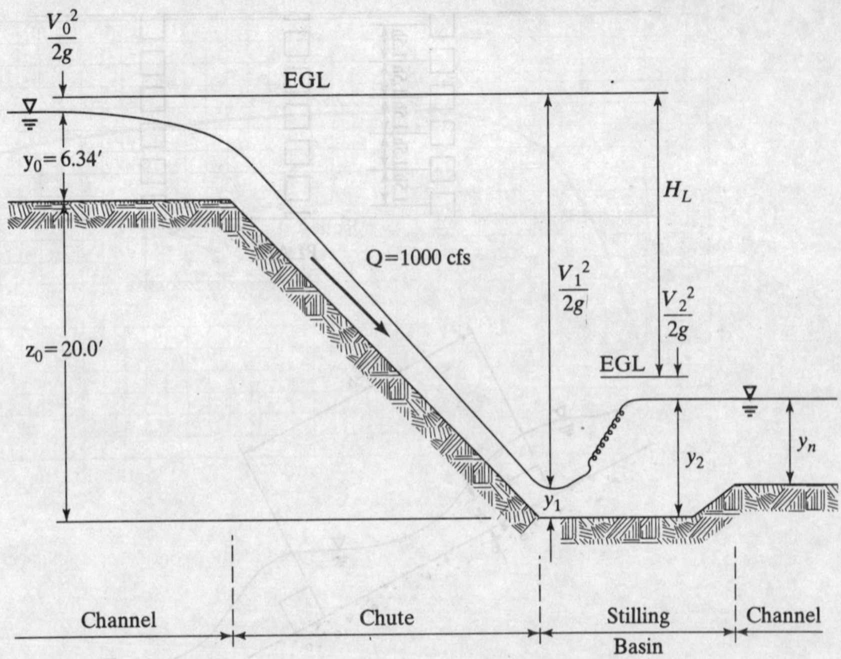

Example 5.25 Hydraulic jump.

$$Fr_1 = \frac{V_1}{\sqrt{gy_1}} = \frac{41.0}{\sqrt{32.2 \times 1.22}} = 6.54$$

From Eq. 5.68

$$y_2 = \frac{y_1}{2}\left(-1 + \sqrt{1 + 8Fr_1^2}\right) = \frac{1.22}{2}\left(-1 + \sqrt{1 + 8 \times 6.54^2}\right) = 10.69 \text{ ft}$$

$$V_2 = \frac{Q}{A_2} = \frac{1,000}{20 \times 10.69} = 4.68 \text{ fps}$$

$$H_L = E_1 - E_2 = 27.31 - 10.69 - \frac{4.68^2}{64.4} = 16.28 \text{ ft}$$

Approximately 60 percent (16.28/27.31 × 100) of the energy in the flow is dissipated in the hydraulic jump. If the normal depth in the downstream channel is less than 10.69 ft, then the floor of the stilling basin would have to be below the downstream channel grade to hold the hydraulic jump at the base of the chute.

5.8.4 Baffled Chute

Roughness elements are placed along the chute to cause energy dissipation with tumbling flow. The flow in a baffled chute (Fig. 5.30) changes from subcritical to

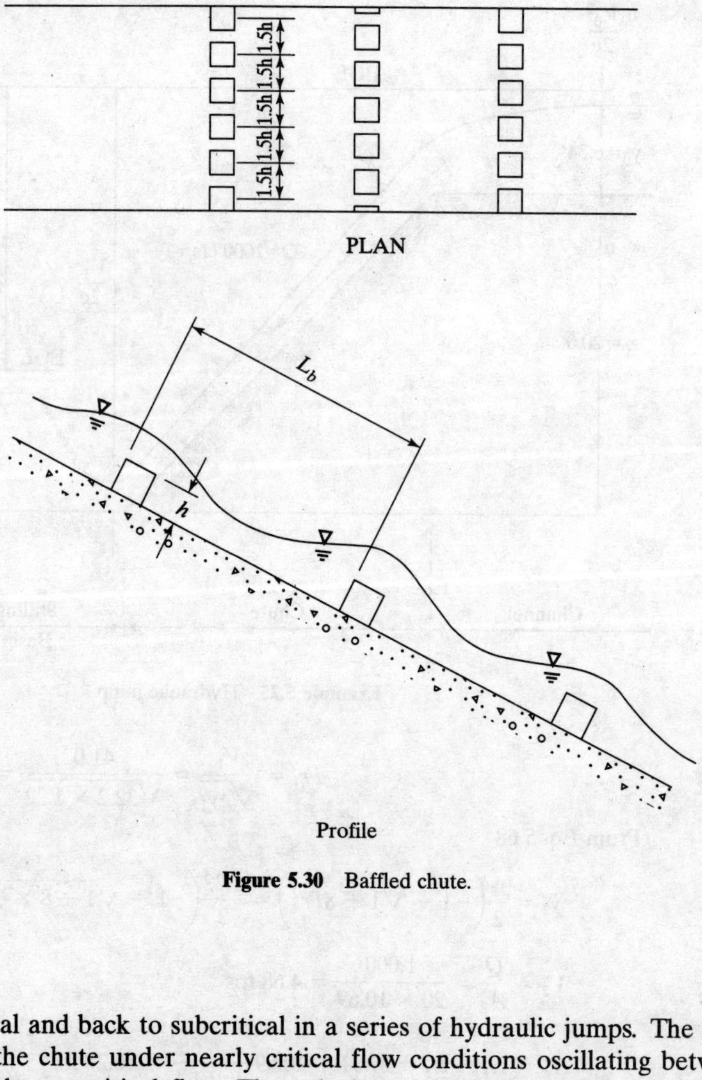

Figure 5.30 Baffled chute.

supercritical and back to subcritical in a series of hydraulic jumps. The flow travels down the chute under nearly critical flow conditions oscillating between subcritical and supercritical flow. The velocity at the outlet of the baffled chute is essentially critical velocity.

The minimum height of cubical roughness element is

$$h = 0.7y_c \tag{5.69}$$

where y_c is the critical depth. The lateral spacing of the roughness elements is 1.5 h with elements in adjacent rows staggered. The longitudinal spacing of adjacent rows of baffles (L_b) is 10 h. The slope of the chute is normally 2H:1V or flatter. Since the velocity in the chute is near critical, erosion protection may be required for a downstream earth channel.

Example 5.26 Baffled Chute

The discharge rate in a rectangular concrete chute 5.0 ft wide is 25 cfs. The chute is 24 ft long at a slope of 2H:1V. Design a baffled chute.

$$y_c = \sqrt[3]{\frac{q^2}{g}} = \sqrt[3]{\frac{5^2}{32.2}} = 0.92 \text{ ft}$$

Size the critical roughness elements

$$h = 0.7y_c = 0.64 \text{ ft}$$

$$\text{Use } h = 8.0 \text{ inches}$$

Lateral spacing (LS) of elements

$$LS = 1.5h = 0.96 \text{ ft}$$

$$\text{Use } LS = 12 \text{ inches}$$

Longitudinal spacing of rows of baffles (LR)

$$LR = 10h = 6.4 \text{ ft}$$

Use 4 rows of baffles spaced 6.0 ft apart.

5.8.5 Vertical Drop

A vertical drop structure can be used to dissipate energy in small channels and spillways. The flow geometry of the vertical drop can be described in terms of a drop number (D_n)

$$D_n = \frac{q^2}{gh^3} \tag{5.70}$$

where q is the discharge per unit width of channel, and h is the height of the drop. The flow geometry is

$$\frac{L_d}{h} = 4.30D_n^{0.27} \tag{5.71}$$

$$\frac{y_1}{h} = 0.54D_n^{0.425} \tag{5.72}$$

$$\frac{y_2}{h} = 1.66D_n^{0.27} \tag{5.73}$$

$$L = 6.9(y_2 - y_1) \tag{5.74}$$

where L_d, y_1, and y_2 are defined in Fig. 5.31. Supercritical flow occurs at y_1 with a hydraulic jump formed between y_1 and y_2. If the water depth in the downstream channel (tailwater) is less than y_2, the jump will move downstream. If the tailwater depth is greater than y_2, the jump will become submerged but still effective. An end sill can be added to ensure that the hydraulic jump remains just downstream of the vertical drop.

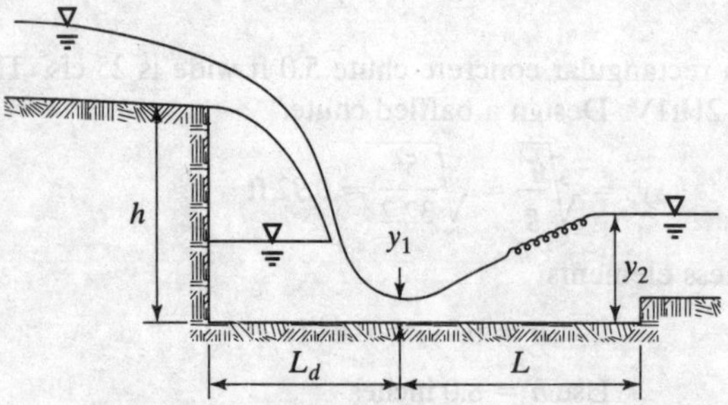

Figure 5.31 Vertical drop.

Example 5.27 Vertical Drop

The discharge rate in a rectangular channel is 5 cfs/ft of width. A vertical drop is used to lower the channel 6.0 ft. The flow is subcritical above and below the drop structure. Determine the dimensions of the drop structure for a tailwater depth of 1.67 ft.

Drop Number

$$D_n = \frac{q^2}{gh^3} = \frac{5^2}{32.2 \times 6^3} = 3.6 \times 10^{-3}$$

$$\frac{L_d}{h} = 4.30 D_n^{0.27} = 4.30(3.6 \times 10^{-3})^{0.27} = 0.94$$

$$L_d = 0.94 \times 6 = 5.6 \text{ ft}$$

$$\frac{y_1}{h} = 0.54 D_n^{0.425} = 0.54(3.6 \times 10^{-3})^{0.425} = 0.049$$

$$y_1 = 6 \times 0.049 = 0.30 \text{ ft}$$

$$\frac{y_2}{h} = 1.66 D_n^{0.27} = 1.66(3.6 \times 10^{-3})^{0.27} = 0.36$$

$$y_2 = 0.36 \times 6 = 2.18 \text{ ft}$$

$$L = 6.9(y_2 - y_1) = 6.9(2.18 - 0.30) = 13.0 \text{ ft}$$

Height of end sill (Δy)

$$\Delta y = y_2 - y_{tw} = 2.18 - 1.67 = 0.51 \text{ ft}$$

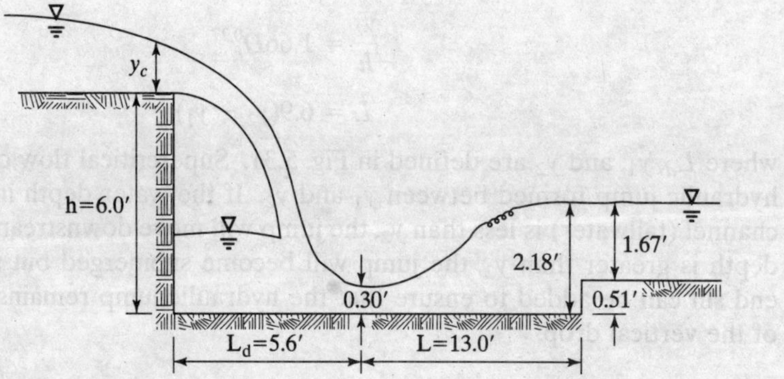

5.8.6 Stilling Basin

Stilling basins are generally reinforced concrete structures designed to contain the turbulent flow of a hydraulic jump. A concrete floor in the stilling basin is provided for the length of the jump. The floor elevation is set to provide the necessary tail-water depth to hold the jump in the stilling basin. Chute blocks, baffle piers, and end sill (Fig. 5.32) are often added to reduce the length of the jump and to ensure efficient operation over a wide range of flows.

Chute blocks increase the thickness of the inflow jet to the jump, increase energy dissipation, and decrease the length of the jump. Baffle piers stabilize the jump and increase energy dissipation. The end sill tends to hold the jump in the stilling basin. The U.S. Bureau of Reclamation (U.S.B.R.) has developed 10 stilling basin designs briefly described as follows.

1. U.S.B.R. Basin I—straight, horizontal, plain rectangular basin.
2. U.S.B.R. Basin II—used on high spillways and large canal structures for $Fr_1 > 4.5$. Contains row of inlet chute blocks at basin inlet and a dentated end sill.

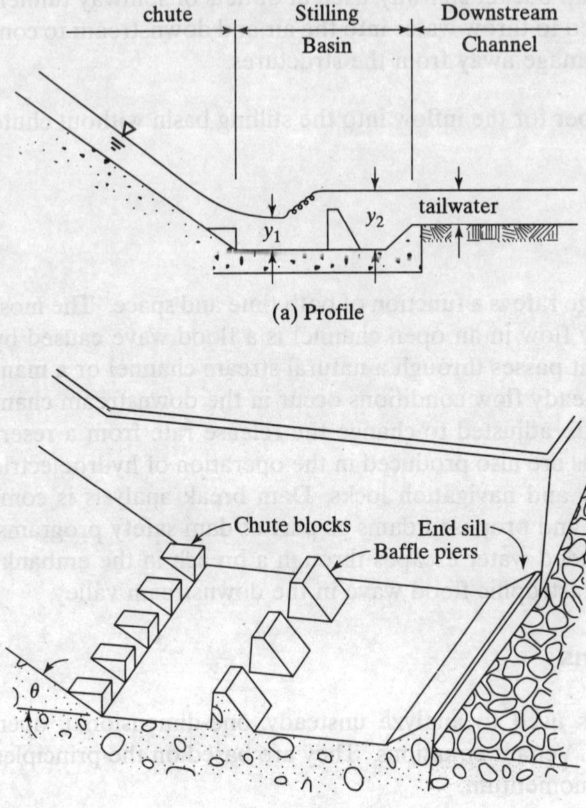

Figure 5.32 Stilling basin.

3. U.S.B.R. Basin III—used on small spillways, outlet works, etc., for $Fr_1 > 4.5$. Contains row of inlet chute blocks, row of baffle piers, and solid end sill.
4. U.S.B.R. Basin IV—used for structures with $2.5 < Fr_1 < 4.5$. Uses chute blocks and solid end sill.
5. U.S.B.R. Basin V—consists mainly of sloping apron, used where economy requires sloping apron, usually on high dam spillways.
6. U.S.B.R. Basin VI—used for pipe or open channel outlets, uses a small box and vertical "hanging baffle wall," with flow energy reduced by impact on wall and flow emerging beneath the baffle.
7. U.S.B.R. Basin VII—used for spillways and other structures with unsubmerged crests. Consists of slotted-end, bucket-type dissipator, with upsloping bed at outlet.
8. U.S.B.R. Basin VIII—the hollow-jet valve stilling basin, a short basin used with an outlet works control structure discharging an annular jet of water.
9. U.S.B.R. Basin IX—baffled apron, with large chute blocks or baffle piers staggered in rows down the chute or spillway drop.
10. U.S.B.R. Basin X—flip bucket spillway, used at outlets of spillway tunnels or large conduits, used to throw water into the air and downstream to concentrate riverbed damage away from the structures.

where Fr_1 is the Froude Number for the inflow into the stilling basin without chute blocks.

5.9 UNSTEADY FLOW

For unsteady flow, the discharge rate is a function of both time and space. The most common example of unsteady flow in an open channel is a flood wave caused by precipitation or snow melt that passes through a natural stream channel or a manmade drainage channel. Unsteady flow conditions occur in the downstream channel when the spillway gates are adjusted to change the release rate from a reservoir. Unsteady flow conditions are also produced in the operation of hydroelectric power plants, pumping plants, and navigation locks. Dam break analysis is commonly performed for existing and proposed dams as part of dam safety programs. When a dam fails, the impounded water escapes through a breach in the embankment causing a potentially catastrophic flood wave in the downstream valley.

5.9.1 Governing Equations

The two governing equations used to analyze unsteady, one-dimensional, open channel flow are called the St. Venant equations. They are based on the principles of conservation of mass and momentum.

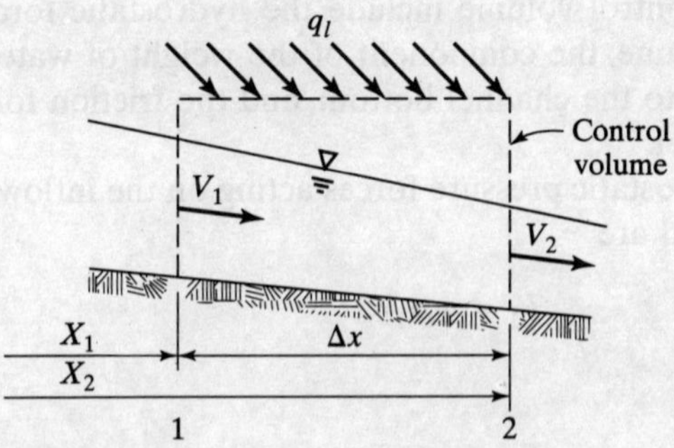

Figure 5.33 Control volume for continuity equation.

Applying the conservation of mass principle to the control volume in Fig. 5.33 gives

$$\frac{d}{dt}\int_{x_1}^{x_2} \rho A dx + \rho A_2 V_2 - \rho A_1 V_1 - \rho q_\ell (x_2 - x_1) = 0 \tag{5.75}$$

where A is the flow area, V is the velocity, and q_ℓ is the lateral inflow per unit length between Sections 1 and 2. Dividing by ρ and substituting Q for VA gives

$$\int_{x_1}^{x_2} \frac{\partial A}{\partial t} dx + Q_2 - Q_1 - q_\ell (x_2 - x_1) = 0 \tag{5.76}$$

This equation reduces to the conservation form of the continuity equation or

$$\frac{\partial A}{\partial t} + \frac{\partial Q}{\partial x} = q_\ell \tag{5.77}$$

where the lateral inflow (q_ℓ) is considered positive for a flow into the channel.

Newton's second law of motion requires that the summation of forces acting on the control volume is equal to the rate of change of momentum or

$$\Sigma F_x = \frac{d}{dt}\int_{x_1}^{x_2} V \rho A dx + V_2 \rho A_2 V_2 - V_1 \rho A_1 V_1 - V_x \rho q_\ell (x_2 - x_1) \tag{5.78}$$

where V_x is the component of the velocity of the lateral inflow that is in the x direction. Substituting Q for VA, Eq. 5.78 becomes

$$\Sigma F_x = \int_{x_1}^{x_2} \rho \frac{\partial Q}{\partial t} dx + \rho Q_2 V_2 - \rho Q_1 V_1 - V_x \rho q_\ell (x_2 - x_1) \tag{5.79}$$

dividing through by $\rho(x_2 - x_1)$ gives

$$\frac{\Sigma F_x}{\rho(x_2 - x_1)} = \frac{\partial Q}{\partial t} + \frac{\partial (QV)}{\partial x} - V_x q_\ell \tag{5.80}$$

The forces acting on the control volume include the hydrostatic forces acting on the ends of the control volume, the component of the weight of water in the control volume that is parallel to the channel bottom, and the friction force acting on the perimeter of the channel.

F_1 and F_2 are the hydrostatic pressure forces acting on the inflow and outflow boundaries, respectively, and are

$$F_1 = \gamma A_1 \bar{y}_1 \tag{5.81}$$

and

$$F_2 = \gamma A_2 \bar{y}_2 \tag{5.82}$$

where $\bar{y}$ is the depth of the centroid of the area (A) below the water surface.

F_3 is the component of the weight of the water in the control volume that is parallel with the channel bottom and is

$$F_3 = \gamma \int_{x_1}^{x_2} A S_o dx \tag{5.83}$$

where S_o is the slope of the channel bottom.

F_4 is the force due to shear between the water and the perimeter of the channel.

$$F_4 = \int_{x_1}^{x_2} \tau P dx \tag{5.84}$$

where P is the wetted perimeter of the channel and τ is the shear stress. Substituting $\tau = \rho f V^2/8$ and

$$S_f = \frac{fV^2}{4R\,2g}$$

gives

$$F_4 = \gamma \int_{x_2}^{x_1} A S_f dx \tag{5.85}$$

Neglecting the wind stress on the water surface

$$\Sigma F_x = F_1 - F_2 + F_3 - F_4 \tag{5.86}$$

Substituting the equations (5.81, 5.82, 5.83, and 5.85) of the forces into Eq. 5.80 and simplifying gives the conservation form of the momentum equation

$$\frac{\partial Q}{\partial t} + \frac{\partial}{\partial x}(QV + gA\bar{y}) = gA(S_o - S_f) + V_x q_\ell \tag{5.87}$$

The conservation form of the St. Venant equations for one-dimensional unsteady flow in vector form are

$$\frac{\partial \mathbf{U}}{\partial t} + \frac{\partial \mathbf{F}}{\partial x} + \mathbf{S} = 0 \tag{5.88}$$

where

$$\mathbf{U} = \begin{pmatrix} A \\ VA \end{pmatrix}$$

$$\mathbf{F} = \begin{pmatrix} VA \\ V^2A + gA\bar{y} \end{pmatrix}$$

$$\mathbf{S} = \begin{pmatrix} -q_\ell \\ -gA(S_o - S_f) - V_x q_\ell \end{pmatrix}$$

V = velocity
A = cross-section area
g = acceleration due to gravity
$\bar{y}$ = distance from the water surface to the centroid of the area
q_l = lateral inflow
V_x = component of the velocity of the lateral inflow that is in the x direction
S_o = slope of the channel and
S_f = slope of the energy line

Finite-difference approximations are made for the partial derivatives in the governing equations. As shown in Fig. 5.34, the distance along the channel is divided

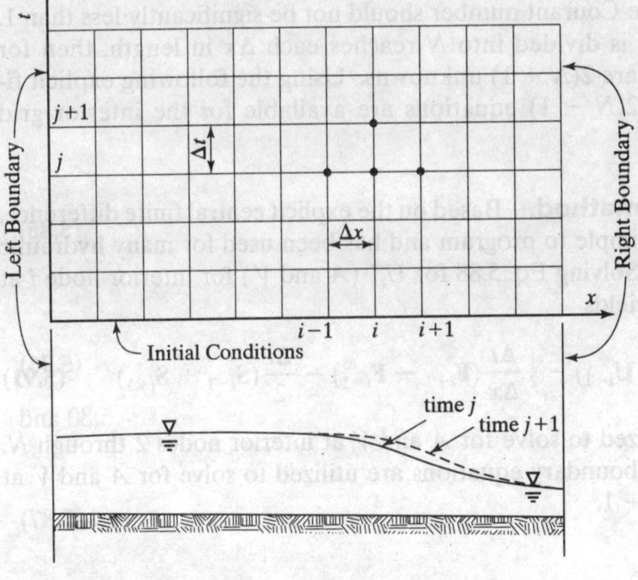

Figure 5.34 Finite difference grid for unsteady flow.

into grid intervals Δx in length and the grid interval along the time axis is Δt. The two independent variables are x and t, while the two dependent variables are A (or y) and V (or Q). All values of the dependent variables are known along the present time line (J) and the values of the dependent variables along the next time line ($J + 1$) are to be computed.

5.9.2 Solution

A closed-form solution of the governing equations is not available except for very simplified examples. However, there are a number of explicit and implicit finite-difference methods available to solve the more complex problems. In the explicit scheme, the spatial partial derivatives are written in terms of the variables at the known time level (J), whereas in the implicit finite-difference scheme, the spatial partial derivatives are written in terms of the variables at both the known (J) and unknown ($J + 1$) time levels. The implicit method requires that all equations for a time level be solved simultaneously. With the explicit method, the equations are solved node by node along the channel for each time step. Because explicit methods are simpler to implement than implicit methods, only explicit methods will be presented in this chapter. An implicit method is presented in the hydraulic routing section of Chapter 6.

Stability for the explicit method requires that the Courant number (C_n) be less than or equal to 1

$$C_n = \frac{\text{actual wave speed}}{\text{numerical wave speed}} = \frac{|V| + a}{\Delta x/\Delta t} \qquad \textbf{(5.89)}$$

where a is the celerity and is a function of the hydraulic depth (D)

$$a = \sqrt{gD} \qquad \textbf{(5.90)}$$

For improved accuracy, the Courant number should not be significantly less than 1.

If a stream channel is divided into N reaches each Δx in length, then for each new time step there are $2(N + 1)$ unknowns. Using the following explicit finite-difference methods, $2(N - 1)$ equations are available for the interior grid nodes.

5.9.2.1 Diffusive method. Based on the explicit central finite difference, the difference method is simple to program and has been used for many hydraulic engineering applications. Solving Eq. 5.88 for U_i^+ (A and V) for interior node i at the next time step $J + 1$ yields

$$\mathbf{U}_i^+ = \tfrac{1}{2}(\mathbf{U}_{i-1} + \mathbf{U}_{i-1}) - \tfrac{1}{2}\frac{\Delta t}{\Delta x}(\mathbf{F}_{i+1} - \mathbf{F}_{i-1}) - \frac{\Delta t}{2}(\mathbf{S}_{i-1} + \mathbf{S}_{i+1}) \qquad \textbf{(5.91)}$$

Equation 5.91 can be utilized to solve for A and V at interior nodes 2 through N. Boundary conditions and boundary equations are utilized to solve for A and V at boundary nodes 1 and $N + 1$.

5.9.2.2 MacCormack method. The MacCormack method is an explicit predictor-corrector procedure for solving the governing equations at interior nodes. It is presented as an alternative to the diffusive method and is second-order accurate in both time and space. In this two-step procedure, the backward difference is utilized for the spacial partial derivative in the predictor step and the forward difference is utilized for the spacial partial derivative in the corrector step.

In the predictor step, the partials are defined

$$\frac{\partial \mathbf{U}}{\partial t} = \frac{\mathbf{U}_i^* - \mathbf{U}_i}{\Delta t} \tag{5.92}$$

and

$$\frac{\partial \mathbf{F}}{\partial x} = \frac{\mathbf{F}_i - \mathbf{F}_{i-1}}{\Delta x} \tag{5.93}$$

where the superscript * indicates the variables to be computed in the predictor step. Substituting the finite-difference approximations into Eq. 5.88 and solving

$$\mathbf{U}_i^* = \mathbf{U}_i - \frac{\Delta t}{\Delta x}(\mathbf{F}_i - \mathbf{F}_{i-1}) - \Delta t \mathbf{S}_i \tag{5.94}$$

The computed value $\mathbf{U}_i^*$ gives A^* (and y^*) and V^* (and Q^*), which are then used to compute $\mathbf{F}^*$ and $\mathbf{S}^*$.

In the corrector step, the partials are defined

$$\frac{\partial \mathbf{U}}{\partial t} = \frac{\mathbf{U}_i^{**} - \mathbf{U}_i}{\Delta t} \tag{5.95}$$

$$\frac{\partial \mathbf{F}}{\partial t} = \frac{\mathbf{F}_{i+1}^* - \mathbf{F}_i^*}{\Delta x} \tag{5.96}$$

Substituting these finite-difference approximations and $S = S_i^*$ into Eq. 5.88 and solving yields

$$\mathbf{U}_i^{**} = \mathbf{U}_i - \frac{\Delta t}{\Delta x}(\mathbf{F}_{i+1}^* - \mathbf{F}_i^*) - \Delta t \mathbf{S}_i^* \tag{5.97}$$

The values of the dependent variables at the next time step $J + 1$ are

$$\mathbf{U}_i^+ = \tfrac{1}{2}(\mathbf{U}_i^* + \mathbf{U}_i^{**}) \tag{5.98}$$

5.9.3 Boundary Conditions

Two additional equations are available for the computation of unsteady flow in a channel by specifying boundary conditions. Boundary conditions are essential in defining the flow conditions in the channel as a function of time. For unsteady flow conditions to exist in the channel, at least one boundary condition must be either a function of time or different from the initial conditions. For subcritical flow, a boundary condition must be specified for both the upstream and downstream boundaries. For supercritical flow, two boundary conditions are required for the upstream boundary and none required at the downstream boundary.

As shown in Fig. 5.35, typical boundary conditions might include an upstream or downstream reservoir where the water surface elevation is either constant or a function of time, an upstream or downstream control gate where the gate opening is either constant or a function of time, an inflow hydrograph where the discharge rate is specified as a function of time, or a rating curve where the discharge rate is a function of water depth and slope of the energy line.

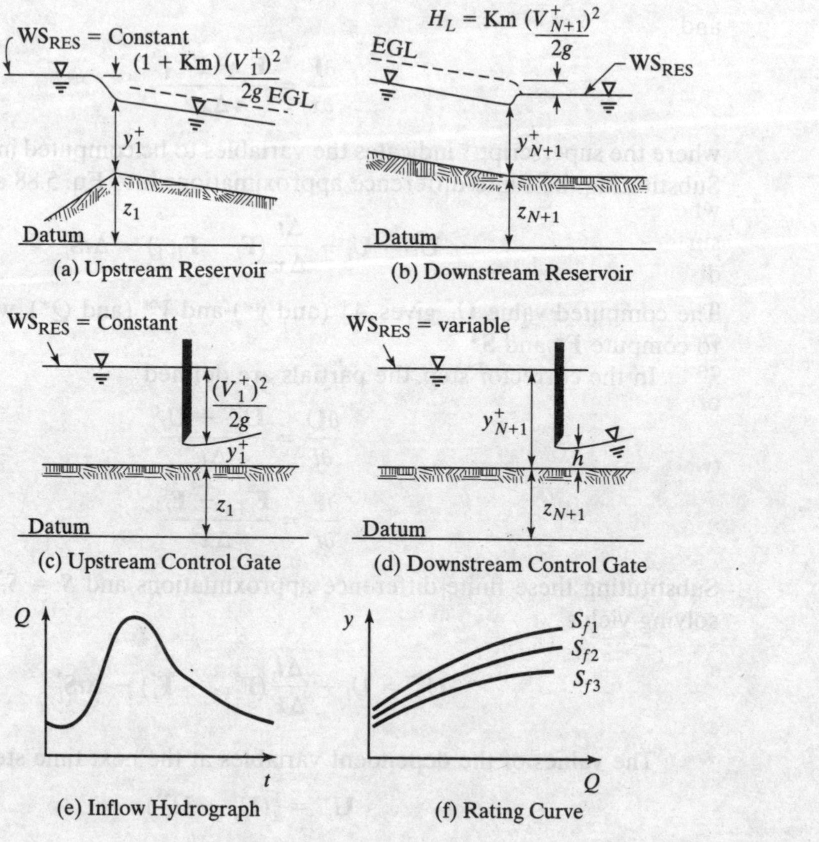

Figure 5.35 Boundary conditions.

The boundary conditions shown in Fig. 5.35 can be expressed by the following equations:

(a) Upstream Reservoir

$$y_1^+ = WS_{\text{Res}} - Z_1 - (1 + K_m)\frac{(V_1^+)^2}{2g} \qquad \textbf{(5.99)}$$

(b) Downstream Reservoir

$$y_{N+1}^{+} = WS_{\text{Res}} - Z_{N+1} - (1 + K_m)\frac{(V_{N+1}^{+})^2}{2g} \tag{5.100}$$

(c) Upstream Control Gate

$$Q_1^{+} = C_o A_g \sqrt{2g(WS_{\text{Res}} - Z_1 - y_1^{+})} \tag{5.101}$$

(d) Downstream Control Gate

$$Q_{N+1}^{+} = C_o A_g \sqrt{2g(y_{N+1}^{+} - h)} \tag{5.102}$$

(e) Inflow Hydrograph

$$Q_1^{+} = f(t) \tag{5.103}$$

(f) Rating Curve

$$Q_{N+1}^{+} = f(y_{N+1}^{+}, S_{fN+1}) \tag{5.104}$$

where K_m is the entrance or exit loss coefficient, WS_{Res} is the reservoir water surface elevation, Z is the channel bottom elevation, C_o is the control gate discharge coefficient, A_g is the area of the gate opening, h is the height of the gate opening, and S_f is the slope of the energy line. The expansion loss coefficient in Eq. 5.100 is negative. Because the slope of the energy line usually changes slowly, the slope of the energy line in Eq. 5.104 can be computed for the present time step.

Channels in series can also be analyzed by including a junction node between the two channels. As shown in Fig. 5.36, the junction node represents the

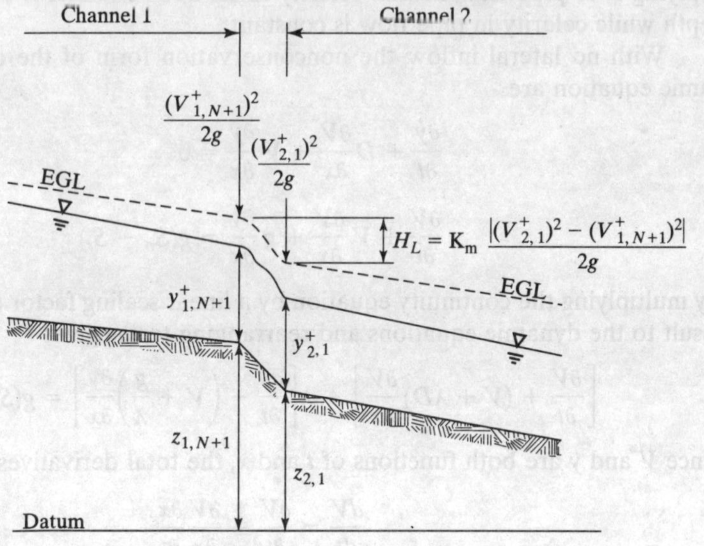

Figure 5.36 Junction node.

downstream boundary of channel 1 and the upstream boundary of channel 2. Writing the continuity and energy equations across the junction yields

$$(V^+_{1,N+1})(A^+_{1,N+1}) = (V^+_{2,1})(A^+_{2,1}) \tag{5.105}$$

$$(1 + K_m)\frac{(V^+_{1,N+1})^2}{2g} + Z_{1,N+1} + y^+_{1,N+1} = (1 + K_m)\frac{(V^+_{2,1})^2}{2g} + Z_{2,1} + y^+_{2,1} \tag{5.106}$$

where the minor loss coefficient (K_m) is equal to K_m for a contraction and equal to $-K_m$ for an expansion.

Unsteady flow in nonprismatic channels may be simulated by dividing the channel into a series of prismatic reaches with different channel cross sections and bottom slopes.

5.9.4 Boundary Equations

In addition to the $2(N - 1)$ finite-difference equations written for the interior nodes and the two boundary conditions, two additional equations are required to solve for the $2(N + 1)$ unknown dependent variables in the channel. The method of characteristics will be presented in this section to provide the remaining two boundary equations. For subcritical flow, the minus characteristic boundary equation will be applied at the upstream boundary, and the plus characteristic boundary equation will be applied at the downstream boundary. For supercritical flow, both the plus and minus characteristic boundary equations will be applied at the downstream boundary.

The method of characteristics presented in this section is very similar to the method of characteristics presented for unsteady flow in pipes and pipe networks. Applying the method of characteristics to open channel flow is more complex than applying it to pipe flow, in that celerity in an open channel is a function of water depth while celerity in pipe flow is constant.

With no lateral inflow, the nonconservation form of the continuity and dynamic equation are

$$\frac{\partial y}{\partial t} + D\frac{\partial V}{\partial x} + V\frac{\partial y}{\partial x} = 0 \tag{5.107}$$

$$\frac{\partial V}{\partial t} + V\frac{\partial V}{\partial x} + g\frac{\partial y}{\partial x} = g(S_o - S_f) \tag{5.108}$$

By multiplying the continuity equation by a linear scaling factor (λ) and adding the result to the dynamic equations and rearranging terms gives

$$\left[\frac{\partial V}{\partial t} + (V + \lambda D)\frac{\partial V}{\partial x}\right] + \lambda\left[\frac{\partial y}{\partial t} + \left(V + \frac{g}{\lambda}\right)\frac{\partial y}{\partial x}\right] = g(S_o - S_f) \tag{5.109}$$

Since V and y are both functions of t and x, the total derivatives are

$$\frac{dV}{dt} = \frac{\partial V}{\partial t} + \frac{\partial V}{\partial x}\frac{\partial x}{\partial t} \tag{5.110}$$

$$\frac{dy}{dt} = \frac{\partial y}{\partial t} + \frac{\partial y}{\partial x}\frac{\partial x}{\partial t} \tag{5.111}$$

The terms inside the brackets of Eq. 5.109 are equal to the total derivatives of V and y, respectively, if

$$\frac{\partial x}{\partial t} = \frac{dx}{dt} = V + \lambda D \tag{5.112}$$

and

$$\frac{\partial x}{\partial t} = \frac{dx}{dt} = V + \frac{g}{\lambda} \tag{5.113}$$

or

$$\lambda = \pm\sqrt{\frac{g}{D}} \tag{5.114}$$

Since celerity (a) is defined as

$$a = \sqrt{gD} \tag{5.115}$$

then

$$\lambda = \pm\frac{g}{a} \tag{5.116}$$

The plus characteristic equation (C^+) is

$$\frac{dV}{dt} + \frac{g}{a}\frac{dy}{dt} = g(S_o - S_f) \tag{5.117}$$

with

$$\frac{dx}{dt} = V + a \tag{5.118}$$

The negative characteristics equation (C^-) is

$$\frac{dV}{dt} - \frac{g}{a}\frac{dy}{dt} = g(S_o - S_f) \tag{5.119}$$

with

$$\frac{dx}{dt} = V - a \tag{5.120}$$

The positive characteristics equation (C^+) is valid along the positive characteristic line where $dx/dt = V + a$, and the negative characteristic equation (C^-) is valid along the negative characteristic line where $dx/dt = V - a$.

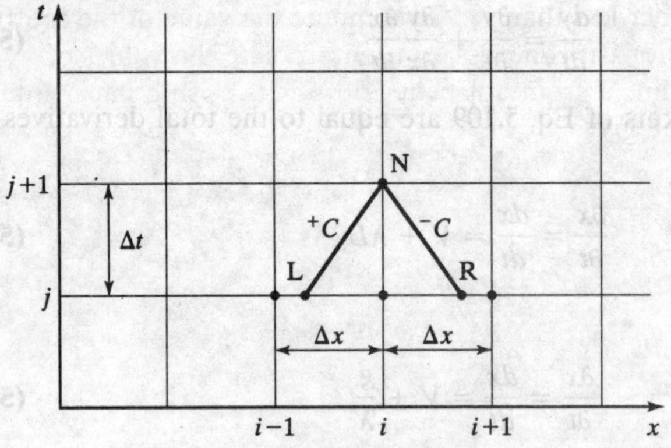

Figure 5.37 Method of characteristics grid.

Fig. 5.37 shows the $+C$ characteristic line (LN) and negative characteristic line (RN). Integrating Eqs. 5.117 and 5.119 along the characteristics LN and RN, respectively yields

$$V_i^+ - V_L + \left(\frac{g}{a}\right)_L (y_i^+ - y_L) = g(S_o - S_f)_L \,\Delta t \tag{5.121}$$

$$V_i^+ - V_R - \left(\frac{g}{a}\right)_R (y_i^+ - y_R) = g(S_o - S_f)_R \,\Delta t \tag{5.122}$$

In short form notation, Eqs. 5.121 and 5.122 are

$$C^+: \quad V_i^+ = C_1 - C_2 y_i^+ \tag{5.123}$$

$$C^-: \quad V_i^+ = C_3 + C_4 y_i^+ \tag{5.124}$$

where

$$C_1 = V_L + C_2 y_L + g(S_o - S_f)_L \,\Delta t$$

$$C_2 = \left(\frac{g}{a}\right)_L$$

$$C_3 = V_R - C_4 y_R + g(S_o - S_f)_R \,\Delta t$$

$$C_4 = \left(\frac{g}{a}\right)_R$$

During the time interval Δt between time line J and time line $J + 1$, the coefficients C_1, C_2, C_3, and C_4 are considered constant, and the characteristic Eqs. 5.123 and 5.124 are simple linear equations.

On time line J, the characteristic lines do not originate at grid points, but instead originate at L and R. For the method to be numerically stable, it is necessary that point L be to the right of grid point $i - 1$ and point R be to the left of grid point $i + 1$ (Courant number less than 1). To compute the value of the coefficients in the characteristic equations, it is necessary to interpolate the values of V, y, and a at L and R along time line J from adjacent grid points. Using linear interpolation along the old time line gives

$$\frac{V_i - V_L}{V_i - V_{i-1}} = \frac{x_i - x_L}{\Delta x} = \frac{V_L + a_L}{\Delta x/\Delta t} \tag{5.125}$$

$$\frac{a_i - a_L}{a_i - a_{i-1}} = \frac{V_L + a_L}{\Delta x/\Delta t} \tag{5.126}$$

and

$$\frac{y_i - y_L}{y_i - y_{i-1}} = \frac{V_L + a_L}{\Delta x/\Delta t} \tag{5.127}$$

Combining equations yields

$$V_L = \frac{V_i - (a_{i-1}V_i - a_i V_{i-1})\,\Delta t/\Delta x}{1 + ((V_i - V_{i-1}) + (a_i - a_{i-1}))\Delta t/\Delta x} \tag{5.128}$$

$$a_L = \frac{a_i - V_L\,(a_i - a_{i-1})\Delta t/\Delta x}{1 + (a_i - a_{i-1})\Delta t/\Delta x} \tag{5.129}$$

$$y_L = y_i - (V_L + a_L)(y_i - y_{i-1})\Delta t/\Delta x \tag{5.130}$$

Similarly for point R

$$V_R = \frac{V_i - (a_{i+1}V_i - a_i V_{i+1})\Delta t/\Delta x}{1 - ((V_i - V_{i+1}) + (a_{i+1} - a_i))\Delta t/\Delta x} \tag{5.131}$$

$$a_R = \frac{a_i + V_R\,(a_i - a_{i+1})\Delta t/\Delta x}{1 + (a_i - a_{i+1})\Delta t/\Delta x} \tag{5.132}$$

$$y_R = y_i + (V_R - a_R)(y_i - y_{i+1})\Delta t/\Delta x \tag{5.133}$$

The method of characteristics can be applied to the interior nodes as well as the boundary nodes. However, its application to interior nodes has been limited because relatively simple explicit finite-difference methods with higher accuracy are available. The accuracy of the boundary equations can generally be less than that of the interior nodes without affecting the overall accuracy of the scheme. Figure 5.38 shows the characteristic lines at the boundaries of an open channel. The channel is divided into N segments each Δx in length. The values of V, y, and a must be interpolated along time line J for point R between nodes 1 and 2 (Eqs. 5.131–5.133) and point L between nodes N and $N + 1$ (Eqs. 5.128–5.130). The C^- characteristic equation (Eq. 5.124) can be written for the left boundary, and the C^+ characteristic equation (Eq. 5.123) can be written for the right boundary.

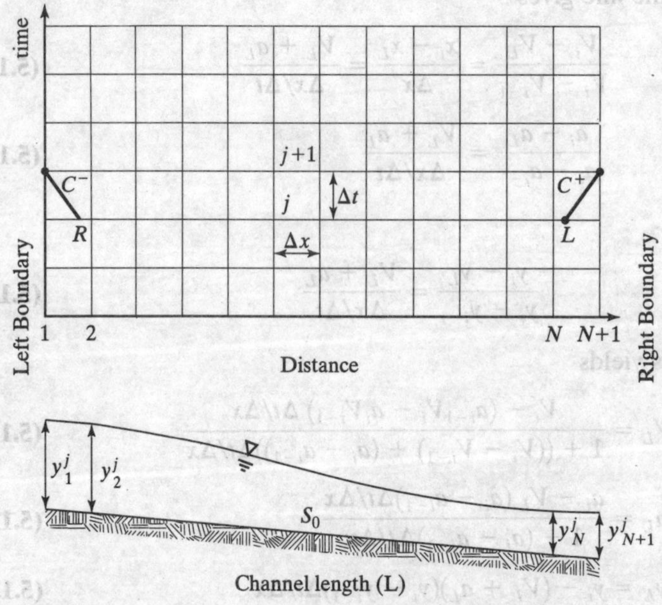

Figure 5.38 Characteristic lines at boundaries.

5.9.5 Initial Conditions

Values of the dependent variables (Q and A or V and y) for all nodes at the start of the simulation ($t = 0$) are referred to as initial conditions and must be known. The initial conditions are assumed to occur before the flow becomes unsteady and can usually be computed using the gradually varied flow equations.

Example 5.28 Unsteady Flow in a Trapezoidal Channel

A trapezoidal earth channel ($n = 0.024$) has a bottom width of 10 m and 3H:1V side slopes. The canal is 5,000 m long with a slope of 0.0001. The canal runs from a reservoir at the upstream end to a sluice gate at the downstream end and carries a discharge of 200 m^3/s. The minor loss coefficient at the entrance is 0.5, and the upstream invert elevation of the channel is 100 m. If the sluice gate is closed in 500 s, compute the depth and velocity in the channel at time 500, 1,000, 1,500, 2,000, and 2,500 seconds. For computational purpose, divide the channel into 50 sections.

Two programs were developed for this example problem. MOC.FOR uses the method of characteristics (MOC) to solve the unsteady flow equations. MACK.FOR uses the MacCormack scheme to solve the unsteady flow equations for the internal nodes and the MOC for the boundary nodes. The input file is listed along with plots of velocity and water depth. Both methods give nearly identical results.

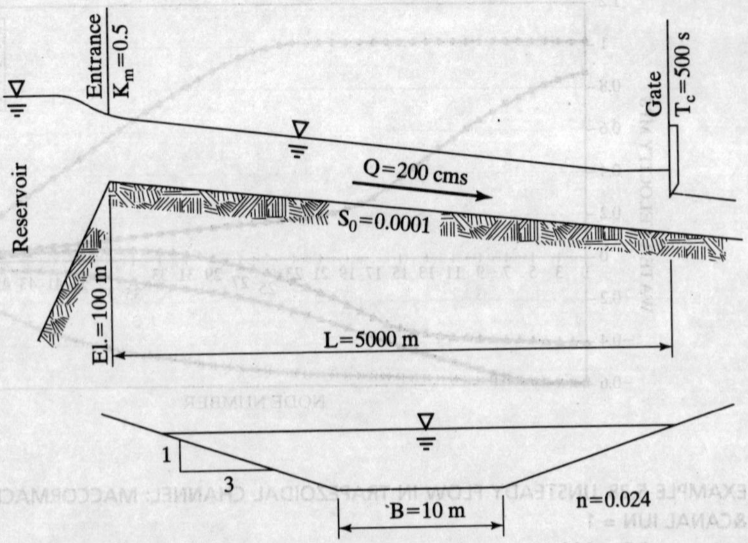

EXAMPLE 5.28 UNSTEADY FLOW TRAPEZOID CHANNEL: METHOD OF CHARACTERISTICS

&CANAL IUN = 1
QO = 200.
ELEV = 100.
RKM = 0.5
BO = 10.0
SS = 3.0
CMN = 0.024
CHL = 5000.
SO = 0.00010
NSEC = 50
TLAST = 2600.
T1 = 500.
AOUT = 500.

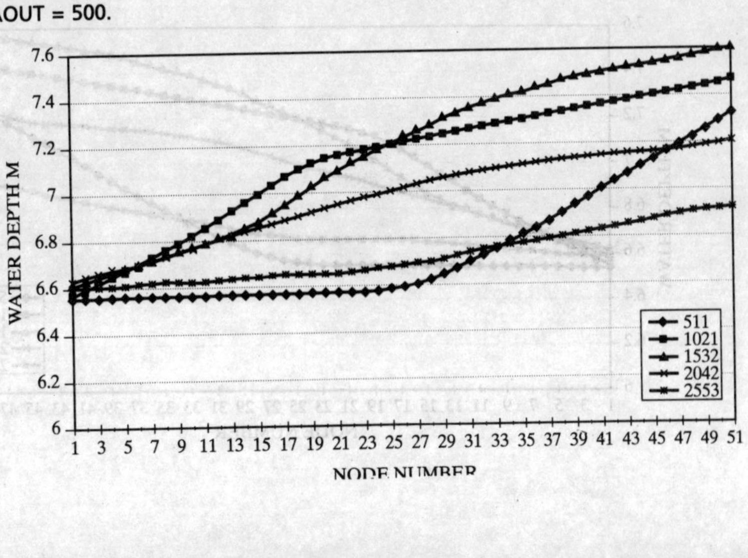

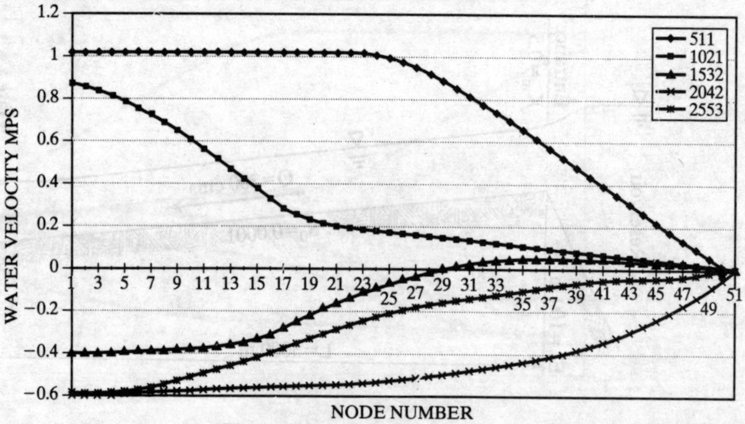

EXAMPLE 5.28 UNSTEADY FLOW IN TRAPEZOIDAL CHANNEL: MACCORMACK SCHEME
&CANAL IUN = 1
QO = 200.
ELEV = 100.
RKM = 0.5
BO = 10.0
SS = 3.0
CMN = 0.024
CHL = 5000.
SO = 0.0001
NSEC = 50
TLAST = 2600.
T1 = 500.
AOUT = 500.

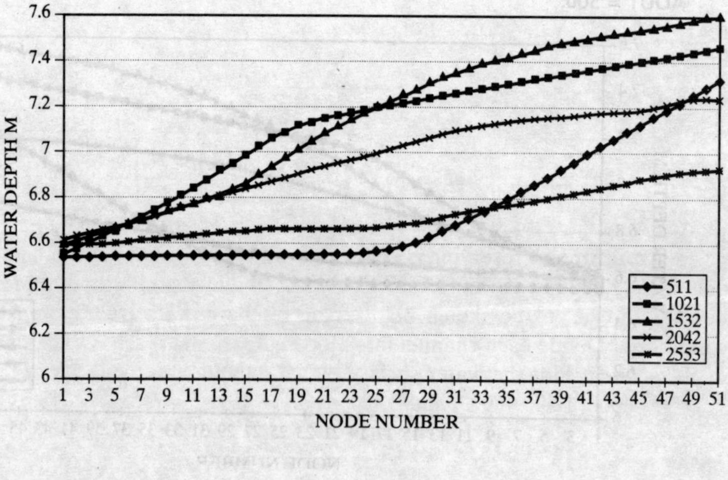

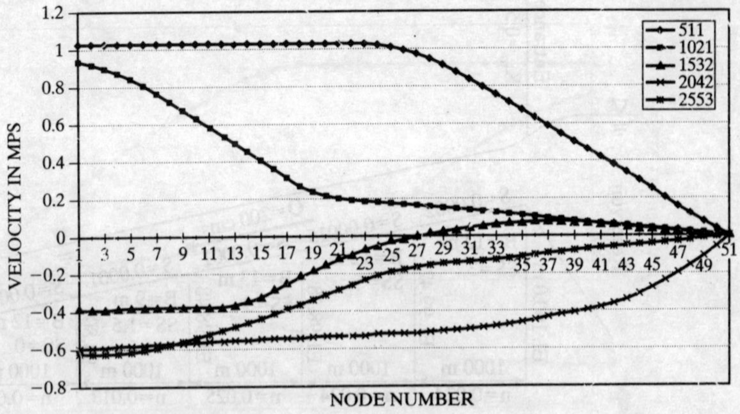

Example 5.29 Unsteady Flow in Channels in Series

Modify the program (MACK.FOR) using the MacCormack scheme to handle channels in series. Neglect the headloss at the junction nodes.

The four equations for the junction node connecting channel j with channel $j+1$ are

$$\text{Continuity:}\quad V^+_{j,N+1}A^+_{j,N+1} = V^+_{j+1,1}A^+_{j+1,1}$$

$$\text{Energy:}\quad \frac{(V^+_{j,N+1})^2}{2g} + y^+_{j,N+1} + Z_{j,N+1} = \frac{(V_{j+1,1})^2}{2g} + y^+_{j+1,1} + Z_{j+1,1}$$

$$C^+:\quad V^+_{j,N+1} = C_1 - C_2\, y^+_{j,N+1}$$

$$C^-:\quad V^+_{j+1,1} = C_3 + C_4\, y^+_{j+1,1}$$

where the first subscript represents the channel and the second subscript represents the node.

Combining into two equations and two unknowns:

$$F_1 = \frac{(C_3 + C_4\, y^+_{j+1,1})^2}{2g} + y^+_{j+1,1} + Z_{j+1,1} - \frac{(C_1 - C_2\, y^+_{j,N+1})^2}{2g} - y^+_{j,N+1} - Z_{j,n+1} = 0$$

$$F_2 = A^+_{j+1,1}(C_3 + C_4\, y^+_{j+1,1}) - A^+_{j,N+1}(C_1 - C_2\, y^+_{j,N+1}) = 0$$

Using matrix notation

$$\begin{bmatrix} \dfrac{\partial F_1}{\partial y_{j,N+1}} & \dfrac{\partial F_1}{\partial y_{j+1,1}} \\ \dfrac{\partial F_2}{\partial y_{j,N+1}} & \dfrac{\partial F_2}{\partial y_{j+1,1}} \end{bmatrix} \begin{bmatrix} \Delta y_{j,N+1} \\ \Delta y_{j+1,1} \end{bmatrix} = - \begin{bmatrix} F_1 \\ F_2 \end{bmatrix}$$

An upstream reservoir is connected to a downstream sluice gate by five channels (L = 1,000 m each) in series. The channel dimensions, slopes, and roughness values are shown on the sketch below. The discharge rate is 200 cms for steady-state conditions. Divide each channel into 10 computational reaches. The sluice gate closes in 50 seconds. Plot the water surface elevation and water velocity in the channel for t = 500, 1,000, 1,500, 2,000, and 2,500 seconds. The modified program for channels in

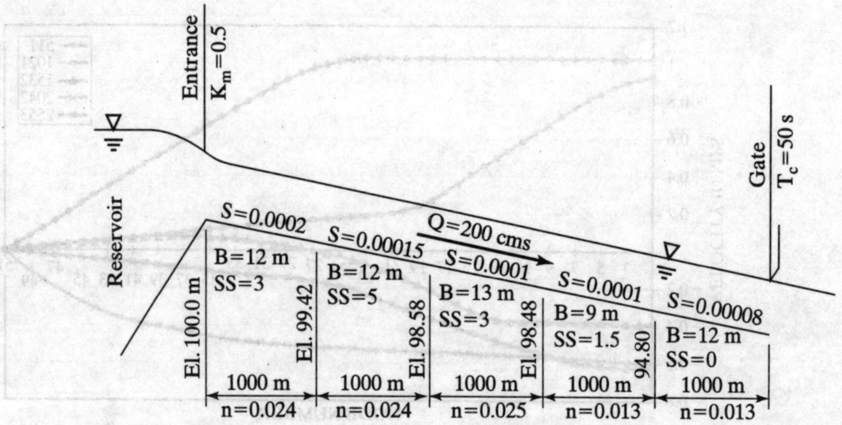

series is called MACKS.FOR and the input file is MACKS.DAT. A listing of the input file is listed below.

```
EXAMPLE 5.29 UNSTEADY FLOW FOR CHANNELS IN SERIES
&CANAL IUN = 1
          QO = 200.
          RKM = 0.5
          NOC = 5
          TLAST = 2600.
          T1 = 50.
          AOUT = 500./
100.00    12.0    3.0    0.024    1000.    0.00020    10
99.42     12.0    5.0    0.024    1000.    0.00015    10
98.58     13.0    3.0    0.025    1000.    0.00010    10
98.48      9.0    1.5    0.013    1000.    0.00010    10
94.80     12.0    0.0    0.013    1000.    0.00008    10
```

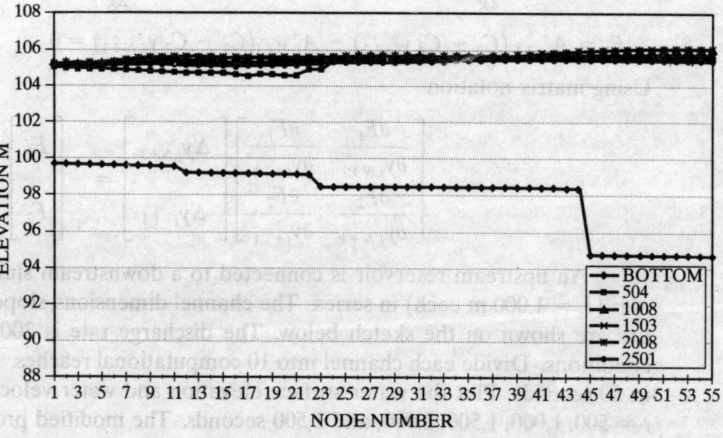

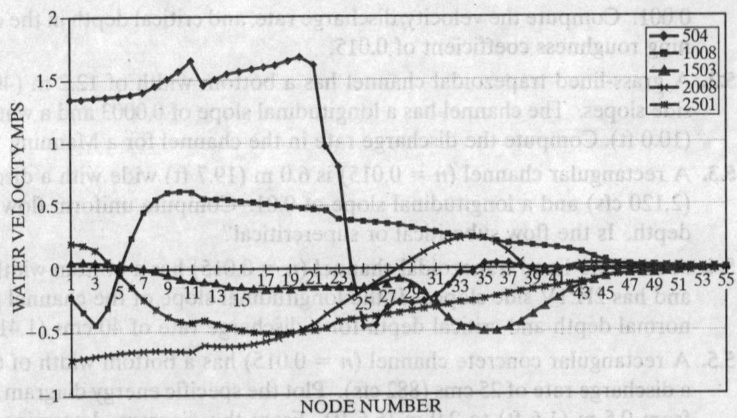

5.10 GENERALIZED OPEN CHANNEL HYDRAULICS MODELS

The River Analysis System (RAS), developed by the Hydrologic Engineering Center (HEC), is the most widely used of the generalized open channel hydraulics models. The public domain HEC-RAS software, users manual, and technical reference manual may be downloaded directly from the HEC website (Table 1.7). HEC-RAS simulates flow in single reaches or complex networks of natural or man-made channels. Bridges, culverts, weirs, encroachments, and channel improvements may be incorporated in the simulation. Either steady or unsteady flow analyses may be performed. Steady-flow water surface profile computations use the standard step method solution of the energy equation. Unsteady flow simulations are based on an implicit finite-difference solution of the St. Venant equations. The methods for modeling the effects of hydraulic structures are essentially the same with either steady or unsteady flow simulations. HEC-RAS also includes capabilities for predicting scour at bridges.

The computer program HEC-2 Water Surface Profiles was a predecessor to HEC-RAS. HEC-2 evolved through various versions dating back to the 1960's and has been widely applied by agencies and consulting firms throughout the U.S. and abroad. The first version of HEC-RAS released in 1997 did not include unsteady flow analysis. Dynamic routing was added in 2001. The HEC-RAS dynamic routing routines are based on methods previously incorporated in the UNET model (one-dimensional unsteady flow through a full network of open channels), which is still maintained by the HEC. Erosion and sediment transport modeling capabilities are being added to future versions of HEC-RAS.

PROBLEMS

5.1. Uniform flow occurs in a rectangular channel 3.05 m (10.0 ft) wide and 1.52 m (5.0 ft) deep. Compute the mean shear stress on the channel perimeter if the channel slope is

0.001. Compute the velocity, discharge rate, and critical depth in the channel for a Manning roughness coefficient of 0.015.

5.2. A grass-lined trapezoidal channel has a bottom width of 12.2 m (40.0 ft) with 4H:1V side slopes. The channel has a longitudinal slope of 0.0003 and a water depth of 3.05 m (10.0 ft). Compute the discharge rate in the channel for a Manning "n" value of 0.035.

5.3. A rectangular channel (n = 0.015) is 6.0 m (19.7 ft) wide with a discharge rate 60 cms (2,120 cfs) and a longitudinal slope of 0.01. Compute uniform flow depth and critical depth. Is the flow subcritical or supercritical?

5.4. A concrete-lined trapezoidal channel (n = 0.015) has a bottom width of 6.0 m (19.7 ft) and has 2H:1V side slopes. If the longitudinal slope of the channel is 0.0004, compute normal depth and critical depth for a discharge rate of 40 cms (1,411 cfs).

5.5. A rectangular concrete channel (n = 0.015) has a bottom width of 6.0 m (19.7 ft) and a discharge rate of 25 cms (882 cfs). Plot the specific energy diagram for depths ranging from 0.5 m (1.6 ft) to 2.0 m (6.6 ft). From the diagram, determine (a) critical depth, (b) minimum specific energy, (c) specific energy for a water depth of 1.9 m (6.2 ft), and (d) water depths when the specific energy is 2.2 m (7.2 ft). What channel slope would result in critical flow?

5.6. Determine the discharge rate in a 1.52-m (60-in.) diameter corrugated metal pipe (CMP) (n = 0.024) on a slope of 0.005 flowing full. What is the discharge rate if the water depth is 0.91 m (3.0 ft)? Compute the water depth for a discharge rate of 1.70 cms (60 cfs).

5.7. Determine the discharge rate in a 1.52-m (60-inch) reinforced concrete pipe (RCP) (n = 0.013) on a slope of 0.005 flowing full. What is the discharge rate if the water depth is 0.91 m (3.0 ft)? Compute the water depth for a discharge rate of 1.70 cms (60 cfs).

5.8. Determine the discharge rate for a 1.52-m × 1.52-m (5-ft × 5-ft) concrete box culvert (n = 0.012) on a slope of 0.005 flowing full. What is the discharge rate if the water depth is 0.91 m (3.0 ft)? Compute the water depth if the discharge is 1.70 cms (60 cfs).

5.9. A street gutter has a side slope of 20:1, a longitudinal slope of 0.02, and a Manning "n" value of 0.017. Compute the discharge rate and velocity if the depth of flow at the curb is 0.12 m (0.4 ft). Compute the water depth and velocity if the discharge rate in the gutter is 0.14 cms (5.0 cfs).

5.10. A trapezoidal concrete channel (n = 0.015) has side slopes of 1.5H:1V, a bottom width of 3.66 m (12.0 ft), and a longitudinal slope of 0.005. Compute the discharge rate and velocity if the depth of flow is 1.52 m (5.0 ft). Compute the normal depth and critical depth for a discharge rate of 5.67 cms (200 cfs). Compute critical slope for a discharge of 5.67 cms (200 cfs).

5.11. A grass-lined earth channel (n = 0.035) has side slopes of 4H:1V, a bottom width of 11.0 m (36.0 ft), and a longitudinal slope of 0.001. Compute the discharge rate and velocity if the water depth is 1.83 m (6.0 ft). Compute normal depth and critical depth for a discharge rate of 14.2 cms (500 cfs).

5.12. A rectangular concrete channel (n = 0.012) has a bottom width of 3.05 m (10.0 ft) and a longitudinal slope of 0.002. Compute the discharge rate and velocity for a water depth of 1.52 m (5.0 ft). Compute normal depth, critical depth, and critical slope for a discharge of 5.67 cms (200 cfs).

5.13. What longitudinal slope is necessary to carry 20 cms (706 cfs) at a depth of 2.0 m (6.56 ft) in a rectangular channel 5.0 m (16.4 ft) wide ($n = 0.015$)? Compute critical slope.

5.14. What longitudinal slope is necessary to carry 0.051 cms (1.8 cfs) at a depth of 0.091 m (0.3 ft) in a street gutter ($n = 0.017$) with a side slope of 20:1?

5.15. What longitudinal slope is necessary to carry 2.83 cms (100 cfs) at a velocity of 0.61 mps (2.0 fps) in a trapezoidal earth canal ($n = 0.03$) with a bottom width of 3.05 m (10.0 ft) and 3H:1V side slopes.

5.16. Determine the water depth at which 1.0 cms (35.3 cfs) will flow in a circular concrete pipe ($n = 0.013$) 2.0 m (6.56 ft) in diameter on a longitudinal slope of 0.0002.

5.17. Determine the bottom width (B) of a rectangular concrete channel ($n = 0.012$) required to carry 50 cms (1,765 cfs) at a water depth $d = B/2$ if the longitudinal slope is 0.001.

5.18. What bottom width (B) trapezoidal grass-lined channel ($n = 0.035$) is required to carry 100 cms (3,530 cfs) at a water depth $d = B/4$? The channel has side slopes of 4H:1V and a longitudinal slope of 0.004.

5.19. What bottom width (B) trapezoidal concrete-lined channel ($n = 0.015$) is required to carry 100 cms (3,530 cfs) at a water depth $d = B/2$? The channel has side slopes of 2H:1V and a longitudinal slope of 0.002.

5.20. A rectangular concrete channel ($n = 0.012$) has a bottom width of 10.0 m (32.8 ft), a water depth of 4.0 m (13.1 ft), and a longitudinal slope of 0.001. There is a horizontal bend with a radius of curvature of 35.0 m (115 ft). Compute the Froude Number and water depth in the bend along the inside and outside walls.

5.21. A rectangular concrete channel ($n = 0.015$) is 3.0 m (9.84 ft) wide, has a slope of 0.009, and a discharge rate of 5.0 cms (176 cfs). Compute normal depth and critical depth. If the depth of flow is 1.5 m (4.9 ft), is the flow subcritical or supercritical? What is the type of profile?

5.22. Compute the flow rate for the composite channel shown below if the longitudinal slope is 0.0003.

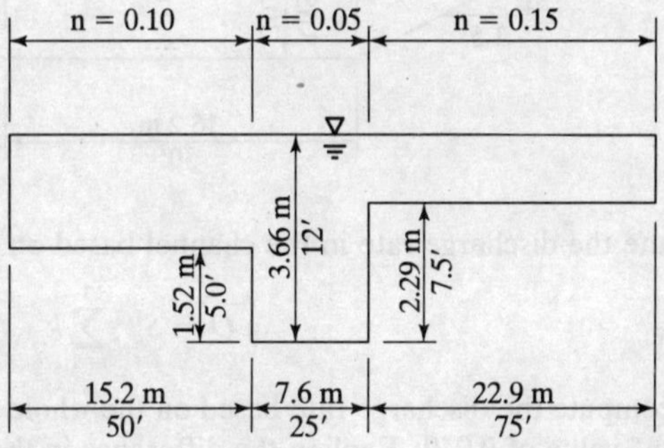

Compute critical depth.

5.23. Compute normal depth and critical depth for the channel in Problem 5.22 if the discharge rate is 56.6 cms (2,000 cfs).

5.24. A trapezoidal channel with a 15.2 m (50 ft) bottom width, 2H:1V side slopes, a Manning "n" value of 0.04, a longitudinal slope of 0.0009, and a water depth of 3.05 m (10.0 ft).

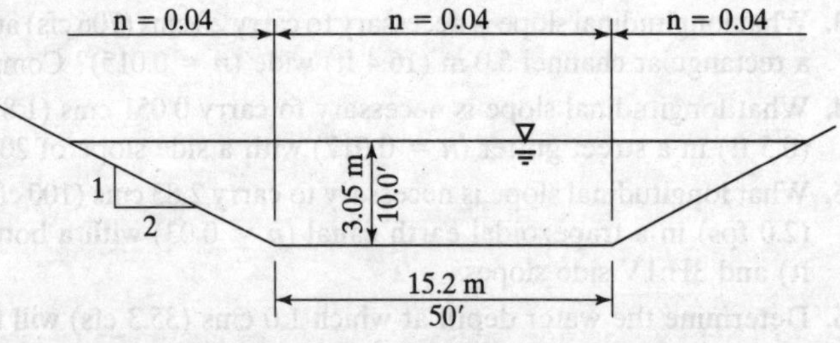

Compute the discharge rate based on single channel segment

$$Q = \frac{C_m}{n} A R^{2/3} S_o^{1/2}$$
$$= K_T S_o^{1/2}$$

and based on three channel segments where

$$Q = S_o^{1/2} \sum_{i=1}^{3} k_i$$

Explain the difference in the two discharge rates.

5.25. A trapezoidal channel shown below with a longitudinal slope of 0.0009 and has channel banks on a side slope of $1\frac{1}{2}$ H:1V with dense vegetation. The estimated Manning "n" value of the channel is shown below.

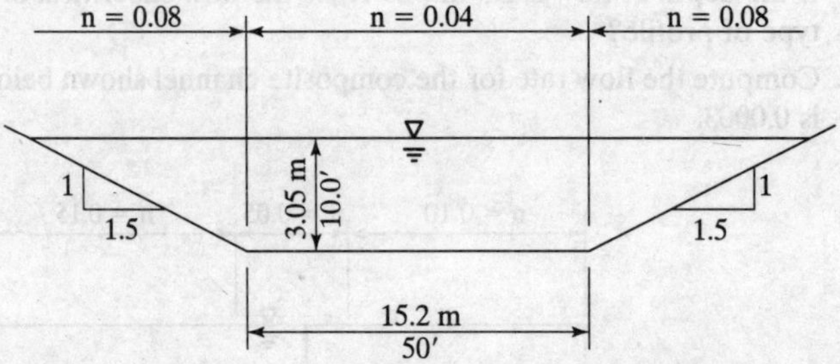

Compute the discharge rate in the channel based on three channel segments where

$$Q = S_o^{1/2} \sum_{i=1}^{3} k_i$$

Also compute the discharge rate based on the whole channel using an estimated Manning "n" value of 0.040. Explain the difference in the two discharge rates.

5.26. Design an earth channel ($n = 0.03$) to carry a discharge of 283.3 cms (10,000 cfs) with 2H:1V side slopes, B/y_n of 6.0, and a velocity of 0.91 mps (3.0 fps). Use 3.6-m (12-ft) wide berms for each bank and the USBR equation for freeboard. Compute the tractive force at design capacity.

5.27. Design a grass-lined drainage channel ($n = 0.035$) to carry 283.3 cms (10,000 cfs) with 4H:1V side slopes, a B/y_n of 10, and a longitudinal slope of 0.0007. Use the USBR equation for freeboard. Compute the velocity, Froude Number, and tractive force at design capacity.

5.28. A grass-lined drainage channel ($n = 0.035$) carries a discharge of 283.3 cms (10,000 cfs) with 4H:1V side slopes, a maximum velocity of 1.52 mps (5.0 fps), and a longitudinal slope of 0.0007. Size the channel and compute the tractive force and Froude Number at design capacity. Use the USBR equation for freeboard.

5.29. Design a concrete-lined drainage channel ($n = 0.015$) to carry 283.3 cms (10,000 cfs) with 1.5H:1V side slopes, $B/y_n = 3$, and a longitudinal slope of 0.0007. Compute the velocity and Froude Number at design capacity.

5.30. Compute water surface profiles in a long rectangular concrete channel ($n = 0.015$). The channel is 10.0 m (32.8 ft) wide with a flow rate of 50 cms (1,765 cfs). The channel slope changes abruptly from 0.001 to 0.02.

5.31. A rectangular concrete channel ($n = 0.015$) is 3.0 m (9.84 ft) wide, has a longitudinal slope of 0.001, and a discharge rate of 5.0 cms (176 cfs). If the channel ends in a free outfall, compute the water surface profile for a distance of 150 m (500 ft) upstream of the outfall.

5.32. If the channel in Problem 5.21 ended in a free outfall, compute the water surface profile for a distance of 150 m (500 ft) upstream.

5.33. A concrete rectangular channel ($n = 0.015$) is 10.0 m (32.8 ft) wide and carries a discharge of 50 cms (1,764 cfs). Using the direct step method of water surface profile computation, compute the following profiles (a) for the M-2 portion of the channel, (b) for the S-2 portion 100 m (328 ft), and (c) for the M-3 portion of the channel.

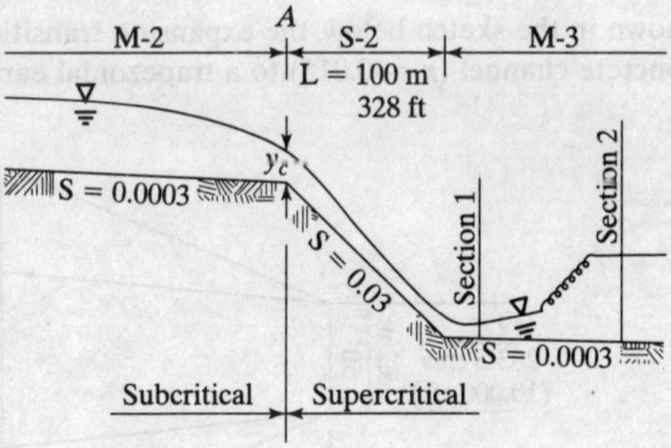

5.34. Compute the water surface profile upstream of a road crossing over a trapezoidal concrete-lined ($n = 0.015$) channel. The channel has a bottom width of 3.0 m (9.8 ft) and side slopes of 2H:1V. The slope of the channel is 0.0005 and the discharge rate is 20.0 cms (706 cfs). The starting water depth (y_1) upstream of the road crossing is 4.0 m (13.1 ft).

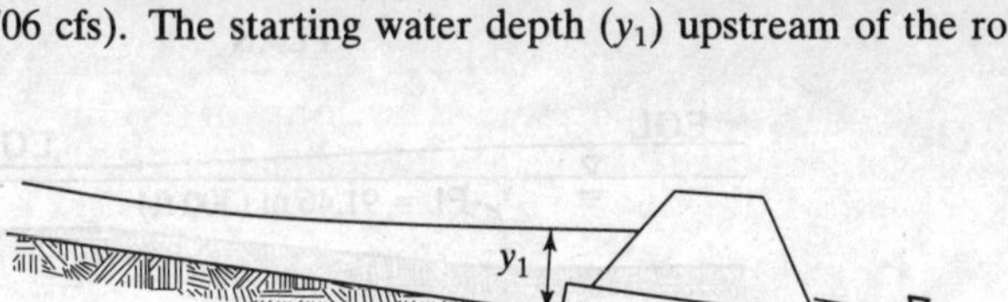

5.35. Water is flowing in a rectangular channel 12.2 m (40 ft) at a velocity of 1.0 mps (3.28 fps) and a water depth of 2.5 m (8.2 ft). What is the water depth in a gradual contraction in the channel to a width of 10.0 m (32.8 ft)? Determine the largest width in the contraction that will just cause critical depth to occur in the constriction.

5.36. A rectangular channel is 91.5 m (300 ft) wide and the normal depth is 2.74 m (9.0 ft) with an average velocity of 1.83 mps (6.0 fps). A roadway is to cross the channel using a bridge with equal encroachments from each side. Determine the water depth at Sections 2 and 3 for bridge lengths (B) of 30.5, 45.7, and 61.0 m (100, 150, and 200 ft). What is the maximum length of bridge that will cause critical depth to occur in the constriction? Neglect headloss and channel slope.

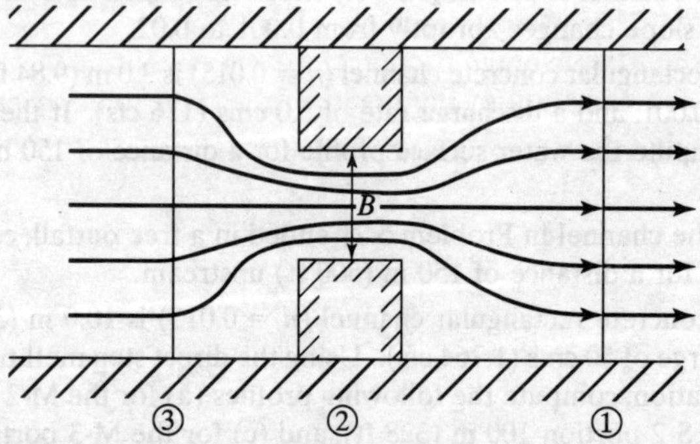

5.37. As shown in the sketch below, the expansion transition (n = 0.015) from a rectangular concrete channel (n = 0.015) to a trapezoidal earth channel (n = 0.035) is 61.0 m

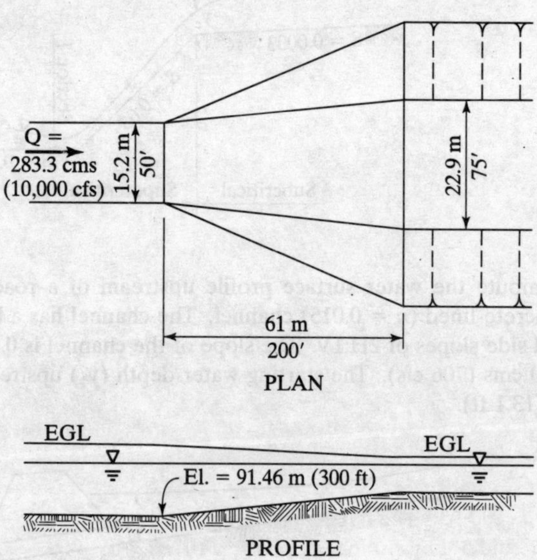

(200 ft) long. The rectangular channel has a bottom width of 15.2 m (50.0 ft) and a slope of 0.0005 while the trapezoidal channel has a bottom width of 22.9 m (75 ft), 4H:1V side slopes, and a longitudinal slope of 0.001. If the discharge in the channel is 283.3 cms (10,000 cfs), compute the water depth in each channel and the headloss in the transition based on the average friction slope and an expansion loss coefficient of 0.4. Compute the elevation of the invert of the trapezoidal channel at the end of the transition.

5.38. A contraction transition (n = 0.015) from a trapezoidal earth channel (n = 0.035) to a rectangular concrete channel (n = 0.015) is 15.2 m (50 ft) long. The trapezoidal channel has a bottom width of 22.9 m (75 ft), 4H:1V side slopes, and a longitudinal slope of 0.001 while the rectangular channel has a bottom width of 15.2 m (50 ft) and a longitudinal slope of 0.0005. If the discharge rate is 283.3 cms (10,000 cfs), compute the water depth in each channel and the headloss in the transition based on the average friction slope and a contraction loss coefficient of 0.2. Compute the elevation of the invert of the trapezoidal channel at the beginning of the transition.

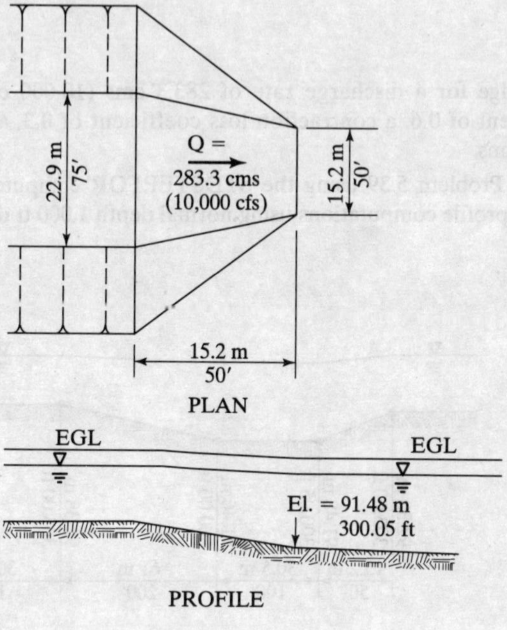

5.39. A bridge extends over a trapezoidal earth channel (n = 0.035) as shown below. The bridge section is represented by a rectangular concrete channel (n = 0.015) with a bottom width of 15.2 m (50 ft), a longitudinal slope of 0.0005, and a length of 30.5 m (100 ft). The earth channel has a bottom width of 22.9 m (75 ft), 4H:1V side slopes, and a longitudinal slope of 0.001. Compute the water surface profile through

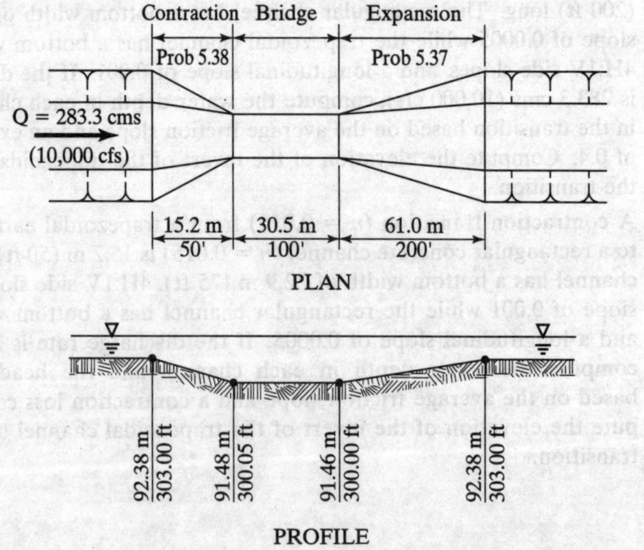

the bridge for a discharge rate of 283.3 cms (10,000 cfs). Use an expansion loss coefficient of 0.6, a contraction loss coefficient of 0.3, and n value of 0.015 for the transitions.

5.40. Repeat Problem 5.39 using the STDSTEP.FOR computer program. Start the water surface profile computations using normal depth 1,000 ft downstream of the expansion section.

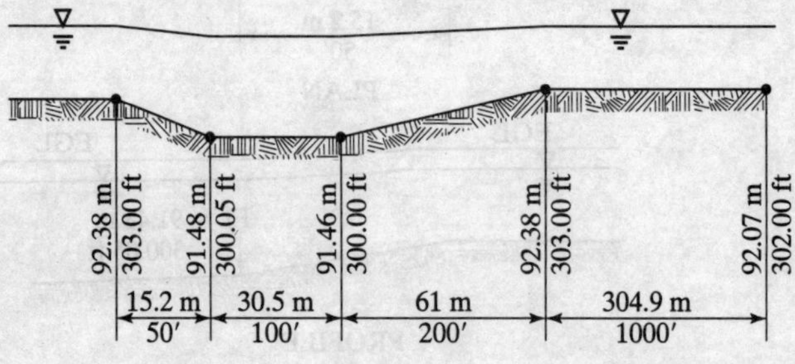

Approximate the rectangular section at the bridge using an eight point cross-section shown below.

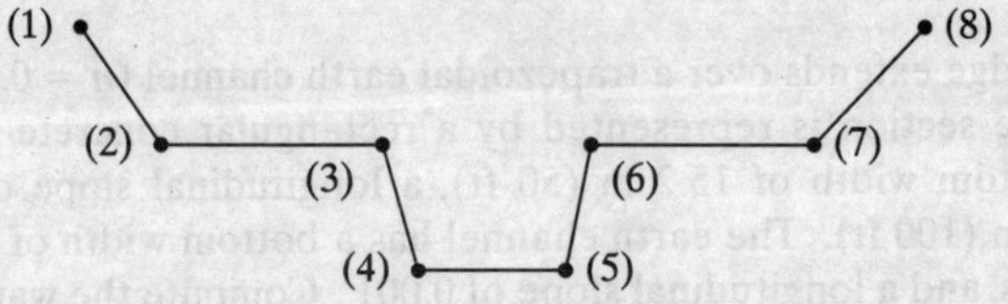

Approximate the rectangular section at the bridge using an eight point cross-section show below.

Point	Distance m	Distance ft	Elevation m	Elevation ft
1	0.00	0.00	103.66	340.00
2	3.05	10.00	100.91	331.00
3	60.97	200.00	100.61	330.00
4	60.98	200.01	91.46	300.00
5	76.22	250.00	91.46	300.00
6	76.23	250.01	100.61	330.00
7	121.95	400.00	100.91	331.00
8	125.00	410.00	103.66	340.00

5.41. A highway bridge extends across a natural stream ($n = 0.035$). The natural channel is approximated as a rectangular channel 61.0 m (200 ft) wide with a longitudinal slope of 0.001, while the bridge section is approximated as a rectangular earth channel ($n = 0.035$) 30.5 m (100 ft) wide. Compute the water surface profile through the bridge using an expansion coefficient of 0.6 and a contraction coefficient of 0.3.

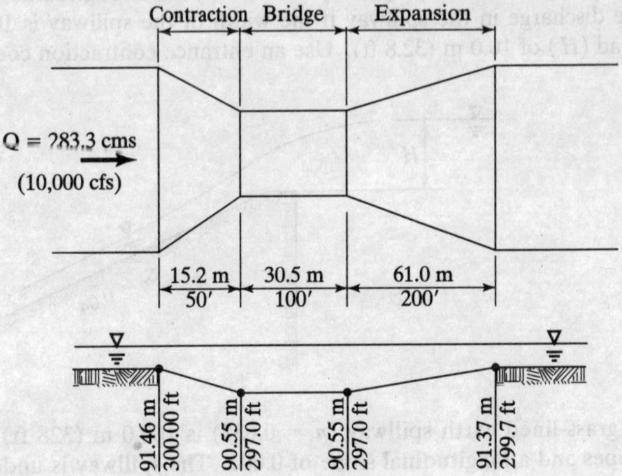

5.42. Repeat Problem 5.41 using the STDSTEP.FOR computer program. Start the profile computation using normal depth 305 m (1,000 ft) downstream of the expansion section where the invert elevation is 91.07 m (298.7 ft).

5.43. Compute the water surface profile at 30.5-m (100-ft) intervals for 305 m (1,000 ft) for discharge rates of 141.6, 283.3, 424.9, and 566.6 cms (5,000, 10,000, 15,000, and 20,000 cfs)

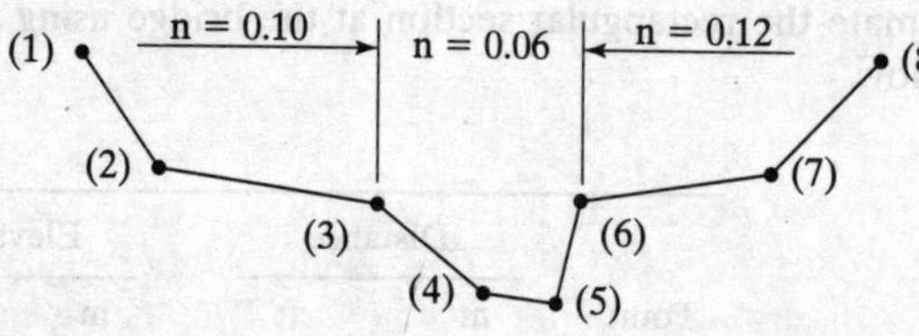

starting with critical depth. The channel slope is 0.001. A typical cross-section of the channel is shown below.

Point	Distance m	Distance ft	Elevation m	Elevation ft
1	0.0	0.0	304.9	1,000.0
2	15.2	50.0	298.8	980.0
3	76.2	250.0	298.2	978.0
4	82.3	270.0	295.7	970.0
5	91.5	300.0	295.7	970.0
6	100.6	330.0	297.9	977.0
7	152.4	500.0	298.5	979.0
8	167.7	550.0	304.9	1,000.0

5.44. A rectangular concrete spillway (n = 0.013) has a longitudinal slope of 0.04. Compute the discharge in the spillway if the width of the spillway is 100.0 m (328 ft) under a head (H) of 10.0 m (32.8 ft). Use an entrance contraction coefficient of 0.1.

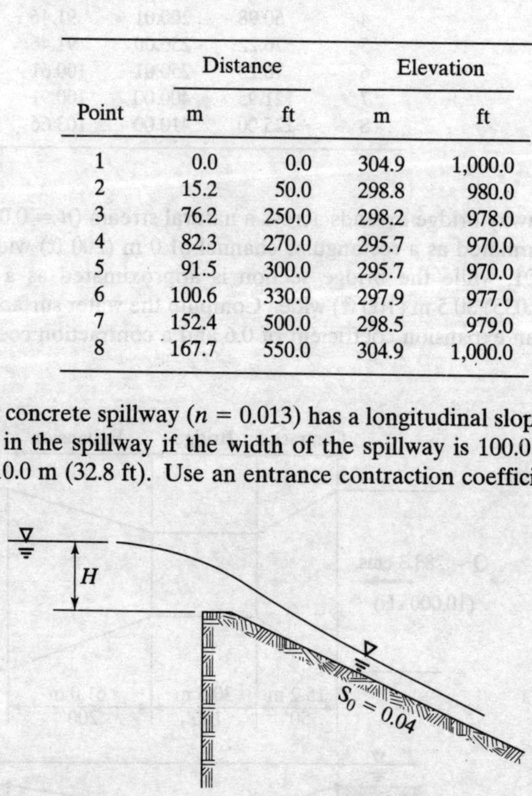

5.45. A grass-lined earth spillway (n = 0.035) is 100.0 m (328 ft) wide with 2H:1V side slopes and a longitudinal slope of 0.001. The spillway is under a head (H) of 3.0 m

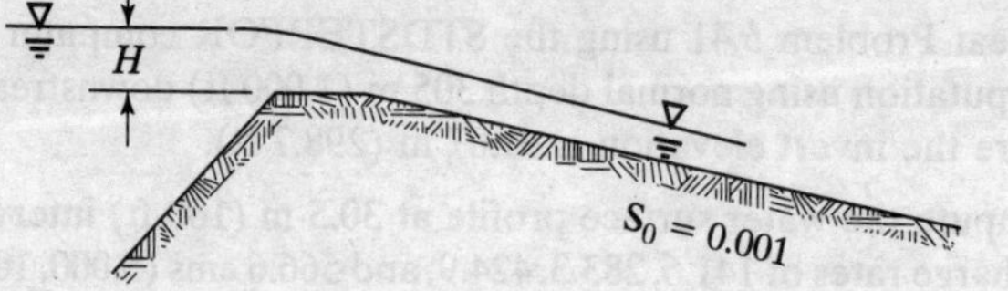

(9.8 ft). Compute the discharge in the spillway if the spillway is hydraulically long (normal depth occurs in the spillway below the entrance). Use an entrance loss coefficient of 0.3.

5.46. A culvert flowing partly full is the same as an open channel connecting two reservoirs. A 3.05 × 3.05 m (10 × 10 ft) box culvert (n = 0.015) 61.0 m (200 ft) long is used for cross-drainage under a highway. The headwater depth (H) is 3.05 m (10.0 ft), and the tailwater depth (y_t) is less than critical depth. Compute the discharge rate through the culvert for a culvert slope (S_o) of 0.02 and 0.002. Use a contraction loss coefficient at the entrance of 0.5 and an expansion loss coefficient at the outlet of 1.0.

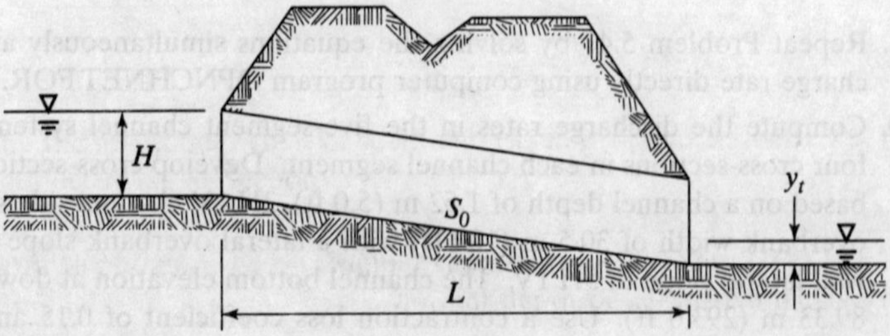

5.47. A grass-lined spillway channel (n = 0.035) extends from a reservoir 100.0 m (328 ft) on a slope of 0.001 under subcritical flow. At 100 m (328 ft) the slope of the channel increases to 0.03 and the flow becomes supercritical. Compute water surface profiles for a range of discharge rates using cross-sections spaced at 20-m (65.6-ft) intervals. Start the computations using critical depth. Plot the rating curve for the spillway and determine the discharge rate for a water surface elevation in the reservoir of 92.99 m (305.0 ft). Use a contraction loss coefficient of 0.3 at the entrance.

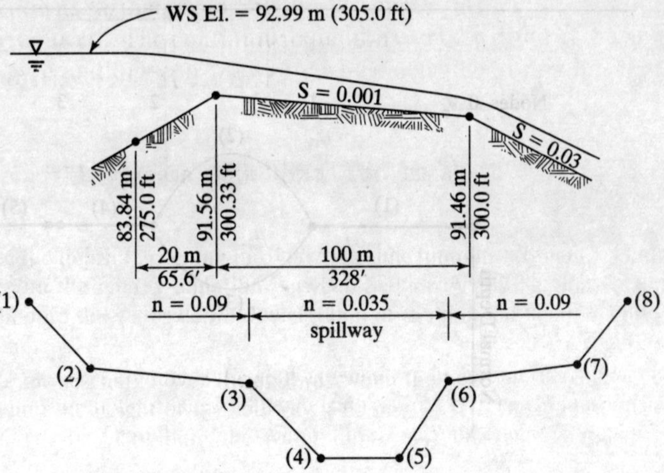

	Distance		Elevation	
Point	m	ft	m	ft
1	0.0	0.0	97.56	320.0
2	6.10	20.0	94.66	310.5
3	30.49	100.0	94.51	310.0
4	36.58	120.0	91.46	300.0
5	67.07	220.0	91.46	300.0
6	73.17	240.0	94.51	310.0
7	91.46	300.0	94.66	310.5
8	97.56	320.0	97.56	320.0

5.48. Repeat Problem 5.47 by solving the equations simultaneously and compute the discharge rate directly using computer program OPNCHNET.FOR.

5.49. Compute the discharge rates in the five-segment channel system shown below. Use four cross-sections in each channel segment. Develop cross-sections for each segment based on a channel depth of 1.52 m (5.0 ft), 2H:1V channel side slope, a left and right overbank width of 30.5 m (100 ft) with a lateral overbank slope of 0.01, and a valley wall lateral slope of 5H:1V. The channel bottom elevation at downstream boundary is 89.33 m (293.0 ft). Use a contraction loss coefficient of 0.15 and an expansion loss coefficient of 0.3.

	Bottom Width		Channel Length		Slope	Manning "n" Values		
Segment	m	ft	m	ft	%	LOB	CH	ROB
1	6.1	20	91.5	300	0.100	0.10	0.04	0.12
2	4.57	15	182.9	600	0.133	0.10	0.04	0.12
3	3.05	10	213.4	700	0.114	0.12	0.05	0.10
4	6.10	20	152.4	500	0.120	0.10	0.05	0.12
5	7.62	25	91.5	400	0.100	0.10	0.04	0.12

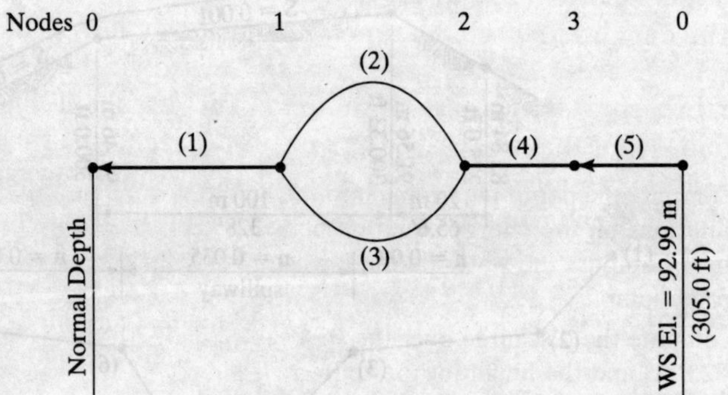

5.50. Repeat Problem 5.49, except the upstream boundary has been changed to a constant discharge of 141.6 cms (5,000 cfs), and the downstream boundary has been changed to critical depth. Compute the water surface elevation at each of the nodes.

5.51. A rectangular concrete channel has a bottom width of 6.0 m (19.7 ft), a water depth (y_1) of 3.0 m (9.8 ft) at Section 1, and a discharge rate of 20.0 cms (706 cfs). There is a rise or hump in the channel bottom (h) of 1.0 m (3.3 ft). Compute depth of flow (y_2) at the top of the rise, neglecting headloss and channel slope between Sections 1 and 2.

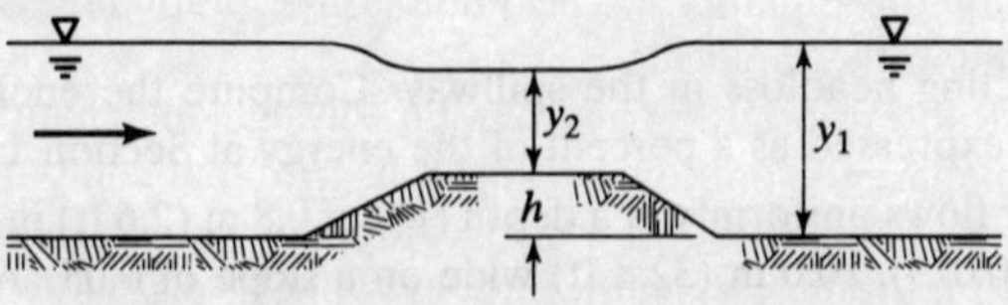

5.52. Compute the minimum height of the hump (h) in Problem 5.51 that will result in critical depth occurring at Section 2.

5.53. Compute the minimum height of a hump in a channel that will cause critical depth to occur over the hump. The rectangular channel (n = 0.015) is 3.0 m (9.8 ft) wide with a longitudinal slope of 0.001 and a normal depth of 2.0 m (6.6 ft). Neglect headloss caused by the hump.

5.54. A roadway crosses a wide, rectangular-shaped channel. The channel has a Manning "n" value of 0.05, a longitudinal slope of 0.001, a bottom width of 100 m (328 ft). The normal depth in the channel is 3.0 m (9.84 ft). Compute the depth of flow at the center of the roadway and at the upstream toe of the embankment for road heights of 0.61, 0.91, and 1.22 m (2.0, 3.0, and 4.0 ft). Neglect headlosses.

5.55. Compute the discharge rate through the gate shown below for y_1 equal to 3.0 m (9.8 ft) and y_2 equal to 1.0 m (3.3 ft). The gate is 2.0 m (6.6 ft) wide. Neglect headloss through the gate.

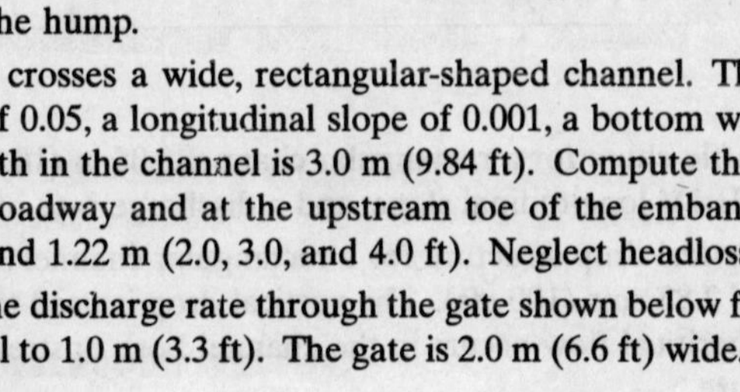

5.56. Compute the depth (y_1) going into the hydraulic jump in Problem 5.33. Base the computations on the energy equation between Section A and Section 1, neglecting headloss. Compute the water depth downstream of the jump at Section 2 and the headloss in the jump.

5.57. Compute the discharge over the spillway when the head on the spillway (h_1) is 10.0 m (32.8 ft) and the height of the spillway is 60 m (197 ft). The spillway discharge coefficient (K_R) is 0.5. Compute the water depth (y_1) upstream of the hydraulic jump,

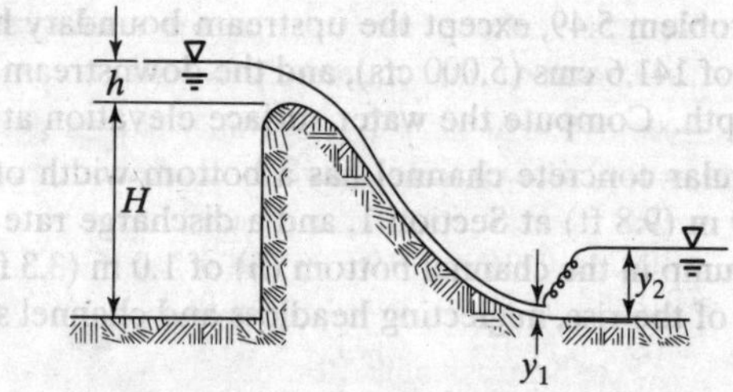

neglecting headloss in the spillway. Compute the energy dissipated in the hydraulic jump expressed as a percent of the energy at Section 1.

5.58. Water flows uniformly at a depth (y_1) of 0.8 m (2.6 ft) in a rectangular concrete channel (n = 0.015), 10.0 m (32.8 ft) wide on a slope of 0.05. A sill is installed in the channel to cause a hydraulic jump to form. Compute the height of the hydraulic jump (y_2) and the energy lost in the jump expressed as a percent of the energy at 1.

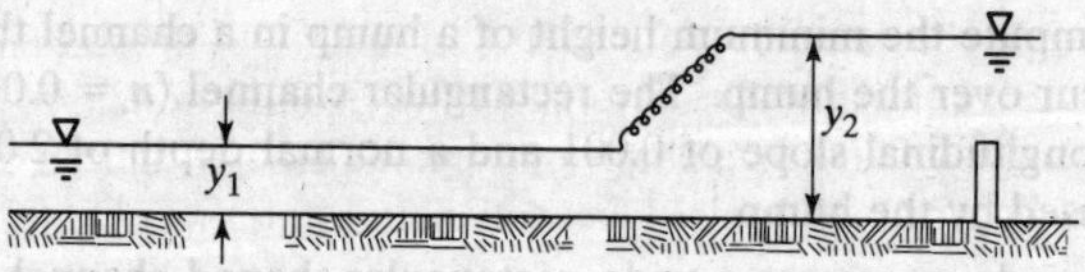

5.59. Design a baffle chute for a rectangular channel 3.05 m (10.0 ft) wide, 12.20 m (40.0 ft) long on a 2H:1V longitudinal slope, and a discharge rate of 2.83 cms (100 cfs).

5.60. Design a vertical drop structure for a rectangular channel 3.05 m (10.0 ft) wide with a discharge of 2.83 cms (100 cfs). The vertical drop is used to lower the channel 1.52 m (5.0 ft). Subcritical flow occurs in the channel both upstream and downstream of the drop structure.

5.61. Compute the water depth (y_2) downstream of the hydraulic jump for a head (h) of 5.0 m (16.4 ft) and a spillway height (H) of 30.0 m (98.4 ft). The spillway discharge coefficient is 0.46. Neglect headloss on the spillway.

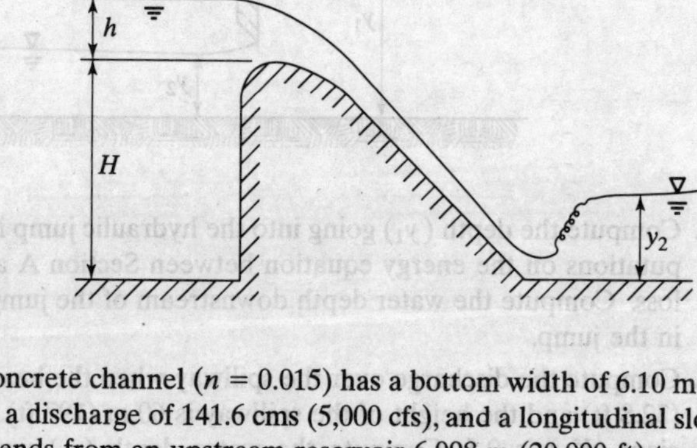

5.62. A trapezoidal concrete channel (n = 0.015) has a bottom width of 6.10 m (20.0 ft), side slopes of 2H:1V, a discharge of 141.6 cms (5,000 cfs), and a longitudinal slope of 0.0001. The channel extends from an upstream reservoir 6,098 m (20,000 ft) to a downstream sluice gate. Use an entrance loss coefficient at the reservoir of 0.5 where the channel invert elevation is 91.46 m (300.0 ft). If the sluice gate is closed in 600 seconds, compute the velocity and water depth along the channel after 1,200, 1,800, and 2,400 seconds.

5.63. The sluice gate in Problem 5.62 is partially closed such that the discharge in the channel is 70.8 cms (2,500 cfs). If the sluice gate is then opened in 600 seconds, compute the velocity and water depth along the channel after 1,200, 1,800, and 2,400 seconds. (Note: Program MACK requires modification for the problem.)

5.64. Modify program MACK.FOR for an upstream inflow hydrograph and a downstream rating curve. At the upstream boundary, the discharge is specified as a function of time and the minus characteristic equation can be used to compute the depth and velocity. At the downstream boundary, use the plus characteristic equation and the Manning equation to compute the depth and velocity. Use the modified program to route the following hydrograph through the trapezoidal channel in Example 5.11. Note that the computational time step (*dt*) is different from that of the inflow hydrograph, and the inflow must be interpolated at any time from the inflow hydrograph.

Time	Discharge		Time	Discharge		Time	Discharge	
hrs	cms	cfs	hrs	cms	cfs	hrs	cms	cfs
0	10	353	10	140	4,940	20	35	1,240
1	20	706	11	120	4,240	21	30	1,060
2	40	1,412	12	110	3,880	22	25	880
3	80	2,824	13	100	3,530	23	20	710
4	140	4,940	14	90	3,180	24	18	640
5	200	7,060	15	80	2,820	25	16	560
6	195	6,880	16	70	2,470	26	14	490
7	`190	6,710	17	60	2,120	27	13	460
8	180	6,350	18	50	1,760	28	12	420
9	160	5,650	19	40	1,410	29	11	390

BIBLIOGRAPHY

CHANSON, H., *The Hydraulics of Open Channel Flow: An Introduction,* Arnold Publishers, London, UK, 1999.

CHAUDHRY, M. H., *Open-Channel Flow,* Prentice-Hall, Upper Saddle River, NJ, 1993.

CHOW, V. T., *Open-Channel Hydraulics,* McGraw-Hill, New York, NY, 1959.

FRENCH, R. H., *Open-Channel Hydraulics,* McGraw-Hill, New York, NY, 1985.

GRAF, W. H., *Fluvial Hydraulics,* John Wiley & Sons, New York, NY, 1998.

HENDERSON, F. M., *Open Channel Flow,* Macmillan, New York, NY, 1966.

JAIN, S. C., *Open Channel Flow,* John Wiley & Sons, New York, NY, 2001.

MAYS, L. W. (Ed.), *Hydraulic Design Handbook,* McGraw-Hill, New York, NY, 1999.

MONTES, S., *Hydraulics of Open Channel Flow,* American Society of Civil Engineers Press, Reston, VA, 1998.

NOVAK, P., A. I. B. MOFFAT, C. NALLURI, and R. NARAYANAN, *Hydraulic Structures,* Chapman & Hall, London, UK, 1996.

STURM, T. W., *Open Channel Hydraulics,* McGraw Hill, New York, NY, 2001.

6

Flood Routing

Routing is a process used to determine the variations in flow rate for a flood wave as it moves through a watercourse. Routing procedures are classified as hydrologic or hydraulic. Either approach may be applied to natural and man-altered streams, but hydraulic routing is generally more accurate. Hydrologic routing works better for reservoirs than for streams and channels. Routing methods may also be applied to precipitation runoff in a watershed.

Hydrologic routing is often called storage routing. It uses a storage form of the continuity equation combined with a relationship between storage and discharge. A variety of hydrologic routing techniques are reported in the literature. Two commonly applied methods are presented in this chapter: (1) storage-outflow and (2) Muskingum. The storage-outflow method is based on developing a relationship between storage and outflow. The Muskingum method is based on the premise that storage is a function of weighted inflow and outflow. Hydrologic routing results in a discharge hydrograph. The stage for the peak discharge is determined separately using methods from Chapter 5.

Hydraulic routing involves simultaneously computing both stage and discharge as a function of location and time. Dynamic routing is based on the complete St. Venant equations (Eq. 5.88) representing conservation of mass and momentum for unsteady flow in open channels. Other variations of hydraulic routing are based on simplified approximations. Kinematic routing is a simplified hydraulic routing procedure often used for overland flow. In kinematic routing, the momentum equation is replaced with the Manning equation (Eq. 5.7). Both dynamic and kinematic routing are covered in this chapter.

6.1 HYDROLOGIC ROUTING

Hydrologic routing consists of computing the outflow hydrograph corresponding to a given inflow hydrograph. Hydrographs may be routed through a reservoir (Fig. 6.1), stream or river reach (Fig. 6.2), or other container. In Section 8.6.6, hydrologic routing is applied to a watershed to convert precipitation runoff to a hydrograph at the outlet. Stream and reservoir routing are also commonly incorporated as components of the watershed models explored in Chapter 8.

Storage (S), inflow (I), and outflow (O) at an instant in time are related by the following form of the continuity equation.

$$\frac{dS}{dt} = I - O \tag{6.1}$$

For a time interval Δt, the continuity equation may be written in terms of average inflow ($\overline{I}$) and outflow ($\overline{O}$) as

$$\frac{\Delta S}{\Delta t} = \overline{I} - \overline{O} \tag{6.2}$$

Referring to Fig. 6.3, the area between the inflow and outflow hydrographs for a given time interval Δt represents a change in storage volume. If the average inflow

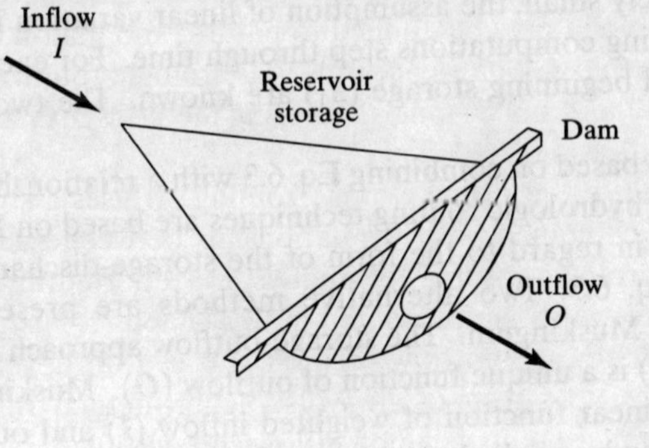

Figure 6.1 Reservoir routing.

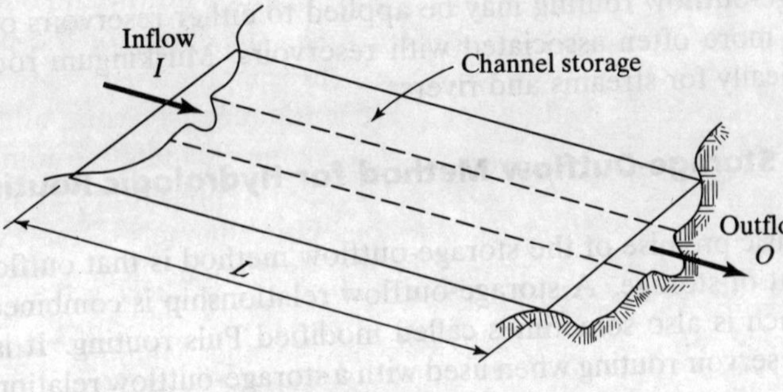

Figure 6.2 Stream routing.

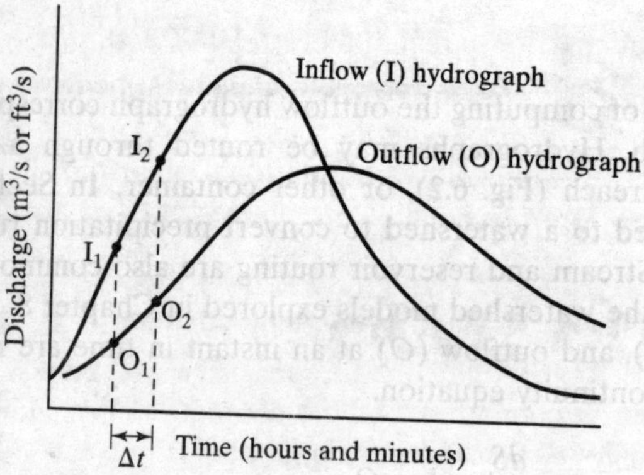

Figure 6.3 Hydrologic routing.

$(\bar{I})$ is greater than the average outflow $(\bar{O})$ during Δt, the change in storage (ΔS) will be positive. S decreases if $\bar{I}$ is less than $\bar{O}$. The effect of storage in decreasing the peak flow and broadening the time base of the hydrograph is called *attenuation*.

For computational purposes, Eq. 6.2 is rewritten as follows.

$$\frac{S_2 - S_1}{\Delta t} = \frac{I_1 + I_2}{2} - \frac{O_1 + O_2}{2} \tag{6.3}$$

where subscripts 1 and 2 refer to the beginning and end of the computational time interval Δt. If Δt is relatively small, the assumption of linear variation in discharge rate is adequate. The routing computations step through time. For each time step, the inflows (I_1 and I_2) and beginning storage (S_1) are known. The two unknowns are O_2 and S_2.

Hydrologic routing is based on combining Eq. 6.3 with a relationship between storage and discharge. All hydrologic routing techniques are based on Eq. 6.3. The alternative methods differ in regard to the form of the storage-discharge relation that is combined with Eq. 6.3. Two alternative methods are presented here: (1) storage-outflow and (2) Muskingum. The storage-outflow approach is based on the premise that storage (S) is a unique function of outflow (O). Muskingum routing relates storage S to a linear function of weighted inflow (I) and outflow (O). Storage-outflow routing may be applied to either reservoirs or streams and rivers, but is more often associated with reservoirs. Muskingum routing was developed specifically for streams and rivers.

6.1.1 Storage-Outflow Method for Hydrologic Routing

The basic premise of the storage-outflow method is that outflow is known for any amount of storage. A storage-outflow relationship is combined with Eq. 6.3. This approach is also sometimes called modified Puls routing. It is often called level-pool reservoir routing when used with a storage-outflow relationship for a reservoir. If reservoir storage is large compared with the outflow rate, the water surface is nearly horizontal, storage and outflow are uniquely related without considering inflow, and the basic premise of storage-outflow routing is most nearly valid.

6.1.1.1 Routing algorithm. The computational algorithm is based on rearranging Eq. 6.3 with the unknowns O_2 and S_2 grouped together as follows.

$$\left[\frac{2S_2}{\Delta t} + O_2\right] = I_1 + I_2 + \left[\frac{2S_1}{\Delta t} - O_1\right] \tag{6.4}$$

At each time step, the terms to the right of the equal sign are known, and the $(2S/\Delta t + O)$ term on the left is computed. As the computational algorithm advances to the next time step, $(2S/\Delta t - O)$ is determined as

$$\left[\frac{2S}{\Delta t} - O\right] = \left[\frac{2S}{\Delta t} + O\right] - 2O \tag{6.5}$$

A relationship between the term on the left of Eq. 6.4 and outflow O

$$\frac{2S}{\Delta t} + O \quad \text{versus} \quad O$$

is required to determine O. This relationship may be in the format of a table, graph, or equation. A table requires interpolation. A graph must be read.

Equation 6.4 is applied in identically the same manner for either reservoir or stream routing. However, procedures for developing the $(2S/\Delta t + O)$ versus O relationship differ between reservoirs and stream reaches. Examples 6.1 and 6.2 illustrate reservoir and stream routing, respectively.

6.1.1.2 Storage-outflow relationship for a reservoir. An elevation versus storage volume relationship is developed from a topographic map of a reservoir site. The horizontal area A enclosed by each contour and the dam is determined. The incremental storage volume ΔS between two contours is computed as

$$\Delta S = \left(\frac{A_{\text{bottom}} + A_{\text{top}}}{2}\right)(\text{contour interval}) \tag{6.6}$$

The total reservoir storage S below a given contour elevation is the sum of all incremental volumes below that elevation. Elevation-storage volume plots or tables are available for most existing major reservoirs and can be developed from a topographic map for proposed new projects.

An elevation versus outflow relationship, often called an outlet rating curve, is developed based on the hydraulics of the outlet structures. Flow through spillways and other outlets can often be related to the water surface elevation in the reservoir by weir and/or orifice equations. For uncontrolled (ungated) outlet structures, there is a single elevation-outflow relationship. For gated outlet structures, the elevation-outflow relationship varies with gate openings.

The S versus elevation and O versus elevation relationships are combined to develop a relationship between S and O. This relationship allows simple computation of a relationship between O and $(2S/\Delta t + O)$. The computations are illustrated by Example 6.1.

Example 6.1 Storage-Outflow Routing through a Reservoir

A dam has an uncontrolled weir spillway 10 m wide with a crest elevation of 548.0 m and a discharge coefficient of (K_R in Eq. 5.57) of 0.45. The reservoir water surface elevation versus area relationship provided in columns 1 and 2 of the table was developed from a topographic map. The elevation versus discharge relationship (columns 1 and 5) is computed using the weir equation (Eq. 5.57). The inflow hydrograph provided in columns 1 and 2 of the routing table is to be routed through the reservoir. The starting water surface elevation in the reservoir is 544.0 m.

Reservoir storage is computed based on Eq. 6.6.

$$S_{i+1} = S_i + (A_{i+1} + A_i)\frac{\Delta h}{2}$$

where $\Delta h = 2$ m.

The discharge rate through the spillway is computed from Eq. 5.57

$$Q = K_R\sqrt{2g}\,L_w H^{3/2} = 0.45\sqrt{19.62} \times 10\,H^{3/2} = 19.9\,H^{3/2}$$

The last column in the storage discharge table is computed as

$$\frac{2S}{\Delta t} + O$$

where Δt is 3,600 seconds and the units of $2S/\Delta t$ are m^3/s.

RESERVOIR STORAGE-OUTFLOW RELATIONSHIP

Elevation (h) m (1)	Area (A) m^2 (2)	Storage (S) 1000 m^3 (3)	Head (H) m (4)	Outflow (O) m^3/s (5)	$\frac{2S}{\Delta t} + O$, m^3/s (6)
530	0	0	0	0	0
532	1,000	1	0	0	0.5
534	2,000	4	0	0	2.2
536	5,000	11	0	0	6.1
538	9,000	25	0	0	13.9
540	20,000	54	0	0	30
542	35,000	109	0	0	60
544	60,000	204	0	0	113
546	110,000	374	0	0	208
548	200,000	684	0	0	380
550	330,000	1,214	2	56	730
552	430,000	1,974	4	159	1,256
554	550,000	2,954	6	292	1,933
556	700,000	4,204	8	450	2,785
558	1,000,000	5,804	10	630	3,854

The routing table is organized in columns representing the terms in the continuity equation (Eq. 6.4) and rows representing time. The water surface elevation at time 0 is given as 544.0 m with a value $2S/\Delta t + O$ of 113 m^3/s and a discharge of 0 (from the storage discharge table). On any time line, column 3 in the routing table is equal to column 4 minus twice the outflow in column 5. Following the continuity equation I_2 (column 2, line 2) plus I_1 (column 2, line 1) plus $2S_1/\Delta t - O_1$ (column 3, line 1) equals $2S_2/\Delta t + O_2$ (column 4, line 2). The value of O_2 is interpolated from the storage/discharge table for the value of $2S_2/\Delta t + O_2$. This process is repeated until the outflow hydrograph approaches zero. The inflow and outflow hydrographs are plotted below.

The Program RESROUTE.FOR performs hydrologic routing computations. An outflow hydrograph computed with RESROUTE is nearly identical to that in the routing table computed using manual computations.

ROUTING TABLE

Time (t) hrs (1)	Inflow (I) m^3/s (2)	$\frac{2S_1}{\Delta t} - O_1$, m^3/s (3)	$\frac{2S_2}{\Delta t} + O_2$, m^3/s (4)	Outflow (O) m^3/s (5)
0	10	113	113	0
1	90	213	213	0
2	300	533	603	35
3	400	925	1,233	154
4	500	1,285	1,825	270
5	450	1,555	2,235	340
6	400	1,645	2,405	380
7	300	1,605	2,345	370
8	250	1,495	2,155	330
9	200	1,365	1,945	290
10	150	1,215	1,715	250
11	125	1,070	1,490	210
12	100	955	1,295	170
13	75	870	1,130	130
14	65	780	1,010	115
15	50	727	895	84
16	40	667	817	75
17	30	625	737	56
18	20	585	675	45
19	10	545	615	35
20	0	505	555	25
21	0	465	505	20
22	0	435	465	15
23	0	415	435	10
24	0	405	415	5
25	0	397	405	4

```
EXAMPLE 6.1 RESERVOIR ROUTING OF HYDROGRAPH
&HYDR
IUN =           1
NH =           25
DT =      1.000000
NR =           15
ELST =    544.000000
1        10.00        .00
2        90.00        .00
3       300.00      35.76
4       400.00     154.51
5       500.00     270.62
6       450.00     347.64
7       400.00     376.34
8       300.00     366.57
9       250.00     332.60
10      200.00     292.68
11      150.00     246.47
12      125.00     203.65
13      100.00     167.83
14       75.00     136.29
15       65.00     110.27
16       50.00      89.56
17       40.00      72.07
18       30.00      57.52
19       20.00      46.83
20       10.00      36.65
21         .00      26.53
22         .00      18.04
23         .00      12.27
24         .00       8.35
25         .00       5.68
26         .00       3.86
27         .00       2.63
28         .00       1.79
29         .00       1.21
30         .00        .83
31         .00        .56
32         .00        .38
33         .00        .26
34         .00        .18
35         .00        .12
TOTAL INFLOW VOLUME =      1.283400E+07 CU M
TOTAL OUTFLOW VOLUME =     1.233647E+07 CU M
```

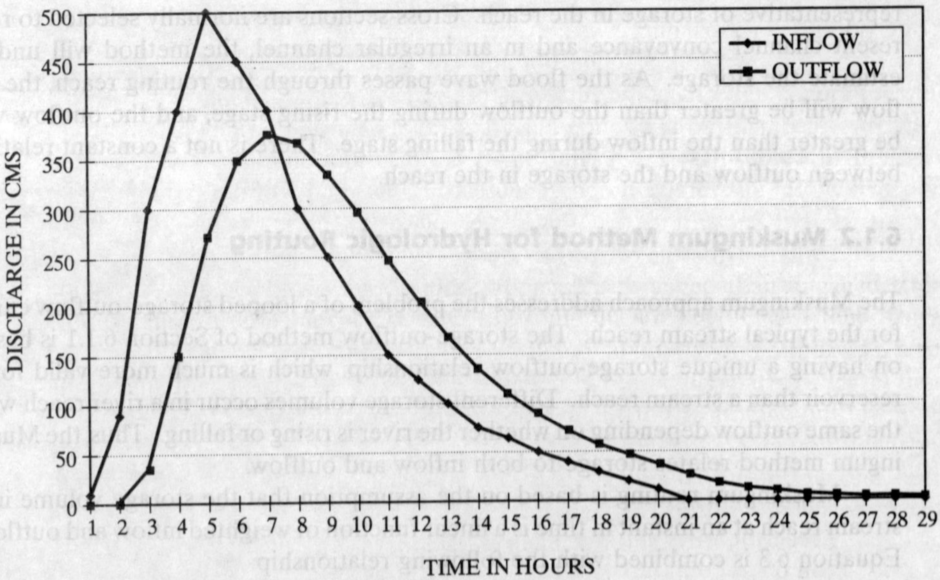

6.1.1.3 Storage-outflow relationship for a river reach. An S versus O relationship for a stream reach is approximated based on developing water surface profiles for various discharges using open channel hydraulics methods of Chapter 5. The standard step method (Section 5.7) is usually applied for water surface profile computations. This general approach can also be applied to reservoir routing with a nonlevel water surface by using a starting water surface elevation from the outlet structure elevation-outflow relationship and computing water surface profiles through the reservoir.

The storage S in a stream reach (Fig. 6.2) can be estimated as

$$S = AL \tag{6.7}$$

where A is the cross-sectional area and L is the reach length. The reach may be subdivided into any number of subreaches with the water surface elevation computed at the cross-sections defining the ends of the subreaches. The storage volume S is determined as

$$S = \sum_{i=1}^{N-1} (A_i + A_{i+1})L_{i+1} \tag{6.8}$$

where the subscript refers to the cross-section locations in the routing reach, A is cross-sectional flow area, and L is the distance to the downstream cross-section. Water surface profiles are computed and storage volumes determined for a number of discharge rates. The discharge incorporated in the water surface profile computations is assumed to be the outflow. The results provide an S-O relationship that is converted to a $(2S/\Delta t + O)$ versus O relationship.

The storage-outflow routing method should be used with caution for streams. The method assumes a uniform discharge rate in the reach, and cross-sections are representative of storage in the reach. Cross-sections are normally selected to represent channel conveyance and in an irregular channel, the method will underestimate the storage. As the flood wave passes through the routing reach, the inflow will be greater than the outflow during the rising stage, and the outflow will be greater than the inflow during the falling stage. There is not a constant relation between outflow and the storage in the reach.

6.1.2 Muskingum Method for Hydrologic Routing

The Muskingum approach addresses the problem of a looped storage-outflow curve for the typical stream reach. The storage-outflow method of Section 6.1.1 is based on having a unique storage-outflow relationship, which is much more valid for a reservoir than a stream reach. Different storage volumes occur in a river reach with the same outflow depending on whether the river is rising or falling. Thus, the Muskingum method relates storage to both inflow and outflow.

Muskingum routing is based on the assumption that the storage volume in a stream reach at an instant in time is a linear function of weighted inflow and outflow. Equation 6.3 is combined with the following relationship

$$S = K(xI + (1 - x)O) \tag{6.9}$$

to obtain the Muskingum routing equation (Eq. 6.10).

$$O_2 = C_1 I_2 + C_2 I_1 + C_3 O_1 \tag{6.10a}$$

where

$$C_1 = \frac{0.5\Delta t - Kx}{K - Kx + 0.5\Delta t} \tag{6.10b}$$

$$C_2 = \frac{0.5\Delta t + Kx}{K - Kx + 0.5\Delta t} \tag{6.10c}$$

$$C_3 = \frac{K - Kx - 0.5\Delta t}{K - Kx + 0.5\Delta t} \tag{6.10d}$$

$$C_1 + C_2 + C_3 = 1$$

The inflow (I) and outflow (O) are the flow rates at the upstream and downstream ends of the reach. Subscripts 1 and 2 for I and O refer to the beginning and end of the time interval Δt. S is the storage volume in the reach at an instant in time. The Muskingum method has two parameters, x and K. The weighting factor x is a dimensionless number between 0.0 and 1.0 representing the relative influence of I versus O in determining S. For most natural channels, x ranges between 0.1 and 0.3. The parameter K has units of time. K and Δt must be in the same units (minutes, hours, and days, etc.).

The flow area (A) in Eq. 6.7 can be replaced with Q/V yielding

$$S = Q\frac{L}{V} \tag{6.11}$$

where L/V is the travel time through the reach. Comparison of Eqs. 6.9 and 6.11 suggests that K represents the travel time through the reach.

The parameters x and K are normally established through calibration studies (Section 8.2.3) based on gaged hydrographs at either end of the reach. A reach may be divided into multiple subreaches with an x and K for each subreach. In this case, the computed outflow hydrograph for a subreach becomes the inflow hydrograph for the next downstream subreach. The number of subreaches may be treated as another parameter to be calibrated. The Muskingum method is limited by the availability of observed flows and is more applicable to gaged than ungaged streams. Rough estimates for ungaged streams are sometimes made. Since x tends to range between 0.1 and 0.3 for natural streams, a value of about 0.2 or 0.25 provides a rough approximation. Values for K may be assigned based on estimates of travel time.

If x is zero, Muskingum routing is equivalent to the storage-outflow method with $S = KO$. This is called a linear reservoir. With the storage-outflow routing method, the peak outflow occurs at the intersection of the inflow and outflow hydrographs. Inflows exceed outflows before the peak outflow and vice versa after. This is not necessarily the case in Muskingum routing with nonzero x.

Example 6.2 Hydrologic Stream Routing

The inflow hydrograph from Example 6.1 is routed through the 4.5-mile reach of No Name Creek as shown in the sketch below using alternatively the (a) storage-outflow and (b) Muskingum routing methods. The storage-outflow routing method requires a

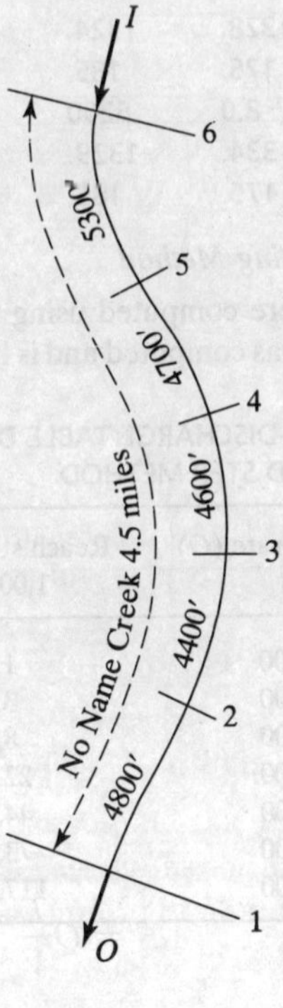

Stream rounting reach.

storage-outflow relationship for the channel reach. The standard step method (Chapter 5) is used to develop water surface profiles for use in developing a storage-outflow table. Six cross-sections were scaled from a topographic map of the stream reach and are included in the input file (STDSTEP.DAT) listed below. The Manning roughness values for the channel were estimated from aerial photographs. For the Muskingum routing method, a value of x of 0.25 is assumed. K is set equal to the travel time for the peak discharge rate computed in the storage-outflow method routing. Δt for the inflow hydrograph is equal to 3,600 seconds.

```
EXAMPLE 6.2 ROUTING HYDROGRAPH THROUGH REACH OF NO NAME CREEK
SC    00      0.001    0.001    0.001    0.001    0.001    0.001    0.001
QT            100.     300.     800.     2000.    5000.    10000.   20000.
MN            0.07     0.04     0.08     .15      0.3
XS            1.0      8.0      0        0        0
XZ    1320.   1310.    1308.    1304.    1304.    1307.    1309.    1320.
XX    0.      50.      175.     185.     215.     225.     350.     400.
XS            2.0      8.0      4800     4800     4800
XZ    1325.   1315.    1313.    1309.    1309.    1312.    1314.    1325.
XX    0.      30.      150.     160.     200.     210.     354.     400.
XS            3.0      8.0      4500     4400     4300
XZ    1330.   1320.    1317.    1315.    1315.    1318.    1319.    1330.
XX    0.      40.      200.     220.     250.     260.     360.     490.
XS            4.0      8.0      4700     4600     4500
XZ    1335.   1324.    1322.    1319.    1319.    1323.    1325.    1335.
XX    0.      50.      175.     185.     215.     225.     350.     400.
XS            5.0      8.0      4700     4700     4700
XZ    1340.   1330.    1328.    1324.    1324.    1329.    1330.    1340.
XX    0.      30.      175.     185.     215.     225.     370.     410.
XS            6.0      8.0      5200     5300     5400
XZ    1345.   1336.    1334.    1329.    1329.    1334.    1335.    1345.
XX    0.      50.      175.     185.     215.     225.     350.     400.
```

(a) *Storage-Outflow Routing Method*

Water surface profiles were computed using the standard step input file listed above. The storage in the reach was computed and is listed in the storage-discharge table below.

STORAGE-DISCHARGE TABLE DEVELOPED USING STANDARD STEP METHOD

Discharge rate (O) cfs	Reach storage (S) 1,000 ft^3	$\frac{2S}{\Delta t} + O$ (cfs)
100	1,600	990
300	3,350	2,160
800	8,830	5,700
2,000	22,250	14,360
5,000	44,660	29,800
10,000	73,050	50,600
20,000	117,520	85,300

The discharge rate in column 1 was plotted against the value of $2S/\Delta t + O$ in column 3 to assist in stream routing using Eq. 6.4. The routing table for the standard step storage method is listed below.

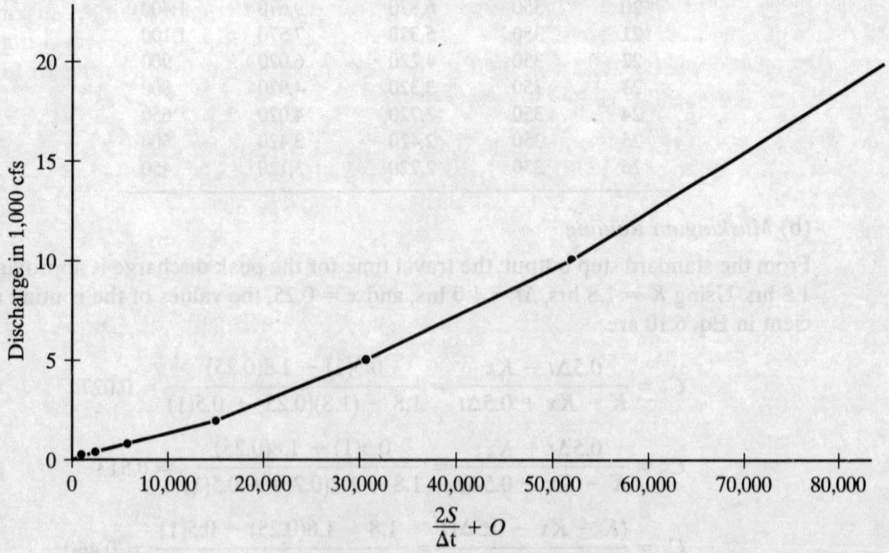

ROUTING TABLE FOR STORAGE-OUTFLOW METHOD FOR EXAMPLE 6.2

Time hrs	Inflow (I) cfs	$\frac{2S_1}{\Delta t} - O_1$ cfs	$\frac{2S_2}{\Delta t} + O_2$ cfs	O cfs
1	350	1,500	2,200	350
2	3,180	3,530	5,030	750
3	10,600	12,110	17,310	2,600
4	14,100	23,400	36,800	6,700
5	17,600	32,500	55,100	11,300
6	15,900	36,600	66,000	14,700
7	14,100	37,000	66,600	14,800
8	10,600	34,900	61,700	13,400
9	8,820	31,920	54,320	11,200
10	7,060	29,000	47,800	9,400
11	5,300	25,760	41,360	7,800
12	4,410	22,670	35,470	6,400
13	3,530	20,410	30,610	5,100
14	2,650	17,590	26,590	4,500
15	2,300	15,140	22,540	3,700
16	1,760	13,200	19,200	3,000
17	1,410	11,570	16,370	2,400
18	1,060	10,040	14,040	2,000
19	710	8,610	11,810	1,600

Time hrs	Inflow (I) cfs	$\frac{2S_1}{\Delta t} - O_1$ cfs	$\frac{2S_2}{\Delta t} + O_2$ cfs	O cfs
20	350	6,870	9,670	1,400
21	350	5,370	7,570	1,100
22	350	4,220	6,020	900
23	350	3,320	4,920	800
24	350	2,720	4,020	650
25	350	2,420	3,420	500
26	350	2,220	3,120	450

(b) *Muskingum Routing*

From the standard step output, the travel time for the peak discharge is approximately 1.8 hrs. Using $K = 1.8$ hrs, $\Delta t = 1.0$ hrs, and $x = 0.25$, the values of the routing coefficient in Eq. 6.10 are

$$C_1 = \frac{0.5\Delta t - Kx}{K - Kx + 0.5\Delta t} = \frac{0.5(1) - 1.8(0.25)}{1.8 - (1.8)(0.25) + 0.5(1)} = 0.027$$

$$C_2 = \frac{0.5\Delta t + Kx}{K - Kx + 0.5\Delta t} = \frac{0.5(1) + 1.8(0.25)}{1.8 - 1.8(0.25) + 0.5(1)} = 0.513$$

$$C_3 = \frac{(K - Kx - 0.5\Delta t)}{K - Kx + 0.5\Delta t} = \frac{1.8 - 1.8(0.25) - 0.5(1)}{1.8 - (1.8)(0.25) + 0.5(1)} = 0.460$$

The Muskingum routing equation is

$$O_2 = C_1 I_2 + C_2 I_1 + C_3 O_1 = 0.027 I_2 + 0.513 I_1 + 0.460 O_1$$

The computations are summarized in the routing table.

MUSKINGUM ROUTING TABLE FOR EXAMPLE 6.2

Time hrs	Inflow (I) cfs	Outflow (O) cfs
1	350	350
2	3,180	435
3	10,600	2,140
4	14,100	6,810
5	17,600	10,850
6	15,900	14,400
7	14,100	15,200
8	10,600	14,500
9	8,820	12,300
10	7,060	10,400
11	5,300	8,540
12	4,410	6,760
13	3,530	5,460
14	2,650	4,390

Time hrs	Inflow (I) cfs	Outflow (O) cfs
15	2,300	3,370
16	1,760	2,780
17	1,410	2,180
18	1,060	1,750
19	710	1,370
20	350	1,000
21	350	650
22	350	490
23	350	410
24	350	380
25	350	360
26	350	350

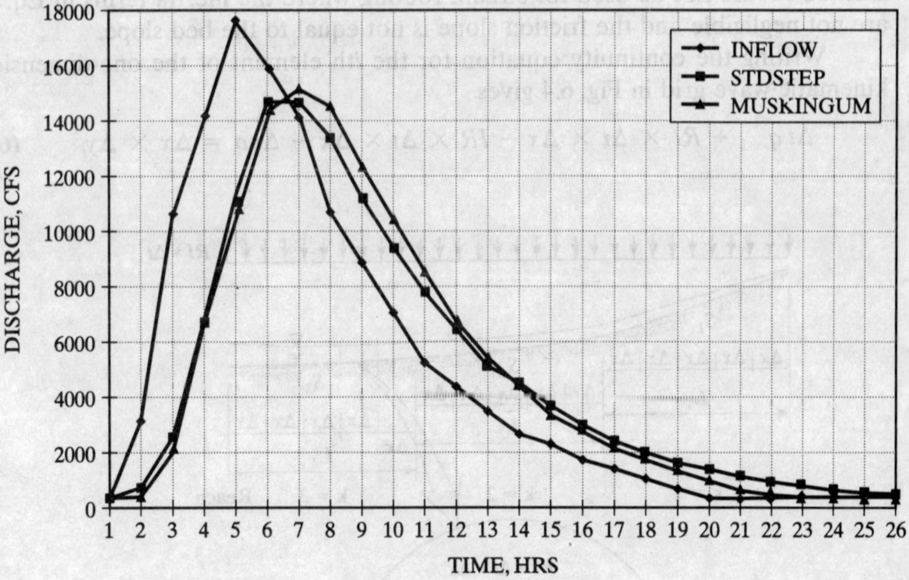

Example 6.2 Inflow and outflow hydrographs.

6.2 KINEMATIC ROUTING

During the precipitation-runoff process, overland flow occurs before the runoff is concentrated in swales or drainage channels. The depth of flow is generally less than the height of the vegetation and the water velocity is low. The Manning roughness values for overland flow are typically much greater than for channel flow. Typical Manning roughness factors for overland flow are listed in Table 6.1.

TABLE 6.1 MANNING ROUGHNESS FACTORS FOR OVERLAND FLOW

Surface	n
Asphalt/concrete	0.05–0.15
Bare smooth soil	0.10
Sparse vegetation	0.15
Light turf	0.20
Dense grass	0.30
Forest litter	0.40

The kinematic wave model is often used for overland flow routing where the friction slope and the bed slope are approximately equal. The kinematic wave method should not be used for stream routing where the inertia terms in Eq. 5.88 are not negligible and the friction slope is not equal to the bed slope.

Writing the continuity equation for the ith element of the one-dimensional kinematic wave grid in Fig. 6.4 gives

$$\Delta t\, q_{i-1} + RI \times \Delta t \times \Delta x - IR \times \Delta t \times \Delta x - \Delta t\, q_i = \Delta x \times \Delta y_i \tag{6.12}$$

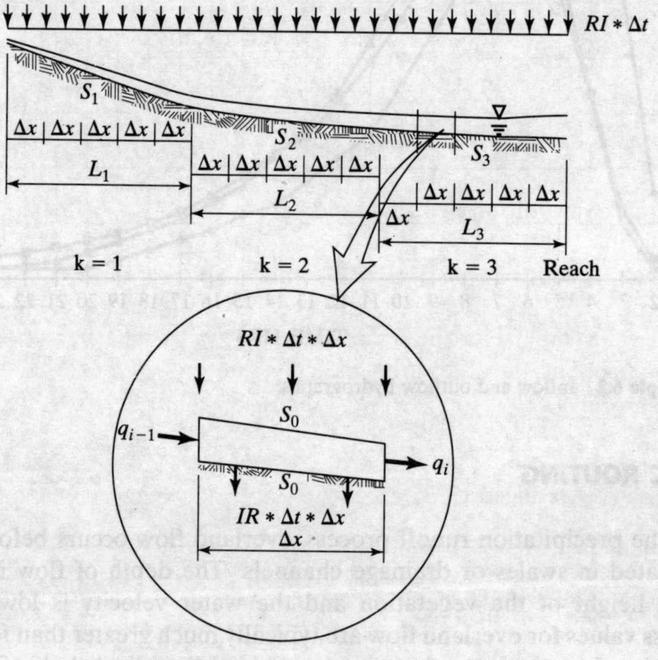

Figure 6.4 Overland flow routing.

where

q is the unit discharge $\left(\frac{L^2}{t}\right)$,

RI is rainfall intensity $\left(\frac{L}{t}\right)$,

IR is infiltration rate $\left(\frac{L}{t}\right)$, and

Δy_i is change in water depth (L).

This equation reduces to

$$\frac{q_{i-1} - q_i}{\Delta x} + RI - IR = \frac{\Delta y}{\Delta t}$$

or

$$\frac{\partial q}{\partial x} + \frac{\partial y}{\partial t} = RE \tag{6.13}$$

where RE is the effective rainfall ($RI - IR$).

In kinematic wave routing, the relation between water depth (y) and unit discharge rate (q) is typically provided by Manning equation. For shallow overland flow, the Manning equation is

$$q = \frac{C_m}{n} y^{5/3} S_o^{1/2} \tag{6.14}$$

or

$$y = \left(\frac{n}{C_m S_o^{1/2}}\right)^{3/5} q^{3/5} \tag{6.15}$$

Using a backward difference finite-element scheme, Eq. 6.13 can be written as

$$\frac{q_i^+ - q_{i-1}^+}{\Delta x_k} + \frac{\alpha_k (q_i^+)^{3/5} - \alpha_k (q_i)^{3/5}}{\Delta t} = RE^+ \tag{6.16}$$

where the plus superscript indicates next time step, the k subscript indicates overland flow segment, the i subscript indicates the grid element, and

$$\alpha_k = \left(\frac{n}{C_m S_o^{1/2}}\right)_k^{3/5}$$

Eq. 6.16 can be reduced to

$$F(q_i) = q_i^+ + A_k (q_i^+)^{3/5} - q_{i-1}^+ - A_k (q_i)^{3/5} - \Delta x_k \times RE^+ = 0 \tag{6.17}$$

where

$$A_k = \frac{\Delta x_k}{\Delta t} \alpha_k$$

Since computations proceed from left to right in Fig. 6.4, the outflow rate (q_{i-1}^{+}) has been computed for grid $i - 1$, and Eq. 6.17 includes only one unknown (q_i^{+}). Equation 6.17 is nonlinear and can be solved using the Newton–Raphson procedure.

$$q^{j+1} = q^j - \frac{F(q^j)}{F'(q^j)} \tag{6.18}$$

where the superscript j refers to the iteration and

$$F'(q^j) = 1 + \tfrac{3}{5}A_k(q_i^j)^{-2/5} \tag{6.19}$$

The iterative process continues until Eq. 6.17 is equal to zero.

Example 6.3(a) Kinematic Wave Routing

Compute the runoff hydrograph for the overland flow model shown in Fig. 6.4 for a 100-yr return period 6-hr duration rainfall. The three-segment overland flow model has the following characteristics:

Segment	Length ft	Slope %	Impervious %	Manning n
1	75	3.0	15	0.2
2	75	2.0	10	0.25
3	100	1.0	0	0.3

The 100-hr return period rainfall intensity (RI) in inches/hr is given by the equation

$$RI = \frac{a}{(t + b)^c}$$

where $a = 96$, $b = 8.0$, $c = 0.73$, and t is time in minutes. Use a balanced triangular rainfall distribution with a computational time step (Δt) equal to 5 minutes.

Losses for the pervious area are to be estimated using Green and Ampt infiltration equation. Neglecting ponding depth, the infiltration rate (IR) is given by

$$IR = K_s + K_s\frac{(\text{POR} - \text{VMC})\text{WFS}}{F}$$

where K_s is the hydraulic conductivity of the soil, POR is the porosity of the soil, VMC is the volumetric moisture content of the soil prior to the storm, WFS is the wetting front suction head, and F is the total infiltrated volume ($F = \Sigma IR$). If the rainfall rate is less than the infiltration rate, the infiltration is equal to rainfall rate. The soil in this example has been classified as silty clay loam with the following Green and Ampt parameters:

$K_s = 0.033$ inches/hr,

POR $= 0.43$,

VMC $= 0.25$, and

WFS $= 11.0$ inches

For interception and depression storage, use an initial abstraction of 0.5 inches.

Solution A FORTRAN program (KINRT.FOR) was written for up to five overland flow segments in series. Comment lines are included in the program listing explaining the program and input requirements. The input (KINRT.DAT) file for the example problem is listed below. The incremental precipitation and runoff hydrograph are plotted below.

```
EXAMPLE 6.3 KINEMATIC WAVE ROUTING OF OVERLAND FLOW
&GRID IUN = 2, NSEG = 3, DT = 0.0833, DUR = 6./
&PRECIP B = 96., D = 8.0, E = .730/
1    75.    3.00   0.20   15.0   0.5   .033   0.43   0.25   11.0
2    75.    2.00   0.25   10.0   0.5   .033   0.43   0.25   11.0
3    100.   1.00   0.30    0.0   0.5   .033   0.43   0.25   11.0
```

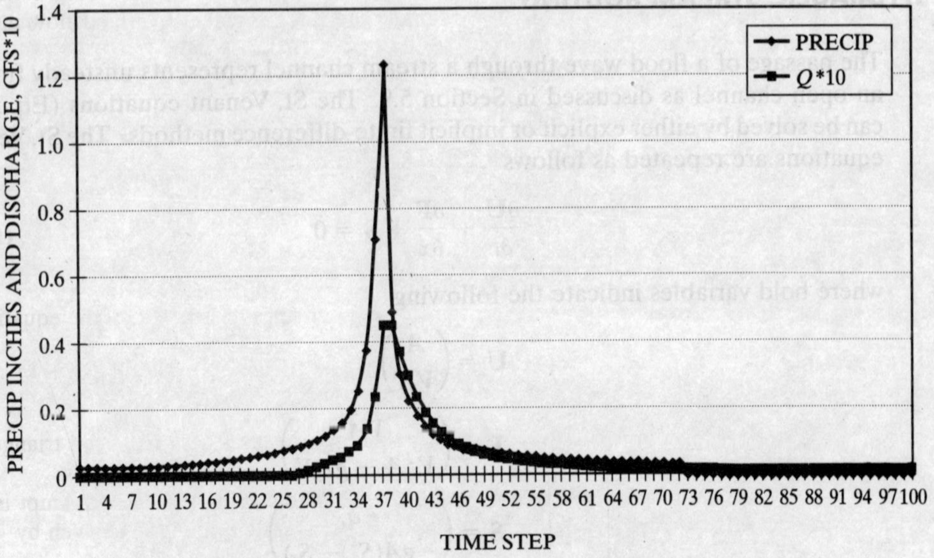

Example 6.3 Kinematic overland flow model. Plot of incremental precipitation and runoff hydrograph for $DT = 0.0833$ hours.

Example 6.3(b) Hydroplaning

Assume that the potential for hydroplaning on a high speed roadway will occur if the water depth exceeds 0.25´inches. Determine the water depth at the outer edge of the pavement for a 2-yr return period storm. The equation for the rainfall intensity (I) in inches per hour is

$$I = \frac{65}{(t + 8.0)^{0.806}}$$

where t is the time in minutes.

The roadway is 90 ft wide. Consider two conditions:

(1) normal section where the length of surface runoff is 45 ft and
(2) roadway curve where the length of overland flow is 90 ft.

For the normal section, the overland flow is divided into three 15-ft sections with slopes of 1 percent, 2 percent, and 3 percent, respectively. For the roadway curve, the overland flow is divided into three 30-ft sections, each with a slope of 3%. Use a Manning "n" value of 0.10 for the roadway.

The kinematic wave model was run for the two conditions and the maximum depth of flow was computed as (1) 0.41 and (2) 0.58 inches, respectively. Results indicate that there is a potential for hydroplaning on the roadway.

6.3 HYDRAULIC STREAM ROUTING

The passage of a flood wave through a stream channel represents unsteady flow in an open channel as discussed in Section 5.9. The St. Venant equations (Eq. 5.88) can be solved by either explicit or implicit finite-difference methods. The St. Venant equations are repeated as follows

$$\frac{\partial \mathbf{U}}{\partial t} + \frac{\partial \mathbf{F}}{\partial x} + \mathbf{S} = 0 \qquad \textbf{(5.88)}$$

where bold variables indicate the following

$$\mathbf{U} = \begin{pmatrix} A \\ VA \end{pmatrix}$$

$$\mathbf{F} = \begin{pmatrix} VA \\ V^2A + gA\overline{Y} \end{pmatrix}$$

$$\mathbf{S} = \begin{pmatrix} -q_\ell \\ -gA(S_o - S_f) \end{pmatrix}$$

In the St. Venant equations, the area in the continuity equation represents storage area, while the area in the dynamic equation represents conveyance area. In a natural channel, the two areas will be different.

6.3.1 Explicit Method

The two-step, predictor/corrector MacCormack method (Section 5.9.2.2) will be used to solve for V and y at the interior nodes. Using the backwards difference for the spatial partial derivative in the predictor steps gives

$$\mathbf{U}_i^* = \mathbf{U}_i - \frac{\Delta t}{\Delta x}(\mathbf{F}_i - \mathbf{F}_{i-1}) - \Delta t\mathbf{S}_i \qquad \textbf{(5.94)}$$

Using the forward difference for the spatial partial derivative in the corrector step gives

$$\mathbf{U}_i^{**} = \mathbf{U}_i - \frac{\Delta t}{\Delta x}\left(\mathbf{F}_{i+1}^{*} - \mathbf{F}_i^{*}\right) - \Delta t\mathbf{S}^{*} \tag{5.97}$$

where the superscript asterisk indicates the terms were evaluated based on the results of the predictor step. The values of A and AV for interior node i at the next time step are

$$\mathbf{U}_i^{+} = \tfrac{1}{2}(\mathbf{U}_i^{*} + \mathbf{U}_i^{**}) \tag{5.98}$$

The method of characteristics will be used at the upstream and downstream boundaries. For hydraulic stream routing, the upstream boundary is an inflow hydrograph and the discharge rate (Q_1) is known. The negative characteristic equation at the upstream boundary ($i = 1$) is

$$C^-: \quad V_1^+ = C_3 + C_4 y_1^+ \tag{5.124}$$

This equation combined with the given discharge rate gives

$$F(y_1^+) = Q - C_3 A_1^+ - C_4 A_1^+ y_1^+ = 0 \tag{6.20}$$

Equation 6.20 can be solved using the Newton–Raphson method or

$$y_1^{j+1} = y_1^j - \frac{F(y_1^j)}{F'(y_1^j)}$$

where the superscript j refers to the iteration and

$$F'(y_1^j) = -T_1^j(C_3 + C_4 y_1^j) - C_4 A_1^j$$
$$T = \frac{dA}{dy} = \text{top width} \tag{6.21}$$

The downstream boundary equations ($i = N + 1$) are the Manning equation and the plus characteristic equation

$$\text{Energy:} \quad V_{N+1}^+ = \frac{C_m}{n}\left(R^{2/3} S_f^{1/2}\right)_{N+1}^{+} \tag{6.22}$$

$$C^+: \quad V_{N+1}^+ = C_1 - C_2 y_{N+1}^+ \tag{5.123}$$

Combining equations gives

$$F(y_{N+1}^+) = \frac{C_m}{n}\left(\frac{A_{N+1}^+}{P_{N+1}^+}\right)^{2/3}(S_{fN+1})^{1/2} - C_1 + C_2 y_{N+1}^+ = 0 \tag{6.23}$$

Using the friction slope at the present time step, Eq. 6.23 can be solved for y_{N+1}^+ using the Newton–Raphson procedure

$$y_{N+1}^{j+1} = y_{N+1}^{j} - \frac{F(y_{N+1}^j)}{F'(y_{N+1}^j)} \tag{6.24}$$

For a trapezoidal channel, the derivative $(F'(y^j_{N+1}))$ can be evaluated directly such as Eq. 6.21. However, for natural channels, it will probably be easier to evaluate the derivative of Eq. 6.23 numerically.

Example 6.4(a) Hydraulic Stream Routing through a Trapezoidal Channel

Route the inflow hydrograph from Example 6.1 through the 4.5-mile stream reach of No Name Creek (Example 6.2) using hydraulic stream routing. Assume that the stream reach can be approximated as a single, trapezoidal channel 23,800 ft long with a longitudinal slope of 0.001. The channel has a bottom width of 320 ft with 4:1 side slopes and a Manning "n" value of 0.07.

Solution Program MACK.FOR was modified to include

1. Upstream boundary Eq. 6.20
2. Downstream boundary Eq. 6.23
3. Read inflow hydrograph (Q_1)

The program listing for this example problem is HYRT1.FOR. Comments are included in the listing explaining input and computation steps. The input file (HYRT1.DAT) is included below.

```
EXAMPLE 6.4(a) HYDRAULIC ROUTING THROUGH A TRAPEZOIDAL CHANNEL
&CANAL IUN = 2, NOH = 40, DTH = 1.0/
1334.00        320.0     4.0   23800.0          0.07     0.0010        60
```

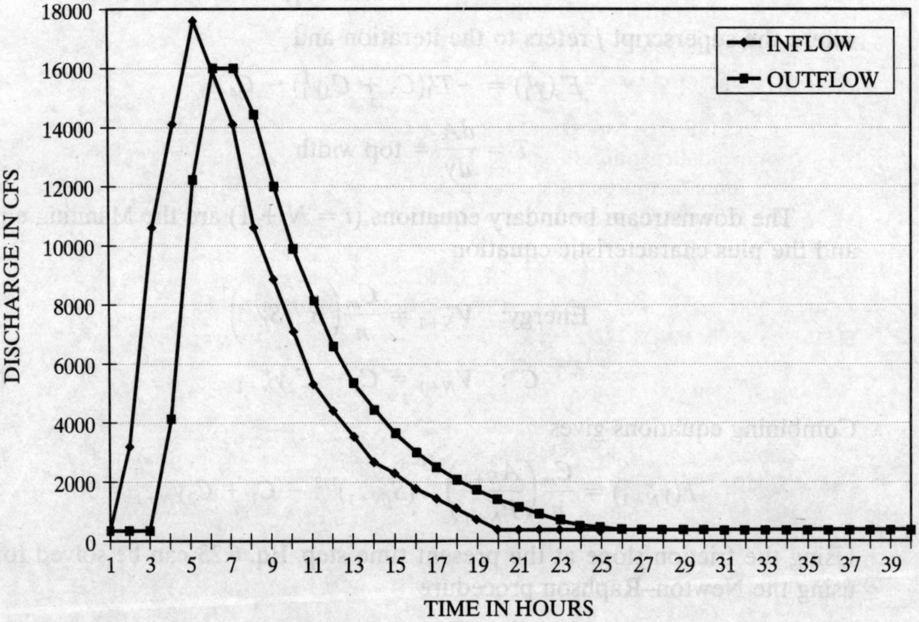

Example 6.4(a) Hydraulic stream routing through a trapezoidal channel using the MacCormack scheme.

Example 6.4(b) Hydraulic Stream Routing through Trapezoidal Channels in Series

Assume that the stream reach can be approximated with six trapezoidal channel segments in series with the following characteristics.

Bottom width ft	Side slopes	Length ft	Manning "n" value	Slope
300*	5.0	2,650	0.07	0.001
340	3.50	5,000	0.07	0.001
300	5.00	4,650	0.07	0.001
320	3.50	4,500	0.07	0.001
324	3.50	4,600	0.07	0.001
300**	5.00	2,400	0.07	0.001

*Upstream segment.

**Downstream segment.

Also include the option of having lateral inflow (q_ℓ in Eq. 5.88) into the channel. Assume that the computed hydrograph from the kinematic wave model in Example 6.3(a) represents one-half the lateral inflow along the channel.

Solution Program MACKS.FOR was modified to include:

1. Upstream boundary Eq. 6.20
2. Downstream boundary Eq. 6.23
3. Reading inflow hydrograph (Q_1)
4. Reading lateral inflow hydrograph
5. Lateral inflow

FORTRAN program HYRT2.FOR was developed for hydraulic routing through trapezoidal channels in series. The program listing includes comment lines explaining the input requirements and program computational procedures. The input file (HYRT2.DAT) is listed below for the example problem.

```
EXAMPLE 6.4(b) HYDRAULIC ROUTING THROUGH TRAPEZOIDAL CHANNELS IN SERIES
&CANAL IUN = 2, LI = 2, NOC = 6, NOH = 40, DTH = 1.0, NOL = 100, DTL = 0.0833,
ELEV = 1334.00, MNOS = 7/
300.0      5.0      2650.0      0.07      0.001
340.0      3.5      5000.0      0.07      0.001
300.0      5.0      4650.0      0.07      0.001
320.0      3.5      4500.0      0.07      0.001
324.0      3.5      4600.0      0.07      0.001
300.0      5.0      2400.0      0.07      0.001
```

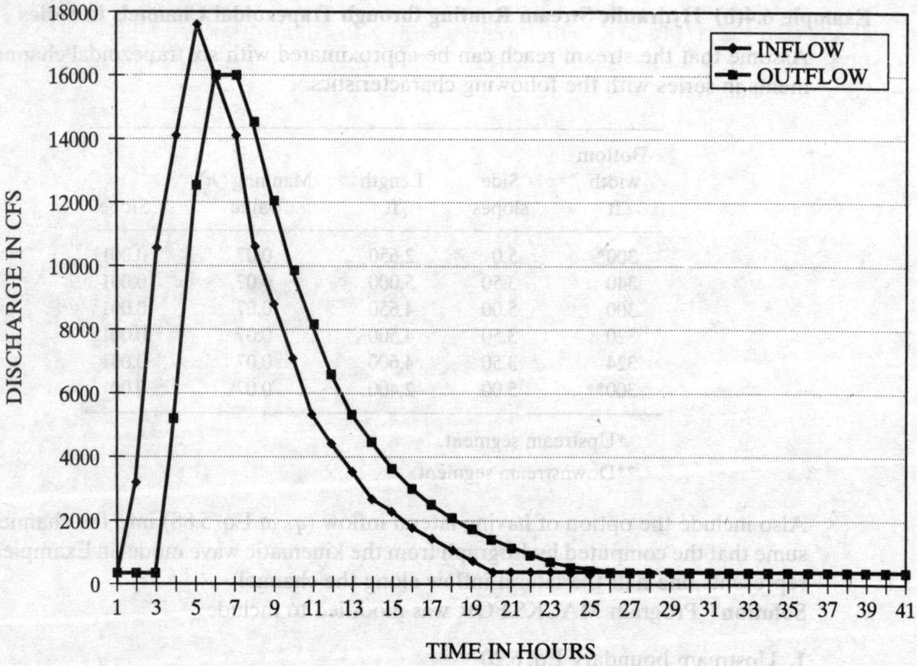

Example 6.4(b) Hydraulic routing through trapezoidal channels in series using the MacCormack scheme.

Example 6.4(c) Hydraulic Stream Routing through a Natural Channel

Modify program developed in Example 6.4(b) (HYRT2.FOR) for a nonuniform, natural channel described by eight-point cross-sections. Each cross-section is assumed to be representative of a stream reach extending from halfway to the downstream cross-section (or downstream boundary) and halfway to the upstream cross-section (or upstream boundary). Stream reaches will be considered channels in series and will be connected with junction nodes as described by Eqs. 5.105 and 5.106.

Solution Subroutine RATET was written to generate a rating table for each stream reach. The rating table includes area, hydraulic radius, top width, wetted perimeter, conveyance, celerity, α coefficient, and depth to the centroid of area for 20 water depths. Subroutine LOOKUP (RX, CA, J) is used to find the values in the table required in the model.

FORTRAN program HYRT3.FOR was developed for hydraulic routing through natural channels. Equation 6.23 was modified for natural channels to

$$F(y_{N+1}) = \frac{K_{N+1}^{+}}{A_{N+1}^{+}}(S_{fN+1})^{1/2} - C_1 + C_2 Y_{N+1}^{+}$$

where K_{N+1}^{+} is the conveyance. The program listing includes comment lines explaining the input requirements and program computational procedures. The input file (HYRT3.DAT) is listed below.

EXAMPLE 6.4(c) HYDRAULIC ROUTING THROUGH A NATURAL CHANNEL
&CHANNEL IUN = 2, LI = 2, NOC = 6, NOH = 40, DTH = 1.0, NOL = 100, DTL = 0.0833,
ELEV = 1334.00, MNOS = 5/

1	0.07	0.04	0.08	2650.	0.001		
1345.	1336.	1334.	1329.	1329.	1334.	1335.	1345.
0.	50.	175.	185.	215.	225.	350.	400.
2	0.07	0.04	0.08	5000.	0.001		
1340.	1330.	1328.	1324.	1324.	1329.	1330.	1340.
0.	30.	175.	185.	215.	225.	370.	410.
3	0.07	0.04	0.08	4650.	0.001		
1335.	1324.	1322.	1319.	1319.	1323.	1325.	1335.
0.	50.	175.	185.	215.	225.	350.	400.
4	0.07	0.04	0.08	4500.	0.001		
1330.	1320.	1317.	1315.	1315.	1318.	1319.	1330.
0.	40.	200.	220.	250.	260.	360.	490.
5	0.07	0.04	0.08	4600.	0.001		
1325.	1315.	1313.	1309.	1309.	1312.	1314.	1325.
0.	30.	150.	160.	200.	210.	354.	400.
6	0.07	0.04	0.08	2400.	0.001		
1320.	1310.	1308.	1304.	1304.	1307.	1309.	1320.
0.	50.	175.	185.	215.	225.	350.	400.

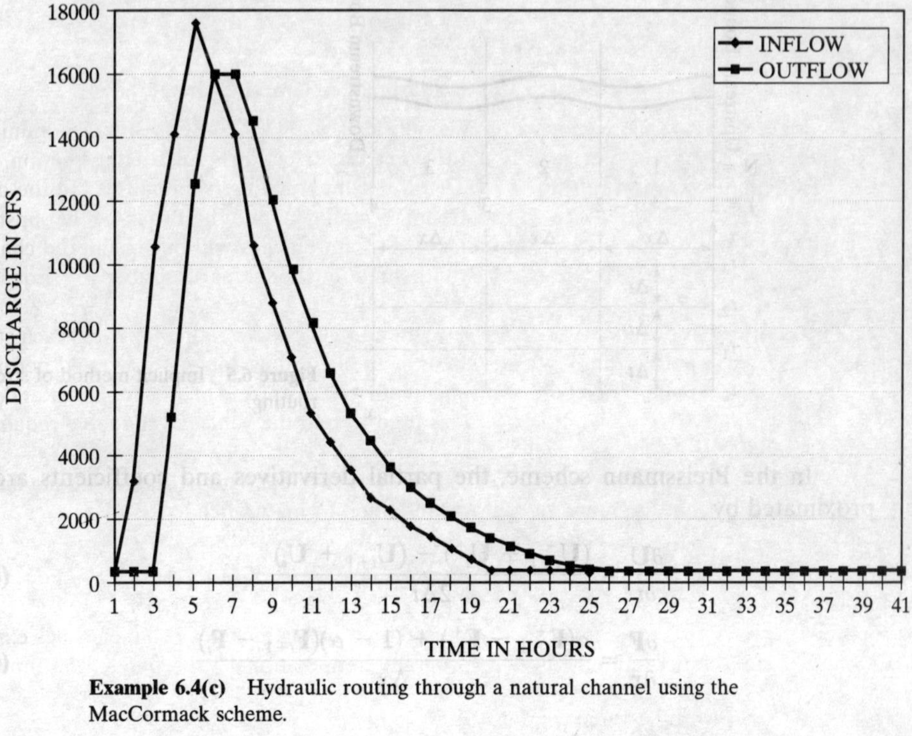

Example 6.4(c) Hydraulic routing through a natural channel using the MacCormack scheme.

6.3.2 Implicit Method

The implicit finite-difference scheme for stream routing has an advantage over explicit methods in that the Courant number does not have to be equal or less than one for stability. The implicit scheme is unconditionally stable, and there are no restrictions on the size of Δx and Δt for stability; however, for accuracy, the Courant number should be close to one.

Implicit finite-difference schemes require that the spatial partial derivatives and coefficients be written in terms of the values at the unknown time level and the system of algebraic equations for the entire routing reach be solved simultaneously. The Preissmann scheme is popular.

A sketch of a routing reach is shown in Fig. 6.5. For example, the routing reach has been divided into N (three) subreaches each Δx in length. There are $N + 1$ computational nodes. The subscript i is used to identify computational nodes, and the subscript N is used to identify subreaches. Node 1 represents the upstream boundary where the inflow hydrograph is specified, and node $N + 1$ represents the outflow boundary where a rating table is specified. The velocity and water depths are unknown at each computational node. There are $2N + 2$ unknowns at each time step.

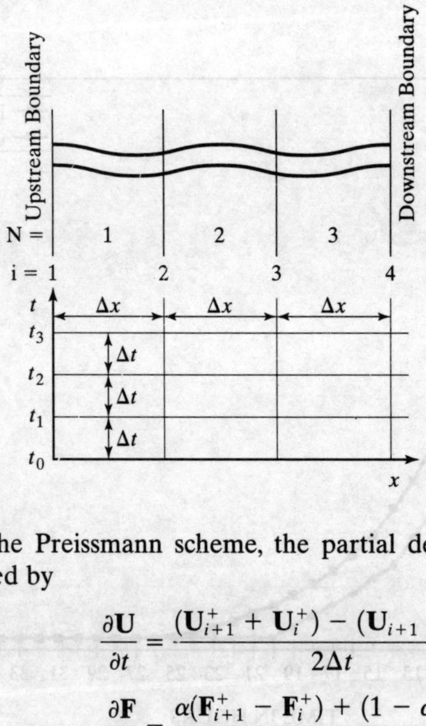

Figure 6.5 Implicit method of stream routing.

In the Preissmann scheme, the partial derivatives and coefficients are approximated by

$$\frac{\partial \mathbf{U}}{\partial t} = \frac{(\mathbf{U}_{i+1}^{+} + \mathbf{U}_{i}^{+}) - (\mathbf{U}_{i+1} + \mathbf{U}_{i})}{2\Delta t} \tag{6.25}$$

$$\frac{\partial \mathbf{F}}{\partial x} = \frac{\alpha(\mathbf{F}_{i+1}^{+} - \mathbf{F}_{i}^{+}) + (1 - \alpha)(\mathbf{F}_{i+1} - \mathbf{F}_{i})}{\Delta x} \tag{6.26}$$

$$\mathbf{S} = \frac{1}{2}\left[\alpha(\mathbf{S}_{i+1}^{+} + \mathbf{S}_{i}^{+}) + (1-\alpha)(\mathbf{S}_{i+1} + \mathbf{S}_{i})\right] \tag{6.27}$$

where α is weighting coefficient. For $\alpha = 0.0$, the scheme is explicit and for $\alpha = 1.0$, the scheme is implicit. The scheme is stable for values of α between 0.6 and 1.0.

Substituting Eqs. 6.25, 6.26, and 6.27 into Eq. 5.88 and multiplying by $2\Delta t$ gives

$$\mathbf{U}_{i+1}^{+} + \mathbf{U}_{i}^{+} - (\mathbf{U}_{i+1} + \mathbf{U}_{i}) + \alpha 2\frac{\Delta t}{\Delta x}(\mathbf{F}_{i+1}^{+} - \mathbf{F}_{i}^{+}) + (1-\alpha)\, 2\frac{\Delta t}{\Delta x}(\mathbf{F}_{i+1} - \mathbf{F}_{i}) + \alpha\, \Delta t(\mathbf{S}_{i+1}^{+} + \mathbf{S}_{i}^{+}) + (1-\alpha)\, \Delta t(\mathbf{S}_{i+1} + \mathbf{S}_{i}) = 0 \tag{6.28}$$

Expanding into the continuity and the dynamic equations yields

$$F_{C_N} = A_{i+1}^{+} + A_{i}^{+} + 2\alpha\frac{\Delta t}{\Delta x}((VA)_{i+1}^{+} - (VA)_{i}^{+}) + C_1 = 0 \tag{6.29}$$

where

F_{C_N} = the continuity equation for reach N

$$C_1 = 2(1-\alpha)\frac{\Delta t}{\Delta x}((VA)_{i+1} - (VA)_{i}) - A_{i+1} - A_{i}$$

and

$$F_{D_N} = (VA)_{i+1}^{+} + (VA)_{i}^{+} + 2\alpha\frac{\Delta t}{\Delta x}((V^2A + gA\bar{y})_{i+1}^{+} - (V^2A + gA\bar{y})_{i}^{+}) - g\alpha\Delta t((A(S_o - S_f))_{i+1}^{+} + (A(S_o - S_f))_{i}^{+}) + C_2 = 0 \tag{6.30}$$

where

F_{D_N} = the dynamic equation for reach N

$$C_2 = -(VA)_{i+1} - (VA)_{i} + 2(1-\alpha)\frac{\Delta t}{\Delta x}((V^2A + gA\bar{y})_{i+1} - (V^2A + gA\bar{y})_{i}) - g(1-\alpha)\Delta t((A(S_o - S_f))_{i+1} + (A(S_o - S_f))_{i})$$

The continuity and dynamic equations are written for N subreaches giving $2N$ equations. The other two equations are boundary equations. For the upstream boundary, the inflow hydrograph is specified and the discharge rate (Q) is known. The upstream boundary equation is

$$F_{B_1} = V_1^{+}A_1^{+} - Q = 0 \tag{6.31}$$

Manning's equation is used at the downstream boundary to compute the discharge rate. The downstream boundary equation is

$$F_{B_2} = \frac{C_m}{n}(A_{N+1}(R_{N+1})^{2/3}S_f^{1/2})^{+} - (AV)^{+}_{N+1} = 0 \tag{6.32}$$

The Newton–Raphson method can be used to solve the system of $2N + 2$ equations. The estimate of the unknown variables V and y at each computation node is taken as the value from the previous time step. This is an iterative procedure and the equations are solved for the corrections (Δ) at each iteration. The procedure continues until the sum of the absolute value of the corrections is within a specified tolerance limit.

In matrix notation

$$\mathbf{A}\boldsymbol{\Delta} = -\mathbf{F} \tag{6.33}$$

where

$$\mathbf{A} = \begin{bmatrix}
\frac{\partial F_{B_1}}{\partial V_1} & \frac{\partial F_{B_1}}{\partial y_1} & & & & & & \\
\frac{\partial F_{C_1}}{\partial V_1} & \frac{\partial F_{C_1}}{\partial y_1} & \frac{\partial F_{C_1}}{\partial V_2} & \frac{\partial F_{C_1}}{\partial y_2} & & & & \\
\frac{\partial F_{D_1}}{\partial V_1} & \frac{\partial F_{D_1}}{\partial y_1} & \frac{\partial F_{D_1}}{\partial V_2} & \frac{\partial F_{D_1}}{2 y_2} & & & & \\
 & & \frac{\partial F_{C_2}}{\partial V_2} & \frac{\partial F_{C_2}}{2 y_2} & \frac{\partial F_{C_2}}{\partial V_3} & \frac{\partial F_{C_2}}{2 y_3} & & \\
 & & \frac{\partial F_{D_2}}{\partial V_2} & \frac{\partial F_{D_2}}{2 y_2} & \frac{\partial F_{D_2}}{2 V_3} & \frac{\partial F_{D_2}}{\partial y_3} & & \\
 & & & & \frac{\partial F_{C_3}}{\partial V_3} & \frac{\partial F_{C_3}}{\partial y_3} & \frac{\partial F_{C_3}}{\partial V_4} & \frac{\partial F_{C_3}}{\partial y_4} \\
 & & & & \frac{\partial F_{D_3}}{\partial V_3} & \frac{\partial F_{D_3}}{\partial y_3} & \frac{\partial F_{D_3}}{\partial V_4} & \frac{\partial F_{D_3}}{\partial y_4} \\
 & & & & & & \frac{\partial F_{B_2}}{\partial V_4} & \frac{\partial F_{B_2}}{\partial y_4}
\end{bmatrix}$$

$$\boldsymbol{\Delta} = \begin{bmatrix} \Delta V_1 \\ \Delta y_1 \\ \Delta V_2 \\ \Delta y_2 \\ \Delta V_3 \\ \Delta y_3 \\ \Delta V_4 \\ \Delta y_4 \end{bmatrix}$$

$$\mathbf{F} = \begin{bmatrix} F_{B_1} \\ F_{C_1} \\ F_{D_1} \\ F_{C_2} \\ F_{D_2} \\ F_{C_3} \\ F_{D_3} \\ F_{B_2} \end{bmatrix}$$

In the following example, the Gauss–Jordan method is used to solve Eq. 6.33, when routing a hydrograph through natural channels in series.

Example 6.5(a) Hydraulic Stream Routing Through a Natural Channel Using Implicit Finite-Difference Method

Modify program HYRT3.FOR used in Example 6.4(c) to set up a system of equations (Eq. 6.33) and solve the equations simultaneously. At the junction of channels in series, use the energy and continuity equations (Eqs. 5.105 and 5.106). Solve the system of equations using the Gauss–Jordan method. Route the inflow hydrograph from Example 6.4 through the natural channel using the cross-sections from Example 6.4(c).

$$\left[\begin{array}{cccccccc|c}
a_{11} & a_{21} & 0 & 0 & 0 & 0 & 0 & 0 & F_1 \\
a_{12} & a_{22} & a_{32} & a_{42} & 0 & 0 & 0 & 0 & F_2 \\
a_{13} & a_{23} & a_{33} & a_{43} & 0 & 0 & 0 & 0 & F_3 \\
0 & 0 & a_{34} & a_{44} & a_{54} & a_{64} & 0 & 0 & F_4 \\
0 & 0 & a_{35} & a_{45} & a_{55} & a_{65} & 0 & 0 & F_5 \\
0 & 0 & 0 & 0 & a_{56} & a_{66} & a_{76} & a_{86} & F_6 \\
0 & 0 & 0 & 0 & a_{57} & a_{67} & a_{77} & a_{87} & F_7 \\
0 & 0 & 0 & 0 & 0 & 0 & a_{78} & a_{88} & F_8
\end{array}\right]$$

$$\Downarrow$$

$$\left[\begin{array}{cccccccc|c}
1 & 0 & 0 & 0 & 0 & 0 & 0 & 0 & \Delta V_1 \\
0 & 1 & 0 & 0 & 0 & 0 & 0 & 0 & \Delta y_1 \\
0 & 0 & 1 & 0 & 0 & 0 & 0 & 0 & \Delta V_2 \\
0 & 0 & 0 & 1 & 0 & 0 & 0 & 0 & \Delta y_2 \\
0 & 0 & 0 & 0 & 1 & 0 & 0 & 0 & \Delta V_3 \\
0 & 0 & 0 & 0 & 0 & 1 & 0 & 0 & \Delta y_3 \\
0 & 0 & 0 & 0 & 0 & 0 & 1 & 0 & \Delta V_4 \\
0 & 0 & 0 & 0 & 0 & 0 & 0 & 1 & \Delta y_4
\end{array}\right]$$

Example 6.5 Representation of the Gauss–Jordan method of solving a system of equations.

The Gauss–Jordan method is a simple and convenient method of solving the system of equations. As shown in the figure on page 383, the A array is converted to an identity matrix using matrix algebra and the F column becomes the solution. For convenience, the partials in the A array have been replaced with a_{ij} elements in the figure.

FORTRAN program HYRTIMP.FOR was written for hydraulic routing using the implicit method. The computational time step (Δt) is read for the input file (HYRTIMP.DAT) along with the weighting coefficient (α) and accuracy for convergence of the equations. The number of computational reaches in each channel is included in the input file.

Route the inflow hydrograph through the channel using Δt = 0.1 and 1.0 hrs and compare the outflow hydrographs with that computed for Example 6.4(c). A plot of the inflow and outflow hydrographs is shown below. The implicit method with DT = 1.0 hr gives a significantly lower peak discharge rate.

```
EXAMPLE 6.5 IMPLICIT HYDRAULIC ROUTING THROUGH A NATURAL CHANNEL
&CHANNEL IUN = 2, NOC = 6, NOH = 40, DTH = 1.0, DT = 1.0, ELEV = 1334.00, AWC = 0.90,
ACC = 0.001/
    1       3     0.07     0.04     0.08    2650.    0.001
 1345.   1336.   1334.    1329.    1329.    1334.    1335.    1345.
    0.     50.    175.     185.     215.     225.     350.     400.
    2       3     0.07     0.04     0.08    5000.    0.001
 1340.   1330.   1328.    1324.    1324.    1329.    1330.    1340.
    0.     30.    175.     185.     215.     225.     370.     410.
    3       3     0.07     0.04     0.08    4650.    0.001
 1335.   1324.   1322.    1319.    1319.    1323.    1325.    1335.
    0.     50.    175.     185.     215.     225.     350.     400.
    4       3     0.07     0.04     0.08    4500.    0.001
 1330.   1320.   1317.    1315.    1315.    1318.    1319.    1330.
    0.     40.    200.     220.     250.     260.     360.     490.
    5       3     0.07     0.04     0.08    4600.    0.001
 1325.   1315.   1313.    1309.    1309.    1312.    1314.    1325.
    0.     30.    150.     160.     200.     210.     354.     400.
    6       3     0.07     0.04     0.08    2400.    0.001
 1320.   1310.   1308.    1304.    1304.    1307.    1309.    1320.
    0.     50.    175.     185.     215.     225.     350.     400.
```

Example 6.5(b) Hydraulic Stream Routing Using Implicit Finite-Difference Method with Pointer Matrix

The program developed in Example 6.5(a) uses a square A array to store the partials used to solve the equations. For a large problem, this would be very inefficient. Modify the FORTRAN program HYRTIMP.FOR and replace the square A array with a pointer matrix. Repeat the example Problem 6.5(a) and compare the results. Program HYRTI.FOR was developed and gives identical results.

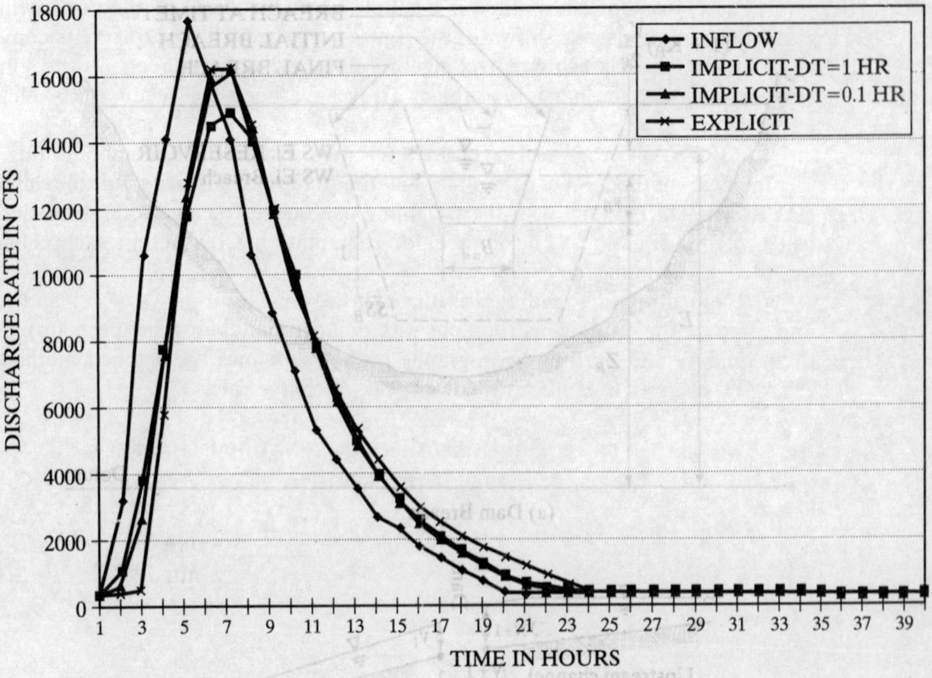

Example 6.5 Routing a hydrograph through a natural channel. A comparison of outflow hydrographs using implicit and explicit routing procedures.

6.4 DAM BREAK ANALYSIS

Dam break analysis is often required in safety studies of major dams where failure might result in loss of life and serious damage to homes, buildings, and public utilities. Dam break analysis is similar to hydraulic routing through natural channels in series except the dam is specified at a junction node. The following discussion applies only to the junction node at the dam.

The dam break problem represents unsteady flow in an open channel where the discharge at the dam is controlled by the size of the breach or weir. The size of the breach will increase with time. The break in Fig. 6.6 is shown as a trapezoidal weir that increases from its initial size to the final size over some time period. If it is assumed that critical depth occurs in the breach, the discharge rate through the break can be estimated. As shown in Fig. 6.6, E is the total energy head and

$$E = (1 + K_B)\frac{V_c^2}{2g} + y_B + Z_B \tag{6.34}$$

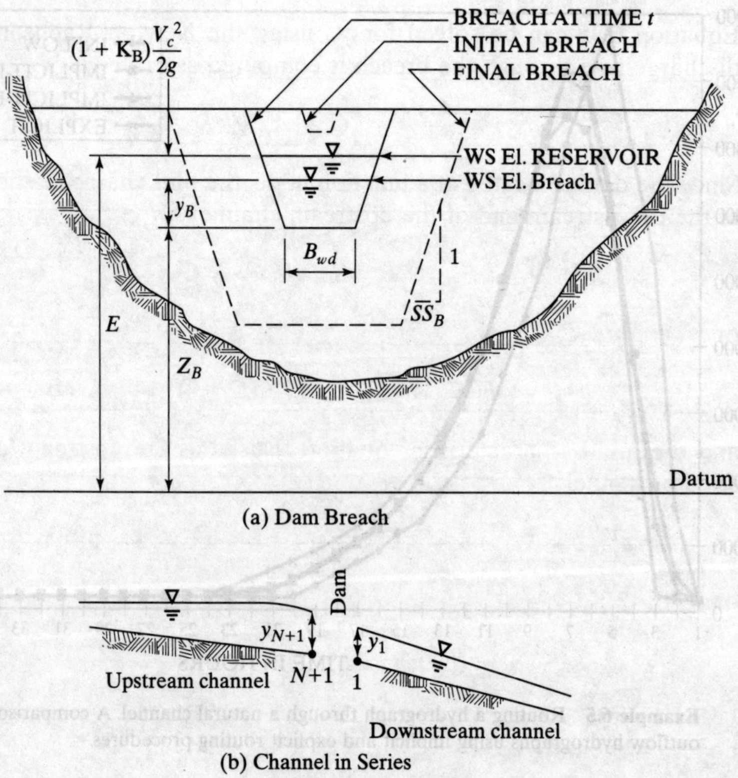

Figure 6.6 Dam break.

where E is the elevation of the energy line in the reservoir, K_B is the contraction loss coefficient plus the friction loss coefficient for the breach, $V_c^2/2g$ is the critical velocity head in breach [equal to one-half the hydraulic depth (D)], y_B is the water depth in the breach, and Z_B is the bottom elevation of the breach. The breach size (B_{WD}, SS_B, and Z_B) is specified as a boundary condition.

6.4.1 Explicit Routing

With the explicit routing of the dam break flood wave, E is computed from the previous time step and Eq. 6.34 becomes

$$F(y) = E - (1 + K_B)\frac{D}{2} + y_B + Z_B = 0 \tag{6.35}$$

where

$$D = \left(\frac{(B_{WD} + SS_B \times y_B)y_B}{B_{WD} + 2 \times SS_B \times y_B}\right)$$

Equation 6.35 can be solved for y_B using the Newton–Raphson procedure. The discharge rate through the breach is computed as

$$Q = A_B\sqrt{gD} \tag{6.36}$$

Since the dam is located at a junction node, the plus characteristic equation is used at the downstream end of the upstream channel

$$C^+: \quad V_{N+1}^+ = C_1 - C_2 y_{N+1}^+ \tag{6.37}$$

or

$$F(y_{N+1}^+) = Q - A_{N+1}^+(C_1 - C_2 y_{N+1}^+) = 0 \tag{6.38}$$

and the minus characteristic equation is used at the upstream end of the downstream channel

$$C^-: \quad V_1^+ = C_3 + C_4 y_1^+ \tag{6.39}$$

or

$$F(y_1^+) = Q - A_1^+(C_3 + C_4 y_1^+) = 0 \tag{6.40}$$

Using Eqs. 6.37–6.40, V_{N+1}^+, y_{N+1}^+, V_1^+, and y_1^+ can be computed for the junction node representing the dam.

6.4.2 Implicit Routing

For the implicit dam break routing procedure, three equations are developed for the junction node representing the dam. The continuity equation (F_1) is written between the last node ($N + 1$) of the upstream channel (J) and the breach (B):

$$F_1 = A_{N+1,j} \times V_{N+1,j} - A_B\sqrt{gD_B} = 0 \tag{6.41}$$

The energy equation (F_2) is written between the last node of the upstream channel and the breach:

$$F_2 = \frac{V_{N+1,j}^2}{2g} + y_{N+1,j} + Z_{N+1,j} - Z_B - (1 + K_B)\frac{D_B}{2} - y_B = 0 \tag{6.42}$$

The final equation for the junction representing the dam is the continuity equation (F_3) between the breach and the first node (1) of the downstream channel ($J + 1$):

$$F_3 = A_{1,J+1} \times V_{1,J+1} - A_B\sqrt{gD_B} = 0 \tag{6.43}$$

Since A_B and D_B are functions of y_B, there is only one additional unknown for the junction node representing the dam, y_B.

For the implicit routing procedure, Eqs. 6.41–6.43 are combined with the reach equations (6.29 and 6.30), the junction node equations (5.105 and 5.106), upstream boundary equation (6.31), and the downstream boundary equation (6.32). All equations are solved simultaneously at each time step.

6.4.3 Submerged Breach

If the capacity of the downstream channel is not adequate to handle the discharge from the breach, the tailwater below the dam will rise and submerge the breach or weir. As the weir becomes submerged, the discharge (Q_s) over the weir is reduced. The approximate effect of weir submergence (H_2/H_1) on the discharge rate (Q_s/Q) is shown in Fig. 6.7. Submergence has considerable affect on the discharge of a sharp-crested weir, but has a smaller affect on the discharge of a broad-crested weir. The discharge over a broad-crested weir remains unaffected by submergence until the submergence (H_2/H_1) exceeds 0.67. The effect of submergence of a broad-crested weir can be approximated by the following simple equation (French, 1995).

$$K_{ws} = 1.0 \quad \text{for} \quad \frac{H_2}{H_1} \le 0.67$$

$$K_{ws} = 1.0 - 2.78\left(\frac{H_2}{H_1} - 0.67\right) \quad \text{for} \quad \frac{H_2}{H_1} > 0.67 \tag{6.44}$$

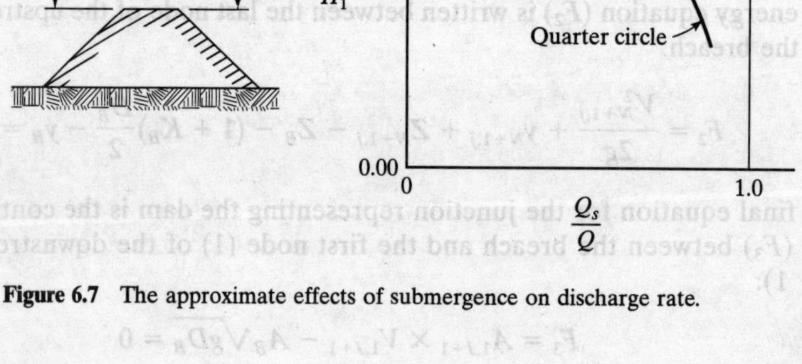

Figure 6.7 The approximate effects of submergence on discharge rate.

Example 6.6(a) Explicit Stream Routing of a Dam Break Flood Wave through a Natural Channel

Modify program (HYRT3.FOR) used in Example 6.4(c) to include a dam at an internal junction node. The discharge through the dam is controlled by a breach (trapezoidal weir) that grows linearly with time.

The dam is located at the downstream end of stream segment 5 of Example 6.4(c). The initial water level in the reservoir is 1,355 ft. The cross-sections in the reservoir were extended to elevation 1,355 ft and modified to represent a reservoir. The breach bottom width increases from 30 ft to 150 ft, and the breach elevation decreases to 1,325 ft. The side slopes of the breach are 0.5:1.

Solution The FORTRAN program (HYRTEDB.FOR) was modified to include Eqs. 6.35–6.40 at an internal junction node specified as the location of the dam. The program will include an inflow hydrograph at the upstream boundary (Eq. 6.20), lateral inflow along the stream, and a rating table at the downstream boundary (Eq. 6.23).

The input data file (HYRTEDB.DAT) is included below. The program listing includes comment lines explaining the program and input requirements. Use the inflow hydrograph from Example 6.4 and run the model for a breach time of 4 and 8 hrs.

```
EXAMPLE 6.6(a) EXPLICIT ROUTING DAM BREAK HYDROGRAPH THROUGH A NATURAL CHANNEL
&CHANNEL IUN = 2, LI = 0, NOC = 6, NOH = 40, DTH = 1.0, NOL = 00, DTL = 0.0833, ELEV = 1334.00,
MNOS = 3, TFT = 0.9/
&BREACH IBSEG = 5, WSELD = 1355., BWDI = 30., BWDF = 150., BELF = 1315., SSB = 0.5, BTST = 0.,
BTDR = 4., BK = 0.2/
      1      0.10     0.10     0.10     2650.    0.001    1.0
1355.    1336.    1334.    1329.    1329.    1334.    1335.    1355.
   0.      50.     175.     185.     215.     225.    1000.    2000.
      2      0.10     0.10     0.10     5000.    0.001    1.0
1355.    1330.    1328.    1324.    1324.    1329.    1330.    1355.
   0.      30.     175.     185.     215.     225.    2000.    3000.
      3      0.10     0.10     0.10     4650.    0.001    1.0
1355.    1324.    1322.    1319.    1319.    1323.    1325.    1355.
   0.      50.     175.     185.     215.     225.    3000.    4000.
      4      0.10     0.10     0.10     4500.    0.001    1.0
1355.    1320.    1317.    1315.    1315.    1318.    1319.    1355.
   0.      40.     200.     220.     250.     260.    4000.    5000.
      5      0.10     0.10     0.10     4600.    0.001    1.0
1355.    1315.    1313.    1309.    1309.    1312.    1314.    1355.
   0.      30.     150.     160.     200.     210.    5000.    6000.
      6      0.07     0.04     0.08     2400.    0.001    1.0
1320.    1310.    1308.    1304.    1304.    1307.    1309.    1320.
   0.      50.     175.     185.     215.     225.     350.     400.
```

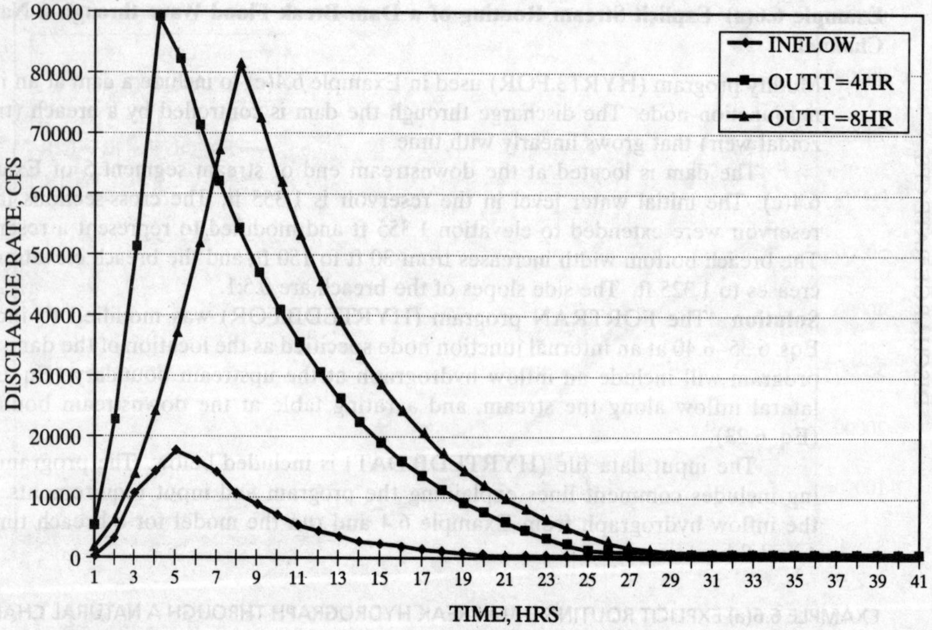

Example 6.6(a) Explicit routing of dam break flood wave through a natural channel for breach times of 4 and 8 hrs.

Example 6.6(b) Implicit Stream Routing of a Dam Break Flood Wave through a Natural Channel

Modify program HYRTI.FOR to include a dam at an internal junction node. Run the same example problem as 6.6(a) and compare the results.

Solution FORTRAN program HYRTIDB.FOR was modified to include Eqs. 6.41, 6.42, and 6.43. The program requires an inflow hydrograph at the upstream boundary, but does not include lateral inflow along the channel. The program listing includes comment lines explaining the program input requirements.

Example 6.6(c) Breach Submergence

Include the weir submergence Eq. 6.44 into the models and repeat [Example 6.6(a) and Example 6.6(b)].

Solution FORTRAN programs HYRTEDB.FOR and HYRTIDB.FOR were modified to include Eq. 6.44 and were renamed HYRTEDBS.FOR and HYRTIDBS.FOR. The programs were rerun for a breach time of 4 hr with the outflow hydrographs plotted below.

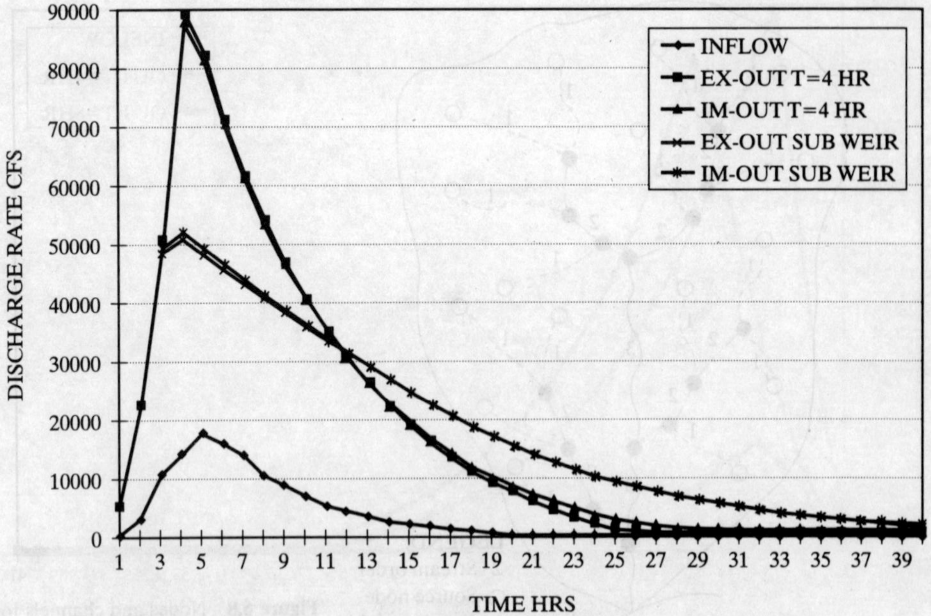

Example 6.6(b and c) Explicit and implicit routing of dam break flood wave with and without weir submergence.

6.5 WATERSHED ROUTING

Watershed routing is a combination of kinematic wave overland flow routing and hydraulic stream routing applied to a branching channel system to form a runoff model. Runoff from the watershed can only get into the channel system by lateral overland flow into any of the channels and overland inflow at the upstream end of first order channels (see Fig. 6.8). At a junction node, more than one channel can flow into the junction but only one channel can flow out of the junction. The following section is presented to show the wide variation in watershed characteristics and channel configurations that can be anticipated when modeling runoff from watersheds.

The drainage basin for any point along a stream is the watershed area as defined by topography that can contribute runoff to the flow in the channel. The drainage basin collects the precipitation excess in a branch channel system that transports the runoff to the basin outlet. Erosion, sediment transportation, and deposition processes also occur in the watershed. In a graded drainage basin, the geological processes have reached near steady-state conditions. The resulting topography and channel configuration depends on the geological processes, surface soils, subsurface formations, and climate.

Characteristics of a drainage basin that affect the hydrologic response include soils, vegetation, surface roughness and slope, basin shape and area, and

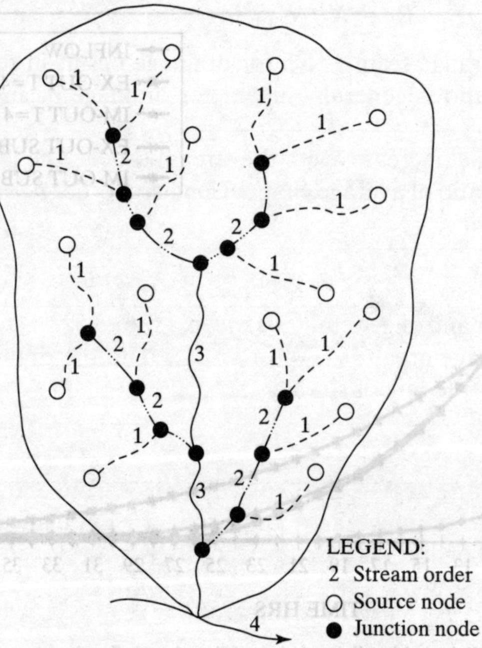

Figure 6.8 Nodes and channels for an idealized drainage system.

channel length, size, slope, roughness, and configuration. Much of this information can be obtained from topographic maps, soil surveys and aerial photography. Studies have shown that there is often geometric similarity between drainage basins of different sizes and stream order can be used to indicate some watershed characteristics.

As shown in Fig. 6.8, the channel system in a drainage basin can be represented by a system of channel segments connecting nodes to form a tree-like network (Strahler, 1964). A first-order channel has no tributaries and extends from a source node to a junction node. The first-order channels receive only overland flow. The junction of two or more first-order channels forms a second-order channel. The second-order channel is larger than the first-order channels because it receives the flow from two or more first-order channels and overland flow. A second-order channel begins at the junction of two or more first-order channels and extends to the intersection with a second-order (or higher) channel. Similarly, a third-order channel begins at the junction of two or more second-order channels and extends to the intersection with a third-order (or higher) channel. A third-order channel receives the flow from at least two second-order channels and overland flow. The order of the drainage basin is the order of its highest order channel.

The number of channel segments decrease with the channel order. The ratio of the number of segments (N) of order n to the number of segments (N) of order $n + 1$ is called the bifurcation ratio (R_b):

$$R_b = N_n/N_{n+1} \tag{6.45}$$

Bifurcation ratios generally range from 3 to 5 for drainage basins in relatively homogeneous material. The ratio is generally higher for drainage basins in dipping rock strata.

The length of the channel increases with the stream order. The law of stream length (Horton, 1945) is the ratio of average channel length of order $n + 1$ to average channel length of order n:

$$R_\ell = L_{n+1}/L_n \tag{6.46}$$

R_ℓ is the stream-length ratio and is generally between 1.5 to 3.5.

R_a is the ratio of drainage area (DA) of order $n + 1$ to the drainage area of order n and is referred to as the law of stream areas:

$$R_a = DA_{n+1}/DA_n \tag{6.47}$$

R_a is often nearly constant for a region and typically ranges from about 3 to 5.

A relation between the drainage area and channel length (measured from the outlet to a point on the drainage divide) has been developed by several researchers:

$$DA_n = CL_n^m \tag{6.48}$$

Hack (1957) noted that if geometrical similarity is to be preserved as the drainage area increases, the exponent m should have a value of 2. Researchers have generally found m to have a value of about 1.7, indicating that most drainage basins elongate as they increase in size.

The average channel slope (CS) tends to decrease as the channel becomes larger. The ratio of average channel slope for channels of order $n + 1$ to the average channel slope for channels of order n is known as the law of stream slopes:

$$R_s = CS_{n+1}/CS_n \tag{6.49}$$

The typical value of R_s is approximately 0.6.

The geology underlying the drainage basin affects the hydrology in several ways. The subsurface rock formations often control the topography and the shape of the drainage basin and channel network. *In situ* soils reflect the character of the parent material. Soils developed from shales, silts, and limestone are fine grain and have low infiltration rates while soils developed from sandstone, conglomerates, and granites are coarse grain and have high infiltration rates. Sometimes the soils are transported into the area by wind (aeolian), water (fluvial), or glaciers. These soils include windblown silts (loess) and sand, glacial till and outwash plains, and fine grain lake deposits (lacustrine plains) and coarse grain terrace deposits.

The drainage pattern is a reflection of the surface soils, subsurface rock formations, and the geological process. Parvis (1950) identified several recurring

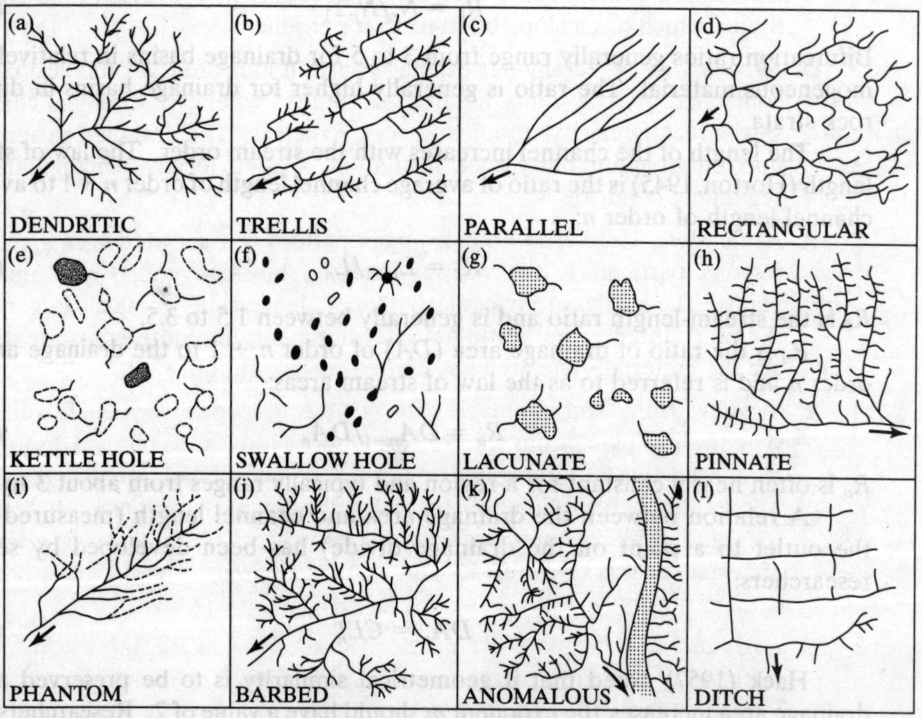

Figure 6.9 Drainage patterns.

drainage patterns of hydrologic significance. The patterns are shown in Fig. 6.9. They are included to give the reader a better understanding of the relation between hydrology and geology and the limitations of the general Eqs. 6.45 through 6.49.

A. *Dendritic.* The most common drainage pattern. It is characteristic of essentially flat-lying relatively homogeneous materials where tributaries were formed in a random pattern. A coarse-textured dendritic pattern is found on geological young landforms while fine-textured dendritic pattern is found on mature landforms. Where there is a thick soil mantle, the local drainage pattern may be dendritic while the regional drainage pattern may be controlled by subsurface formations.

B. *Trellis.* This drainage pattern is found in regions of tilted sedimentary rocks where the soft rocks form the valleys and harder rocks form the parallel ridges. The ridges are cut by water gaps that give the characteristic trellis shape.

C. *Parallel.* A parallel drainage pattern is indicative of a regional sloping terrain or a system of parallel faults or rock joints. It may occur on a regional basis on an alluvial apron such as the Great Plains Province of the United States, which has a slope of 10 to 15 ft per mile.

D. *Rectangular.* A rectangular drainage pattern indicates that the drainage is controlled by rock joints, fissures, and faults. Whenever the main channel shows rectangularity, it indicates rock structure. The drainage pattern for horizontal sedimentary rock formation is often locally rectangular, but regionally dendritic.

E. *Kettle Hole.* The kettle hole drainage pattern is associated with glaciated regions, particularly the various forms of morainic ridges and outwash plains. The kettles are irregularly shaped depressions or lakes. In general, the more intense the kettle-hole development, the more coarse textured are the material. The area has limited surface runoff.

F. *Swallow Hole.* This drainage pattern often includes very small solution-type swallow hole basins where channels lead into the basins and disappear. It is indicative of either limestone rock or underlying granular material. The swallow holes (solution basins) will be very numerous on limestone rock and swallow holes (infiltration basins) less pronounced in granular material such as river terraces and outwash plains. The area has limited surface runoff.

G. *Lacunate.* The lacunate drainage pattern occurs in the tertiary outwash material of the high plains section of the Great Plains Province of the United States. The depressions are a form of swallow hole, but on a grander scale. The area has limited surface runoff.

H. *Pinnate.* Pinnate drainage pattern is the local drainage pattern in deep aeolian silt or loess deposits. The regional drainage pattern will depend on the subsurface formations.

I. *Phantom.* Phantom drainage pattern occurs where an internally well-drained soil, such as a thin layer of aeolian silt material, overlies an impervious subsoil.

J. *Barbed.* The barbed drainage pattern is found on clay-shale or sandy-shale plains where channel piracy has occurred.

K. *Anomalous.* The anomalous pattern is a local drainage pattern whereby the upland drainage is deflected through the slackwater area of a granular terrace associated with a river.

L. *Ditch.* The ditch drainage pattern is a system of open ditch drainage channels and is generally associated with irrigated areas.

Example 6.7 Watershed Routing

Compute the runoff hydrograph for the watershed shown below for natural watershed, partially urbanized watershed, and fully developed watershed. Kinematic wave overland flow routing is to be used to get the precipitation excess into the channel system, and hydraulic channel routing is to be utilized to route the runoff to the watershed outlet. The rainfall intensity (I) in inches per hour is given by

$$I = \frac{96.0}{(t + 8.0)^{0.730}}$$

where t is time in minutes. Use a 6-hr duration storm with a 5-minute computational time step. Soils in the watershed range from loam along the stream channels to sandy clay loam in the upland areas.

The watershed area is 0.60 square miles and includes 10 first-order channels, 5 second-order channels, and 2 third-order channels. The first-order channels were assumed to have a slope of 0.0025, the second-order channels were assumed to have a slope of 0.0015, and the third-order channels were assumed to have a slope of 0.001. An eight-point cross-section was developed for each of the 17 channel segments. The following watershed and channel characteristics were used in the models.

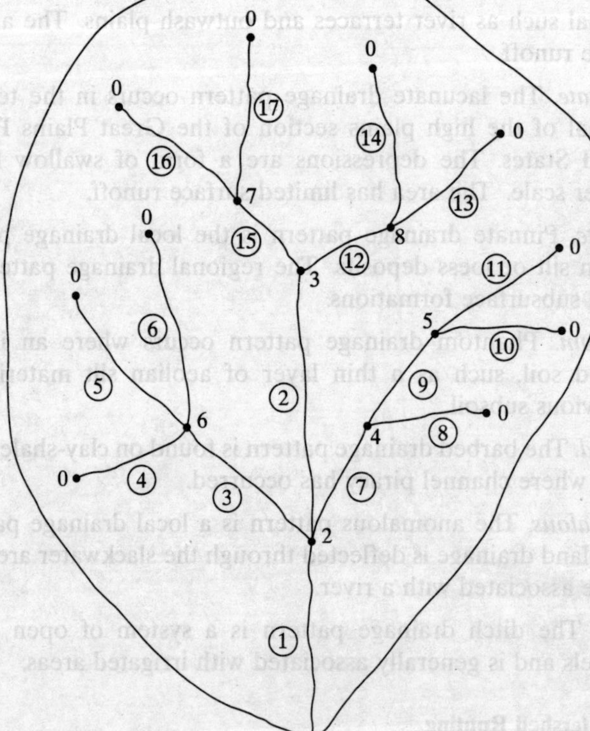

Watershed with nodes and channels numbered.

	Natural	Partially developed	Fully developed
Percent watershed impervious	0	25	50
Manning "n" value			
overland flow	0.30	0.15	0.10
Channel	0.08	0.04	0.04
Channel overbank	0.16	0.08	0.04

The channels and nodes in the watershed have been numbered. All source nodes (located at the upper end of first order channels) are labeled with a node number of 0. The node representing the outlet of the watershed has a node number of 1 and the watershed outlet channel is labeled as channel 1.

Two programs were developed for this problem. HYRTWD1.FOR computes the rating tables for each channel segment and the runoff hydrographs using the kinematic wave model. The second program HYRTWD2.FOR does the hydraulic channel routing.

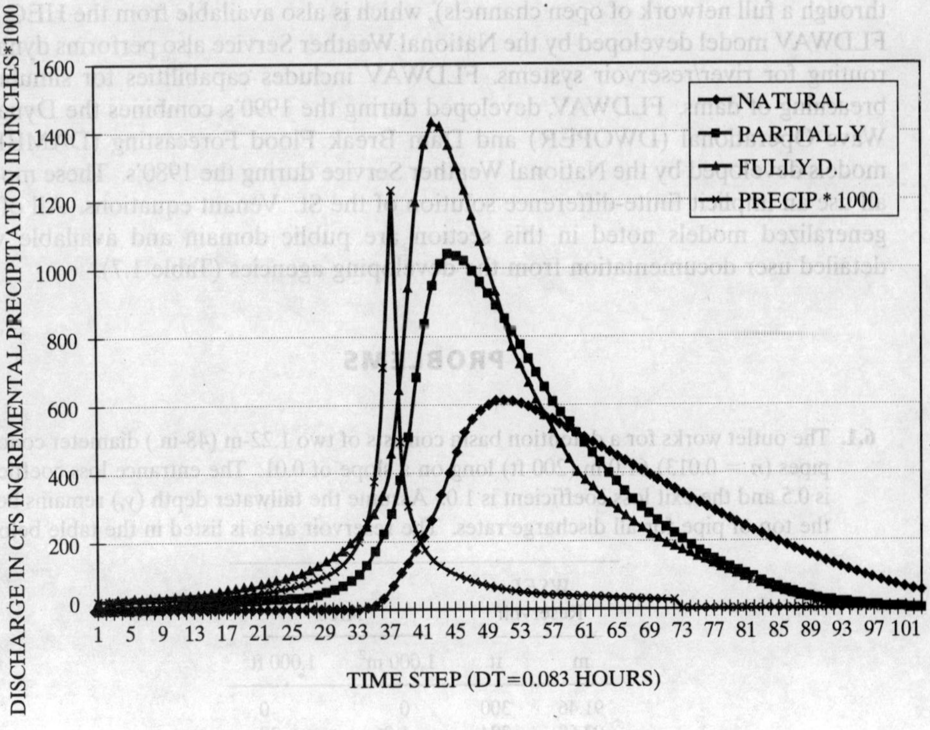

Example 6.7 Watershed routing using kinematic overland flow routing and hydraulic stream routing. A comparison of outflow hydrographs for natural, partly and fully developed watersheds.

A plot of the computed hydrographs and the incremental precipitation is shown on the previous page. The peak discharge was 600 cfs for the natural watershed, 1,000 cfs for the partially urbanized watershed, and 1,400 cfs for the fully urbanized watershed.

The primary disadvantage of the model is the relatively large amount of input data required to model a watershed. As a result, the model is limited to relatively small watersheds.

6.6 GENERALIZED FLOOD ROUTING MODELS

Hydrologic methods for stream and reservoir routing are components of many watershed models, such as the HEC-HMS Hydrologic Modeling System discussed in Section 8.9.1. Hydrologic storage routing is also incorporated in other types models for modeling reservoir storage. HEC-HMS also has kinematic wave watershed and channel routing capabilities, which are designed primarily for modeling urban watersheds. Dynamic routing is incorporated in the HEC-RAS River Analysis System discussed in Section 5.10. The HEC-RAS dynamic routing routines are based on methods originally developed for the UNET model (one-dimensional unsteady flow through a full network of open channels), which is also available from the HEC. The FLDWAV model developed by the National Weather Service also performs dynamic routing for river/reservoir systems. FLDWAV includes capabilities for simulating breaching of dams. FLDWAV, developed during the 1990's, combines the Dynamic Wave Operational (DWOPER) and Dam Break Flood Forecasting (DAMBRK) models developed by the National Weather Service during the 1980's. These models all use an implicit finite-difference solution of the St. Venant equations. All of the generalized models noted in this section are public domain and available with detailed user documentation from the developing agencies (Table 1.7).

PROBLEMS

6.1. The outlet works for a detention basin consists of two 1.22-m (48-in.) diameter concrete pipes (n = 0.013) 61.0 m (200 ft) long on a slope of 0.01. The entrance loss coefficient is 0.5 and the exit loss coefficient is 1.0. Assume the tailwater depth (y_t) remains below the top of pipe for all discharge rates. The reservoir area is listed in the table below.

WSEL Reservoir		Area	
m	ft	1,000 m^2	1,000 ft^2
91.46	300	0	0
92.68	304	1.86	20
93.90	308	9.30	100
95.12	312	18.59	200
96.34	316	46.48	500
97.56	320	92.95	1,000
98.78	324	185.90	2,000

Route the following hydrograph through the detention basin and determine the peak discharge rate from the reservoir and the maximum water surface elevation in the reservoir. The detention basin is empty at the start of the storm.

Time	Discharge		Time	Discharge		Time	Discharge	
hrs	cms	cfs	hrs	cms	cfs	hrs	cms	cfs
0	0.28	10	17	7.37	260	34	3.68	130
1	0.57	20	18	6.80	240	35	3.40	120
2	1.13	40	19	6.23	220	36	3.12	110
3	2.83	100	20	5.67	200	37	2.83	100
4	7.04	250	21	5.52	195	38	2.55	90
5	14.16	500	22	5.38	190	39	2.27	80
6	28.32	1,000	23	5.24	185	40	1.98	70
7	26.91	950	24	5.10	180	41	1.70	60
8	25.50	900	25	4.96	175	42	1.42	50
9	22.66	800	26	4.82	170	43	1.13	40
10	19.83	700	27	4.67	165	44	0.85	30
11	17.00	600	28	4.53	160	45	0.57	20
12	14.16	500	29	4.39	155	46	0.28	10
13	11.33	400	30	4.25	150	47	0.28	10
14	9.92	350	31	4.11	145	48	0.28	10
15	8.52	300	32	3.97	140	49	0.28	10
16	7.93	280	33	3.82	135	50	0.28	10

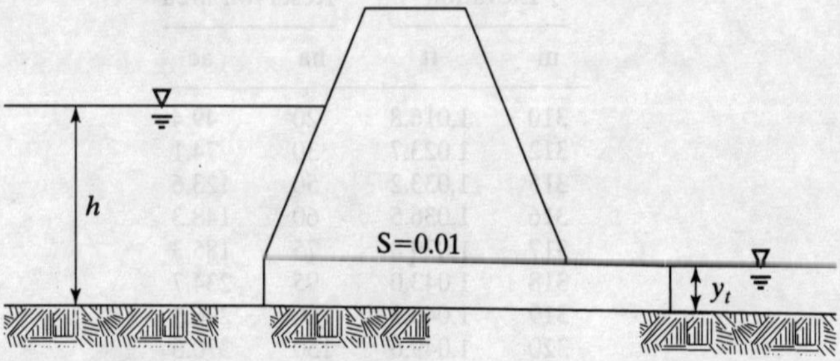

6.2. Outlet works for a detention basin is a spillway with a width (L) of 9.15 m (30.0 ft) where the discharge is given by the weir equation.

$$Q = 0.40\sqrt{2g}\,Lh^{3/2}$$

The reservoir area is listed below.

Elevation		Area	
m	ft	1,000 m^2	1,000 ft^2
152.4	500	93.0	1,000
153.0	502	139.4	1,500
153.6	504	185.9	2,000
154.3	506	278.8	3,000
154.9	508	464.8	5,000
155.5	510	836.6	9,000

Route the inflow hydrograph from Problem 6.1 through the detention basin and determine the maximum discharge rate from the basin and the maximum water level in the basin. The starting water surface elevation in the basin is 152.4 m (500 ft).

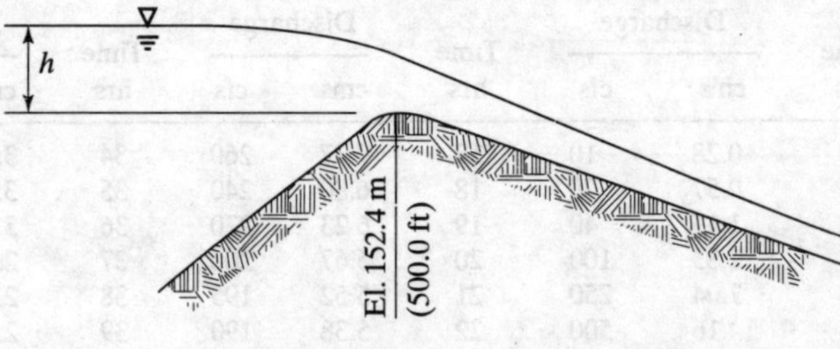

6.3. The outlet works for a reservoir include both a pipe through the embankment and a spillway. The spillway is 10.0 m (32.8 ft) wide (L) and the discharge is given by the weir equation

$$Q = 0.40\sqrt{2g}\,Lh^{3/2}$$

The outlet pipe is 1.0 m (3.28 ft) in diameter with a total length of 75 m (246 ft). The Manning "n" value for the outlet pipe is 0.024 with a minor loss coefficient totaling 2.8 (entrance loss 0.8, bend loss 1.0, and exit loss 1.0). Assume the tailwater elevation remains below the top of pipe for the range in discharge rates expected. The water surface area in the reservoir is listed below.

Elevation		Reservoir area	
m	ft	ha	ac
310	1,016.8	20	49.4
312	1,023.7	30	74.1
315	1,033.2	50	123.6
316	1,036.5	60	148.3
317	1,039.8	75	185.3
318	1,043.0	95	234.7
319	1,046.3	120	296.5
320	1,049.6	150	370.6

Route the following hydrograph through the reservoir. The water surface elevation in the reservoir at the start of the storm is 310.0 m (1,016.8 ft).

Time	Discharge		Time	Discharge	
hrs	cms	cfs	hrs	cms	cfs
0	0	0	4.5	100	3,530
0.5	8	282	5.0	80	2,824
1.0	30	1,059	5.5	65	2,294
1.5	100	3,530	6.0	50	1,765
2.0	170	6,000	6.5	35	1,236
2.5	165	5,824	7.0	20	706
3.0	155	5,472	7.5	10	353
3.5	140	4,942	8.0	5	176
4.0	120	4,236	8.5	0	0

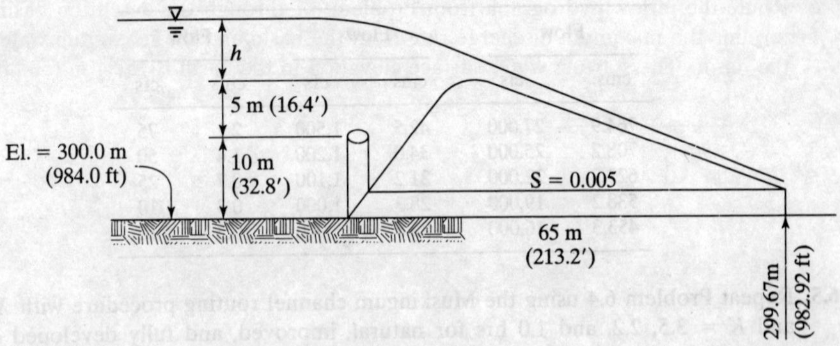

6.4. The following cross-section is representative of a channel routing reach 4,573 m (15,000 ft) long. Estimate the relation between storage and discharge using the standard step water surface profile computations for a range in discharge rates.

Cross-section Elevation								
m	304.9	298.8	298.2	295.7	295.7	298.5	299.1	304.9
ft	1,000	980	978	970	970	979	981	1,000

Distance								
m	0	15.2	76.2	83.8	91.5	97.6	289.6	304.9
ft	0	50	250	275	300	320	950	1,000

The Manning "n" values and channel slope listed below indicate the degree of channel improvements.

Manning "n"

Channel	LOB	CH	ROB	Slope
Natural	0.16	0.08	0.16	0.0006
Improved	0.08	0.04	0.08	0.0006
Developed	0.04	0.03	0.04	0.001

Route the following hydrograph (tabulated at 1-hr intervals) through the routing reach for the natural, improved, and fully developed channel using the storage routing procedure.

Flow		Flow		Flow	
cms	cfs	cms	cfs	cms	cfs
5.7	200	368.3	13,000	22.7	800
14.2	500	283.3	10,000	19.8	700
42.5	1,500	226.6	8,000	17.0	600
85.0	3,000	198.3	7,000	14.2	500
170.0	6,000	170.0	6,000	11.3	400
339.9	12,000	141.6	5,000	8.5	300
679.9	24,000	113.3	4,000	7.1	250
849.9	30,000	85.0	3,000	5.7	200
821.5	29,000	70.8	2,500	4.2	150
793.2	28,000	56.7	2,000	2.8	100

Flow		Flow		Flow	
cms	cfs	cms	cfs	cms	cfs
764.9	27,000	42.5	1,500	2.1	75
708.2	25,000	34.0	1,200	1.4	50
623.2	22,000	31.2	1,100	0.7	25
538.2	19,000	28.3	1,000	0.3	10
453.3	16,000	25.5	900	0.3	10

6.5. Repeat Problem 6.4 using the Muskingum channel routing procedure with $X = 0.20$ and $K = 3.5$, 2.2, and 1.0 hrs for natural, improved, and fully developed channel, respectively.

6.6. Use the kinematic wave model to compute the time of concentration, the maximum runoff rate, and the maximum water depth from an area 73.2 m (240 ft) long, 1 unit wide with a longitudinal slope of 2.0 percent, and a Manning "n" value of 0.25. The soil for the pervious area has been classified as silty clay loam [see Example 6.3(a)] with an initial abstraction of 12.7 mm (0.5 in.). The rainfall intensity (RI) in mm/hr (in./hr) is given by

$$RI = \frac{b}{(t + d)^e}$$

where $b = 127$ for RI in mm/hr (5.0, RI in in./hr), $d = 8.0$, $e = 0.0$, and t is time in minutes. Use a 1-hr duration rainfall with $dt = 1.0$ minute. Assume the area is 100 percent pervious.

6.7. Repeat Problem 6.6 with the area 100 percent impervious. Compare the peak discharge rate with the steady state discharge rate (q)

$$q = RI \times \text{Area}$$

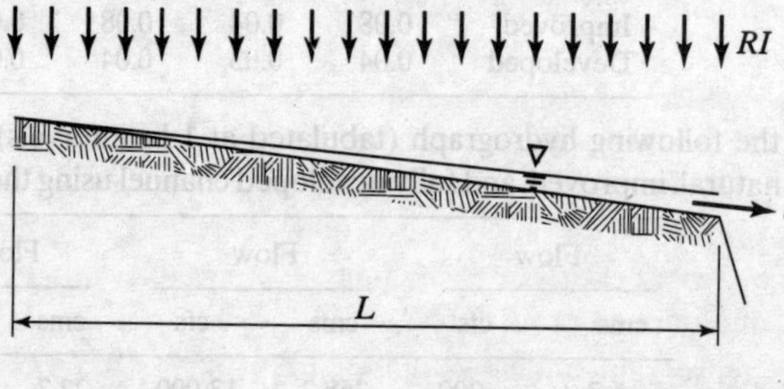

6.8. Compute the overland flow into a stream channel where the cross-section is shown on the following page. The rainfall intensity (RI) in mm/hr (in./hour) is given by

$$RI = \frac{b}{(t + d)^e}$$

where $b = 2{,}438$ for RI in mm/hour (96, in./hour), $d = 8.0$, $e = 0.73$, and t is time in minutes. Use a 6-hour duration storm with $dt = 5$ minutes. Use an initial abstraction of 12.7 mm (0.5 in.) and a balanced triangular rainfall distribution. The soil and infiltration characteristics are listed as follows:

	Soil 1	Soil 2	Soil 3
Type	Clay loam	Silty loam	Silty clay
K, mm/hr (in./hr)	1.3 (0.05)	3.6 (0.14)	0.5 (0.02)
Porosity	0.39	0.49	0.43
VMC	0.25	0.20	0.30
WFS mm (in.)	208 (8.2)	168 (6.6)	292 (11.5)

What is the maximum water depth and discharge rate at the channel?

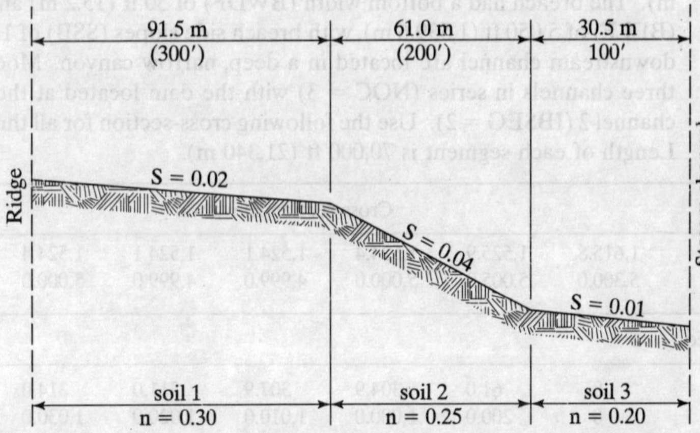

6.9. Repeat Problem 6.4 using the explicit hydraulic routing procedure for a natural channel (HYRT3.FOR). Use one channel with 30 computational segments.

6.10. Repeat Problem 6.4 using the implicit hydraulic routing procedure for a natural channel (HYRTI.FOR). Use one channel with 5 computational segments, Dt = 0.1 hr, AWC = 0.90, and ACC = 0.001.

6.11. Repeat problem 6.4 using three channels in series, each 1,524 m (5,000 ft) long. Use the same cross-section for each channel. Route the inflow hydrograph using both the explicit and implicit procedures.

6.12. Use the hydrograph and cross-section in Problem 6.4 for a dam break analysis. Divide the routing reach into three segments, each 1,524 m (5,000 ft) long using the same cross-section (natural conditions) for each segment. The dam is to be located at the downstream end of channel 2 (IBSEG = 2). The initial water surface elevation in the reservoir (WSELD) is 330.0 m (1,082.4 ft), initial breach width (BWDI) is 10.0 m (32.8 ft), final breach width (BWDF) is 40.0 m (132.2 ft), final breach elevation (BELF) is 305.0 m (1,000 ft), side slopes of the breach (SSB) are 0.5, start time of the breach (BTST) is 2.0 hrs, and time for the breach to develop (BTDR) is 6.0 hrs. The cross-sections above the dam are to be modified to represent the reservoir as follows:

Cross-section segment	Distance Pt 7 m	Distance Pt 7 ft	Distance Pt 8 m	Distance Pt 8 ft	Elevation Pts 1 & 8 m	Elevation Pts 1 & 8 ft
1	500	1,640	1,000	3,280	330	1,082.4
2	1.000	3,280	2,000	6,560	330	1,082.4

6.13. The Teton Dam on the Teton River in Idaho failed in 1976. The reservoir contained 250,000 ac-ft (309 million m^3) of water when the 300-ft (91.5-m) high earthfill dam failed in approximately 2 hrs (BTDR = 2.0), resulting in a peak discharge rate of 2,300,000 cfs (65,200 cms) downstream of the dam and a peak discharge rate of 1,060,000 cfs (30,000 cms) in the channel approximately 50,000 ft (15,200 m) downstream of the dam. The initial water surface elevation in the reservoir (WSELD) was 5,300 ft (1,615.8 m). The breach had a bottom width (BWDF) of 50 ft (15.2 m) and a bottom elevation (BELF) of 5,050 ft (1,539.6 m), with breach side slopes (SSB) of 1.5. The reservoir and downstream channel are located in a deep, narrow canyon. Model the dam break as three channels in series (NOC = 3) with the dam located at the downstream end of channel 2 (IBSEG = 2). Use the following cross-section for all three channel segments. Length of each segment is 70,000 ft (21,340 m).

	Cross-section Elevation							
m	1,615.8	1,525.9	1,524.4	1,524.1	1,524.1	1,524.4	1,525.9	1,615.8
ft	5,300.0	5,005.0	5,000.0	4,999.0	4,999.0	5,000.0	5,005.0	5,300.0
Distance								
m	0	61.0	304.9	307.9	311.0	314.0	457.3	609.8
ft	0	200.0	1,000.0	1,010.0	1,020.0	1,030.0	1,500.0	2,000.0

Channel segment	Manning "n" values Lt	Ch	Rt	ASF	Slope
1	0.20	0.20	0.20	1.1	0.001
2	0.20	0.20	0.20	1.1	0.002
3	0.08	0.06	0.08	1.3	0.002

Model the dam break for 24 hrs (NOH = 24) with no lateral inflow (LI = 0 and NOL = 0). The invert elevation of the upstream end of channel 1 (ELEV) is 5,210 ft (1,588.4 m). At the start of the break (BTST = 0), the width of the breach (BWDI) is 10.0 ft (3.05 m). Use the explicit routing procedure with 30 computational segments in each channel (MMOS = 30) and TFT = 0.7. An inflow hydrograph file will be required (INFLOWHY.DAT). Use 24 values of 1,000 cfs (28.3 cms) for the inflow hydrograph. Modify the cross-section for the downstream channel three by lowering the end elevations (Pts 1 and 8) to 5,150 ft (1,570.1 m).

6.14. The dam break program requires large Manning "n" values in the reservoir for stability at the junction nodes. To eliminate the large change in cross-sections at the junction nodes, modify subroutine LOOKUP to interpolate cross-section values along the length of the channel. Assume each cross-section is representative of channel (J) at node 1 and at the last node (N) of the upstream channel ($J - 1$). No interpolation will be required for the last channel above the reservoir and for the last channel downstream of the dam.

6.15. The two watersheds shown below have the same area (AW) 0.32 sq km (0.125 sq mi) but have different shapes and channel configurations. Compute the runoff hydrographs from each watershed for a 6-hr duration storm (DUR = 6.0), with a computational time step of 5 minutes (DT = 0.0833) using the rainfall intensity equation from Problem 6.8. Both watersheds have sandy clay loam soils (soil code = 6) with 10 percent impervious. Use an initial rainfall abstraction of 12.7 mm (0.5 in.) and a volumetric soil moisture content of 0.2. Watershed (a) has 12 channels (NOC = 12), including 8 first-order channels and 5 junctions (NOJ = 5). Watershed (b) has 13 channels (NOC = 13), including 8 first-order channels and 6 junctions (NOJ = 6).

Use the following cross-section for channels 1a, 2a, and 1b with a slope of 0.002 and a length of 457.3 m (1,500 ft).

	Elevation							
m	30.5	29.0	28.7	27.4	27.4	28.7	29.0	30.5
ft	100.0	95.0	94.0	90.0	90.0	94.0	95.0	100.0

	Distance							
m	0	3.0	12.2	15.2	16.8	19.8	27.4	30.5
ft	0	10.0	40.0	50.0	55.0	65.0	90.0	100.0

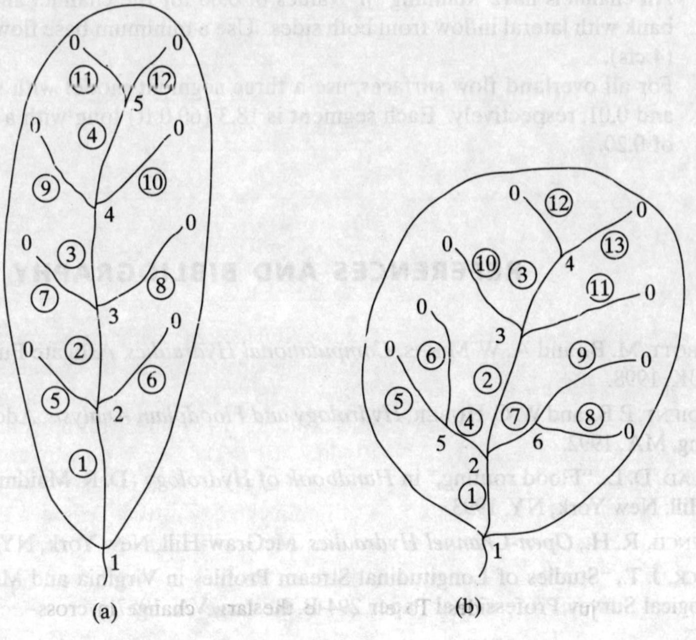

Use the following cross-section for channels 3a, 4a, 2b, 3b, 4b, and 7b with a channel slope of 0.003 and a length of 274.4 m (900 ft).

Elevation								
m	30.5	29.0	28.7	27.4	27.4	28.7	29.0	30.5
ft	100.0	95.0	94.0	90.0	90.0	94.0	95.0	100.0
Distance								
m	0	9.1	12.2	13.7	14.9	16.8	19.8	29.0
ft	0	30.0	40.0	45.0	49.0	55.0	65.0	95.0

For all first-order channels, use the following cross-section with a channel slope of 0.004 and a channel length of 152.4 m (500 ft).

Elevation								
m	30.5	29.0	28.7	27.4	27.4	28.7	29.0	30.5
ft	100.0	95.0	94.0	90.0	90.0	94.0	95.0	100.0
Distance								
m	0	9.1	10.7	12.2	13.1	15.2	16.8	25.9
ft	0	30.0	35.0	40.0	43.0	50.0	55.0	85.0

All channels have Manning "*n*" values of 0.06 for the channel and 0.10 for the overbank with lateral inflow from both sides. Use a minimum base flow (BFM) of 0.11 cms (4 cfs).

For all overland flow surfaces, use a three-segment model with slopes of 0.01, 0.02, and 0.01, respectively. Each segment is 18.3 (60.0 ft) long with a Manning "*n*" value of 0.20.

REFERENCES AND BIBLIOGRAPHY

ABBOTT, M. B., and A. W. MINNS, *Computational Hydraulics,* Ashgate Publishing, Aldershot, UK, 1998.

BEDIENT, P. B., and W. C. HUBER, *Hydrology and Floodplain Analysis,* Addison-Wesley, Reading, MA, 1992.

FREAD, D. L., "Flood routing," in *Handbook of Hydrology* (D. R. Maidment, Ed.), McGraw-Hill, New York, NY, 1993.

FRENCH, R. H., *Open-Channel Hydraulics,* McGraw-Hill, New York, NY, 1985.

HACK, J. T., "Studies of Longitudinal Stream Profiles in Virginia and Maryland," U.S. Geological Survey Professional Paper 294-B, Reston, VA, 1957.

HORTON, R. E., "Erosional Development of Streams and Their Drainage Basins: Hydrophysical Approach to Quantitative Morphology," *Bull. Geol. Soc. Am.*, Vol. 56, pp. 275–370, 1945.

MCCUEN, R. H., *Hydrologic Analysis and Design,* Prentice Hall, Upper Saddle River, NJ, 1998.

PARVIS, M., "Drainage Pattern Significance in Airphoto Interpretation of Soils and Rocks," *Soil Exploration and Mapping,* Highway Research Board, Bulletin 28, November 1950.

ROBERSON, J. A., J. J. CASSIDY, and M. H. CHAUDHRY, *Hydraulic Engineering,* 2nd Ed., John Wiley & Sons, New York, NY, 1998.

SINGH, V. P., *Kinematic Wave Modeling in Water Resources,* John Wiley & Sons, New York, NY, 1996.

STURM, T. W., *Open Channel Hydraulics,* McGraw-Hill, New York, NY, 2001.

STRAHLER, A. N., "Quantitative Geomorphology of Drainage Basins and Channel Networks," in *Handbook of Applied Hydrology* (V. T. Chow, Ed.), McGraw-Hill, New York, NY, 1964.

VENNARD, J. K., and R. L. STREET, *Elementary Fluid Mechanics,* John Wiley & Sons, New York, NY, 1975.

7

Hydrologic Frequency Analysis

Hydrologic phenomena are characterized by great variability, randomness, and uncertainty. Precipitation, streamflow, and other quantities of importance in water resources engineering must be treated as random variables, with associated measures of frequency that represent likelihood, percentage of time, or probability. Chapter 7 presents techniques for associating frequencies with hydrologic variables.

7.1 HYDROLOGIC RANDOM VARIABLES AND DATA

The risks that the flow capacity of hydraulic structures will be exceeded, water supply systems will fail to meet demands, water quality requirements will be violated, and flooding streams will endanger life and property are fundamental to water resources engineering. Frequency analysis methods are essential to hydrologic design and assessment.

Hydrologic design of hydraulic structures is based on adopting acceptable levels of risk, which are often specified in design criteria manuals developed by federal and state water agencies, transportation agencies, and cities. Bridges for major highways may be designed to pass a flood with an annual exceedance frequency of 1 percent without overtopping the roadway. Bridges and culverts for streets with lower traffic volumes may be designed based on less stringent criteria, perhaps a 2 or 4 percent exceedance frequency design flood. Design criteria for storm sewers, drainage ditches, detention basins, and other components of urban stormwater management systems are likewise based on specified exceedance probabilities.

Frequency analysis is a key aspect of the economic analysis of flood mitigation plans. The 100-year recurrence interval flood has been adopted as the base flood for floodplain management regulations under the National Flood Insurance Program.

Municipal water supply systems must maintain very high reliabilities. Agricultural irrigation may be associated with greater risks of failing to supply demands. Acceptable reliabilities for generating hydroelectric power are dependent on alternative thermal energy sources available to meet energy requirements during periods of low streamflow. Low-flow frequencies associated with environmental instream flow requirements are an important aspect of river basin management. Water quality is often assessed in terms of the frequency in which constituent concentrations fall within specified ranges.

7.1.1 Random Variables

A random variable is a quantity that depends on chance. The value or range of values can be predicted only with an associated probability, not with certainty. Random variables are formulated to fit a particular water resources engineering application. Examples of common hydrologic random variables include:

- mean daily, monthly, or annual stream discharge (m^3/s), precipitation (mm), evaporation (mm), net evaporation less precipitation (mm), water quality constituent concentration (mg/l) or load (kg), aquifer water table elevation (m), or reservoir water surface elevation (m)
- annual maximum or minimum mean streamflow (m^3/s) or water quality constituent concentration (mg/l) for a specified time duration, such as 1 day, 7 days, or 30 days
- maximum rainfall depth (mm) or mean intensity (mm/hr) for a specified time duration
- maximum annual instantaneous flood stage (m) or discharge (m^3/s)

A frequency relationship or probability distribution function represents the likelihood of occurrence of values of a random variable. Frequency relationships are developed based on observed and/or simulated data.

7.1.2 Hydrologic Data

Hydrologic processes and associated measured data are covered in Chapter 2. Data sources include those discussed in Section 2.11. A data set is provided in Table 7.1 for use in illustrating the concepts discussed later in Sections 7.4–7.7. Examples 7.4–7.8 use the data set. Table 7.1 is a tabulation of the maximum annual discharge for each year from 1933 through 1998 observed at a U.S. Geological Survey (USGS) gaging station on the Mississippi River at St. Louis, Missouri. The river basin is

TABLE 7.1 MAXIMUM ANNUAL DISCHARGE IN THE MISSISSIPPI RIVER AT ST. LOUIS

Year	Flow (m^3/s)	Year	Flow (m^3/s)	Year	Flow (m^3/s)
1933	12,400	1955	8,800	1977	11,000
1934	6,260	1956	5,860	1978	16,200
1935	18,500	1957	9,620	1979	19,500
1936	9,450	1958	14,300	1980	9,930
1937	10,600	1959	10,300	1981	14,400
1938	12,300	1960	19,000	1982	20,700
1939	15,100	1961	16,700	1983	20,300
1940	5,240	1962	16,700	1984	16,400
1941	14,000	1963	8,510	1985	19,500
1942	18,900	1964	8,710	1986	20,500
1943	23,700	1965	15,600	1987	11,900
1944	23,700	1966	10,500	1988	8,850
1945	17,400	1967	15,000	1989	9,280
1946	14,200	1968	9,790	1990	17,000
1947	22,300	1969	17,500	1991	12,400
1948	17,900	1970	15,300	1992	14,600
1949	12,000	1971	11,900	1993	30,600
1950	13,100	1972	11,500	1994	17,000
1951	22,200	1973	24,200	1995	22,500
1952	19,400	1974	16,500	1996	17,400
1953	10,400	1975	13,700	1997	15,400
1954	8,230	1976	12,700	1998	15,500

discussed in Chapter 2, and the hydrograph for this station is plotted in Fig. 2.14. The watershed above the gage has a drainage area of 1,810,000 km^2.

Data sets used in frequency analyses may be developed in various forms. The hydrograph for the Mississippi River at St. Louis plotted in Fig. 2.14 is a complete series of 24,100 mean daily flows covering the 1933–1998 period of record. This general type of data set is used in Section 7.9 to develop flow-duration relations for use in water supply and low flow studies. Flood peak discharge frequency analyses may be based on two types of data sets: an annual series or partial duration series. Table 7.1 is an annual series with the one peak flow for each of 66 years. A partial duration series includes all floods with peaks above a base flood magnitude, which could include multiple floods in the same year. The base flood magnitude for a partial duration series could be the channel capacity, level at which significant damages occur, or any other selected level. Probability distribution functions discussed in Section 7.5 are applicable to annual series, but not to partial duration series. The plotting position formulas in Section 7.4 are applicable for either annual or partial duration series.

Data sets used in frequency analyses may be gage observations, adjusted gage observations, simulation model results, or combinations thereof. Homogeneity is an important consideration in using gaged streamflow data. Construction of reservoir projects, water supply diversions and return flows, and watershed land use changes during the period of record affect the flows at the gage and introduce nonhomogeneities. Frequency analyses should represent a specified condition of watershed development. In many cases, gaged flows are adjusted for the historical effects of reservoirs and other watershed development activities. In many other cases, watershed simulation models (Chapter 8) are used instead of applying frequency analyses methods directly to observed gaged flows. Construction of reservoir projects and other activities in the Mississippi River Basin above St. Louis during 1933–1998 have affected to some degree the observed flows tabulated in Table 7.1. However, major long-term trends of changing characteristics of high flows are not evident from the hydrograph of Fig. 2.14. We consider the data set to be homogeneous enough for purposes of the examples in Sections 7.4–7.6.

7.2 PROBABILITY RELATIONSHIPS

Probability theory and methods are covered in depth by many probability and statistics books (Ang and Tang, 1991; Kottegoda and Rosso, 1997; Devore, 2000) and hydrology books (Helsel and Hirsch, 1992; Viessman and Lewis, 1996; McCuen, 1998; Rao and Hamed, 2000). Our treatment of probability is limited to basic concepts required to apply hydrologic frequency analysis techniques.

Random variables may be measured either as discrete integer values (0, 1, 2, ...) or over a continuous scale. Of the many discrete distribution functions covered in probability and statistics books, our attention will be limited to the binomial distribution described in Section 7.3.1. The analytical probability distribution functions discussed in Section 7.5 model continuous random variables. Peak flood flows are used throughout Sections 7.3–7.6 to illustrate the methods presented. However, the concepts are applicable to a myriad of other random variables in water resources engineering and various other fields.

The cumulative probability $F(x)$ and exceedance probability $P(x)$ for a random variable X are complements of each other defined as

$$F(x) = \text{Probability}\,(X \le x) = 1 - P(x) \tag{7.1}$$

$$P = P(x) = \text{Probability}\,(X > x) = 1 - F(x) \tag{7.2}$$

Probabilities are dimensionless numbers between zero and one ($0.0 \le$ probability ≤ 1.0). Although probabilities are sometimes also expressed as percentages, percentages are more commonly associated with the equivalent term *frequency* ($0\% \le$ frequency $\le 100\%$).

For continuous random variables,

$$\text{Probability}\,(X = x) = 0$$

since x is just one of an infinite number of possible values for X. Thus, probabilities are assigned to ranges of X rather than single values. Since for continuous random variables,

$$\text{Probability}\,(X \ge x) = \text{Probability}\,(X > x)$$

exceedance probability P may be defined equivalently as the probability either that a specified magnitude is exceeded or that the specified magnitude is equaled or exceeded.

The *annual exceedance probability* P is the probability that a specified magnitude will be equaled or exceeded in any given year. The *recurrence interval* or *return period* T is the average interval, in years, between successive occurrences of events equaling or exceeding a specified magnitude. The recurrence interval T and annual exceedance probability P are reciprocals of each other.

$$T = \frac{1}{P} \quad \text{and} \quad P = \frac{1}{T} \tag{7.3}$$

For example, the 100-year recurrence interval event has an annual exceedance probability of 0.01 or annual exceedance frequency of 1 percent.

The following alternative methods for assigning probabilities to a random variable based on sample data are presented in Sections 7.4 and 7.5.

- Empirical relative frequency relations (Section 7.4)
- Analytical probability distribution functions (Section 7.5)
  — Normal distribution
  — Log-normal distribution
  — Log-Pearson type III distribution
  — Gumbel distribution

The probability distribution functions described in Section 7.5 may be used with observed or computed data to develop relationships between a random variable and annual exceedance probability P, which is the probability that a specified magnitude will be exceeded at least once in any single year. Given a P estimated based on data using Section 7.5 methodologies, the binomial distribution covered in Section 7.3.1 allows us to estimate the probability that the specified magnitude will be exceeded any specified number of times in any specified number of years. The risk formula presented in Section 7.3.2 is developed from the binomial distribution. Given the annual exceedance probability P, the risk formula allows us to determine the probability that the specified magnitude is exceeded at least once in any specified number of years.

7.3 BINOMIAL DISTRIBUTION AND RISK FORMULA

The binomial probability function is a discrete ($X = 0, 1, 2, 3, \ldots$) distribution that is applied in many fields for many different purposes. It has a variety of applications in water resources engineering. The risk formula presented in Section 7.3.2 is based on the same premises and derives directly from the binomial distribution.

7.3.1 Binomial Distribution

In the binomial probability distribution, the random variable X is the integer number of occurrences of an event in N discrete trials. The following conditions must be satisfied for the binomial distribution to be applicable.

- There are N trials or possible occurrences of an event.
- In each trial, the event either occurs or does not occur. The probabilities of occurrence P and nonoccurrence $(1 - P)$ sum to 1.0.
- The probability of occurrence of the event is constant from trial to trial. The trials are independent, meaning occurrence of the event in any trial does not affect the probability of its occurrence in any other trial.

The binomial distribution is

$$P_x(X = x) = \frac{N!}{x!(N - x)!} P^x(1 - P)^{N-x} \tag{7.4}$$

$P_x(X = x)$ denotes the probability that the event will occur x times in N trials. P is the probability that the event occurs in any single trial, and $(1 - P)$ is the probability that the event does not occur in any single trial. The expression

$$\frac{N!}{x!(N - x)!}$$

is called the binomial coefficient and represents the number of different sequences of N trials with x occurrences that are possible. $N!$ is read N factorial and defined as

$$N! = N(N - 1)(N - 2) \dots (2)(1) \tag{7.5}$$

For example, $4! = (4)(3)(2)(1) = 24$. The factorial $0!$ is defined as 1.

Example 7.1

Compute the probability of getting exactly 5 heads when a coin is flipped 10 times. In each of the 10 tosses of the coin, the probabilities of obtaining a head and tail are each 0.5.

$$P_x(X = 5) = \frac{10!}{5!(10 - 5)!}(0.5)^5(1 - 0.5)^{10-5} = 0.246$$

Example 7.2

A culvert is designed to just pass the 20-year recurrence interval discharge without overtopping the road. Estimate the probability that the road will be overtopped exactly once in the next 30 years. (Note that the binomial coefficient is very easy to compute recognizing that 30! divided by 29! is 30.)

$$P = \frac{1}{T} = \frac{1}{20 \text{ years}} = 0.05$$

$$P_x(X = 1) = \frac{30!}{1!(30 - 1)!} 0.05^1(1 - 0.05)^{30-1} = 0.339$$

Estimate the probability that the peak flow in each of exactly 2 of the next 30 years will overtop the bridge.

$$P_x(X = 2) = \frac{30!}{2!(30 - 2)!}(0.05)^2(1 - 0.05)^{30-2} = 0.259$$

If we continued this example by computing $P(X = 0)$, $P(X = 1)$, $P(X = 2)$, $P(X = 3)$, $P(X = 4), \ldots, P(X = 29)$, $P(X = 30)$, for $N = 30$, the 31 probabilities would sum to 1.0. The discrete random variable X is the number of years during a 30-year period in which the peak flow exceeds the 20-year recurrence interval flow.

7.3.2 Risk Formula

The random variable X is again the integer number of occurrences of an event in N discrete trials, with associated probability $P_x(X = x)$. The probability of at least one occurrence of the event in N trials is

$$P_x(X > 0) = 1 - P_x(X = 0)$$

that may be combined with the binomial distribution expression for $P_x(X = 0)$

$$P_x(X = 0) = \frac{N!}{x!(N - x)!}P^N(1 - P)^{N-x} = \frac{N!}{0!(N - 0)!}P^0(1 - P)^{N-0} = (1 - P)^N$$

to obtain

$$P_x(X > 0) = 1 - (1 - P)^N \qquad \textbf{(7.6)}$$

Equation 7.6 is commonly used as an expression for risk written as

$$R = 1 - (1 - P)^N \qquad \textbf{(7.7)}$$

where the risk R is the probability that a specified magnitude will be equaled or exceeded at least once in a series of N years, and P is the annual exceedance probability. The expression $(1 - P)^N$ is the probability that the specified magnitude will not be equaled or exceeded in a series of N years.

Example 7.3

A hydraulic structure is sized for a 50-year recurrence interval design discharge. What is the risk that the flow capacity will be exceeded during any future 20-year period?

$$P = 1/50 \text{ years} = 0.02$$

$$R = 1 - (1 - P)^N = 1 - (1 - 0.02)^{20} = 0.332$$

What is the probability that the 50-year recurrence interval peak flow rate will be exceeded in the next 50 years?

$$R = 1 - (1 - P)^N = 1 - (1 - 0.02)^{50} = 0.636$$

A recurrence interval T or annual exceedance probability P was given in Examples 7.2 and 7.3. In the following sections, we explore methods for estimating P and T using sets of observed or simulated data.

7.4 EMPIRICAL RELATIVE FREQUENCY RELATIONS

A series of N observations may be ranked in descending order with the highest value assigned a rank m of 1 and the smallest assigned a rank m of N. The probability P_m that the observation with rank m is equaled or exceeded becomes

$$P_m = \left(\frac{m}{N}\right)_{N\to\infty} \quad (7.8)$$

as the number of observations (sample size) N approaches infinity. Without N approaching infinity, the relative frequency relation

$$P_m = \frac{m}{N} \quad (7.9)$$

provides an estimate of the probability of observation m being equaled or exceeded, with the accuracy improving with increasing sample size. Equation 7.9 will assign an exceedance probability of 1.0 to the smallest of the N observations, indicating a zero probability of obtaining a value less than those observed, which is usually not correct. Other frequency relations (Eqs. 7.10 and 7.11) have been formulated that eliminate assigning an exceedance probability of 1.0 to an observation. Empirical frequency relations are often called plotting position formulas because they are used to plot observations on probability graph paper.

The general form of most plotting position formulas is as follows:

$$P_m = \frac{m-a}{N+b} \quad (7.10)$$

Equation 7.9 with $a = b = 0$ and the Weibull formula (Eq. 7.11) with $a = 0$ and $b = 1$ are the most commonly used forms of Eq. 7.10.

$$P_m = \frac{m}{N+1} \quad (7.11)$$

The Weibull formula may be expressed in terms of either annual exceedance probability or recurrence interval T for rank m and number of years of observation N.

$$P = \frac{m}{N+1} \quad \text{and} \quad T = \frac{N+1}{m} \quad (7.12)$$

The exceedance probability may be expressed as an exceedance frequency in percent by multiplying P_m and P from Eqs. 7.8–7.12 by 100 percent.

Example 7.4

The Weibull formula is used to develop a frequency relationship for peak flows on the Mississippi River at St. Louis. The observations of peak annual flows in Table 7.1 are rearranged in ranked order in Table 7.2. Annual exceedance probabilities for each observed flow are assigned using the Weibull formula. The flows are plotted with their assigned exceedance frequencies on normal probability paper in Fig. 7.1 and on log-normal probability paper in Fig. 7.2. These plots and the other information included in Figs. 7.1 and 7.2 are discussed in Sections 7.6 and 7.7.

TABLE 7.2 FLOWS FROM TABLE 7.1 IN RANKED ORDER WITH *P* AND *T* FROM WEIBULL FORMULA (EXAMPLE 7.4)

Rank m	$P = m/67$	$T = 67/m$	Flow (m^3/s)	Year	Rank m	$P = m/67$	$T = 67/m$	Flow (m^3/s)	Year
1	0.0149	67.0	30,600	1993	34	0.508	1.97	14,600	1992
2	0.0299	33.5	24,200	1973	35	0.522	1.91	14,400	1981
3	0.0448	22.3	23,700	1943	36	0.537	1.86	14,300	1958
4	0.0597	16.8	23,700	1944	37	0.552	1.81	14,200	1946
5	0.0746	13.4	22,500	1995	38	0.567	1.76	14,000	1941
6	0.0896	11.2	22,200	1951	39	0.582	1.72	13,700	1975
7	0.1045	9.6	20,700	1982	40	0.597	1.68	13,100	1950
8	0.119	8.4	20,500	1986	41	0.612	1.63	12,700	1976
9	0.134	7.4	20,300	1947	42	0.627	1.60	12,400	1933
10	0.149	6.7	20,300	1983	43	0.642	1.56	12,400	1991
11	0.164	6.1	19,500	1979	44	0.657	1.52	12,300	1938
12	0.179	5.6	19,500	1985	45	0.672	1.49	12,000	1949
13	0.194	5.2	19,400	1952	46	0.687	1.46	11,900	1971
14	0.209	4.8	19,400	1960	47	0.702	1.43	11,900	1987
15	0.224	4.5	18,900	1942	48	0.716	1.40	11,500	1972
16	0.239	4.2	18,500	1935	49	0.731	1.37	11,000	1977
17	0.254	3.9	17,900	1948	50	0.746	1.34	10,600	1937
18	0.269	3.7	17,500	1969	51	0.761	1.31	10,500	1966
19	0.284	3.5	17,400	1945	52	0.776	1.29	10,400	1953
20	0.299	3.4	17,400	1996	53	0.791	1.26	10,300	1959
21	0.313	3.2	17,000	1990	54	0.806	1.24	9,930	1980
22	0.328	3.0	17,000	1994	55	0.821	1.22	9,790	1968
23	0.343	2.9	16,700	1961	56	0.836	1.20	9,620	1957
24	0.358	2.8	16,700	1962	57	0.851	1.18	9,450	1936
25	0.373	2.7	16,500	1974	58	0.866	1.16	9,280	1989
26	0.388	2.6	16,400	1984	59	0.881	1.14	8,850	1988
27	0.403	2.5	16,200	1978	60	0.896	1.12	8,800	1955
28	0.418	2.4	15,600	1965	61	0.910	1.10	8,710	1964
29	0.433	2.3	15,500	1998	62	0.925	1.08	8,510	1963
30	0.448	2.2	15,400	1997	63	0.940	1.06	8,230	1954
31	0.463	2.2	15,300	1970	64	0.955	1.05	6,260	1934
32	0.478	2.1	15,100	1939	65	0.970	1.03	5,860	1956
33	0.493	2.0	15,000	1967	66	0.985	1.02	5,240	1940

With 66 years of observations, the recurrence interval assigned to the highest observed discharge is

$$T = \frac{N+1}{m} = \frac{66+1}{1} = 67 \text{ years}$$

with an associated exceedance probability of

$$P = \frac{m}{N+1} = \frac{1}{66+1} = 0.0149$$

Occurrence of extreme flood events, with recurrence intervals much greater than 67 years, during the 66-year period of record is certainly possible, as demonstrated by Eq. 7.7. With the Weibull formula, any such extreme event would be inaccurately assigned a recurrence interval of 67 years. The empirical relative frequency relation provides reasonably accurate estimates of probabilities for frequent events well within the range covered by the observations. However, the estimates of exceedance probability assigned to the largest floods in the observed data may be highly inaccurate. The plots of observed data should not be extrapolated. An analytical probability distribution function, such as the log-Pearson type III discussed in the next section, may be used for a full range of values, including events with recurrence intervals greater than the number of years of observation.

Graphs using plotting position formulas provide a visual display of the closeness of fit of an analytical probability distribution to the observed annual data series. Plotting position formulas are also used with partial duration series to analyze relatively frequent events with recurrence intervals ranging from less than a year to several years. The analytical probability distribution functions discussed next are limited to annual data series and recurrence intervals greater than a year.

7.5 ANALYTICAL PROBABILITY DISTRIBUTIONS

Numerous probability distribution functions have been used to model phenomena characterized by significant variability not deterministically explained by physical principles. Many probability distribution functions for continuous random variables can be expressed as either

$$X = \overline{X} + KS \tag{7.13}$$

or

$$\log X = \overline{\log X} + KS_{\log X} \tag{7.14}$$

The sample mean $\overline{X}$ and standard deviation S are estimates of the population mean μ and standard deviation σ of the random variable X. Frequency factors K are read from published tables previously developed by integrating the appropriate probability density function. $\overline{\text{Log}\, X}$ and $S_{\log X}$ are the mean and standard deviation of the logarithm of the random variable X. Application of Eq. 7.14 consists of transforming the data to their logarithms and applying Eq. 7.13 to the logarithms. The normal and Pearson type III distributions are applied using Eq. 7.13. The log-normal and log-Pearson type III distributions are applied using Eq. 7.14.

A distribution function provides a probabilistic model of the phenomena represented by a particular random variable. Model parameters are computed from sample observations. The parameters mean $\overline{X}$, standard deviation S, and skew coefficient G are computed from n observations X_i with the following formulas.

$$\overline{X} = \frac{1}{n} \sum_{i=1}^{n} X_i \tag{7.15}$$

$$S = \left[\frac{1}{n-1} \sum_{i=1}^{n} (X_i - \overline{X})^2 \right]^{0.5} \tag{7.16}$$

$$G = \frac{n \sum_{i=1}^{n} (X_i - \overline{X})^3}{(n-1)(n-2)S^3} \quad \textbf{(7.17)}$$

Probability distributions for continuous random variables are typically derived by statisticians and mathematicians as a probability density function that is integrated between limits to obtain probabilities associated with ranges of values of the random variable. Since the density functions may be difficult to integrate, tables are developed and published in statistics books and other references to facilitate application of the probability distribution functions. Values for the frequency factor K in Eqs. 7.13 and 7.14 are obtained from the tables. Standard probability distributions commonly used in water resources engineering include the normal, log-normal, log-Pearson type III, and Gumbel. The Gumbel probability function may be applied directly without use of Eqs. 7.13 and 7.14 and frequency factor tables.

7.5.1 Normal and Log-Normal Distributions

The normal probability distribution function, also known as the Gaussian distribution, is described in essentially all probability and statistics textbooks and many other references. It models many different processes that are subject to random and independent variations. The normal distribution has two parameters: the mean and standard deviation. Its probability density function is bell-shaped and symmetrical about the mean. For purposes of practical applications, the normal distribution is represented by Eq. 7.13, with the frequency factor K being the standard ($\mu = 0$ and $\sigma = 1$) normal variant z that is related to cumulative probability in tables provided in all statistics and many hydrology books.

The log-normal distribution of X is equivalent to applying the normal distribution to the transformed random variable log X. It is represented by Eq. 7.14 with $\overline{\log X}$ and $S_{\log X}$ computed from the logarithms of the data. The frequency factor K is the standard normal variant from the normal probability table.

Many hydrologic variables exhibit a marked skewness, largely because physically they cannot be negative. Whereas the normal distribution allows the random variable to range without limit from negative infinity to positive infinity, the log-normal distribution has a lower limit of zero. Thus, for random variables such as streamflow and precipitation that should have zero probability of being negative, the log-normal distribution provides this advantage over the normal distribution while still preserving most properties of the normal distribution.

7.5.2 Pearson Type III and Log-Pearson Type III Distributions

Pearson (1930) proposed a general formulation that fits many probability distributions, including the normal, beta, and gamma distributions. A form of the Pearson probability distribution, called the Pearson type III, has three parameters that include the skew coefficient (Eq. 7.17), as well as the mean and standard deviation. The Pearson type III distribution is represented by Eq. 7.13 with K determined from Table 7.3. The Interagency Committee on Water Data (1982) provides a larger table with more values than Table 7.3. If the skew coefficient G has a value of zero, the Pearson type III distribution reduces to the normal distribution. Thus, the values for K in Table 7.3 for a G of zero are the same as the values in the normal probability tables found in many other books.

TABLE 7.3 *K* **VALUES FOR THE PEARSON TYPE III AND LOG-PEARSON TYPE III DISTRIBUTIONS**

Skew	Recurrence interval, years							
Coefficient	1.0101	2	5	10	25	50	100	200
G	Exceedance frequency, percent							
	99	50	20	10	4	2	1	0.5
3.0	−0.667	−0.396	0.420	1.180	2.278	3.152	4.051	4.970
2.8	−0.714	−0.384	0.460	1.210	2.275	3.114	3.973	4.847
2.6	−0.769	−0.368	0.499	1.238	2.267	3.071	3.889	4.718
2.4	−0.832	−0.351	0.537	1.262	2.256	3.023	3.800	4.584
2.2	−0.905	−0.330	0.574	1.284	2.240	2.970	3.705	4.444
2.0	−0.990	−0.307	0.609	1.302	2.219	2.912	3.605	4.298
1.8	−1.087	−0.282	0.643	1.318	2.193	2.848	3.499	4.147
1.6	−1.197	−0.254	0.675	1.329	2.163	2.780	3.388	3.990
1.4	−1.318	−0.225	0.705	1.337	2.128	2.706	3.271	3.828
1.2	−1.449	−0.195	0.732	1.340	2.087	2.626	3.149	3.661
1.0	−1.588	−0.164	0.758	1.340	2.043	2.542	3.022	3.489
0.8	−1.733	−0.132	0.780	1.336	1.993	2.453	2.891	3.312
0.6	−1.880	−0.099	0.800	1.328	1.939	2.359	2.755	3.132
0.4	−2.029	−0.066	0.816	1.317	1.880	2.261	2.615	2.949
0.2	−2.178	−0.033	0.830	1.301	1.818	2.159	2.472	2.763
0.0	−2.326	0.000	0.842	1.282	1.751	2.054	2.326	2.576
−0.2	−2.472	0.033	0.850	1.258	1.680	1.945	2.178	2.388
−0.4	−2.615	0.066	0.855	1.231	1.606	1.834	2.029	2.201
−0.6	−2.755	0.099	0.857	1.200	1.528	1.720	1.880	2.016
−0.8	−2.891	0.132	0.856	1.166	1.448	1.606	1.733	1.837
−1.0	−3.022	0.164	0.852	1.128	1.366	1.492	1.588	1.664
−1.2	−3.149	0.195	0.844	1.086	1.282	1.379	1.449	1.501
−1.4	−3.271	0.225	0.832	1.041	1.198	1.270	1.318	1.351
−1.6	−3.388	0.254	0.817	0.994	1.116	1.166	1.197	1.216
−1.8	−3.499	0.282	0.799	0.945	1.035	1.069	1.087	1.097
−2.0	−3.605	0.307	0.777	0.895	0.959	0.980	0.990	0.995
−2.2	−3.705	0.330	0.752	0.844	0.888	0.900	0.905	0.907
−2.4	−3.800	0.351	0.725	0.795	0.823	0.830	0.832	0.833
−2.6	−3.889	0.368	0.696	0.747	0.764	0.768	0.769	0.769
−2.8	−3.973	0.384	0.666	0.702	0.712	0.714	0.714	0.714
−3.0	−4.051	0.396	0.636	0.660	0.666	0.666	0.667	0.667

The log-Pearson type III distribution of X is equivalent to applying the Pearson type III distribution to the transformed random variable log X. It is represented by Eq. 7.14 with $\overline{\log X}$ and $S_{\log X}$ computed from the logarithms of the X_i. The frequency factor K values provided in Table 7.3 are applicable for either the Pearson type III or log-Pearson type III distributions. If the skew coefficient is zero, the log-Pearson type III distribution is equivalent to the log-normal distribution.

As discussed in Section 7.7, the log-Pearson type III distribution is used by the federal water agencies for performing flood frequency analyses. This probability distribution is recommended, and guidelines for its application are provided in Bulletin 17B (Interagency Committee on Water Data, 1982). The log-Pearson III distribution is advantageous over the log-normal partly because use of the skew coefficient provides greater flexibility for fitting the model (probability distribution) to the real-world phenomena (streamflow) using observed data.

7.5.3 Gumbel Distribution

The theory of extreme values considers the distribution of the largest or smallest observations occurring in each group of repeated samples. Based on extreme value theory treating each year as a sample, Gumbel (1958) applied the extreme value type I function, now commonly called the Gumbel distribution, to flood flows. The probability distribution is

$$P = 1 - e^{-e^{-b}} \tag{7.18}$$

$$b = \frac{1}{0.7797S}(X - \overline{X} + 0.45S) \tag{7.19}$$

where the constant $e = 2.71828$ is the base of the natural logarithms. Thus, the annual exceedance probability P is determined as a function of two parameters, the mean $\overline{X}$ and standard deviation S. The Gumbel distribution may be used to model a variety of phenomena involving extreme events. It is used in Europe to model flood flows and has been applied by the U.S. National Weather Service (NWS) in analyzing precipitation. Equations 7.18 and 7.19 can be applied directly without needing a table of frequency factors. Another alternative, the log-Gumbel distribution, consists of applying Eqs. 7.18 and 7.19 to the logarithms of X.

7.5.4 Example Problems Applying the Alternative Distributions

Examples 7.5 and 7.6 consist of applying the normal, log-normal, log-Pearson type III, and Gumbel distributions with the data from Table 7.1 to model flood flows for the Mississippi River at St. Louis. Following the guidelines of the federal water agencies discussed in Section 7.7, the log-Pearson type III distribution would typically be adopted for a frequency analysis of peak flood flows. However, the examples provide an opportunity to illustrate application of the alternative distributions.

The probability distributions may be applied in the same manner to many other types of random variables as well as peak flood flows.

Example 7.5

Estimate the 10-year and 100-year recurrence interval peak flows on the Mississippi River at St. Louis using the data from Table 7.1. Model the flows alternatively with the normal, log-normal, log-Pearson type III, and Gumbel probability distributions.

Solution Equations 7.15 and 7.16 are applied to compute the mean and standard deviation for the 66 flows for use in the normal and Gumbel distributions.

$$\overline{X} = 14{,}776 \text{ m}^3/\text{s}$$

$$S = 5{,}242 \text{ m}^3/\text{s}$$

The base 10 logarithms of each of the 66 flows are computed and substituted into Eqs. 7.15, 7.16, and 7.17 to obtain values for the parameters of the log-normal and log-Pearson type III distributions.

$$\overline{\log X} = 4.149$$

$$S_{\log X} = 0.1511$$

$$G_{\log X} = -0.427$$

Since this textbook does not provide a normal probability table, Table 7.3 is used with $G = 0.0$ to obtain the frequency factor K for the normal and log-normal distributions. The K for T of 10 and 100 years are 1.282 and 2.326, respectively. Values of K for the log-Pearson type III distribution obtained from linear interpolation of Table 7.3 for G of -0.427 and T of 10 and 100 years are 1.227 and 2.009.

Normal Distribution

$$Q_{10\text{ years}} = \overline{Q} + KS = 14{,}776 + (1.282)(5{,}242) = 21{,}500 \frac{\text{m}^3}{\text{s}}$$

$$Q_{100\text{ years}} = \overline{Q} + KS = 14{,}776 + (2.326)(5{,}242) = 27{,}000 \frac{\text{m}^3}{\text{s}}$$

Log-Normal Distribution

$$\log Q_{10\text{ years}} = \overline{\log Q} + KS_{\log Q} = 4.149 + (1.282)(0.1511) = 4.343$$

$$Q_{10\text{ years}} = 10^{4.343} = 22{,}000 \frac{\text{m}^3}{\text{s}}$$

$$\log Q_{100\text{ years}} = \overline{\log Q} + KS_{\log Q} = 4.149 + (2.326)(0.1511) = 4.5005$$

$$Q_{100\text{ years}} = 10^{4.5005} = 31{,}700 \frac{\text{m}^3}{\text{s}}$$

Log-Pearson Type III Distribution

$$\log Q_{10\text{ years}} = \overline{\log Q} + KS_{\log Q} = 4.149 + (1.227)(0.1511) = 4.334$$

$$Q_{10\text{ years}} = 21{,}600 \frac{\text{m}^3}{\text{s}}$$

$$\log Q_{100\text{ years}} = \overline{\log Q} + KS_{\log Q} = 4.149 + (2.009)(0.1511) = 4.453$$

$$Q_{100\text{ years}} = 28{,}300\,\frac{\text{m}^3}{\text{s}}$$

Gumbel Distribution

$$P = 1 - e^{-e^{-b}}$$

$$b = \frac{1}{0.7797S}(X - \overline{X} + 0.45S)$$

Gumbel, T = *10 Years*

$$0.10 = 1 - e^{-e^{-b}}$$

$$e^{-e^{-b}} = 0.9$$

$$\ln(e^{-e^{-b}}) = \ln(0.9)$$

$$-e^{-b} = -0.1054$$

$$\ln(e^{-b}) = \ln(0.1054)$$

$$b = 2.2504$$

$$b = \frac{1}{0.7797S}(X - \overline{X} + 0.45S)$$

$$2.2504 = \frac{1}{0.7797(5242)}(Q_{10\text{yr}} - 14{,}776 + 0.45(5{,}242))$$

$$Q_{10\text{ years}} = 21{,}600\,\frac{\text{m}^3}{\text{s}}$$

Gumbel, T = *100 Years*

$$0.01 = 1 - e^{-e^{-b}}$$

$$b = 4.600$$

$$4.600 = \frac{1}{0.7797(5{,}242)}(Q_{100\text{yr}} - 14{,}776 + 0.45(5{,}242))$$

$$Q_{100\text{ years}} = 31{,}200\,\frac{\text{m}^3}{\text{s}}$$

EXAMPLE 7.5 SOLUTION SUMMARY

Probability distribution	T = 10 years Q, m³/s	T = 100 years Q, m³/s
Normal	21,500	27,000
Log-normal	22,000	31,700
Log-Pearson type III	21,600	28,300
Gumbel	21,600	31,200

Example 7.6

Estimate the annual exceedance probability and recurrence interval for a flow of 25,000 m^3/s alternatively using the normal, log-normal, log-Pearson type III, and Gumbel probability distributions.

Normal Distribution

$$X = \overline{X} + KS$$

$$25{,}000 = 14{,}776 + K(5{,}242)$$

$$K = 1.9504$$

Linear interpolation of Table 7.3 with $G = 0.0$ yields: $T = 31$ years and $P = 3.2\%$

Log-Normal

$$\log X = \overline{\log X} + KS_{\log X}$$

$$\log(25{,}000) = 4.149 + K(0.1511)$$

$$K = 1.6475$$

Linear interpolation of Table 7.3 with $G = 0.0$ yields: $T = 22$ years and $P = 4.6\%$

Log-Pearson Type III

$$\log X = \overline{\log X} + KS_{\log X}$$

$$\log(25{,}000) = 4.149 + K(0.1511)$$

$$K = 1.6475$$

Linear interpolation of Table 7.3 with $G = -0.427$ yields: $T = 31$ years and $P = 3.2\%$

Gumbel

$$b = \frac{1}{0.7797S}(X - \overline{X} + 0.45S) = \frac{1}{0.7797(5{,}242)}(25{,}000 - 14{,}776 + 0.45(5{,}242))$$

$$= 3.0786$$

$$P = 1 - e^{-e^{-b}} = 1 - e^{-e^{-3.0786}} = 0.0450$$

$$T = \frac{1}{P} = \frac{1}{0.045} = 22 \text{ years}$$

EXAMPLE 7.6 SOLUTION SUMMARY

Probability distribution	25,000 m^3/s T	P
Normal	31 years	0.032
Log-normal	22 years	0.046
Log-Pearson type III	31 years	0.032
Gumbel	22 years	0.045

7.6 FREQUENCY GRAPHS

The frequency analysis for the Mississippi River at St. Louis is presented graphically in Figs. 7.1 and 7.2. These graphs were printed from the Hydrologic Engineering Center-Flood Frequency Analysis (HEC-FFA) (Hydrologic Engineering Center, 1992) computer program discussed in Section 7.7.1. The confidence limits on the frequency curves are discussed in Section 7.7.2. The Weibull plotting positions from Section 7.4 and analytical flow frequency curves from Section 7.5 are discussed in the following paragraphs.

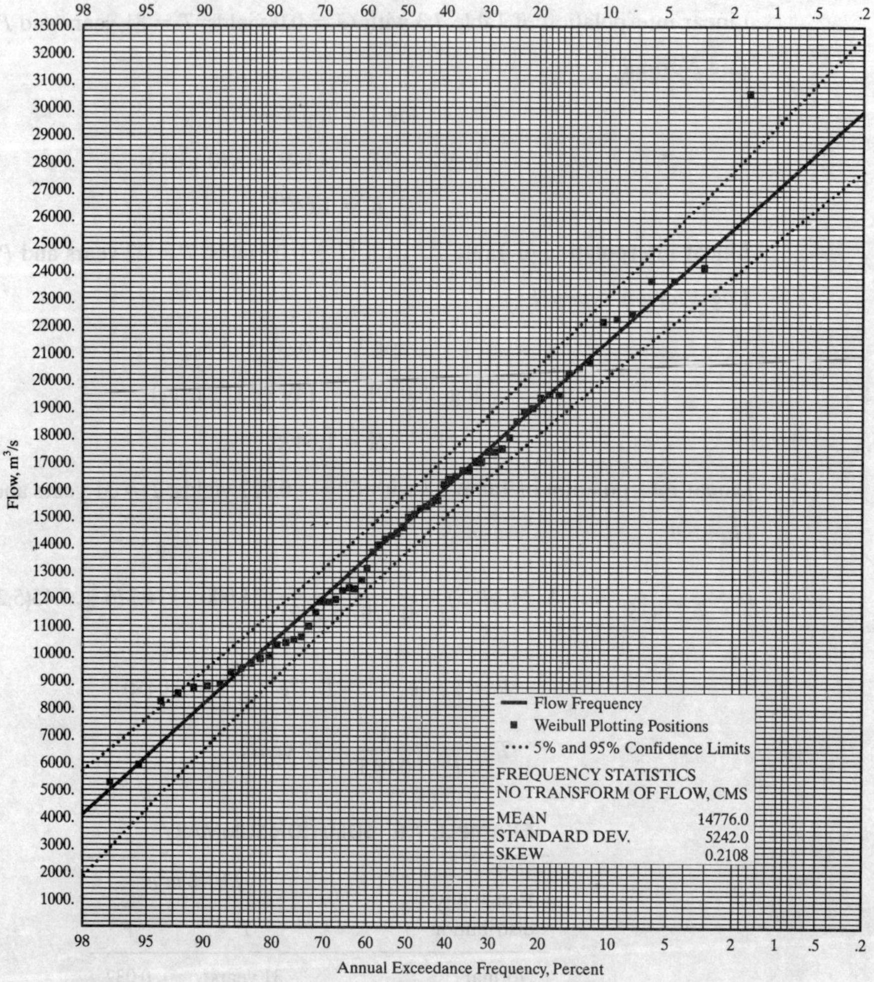

Figure 7.1 The normal frequency curve and Weibull plotting positions for peak annual flows in the Mississippi River at St. Louis are graphed on normal probability paper.

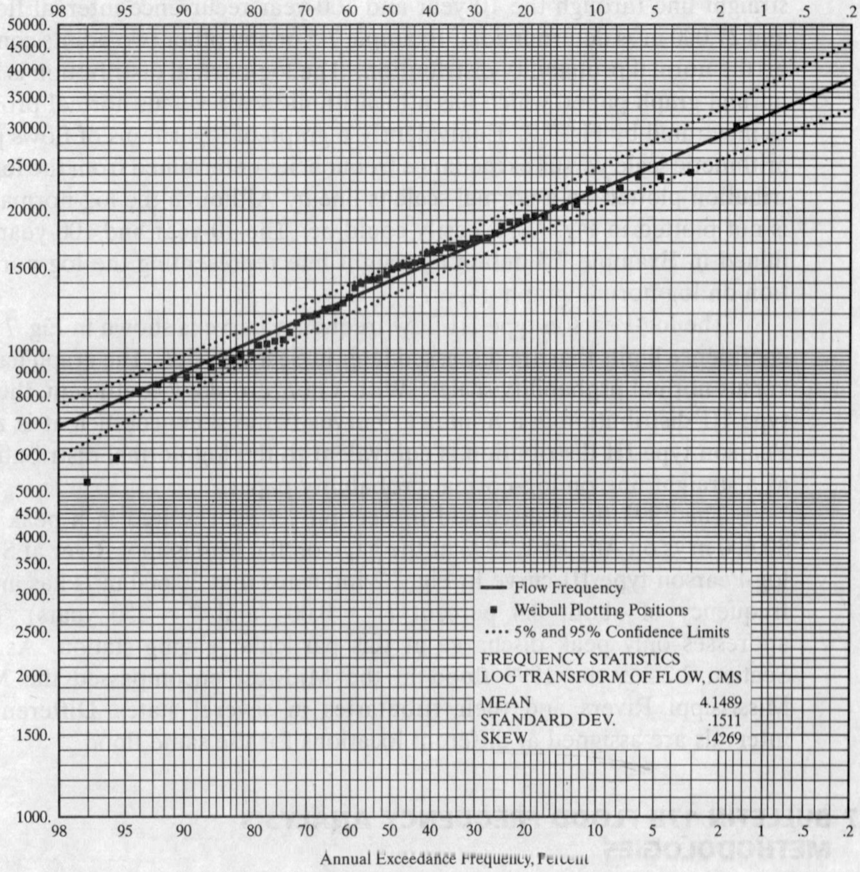

Figure 7.2 The log-Pearson type III frequency curve and Weibull plotting positions for peak annual flows in the Mississippi River at St. Louis are graphed on log-normal probability paper.

In Fig. 7.1, flows are on an arithmetic scale versus exceedance frequency on a normal probability scale. In the log-normal graph of Fig. 7.2, the flows are on a logarithmic scale. The Weibull plotting positions from Table 7.2 are plotted on both graphs as discussed in Section 7.4. A curve could be drawn through the 66 data points manually based on judgment regarding the best fit. Different people might draw the line somewhat differently. However, the frequency curve lines actually included on the two graphs are based on the analytical probability functions discussed in Section 7.5, not Weibull plotting positions. The frequency curves are fixed precisely by the analytical distributions with parameters computed from the data. The normal distribution and log-Pearson type III distribution are graphed in Figs. 7.1 and 7.2, respectively.

The normal distribution is a straight line on graph paper with an arithmetic scale versus normal probability scale. Thus, the frequency curve in Fig. 7.1 is a straight line through the 10-year and 100-year recurrence interval flows of 21,500 and 27,000 m^3/s determined in Example 7.5 or any other two points computed based on the normal probability distribution. The log-normal distribution is linear on log-normal graph paper, which has a logarithmic scale versus normal probability scale as illustrated by Fig. 7.2. Equivalently, a graph of logarithms of flows plotted on an arithmetic scale versus exceedance frequencies determined from the log-normal distribution plotted on a normal scale is linear. Although the log-normal distribution is not plotted in Fig. 7.2, it easily could be. The 10-year and 100-year flows determined in Example 7.5 define a straight line representing the log-normal distribution on log-normal graph paper.

The log-Pearson type III flow frequency curve is shown in Fig. 7.2, along with confidence limits that are discussed later in Section 7.7.2. The graph has logarithmic versus normal probability scales. With a nonzero skew coefficient, the log-Pearson type III distribution is a nonlinear curve. If the skew coefficient is zero, the log-Pearson type III distribution is equivalent to the log-normal distribution and plots as a straight line on log-normal probability paper.

The 1993 flood discussed in Section 2.2.3.2 resulted in a peak discharge of 30,600 m^3/s on August 1, 1993 at the gage on the Mississippi River at St. Louis. The log-Pearson type III curve in Fig. 7.2 indicates that 30,600 m^3/s has an exceedance frequency of about 0.4 percent ($P = 0.004$ and $T = 250$ years). This analysis addresses only peak discharge at this particular gaging station. As discussed in Section 2.2.3.2, the 1993 flood in the Midwest encompassed the Missouri and Mississippi Rivers and their tributaries in several states. Different recurrence intervals are assigned at different locations for the same flood.

7.7 BULLETIN 17B FLOOD FREQUENCY ANALYSIS METHODOLOGIES

The Hydrology Committee of the former U.S. Water Resources Council developed guidelines for flood frequency analysis to be followed consistently by the federal water agencies. Each of the major water agencies was represented on the committee. The guidelines were published as Bulletin 17 in 1976 and in revised form as Bulletin 17B in 1982 (Interagency Advisory Committee on Water Data, 1982) and continue to be followed by the federal water agencies and professional water resources engineering community. The log-Pearson type III distribution is adopted for modeling peak flows. Although Bulletin 17B was developed specifically for peak annual discharge, the general methodology is applied to other random variables as well.

7.7.1 Hydrologic Engineering Center (HEC)-Flood Frequency Analysis (FFA) Model

Flood frequency analysis procedures outlined in Bulletin 17B and reiterated in EM 1110-2-1415 (U.S. Army Corps of Engineers, 1993) are coded in the computer

TABLE 7.4 FLOW FREQUENCY RELATIONSHIP FOR MISSISSIPPI RIVER AT ST. LOUIS

Exceedance Frequency (percent)	Recurrence Interval (years)	Discharge (m^3/s)	95% Confidence Limits Upper (m^3/s)	95% Confidence Limits Lower (m^3/s)
0.2	500.00	32,100	37,600	28,400
0.5	200	30,000	34,800	26,700
1.0	100	28,300	32,600	25,400
2.0	50	26,500	30,200	23,900
5.0	20	23,900	26,800	21,700
10.0	10	21,600	24,000	19,800
20.0	5	19,000	20,800	17,600
50.0	2	14,400	15,500	13,500
90.0	1.11	8,910	9,730	8,000

program HEC-FFA Flood Frequency Analysis developed by the HEC of the U.S. Army Corps of Engineers (Hydrologic Engineering Center, 1992). Log-Pearson type III is the standard distribution coded into the HEC-FFA model. However, user-specified options include (1) either a logarithmic transform of the data or no transform and (2) computation of the skew coefficient or a user-specified value. Thus, the normal, log-normal, Pearson type III, and log-Pearson type III distributions may be adopted in applying HEC-FFA, although the log-Pearson type III is the Bulletin 17B recommended choice.

The HEC-FFA computations are illustrated in Examples 7.4 and 7.5. The flows from Table 7.1 are input to the HEC-FFA program. Figures 7.1 and 7.2 and Table 7.4 are reproduced from the HEC-FFA output. The discharge versus annual exceedance frequency relationship for the Mississippi River at St. Louis, based on the log-Pearson type III distribution and 1933–1998 annual series of observed peak flows, is tabulated in Table 7.4 and presented graphically in Fig. 7.2.

7.7.2 Confidence Limits

Bulletin 17B provides procedures for estimating confidence limits based on describing errors with the noncentral t probability distribution. This procedure provides the 95 percent upper and lower confidence limits shown in Table 7.4 and Figs. 7.1 and 7.2.

Our Mississippi River frequency analysis is based on the 66-year 1933–1998 data series. As additional years of data since 1998 accumulate, the analysis should be periodically updated. The frequency curve will change, particularly if extreme flood events are added. The results of the frequency estimates are dependent on sample size, with accuracy improving with additional data. The width of the confidence band decreases with increasing sample size. Confidence limits highlight our dependence on data series of limited length. However, the accuracy of the flow frequency estimates is also dependent on other uncertainties not reflected in the confidence limits, such as inaccuracies in streamflow measurements, errors in storing and retrieving data, computational blunders, and approximations in modeling real-world phenomena with the log-Pearson type III model.

Although physically impossible, hypothetically assume that we have millions of years of recorded flows at our gaging station. This data set is subdivided into many thousands of independent 66-year sequences of peak annual flows. Applying the log-Pearson type III methodology, each 66-year sequence would yield a different flow frequency relationship. For a specified annual exceedance probability, the computed discharge will be less than the 95 percent upper confidence limit for 95 percent of the many thousands of 66-year sequences. The computed discharge will be greater than the 95 percent lower confidence limit for 95 percent of the sequences. Also, assume the true flow-frequency relationship could be developed based on the complete series of millions of years. The true flow for a given exceedance frequency has a 90 percent chance of falling within the 90 percent confidence band defined by the 95 percent upper and lower confidence limits developed based on any one of the 66-year sequences.

From Table 7.4, the confidence limits indicate a 95 percent probability that the true 100-year recurrence interval discharge is less than 32,600 m^3/s. There is also a 95 percent probability that the true 100-year discharge is greater than 25,400 m^3/s. Based on 66 years of observed data, the probability is 90 percent that the true 100-year peak flow falls between 25,400 m^3/s and 32,600 m^3/s. The confidence limits are a function of sample size (66 years) with the width of the band decreasing with increasing sample size.

The noncentral *t*-distribution and the statistical basis for the derivation of confidence limit procedures based on it will not be explored in this textbook. However, we will present the basic equations adopted in Bulletin 17B and HEC-FFA and work an example problem. Although confidence limits may also be applied to frequency relationships developed with the normal and other distributions, terms in the following equations are defined from the perspective of the log-Pearson type III distribution that uses the logarithms of the flows.

Let X_P^* denote the true or population logarithmic discharge that has exceedance probability P. Upper and lower confidence limits for X_P^*, with confidence level C, are defined to be numbers $U_{P,C}$ and $L_{P,C}$ such that based on observed records

$$\text{Probability}\,(U_{P,C} \geq X_P^*) = C \tag{7.20}$$

$$\text{Probability}\,(L_{P,C} \leq X_P^*) = C \tag{7.21}$$

$U_{P,C}$ and $L_{P,C}$ are called one-sided confidence limits because each describes a bound on just one side of the P-probability discharge. A two-sided confidence interval can be formed from the overlap or union of the two one-sided intervals as follows.

$$\text{Probability}\,(L_{P,C} \leq X_P^* \leq U_{P,C}) = 2C - 1 \tag{7.22}$$

Thus, the union of two one-sided 95 percent confidence intervals is a two-sided 90 percent confidence interval.

Confidence limits are computed with the following equations.

$$U_{P,C} = \overline{X} + SK^{U}_{P,C} \tag{7.23}$$

$$L_{P,C} = \overline{X} + SK^{L}_{P,C} \tag{7.24}$$

$$K^{U}_{P,C} = \frac{K_P + (K_P^2 - ab)^{0.5}}{a} \tag{7.25}$$

$$K^{L}_{P,C} = \frac{K_P - (K_P^2 - ab)^{0.5}}{a} \tag{7.26}$$

$$a = 1 - \frac{Z_C^2}{2(N-1)} \tag{7.27}$$

$$b = K_P^2 - \frac{Z_C^2}{N} \tag{7.28}$$

where

- P specified annual exceedance probability
- C specified confidence level
- Z_C standard normal deviate from normal probability table
- $U_{P,C}$ upper confidence limit as a logarithm of discharge
- $L_{P,C}$ lower confidence limit as a logarithm of discharge
- $K^U_{P,C}$ upper confidence limit coefficient
- $K^U_{P,C}$ lower confidence limit coefficient
- $\overline{X}$ mean of logarithms of annual peak flows
- S standard deviation of logarithms of annual peak flows
- K_P frequency factor from Table 7.3 for exceedance probability P
- N record length

As previously discussed, the flow frequency relationship presented in Table 7.4 was developed with the HEC-FFA computer model that is based on Bulletin 17B procedures. The frequency curve and confidence limits are plotted in Fig. 7.2. The $C = 95$ percent confidence limits indicate a 90 percent chance that the $P = 0.01$ flood is between 25,400 m^3/s and 32,600 m^3/s. Example 7.7 illustrates the application of Eqs. 7.23–7.28 to compute the confidence limits of 25,400 m^3/s and 32,600 m^3/s found in Table 7.4.

Example 7.7

A 0.01 exceedance probability discharge of 28,300 m^3/s was determined in Example 7.5. We will now extend the analysis to determine the corresponding 95 percent upper and lower confidence limits. In Example 7.5, application of Eqs. 7.15–7.17 and Table 7.3 to the logarithms of the 1993–1998 flows in Table 7.1 resulted in the following statistics for the 0.01 exceedance probability flow.

$$\overline{\log X} = 4.149$$

$$S_{\log X} = 0.1511$$

$$G_{\log X} = -0.427$$

$$K = 2.009 \text{ from Table 7.3}$$

For a confidence level C of 95 percent, $Z_c = 1.645$ is read from a normal probability table that also can be approximated by interpolation of Table 7.3.

$$a = 1 - \frac{Z_{0.95}^2}{2(N-1)} = 1 - \frac{(1.645)^2}{2(66-1)} = 0.9792$$

$$b = K_{0.01}^2 - \frac{Z_{0.95}^2}{N} = (2.009)^2 - \frac{(1.645)^2}{66} = 3.995$$

$$K_{0.01,0.95}^U = \frac{K_{0.01} + (K_{0.01}^2 - ab)^{0.5}}{a} = \frac{2.009 + ((2.009)^2 - (0.9792)(3.995))^{0.5}}{0.9792} = 2.412$$

$$K_{0.01,0.95}^L = \frac{K_{0.01} - (K_{0.01}^2 - ab)^{0.5}}{a} = \frac{2.009 - ((2.009)^2 - (0.9792)(3.995))^{0.5}}{0.9792} = 1.692$$

$$U_{0.01,0.95} = \overline{X} + SK_{0.01,0.95}^U = 4.149 + (0.1511)(2.412) = 4.513$$

$$L_{0.01,0.95} = \overline{X} + SK_{0.01,0.95}^L = 4.149 + (0.1511)(1.692) = 4.405$$

$$\text{Upper Limit} = 10^{4.513} = 32{,}600 \frac{\text{m}^3}{\text{s}}$$

$$\text{Lower Limit} = 10^{4.405} = 25{,}400 \frac{\text{m}^3}{\text{s}}$$

7.7.3 Skew Coefficient

The sample mean, standard deviation, and skew coefficient are computed for the logarithms of the observed annual peak flows based on Eqs. 7.15, 7.16, and 7.17. The skew coefficient is particularly sensitive to extreme flood events due to the cube term in Eq. 7.17. Skew coefficient estimates from small samples may be highly inaccurate. Therefore, Bulletin 17B outlines a procedure for developing regionalized skew coefficients based on at least 40 gaging stations located within a 160-km (100-mi.) radius of the site of concern, with each having at least 25 years of data. If this procedure is not feasible, the generalized skew map reproduced as Fig. 7.3 is provided as an easier but less accurate alternative. This map of generalized logarithmic skew coefficients for peak annual flows was developed from skew coefficients computed for 2,972 gaging stations, all having at least 25 years of record, following procedures outlined in Bulletin 17B. Depending on the number of years

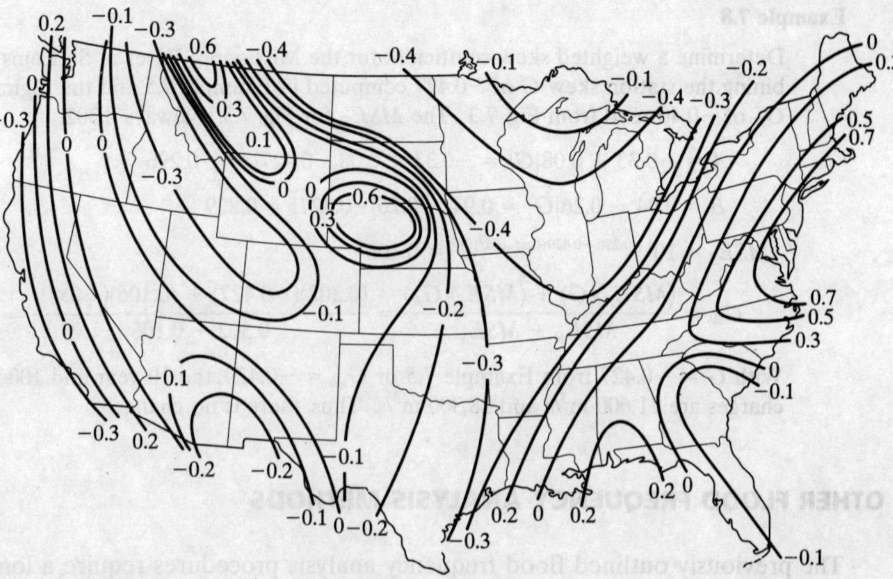

Figure 7.3 Generalized skew coefficients of the logarithms of annual maximum streamflow were developed by the Interagency Advisory Committee on Water Data (1982).

of gage record, regionalized skew coefficients are used either in lieu of or in combination with values computed from observed flows at the particular stream gage of concern.

Equations 7.29 and 7.30 allow a weighted skew coefficient G_W to be computed by combining a regionalized skew coefficient G_R and station skew coefficient G.

$$G_W = \frac{(MSE_R)(G) + (MSE_S)(G_R)}{MSE_R + MSE_S} \tag{7.29}$$

$$MSE_S = 10^{[A - B(\text{Log}_{10}(N/10))]} \tag{7.30}$$

$$A = -0.33 + 0.08|G| \quad \text{if} \quad |G| \leq 0.90$$

$$A = -0.52 + 0.30|G| \quad \text{if} \quad |G| > 0.90$$

$$B = 0.94 - 0.26|G| \quad \text{if} \quad |G| \leq 1.50$$

$$B = 0.55 \quad \text{if} \quad |G| > 1.50$$

The station skew G is computed from observed flows at the station of interest using Eq. 7.17. The regional skew G_R is either developed from multiple stations following procedures outlined in Bulletin 17B or read from Fig. 7.3 also supplied by Bulletin 17B. MSE_S denotes the mean square error of the station skew. MSE_R is the mean square error of the regional skew. If G_R is taken from Fig. 7.3, the MSE_R is 0.302.

Example 7.8

Determine a weighted skew coefficient for the Mississippi River at St. Louis by combining the station skew G of -0.427 computed in Example 7.5 and the regional skew G_R of -0.40 read from Fig. 7.3. The MSE_R for Fig. 7.3 is always 0.302.

$$A = -0.33 + 0.08|G| = -0.33 + 0.08|-0.427| = -0.296$$

$$B = 0.94 - 0.26|G| = 0.94 - 0.26|-0.427| = 0.829$$

$$MSE_S = 10^{[-0.296 - 0.829(\mathrm{Log}_{10}(66/10))]} = 0.106$$

$$G_W = \frac{(MSE_R)(G) + (MSE_S)(G_R)}{MSE_R + MSE_S} = \frac{(0.302)(-0.427) + (0.106)(-0.40)}{0.302 + 0.106} = -0.420$$

With $G = -0.427$ from Example 7.5 or $G_W = -0.420$, the 10-year and 100-year discharges are 21,600 m^3/s and 28,300 m^3/s. Thus, there is no change.

7.8 OTHER FLOOD FREQUENCY ANALYSIS METHODS

The previously outlined flood frequency analysis procedures require a long series of unregulated, homogeneous observed peak annual flows at the location of concern. Gaged flows are often adjusted to remove nonhomogeneities caused by reservoir regulation and other river basin development activities. The availability of streamflow data is contingent on gaging stations having been maintained at pertinent locations for a significantly long period of time. Extensive adjustments may be required to remove nonhomogeneities caused by river basin development. Consequently, alternative flood flow frequency approaches are often necessary.

Flood frequency analysis methods vary depending on (1) the availability of streamflow and precipitation data and (2) whether the application requires peak discharges, volumes, or entire flood hydrographs. Working directly with gaged streamflow observations provides important advantages. However, watershed (precipitation-runoff) modeling is often either advantageous or an absolute necessity. Precipitation gages are much more abundant than streamflow gages, and watershed development does not affect the homogeneity of precipitation data like it does streamflow data. Sizing hydraulic structures and delineating floodplains are typically based on peak discharges. Many applications, such as those involving reservoir storage, require volumes and/or complete hydrographs. Alternative approaches include the following.

1. If a sufficiently long, homogeneous stream gage record is available at the pertinent location, the log-Pearson type III distribution may be applied following guidelines provided by Bulletin 17B as discussed in the preceding Section 7.7.
2. As discussed in Section 7.8.1, rainfall intensity-duration-frequency (IDF) relationships and associated design storms outlined in Section 7.11 may be combined with the watershed models of Chapter 8.

3. Statistical regression equations based on watershed parameters may be used to transfer flow frequency relations developed at a number of stream gaging stations to pertinent ungaged sites as discussed in Section 7.8.2.
4. As discussed in Section 7.8.3, variations of the frequency analysis techniques applied to discharges may also be applied to volumes.

7.8.1 Watershed (Precipitation-Runoff) Models

Chapter 8 explores watershed models used to simulate the hydrologic processes by which precipitation is transformed to streamflow. The variety of roles played by watershed models includes various flood frequency analysis approaches. Watershed models have the advantage of simulating specified conditions of watershed development. Complete runoff hydrographs may be developed as well as peak flows and volumes. Hydrographs may be developed for ungaged as well as gaged locations. Several strategies for applying watershed (precipitation-runoff) models for flood frequency analyses are outlined as follows.

The rational method described in Section 8.4 is often used to determine peak flows from small, relatively impervious watersheds for use in designing storm drainage facilities. Rainfall IDF relationships discussed in Section 7.11.1 provide rainfall intensities associated with specified exceedance frequencies. The rational method (Eq. 8.11) provides the corresponding peak discharge.

Rainfall IDF relationships are commonly used to synthesize design storms associated with specified exceedance probabilities as outlined in Section 7.11.2. A design storm, representing a precipitation event, is input to a watershed model to determine the associated flow hydrograph, which is assumed to have the same exceedance probability as the design storm. The watershed model is executed with several design storms covering a range of exceedance probabilities to develop a flow-frequency relationship.

Approximations in the flow-frequency relationship result from assumptions regarding rainfall duration and temporal distribution inherent in design storms. Thus, the approach outlined next may result in a more accurate flow-frequency relationship, but more work is required.

A flood flow-frequency analysis can be performed based on gaged rainfall as follows. The most severe rain storm in each year is selected from the records. The corresponding runoff hydrographs are computed using a watershed model. Most severe rain storm is defined in terms of producing the greatest peak streamflow discharge. Multiple storms in the same year may be compared by inputting each to the watershed model to determine which produces the highest peak flow. Thus, if for example, there is 90 years of recorded precipitation data, the 90 annual most severe precipitation events are input in turn to a watershed model to obtain the corresponding series of 90 annual peak flows. The log-Pearson type III distribution is then fitted to this annual peak flow series following the procedures discussed in Section 7.7.

7.8.2 Regional Regression Equations

A variety of regional frequency analysis approaches have been developed and applied for developing flow-frequency relationships for ungaged locations. Statistical regression and frequency analyses techniques are combined to develop equations for predicting peak discharges at ungaged sites based on analysis of observed streamflow records at gaged sites in the same hydrologic region as follows.

Step 1. A number of streamflow gaging stations located within a hydrologically homogeneous region of reasonably uniform land use, topography, and climate are selected.

Step 2. For each individual gaging station, flows for specified exceedance frequencies are determined by applying the frequency analysis techniques of Sections 7.5–7.7.

Step 3. The flows are related to watershed and climatic characteristics using statistical regression analysis techniques to develop equations that may be applied to ungaged locations in the region.

In many cases, the existing gaging stations provide too few years of record to reliably develop flow-frequency relationships directly using the log-Pearson type III or other probability distribution functions. The limited years of flow record may still be adequate to calibrate a watershed model. Various strategies are adopted from incorporating watershed modeling in step 2 to generate flow-frequency relationships for the gaging stations.

Regression techniques are covered in most statistics books (Helsel and Hirsch, 1992; Kottega and Rosso, 1997). In this book, we are interested in applying the equations resulting from regression analyses, but do not present the standard statistical techniques used to develop the equations. Peak discharge regression equations are typically of the form

$$Q_T = aB^bC^cD^d \tag{7.31}$$

where Q_T is the peak discharge associated with recurrence interval T, and a, b, c, and d are coefficients determined from regression analyses. B, C, and D are explanatory or predictor variables for which values can be estimated from maps, field surveys, or other available sources of information. Drainage area is essentially always included. Other typical explanatory variables include watershed slope or channel slope, percent of the watershed with impervious land cover (streets, parking lots, buildings, etc.), percent of watershed covered by lakes and ponds, and mean annual precipitation or precipitation for a specified recurrence interval and duration. Typical applications include hydrologic sizing of highway culverts.

Since 1973, the USGS, in cooperation with state highway departments and other entities, has developed and published regional regression equations for estimating flood flow frequency relations for rural, unregulated watersheds in every state, at least once. Equations have also been developed for many specific metropolitan areas. Nationwide urban equations (Eqs. 7.39–7.45) are presented later in this section. The USGS, in cooperation with the Federal Highway Administration and Federal Emergency Management Agency, compiled the regression equations into a computer program called the National Flood Frequency Program and an accompanying report (Jennings, Thomas, and Riggs, 1994). The nationwide summary report presents the equations for each state with references to the reports documenting the original statewide frequency and regression analyses. The rural equations for Illinois are summarized here as an example.

The USGS applied the log-Pearson type III distribution following *Bulletin 17B* guidelines to develop discharges corresponding to recurrence intervals of 2, 5, 10, 25, 50, 100, and 500 years for 268 stream gaging stations with unregulated rural watersheds. Statistical regression analyses methods were then applied to relate the discharges to watershed and climatic parameters, resulting in Eqs. 7.32–7.38, which are applicable for rural watersheds in Illinois with drainage areas ranging from 0.2 to 10,000 square miles (0.5–25,900 km). Standard errors of estimate for the regression equations range from 35 to 50 percent.

$$Q_2 = 38.1A^{0.790}S^{0.481}(I - 2.5)^{0.677}R_f \tag{7.32}$$

$$Q_5 = 63.0A^{0.786}S^{0.513}(I - 2.5)^{0.719}R_f \tag{7.33}$$

$$Q_{10} = 78.9A^{0.785}S^{0.532}(I - 2.5)^{0.742}R_f \tag{7.34}$$

$$Q_{25} = 98.2A^{0.786}S^{0.552}(I - 2.5)^{0.768}R_f \tag{7.35}$$

$$Q_{50} = 112A^{0.786}S^{0.566}(I - 2.5)^{0.786}R_f \tag{7.36}$$

$$Q_{100} = 125A^{0.787}S^{0.578}(I - 2.5)^{0.803}R_f \tag{7.37}$$

$$Q_{500} = 155A^{0.789}S^{0.601}(I - 2.5)^{0.838}R_f \tag{7.38}$$

where

Q_T T-years recurrence interval peak discharge in ft^3/s

A drainage area in square miles

S main channel slope in feet per mile

I rainfall depth in inches for 2-year recurrence interval and 24-hour duration

R_f dimensionless regional factor from Table 7.5

The channel slope S is based on the difference in elevation divided by distance between points 10 percent and 85 percent of the total distance measured along the low-water channel from the site to the channel divide. Illinois is divided into the four regions shown in Fig. 7.4, and the regional factors R_f for Eqs. 7.32–7.38 for each region are tabulated in Table 7.5. The 2-year recurrence interval, 24-hour duration rainfall depth I is provided in Fig. 7.5.

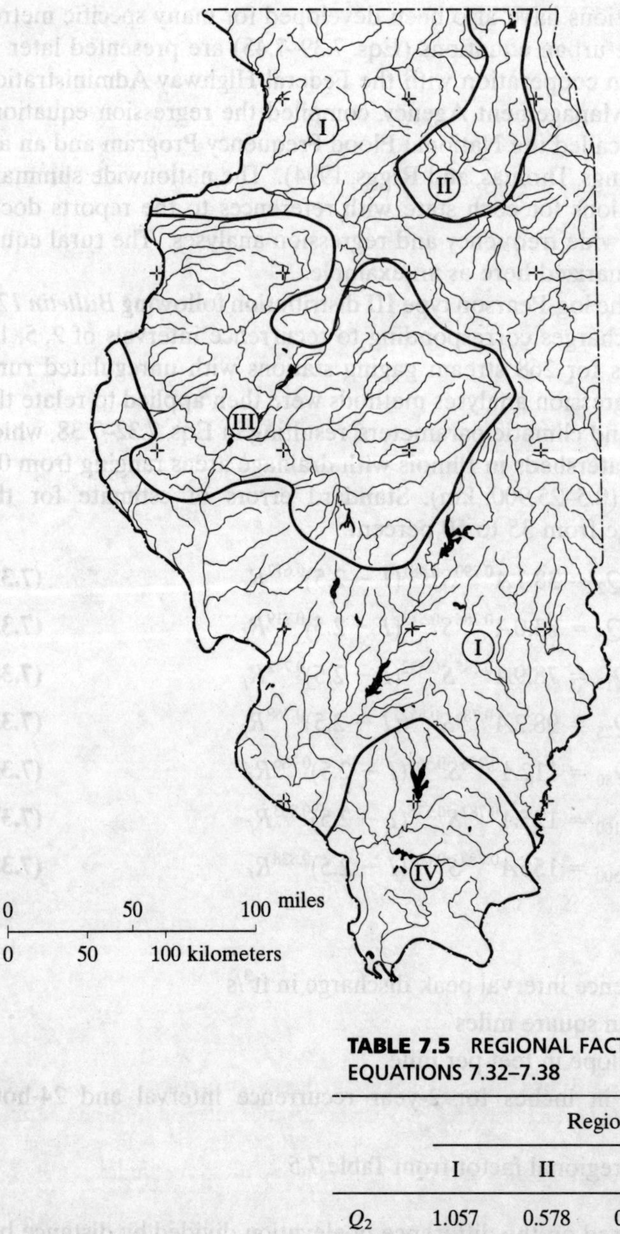

Figure 7.4 Flood frequency region map for Illinois (Jennings et al., 1994).

TABLE 7.5 REGIONAL FACTORS R_f IN EQUATIONS 7.32–7.38

	Region			
	I	II	III	IV
Q_2	1.057	0.578	0.805	0.983
Q_5	1.053	0.576	0.822	0.894
Q_{10}	1.053	0.574	0.837	0.859
Q_{25}	1.051	0.570	0.853	0.826
Q_{50}	1.050	0.567	0.862	0.806
Q_{100}	1.048	0.563	0.870	0.790
Q_{500}	1.044	0.555	0.886	0.759

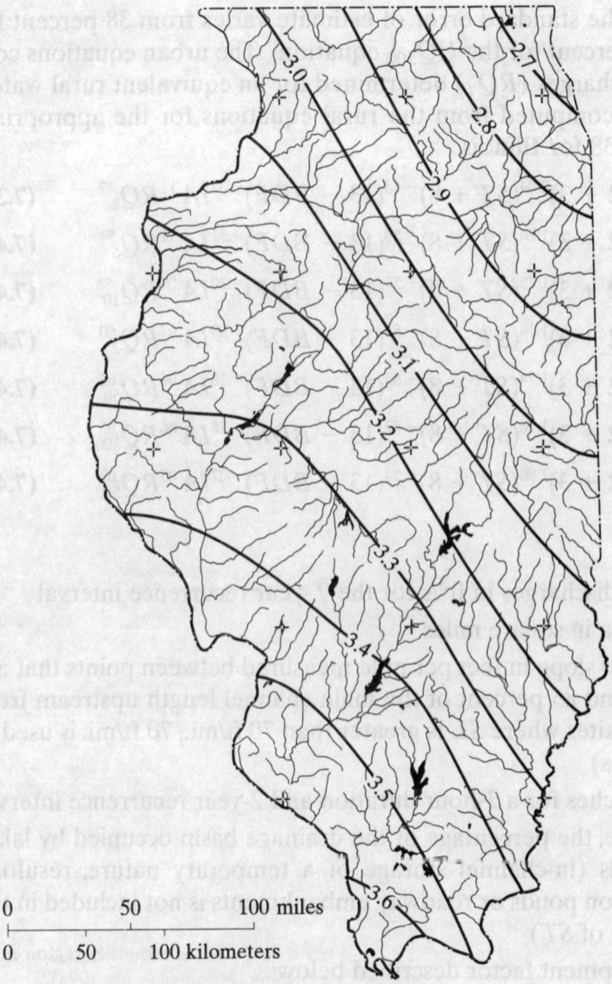

Figure 7.5 Precipitation depth in inches for 2-year recurrence interval, 24-hour duration, in Illinois (Jennings et al., 1994).

Example 7.9

Estimate the 50-year recurrence interval peak flood discharge at a location on a stream in Region I of Illinois (Fig. 7.4) that has a 2-year recurrence interval, 24-hour duration rainfall depth of 3.1 inches (Fig. 7.5). From a topographic map, the drainage area A is determined to be 625 square miles and the channel slope is 2.5 feet/mile.

$$Q_{50} = 112\,A^{0.786} S^{0.566} (I - 2.5)^{0.786} R_f = 112(625)^{0.786}(2.5)^{0.566}(3.1 - 2.5)^{0.786}(1.050)$$

$$Q_{50} = 21{,}000 \text{ ft}^3/\text{s}$$

The USGS flood frequency regression equation report and computer program include equations for a number of specific metropolitan areas. Also, the following nationwide urban equations may be applied for urban watersheds throughout the nation. The standard error of estimate varies from 38 percent for the UQ_2 equation to 49 percent for the UQ_{500} equation. The urban equations consist of adjustments to discharges (RQ_T) determined for an equivalent rural watershed in the same region computed from the rural equations for the appropriate state such as Eqs. 7.32–7.38 for Illinois.

$$UQ_2 = 2.35A^{.41}SL^{.17}(RI2 + 3)^{2.04}(ST + 8)^{-.65}(13 - BDF)^{-.32}IA^{.15}RQ_2^{.47} \quad (7.39)$$

$$UQ_5 = 2.70A^{.35}SL^{.16}(RI2 + 3)^{1.86}(ST + 8)^{-.59}(13 - BDF)^{-.31}IA^{.11}RQ_5^{.54} \quad (7.40)$$

$$UQ_{10} = 2.99A^{.32}SL^{.15}(RI2 + 3)^{1.75}(ST + 8)^{-.57}(13 - BDF)^{-.30}IA^{.09}RQ_{10}^{.58} \quad (7.41)$$

$$UQ_{25} = 2.78A^{.31}SL^{.15}(RI2 + 3)^{1.76}(ST + 8)^{-.55}(13 - BDF)^{-.29}IA^{.07}RQ_{25}^{.60} \quad (7.42)$$

$$UQ_{50} = 2.67A^{.29}SL^{.15}(RI2 + 3)^{1.74}(ST + 8)^{-.53}(13 - BDF)^{-.28}IA^{.06}RQ_{50}^{.62} \quad (7.43)$$

$$UQ_{100} = 2.50A^{.29}SL^{.15}(RI2 + 3)^{1.76}(ST + 8)^{-.52}(13 - BDF)^{-.28}IA^{.06}RQ_{100}^{.63} \quad (7.44)$$

$$UQ_{500} = 2.27A^{.29}SL^{.16}(RI2 + 3)^{1.86}(ST + 8)^{-.54}(13 - BDF)^{-.27}IA^{.05}RQ_{500}^{.63} \quad (7.45)$$

where

- UQ_T urban peak discharges in ft^3/s for the T-year recurrence interval
- A drainage area in square miles
- SL main channel slope in feet per mile measured between points that are 10 percent and 85 percent of the main channel length upstream from the site (for sites where SL is greater than 70 ft/mi., 70 ft/mi. is used in the equations)
- $RI2$ rainfall in inches for a 2-hour duration and 2-year recurrence interval
- ST basin storage, the percentage of the drainage basin occupied by lakes and wetlands (in-channel storage of a temporary nature, resulting from detention ponds or roadway embankments is not included in the computation of ST)
- BDF basin development factor described below
- IA percentage of watershed occupied by impervious surfaces, such as buildings, streets, and parking lots
- RQ_T peak discharge in ft^3/s for recurrence interval T for a rural watershed in the same hydrologic region

The basin development factor BDF is an index of the prevalence of urban drainage improvements. The procedure for determining BDF starts with dividing the basin into upper, middle, and lower thirds on a drainage map, with each third containing about one-third of the drainage area. Within each third of the basin, the following four characteristics are assigned a code of zero or one.

1. Channel improvements—If more than 50 percent of the main drainage channels and principal tributaries are improved over natural conditions, a code of 1 is assigned. Otherwise, zero is assigned.

2. Channel linings—If more than 50 percent of the length of the main channels and principal tributaries are lined with an impervious surface, such as concrete, a code of 1 is assigned. Otherwise, zero is assigned.
3. Storm sewers—If more than 50 percent of the secondary tributaries consists of storm sewers (pipes), a code of 1 is assigned. Otherwise, zero is assigned.
4. Curb-and-gutter streets—If more than 50 percent of the subarea is urbanized and more than 50 percent of the streets are constructed with curbs and gutters, a code of 1 is assigned. Otherwise, zero is assigned.

The *BDF* is the summation of the 12 codes representing four characteristics in each of the three subbasins.

7.8.3 Flood Volume Frequency Analysis

Flood volume-duration-frequency analyses are performed for reservoir design and operation studies and other applications. Annual maximum flood volumes for various time intervals may be compiled from stream gage records of daily flows. The USGS database discussed in Section 2.11 includes tabulations of annual maximum volumes for periods of 1, 3, 7, 15, 30, 60, 120, and 183 days at numerous gaging stations.

The log-Pearson type III distribution is recommended for flood volume as well as peak discharge (U.S. Army Corps of Engineers, 1993). The log-Pearson type III distribution is applied to an annual series of maximum streamflow volumes for a specified duration. For example, for each year of the stream gage record, the maximum volume to occur during any 7-day period is selected to develop the annual maximum 7-day volume series. The mean, standard deviation, and skew coefficient of the logarithms of the 7-day volumes are used with the log-Pearson type III distribution to develop an exceedance frequency versus volume relationship. The procedure is then repeated for other durations, such as 1 day or 30 days. Regional regression equations may be developed to transfer flood volume frequency relationships from gaged to ungaged sites.

Balanced hydrographs developed from volume-duration-frequency relationships are used as reservoir inflows in planning studies to size storage capacities and evaluate operating plans. A balanced hydrograph has volumes for specified durations consistent with established volume-duration-frequency relations. For example, a balanced hydrograph for an exceedance frequency of 2 percent is developed so that the peak 6-hour, 12-hour, 24-hour, and 72-hour volumes each corresponds to a 2 percent exceedance frequency. This balanced hydrograph concept is analogous to the balanced triangular precipitation concept outlined in Section 7.11.2.

7.9 FLOW-DURATION, CONCENTRATION-DURATION, AND LOW-FLOW FREQUENCY RELATIONSHIPS

These relationships are important in assessing streamflow capabilities for meeting water supply, water quality, and environmental instream flow requirements. Low flows govern the dilution and transport of pollutant loads, vitality of fish and ecosystems, and water supply reliabilities. Flow-duration and concentration-duration relationships are based on Eq. 7.9. Developing low flow-frequency relationships is another application of Eqs. 7.10–7.19.

7.9.1 Flow-Duration and Concentration-Duration Relationships

Flow-duration relationships show the frequency or percentage-of-time that stream discharge falls within various ranges. They are developed by counting the number of periods, typically days or months, during which the mean discharge equaled or exceeded specified levels during a period of analysis at a stream gage. Alternatively, the flow-duration relationship may be expressed in terms of flows being less than specified levels. The duration or frequency associated with a given discharge is computed by dividing the number of periods it is equaled or exceeded by the total number of periods. Thus, Eq. 7.9 is applied.

A hydrograph and flow-duration curve for the Brazos River near Richmond, Texas, are presented in Figs. 7.6 and 7.7. This location is about 100 km upstream of the river's mouth and has a drainage area of 117,000 km^2. The data consist of

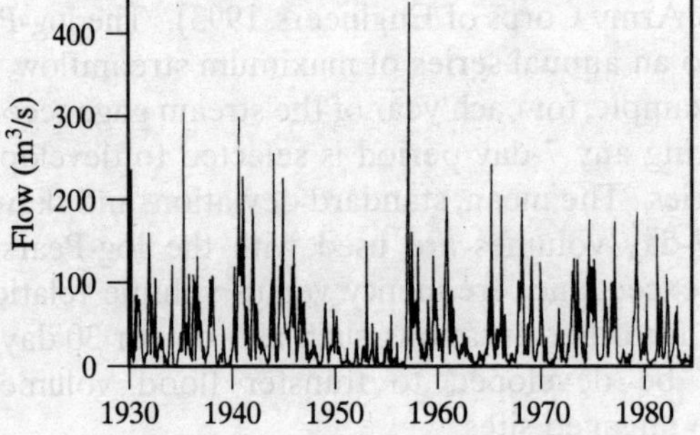

Figure 7.6 Hydrograph for the Brazos River near Richmond.

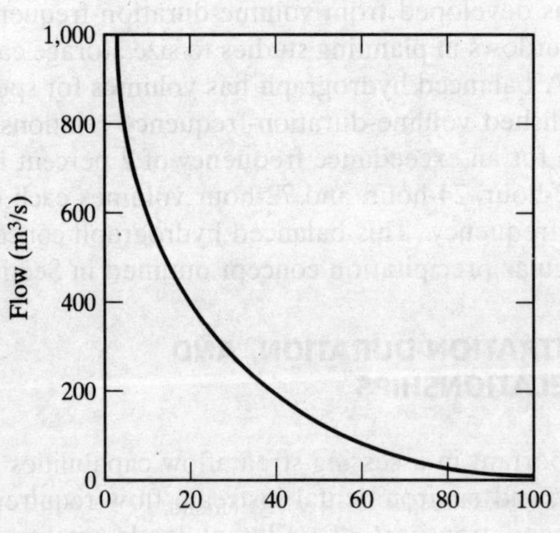

Figure 7.7 Flow-duration curve for the Brazos River near Richmond.

mean monthly gaged flows from October 1922 through December 1984 adjusted to represent unregulated conditions. The adjustments consisted of removing the effects of reservoirs and water supply diversions. The hydrograph shows the great variation in streamflow, including the 1950–1957 drought of record, which ended with the greatest flow on record in April 1957. The flow-duration curve shows the percentage of the time that the discharge exceeds specified rates or equivalently the probability that the flow will exceed a specified rate at a randomly selected time.

Salinity is a concern to entities that use the Brazos River as a water supply source. Municipal supplies normally should have total dissolved solids (TDS) concentrations of less than 500 mg/liter. Mean monthly observed TDS concentrations at the Richmond gage on the Brazos River are plotted in Fig. 7.8. The corresponding TDS concentration-duration curve is plotted as Fig. 7.9. Figures 7.7 and 7.9 are developed in the same manner using Eq. 7.9.

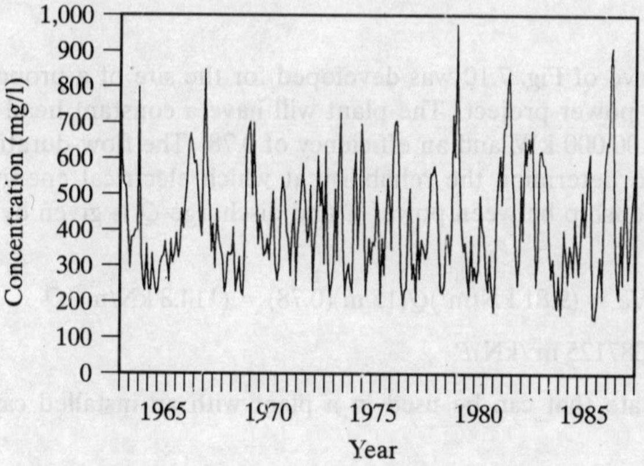

Figure 7.8 TDS concentrations for the Brazos River near Richmond.

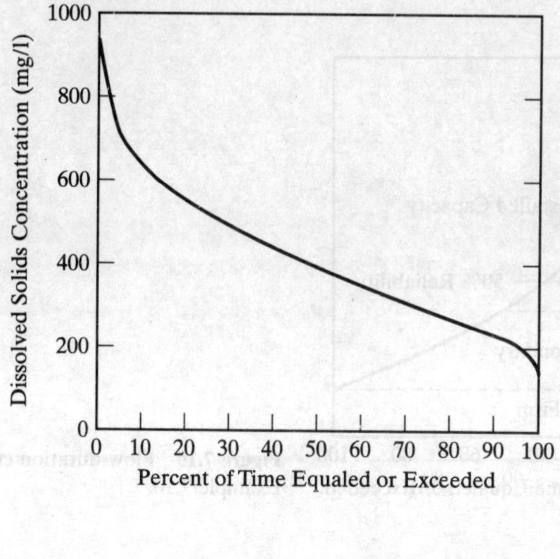

Figure 7.9 TDS concentration-duration curve for the Brazos River near Richmond.

7.9.2 Run-of-River Hydroelectric Energy Reliability

A flow-duration relationship provides a measure of the amount of water supplied at various levels of reliability for various water uses. Example 7.10 illustrates the application of a flow-duration curve in analyzing the amount of electrical energy that can be produced by a run-of-river hydroelectric power plant. The power equation

$$P = \gamma Q H e \tag{7.46}$$

relates electrical power P generated to discharge Q through the turbines. Head H is the difference in water surface elevation upstream and downstream of the turbines. The efficiency e is the ratio of electrical energy to hydraulic energy. The unit weight of water is denoted by γ. Energy and power are related as follows:

$$\text{Power} = \frac{\text{Energy}}{\text{Time}} \tag{7.47}$$

Example 7.10

The flow-duration curve of Fig. 7.10 was developed for the site of a proposed run-of-river hydroelectric power project. The plant will have a constant head of 15 m, installed capacity of 100,000 kW, and an efficiency of 0.78. The flow-duration curve of Fig. 7.10 is used to determine the reliability at which electrical energy can be generated. The relationship between power P and discharge Q is given by Eq. 7.46 as follows.

$$P = \gamma Q H e = (9.81 \text{ kN/m}^3)Q(15 \text{ m})(0.78) = (114.8 \text{ kN/m}^2)Q$$

$$Q = (0.0087125 \text{ m}^2/\text{kN})P$$

The maximum flow rate that can be used in a plant with an installed capacity of 100,000 kW is:

$$Q = (0.0087125 \text{ m}^2/\text{kN})(100{,}000 \text{ kN}\cdot\text{m/s}) = 871 \text{ m}^3/\text{s}$$

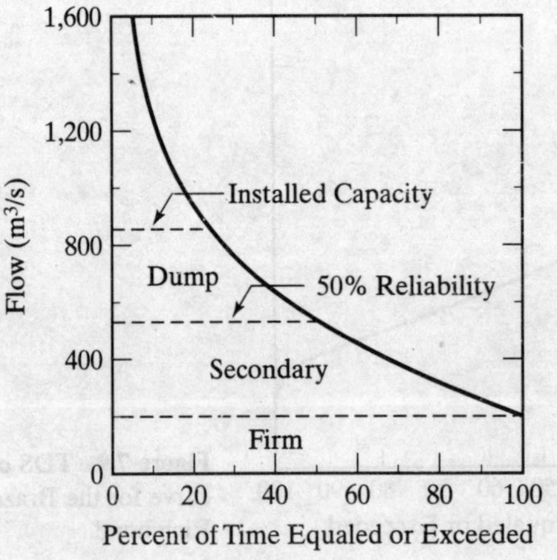

Figure 7.10 Flow-duration curve for Example 7.10.

From Fig. 7.10, a flow of 871 m^3/s is equaled or exceeded about 23 percent of the time. Thus, 100,000 kW can be generated about 23 percent of the time. Firm power is the maximum power that can be provided 100 percent of the time for the given streamflow record. From Fig. 7.10, a flow of 200 m^3/s is available 100 percent of the time. Thus firm power is:

$$P = (114.8 \text{ kN/m}^2)(200\text{m}^3/\text{s}) = 22{,}960 \text{ kN} \cdot \text{m/s} = 22{,}960 \text{ kW}$$

Since power is the time rate of producing energy, the firm energy is:

$$\text{firm energy} = (22{,}960 \text{ kW})(8{,}760 \text{ hrs/yr}) = 2.01 \times 10^8 \text{ kW hrs/yr}$$

From Fig. 7.10, a flow of 520 m^3/s or greater is available 50 percent of the time. The area under the flow-duration curve between the 100 percent reliability flow of 200 m^3/s and the 50 percent reliability flow of 520 m^3/s is approximately:

$$(520 - 200 \text{ m}^3/\text{s})(72\%) = 230 \text{ m}^3/\text{s}$$

The secondary energy that can be provided at least 50 percent of the time, in addition to the firm energy, is:

$$\text{energy} = (114.8 \text{ kN/m}^2)(230 \text{ m}^3/\text{s})(8{,}760 \text{ hrs/yr}) = 2.31 \times 10^8 \text{ kW} \cdot \text{hrs/yr}$$

7.9.3 Low Flow-Frequency Relationships

Low flow-frequency curves are among the various alternative methods used to quantify the risk of drought conditions occurring. Nonexceedance frequency relationships are developed by determining the minimum streamflow during periods of various lengths that occurred during each year of the gage record. For example, for each year, the minimum 1-day, 7-day, 30-day, and/or 90-day mean flows are tabulated from the record. Flow nonexceedance frequency relationships are developed for each time duration.

The low flow-frequency relationship for annual minimum 7-day flows for the Choctawhatchee River near Newton, Alabama, is presented in Fig. 7.11 (Atkins and Pearman, 1994). The gage record covers 60 years, 1923–1927 and 1936–1990. The 60-year mean discharge is 958 ft^3/s. The data set consists of the minimum 7-day mean discharge in each of 60 years. The Weibull plotting position formula (Eq. 7.12) was used to compute the points plotted as the circular symbol in Fig. 7.11. The log-Pearson type III distribution is plotted as a solid line in Fig. 7.11.

7.10 RESERVOIR/RIVER SYSTEM RELIABILITY

Because streamflow and other variables are characterized by great variability and uncertainty, reservoir/stream system capabilities for satisfying water management requirements must be evaluated from a reliability or risk perspective. Reliability is a measure of the level of dependability at which water supply, hydroelectric power generation, environmental requirements, and other needs can be met. Conversely, risk is a measure of the likelihood of failures in providing these services.

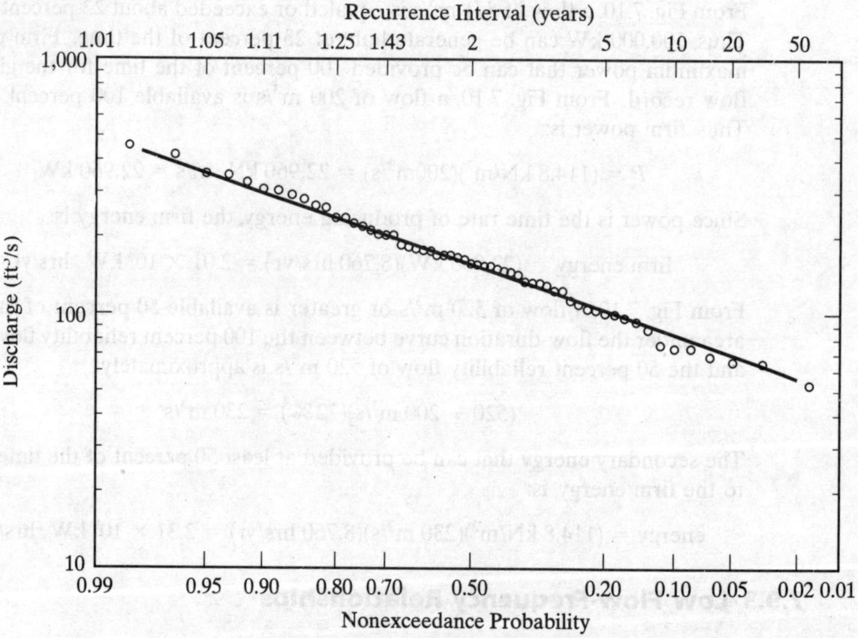

Figure 7.11 Low-flow frequency curve of annual minimum 7-day low flows for the Choctawhatchee River near Newton, Alabama (Atkins and Pearman, 1994).

Reservoir/river system reliability studies are usually based on a simulation model that combines a specified set of water use requirements with adjusted historical streamflows, such as those plotted in Fig. 7.6. Chapters 11 and 12 present the concept of a simulation model consisting of sequential period-by-period water balance accounting as specified water use demands are placed on reservoir storage and streamflow sequences. Streamflows representing a specified condition of river basin development are typically developed by adjusting gaged flows to remove nonhomogeneities.

Various reliability indices can be formulated to measure the capabilities of the simulated reservoir/river system to satisfy specified water use requirements during a postulated repetition of historical hydrology. Water use requirements may involve water supply diversions, instream flow needs, hydroelectric power production, or maintenance of reservoir storage. Variations of the concepts of volume reliability R_V and period reliability R_P are commonly adopted in modeling studies (Wurbs, 1996). These indices are computed from the results of a simulation as follows.

$$R_V = \frac{v}{V}(100\%) \tag{7.48}$$

$$R_P = \frac{n}{N}(100\%) \tag{7.49}$$

Volume reliability R_V is the ratio of the volume of water supplied v to the volume needed V, expressed as a percentage. Unlike period reliability, volume reliability reflects the shortage magnitude as well as frequency. The shortage volume is the demand target V less the volume supplied v within the constraints of water availability. Period reliability R_P is the ratio of the number of time periods n during which the target is met to the total number of time periods N in the simulation. The period reliability R_P represents the percentage of time that a specified water use requirement is met or equivalently the probability of the requirement being met in any randomly selected period.

Firm yield (also called safe or dependable yield) is a commonly used measure of water supply and hydroelectric power generation capabilities. Firm yield is the estimated maximum release or diversion rate or hydroelectric energy production rate that has a R_V and R_P of 100 percent. Reliability estimates are indices reflecting modeling premises and imperfect limited data. A reliability of 100 percent certainly does not mean that the corresponding yield can be provided with certainty. The reliability estimate is based on historical hydrology, as well as other modeling assumptions. A drought more severe than the drought of record will occur at some unknown future time.

7.11 PRECIPITATION FREQUENCY ANALYSIS

Precipitation frequency analysis is explored in this section from the perspective of developing precipitation input for the watershed models covered in Chapter 8. Our attention is focused on assessment of flooding problems and hydrologic design of drainage, stormwater management, and flood mitigation facilities and other hydraulic structures.

7.11.1 Intensity-Duration-Frequency (IDF) Relationships

Rainfall characteristics for a location may be expressed as a relationship between (1) rainfall intensity or depth; (2) duration; and (3) annual exceedance probability, exceedance frequency, or recurrence interval. This is commonly called an intensity-duration-frequency (IDF) relationship. In previous sections, we recognized that instantaneous peak flow rate is a very meaningful, useful random variable. However, with rainfall, duration is an essential component; an instantaneous peak intensity is not a useful random variable in our analyses. Thus, we work with rainfall depth or mean intensity over a specified duration.

IDF relationships are developed from precipitation gage records. A set of durations is selected that might include 5, 15, and 30 minutes, and 1, 3, 6, 12, 18, 24, 48, and 72 hours. For each duration, an annual series is compiled that consists of the maximum rainfall depth occurring over that duration in each year. An analytical probability distribution is applied to the annual series for a specified duration. The process is repeated for each duration. The Gumbel distribution

(Eqs. 7.18 and 7.19) has often been adopted, but other alternative probability distributions are used as well.

Water resources engineers may be called on to develop IDF relationships for a particular city or region in conjunction with various types of investigations. However, most often, engineers apply IDF relationships that have been developed by others. The NWS has published IDF atlases for major regions of the United States as sets of isohyetal maps (Hershfield, 1961; Miller, 1964; Miller, Frederick, and Tracey, 1973; Frederick, Myers, and Auciello, 1977). The USGS has developed IDF relationships for several state transportation departments (Parrett, 1997; Asquith, 1998; Tortorelli, Rea, and Asquith, 1999). Other state and local agencies and their consultants have developed IDF relationships for application in particular cities or regions (Huff and Angel, 1992).

Rainfall IDF relationships are expressed in various formats. For example, Fig. 7.5 is an isohyetal map of Illinois prepared by the USGS showing precipitation depths for a recurrence interval of 2 years and duration of 24 hours. NWS atlases consist of sets of maps covering particular regions in this general format, with a map for each combination of recurrence interval and duration. Fig. 7.12 illustrates the common approach of expressing the IDF relationship graphically for a particular location, which is often adopted in drainage and stormwater management manuals developed by cities. Various entities have expressed IDF relationships as equations with variations of the following general form.

$$i = \frac{a}{(t + b)^c} \tag{7.50}$$

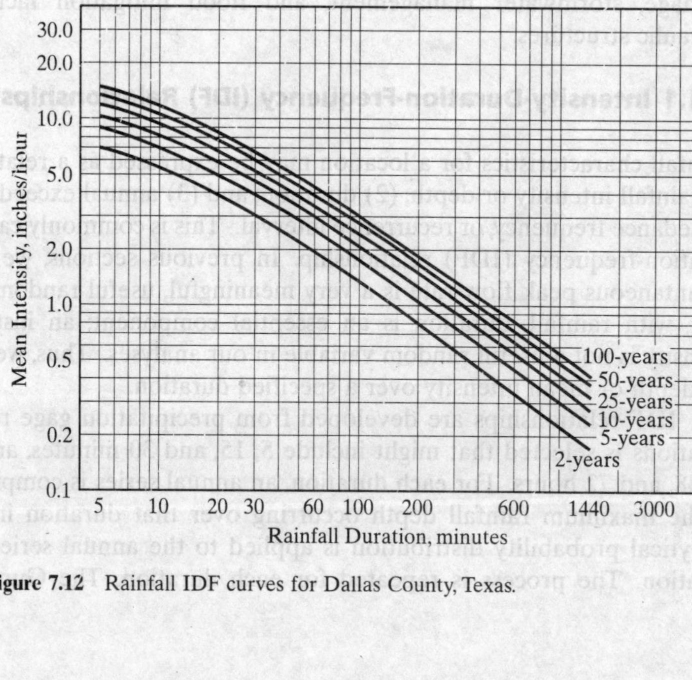

Figure 7.12 Rainfall IDF curves for Dallas County, Texas.

TABLE 7.6 COEFFICIENTS FOR EQ. 7.50 FOR FOUR TEXAS COUNTIES

T	Brazos			Dallas			El Paso			Harris		
years	*a*	*b*	*c*	*a*	*b*	*c*	*a*	*b*	*c*	*a*	*b*	*c*
2	65	8.0	0.806	54	8.3	0.791	24	9.5	0.797	68	7.9	0.800
5	76	8.5	0.785	68	8.7	0.782	34	12.0	0.802	70	7.7	0.749
10	80	8.5	0.763	78	8.7	0.777	42	12.0	0.795	81	7.7	0.753
25	89	8.5	0.754	90	8.7	0.774	60	12.0	0.843	81	7.7	0.724
50	98	8.5	0.745	101	8.7	0.771	90	12.0	0.900	91	7.7	0.728
100	96	8.0	0.730	106	8.3	0.762	65	9.5	0.825	91	7.9	0.706

The Texas Department of Transportation (1986) hydraulic design manual includes Eq. 7.50, which relates rainfall intensity i in inches/hour to rainfall duration t in minutes, along with a table of the coefficients a, b, and c tabulated as a function of recurrence interval for each of the 254 counties of the state. Coefficients are provided in Table 7.6 for the four Texas counties that contain the cities of College Station, Dallas, El Paso, and Houston. The IDF curves for Dallas County presented in Fig. 7.12 are graphs of Eq. 7.50 with appropriate coefficient values from Table 7.6. Wenzel (1982) provides intensity-duration relationships for a 10-year recurrence interval for the cities listed in Table 7.7, based on Eq. 7.50 where again rainfall intensity i is in inches/hour and duration t is in minutes.

7.11.2 Synthetic Design Storms

Many hydrologic engineering applications involve developing streamflow hydrographs for a specified recurrence interval by providing a design storm as input to

TABLE 7.7 COEFFICIENTS FOR EQ. 7.50 FOR A RECURRENCE INTERVAL *T* OF 10 YEARS

City	*a*	*b*	*c*
Atlanta, Georgia	64.1	8.16	0.76
Chicago, Illinois	60.9	9.56	0.81
Cleveland, Ohio	47.6	8.86	0.79
Denver, Colorado	50.8	10.50	0.84
Helena, Montana	30.8	9.56	0.81
Los Angeles, California	10.9	1.15	0.51
Miami, Florida	79.9	7.24	0.73
New York, New York	51.4	7.85	0.75
Olympia, Washington	6.3	0.60	0.40
Santa Fe, New Mexico	32.2	8.54	0.76
St. Louis, Missouri	61.0	8.96	0.78

a watershed model. A design storm is a synthetically derived rainfall event with the following characteristics: annual exceedance probability or recurrence interval, rainfall duration, rainfall depth, computational time interval, spatial distribution, and temporal distribution.

The exceedance probability is selected based on the particular application. For example, the design criteria for sizing a highway culvert might specify a 50-year recurrence interval. Floodplain management activities may be based on delineating the area inundated by the 100-year recurrence interval flood. Selection of a rainfall duration is somewhat arbitrary. A 24-year duration has been adopted in many applications as being both reasonable and convenient. As discussed in Chapter 8, in some situations, setting the rain duration at about the watershed time of concentration has been judged to be appropriate.

For a specified exceedance frequency and rainfall duration, the rainfall depth is determined from an IDF relationship. A uniform spatial distribution of rainfall over the watershed or subbasin is typically assumed, except for very large watersheds. Modeling of spatial variations over large river basins significantly complicate analyses.

A hyetograph of rainfall measurements at a gage is presented in Fig. 7.13 to illustrate the great random variability of rainfall intensity over time that occurs during a typical storm. The synthetic design storm hyetograph of Fig. 7.14 has a triangular shape that is clearly different from the sporadic intensity variations of the actual storm of Fig. 7.13. The synthetic design storm incorporates the concept of a balanced or nested temporal distribution. For a specified exceedance probability, the rainfall depth for a specified duration is contained within the depth for a longer duration. This concept is discussed further after presenting Example 7.11. This example illustrates development of a synthetic design storm with a triangular balanced (nested) temporal distribution. This rainfall distribution is also called the alternating block method.

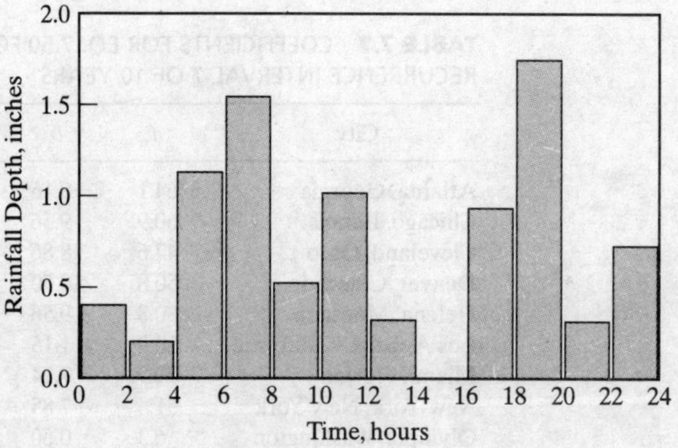

Figure 7.13 The distribution of rainfall during an actual storm is typically very random and sporadic.

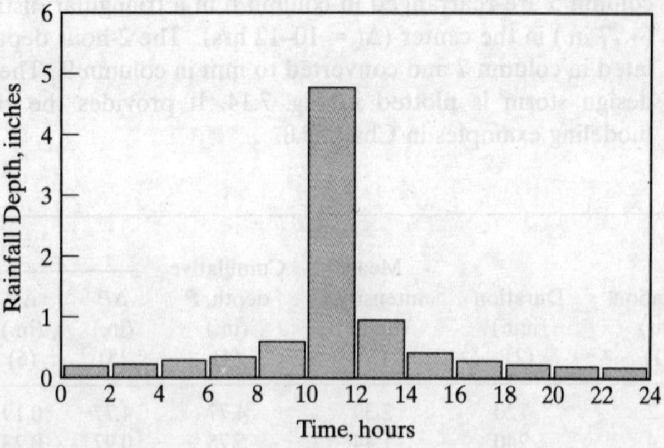

Figure 7.14 The hyetograph for the synthetic design storm developed in Example 7.11 illustrates the balanced triangular (alternating block) distribution of rainfall over time.

Example 7.11

Develop a design storm for a watershed in Dallas, Texas, for an annual exceedance probability of 0.02 and duration of 24 hours, using a computational time interval of 2.0 hours.

Solution An annual exceedance probability of 0.02 is equivalent to a recurrence interval of 50 years.

$$T = \frac{1}{P} = \frac{1}{0.02} = 50 \text{ years}$$

Equation 7.50 is used with coefficients from Table 7.6 for a 50-year storm in Dallas County. The city of Dallas is located within Dallas County.

$$i = \frac{a}{(t + b)^c} = \frac{101}{(t + 8.7)^{0.771}}$$

For a rainfall duration t of 24 hours or 1,440 minutes,

$$i = \frac{101}{(1{,}440 + 8.7)^{0.771}} = 0.369 \text{ inches/hour} \quad (9.38 \text{ mm/hr})$$

The total rainfall depth P for this 50-year, 24-hour rainfall event is

$$P = i\Delta t = (0.369 \text{ in./hr})(24 \text{ hrs}) = 8.86 \text{ inches } (225 \text{ mm})$$

The total rainfall durations t from column 2 of the table are input into the IDF equation to obtain the intensities i recorded in column 3 that are transformed to depths P in column 4. The incremental increase in precipitation depth for each 2-hour incremental increase in duration is tabulated in column 5 ($\Delta P = P_t - P_{t-1}$). The ΔP's of column 5 are rearranged in column 6 in a triangular distribution with the largest ΔP (4.77 in.) in the center (Δt = 10–12 hrs). The 2-hour depths of column 6 are accumulated in column 7 and converted to mm in column 9. The 50-year recurrence interval design storm is plotted in Fig. 7.14. It provides the rainfall input for watershed modeling examples in Chapter 8.

				Design storm				
Duration (hr) (1)	Duration (min) (2)	Mean intensity, i (in./hr) (3)	Cumulative depth, P (in.) (4)	ΔP (in.) (5)	ΔP (in.) (6)	P (in.) (7)	ΔP (mm) (8)	P (mm) (9)
2	120	2.39	4.77	4.77	0.19	0.19	4.8	4.8
4	240	1.44	5.75	0.97	0.23	0.42	5.8	10.6
6	360	1.06	6.36	0.62	0.28	0.70	7.1	17.8
8	480	0.85	6.83	0.46	0.38	1.08	9.7	27.3
10	600	0.72	7.20	0.38	0.62	1.70	15.7	43.0
12	720	0.63	7.52	0.32	4.77	6.47	121.0	164.0
14	840	0.56	7.81	0.28	0.97	7.44	24.6	189.0
16	960	0.50	8.06	0.25	0.46	7.90	11.6	201.0
18	1,080	0.46	8.28	0.23	0.32	8.22	8.1	210.0
20	1,200	0.42	8.49	0.21	0.25	8.47	6.4	215.0
22	1,320	0.39	8.68	0.19	0.21	8.68	5.3	221.0
24	1,440	0.37	8.86	0.18	0.18	8.86	4.6	225.0

The balanced triangular distribution of rainfall over time reflected in the synthetic design storm is not intended to represent the temporal distribution of an actual rainfall event. Rather rainfall depths for the specified recurrence interval for various durations are nested. In Example 7.11, the 50-year, 2-hour depth of 4.77 inches is contained within the 50-year, 4-hour depth of 5.75 inches that is nested within the 50-year, 6-hour depth of 6.36 inches and so forth. Selection of a duration for a design storm is somewhat arbitrary. Combining a reasonably long duration with a balanced (nested) triangular temporal distribution provides a conservative design storm.

The Natural Resource Conservation Service (NRCS) developed the 24-hour rainfall distributions provided in Table 7.8 based on the NWS IDF information. The types I, IA, II, and III distributions are representative of the regions shown in Fig. 7.15. These distributions reflect the concept of shorter duration depths being embedded within longer duration depths, for a specified recurrence interval, as described in the previous paragraph. However, the most intense portion of the storm occurs earlier in the types I and IA, for the Pacific coastal regions, than in the types II and III distributions. The type III distribution for the Atlantic and Gulf coastal regions is almost identical to the type II. Types II and III represent all of the contiguous United States except the west coast. Application of the NRCS methodology is illustrated by Example 7.12.

TABLE 7.8 NRCS RAINFALL DISTRIBUTIONS

Time hours	I	IA	II	III	Time hours	I	IA	II	III
0.5	0.008	0.010	0.0053	0.0050	12.5	0.706	0.683	0.7351	0.7020
1.0	0.017	0.020	0.0108	0.0100	13.0	0.728	0.701	0.7724	0.7500
1.5	0.026	0.035	0.0164	0.0150	13.5	0.748	0.719	0.7989	0.7835
2.0	0.035	0.050	0.0223	0.0200	14.0	0.766	0.736	0.8197	0.8110
2.5	0.045	0.067	0.0284	0.0252	14.5	0.783	0.753	0.8380	0.8341
3.0	0.055	0.082	0.0347	0.0308	15.0	0.799	0.769	0.8538	0.8542
3.5	0.065	0.098	0.0414	0.0367	15.5	0.815	0.785	0.8676	0.8716
4.0	0.076	0.116	0.0483	0.0430	16.0	0.830	0.800	0.8801	0.8860
4.5	0.087	0.135	0.0555	0.0497	16.5	0.844	0.815	0.8914	0.8984
5.0	0.099	0.156	0.0632	0.0568	17.0	0.857	0.830	0.9019	0.9095
5.5	0.112	0.180	0.0712	0.0642	17.5	0.870	0.844	0.9115	0.9194
6.0	0.126	0.206	0.0797	0.0720	18.0	0.882	0.858	0.9206	0.9280
6.5	0.140	0.237	0.0887	0.0806	18.5	0.893	0.871	0.9291	0.9358
7.0	0.156	0.268	0.0984	0.0905	19.0	0.905	0.844	0.9371	0.9432
7.5	0.174	0.310	0.1089	0.1016	19.5	0.916	0.896	0.9446	0.9503
8.0	0.194	0.425	0.1203	0.1140	20.0	0.926	0.908	0.9519	0.9570
8.5	0.219	0.480	0.1328	0.1284	20.5	0.936	0.920	0.9588	0.9634
9.0	0.254	0.520	0.1467	0.1458	21.0	0.946	0.932	0.9653	0.9694
9.5	0.303	0.550	0.1625	0.1659	21.5	0.956	0.944	0.9717	0.9752
10.0	0.515	0.577	0.1808	0.1890	22.0	0.965	0.956	0.9777	0.9808
10.5	0.583	0.601	0.2042	0.2165	22.5	0.974	0.967	0.9836	0.9860
11.0	0.624	0.624	0.2351	0.2500	23.0	0.983	0.978	0.9892	0.9909
11.5	0.655	0.645	0.2833	0.2980	23.5	0.992	0.989	0.9947	0.9956
12.0	0.682	0.664	0.6632	0.5000	24.0	1.000	1.000	1.0000	1.0000

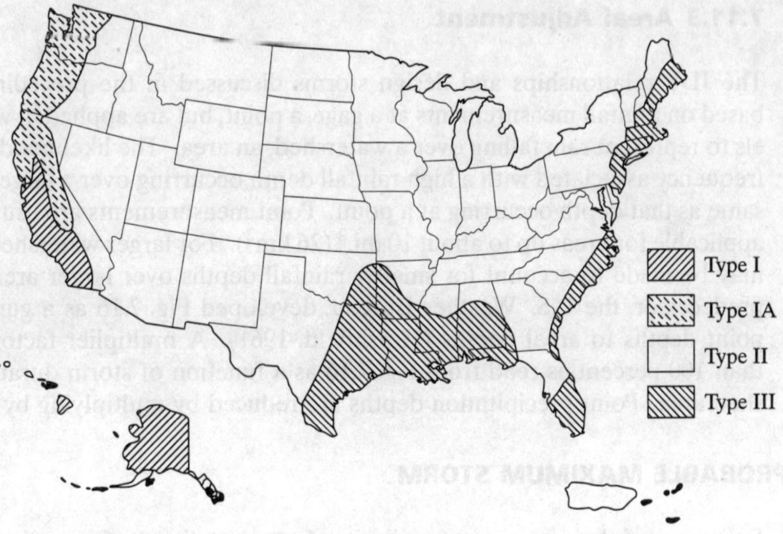

Figure 7.15 Regions of applicability for the NRCS rainfall distributions (NRCS, 1986).

Example 7.12

Repeat Example 7.11 using the NRCS type III distribution

Solution From Example 7.11, the 50-year, 24-hour rainfall depth is 8.86 inches (225 mm). This amount is distributed over the 24-hour duration using a 2-hour time step as follows.

Duration (hr)	Ratio from Table 7.8	Cumulative depth, P (in.)	Incremental depth, P (in.)	Cumulative depth, ΔP (mm)	Incremental depth, ΔP (mm)
2	0.0200	0.18	0.18	4.5	4.5
4	0.0430	0.38	0.20	9.7	5.2
6	0.0720	0.64	0.26	16.2	6.5
8	0.1140	1.01	0.37	25.7	9.5
10	0.1890	1.67	0.66	42.5	16.8
12	0.5000	4.43	2.76	112.5	70.0
14	0.8110	7.19	2.76	182.5	70.0
16	0.8860	7.85	0.66	199.4	16.9
18	0.9280	8.22	0.37	208.8	9.4
20	0.9570	8.48	0.26	215.3	6.5
22	0.9808	8.69	0.21	220.6	5.3
24	1.0000	8.86	0.17	225.0	4.4

Differences between these two similar alternative methods are as follows. The balanced triangular methodology illustrated by Example 7.11 is applicable for any duration and results in a temporal rainfall distribution shaped by the IDF relationship for the particular location. The four 24-hour NRCS rainfall distributions are generically applied to all locations within the regions delineated in Fig. 7.15.

7.11.3 Areal Adjustment

The IDF relationships and design storms discussed in the preceding sections are based on rainfall measurements at a gage, a point, but are applied in watershed models to represent rain falling over a watershed, an area. The likelihood or exceedance frequency associated with a high rainfall depth occurring over a large area is not the same as that depth occurring at a point. Point measurements are usually considered applicable for areas up to about 10 mi.^{2} (26 km^2). For larger watersheds, adjustments may be made to account for smaller rainfall depths over larger areas. The NWS's predecessor, the U.S. Weather Bureau, developed Fig. 7.16 as a guide in reducing point depths to areal depths (Hershfield, 1961). A multiplier factor, which is less than 100 percent, is read from Fig. 7.16 as a function of storm duration and watershed area. Point precipitation depths are reduced by multiplying by this factor.

7.12 PROBABLE MAXIMUM STORM

Estimates of the extreme upper limits of precipitation and runoff are required for design situations in which failure could result in catastrophic consequences. For ex-

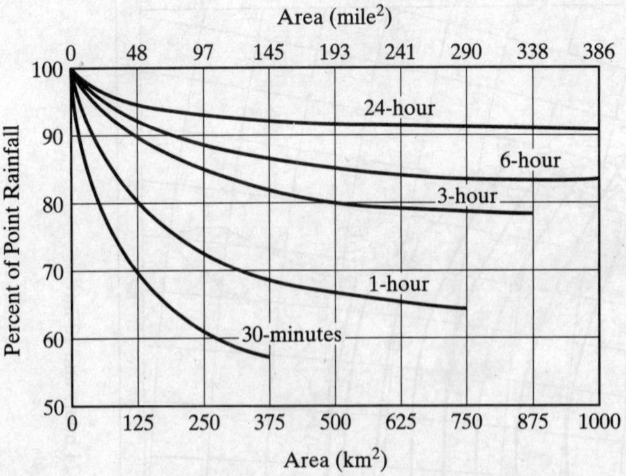

Figure 7.16 Areal depth adjustment factor.

ample, a flood over-topping an embankment dam could breach the dam and result in severe downstream flooding that is much worse than if the dam did not exist. Thus, hydrologic design of a dam and emergency spillway is based on the probable maximum flood. Likewise, federal and state dam safety programs include evaluation of capabilities of existing dams for handling the probable maximum flood without over-topping.

The probable maximum precipitation (PMP) is the estimated greatest depth of precipitation for a given duration that is physically possible over a given size storm area at a particular geographical location for a certain time of the year. The probable maximum storm (PMS) is the PMP with appropriate temporal and spatial distribution. The PMS is provided as input to a watershed model to develop the probable maximum flood (PMF) hydrograph.

The NWS develops PMP depth-duration-area relationships by increasing observed precipitation to physical limits. The process includes (1) storm transposition, (2) moisture maximization, and (3) envelopment of the adjusted transposed storms. Transposition enhances observed precipitation records by considering not only storms that have occurred at the location of interest, but also those that actually occurred elsewhere but could have occurred there. Moisture maximization consists of increasing precipitation observed for an extreme past storm event by a factor that reflects the maximum amount of moisture that physically could have existed in the atmosphere for the storm location and season of the year. Envelopment involves construction of smooth curves that envelope precipitation maxima for various durations and area sizes. Geographic smoothing is performed to ensure regional consistency.

Generalized estimates of PMP for the U.S. developed by the NWS are reported in its NOAA Hydrometeorological Report (HMR) series. HMR 51 (NWS, 1978) provides PMP information in the format of Fig. 7.17 for the U.S. east of the 105th meridian. Other reports cover the western states. The World Meteorological Organization (1986) addresses PMP estimates for regions throughout the world.

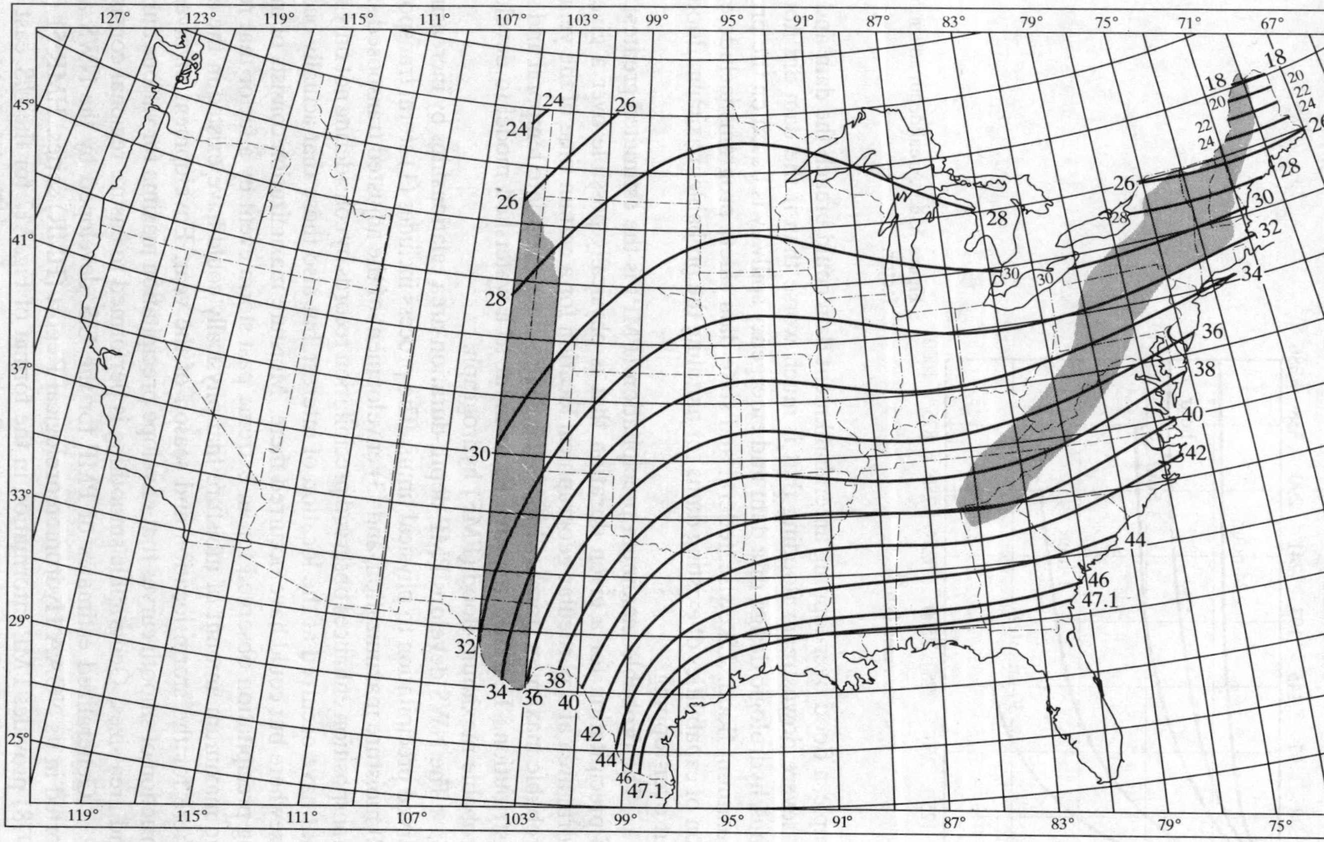

Figure 7.17 All-season probable maximum precipitation PMP for 24 hours and 10 mi^2 (NWS, 1978).

The PMP at a location is a function of rainfall duration, areal extent, and season of the year. The NWS PMP atlases provide maps for various combinations of these variables. The all-season PMP of Fig. 7.17 is for 24 hours and 10 mi^2 (26 km^2).

HMR 52 (NWS, 1982) provides criteria and a step-by-step procedure for developing a PMS using PMP estimates from HMR 51. The USACE Hydrologic Engineering Center developed a computer program called HMR52 that performs the computations outlined in the NWS HMR 52. The following characteristics of the synthetic design storm are reflected in the procedure for developing a PMS: isohyetal pattern, location of the storm center, orientation of the storm, storm area size, spatial variability, duration, temporal distribution, and antecedent storms. Watershed modeling methods described in Chapter 8 are applied to compute the PMF hydrograph to result from the PMS.

PROBLEMS

7.1. A highway bridge is designed to just pass a discharge with an annual exceedance probability of 0.020 without overtopping the roadway.

(a) Compute the probability that the bridge will be overtopped at least once during the next 10 years.

(b) Compute the probability that the bridge will be overtopped exactly one time during the next 10 years.

(c) Compute the probability that the bridge will be overtopped in 2 of the next 50 years.

7.2. What are the probabilities that a flood greater than the 100-year flood will occur (a) during the next 100 years and (b) during any single future year?

7.3. What are the probabilities of having at least one flood at a given location equal to or greater than the 25-year flood during (a) the next year, (b) the next 25 years, and (c) any 5-year period?

7.4. A cofferdam is constructed to protect a construction project in a floodplain. The cofferdam is sized based on a 10-year recurrence interval design flood. What is the probability that the cofferdam will be overtopped during the 3-year duration of the construction project?

7.5. How long must a gaging station be maintained so that the probability of observing a flood equal to or greater than the 10-year flood is 0.80?

7.6. What annual exceedance frequency should be adopted in the design of a hydraulic structure if the engineer accepts a 5 percent risk that the flow capacity will be exceeded during the projected 30-year life of the structure?

7.7. An agricultural levee is designed to just contain a 50-year flood without overtopping. Determine the probabilities that:

(a) the levee will be overtopped in any given year

(b) the levee will be overtopped during any 25-year period

(c) the levee will be overtopped at least once during the next 50 years

(d) the levee will not be overtopped during the next 50 years

(e) the levee will be overtopped one time during the next 50 years

(f) the levee will be overtopped in three or more years during the next 50 years

7.8. Use Eq. 7.50 and Table 7.6 to determine the rainfall intensity for a 24-hour duration in the city of Houston in Harris County, Texas, that has a probability of 0.04 of being equaled or exceeded in any year. What is the probability that the peak 24-hour rainfall depth in any 10-year period will exceed the 4 percent exceedance frequency, 24-hour duration depth?

7.9. The annual precipitation for the city of Houston is plotted in Fig. 2.7. The annual precipitation has a mean of 47.5 inches and standard deviation of 11.6 inches. Assume annual precipitation may be modeled by the normal (Gaussian) probability distribution. Estimate the following:

(a) the annual precipitation amount for Houston for any future year that has a 50 percent chance of being equaled or exceeded
(b) the annual amount that has a 20 percent chance of being equaled or exceeded
(c) the probability that the annual precipitation in any year will equal or exceed 62 inches
(d) the probability that the precipitation in any year will be less than 38 inches
(e) the probability that the precipitation in any year will be between 38 and 62 inches

7.10. The annual rainfall in Everglades National Park in Florida has a mean of 1,410 mm and a standard deviation of 280 mm. Assuming that the annual rainfall is normally distributed, estimate the annual rainfall amounts that have the following chances of being exceeded in any year: (a) 50 percent, (b) 20 percent, and (c) 10 percent.

7.11. The series of maximum 24-hour duration rainfall depths observed each year during a long period of record at a precipitation gage has a mean of 125 mm and a standard deviation of 37 mm. Using the Gumbel probability function, estimate the probability that the maximum 24-hour rainfall depth in any year will exceed 160 mm. What is the probability that the maximum 24-hour rainfall depth will exceed 125 mm?

7.12. Annual precipitation amounts for a 100-year period of record at a gage are distributed as follows.

Less than 150 mm	7 years
Between 150 and 249 mm	15 years
Between 250 and 349 mm	34 years
Between 350 and 449 mm	35 years
450 mm or greater	9 years
Total	100 years

Estimate the probability of annual precipitation in any year:

(a) being less than 250 mm
(b) falling between 250 and 449 mm
(c) equaling or exceeding 450 mm

7.13. The rainfall during July has been at least 5 mm in 21 of the past 89 years at a gage. Estimate the probability of having at least 5 mm of rainfall at this location next July.

7.14. A reservoir/river system is simulated in Example 11.7 of Chapter 11. Supplying water for irrigation is one of the services provided by the multiple-purpose system. Irrigation demands and the corresponding portions of the demands supplied in the simulation are tabulated in Table 11.7. Compute the period reliability and volume reliability for the irrigation water supply based on the simulation results presented in Table 11.7.

7.15. Explain why the Weibull plotting position formula (Eq. 7.12) uses $N + 1$ rather than N.

7.16. A 76-year peak annual flow series for USGS gage 08110500 on the Navasota River near Easterly in south central Texas is provided below.

Year	Flow (ft^3/s)	Year	Flow (ft^3/s)	Year	Flow (ft^3/s)	Year	Flow (ft^3/s)
1925	2,540	1944	60,300	1963	356	1982	13,600
1926	17,600	1945	15,600	1964	740	1983	11,200
1927	15,200	1946	23,200	1965	43,200	1984	1,580
1928	16,800	1947	8,800	1966	39,600	1985	11,400
1929	49,400	1948	4,010	1967	1,410	1986	32,300
1930	30,100	1949	5,350	1968	20,100	1987	6,470
1931	8,100	1950	19,000	1969	12,200	1988	1,120
1932	58,100	1951	710	1970	9,860	1989	4,560
1933	5,110	1952	6,800	1971	1,610	1990	25,200
1934	16,600	1953	25,700	1972	15,000	1991	23,800
1935	18,000	1954	8,300	1973	17,700	1992	61,800
1936	37,200	1955	1,930	1974	11,100	1993	10,600
1937	7,290	1956	4,080	1975	24,900	1994	11,800
1938	13,100	1957	37,700	1976	14,200	1995	16,400
1939	1,620	1958	24,800	1977	23,500	1996	374
1940	1,880	1959	12,600	1978	6,590	1997	23,300
1941	34,300	1960	9,100	1979	29,900	1998	28,500
1942	18,800	1961	33,000	1980	11,900	1999	37,700
1943	4,900	1962	3,780	1981	12,900	2000	4,020

(a) Plot the flows on normal probability paper using the Weibull plotting position formula. Estimate the 10-year recurrence interval discharge from your plot.

(b) Estimate the 10-year and 100-year recurrence interval flows using the log-Pearson type III probability distribution with a station skew coefficient computed directly from the data.

(c) Repeat part b above using a weighted skew coefficient ($G_R = -0.28$ from Fig. 7.3).

(d) Estimate the 10-year and 100-year recurrence interval flows using the log-normal probability distribution.

(e) Estimate the 10-year and 100-year recurrence interval flows using the Gumbel distribution.

(f) Compare the 10-year and 100-year peak flows computed using the alternative methods. Which values do you recommend for adoption? Why?

7.17. Determine the upper and lower 95 percent confidence limits for the 100-year recurrence interval peak flow determined in Problem 7.16(c).

7.18. Determine the upper and lower 95 percent confidence limits for the 10-year recurrence interval peak flow determined in Problem 7.16(c).

7.19. Develop a 90 percent confidence band for the frequency curve of Problem 7.16(c).

7.20. Repeat Problem 7.16 using the 38-year sequence of peak annual flows for 1925–1962.

7.21. Repeat Problem 7.16 using the 44-year annual peak flow series covering from 1933 through 1976 from Table 7.1 for the Mississippi River at St. Louis.

7.22. Repeat Problem 7.16 using the 44-year period 1955–1998 from Table 7.1 for the Mississippi River at St. Louis.

7.23. Determine the upper and lower 95 percent confidence limits for the 100-year recurrence interval peak flow determined in Problem 7.21(c).

7.24. What are the probabilities that the 10-year and 100-year recurrence interval flows from Problems 7.16–7.19 will be equaled or exceeded in any 25-year period?

7.25. Estimate the probability that the 10-year discharge from Problems 7.16 and 7.20–7.22 will be exceeded during the next (a) 10 years and (b) 100 years. Also, compute the probability that the 100-year discharge estimates will be exceeded during the next (a) 10 years and (b) 100 years.

7.26. Select a USGS stream gaging station on a river in your state that has at least 40 years of record. Obtain the record of peak flows through the USGS web site. Determine the 10-year and 100-year recurrence interval peak flows based on the log-Pearson type III distribution.

7.27. Determine the upper and lower 90 percent confidence limits for the 10-year and 100-year recurrence interval peak flows determined in Problem 7.26.

7.28. The flow-duration curve of Fig. 7.10 was developed for the proposed run-of-river hydroelectric power project analyzed in Example 7.10. The present problem consists of reliability analyses for an alternative hydropower plant design at the same location. The flow-duration curve of Fig. 7.10 is still valid. However, the proposed alternative plant has a constant head of 18 m, an installed capacity of 125,000 kW, and an efficiency of 0.80. Determine the (a) firm energy and (b) the secondary energy that can be provided at least 60 percent of the time.

7.29. Repeat Problem 7.28 for a head of 8 m, an installed capacity of 75,000 kW, and an efficiency of 0.75.

7.30. Obtain the mean annual flow for each year of the period of record for a gage included in the National Water Information System accessible through the USGS web site. Choose a gage with at least 40 years of record. The USGS determines mean annual flows by averaging the daily flows recorded throughout the year. Assuming that mean annual flows are modeled by the normal probability distribution, estimate the mean annual flows that have exceedance frequencies of (a) 50 percent, (b) 20 percent, and (c) 10 percent.

7.31. Develop a flow-frequency curve for the mean annual flows of Problem 7.30 using Eq. 7.9.

7.32. Daily flows for USGS gaging station 07010000 are plotted in Fig. 2.14. Obtain the daily flows for the period of record at this station through the USGS web site. Develop and apply a computer program to determine a flow-frequency relationship for this station. Plot a flow-frequency curve.

7.33. Repeat Problem 7.32 for another gaging station of interest.

7.34. The annual series of maximum rainfall depths in millimeters (mm) for durations of 5, 10, 15, 30, and 60 minutes obtained from a recording rain gage are tabulated on the next page. Develop a set of rainfall intensity-duration-frequency curves in the general format of Fig. 7.12. Plot intensity in millimeters/hour versus duration in minutes. Use the Gumbel probability distribution to develop mean intensities (mm/hr) for recurrence intervals of 2, 5, 10, 25, 50, and 100 years for each of the five alternative rainfall durations.

Year	5 min	10 min	15 min	30 min	60 min
1951	2.2	4.2	4.8	7.7	11.5
1952	2.2	3.8	4.5	6.7	9.6
1953	2.9	5.1	6.1	9.0	16.0
1954	4.5	6.7	8.0	9.9	16.3
1955	4.5	7.7	10.6	16.6	25.3
1956	2.9	3.5	4.2	6.1	8.6
1957	2.6	3.5	4.2	5.4	9.6
1958	9.6	10.2	10.9	12.8	18.2
1959	11.2	21.4	28.2	44.5	49.0
1960	6.4	7.4	7.7	8.3	12.2
1961	4.8	6.1	7.7	11.2	21.1
1962	3.2	3.5	4.2	5.1	6.4
1963	5.1	9.6	13.1	20.2	22.4
1964	4.5	8.0	10.6	11.2	12.8
1965	2.6	3.5	4.8	7.4	13.1
1966	5.1	7.7	8.3	11.8	14.1
1967	2.9	5.1	5.8	6.4	9.9
1968	3.2	5.1	6.7	8.3	9.9
1969	2.2	3.2	3.8	6.1	7.7
1970	9.6	10.9	11.2	12.2	12.8
1971	3.8	6.4	6.4	8.3	10.6
1972	8.0	9.0	9.9	11.5	13.4
1973	3.5	5.4	8.0	11.5	13.1
1974	5.1	7.7	12.2	14.7	15.0
1975	2.9	5.4	7.7	11.5	16.6
1976	4.2	6.7	8.6	12.8	18.6
1977	8.0	9.0	12.2	11.5	21.4
1978	2.2	4.5	5.1	7.0	11.5
1979	4.2	6.1	7.7	8.3	9.9
1980	3.5	4.2	5.1	7.0	11.2
1981	3.2	4.5	5.4	6.7	10.2
1982	4.8	7.0	8.6	9.6	10.9
1983	4.5	7.0	9.0	13.4	16.0
1984	3.8	5.8	6.4	8.0	11.5
1985	2.9	4.5	6.1	8.6	9.6
1986	4.5	6.7	8.3	10.6	11.5
1987	3.2	4.5	5.1	6.1	9.3
1988	7.0	13.1	15.0	18.2	21.4
1989	2.6	4.2	5.4	9.0	13.1
1990	2.9	4.8	6.1	8.0	11.2
1991	3.2	5.4	7.4	10.9	15.4
1992	5.4	7.7	8.6	9.9	11.8
1993	2.6	3.8	4.2	5.8	8.3
1994	4.8	5.8	7.4	10.6	16.0
1995	5.8	7.7	9.3	11.2	12.2
1996	4.2	8.6	10.6	16.0	27.2
1997	5.8	7.7	9.3	11.2	13.1
1998	4.5	5.8	6.4	9.6	15.0
1999	2.6	3.5	4.2	6.1	10.9
2000	1.6	3.2	4.8	4.8	8.0

7.35. Repeat Problem 7.34 using the log-Pearson III probability distribution.

7.36. Repeat Problem 7.34 using just the data for the 25-year series covering years 1976–2000. Use the Gumbel distribution.

7.37. Repeat Problem 7.34 using just the data for the 25-year series covering years 1976–2000. Use the log-Pearson III distribution.

7.38. Develop a synthetic design storm for Harris County, Texas, for a 100-year recurrence interval. Use a rainfall duration of 24 hours and a computational time interval of 1.0 hour. Use a balanced triangular distribution of rainfall over time (alternating block method). This design storm is used in Problems 8.16 and 8.31.

7.39. Repeat Problem 7.38 using a NRCS type III temporal distribution of rainfall.

7.40. Develop a design storm for Dallas, Texas, for an annual exceedence frequency of 1.0 percent, a rainfall duration of 24 hours, a balanced triangular distribution over time, and a computational time step of 1.0 hour. Use Fig. 7.16 to adjust the design rainfall for application to a watershed with a drainage area of 275 mi^2.

7.41. Develop a design storm for New York City for an annual exceedence frequency of 10 percent, a rainfall duration of 24 hours, a balanced triangular distribution over time, and a computational time step of 1.0 hour. An intensity-duration-frequency relationship is provided by Eq. 7.50 and Table 7.7.

7.42. Repeat Problem 7.41 using a NRCS type III temporal distribution of rainfall.

REFERENCES

ANG, A. H., and W. H. TANG, *Probability Concepts in Engineering Planning and Design,* John Wiley, New York, NY, 1991.

ASQUITH, W. H., *Depth-Duration Frequency of Precipitation for Texas,* Water Resources Investigations Report 98-4044, U.S. Geological Survey, Austin, TX, 1998.

ATKINS, J. B., and J. L. PEARMAN, *Low-Flow and Flow-Duration Characteristics of Alabama Streams,* Water Resources Investigations Report 93-4186, U.S. Geological Survey, Tuscaloosa, AL, 1994.

DEVORE, J., *Probability and Statistics for Engineering and the Sciences,* 5th Ed., Brooks/Cole, Pacific Grove, CA, 2000.

FREDERICK, R. H., V. A. MEYERS, and E. P. AUCIELLO, *Five- to 60-Minute Precipitation Frequency for the Eastern and Central United States,* NOAA Technical Memorandum NWS HYDRO-35, National Weather Service, Office of Hydrology, Silver Spring, MD, 1977.

GUMBEL, E. J., *Statistics of Extremes,* Columbia University Press, New York, NY, 1958.

HELSEL, D. R., and R. M. HIRSCH, *Statistical Methods in Water Resources,* Studies in Environmental Science 49, Elsevier, New York, NY, 1992.

HERSHFIELD, D. M., *Rainfall Frequency Atlas of the United States for Durations from 30 Minutes to 24 Hours and Return Periods from 1 to 100 Years,* U.S. Weather Bureau Technical Paper 40, Washington, D.C., 1961.

HUFF, F. A., and J. R. ANGEL, *Rainfall Frequency Atlas of the Midwest,* Bulletin 17, Illinois State Water Survey, Urbana, IL, 1992.

Hydrologic Engineering Center, *HEC-FFA Flood Frequency Analysis, Users Manual,* CPD-13, U.S. Army Corps of Engineers, Davis, CA, 1992.

Interagency Committee on Water Data, *Guidelines for Determining Flood Flow Frequency,* Bulletin 17B of the Hydrology Subcommittee, Office of Water Data Coordination, U.S. Geological Survey, Reston, VA, 1982.

JENNINGS, M. E., W. O. THOMAS, and H. C. RIGGS, *Nationwide Summary of U.S. Geological Survey Regional Regression Equations for Estimating Magnitude and Frequency of Floods for Ungaged Sites,* Water Resources Investigations Report 94-4002, Reston, VA, 1994.

KOTTEGODA, N. T., and R. ROSSO, *Probability, Statistics, and Reliability for Civil and Environmental Engineers,* McGraw-Hill, New York, NY, 1997.

MCCUEN, R. H., *Hydrologic Analysis and Design,* 2nd Ed., Prentice Hall, Upper Saddle River, NJ, 1998.

MILLER, J. F., *Two- to Ten-Day Precipitation for Return Periods of 2 to 100 Years in the Contiguous United States,* U.S. Weather Bureau Technical Paper 49, Washington, D.C., 1964.

MILLER, J. F., R. H. FREDERICK, and R. J. TRACEY, *Precipitation Frequency Atlas for the Western United States,* NOAA Atlas 2, National Weather Service, Office of Hydrology, Silver Spring, MD, 1973.

Natural Resource Conservation Service, *Urban Hydrology for Small Watersheds,* Technical Report 55, Washington, D.C. 1986.

National Weather Service, *Probable Maximum Precipitation, United States East of the 105th Meridian,* Hydrometeorological Report No. 51, Silver Spring, MD, 1978.

National Weather Service, *Application of Probable Maximum Precipitation Estimates, United States East of the 105th Meridian,* Hydrometeorological Report No. 52, Silver Spring, MD, 1982.

PARRETT, C. P., *Regional Analysis of Annual Precipitation Maxima in Montana,* Water Resources Investigations Report 97-4004, U.S. Geological Survey, Reston, VA, 1997.

PEARSON, K., *Tables for Statisticians and Biometricians,* 3rd Ed., Columbia University Press, New York, NY, 1930.

RAO, A. R., and K. H. HAMED, *Flood Frequency Analysis,* CRC Press, Boca Raton, FL, 2000.

Texas Department of Transportation, *Hydraulic Manual,* Bridge Division, Austin, TX, 1986.

TORTORELLI, R. L., A. REA, and W. H. ASQUITH, *Depth-Duration Frequency of Precipitation for Oklahoma,* Water Resources Investigations Report 99-4232, U.S. Geological Survey, Oklahoma City, OK, 1999.

U.S. Army Corps of Engineers, *Hydrologic Frequency Analysis,* EM 1110-2-1415, Washington, D.C., 1993.

VIESSMAN, W., and G. L. LEWIS, *Introduction to Hydrology,* 4th Ed., Harper-Collins, New York, NY, 1996.

WENZEL, H. G., "Chapter 2 Rainfall for Urban Stormwater Design," *Urban Stormwater Hydrology* (D. F. Kibler, Ed.), Water Resources Monograph 7, American Geophysical Union, Washington, D.C., 1982.

World Meteorological Organization, *Manual for Estimation of Probable Maximum Precipitation,* WMO 332, 2nd Ed., Geneva, Switzerland, 1986.

WURBS, R. A., *Modeling and Analysis of Reservoir System Operations,* Prentice Hall, Upper Saddle River, NJ, 1996.

8

Modeling Watershed Hydrology

Watershed modeling consists of computing streamflows and in some cases associated sediment or other water quality constituent loads that result from precipitation runoff. The watershed is the system being modeled, precipitation is the input, and flows and loads are the output. Watershed models provide the design hydrographs used to size hydraulic structures, such as dams, spillways, levees, channel improvements, storm sewers, detention basins, culverts, and bridges; and to delineate floodplains in support of floodplain management programs. The models are used to quantify the impacts of land use changes and watershed development activities. Urban stormwater management and control of erosion and pollution from agricultural activities involve modeling of sediment yields and pollutant loads, as well as flows.

8.1 WATERSHED HYDROLOGY

Watershed hydrology is introduced in Section 2.7. As illustrated in Fig. 8.1, the precipitation falling on a watershed drains to its outlet as runoff. Hydrologic processes transform the precipitation into abstractions and runoff to streams. Hydrologic abstractions are losses of precipitation runoff by interception, depression storage, infiltration, and evapotranspiration (Fig. 2.12). Direct runoff is the precipitation that reaches the watershed outlet relatively quickly as a direct response to the storm event. Direct runoff may include interflow, as well as surface flow.

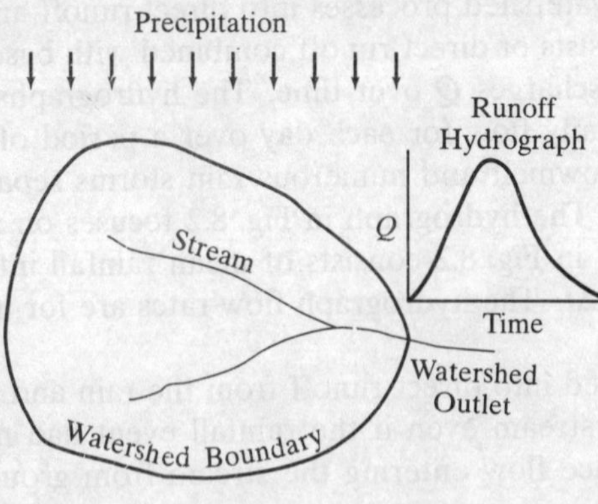

Figure 8.1 The runoff at a location results from precipitation falling on the watershed above that location.

Interflow is flow through relatively pervious soil near the ground surface that reaches the outlet about as quickly as surface flow. Precipitation that infiltrates into the ground may eventually contribute to stream baseflow, but the retention time in the ground is great relative to the time required for the direct runoff to reach the watershed outlet.

Figure 8.2 consists of the hyetograph of spatially averaged rainfall that falls on a watershed and the resulting hydrograph at the watershed outlet. A hyetograph shows precipitation intensities i or depths P over time t. In the hyetograph of Fig. 8.2, mean intensities in centimeters/hour are plotted for each 1.0 hour time interval Δt during the rainfall event. Precipitation depth P is represented by the area under the intensity i hyetograph, recognizing that

$$i = \frac{P}{\Delta t} \quad \text{or} \quad P = i \times \Delta t \tag{8.1}$$

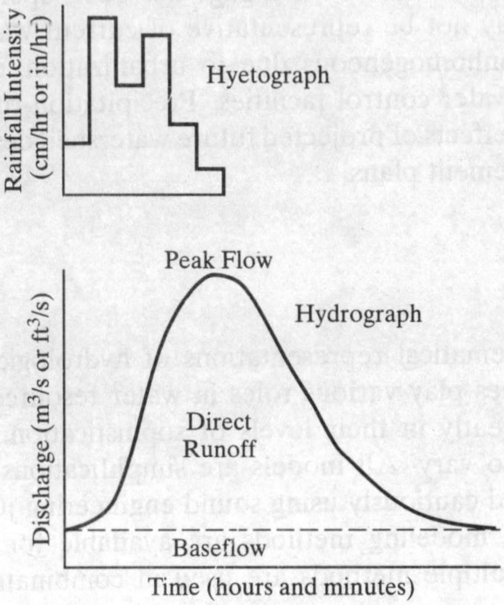

Figure 8.2 The direct runoff portion of the streamflow hydrograph results from the rainfall hyetograph.

The rainfall is separated by watershed processes into direct runoff and hydrologic abstractions. Streamflow consists of direct runoff combined with baseflow.

A hydrograph shows discharges Q over time. The hydrographs in Figs. 2.14 and 2.15 are plots of mean daily flow for each day over a period of many years. These hydrographs reflect snowmelt and numerous rain storms separated by dry periods of varying durations. The hydrograph in Fig. 8.2 focuses on an individual storm event. The hyetograph in Fig. 8.2 consists of mean rainfall intensities over each discrete time increment Δt. The hydrograph flow rates are for instantaneous points in time t.

The hydrograph is divided into direct runoff from the rain and the baseflow that would have been in the stream even if the rainfall event had not occurred. Baseflow is typically subsurface flow entering the stream from groundwater, but may also include reservoir releases, wastewater treatment plant effluent, irrigation return flows, and other inflows. Gaging stations measure the total flow without regard to its origin. Various methods have been adopted for computationally separating the direct runoff and base flow components of observed hydrographs (Singh, 1992; Tallaksen, 1995; McCuen, 1998). The simplest approach is to assume a constant baseflow equal to the discharge just before the rain began, as is done in Fig. 8.2. The direct runoff volume is reflected in Fig. 8.2 as the area between the streamflow and baseflow.

Recorded data from streamflow gaging stations are sometimes used in hydrologic engineering applications, without the need for watershed models. Runoff data from both field measurements and simulation models may be used in combination. Observed data should be used to the fullest extent possible. However, watershed models are commonly required due to insufficient gaged streamflow measurements. Typically, there simply are no gages at the locations of concern. Recently installed gaging stations may have insufficient length of record for hydrologic engineering applications. Even if a gage has been operated for many years, the observed flows may not be representative of current watershed conditions. Past flows may be nonhomogeneous due to urbanization, other land use changes, or construction of water control facilities. Precipitation-runoff modeling also facilitates predicting the effects of projected future watershed development and proposed stormwater management plans.

8.2 WATERSHED MODELS

Watershed models are mathematical representations of hydrologic processes. A variety of modeling techniques play various roles in water resources engineering applications. Models vary greatly in their levels of sophistication. The types of information they provide also vary. All models are simplifications of real-world processes and must be applied cautiously using sound engineering judgment.

An array of watershed modeling methods are available for performing a variety of different tasks. Multiple methods are used in combination to build a model for a particular application. Alternative techniques are available for performing the same task. Hydrologic engineers must have a thorough understanding of:

- each individual method, including its conceptual and empirical basis, premises and assumptions, procedures for applying the method, and its validity for various situations
- comparative advantages and disadvantages of alternative methods for different applications
- relationships between different techniques and approaches for using them in combination
- general strategies for constructing and applying watershed models

8.2.1 Precipitation

Although models may incorporate snowmelt as well as rainfall, this chapter focuses specifically on rainfall. Snow is an important component of streamflow in many regions. However, rainfall is the prevalent form of precipitation in most watershed modeling applications. The rainfall input to watershed models may be in the following forms.

- actual observed rainfall
  — past storm events
  — long continuous sequences of observations
  — real-time observations
- synthetic computed rainfall
  — design storms associated with specified exceedance probabilities (Section 7.11)
  — probable maximum storm (Section 7.12)
  — synthetically generated rainfall sequences

Past storm events are often of interest in watershed modeling studies. An analysis might involve, for example, evaluating the capabilities of a proposed stormwater management plan to handle the flood of May 1986, which is the most severe flood on record for the city. Real-time measurements of rain falling over the past several minutes and hours are used in flood forecasting systems to predict river flows to occur several hours or days in the future. Design of water control facilities and delineation of floodplains are based on synthetic design storms, which are developed in Section 7.11. Continuous models for simulating streamflows and water quality constituents over long periods of time may incorporate either long sequences of observed precipitation or synthetically generated precipitation sequences that preserve the statistical characteristics of the historical observations.

Rainfall intensities vary greatly over short distances. The average rainfall over the watershed or subbasin is input for most models. As discussed in Section 2.4.4, the Thiessen method or other methods combine rainfall measurements at several gages to obtain a spatial average over the watershed.

8.2.2 Modeling Approaches and Techniques

This chapter covers the following watershed modeling techniques:

- Watershed time of concentration and lag time estimation methods (Section 8.3)
- Peak discharge from the intensity-duration-frequency (IDF) relationship: rational method (Section 8.4)
- Runoff volumes for given rainfall depths: the National Resource Conservation Service (NRCS) curve number (CN), Green and Ampt, and Holtan methods (Section 8.5)
- Unit hydrograph approach for transforming runoff volumes to streamflow hydrographs
  - — unit hydrograph from gaged flows (Section 8.6.2)
  - — NRCS, Synder, and Clark synthetic unit hydrographs (Sections 8.6.4–8.6.6)
- Erosion and sediment yield: universal soil loss equation and sediment delivery ratios (Section 8.7)
- Pollutant loadings: the Environmental Protection Agency (EPA) rural loading functions and the U.S. Geological Survey (USGS) urban regression equations (Section 8.8)
- Computer models: Hydrologic Engineering Center (HEC)-Hydrologic Modeling System (HMS), EPA-Stormwater Management Model (SWMM), and the Agricultural Research Service (ARS)-Soil and Water Assessment Tool (SWAT) (Section 8.9)

Watershed models are constructed by selectively interconnecting the various methods covered in this chapter. All are integrally related. The rational method is an example of techniques that are typically applied manually. Most of the methods are applied in combination in generalized computer simulation packages, such as the HEC-HMS, the EPA-SWMM, and the ARS-SWAT.

Watershed models can be classified as single-event or continuous. Single-event models are designed to simulate individual storm events and have no capabilities for replenishing soil infiltration capacity and other watershed abstraction capabilities during dry periods. Continuous models simulate long periods of time that include multiple precipitation events separated by dry periods with no precipitation. This book focuses on single-event modeling techniques, but also introduces the general approach of continuous watershed models.

Models can be classified as dealing with only water quantities versus also dealing with sediment and/or other water quality constituents. The chapter covers water quantity considerations in detail prior to addressing erosion and water quality in later sections.

Watershed modeling techniques can be characterized as being, to various degrees, empirical or conceptual. The term *empirical* implies that a technique is based on observations or gaged data rather than on theoretical considerations. *Conceptual* or theoretical models are based on general principles or ideas regarding the basic processes of concern. Some of the models of this chapter consist of conceptually based equations with values for various parameters estimated from empirical information. Others are purely empirical, having been developed from analyses of field observations.

8.2.3 Model Calibration Studies

The governing equations of mathematical models incorporate parameters for which values must be provided to represent the real-world system of concern. For example, a watershed model may include loss rate parameters, such as the CN, hydraulic conductivity, suction head, and initial moisture deficit discussed in Section 8.5 and runoff timing parameters, such as basin lag, time of concentration, and Clark storage coefficient discussed in Sections 8.3 and 8.6.

Calibration is the process of estimating values for model parameters that achieve simulation results that best reproduce observed data. Calibration of a watershed model requires gaged rainfall and streamflow observations. For given rainfall input, the model parameters are adjusted until computed streamflows are as close as possible to the actual observed streamflows. This is the inverse problem of knowing the correct solution (streamflow hydrograph) and determining by trial-and-error the parameter values that allow the model to closely reproduce this solution. The calibrated model is then applied in other situations in which the hydrograph is unknown and must be computed.

Calibration and regression analyses may be combined to develop relationships between values of parameters obtained by calibration and other parameters that can be estimated based on topographic maps or field surveys. Such analyses are typically confined to selected geographical regions having somewhat uniform hydrologic characteristics. Calibration studies are performed to determine parameter values for a number of gaged watersheds in the region. For example, the Clark storage coefficient R discussed in Section 8.6.6 may be obtained for a number of gaged watersheds by calibration studies. The calibrated values of R for each watershed are then related to area, slope, land use, and/or other factors using regression analysis techniques. The resulting relationships are then used to estimate R for ungaged watersheds in the region using information available from maps and other sources.

Models should be calibrated whenever gaged observations are available for the watershed. However, in the common situation of working with ungaged watersheds, the engineer must either perform regional calibration/regression studies or rely on more generic information developed by others. Parameters are often estimated based on judgment and published empirical data.

Availability of observed data for calibration studies is a key consideration in selecting among alternative modeling techniques. Most of the techniques covered in this chapter can be applied to ungaged watersheds without calibration. However, some of the methods are highly dependent on having the data, time, and expertise required to perform calibration studies. The improved accuracy provided by more complex techniques is often dependent on calibration.

8.2.4 Multiple-Component Models

Watersheds include streams, drainage improvements, reservoirs, and other water management facilities, as well as the land and land cover on which the precipitation falls. Larger watersheds are divided into smaller more hydrologically homogeneous subwatersheds. A watershed model may have numerous, perhaps hundreds, of subwatersheds. Subdivision of watersheds serves various modeling purposes. Hydrographs are developed for various locations of concern. Subdivision into smaller more homogeneous subbasins improves modeling accuracy. *Homogeneity* means uniformity in land use, soil type, vegetation, and other characteristics. Modeling techniques developed for smaller watersheds can still be applied to each subbasin of an extremely large river basin. Runoff hydrographs from the individual subwatersheds are routed through stream reaches and reservoirs using routing methods covered in Chapter 6 and combined at appropriate locations.

The watershed of Fig. 8.3 includes five subbasins, two reservoirs, and three stream routing reaches. A hydrograph at each subbasin outlet is developed by combining rainfall (Section 7.11), hydrologic abstraction (Section 8.5), unit hydrograph (Section 8.6), or perhaps kinematic routing (Section 6.5) modeling methods. The hydrographs from subbasins 1 and 2 are routed through reservoirs (Section 6.1.1). The reservoir outflow hydrographs at stream locations A and B are

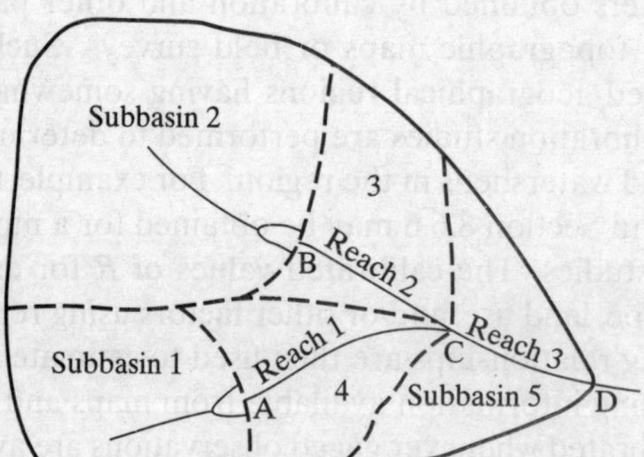

Figure 8.3 The watershed is modeled as a system of subbasins, stream reaches, and reservoirs.

routed (Section 6.1.2) to location C and combined with hydrographs from subbasins 3 and 4. The combined hydrograph at C is routed to D and combined with the hydrograph from subbasin 5. The computations are incorporated in generalized computer models (Sections 1.4 and 8.9).

On completion of this chapter, you should understand the hydrologic modeling techniques required to build each component of this watershed model. Although they may be applied to simpler problems manually or with spreadsheet software, the methods covered are most often associated with computer simulation models, such as those described in Section 8.9, which are applicable to problems that may be quite complex. A thorough understanding of the material in this chapter is absolutely essential to competently apply available generalized computer models to real-world problems. Now we address the details of these various techniques.

8.3 WATERSHED CHARACTERISTICS

Watersheds vary greatly in size, shape, topography, land use, geology, soils, vegetation, stream configuration, water control improvements, and other characteristics. As discussed throughout this chapter, various means are adopted for capturing pertinent characteristics in watershed models. This section focuses on drainage area, lag time, and time of concentration. Delineating the watershed and determining its size is a first step in modeling studies. Time of concentration and lag are parameters that reflect characteristics influencing how quickly precipitation flows off of the watershed.

8.3.1 Drainage Area

A first step in watershed modeling studies is to delineate the boundaries of the watersheds above pertinent sites. We select a point on a river, stream, or drainageway for which flow information is needed. This point of interest is the outlet for our watershed. The boundary is delineated on a topographic map by tracing the drainage divide between the watershed and adjacent watersheds. The drainage area is normally defined by the map contours as illustrated in Fig. 8.4, with the drainage divide following the ridge around the watershed and crossing the stream only at the outlet. However, storm sewers may sometimes cross under drainage divides. Thus, constructed drainage facilities, as well as topography, are considered in delineating watersheds. A watershed may also encompass areas that do not contribute runoff at the outlet. These closed subareas drain to lakes or low terrain not hydraulically connected to the watershed outlet.

The area A encompassed within the watershed boundary is measured from a map considering the map scale. The area is typically expressed in units of square kilometers (km^2), hectares (ha), square miles ($mi.^2$), or acres (ac) (Tables 2.6 and 2.7). The watershed of Fig. 8.4 has a drainage area of 2.4 $mi.^2$, 1,530 ac, 6.2 km^2, or 620 ha.

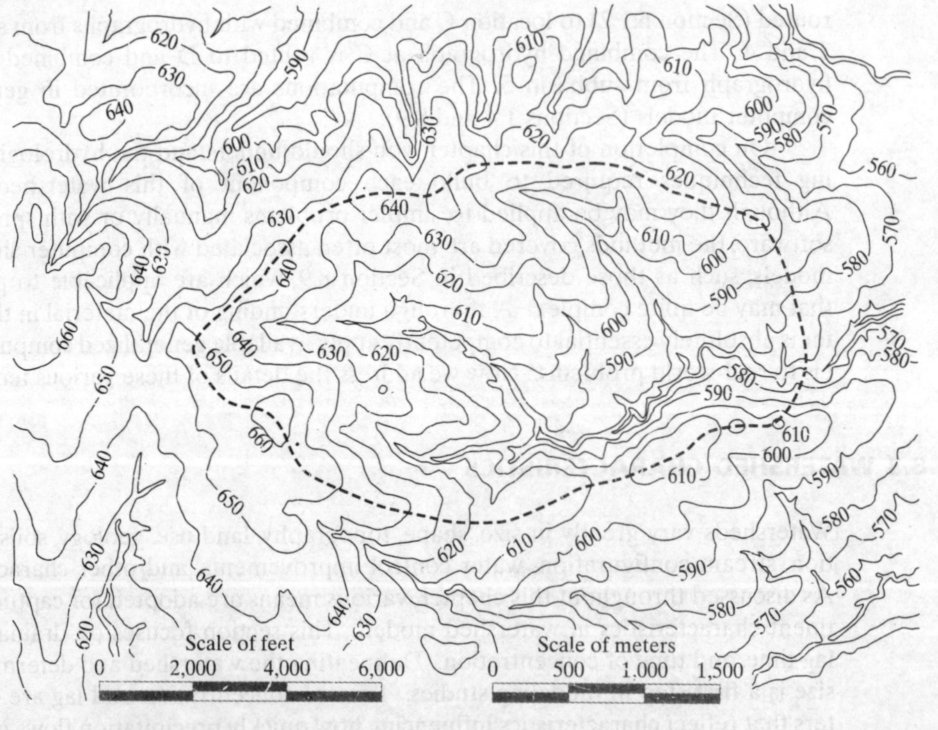

Figure 8.4 The watershed is delineated on a contour map (contour interval = 10 ft).

8.3.2 Lag Time

Lag t_L is sometimes viewed as the time between the center of mass of rainfall and center of mass of the runoff hydrograph. However, more typically, t_L is defined as the time between the center of mass of the rainfall and peak of the hydrograph, as illustrated in Fig. 8.5. The t_L can be measured directly from gaged precipitation and streamflow data. However, the primary application of t_L is as a parameter in synthetic unit hydrographs for ungaged watersheds as discussed in Section 8.6. The t_L characterizing an ungaged watershed must be estimated using information available from maps and field surveys.

Two alternative general approaches are commonly used to estimate t_L for ungaged watersheds. Many empirical lag equations such as Eqs. 8.3–8.6 have been developed based on analyses of data from gaged watersheds (McCuen, 1998). Alternatively, t_L may be estimated by combining time of concentration t_C estimates developed by methods discussed in Section 8.3.3 with a relationship between t_L and t_C, such as Eq. 8.2. The NRCS developed the following relationship based on

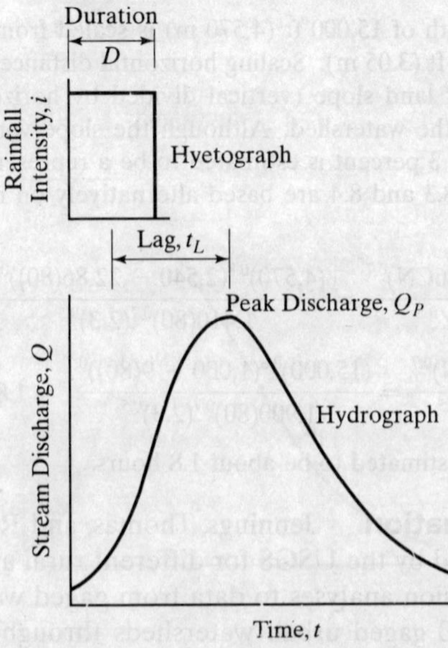

Figure 8.5 The basin lag t_L is the time from the center of mass of the rainfall to the peak of the hydrograph.

empirical analyses of numerous gaged watersheds.

$$t_L = 0.6\, t_C \tag{8.2}$$

Regardless of the method adopted, estimates of t_L and t_C for ungaged watersheds are necessarily approximate and require considerable engineering judgment.

8.3.2.1 NRCS lag equation. The NRCS developed the following equation for watersheds with areas of less than about 8 km^2 (2,000 ac) and CN between 50 and 95 (NRCS, 1985; Haan, Barfield, and Hayes, 1994; McCuen, 1998). Equations 8.3 and 8.4 are English (l in ft) and metric (l in m) versions of the NRCS lag formula.

$$t_L = \frac{l^{0.8}(1{,}000 - 9\mathrm{CN})^{0.7}}{1{,}900\mathrm{CN}^{0.7}Y^{0.5}} \tag{8.3}$$

$$t_L = \frac{l^{0.8}(2{,}540 - 22.86\mathrm{CN})^{0.7}}{1{,}410\mathrm{CN}^{0.7}Y^{0.5}} \tag{8.4}$$

The lag t_L is in hours. The hydraulic length l from the outlet to the most hydraulically remote point in the watershed is in feet (Eq. 8.3) or meters (Eq. 8.4). CN is discussed in Section 8.5.1. Y is the average land slope of the watershed in percent.

Example 8.1

Estimate the lag time for the watershed of Fig. 8.4. The soil and vegetative characteristics of the watershed are represented by a CN of 80.

Solution A hydraulic length of 15,000 ft (4,570 m) is scaled from Fig. 8.4. The map has a contour interval of 10 ft (3.05 m). Scaling horizontal distances between the contours provides estimates of land slope (vertical divided by horizontal distances) at various points throughout the watershed. Although the slope varies significantly at different locations, a Y of 2.3 percent is estimated to be a representative average for the watershed. Equations 8.3 and 8.4 are based alternatively on metric and English units, respectively.

$$t_L = \frac{l^{0.8}(2{,}540 - 22.86\text{CN})^{0.7}}{1{,}410\text{CN}^{0.7}Y^{0.5}} = \frac{(4{,}570)^{0.8}(2{,}540 - 22.86(80))^{0.7}}{1{,}410(80)^{0.7}(2.3)^{0.5}} = 1.83 \text{ hrs}$$

$$t_L = \frac{l^{0.8}(1{,}000 - 9\text{CN})^{0.7}}{1{,}900\text{CN}^{0.7}Y^{0.5}} = \frac{(15{,}000)^{0.8}(1{,}000 - 9(80))^{0.7}}{1{,}900(80)^{0.7}(2.3)^{0.5}} = 1.83 \text{ hrs}$$

The watershed lag time is estimated to be about 1.8 hours.

8.3.2.2 USGS lag equation. Jennings, Thomas, and Riggs (1994) summarize lag equations developed by the USGS for different rural and urban regions of the U.S. by applying regression analyses to data from gaged watersheds. Based on regression analyses for 170 gaged urban watersheds throughout the U.S., the USGS developed the following equation for estimating t_L for urban watersheds nationwide.

$$t_L = 0.003L^{0.71}(13 - BDF)^{0.34}(ST + 10)^{2.53}\, RI2^{-0.44}\, IA^{-0.20}\, SL^{-0.14} \tag{8.5}$$

Lag t_L is in hours. L in miles is the length from the outlet along the main channel and its extension to the watershed divide. Storage ST is the percentage of the watershed occupied by lakes, reservoirs, and wetlands. $RI2$ is the rainfall in inches for a duration of 2 hours and recurrence interval of 2 years. IA is the percentage of the watershed covered by impervious surfaces. SL is the slope of the main channel in feet per mile. The basin development factor (BDF) in Eq. 8.5 is also incorporated in Eqs. 7.39–7.45 and is defined in Section 7.8.2. BDF is an index of the prevalence of urban drainage improvements.

8.3.2.3 Snyder lag equation. As discussed in Section 8.6.5, Snyder (1938) developed a synthetic unit hydrograph based on a study of watersheds in the Appalachian Mountains region with drainage areas ranging from 26 to 26,000 km^2 (10 to 10,000 mi.^{2}). Snyder's unit hydrograph procedure includes the lag equation

$$t_L = C_t(LL_c)^{0.3} \tag{8.6}$$

where lag t_L is in hours, L is the length of the main stream from outlet to divide, and L_c is the length of the main stream from outlet to a point nearest the watershed centroid. C_t is a coefficient that accounts for the slope, land use, and associated storage characteristics of the river basin. With distances L and L_c in miles, Snyder found values of C_t to range from 1.8 to 2.2, with a mean of 2. With distances L and L_c in kilometers, the corresponding range of C_t is 1.35–1.65, with a mean of 1.5.

Values of C_t have since been found to vary significantly outside of these ranges in very mountainous or very flat terrain. Regional studies of gaged watersheds are performed to relate C_t to various parameters that can be measured from maps and field surveys, such as slope and percent impervious.

8.3.3 Time of Concentration

Time of concentration t_C is a parameter used in the rational method (Section 8.4) and Clark unit hydrograph (Section 8.6.6). The t_C is defined as the time required for rain falling at the hydraulically most remote location in the watershed to reach the outlet. It represents the time required after the beginning of a rainfall event for the entire watershed to be contributing runoff at the outlet. Time of concentration is a somewhat abstract concept and is not subject to direct measurement. Estimates of t_C are approximate and require subjective judgement.

Common alternative strategies for estimating t_C involve (1) transforming a t_L to t_C and (2) determining travel times along a flow path. As previously noted, the NRCS (1986) developed the following approximate relationship based on analyses of data from gaged watersheds.

$$t_C = \frac{5}{3} t_L \tag{8.7}$$

Thus, t_C can be estimated by combining Eq. 8.7, with the t_L determined using empirical equations such as those in Section 8.3.2. For example, in Example 8.1, a lag time of 1.8 hours was estimated for the watershed of Fig. 8.4. The corresponding t_C is 5/3 of 1.8 hrs = 3.0 hours.

The other approach for determining t_C is based on travel times along a flow path traced from the outlet to the most hydraulically remote location in the watershed. This is the point from which runoff requires the longest time to reach the outlet. The flow path may include overland flow and concentrated flow in various types of conveyance. For example, runoff from rainfall may begin as overland flow, then concentrate into rills and swales before entering a storm sewer pipe that discharges into a stream that then flows to the watershed outlet. The t_C is the sum of travel time T_i in each component segment of the overall flow path.

$$t_C = \sum T_i \tag{8.8}$$

Travel time T is a function of flow length L and velocity V.

$$T = \frac{L}{V} \tag{8.9}$$

The flow path is drawn on a topographic map, and L for each segment is measured from the map. The method for estimating V depends on the type of flow. V for channelized flow is estimated using hydraulic analysis techniques from Chapter 5, based on some assumption regarding flow depth. For example, the average velocity for a length of stream may be estimated using the Manning equation with a representative cross-section assuming the channel flowing full to the top of banks.

TABLE 8.1 COEFFICIENT k IN EQ. 8.10

Land use	n	R, ft	k
Forest			
Dense underbrush	0.8	0.25	0.7
Light underbrush	0.4	0.22	1.4
Heavy ground liter	0.2	0.20	2.5
Grass			
Bermuda grass	0.41	0.15	1.0
Dense	0.24	0.12	1.5
Short	0.15	0.10	2.1
Rangeland	0.13	0.04	1.3
Grassed waterway	0.095	1.0	15.7
Small upland gullies	0.04	0.5	23.5
Paved area (sheet flow)	0.011	0.06	20.8
Paved area (sheet flow)	0.025	0.2	20.4
Paved gutter	0.011	0.2	46.3

Many empirical equations have been developed for estimating either T or V for overland or sheet flow (McCuen, 1998).

The Manning equation (Section 5.3) may also be applied to estimate velocities for overland flow. The Manning equation may be simplified to

$$V = kS^{0.5} \tag{8.10}$$

where V is the velocity in ft/s, S is the slope in percent, and $k = 1.486R^{2/3}/n$. The hydraulic radius R and roughness coefficient n are defined in Section 5.3. The coefficient k for various types of land cover are provided in Table 8.1 based on estimates of typical values for R and n (NRCS, 1986; McCuen, 1998).

8.4 RATIONAL METHOD FOR ESTIMATING PEAK FLOW

Many applications require estimates of peak discharge only, rather than an entire hydrograph. For example, storm sewers, drainage ditches, and highway culverts are sized for a peak discharge associated with a specified recurrence interval. The rational method proposed over a century ago (Kuichling, 1889) continues to be the most commonly used method for determining peak discharges for designing drainage facilities for small watersheds. The rational formula is

$$Q_p = CiA \tag{8.11}$$

where Q_p, i, and A denote peak discharge, rainfall intensity, and drainage area, and C is a dimensionless runoff coefficient ($0 \le C \le 1.0$). Q_p, i, and A can be expressed in any consistent set of units, or conversion factors can be incorporated as appropriate. For example, with Q_p in m^3/s, i in cm/hr, and A in m^2,

$$Q_p = CiA\left(\frac{1\text{ hr}}{3{,}600\text{ s}}\right)\left(\frac{1\text{ m}}{100\text{ cm}}\right) \tag{8.12}$$

With Q_p in ft³/s, i in inches/hour, and A in acres, the formula is

$$Q_p = CiA\left(\frac{1.0083\ \text{ft}^3/\text{s}}{\text{acre-in./hr}}\right) \tag{8.13}$$

With this common set of units, the conversion factor 1.0083 is often omitted because it is so close to one.

The rainfall intensity is typically determined from an IDF relationship for a specified exceedance frequency and duration equal to the time of concentration for the watershed. The coefficient C is estimated based on tables, such as Table 8.2, which gives ranges of C alternatively (1) for particular types of land use or (2) by type of surface that can be used to develop an area-weighted composite C (American Society of Civil Engineers/Water Environment Federation, 1992; American Society of Civil Engineers, 1996). The value for C increases with rainfall intensity (recurrence interval) and watershed slope and decreases with infiltration capacity. Table 8.2 provides typical values for C for storms with recurrence intervals of 2 to 10 years. C may reasonably be increased by 10 percent, 20 percent, and 25 percent, respectively, for 20-, 50-, and 100-year storms. Similar information is also provided in Table 10.1. C is a ratio of the peak runoff rate to the rainfall rate and has an upper limit of 1.0. A C of 1.0 means that all of the rainfall runs off the watershed with zero hydrologic abstraction.

The rational method is based on the following premises. The rainfall has a constant intensity i for a duration equal to the time of concentration t_C of the watershed. An equilibrium condition is reached with the outflow CiA from the watershed being equal to a proportion C of the inflow iA. The rationale is as follows. After a uniform rainfall intensity continues for a duration of time equal to t_C, the

TABLE 8.2 RATIONAL METHOD COEFFICIENT C FOR 2- TO 10-YEAR RECURRENCE INTERVALS

Composite area	C	Character of surface	C
Business		Pavement	
Downtown	0.70–0.95	Asphalt and brick	0.70–0.95
Neighborhood	0.50–0.70	Brick	0.70–0.85
Residential		Roofs	0.75–0.95
Single-family	0.30–0.50	Lawns, sandy soil	
Multi-units, detached	0.40–0.60	Flat, 2% slope	0.05–0.10
Multi-units, attached	0.60–0.75	Average, 2–7%	0.10–0.15
Residential (suburban)	0.25–0.40	Steep, 7%	0.15–0.20
Apartment	0.50–0.70	Lawns, heavy soil	
Industrial		Flat, 2% slope	0.13–0.17
Light	0.50–0.80	Average, 2–7%	0.18–0.22
Heavy	0.60–0.90	Steep, 7%	0.25–0.35
Parks, cemeteries	0.10–0.25	Water impoundment	1.00
Playgrounds	0.20–0.35		
Railroad yard	0.20–0.35		
Unimproved	0.10–0.30		

entire watershed is contributing runoff at the outflow. The volumetric inflow rate of the rain of intensity i falling on an area A is iA. At equilibrium, the outflow CiA at the outlet is a fraction C of the inflow iA.

The assumptions of constant rainfall intensity and equilibrium between inflow and outflow are reasonably valid only for small watersheds with a short t_C. The assumption of a constant C is most realistic for a C close to one. The runoff from a concrete or asphalt parking lot is close to being a constant percentage of the rainfall. However, for a watershed with sandy soil covered with grass and trees, the percentage of the rainfall that runs off will increase greatly during the duration of the rainfall event as the ground becomes more saturated. The maximum drainage area for which the rational method is applicable is subject to judgment. Drainage criteria manuals and hydrology books provide varying guidelines with the upper limit on drainage area ranging from less than 0.04 km^2 (10 ac) to 12 km^2 (4.6 mi.^{2}).

Example 8.2

Estimate the 25-year recurrence interval peak discharge for the watershed in Fig. 8.4 that is located in Dallas County, Texas. The watershed has an area A of 620 hectares (1,530 ac) and a runoff coefficient C of 0.60.

Solution In Example 8.1, the t_L was estimated to be 1.8 hours using Eq. 8.3, which can be transformed to a t_C of 3.0 hours using Eq. 8.7. The rainfall duration t is set equal to the t_C of 3.0 hours or 180 minutes. Eq. 7.50 and Table 7.6 are used to determine the rainfall intensity for a 25-year recurrence interval.

$$i = \frac{a}{(t + b)^c} = \frac{90}{(180 + 8.7)^{0.774}} = 1.56 \text{ in./hr } (3.96 \text{ cm/hr})$$

The rational formula is applied alternatively with metric and English units.

$$Q_p = CiA \text{ (conversion factors)}$$

$$Q_p = 0.60(3.96 \text{ cm/hr})(620 \text{ ha})\left(\frac{\text{m}}{100 \text{ cm}}\right)\left(\frac{10{,}000 \text{ m}^2}{\text{ha}}\right)\left(\frac{\text{hr}}{3{,}600 \text{ s}}\right) = 41 \text{ m}^3/\text{s}$$

$$Q_p = 0.60(1.56 \text{ in./hr})(1{,}530 \text{ ac})\left(\frac{\text{ft}}{12 \text{ in.}}\right)\left(\frac{43{,}560 \text{ ft}^2}{\text{acre}}\right)\left(\frac{\text{hr}}{3{,}600 \text{ s}}\right) = 1{,}400 \text{ ft}^3/\text{s}$$

Example 8.3

Use the rational method to determine the 10-year Q_P for the watershed in Fig. 8.6, which is located in Houston (Harris County), Texas. The parking lot and park both slope toward a 100-meter-long swale running between them. Rain falling on the parking lot and park drains as overland flow to the swale and then flows to the watershed outlet. Adjacent land drains elsewhere. Mean flow velocities are 0.8, 0.3, and 1.0 m/s for the parking lot, park, and swale, respectively. Runoff coefficients C are 0.9 and 0.25 for the concrete parking lot and grass park, respectively.

Solution The watershed area is

$$A = (100 \text{ m})(100 \text{ m}) = 10{,}000 \text{ m}^2$$

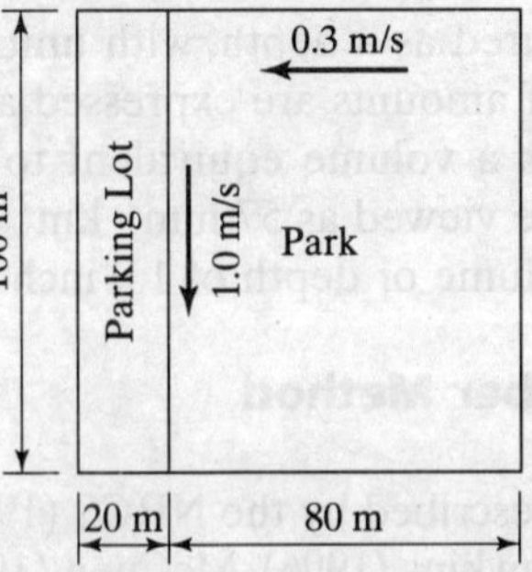

Figure 8.6 The watershed for Example 8.3 consists of a parking lot and a park.

that includes 2,000 m^3 for the parking lot (20 percent of total) and 8,000 m^3 of park. A composite C is computed as an average weighted in proportion to area.

$$C = 20\%(0.9) + 80\%(0.25) = 0.38$$

The most hydraulically remote point in the watershed is the northeast corner of the park. The flow path for determining t_C includes 80 m of overland flow across the park and flow for 100 m in the swale.

$$t_C = T_{\text{park}} + T_{\text{swale}} = \frac{80 \text{ m}}{0.3 \text{ m/s}} + \frac{100 \text{ m}}{1.0 \text{ m/s}} = 367 \text{ s} = 6.1 \text{ min}$$

Eq. 7.50 and Table 7.6 are used to determine the rainfall intensity for a 10-year recurrence interval.

$$i = \frac{a}{(t+b)^c} = \frac{81}{(6.1 + 7.7)^{0.753}} = 11.2 \frac{\text{in.}}{\text{hr}} \left(28.5 \frac{\text{cm}}{\text{hr}}\right)$$

The rational formula is applied to determine Q_p.

$$Q_p = CiA \text{ (conversion factors)}$$

$$Q_p = (0.38)\left(28.5 \frac{\text{cm}}{\text{hr}}\right)\left(10{,}000 \text{ m}^2\right)\left(\frac{\text{m}}{100 \text{ cm}}\right)\left(\frac{\text{hr}}{3{,}600 \text{ s}}\right) = 0.30 \frac{\text{m}^3}{\text{s}}$$

8.5 SEPARATING PRECIPITATION INTO ABSTRACTIONS AND RUNOFF

Hydrologic abstractions include interception, surface storage, infiltration, and evapotranspiration. The terms runoff and rainfall excess are used synonymously to refer to that portion of the precipitation that reaches the watershed outlet. Precipitation P is the input to the watershed. Rainfall excess or runoff is the water remaining after accounting for losses due to hydrologic abstractions.

$$\text{runoff} = \text{precipitation} - \text{abstractions} \tag{8.14}$$

Precipitation is measured as a depth, with units of mm, cm, or inches. Likewise, abstraction and runoff amounts are expressed as depths. A rainfall excess or runoff volume of 57 mm is a volume equivalent to covering the watershed to a depth of 57 mm and may be viewed as 57 mm · km^2 of runoff per km^2 of drainage area. Likewise, a runoff volume or depth of 1.0 inch is equivalent to 1.0 in. · ac/ac.

8.5.1 NRCS Curve Number Method

The NRCS CN method is described by the NRCS (1985, 1986), Haan, Barfield, and Hayes (1994), Ponce and Hawkins (1996), McCuen (1998), and others. A watershed is characterized by a single parameter called the curve number CN. Information developed by the NRCS based on analyses of gaged watersheds facilitate CN estimates. The method is widely used due to its simplicity and the availability of empirical information on which to base estimates of the CN. The basic set of equations is

$$V_R = \frac{(P - 0.2S)^2}{P + 0.8S} \quad \text{for} \quad P \geq I_A = 0.2S \tag{8.15}$$

$$V_R = 0 \quad \text{for} \quad P \leq I_A = 0.2S \tag{8.16}$$

$$S = \frac{2{,}540}{\text{CN}} - 25.4 \quad \text{for} \quad V_R, P, S \text{ in cm} \tag{8.17}$$

$$S = \frac{1{,}000}{\text{CN}} - 10 \quad \text{for} \quad V_R, P, S \text{ in inches} \tag{8.18}$$

V_R is the runoff volume (rainfall excess) to result from precipitation P. S is the maximum potential abstraction after runoff begins. I_A is the initial abstraction before runoff begins. V_R, P, S, and I_A have units of cm or inches. CN is a dimensionless number between zero and 100. Equations 8.15 and 8.16 are valid for any consistent units of depth. Equations 8.17 and 8.18 are in terms of cm and inches, respectively. With a watershed characterized by a value for CN, Eqs. 8.15–8.18 allow V_R to be computed for a known P.

Example 8.4

Estimate the runoff that would result from 3.7 inches (9.4 cm) of rain falling on a watershed characterized by a CN of 78.

Solution

$$S = \frac{1{,}000}{\text{CN}} - 10 = \frac{1{,}000}{78} - 10 = 2.82 \text{ in.}$$

$$V_R = \frac{(P - 0.2S)^2}{P + 0.8S} = \frac{(3.7 - 0.2(2.82))^2}{3.7 + 0.8(2.82)} = 1.7 \text{ in.}$$

$$S = \frac{2{,}540}{\text{CN}} - 25.4 = \frac{2{,}540}{78} - 25.4 = 7.16 \text{ cm}$$

$$V_R = \frac{(P - 0.2S)^2}{P + 0.8S} = \frac{(9.4 - 0.2(7.16))^2}{9.4 + 0.8(7.16)} = 4.2 \text{ cm}$$

The NRCS CN method plays roles in various types of watershed models. Examples 7.11, 8.5, 8.8, and 8.9 illustrate a series of component tasks involved in developing a design hydrograph. In combination, these examples consist of developing a 50-year recurrence interval hydrograph for a particular watershed. A 50-year recurrence interval, 24-hour duration design storm was developed in Example 7.11. In Example 8.5, the runoff for each increment of rainfall is determined. As discussed in Section 8.6, in Example 8.9, these increments of runoff are combined with a unit hydrograph developed in Example 8.8 to develop a discharge hydrograph associated with a 50-year recurrence interval.

Example 8.5

A design storm was developed in Example 7.11. This storm occurs over a watershed characterized by a CN of 80. Estimate the runoff that results from each of the twelve 2-hour increments of rainfall.

Solution The design storm from Example 7.11 is reproduced in the first three columns of the following table. Given P in column 3, V_R is determined using Eqs. 8.15 and 8.16 and recorded in column 4. The P and V_R in Eqs. 8.15 and 8.16 are the cumulative total rainfall depth since the storm began and the corresponding runoff. The ΔV_R in column 5 represents the runoff from each 2-hour increment of rainfall and is computed as the difference between successive depths in column 4. The ΔV_R in column 5 are required for the unit hydrograph computations in Example 8.9.

$$S = \frac{2{,}540}{\text{CN}} - 25.4 = \frac{2{,}540}{80} - 25.4 = 6.35 \text{ cm}$$

$$V_R = \frac{(P - 0.2S)^2}{P + 0.8S} = \frac{(P - 0.2(6.35))^2}{P + 0.8(6.35)} = \frac{(P - 1.27)^2}{P + 5.08}$$

Time (hr) (1)	Rainfall depth, ΔP (cm) (2)	Rainfall depth, P (cm) (3)	Runoff depth, V_R (cm) (4)	Incremental ΔV_R (cm) (5)
0		0.00	0.00	
2	0.48	0.48	0.00	0.00
4	0.58	1.06	0.00	0.00
6	0.72	1.78	0.04	0.04
8	0.95	2.73	0.27	0.24
10	1.57	4.30	0.98	0.70
12	12.13	16.43	10.68	9.70
14	2.46	18.89	12.96	2.27
16	1.18	20.07	14.06	1.10
18	0.82	20.89	14.82	0.76
20	0.63	21.52	15.42	0.60
22	0.53	22.05	15.92	0.50
24	0.45	22.50	16.35	0.43

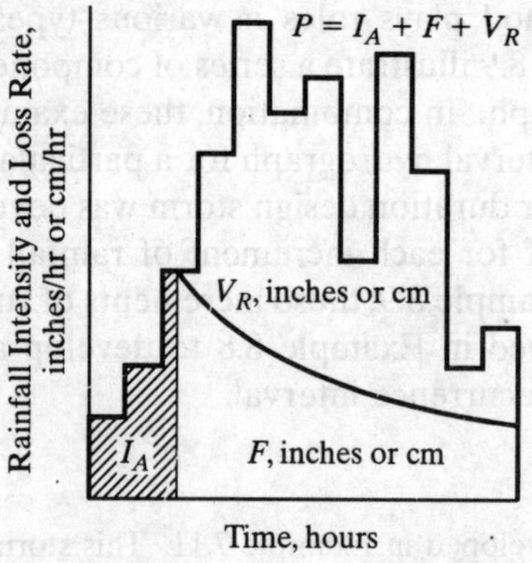

Figure 8.7 The NRCS rainfall–runoff relationship divides precipitation into runoff and abstractions.

The terms in the NRCS relation are illustrated by Fig. 8.7. Rainfall and loss rates are plotted versus time for a rainfall event. The total precipitation depth P is the area under the plot of rainfall intensity. P is divided into I_A, F, and V_R. The initial abstraction I_A is the area under the rainfall intensity curve at the beginning of the rainfall event when all the precipitation is lost through interception, surface storage, infiltration, and other abstractions. I_A ends when runoff begins. F is the area under the loss rate curve after runoff begins. The total abstraction is the sum of I_A and F. The rainfall excess V_R is the area under the rainfall intensity plot remaining after accounting for I_A and F. The rate of infiltration and other losses decreases over time with increasing soil moisture. The watershed parameter S is a somewhat abstract concept. S is the value for F assuming the rainfall continues forever. S is the loss, other than I_A, that would occur if rain fell long enough to saturate the watershed such that losses no longer occur.

The NRCS rainfall–runoff relationship is expressed conceptually as

$$\frac{F}{S} = \frac{V_R}{P - I_A} \tag{8.19}$$

Though not subject to theoretical derivation, this equation has a degree of intuitive appeal. At the beginning of the rainfall event, both F/S and $V_R/(P - I_A)$ are zero. For a long duration of rainfall, F/S and $V_R/(P - I_A)$ both approach one. Substituting $F = P - I_A - V_R$ into Eq. 8.19 and rearranging yields

$$V_R = \frac{(P - I_A)^2}{P - I_A - S} \tag{8.20}$$

Based on analyses of data from numerous gaged watersheds, the NRCS determined that I_A typically is about 20 percent of S.

$$I_A = 0.2S \tag{8.21}$$

Equation 8.21 is substituted into Eq. 8.20 to obtain Eq. 8.15. Alternatively, an engineer may obtain I_A from a calibration study and work directly with Eq. 8.20.

Working with a dimensionless CN with $0 \leq \text{CN} \leq 100$ is more convenient than working directly with the watershed parameter S in cm or inches. Thus, Eq. 8.18 was developed as the definition of CN. A CN of 100 corresponds to S of zero, meaning all precipitation runs off with no watershed retention. A small concrete parking lot may have a CN near 100. The NRCS adopted the term *curve number* because the rainfall–runoff equation may be expressed graphically with curves for different values of CN.

For gaged watersheds, the parameters CN (or S) and I_A may be determined by calibration. However, Eqs. 8.15–8.18 are commonly applied to ungaged watersheds with the single parameter CN being estimated using information from tables like Table 8.3 developed by the NRCS. Tables covering a much broader range of agricultural land use are also available (NRCS, 1985; McCuen, 1998).

The CN depends on soil characteristics, land cover, and antecedent moisture conditions. Information on local soils is available from various sources, including published NRCS county soil surveys. Table 8.3 reflects the standard NRCS soil classification system consisting of four groups (A, B, C, and D).

- *Group A* soils have low runoff potential and high infiltration rates (greater than 0.76 cm/hr) and consist primarily of deep well-drained sands and gravel.
- *Group B* soils have moderate infiltration rates (0.38–0.76 cm/hr) and consist primarily of moderately fine to moderately coarse textured soils, such as loess and sandy loam.
- *Group C* soils have low infiltration rates (0.127–0.38 cm/hr) and consist of clay loam, shallow sandy loam, and clays.
- *Group D* soils have high runoff potential and low infiltration rates (less than 0.127 cm/hr) and consist primarily of clays with a high swelling potential, soils with a permanent high water table, or shallow soils over nearly impervious material.

CNs are provided in Table 8.3 for each soil group for various types of land use. The CN tables are developed from analyses of data from gaged rural watersheds. CNs for urban areas are computed based on average percent impervious. For example, Table 8.3 assigns a CN of 80 for a residential area with an average lot size of 0.5 acre (0.2 ha), which corresponds to 25 percent impervious cover (roofs, driveways, and streets), for hydrologic soil group C. The composite CN of 80 is computed as a weighted average with 25 percent of the watershed being impervious with CN of 98 and the 75 percent pervious area having a CN of 74 given in Table 8.3.

$$\text{composite CN} = 0.25(98) + 0.75(74) = 80$$

The runoff from a particular rainfall event depends on the moisture already in the soil from previous rainfall. The NRCS defines three antecedent moisture conditions.

TABLE 8.3 RUNOFF CN FOR ANTECEDENT MOISTURE CONDITION II

Land use description	Hydrologic condition	Hydrologic soil group A	B	C	D
Fallow, straight row, or bare soil		77	86	91	94
Pasture or range	Poor—less than 25% ground cover density	68	79	86	89
	Fair—between 25% and 50% ground cover density	49	69	79	84
	Good—more than 50% ground cover density	39	61	74	80
Brush	Poor—less than 25% ground cover density	48	67	77	83
	Fair—between 25% and 50% ground cover density	35	56	70	77
	Good—more than 50% ground cover density	30	48	65	73
Woods	Poor—less than 25% ground cover density	45	66	77	83
	Fair—between 25% and 50% ground cover density	36	60	73	79
	Good—more than 50% ground cover density	25	55	70	77
Farmsteads		59	74	82	86
Fully developed urban areas (vegetation established)					
Lawns, open spaces, parks, golf courses, cemeteries, etc.					
Good condition; grass cover on 75% or more of the area		39	61	74	80
Fair condition; grass cover on 50%–75% of the area		49	69	79	84
Poor condition; grass cover on 50% or less of the area		68	79	86	89
Paved parking lots, roofs, driveways		98	98	98	98
Streets and roads					
Paved with curbs and storm sewers		98	98	98	98
Gravel		76	85	89	91
Dirt		72	82	87	89
Paved with open ditches		83	89	92	93

Land use description	Average % impervious	A	B	C	D
Commercial and business areas	85	89	92	94	95
Industrial districts	72	81	88	91	93
Residential with average lot size of					
1/8 acre or less	65	77	85	90	92
1/4 acre	38	61	75	83	87
1/3 acre	30	57	72	81	86
1/2 acre	25	54	70	80	85
1 acre	20	51	68	79	84
2 acres	12	46	65	77	82
Developing urban area, newly graded area with no vegetation established		77	86	91	94
Western desert urban areas					
Natural desert landscaping (pervious area only)		63	77	85	88
Artificial desert landscaping		96	96	96	96

Condition I—Soils are dry but not to wilting point

Condition II—Average conditions

Condition III—Heavy rainfall or light rainfall with low temperatures have occurred within the last 5 days saturating the soil.

Table 8.3 provides CN(II) for average antecedent moisture conditions (condition II). CN(I) and CN(III) for antecedent conditions I and III can be estimated by

$$CN(I) = \frac{4.2CN(II)}{10 - 0.058(II)} \tag{8.22}$$

$$CN(III) = \frac{23CN(II)}{10 + 0.13CN(II)} \tag{8.23}$$

8.5.2 General Framework for Using Infiltration Formulas to Convert Precipitation to Runoff

Prior to discussing the Green and Ampt (1911) and Holtan (1965) infiltration equations, we will outline a general strategy for determining the runoff to result from a specified precipitation amount. The Green and Ampt, Holtan, or other alternative loss rate equations may be incorporated into the general framework.

Precipitation P, infiltration F, and precipitation excess V_R are cumulative depths with typical units of cm, mm, or inches. Incremental infiltration ΔF during time interval Δt has the same units. Rainfall intensity i, infiltration capacity f_C, and the actual infiltration rate f are rates in depth/time units, such as cm/hr, mm/hr, or in./hr.

If the rainfall intensity i is greater than the capacity f_C of the ground to accept water, surface ponding occurs and the actual infiltration f is at capacity f_C. If i is less than f_C, infiltration f is limited by the available supply of rainfall i. Thus, the rate f at which water enters the soil is the lesser of i or f_C.

$$\text{If} \quad i > f_C \quad \text{then} \quad f = f_C \tag{8.24}$$

$$\text{If} \quad i \le f_C \quad \text{then} \quad f = i \tag{8.25}$$

An algorithm for converting P to V_R steps through time with iterative computations being performed for each time step. The increase in F or increment ΔF during a time interval Δt is determined as a function of the mean infiltration rate f during Δt.

$$\Delta F = f\Delta t \tag{8.26}$$

$$F_{t+\Delta t} = F_t + \Delta F \tag{8.27}$$

However, the mean infiltration depends on f_C and F at both the beginning and end of the Δt. Thus, an iterative algorithm is required.

Several infiltration equations express infiltration capacity f_C (depth/time) as a function of cumulative infiltration depth F. The Green and Ampt equation is of the form

$$f_C = A + \frac{B}{F} \tag{8.28}$$

where A and B are computed as a function of the model parameters. In the next section, we use the Green and Ampt equation to determine the constants A and B in Eq. 8.28 and then combine Eqs. 8.24 through 8.28 into an algorithm for converting P to V_R. The Holtan and other alternative infiltration equations may also be used for Eq. 8.28 in the general procedure.

8.5.3 Green and Ampt Infiltration Model

Green and Ampt (1911) developed an infiltration model based on Darcy's law that continues to be applied in various situations including watershed modeling. The Green and Ampt equation is advantageous compared with the NRCS CN method from the perspective of being theoretically derived with physically based parameters. However, estimating accurate parameter values for ungaged watersheds for the Green and Ampt equation is more difficult than estimating CN values. Also, the Green and Ampt equation is limited to infiltration, whereas the CN method includes other abstractions as well.

The Green and Ampt equation is

$$f_C = K\left(\frac{\Psi\Delta\theta}{F} + 1\right) \tag{8.29}$$

where f_C is the infiltration capacity, K is the saturated hydraulic conductivity, Ψ is the suction head at the wetting front, $\Delta\theta$ is the change in moisture content across the wetting front, and F is the cumulative infiltration depth. Any consistent set of units can be used, such as cm/hr, cm/hr, cm, and cm for f, K, Ψ, and F, respectively. $\Delta\theta$ is dimensionless. The hydraulic conductivity K is the flow rate through a unit area under a unit hydraulic gradient. The suction head Ψ is the part of the total energy head due to suction forces binding water to soil particles through surface tension.

Rawls, Brackensick, and Miller (1983) present typical values for n, θ_e, Ψ, and K for various soil types that are summarized in Table 8.4. These and similar

TABLE 8.4 TYPICAL VALUES FOR THE GREEN AND AMPT MODEL (RAWLS, BRACKENSICK, AND MILLER, 1983)

Soil texture	Effective Porosity n	Suction porosity θ_e	Hydraulic head Ψ (cm)	Hydraulic conductivity K (cm/hr)
Sand	0.437	0.417	4.95	11.78
Loamy sand	0.437	0.401	6.13	2.99
Sandy loam	0.453	0.412	11.01	1.09
Loam	0.463	0.434	8.89	0.34
Silt loam	0.501	0.486	16.68	0.65
Sandy clay loam	0.398	0.330	21.85	0.15
Clay loam	0.464	0.309	20.88	0.10
Silty clay loam	0.471	0.432	27.30	0.10
Sandy clay	0.430	0.321	23.90	0.06
Silty clay	0.479	0.423	29.22	0.05
Clay	0.475	0.385	31.63	0.03

published information may be used to estimate approximate values for model parameters. More accurate parameters values may be obtained by calibration studies subject to availability of gaged rainfall and runoff data.

Soil consists of a matrix of particles and voids or pore spaces occupied by air and water or possibly completely saturated with water. The following dimensionless terms are pertinent to the derivation of the Green and Ampt model.

n = porosity = volume of voids/total volume

θ = soil moisture content = volume of water/total volume $\quad \Delta\theta \le \theta \le n$

θ_i = initial moisture content prior to rain

θ_r = residual moisture content after soil has been freely drained

θ_e = effective porosity = $n - \theta_r$

S_e = effective saturation = ratio of available moisture content to maximum possible = $(\theta - \theta_r)/(n - \theta_r) \quad 0 \le S_e \le 1.0$

$\Delta\theta$ = change in moisture content across wetting front = $n - \theta_i$

$$\Delta\theta = (1 - S_e)\theta_e \tag{8.30}$$

The Green and Ampt equation is based on the following assumptions: (1) vertical *piston* or *slug* flow, (2) a wetting front or sharp moving boundary between saturated soil and soil yet unaffected by infiltration, and (3) a uniform moisture content in the soil below the wetting front. These assumptions are significant approximations. Flow through saturated soil is governed by Darcy's law, which is explored further in Chapter 9.

$$q = \frac{Q}{A} = K\frac{\Delta h}{L} \tag{8.31}$$

where q is the flow velocity; Q is the discharge rate through area A; K is the hydraulic conductivity; and $\Delta h/L$ is the hydraulic gradient, with Δh being the change in head across the flow length L.

The derivation of the Green and Ampt equation is illustrated by Fig. 8.8. A sharp wetting front is assumed to move downward as the soil is saturated by infiltration. The soil above the wetting front is saturated ($\theta = n$). The soil below the wetting front is still at the same moisture content as before the rainfall began

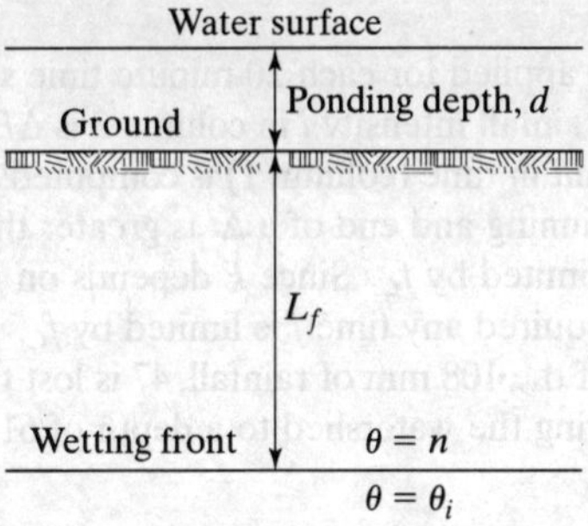

Figure 8.8 The Green and Ampt model is based on the concept of a downward moving wetting front.

($\theta = \theta_i$). Assuming ponding at the surface ($i \geq f$), Darcy's law (Eq. 8.31) can be expressed as

$$f = K\left(\frac{d + L_f + \Psi}{L_f}\right) \tag{8.32}$$

where $(d + L_f + \Psi)$ is the head difference Δh between the water surface and wetting front, and L_f is the depth to the wetting front. If the ponding depth d is assumed negligible, Δh is $L_f + \Psi$.

$$f = K\left(\frac{L_f + \Psi}{L_f}\right) \tag{8.33}$$

The cumulative amount of water that has infiltrated into the soil by time t is F.

$$F = L_f\Delta\theta \quad \text{or} \quad L_f = \frac{F}{\Delta\theta} \tag{8.34}$$

where $\Delta\theta$ is the change in moisture content across the wetting front. Equations 8.33 and 8.34 are combined and rearranged to obtain the Green and Ampt equation.

$$f = K\left(\frac{\Psi\Delta\theta}{F} + 1\right) \tag{8.35}$$

Example 8.6

Use the Green and Ampt model with a computational time step of 20 minutes to determine the direct runoff volume to result from the observed rainfall amounts tabulated as columns 1 and 2 of the table. The rain began at 6:00 a.m. on May 28, 2001 and continued for 3 hours with a total cumulative depth of 108 mm. The watershed has a silt loam soil with an initial effective saturation S_e of 30 percent. We will use the parameter values provided in Table 8.4: $\theta_e = 0.486$, $\Psi = 166.8$ mm, and $K = 6.5$ mm/hr.

Solution Eq. 8.30 converts S_e to $\Delta\theta$. Equation 8.35 provides a relationship between f_C and F.

$$\Delta\theta = (1 - S_e)\theta_e = (1 - 0.30)(0.486) = 0.340$$

$$f_C = K\left(\frac{\Psi\Delta\theta}{F} + 1\right) = 6.5 \text{ mm/hr}\left(\frac{166.8 \text{ mm } (0.340)}{F} + 1\right)$$

$$f_C = \frac{368.6}{F} + 6.5$$

Equations 8.24–8.28 are applied for each 20-minute time step with the results tabulated in the table. The mean rainfall intensity i in column 3 is $\Delta P/\Delta t$. The infiltration capacity f_C (column 4) at an instant in time (column 1) is computed as a function of F (column 5). As long as f_C at the beginning and end of a Δt is greater than i, all the rainfall infiltrates ($f = i$). Otherwise, f is limited by f_C. Since F depends on f_C and vice versa, an iterative solution for f and F is required any time f is limited by f_C. The runoff volume ΔV_R (column 8) is ΔP less ΔF. Of the 108 mm of rainfall, 47 is lost through infiltration, and a volume equivalent to covering the watershed to a depth of 61 mm runs off to the outlet.

Time (a.m.) (1)	ΔP (mm) (2)	i (mm/hr) (3)	f_C (mm/hr) (4)	F (mm) (5)	f (mm/hr) (6)	ΔF (mm) (7)	V_R (mm) (8)
6:00				0			
	5	15			15.0	5.0	0
6:20			80.2	5.0			
	3	9			9.0	3.0	0
6:40			52.6	8.0			
	6	18			18.0	6.0	0
7:00			32.8	14.0			
	13	39			26.6	8.9	4.1
7:20			22.6	22.9			
	26	78			20.6	6.8	19.2
7:40			18.9	29.7			
	17	51			17.8	5.9	11.1
8:00			16.8	35.6			
	29	87			16.1	5.4	23.6
8:20			15.5	41.0			
	8	24			15.0	5.0	3.0
8:40			14.5	46.0			
	1	3			3.00	1.0	0
9:00			14.3	47.0			
Total	108					47.0	61.0

8.5.4 Holtan Model

Holtan (1965) and Holtan, Stiltner, Henson, and Lopez (1975) developed the following infiltration model, which has been used primarily for agricultural watersheds.

$$f_C = GI \cdot A \cdot S_a^{1.4} + f_P \tag{8.36}$$

$$S_a(t + \Delta t) = S_a(t) - f\Delta t + f_P\Delta t + ET\Delta t \tag{8.37}$$

The infiltration capacity f_C, mean actual infiltration rate f during the time interval Δt, and constant rate of percolation f_P through the soil profile below the surface layer have dimensions of depth/time. The percolation rate f_P is the limiting infiltration rate after long wetting, with ranges of values for NRCS soil groups A, B, C, and D cited in Section 8.5.1. The dimensionless growth index GI ($0 < GI \leq 1.0$) represents crop maturity. GI may be treated as the ratio ($ET/ET_{\max}$) of the evapotranspiration (ET) rate at that stage of crop development to the maximum evapotranspiration rate ($ET_{\max}$) for the crop. The term A is the infiltration capacity per depth$^{1.4}$ of available storage [(in./hr)/in.$^{1.4}$ or (cm/hr)/cm$^{1.4}$], which is an index of surface-connected porosity dependent on vegetation. Typical values of A are presented in Table 8.5. The available storage S_a in the surface layer that controls infiltration represents the volume of unused soil moisture storage or pore space, expressed as an equivalent depth of water.

TABLE 8.5 TYPICAL VALUES FOR PARAMETER *A* IN THE HOLTAN MODEL

Land use or cover	Poor condition	Good condition
Fallow	0.10	0.30
Row crops	0.10	0.20
Small grains	0.20	0.30
Hay (legumes)	0.20	0.40
Hay (sod)	0.40	0.60
Pasture (bunchgrass)	0.20	0.40
Temporary pasture (sod)	0.40	0.60
Permanent pasture (sod)	0.80	1.00
Woods and forests	0.80	1.00

The parameters for the Holtan equation are GI, A, f_p, and an initial value of S_a. Equations 8.36 and 8.37 are combined with Eqs. 8.24–8.25 to develop a computational algorithm. Equation 8.37 tracks the amount of available storage S_a for each time interval Δt. At each time step, S_a is depleted by the water infiltrating into the ground, $f\Delta t$. S_a is increased by the water leaving the controlling surface layer or zone of the soil profile, which includes the water percolating deeper, $f_P\Delta t$, and the water lost through evapotranspiration, $ET\Delta t$. The ET rate may be negligible during a rainfall event and is typically neglected, but an ET model may be incorporated if appropriate.

In the NRCS CN and the Green and Ampt methods, the infiltration capacity f_C is related to cumulative infiltration. In the Holtan method, f_C is related to available storage that is a function of both cumulative infiltration and storage capacity recovery mechanisms. Thus, unlike the NRCS and Green and Ampt methods, the Holtan model allows f_C to increase during periods of little or no rainfall.

8.6 UNIT HYDROGRAPH APPROACH FOR ESTIMATING FLOW RATES

Our objective is to compute the hydrograph to result from a given rainfall event. As illustrated by Fig. 8.2, a hydrograph consists of flow rates (m^3/s or ft^3/s) as a function of time. The direct runoff hydrograph at a stream location or watershed outlet results as a direct response to a rainfall event. The total hydrograph includes the baseflow that would have occurred even without the rainfall event, along with the precipitation runoff. The direct runoff volume is the integral of the direct runoff hydrograph and is represented graphically as the area under the plotted hydrograph. A unit hydrograph is a direct runoff hydrograph with a runoff volume of 1 unit, typically 1 cm or 1 inch. Unit volume refers to a volume equivalent to covering the watershed to a depth of 1 unit (cm or in.).

8.6.1 Framework for the General Approach

The concept of modeling the runoff characteristics of a watershed as a representative unit hydrograph proposed by Sherman (1932) continues to be widely applied.

The basic premise of the unit hydrograph approach is that for a particular watershed and rainfall duration, the shape and time base for a runoff hydrograph are the same regardless of the rainfall amount. The magnitude of the hydrograph ordinates (flow rates) vary between rainfall events in direct proportion to runoff volume. All rainfall events are assumed to have the same temporal and spatial distribution. Although the magnitude of runoff volume and flow rates are dependent on antecedent moisture conditions, the time distribution of flows or shape of the unit hydrograph is not. These assumptions are significant approximations, but have been found to be reasonably realistic for practical applications.

Example 8.7

The basic premise of discharge at the outlet being a linear response to precipitation excess (runoff volume) is illustrated by comparing three hydrographs representative of a particular 2,975-acre (1,208-ha) watershed for a specific rainfall duration. Simple triangular-shaped hydrographs are used for this hypothetical illustration. Hydrographs A, B, and C in the table and Fig. 8.9 have a time base of 5.0 hrs. There is no base flow for this watershed, since flows are zero at the beginning and end.

DISCHARGES FOR HYDROGRAPHS

Time (hrs)	Hydrograph A (ft^3/s)	Hydrograph B (ft^3/s)	Hydrograph C (ft^3/s)
0	0	0	0
1	600	900	1,200
2	1,200	1,800	2,400
3	800	1,200	1,600
4	400	600	800
5	0	0	0

The runoff volume V_R is represented by the area under the hydrograph. For these hydrographs with a simple triangular shape, the runoff volume may be computed as the area of a triangle.

$$\text{Area} = 0.5\ (\text{base})\ (\text{height})$$

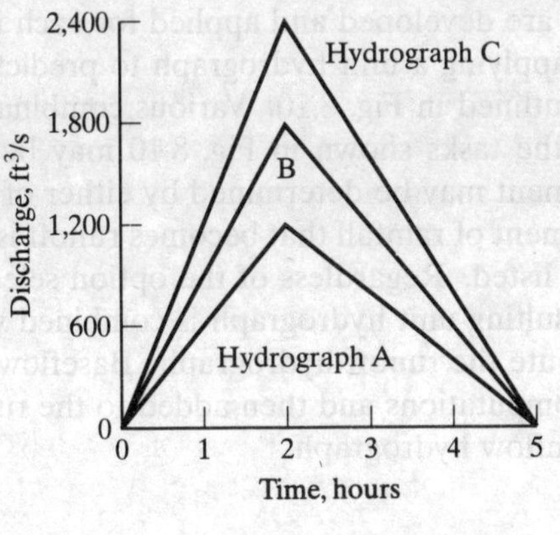

Figure 8.9 The discharges for these hydrographs from Example 8.7 vary in direct proportion to the runoff volume.

Hydrograph A: $V_R = 0.5(1{,}200 \text{ ft}^3/\text{s})(5 \text{ hrs})(3{,}600 \text{ s/hr}) = 10{,}800{,}000 \text{ ft}^3$

Hydrograph B: $V_R = 0.5(1{,}800 \text{ ft}^3/\text{s})(5 \text{ hrs})(3{,}600 \text{ s/hr}) = 16{,}200{,}000 \text{ ft}^3$

Hydrograph C: $V_R = 0.5(2{,}400 \text{ ft}^3/\text{s})(5 \text{ hrs})(3{,}600 \text{ s/hr}) = 21{,}600{,}000 \text{ ft}^3$

Alternatively, the hydrographs can be numerically integrated as

$$V_R = \text{area} = \Sigma Q_t \Delta t = \Delta t \Sigma Q_t$$

where Q_t are the hydrograph ordinates and Δt is the 1.0-hr time interval. For hydrograph A,

$$V_R = (600 + 1{,}200 + 800 + 400 \text{ ft}^3/\text{s})(3{,}600 \text{ s}) = 10{,}800{,}000 \text{ ft}^3$$

The volumes are converted to an equivalent depth covering the 2,975-acre watershed. For hydrograph A,

$$V_R = \left[\frac{10{,}800{,}000 \text{ ft}^3}{(2{,}975 \text{ ac})(43{,}560 \text{ ft}^2/\text{ac})}\right](12 \text{ in./ft}) = 1.00 \text{ in.}$$

Likewise, the V_R for hydrographs B and C are computed to be 1.5 and 2.0 inches, respectively. Notice that the ordinates Q_t of the hydrographs vary in direct proportion to runoff volume. For example, hydrographs A and C have runoff volumes of 1.0 and 2.0, respectively. Thus, for any time t, the Q_t for hydrograph C is twice the value for hydrograph A.

A representative hydrograph can be used to determine other hydrographs associated with different direct runoff volumes. For convenience, the standard of 1 unit (cm, mm, inch) of runoff is adopted for the representative hydrograph. Because rainfall duration affects the shape of the hydrograph, the representative unit hydrograph is for a specific rainfall duration. However, it may be applied to a rainfall event of longer duration subdivided into increments of rainfall. In Example 8.9, a 2-hour unit hydrograph is used to compute the runoff hydrograph for a 24-hour storm by applying the unit hydrograph 12 times, to each of twelve 2-hour increments of rainfall. In typical applications, synthetic unit hydrographs are developed for very short rainfall durations. For example, a unit hydrograph for a 10-minute rainfall duration may be applied to one hundred forty-four 10-minute increments of rainfall to determine the streamflow to result from a 24-hour rainfall event.

As illustrated in Fig. 8.3, a watershed is typically divided into multiple subbasins. Unit hydrographs are developed and applied for each individual subbasin. The general strategy for applying a unit hydrograph to predict the flows to result from a rainfall event is outlined in Fig. 8.10. Various combinations of alternative methods for performing the tasks shown in Fig. 8.10 may be adopted. The rain falling in each time increment may be determined by either of the methods listed. The portion of each increment of rainfall that becomes runoff is determined by any of the alternative options listed. Regardless of the option selected for developing a unit hydrograph, the resulting unit hydrograph is combined with the increments of runoff volume to compute the runoff hydrograph. Baseflow is determined separately from the runoff computations and then added to the runoff hydrograph to determine the total streamflow hydrograph.

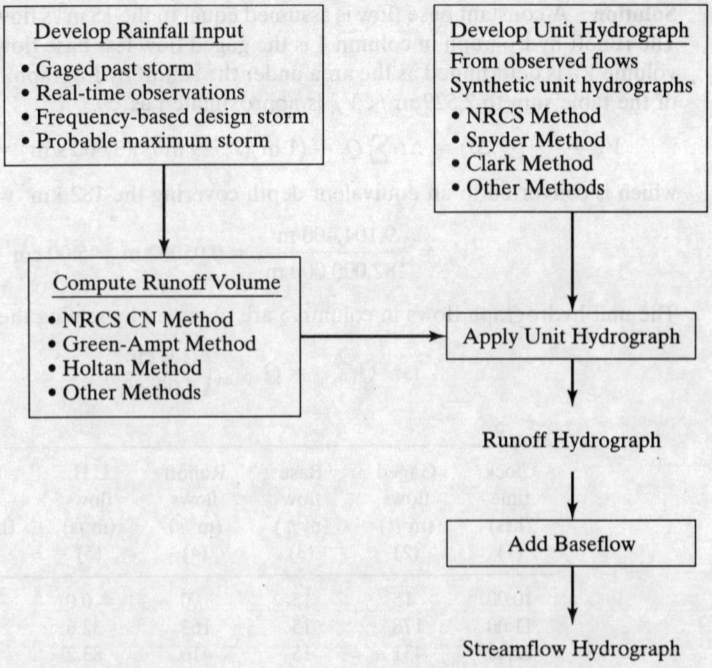

Figure 8.10 Several tasks are combined to develop a model based on the unit hydrograph concept.

8.6.2 Developing a Unit Hydrograph from Gaged Flows

If a stream gaging station has been maintained at the location of interest, a unit hydrograph may be developed from observed flows. Otherwise, the synthetic unit hydrograph methods described in Section 8.6.4–8.6.6 are used. With gaged flows, a rainfall event is selected that is considered to be representative of the storms that occur. Hydrographs from several different flood events with about the same rainfall duration may be averaged together to develop a composite hydrograph that better represents the runoff characteristics of the watershed.

Developing a unit hydrograph from an observed hydrograph consists of the following steps: (1) remove base flow, (2) determine the runoff volume by integrating the hydrograph, and (3) develop the unit hydrograph by scaling discharges in proportion to runoff volume. The rainfall duration for the observed hydrograph becomes the rainfall duration for the unit hydrograph.

Example 8.8

Beginning at 10:00 a.m. on October 18, rain fell for 2 hours on the 182 km^2 watershed above a stream gaging station. The flows recorded in column 2 of the following table were observed. The discharge in the stream prior to the storm was 15 m^3/s. Develop a unit hydrograph (U.H.).

Solution A constant base flow is assumed equal to the 15 m^3/s flow prior to the storm. The runoff hydrograph in column 4 is the gaged flow less base flow. The direct runoff volume V_R is determined as the area under the runoff hydrograph. The Q_t in column 4 of the table sum to 2,529 m^3/s. V_R is approximated as

$$V_R = \sum (Q_t \Delta t) = \Delta t \sum Q_t = (1 \text{ hr})(2{,}529 \text{ m}^3/\text{s})(3{,}600 \text{ s/hr}) = 9{,}104{,}400 \text{ m}^3$$

which is converted to an equivalent depth covering the 182 km^2 watershed.

$$V_R = \frac{9{,}104{,}400 \text{ m}^3}{182{,}000{,}000 \text{ m}^2} = 0.05002 \text{ m} = 5.00 \text{ cm}$$

The unit hydrograph flows in column 5 are obtained by scaling the flows in column 4.

$$Q_{1.0\text{ cm}} = Q_{5.00\text{ cm}}\left(\frac{1.0 \text{ cm}}{5.00 \text{ cm}}\right)$$

Clock time (hrs) (1)	Gaged flows (m^3/s) (2)	Base flow (m^3/s) (3)	Runoff flows (m^3/s) (4)	U.H. flows (m^3/s) (5)	U.H. time (hours) (6)
10:00	15	15	0	0.0	0
11:00	178	15	163	32.6	1
12:00	431	15	416	83.2	2
13:00	562	15	547	109.4	3
14:00	503	15	488	97.6	4
15:00	347	15	332	66.4	5
16:00	245	15	230	46.0	6
17:00	169	15	154	30.8	7
18:00	104	15	89	17.8	8
19:00	75	15	60	12.0	9
20:00	52	15	37	7.4	10
21:00	28	15	13	2.6	11
22:00	15	15	0	0.0	12
23:00	15	15	0		

8.6.3 Applying the Unit Hydrograph

The purpose of a unit hydrograph is to develop hydrographs for selected storms. A relatively small unit hydrograph rainfall duration D is selected to reflect the variation of rainfall and rainfall excess over the duration of the storm. Discharge ordinates are tabulated at a small enough time step to adequately define the variation in flows over time. The process of combining the rainfall excess of each D-hour increment of rainfall with the D-hour unit hydrograph is called convolution. The convolution process consists of scaling the unit hydrograph in proportion to runoff volume and lagging incremental hydrographs to appropriately reflect the time sequencing of each D-hour increment of rainfall. The resulting hydrograph reflects the runoff from the rainfall event. Baseflow is added to determine the total streamflow hydrograph.

Example 8.9

A unit hydrograph (U.H.) for a 182 km^2 watershed in Dallas County, Texas, for a rainfall duration of 2 hours was developed in Example 8.8. A design storm for Dallas County for a recurrence interval of 50 years was developed in Example 7.11. The direct runoff depth associated with each 2-hour increment of rainfall was determined in Example 8.5. Combine the information developed in these previous examples, following the procedure outlined in Fig. 8.10, to develop a 50-year recurrence interval flood hydrograph.

Solution The 2-hour unit (1.0 cm) hydrograph from Example 8.8 is reproduced in columns 1 and 2 of the following table. The direct runoff volume from Example 8.5, in cm, for each of the 2-hour increments of rainfall included in the 50-year design storm is reproduced across the top of columns 3–12. The incremental hydrographs in these columns are computed by multiplying the unit hydrograph ordinates by the runoff volume. The total storm hydrograph in column 13 is the summation of columns 3–12.

Time	U.H.	Hydrographs (m^3/s) for each increment of runoff volume (cm)										Hydrograph
(hrs)	(m^3/s)	0.04	0.24	0.70	9.70	2.27	1.10	0.76	0.60	0.50	0.43	(m^3/s)
(1)	(2)	(3)	(4)	(5)	(6)	(7)	(8)	(9)	(10)	(11)	(12)	(13)
0	0.0	0.0										0.0
1	32.6	1.3										1.3
2	83.2	3.3	0.0									3.3
3	109.4	4.4	7.8									12.2
4	97.6	3.9	20.0	0.0								23.9
5	66.4	2.7	26.3	22.8								51.7
6	46.0	1.8	23.4	58.2	0.0							83.5
7	30.8	1.2	15.9	76.6	316.2							410
8	17.8	0.7	11.0	68.3	807.0	0.0						887
9	12.0	0.5	7.4	46.5	1,061.2	74.0						1,190
10	7.4	0.3	4.3	32.2	946.7	188.9	0.0					1,170
11	2.6	0.1	2.9	21.6	644.1	248.3	35.9					953
12	0.0	0.0	1.8	12.5	446.2	221.6	91.5	0.0				774
13			0.6	8.4	298.8	150.7	120.3	24.8				604
14			0.0	5.2	172.7	104.4	107.4	63.2	0.0			453
15				1.8	116.4	69.9	73.0	83.1	19.6			364
16				0.0	71.8	40.4	50.6	74.2	49.9	0.0		287
17					25.2	27.2	33.9	50.5	65.6	16.3		219
18					0.0	16.8	19.6	35.0	58.6	41.6	0.0	172
19						5.9	13.2	23.4	39.8	54.7	14.0	151
20						0.0	8.1	13.5	27.6	48.8	35.8	134
21							2.9	9.1	18.5	33.2	47.0	111
22							0.0	5.6	10.7	23.0	42.0	81.3
23								2.0	7.2	15.4	28.6	53.1
24								0.0	4.4	8.9	19.8	33.1
25									1.6	6.0	13.2	20.8
26									0.0	3.7	7.7	11.4
27										1.3	5.2	6.5
28										0.0	3.2	3.2
29											1.1	1.1
30											0.0	0.0

8.6.4 NRCS Dimensionless Unit Hydrograph

The majority of hydrologic engineering applications deal with ungaged watersheds. Unit hydrographs must be synthesized from information available from maps or field inspection of the watershed. Various entities, based on analyses of data from gaged watersheds, have devised empirical methods for synthesizing hydrographs for ungaged watersheds.

The NRCS developed a dimensionless unit hydrograph in the 1950's based on analyses of many unit hydrographs for gaged watersheds varying widely in size and location. The NRCS dimensionless unit hydrograph continues to be widely applied throughout the United States and the world. Its popularity is due largely to its ease of use. The watershed area A and lag t_L are the only parameters required, and they can be estimated for ungaged watersheds using the methods described in Section 8.3.

Curvilinear and triangular versions of the NRCS synthetic unit graph are shown in Fig. 8.11. The curvilinear version is more realistic, but the triangular is a little simpler. Ordinates for the curvilinear version are tabulated in Table 8.6. The unit hydrograph for a watershed is developed by combining either version of the dimensionless unit graph with Eqs. 8.38–8.40. The time to peak T_P is estimated as a function of rainfall duration D associated with the unit hydrograph and the watershed lag t_L, all in hours.

$$T_P = \frac{D}{2} + t_L \tag{8.38}$$

The peak of the unit hydrograph is given by Eq. 8.39 (English units) and Eq. 8.40 (metric units).

$$Q_p = \frac{484A}{T_P} \tag{8.39}$$

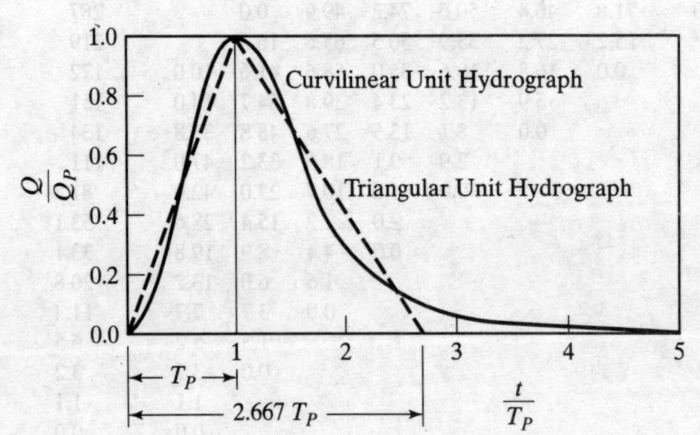

Figure 8.11 Either the triangular or curvilinear version of the NRCS dimensionless unit hydrograph is combined with Eqs. 8.38–8.40.

TABLE 8.6 TIME AND DISCHARGE RATIOS FOR NRCS CURVILINEAR DIMENSIONLESS UNIT HYDROGRAPH

t/T_P	Q/Q_P	t/T_P	Q/Q_P	t/T_P	Q/Q_P
0	0.000	1.1	0.990	2.4	0.147
0.1	0.030	1.2	0.930	2.6	0.107
0.2	0.100	1.3	0.860	2.8	0.077
0.3	0.190	1.4	0.780	3.0	0.055
0.4	0.310	1.5	0.680	3.2	0.040
0.5	0.470	1.6	0.560	3.4	0.029
0.6	0.660	1.7	0.460	3.6	0.021
0.7	0.820	1.8	0.390	3.8	0.015
0.8	0.930	1.9	0.330	4.0	0.011
0.9	0.990	2.0	0.280	4.5	0.005
1.0	1.000	2.2	0.207	5.0	0.000

$$Q_p = \frac{2.08A}{T_P} \tag{8.40}$$

The peak discharge Q_P is in ft^3/s for Eq. 8.39 and m^3/s for Eq. 8.40. The watershed area A is in mi.^{2} or km^2. The time to peak T_P is in hours.

Example 8.10

The direct runoff from observed rainfall was computed in Example 8.6 for the 6.2 km^2 watershed of Fig. 8.4. A t_L of 1.83 hours was estimated in Example 8.1. Develop a unit hydrograph (U.H.) for this watershed for a rainfall duration D of 20 minutes using the NRCS triangular dimensionless unit hydrograph. Use the unit hydrograph to predict the flows to result from the rainfall given in Example 8.6.

Solution

$$T_P = \frac{D}{2} + t_L = \frac{1/3 \text{ hr}}{2} + 1.83 \text{ hr} = 2.0 \text{ hrs}$$

$$Q_p = \frac{2.08A}{T_P} = \frac{2.08(6.2 \text{ km}^2)}{2 \text{ hrs}} = 6.45 \frac{\text{m}^3}{\text{s}}$$

The unit hydrograph is tabulated in columns 1–3 at a time step of 20 minutes. Column 2 is column 1 divided by the T_P of 120 minutes. With the triangular unit hydrograph, the discharge ordinates increase linearly from zero at time zero and 320 minutes (t/T_P of 0 and 2.667) to Q_P = 6.45 m^3/s at T_P = 120 minutes (t/T_P = 1). In Example 8.6, runoff depths ΔV_R for each of five 20-minute increments of rainfall ΔP were estimated to be 4.1, 19.2, 11.1, 23.6, and 3.0 cm, respectively. The results of convoluting the unit hydrograph with these runoff depths are presented in columns 4–8. The resulting storm hydrograph ordinates are tabulated in column 9.

Time (min) (1)	Ratio t/T_P (2)	U.H. (m^3/s) (3)	Incremental hydrographs (m^3/s) 4.1 cm (4)	19.2 cm (5)	11.1 cm (6)	23.6 cm (7)	3.0 cm (8)	Hydrograph (m^3/s) (9)
0	0.00	0.00	0					0
20	0.17	1.07	4	0				4
40	0.33	2.15	9	21	0			29
60	0.50	3.22	13	41	12	0		66
80	0.67	4.30	18	62	24	25	0	129
100	0.83	5.37	22	83	36	51	3	194
120	1.00	6.45	26	103	48	76	6	260
140	1.17	5.80	24	124	60	101	10	318
160	1.33	5.16	21	111	72	127	13	344
180	1.50	4.51	19	99	64	152	16	350
200	1.67	3.87	16	87	57	137	19	316
220	1.83	3.22	13	74	50	122	17	277
240	2.00	2.58	11	62	43	107	15	237
260	2.17	1.93	8	50	36	91	14	198
280	2.33	1.29	5	37	29	76	12	159
300	2.50	0.64	3	25	21	61	10	119
320	2.67	0.00	0	12	14	46	8	80
340				0	7	30	6	43
360					0	15	4	19
380						0	2	2
400							0	0

In analyzing gaged watersheds to develop a dimensionless unit hydrograph, the NRCS found that typically about 0.375 and 0.625 of the flow volume (area under the hydrograph) is before and after the peak. The constant 484 or 2.08 in Eqs. 8.39 and 8.40 is derived directly from this empirical observation. The unit hydrograph volume is a unit depth of runoff V_R multiplied by the watershed area A. The runoff volume is also represented as the integral or the area under the unit hydrograph. For the triangular unit hydrograph, the runoff volume is the area of a triangle with base 2.667 T_P and height Q_P.

$$\text{runoff volume} = V_R\, A = 0.5(2.667\, T_P)Q_P$$

$$Q_P = \frac{0.75 V_R\, A}{T_P}[\text{conversion factors}] \tag{8.41}$$

With Q_P in m^3/s, V_R of 1.0 cm, A in km^2, and T_P in hours, the conversion factors are as follows.

$$Q_P = \frac{0.75 V_R\, A}{T_P}\left[\left(\frac{\text{m}}{100\text{ cm}}\right)\left(\frac{1{,}000{,}000\text{ m}^2}{\text{km}^2}\right)\left(\frac{\text{hr}}{3{,}600\text{ s}}\right)\right]$$

$$Q_P = \frac{2.08A}{T_P} \tag{8.40}$$

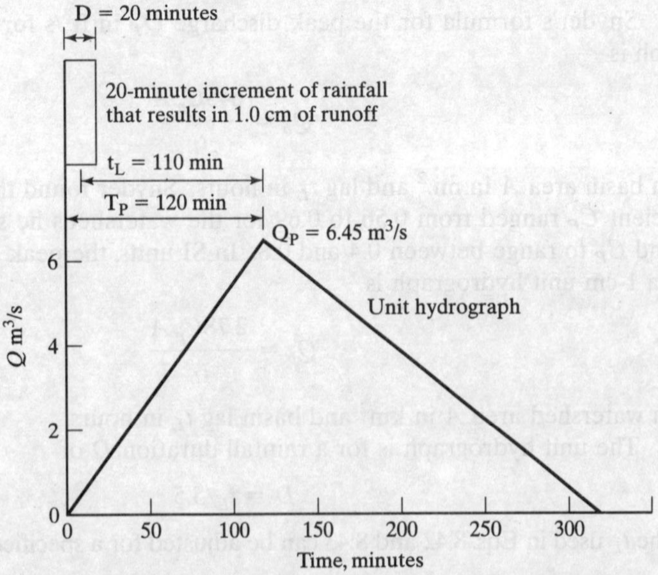

Figure 8.12 This unit hydrograph was developed in Example 8.10 using the NRCS triangular dimensionless unit hydrograph method.

With Q_P in ft^3/s, V_R of 1.0 inch, A in mi.^{2}, and T_P in hours, the resulting constant in Eq. 8.39 is 484.

The constants 484 and 2.08 and associated Q_P in Eqs. 8.39 and 8.40 representing *average* watersheds should logically be greater for a steep quickly peaking watershed than for a flat slow runoff watershed. The Snyder method discussed next replaces this constant with a variable watershed parameter.

8.6.5 Snyder Synthetic Unit Hydrograph

Snyder (1938) pioneered the concept of developing a generic unit hydrograph based on analysis of gaged watersheds that can be applied to ungaged basins using measurements of basin characteristics. He developed a set of formulas relating the physical geometry of the watershed to parameters of the unit hydrograph. The method is based on a study of 20 watersheds located mainly in the Appalachian Highlands of the eastern U.S., with areas ranging from 10 to 10,000 mi.^{2} (25.9–25,900 km^2). Equation 8.6 is Snyder's lag equation. The time to peak T_P is provided by Eq. 8.38. Equations 8.42–8.48 provide Q_P and information regarding the width of the unit hydrograph. The area under the unit hydrograph is set by the unit depth of runoff. The method leaves some flexibility for shaping the unit hydrograph while satisfying Eqs. 8.42–8.48 and the unit volume criterion.

Snyder's formula for the peak discharge Q_P in ft³/s for a 1-inch unit hydrograph is

$$Q_p = \frac{645 C_p A}{t_L} \tag{8.42}$$

with basin area A in mi.^{2} and lag t_L in hours. Snyder found that values for the coefficient C_P ranged from 0.56 to 0.69 for the watersheds he studied. Others have found C_P to range between 0.4 and 0.8. In SI units, the peak discharge Q_P in m³/s for a 1-cm unit hydrograph is

$$Q_p = \frac{2.78 C_p A}{t_L} \tag{8.43}$$

with watershed area A in km^2 and basin lag t_L in hours.

The unit hydrograph is for a rainfall duration D of

$$D = t_L/5.5 \tag{8.44}$$

or the t_L used in Eqs. 8.42 and 8.43 can be adjusted for a specified rainfall duration D'

$$t'_L = t_L + 0.25(D' - D) \tag{8.45}$$

This allows the user of the model to set the rainfall duration D' and adjust the lag t_L accordingly rather than assigning the rainfall duration D from Eq. 8.44 to the unit hydrograph.

The time base t_B of the unit hydrograph in hours is estimated as

$$t_B = 72 + 3\,t_L \tag{8.46}$$

for larger watersheds. A smaller t_B is typically adopted for watersheds with a smaller t_L. The Corps of Engineers developed the following empirical equations to help define the shape.

$$W_{50} = \frac{5.87}{(Q_p/A)^{1.08}} \tag{8.47}$$

$$W_{75} = \frac{3.35}{(Q_p/A)^{1.08}} \tag{8.48}$$

W_{50} and W_{75} are the widths in hours of the unit hydrograph at 50 and 75 percent of Q_P. Q_P is the peak discharge in m³/s. A is the watershed area in km^2. These time widths are proportioned such that one-third is before the peak and two-thirds after the peak.

8.6.6 Clark Synthetic Unit Hydrograph

The Clark (1945) method for developing a unit hydrograph is based on the concept of routing a time–area relationship through a linear reservoir. Other watershed modeling techniques reported in the literature are based on variations of this basic concept. The method is conceptualized as follows. Water covering the basin to a depth of 1 unit (cm, mm, and in.) is released instantaneously and allowed to run off the watershed. A time–area relationship represents the translation hydrograph of this runoff as influenced by watershed characteristics, such as size, shape, and surface roughness. The translation hydrograph is routed through a linear reservoir to capture additional storage effects of the watershed. The resulting unit hydrograph represents the runoff from a rainfall of zero duration. To obtain a unit hydrograph for a rainfall duration D, two instantaneous unit hydrographs lagged by D are combined and then divided by 2 to reduce the resulting two-unit, D-duration hydrograph to a unit hydrograph.

The isochrone lines drawn on the watershed map in Fig. 8.13 are the loci of points of equal travel time to the outlet. For any point in the basin, the travel time refers to the time required for a parcel of water to travel from that point to the outlet. The time of concentration t_C is the travel time from the most hydraulically remote point in the basin to the outlet. A travel time versus watershed–subarea relationship indicates the contributing area as a function of time, with the entire watershed contributing flow at the outlet at t_C. A relationship between travel time and contributing watershed area may be developed based on the travel-time methods of Section 8.3.3. However, whereas estimating t_c is difficult enough, developing an isochrone map and complete time–area relationship is much more difficult. To simplify application of the Clark method, the Hydrologic Engineering Center (1998, 2000a,b) has developed the following synthetic time–area relationship based on empirical analysis of numerous watersheds.

$$A_c/A = 1.414\,T^{1.5} \quad \text{for} \quad 0 \le T \le 0.5$$
$$A_c/A = 1 - 1.414(1-T)^{1.5} \quad \text{for} \quad 0.5 \le T \le 1.0 \qquad \textbf{(8.49)}$$

A_c/A is the contributing area A_c at time T as a fraction of the total watershed area A. T is a fraction of the time of concentration t_C.

A translation hydrograph based on the time–area relationship is routed through a linear reservoir. Hydrologic routing is covered in Section 6.1 from the

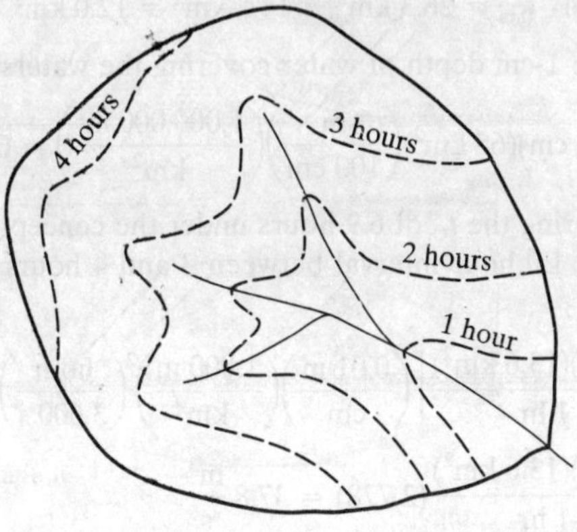

Figure 8.13 The isochrone map shows travel times from the watershed outlet.

perspective of routing hydrographs through reservoirs and stream reaches. For purposes of the Clark method, the continuity equation is written as

$$\bar{I} - \left(\frac{O_1 + O_2}{2}\right) = \frac{S_2 - S_1}{\Delta t} \tag{8.50}$$

where $\bar{I}$ is the mean inflow during the time interval Δt, O is outflow, and S is storage. The subscripts 1 and 2 refer to the beginning and end of the time interval Δt. The concept of a linear reservoir refers to a linear relationship between storage S and outflow O

$$S = RO \tag{8.51}$$

where R is a storage constant. By substituting Eq. 8.51 into Eq. 8.50 and rearranging, the following routing equation is derived

$$O_2 = C\bar{I} + (1 - C)O_1 \tag{8.52}$$

where

$$C = \frac{2\Delta t}{2R + \Delta t} \tag{8.53}$$

Example 8.11

Develop a unit hydrograph (U.H.) for a rainfall duration of 1.0 hour for a watershed characterized by a drainage area A of 65 km^2, a time of concentration t_C of 6.9 hours, and a Clark storage coefficient R of 2.8 hours. Use the Clark method with the HEC synthetic time–area relationship.

Solution Equation 8.49 provides the time–area relationship tabulated in columns 2 and 3. For example, for time t of 4 hours

$$T = t/t_C = 4 \text{ hrs}/6.9 \text{ hrs} = 0.580$$

$$A_c/A = 1 - 1.414(1 - 0.580)^{1.5} = 0.615$$

$$A_c = 0.615(65 \text{ km}^2) = 40.0 \text{ km}^2$$

Incremental areas between isochrones are tabulated in column 5. The incremental area between travel times of 3 and 4 hours is

$$\Delta A_{3\text{–}4\text{hrs}} = 26.3 \text{ km}^2 - 14.3 \text{ km}^2 = 12.0 \text{ km}^2$$

The total volume of the 1-cm depth of water covering the watershed

$$\text{Volume} = (1 \text{ cm})(65 \text{ km}^2)\left(\frac{\text{m}}{100 \text{ cm}}\right)\left(\frac{1{,}000{,}000 \text{ m}^2}{\text{km}^2}\right) = 650{,}000 \text{ m}^3$$

flows past the outlet during the t_C of 6.9 hours under the conceptual scenario of pure translation. During the 1.0-hour interval between 3 and 4 hours, the mean flow rate Q at the outlet is

$$Q = \frac{(1 \text{ cm})(13.6 \text{ km}^2)}{1 \text{ hr}}\left[\left(\frac{0.01 \text{ m}}{\text{cm}}\right)\left(\frac{1{,}000 \text{ m}}{\text{km}}\right)^2\left(\frac{\text{hour}}{3{,}600 \text{ s}}\right)\right]$$

$$= \frac{(1 \text{ cm})(13.6 \text{ km}^2)}{1 \text{ hr}}(2.778) = 37.8 \frac{\text{m}^3}{\text{s}}$$

Column 6 is column 5 multiplied by 2.778. The mean inflow $\bar{I}$ in column 6 is the translation hydrograph that is routed through a linear reservoir to model the storage effects of the watershed using Eq. 8.52.

$$C = \frac{2\Delta t}{2R + \Delta t} = \frac{2(1 \text{ hr})}{2(2.8 \text{ hr}) + 1 \text{ hr}} = 0.303$$

$$O_2 = C\bar{I} + (1 - C)O_1 = 0.303\,\bar{I} + 0.697\,O_1$$

The computations step through time with 0_1 and 0_2 being the outflow at the beginning and end of the Δt. For example, the outflow at time 4 hours is

$$0_2 = 0.303(37.8 \text{ m}^3/\text{s}) + 0.697(17.6 \text{ m}^3/\text{s}) = 23.7 \text{ m}^3/\text{s}$$

The resulting hydrograph tabulated in column 7 represents the runoff from an instantaneous (zero duration) rainfall event. The 1 cm of runoff covering the watershed was released instantaneously. The instantaneous hydrograph of column 7 is converted to the 1-hour rainfall duration unit hydrograph of column 8 by lagging 1-hour, combining, and dividing by 2, which is equivalent to averaging flows at the beginning and end of each Δt. For example, at $t = 2$ hrs, $Q = (4.3 + 10.8)/2 = 7.5$ m^3/s. The unit hydrograph for a 1-hour rainfall duration is presented in columns 1 and 8.

Time, t (hrs) (1)	$T = t/t_c$ (2)	A_c/A (3)	A_c (km^2) (4)	ΔA (km^2) (5)	Translated (m^3/s) (6)	0-hr U.H. (m^3/s) (7)	1-hr U.H. (m^3/s) (8)
0	0	0	0	0	0	0	0
1	0.145	0.078	5.1	5.1	14.1	4.3	2.1
2	0.290	0.221	14.3	9.3	25.8	10.8	7.5
3	0.435	0.405	26.3	12.0	33.4	17.6	14.2
4	0.580	0.615	40.0	13.6	37.8	23.7	20.7
5	0.725	0.796	51.7	11.8	32.7	26.4	25.1
6	0.870	0.933	60.7	9.0	24.9	26.0	26.2
7	1.000	1.000	65.0	4.3	12.0	21.7	23.9
8					0.0	15.2	18.4
9						10.6	12.9
10						7.4	9.0
11						5.1	6.2
25						0.0	0.0

The Hydrologic Modeling System (HMS) model (HEC, 2000, 2001) discussed in Section 8.9 includes a modified Clark option, as well as the conventional Clark method just described. The Clark and other unit hydrograph methods are lumped-parameter techniques, with the entire subbasin represented by a set of parameters, such as CN, t_C, and R, and the rainfall assumed to be evenly distributed spatially over the subbasin. In the HMS-modified Clark method, the subbasin is divided into a set of grid cells. Rainfall excess from each individual grid cell is translated to the outlet based on travel time. The hydrograph at the outlet is computed based on combining the contributions of runoff from each cell. Development of the quasi-distributed modified Clark approach was motivated by applications that use radar to measure rainfall. Rainfall observations are assigned to each individual grid cell.

8.7 EROSION AND SEDIMENT YIELD

Sediment eroded from watersheds is delivered to streams and rivers. The sediment yield from land erosion along with materials from gully and stream bank erosion is transported by streamflow either in suspension or by rolling and sliding along the bed. Sediment is transported by overland and channel flow until flow velocities diminish and sediment deposition occurs. Our focus here is on erosion of land and the resulting sediment yield rather than streambank erosion and sediment transport in rivers.

Soil is detached both by raindrop impact and the shearing force of flowing water. Sediment is transported downslope primarily by flowing water, although there is a small amount of downslope transport by raindrop splash. Runoff and resulting downslope transport do not occur until the rainfall intensity exceeds the infiltration rate. Thus, soil erodibility increases as the infiltration rate decreases. Overland flow includes sheet flow and rill flow. Rills are small channels of flow formed as the sheet flow concentrates. The quantity and size of material transported increase with the velocity of runoff water. At some point downslope, slopes may decrease, resulting in a decreased velocity and transport capacity. Sediment is deposited, starting with the larger particles and aggregates. Smaller particles and aggregates are carried further downslope, resulting in what is known as enrichment of fines. For this reason, the size distribution of eroded sediments has a major impact on soil erosion-deposition processes. Erosion is also greatly influenced by vegetative cover. Agricultural and construction activities involving removal of vegetation often result in a several thousand-fold increase in erosion.

Modeling erosion and sedimentation processes is highly approximate and necessarily empirical, being based mainly on field observations rather than on theoretical considerations. The widely applied techniques presented in this section provide rough estimates of erosion and sedimentation quantities. The universal soil loss equation (USLE) methodology and its later revision (RUSLE) predict soil loss from sheet and rill erosion for alternative land management practices. The USLE was developed originally for estimating average annual soil erosion, but is also used for monthly or seasonal amounts and losses for individual rainfall events. Because much of the eroded sediment is redeposited prior to reaching the watershed outlet, a sediment delivery ratio is combined with the soil loss erosion amounts to obtain the sediment yield at the outlet. The modified USLE (MUSLE) predicts sediment yields at the watershed outlet for individual rainfall events. Reservoir trap efficiencies are used to predict the proportion of the sediment yield inflow that is deposited in reservoirs.

8.7.1 Universal Soil Loss Equation (USLE)

The RUSLE methodology described by the U.S. Department of Agriculture (USDA) Handbook 703 (Renard, Foster, Weesies, McCool, and Yoder, 1997) is an updated and improved version of the USLE documented by USDA Handbook 537 (Wischmeier and Smith, 1978). Many others, such as the Transportation Research Board (1980), have developed additional information for applying the USLE in evaluating the impacts of construction, mining, and other activities, as well as agricultural practices. Textbooks that cover the USLE and RUSLE include Haan, Barfield, and Hayes (1994), Ward and Elliot (1995), and McCuen (1998). Both the RUSLE and original USLE model are based on the following equation.

$$A = RKLSCP \tag{8.54}$$

A is the average soil loss per unit of area, expressed in units selected for K and the time period specified by R. A is the product of factors that account for rainfall-runoff erosivity R, soil erodibility K, flow length L, slope S, cover C, and erosion control practice P. RUSLE and USLE methodology are similar with the RUSLE providing updated and expanded information for estimating parameter values. The RUSLE is designed for either computer or manual application. In the following brief summary of the general methodology, USLE refers to both the RUSLE and USLE.

The USLE predicts the amount of soil eroded by rainfall, sheet flow, and rill flow. The methodology is based on extensive analyses of measurements from numerous field plots. Equation 8.54 and associated empirical information for the factors were developed in U.S. customary units. The soil erosion (A) is typically expressed in tons/acre, which can be converted to SI units of metric tons/hectare by multiplying by 2.242. Renard *et al.* (1997) provide instructions for estimating R, K, LS, C, and P in either English or metric units. The empirical USLE (Eq. 8.54) reflects these major factors:

- Climate is reflected in the rainfall-runoff erosivity index R (Fig. 8.14).
- Soil characteristics are represented by the erodibility factor K (Eq. 8.55).
- Topography is represented by the length-slope factor LS (Fig. 8.15).
- Land cover and soil conservation practices are represented by CP (Table 8.7).

Minimal information is reproduced here to define and illustrate estimation of R, K, LS, C, and P. Renard *et al.* (1997) should be referenced for background information and alternative methods for estimating the factors based on much more detailed considerations.

Values for the rainfall and runoff erosivity index R are provided in Fig. 8.14 for the contiguous U.S. Soil losses are proportional to the product of a storm's kinetic energy E and its maximum 30-minute intensity I. The product EI reflects the combined potential of raindrop impact and runoff turbulence to transport dislodged soil particles. The sum of the EI products for a given year at a particular location is an index of the erosivity of all rainfall for that year. The factor R in Fig. 8.14 is the average value of the series of annual sums of EI products, divided by 100. The units for R are hundreds of (ft · tons · in.)/(ac · hr · yr).

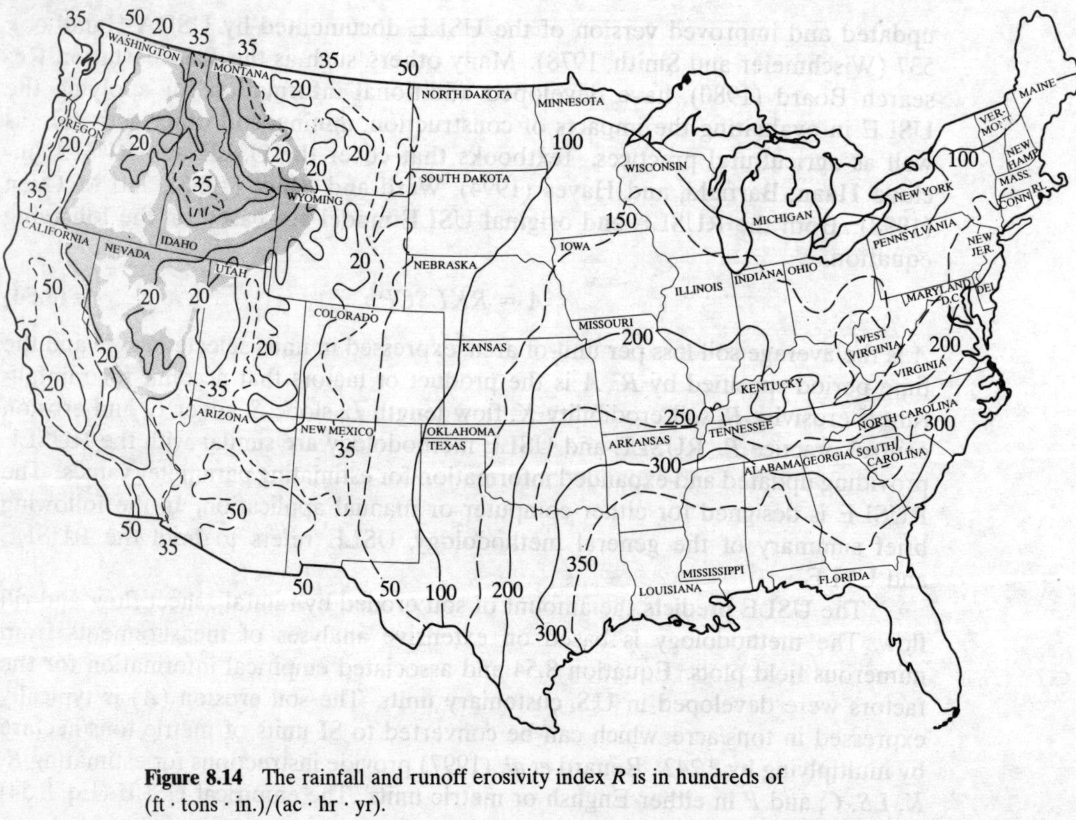

Figure 8.14 The rainfall and runoff erosivity index R is in hundreds of (ft · tons · in.)/(ac · hr · yr).

Although the USLE was developed for predicting long-term average annual soil erosion, shorter time periods may also be considered. Available information regarding the proportion of the *EI* occurring in any month or season provides a value for R in Eq. 8.54 that results in an average soil loss estimate A for the specified months. Information is also available for determining R for single rainfall events.

The soil erodibility factor K is the average soil loss, in tons per acre per unit of the rainfall factor R, from a particular soil in cultivated continuous fallow with standard values of both the plot length and slope, which were selected somewhat arbitrarily as 72.6 ft and 9 percent, respectively. Various alternative methods for estimating K include the following formula developed based on field measurements of losses from fallow plots of different types of soils.

$$K = \frac{2.1(10^{-4})(12 - \%OM)M^{1.14} + 3.25(SI - 2) + 2.5(PI - 3)}{100} \qquad \textbf{(8.55)}$$

$$M = (\%MS + \%VFS)(100 - \%CL)$$

M is a function of the primary particle size fractions. $\%MS$ is the percentage silt (0.002–0.05 mm), VFS is the percentage of very fine sand (0.05–0.1 mm), and $\%CL$

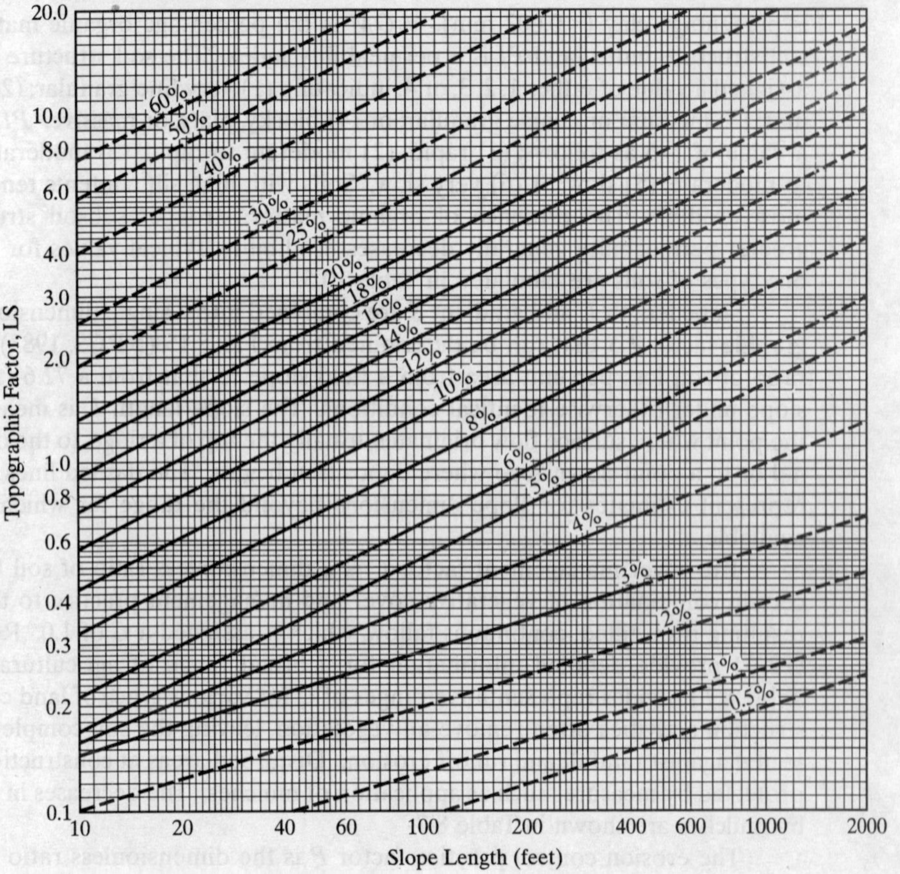

Figure 8.15 Dimensionless topographic factor *LS* represents length and slope.

TABLE 8.7 C FACTOR IN USLE

Land cover	C	Land cover	C
Continuous fallow	1.0	Bare soil	
Undisturbed forest land		Undisturbed except scraped	0.66–1.30
100–75% canopy cover, 100–90% duff cover	0.0001–0.001	Compacted, smooth	1.00–1.40
70–45% canopy cover, 85–75% duff cover	0.002–0.004	Compacted, root racked	0.90–1.20
35–20% canopy cover, 70–40% duff cover	0.003–0.009	Disk tillage, fresh	1.00
Permanent pasture and brush cover		Disk tillage, after one rain	0.89
0% canopy, 80% ground cover		Straw mulch, 0.5 tons/acre	0.30
Grass	0.013	Straw mulch, 1.0 ton/acre	0.18
Weeds	0.043	Straw mulch, 2.0 tons/acre	0.09
50% brush, 80% grass cover	0.012	Straw mulch, 4.0 tons/acre	0.02

is percentage clay (<0.002 mm). $\%OM$ is the percentage organic matter, SI is a soil structure index, and PI is a permeability index. The soil structure index SI is assigned a value of either 1, 2, 3, or 4, indicating (1) very fine granular; (2) fine granular; (3) medium or coarse granular; or (4) blocky, platy, or massive. PI is assigned a value of 1–6 as follows: (1) rapid, (2) moderate to rapid, (3) moderate, (4) slow to moderate, (5) slow, or (6) very slow. Soils with high silt contents tend to be the most erodible. The presence of organic matter, stronger subsoil structure, and greater permeability generally decreases erodibility. Typical values for K for various types of soils are tabulated in Table 10.5.

Slope length L and gradient S are combined into a single dimensionless topographic factor LS that can be determined from Fig. 8.15 (NRCS, 1983). LS is the ratio of soil loss per unit area from a field slope to that from a 72.6-ft, 9 percent slope under otherwise identical conditions. The slope length L is measured from the point where surface flow originates, usually the top of a ridge, to the outlet channel or to a point downslope where deposition begins. The dashed lines in Fig. 8.15 represent estimates for slope dimensions beyond the range for which data were available.

The cover management factor C is a dimensionless ratio of soil loss from a certain combination of vegetative cover and management practice to the soil loss resulting from tilled, continuous fallow, which is assigned a C of 1.0. Renard *et al.* (1997) provide detailed information for a broad range of agricultural practices. Table 8.7 provides representative ranges of C for various types of land cover. Construction activities often remove all vegetation, leaving the soil completely unprotected against rainfall and runoff erosion. Disturbed areas at construction sites are protected by measures such as application of mulches. The decreases in C achieved by mulches are shown in Table 8.7.

The erosion control practice factor P is the dimensionless ratio of soil loss under a certain erosion control practice to the soil loss resulting from straight-row farming. A typical application of P is to model contour farming, strip cropping, and terraces. Contouring consists of performing field operations, such as tillage, planting, and harvesting, approximately on the contour. Contouring reduces surface runoff by impounding water in small depressions and decreasing the development of rills. Strip cropping is the practice of growing alternate strips of different crops in the same field. Strips of grass or small grains, that act as sediment filters by decreasing the overland flow velocity, are alternated with crops that allow greater rates of erosion. Constructing terraces on eroding slopes is another common erosion control practice. Terraces reduce the slope length and pond runoff.

Example 8.12

Compare annual soil losses before and during a construction project in central Ohio. Before the construction activity began, the 448-acre (0.7-mi.^{2}) site was undisturbed forest, with the tree canopy cover extending over about 50% of the ground and duff (decaying organic matter) covering about 80% of the forest floor ($C = 0.003$, Table 8.7). For a short time after clearing the land, the bare soil is characterized by a C of 1.0. Straw mulch applied at a rate of 1.0 ton/acre reduces C to 0.18 (Table 8.7). The ground has an average slope of 4 percent and typical slope lengths of about 600 feet both before and after the construction project. The soil is characterized as 65 percent silt and fine sand (0.002–0.1 mm), 5 percent clay (<0.002 mm), and 3% organic matter, with a fine granular structure ($SI = 2$) and moderate permeability ($PI = 4$).

Solution From Fig. 8.14, R for central Ohio is 150. K is determined using Eq. 8.55.

$$K = \frac{2.1(10^{-4})(12 - 3)[65(100 - 5)]^{1.14} + 3.25(2 - 2) + 2.5(4 - 3)}{100} = 0.42$$

LS is determined from Fig. 8.15 to be 0.8. P is 1.0, since no erosion reduction measures are being adopted. These factors are substituted into Eq. 8.54 to obtain the average annual soil loss. Before the forest is cleared for the construction project,

$$A = RKLSCP = (150)(0.42)(0.8)(0.003)(1.0) = 0.15 \text{ tons/acre/year}$$

or 68 tons/year

After clearing the land of all trees and vegetation, the bare soil is characterized by a C of 1.0 and the following erosion rate.

$$A = RKLSCP = (150)(0.42)(0.8)(1.0)(1.0) = 50 \text{ tons/acre/year}$$

or 23,000 tons/year

Application of 1.0 ton/acre of straw mulch to protect the soil reduces the erosion rate to

$$A = RKLSCP = (150)(0.42)(0.8)(0.18)(1.0) = 9.1 \text{ tons/acre/year}$$

or 4,100 tons/year

8.7.2 Sediment Yield

Much of the soil displaced in the erosion process is subsequently deposited in flatter and/or more densely vegetated areas where flow velocities and sediment transport capabilities are diminished. The USLE predicts the gross amount of soil eroded, not the sediment yield that actually reaches the watershed outlet. One common method for estimating sediment yield Y is based on a delivery ratio D defined as

$$D = \frac{Y}{(\text{gross erosion rate})(\text{watershed area})} \tag{8.56}$$

For small watersheds, the gross erosion rate may be computed as A in tons/acre/year using the USLE. For larger watersheds, gully and streambank erosion may be

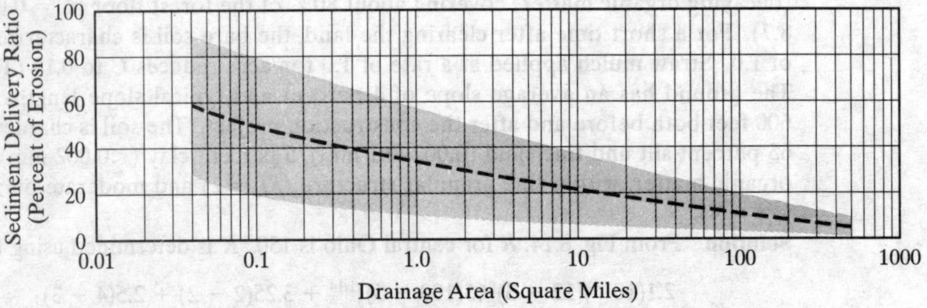

Figure 8.16 The sediment delivery ratio (Eq. 8.56) may be approximated as a function of drainage area (NRCS, 1983).

added. Figure 8.16 provides a rough estimate of the delivery ratio D as a function of drainage area. The NRCS (1983) analyzed data from studies of many watersheds to develop the relation reproduced as Fig. 8.16. The shaded area represents the wide variation in the data, and the dashed line is the median value.

Example 8.13

Determine the average annual sediment yield from the 0.7 mile^2 site of Example 8.12, before and after the clearing and mulching operations. The erosion rate is estimated in Example 8.12 to be 68 and 4,100 tons/year, respectively, before and after clearing and mulching. A sediment delivery ratio D of 37.5 percent is estimated from Fig. 8.16.

$$Y = DA = 37.5\%(68 \text{ tons/yr}) = 26 \text{ tons/yr}$$

$$Y = DA = 37.5\%(4{,}100 \text{ tons/yr}) = 1{,}500 \text{ tons/yr}$$

8.7.3 Modified Universal Soil Loss Equation (MUSLE)

The MUSLE

$$Y = 95(V_R Q_P)^{0.56}\ KLSCP \tag{8.57}$$

determines sediment yield Y in tons for individual storm events (Williams and Berndt, 1977). V_R is the runoff volume in acre-feet. Q_P is the peak discharge in ft^3/s. The other terms are the standard USLE factors. V_R and Q_P may be determined by any of the methods presented earlier in this chapter.

Example 8.14

A watershed is characterized as follows: drainage area A of 320 acres; time of concentration t_C of less than the 3-hour rainfall duration; rational method runoff coefficient C of 0.40; CN of 74; rainfall erosivity factor R of 200; soil erodibility factor K of 0.38; topographic factor LS of 0.75; cover management factor C of 0.40; and erosion control practice factor P of 1.0. Determine the sediment yield to result from a storm with 2.5 inches of rain falling in 3.0 hours with an approximately uniform temporal and spatial distribution.

Solution Various methods may be adopted for estimating V_R and Q_P. The rational (Eq. 8.11) and NRCS *CN* (Eqs. 8.15–8.18) methods are used. Because t_C is less than the rainfall duration, the entire watershed in contributing runoff at the outlet in the rational method.

$$S = \frac{1{,}000}{\text{CN}} - 10 = \frac{1{,}000}{74} - 10 = 3.51 \text{ in.}$$

$$V_R = \frac{(P - 0.2S)^2}{P + 0.8S} = \frac{(2.5 - 0.2(3.51))^2}{2.5 + 0.8(3.51)} = 0.61 \text{ in.}$$

$$V_R = (0.61 \text{ in.})\left(\frac{\text{ft}}{12 \text{ in.}}\right)(320 \text{ ac}) = 16 \text{ ac} \cdot \text{ft}$$

$$Q_P = CiA[\text{conversion factors}]$$

$$Q_P = 0.40\left(\frac{2.5 \text{ in.}}{3.0 \text{ hr}}\right)(320 \text{ ac})\left[\left(\frac{\text{ft}}{12 \text{ in.}}\right)\left(\frac{\text{hr}}{3{,}600 \text{ s}}\right)\left(\frac{43{,}560 \text{ ft}^2}{\text{ac}}\right)\right] = 108 \text{ ft}^3/\text{s}$$

The MUSLE (Eq. 8.57) is used to predict the sediment yield for the 2.5-inch rainfall event.

$$Y = 95(V_R\, Q_P)^{0.56}\, KLSCP$$

$$Y = 95[(16)(108)]^{0.56}\,(0.38)(0.75)(0.40)(1.0)$$

$$Y = 700 \text{ tons (630 metric tons)}$$

Example 8.15

Estimate the average annual sediment yield for the watershed of Example 8.14.

$$Y = DRKLSCP = (0.39)(200)(0.38)(0.75)(0.40)(1.0) = 8.9 \text{ tons/acre/year}$$

$$Y = (8.9 \text{ tons/acre/year})(320 \text{ acres}) = 2{,}800 \text{ tons/year} \quad (2{,}600 \text{ metric tons/year})$$

8.7.4 Sediment Deposition in Reservoirs

Sedimentation is an important consideration in the design and management of reservoirs for water supply, hydroelectric power, flood control, and other purposes. A dam and reservoir project on a stream results in deposition of sediment, which over time may significantly decrease its storage capacity. Typically, storage capacity is lost to sedimentation gradually over many decades. However, in extreme cases, small reservoirs have been almost completely filled with sediment during a single major flood event.

Velocities decrease as streamflow approaches the upstream limits of a reservoir. A delta is formed near the entrance to the reservoir and grows toward the dam over time. Coarser-size materials tend to be deposited first as the flow velocities decrease in upstream reaches. Finer particles are deposited throughout the reservoir. Some of the sediment continues to be suspended in the water released through the outlet structures.

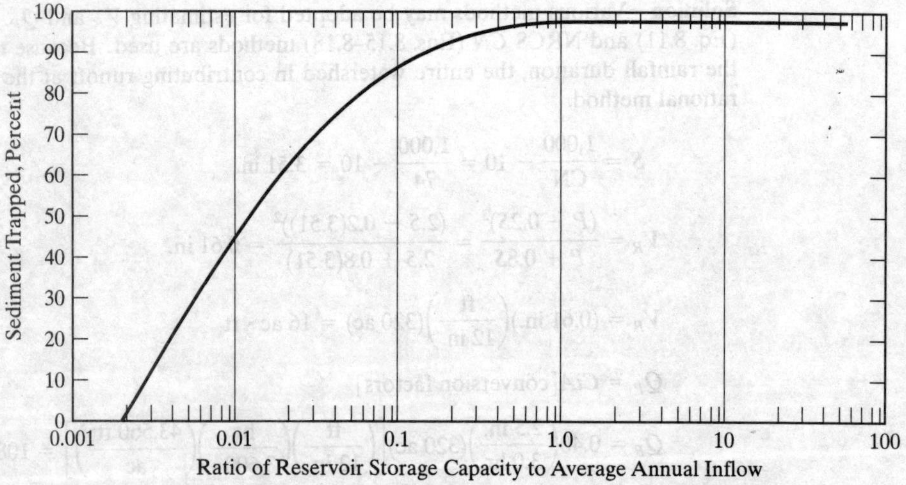

Figure 8.17 Reservoir trap efficiency may be approximated as a function of the ratio of storage capacity to mean annual inflow (U.S. Bureau of Reclamation, 1987).

The amount of sediment deposited in a reservoir is the difference between the incoming sediment yield from the upstream watershed and the materials transported with releases through the outlet structures. Accumulated sediment deposits are inflows less outflows. Reservoir sedimentation amounts are predicted as the sediment yield entering the reservoir multiplied by a trap efficiency. The reservoir trap efficiency is a measure of the proportion of the inflowing sediment that is deposited.

$$\text{trap efficiency} = \frac{\text{sediment amount deposited}}{\text{sediment amount entering}}(100\%) \tag{8.58}$$

Analyses of sediment measurements for a number of reservoirs resulted in Fig. 8.17, which may be used to estimate trap efficiency as a function of the ratio C/I of reservoir storage capacity C to mean annual inflow I for normal ponded reservoirs (U.S. Bureau of Reclamation, 1987). For example, for a reservoir with a storage capacity of 2,000 acre-feet and a mean inflow of 4,000 acre-feet/year, the C/I is 0.5, and the trap efficiency from Fig. 8.17 is 95 percent.

8.8 WATER QUALITY MODELING

As precipitation falls on and runs off the watershed, the water interacts physically, chemically, and biologically with soils, rocks, and biota. Soil particles and other suspended or dissolved substances are transported along with the flowing water. *Sediment* is the larger suspended particles entrained by moving water that will settle out as flow velocities are decreased. Sediment-bound substances are elements or compounds that are physically and chemically associated with sediment. *Particulate* is a more general term that can include settleable sediment, as well as finer colloidal particles that can only be removed from the water by filtration, centrifugation, or coagulation. *Dissolved* matter refers to ions that have disassociated and are coordinated with water molecules. Dissolved matter is operationally defined as that material that passes through a 0.45-μm diameter pore filter. Sediment yield is discussed in the previous Section 8.7. Other water quality constituents transported with the water and sediment are addressed now.

Watershed runoff carries pollutants to receiving water bodies, such as streams, rivers, lakes, reservoirs, bays, and estuaries. The watershed is considered to be a nonpoint, diffuse, or distributed source of pollution. Water quality results from both natural processes and human activities. Streams naturally carry suspended and dissolved matter from their watersheds even if uninfluenced by human development. Increased pollutant loads from human activities include eroded soil from agricultural fields and construction sites, chemical and organic fertilizers, and pesticides from crops and lawns, animal wastes, litter, land fill wastes, septic tank effluents, industrial wastes, substances washed from streets and other pavements, and many other types of pollution.

8.8.1 Flow, Load, and Concentration

The pollutant load L is the mass flux of a given pollutant. In the context of watershed modeling, L is the mass of the pollutant that flows from the watershed during a specified period of time. Typical units are kg/day or kg/year, metric or English tons/year, and pounds/day or pounds/year. Loads are often expressed as mass per time per unit area of watershed in units such as kg/(ha-yr) or lbs/(ac-yr). Various time frames may be pertinent. Average annual load and loads for specific rainfall events are two common expressions of the amount of a water quality constituent flowing past a watershed outlet or site on a stream. Loads at an instant in time, seasonal loads, or other time periods may also be of interest.

Concentration C is the mass of a substance per unit volume of water. For a substance dissolved in flowing water, the mean concentration $\overline{C}$ during a time interval is the mean load $\overline{L}$ divided by the mean discharge $\overline{Q}$ during the time interval.

$$\overline{C} = \frac{\overline{L}}{\overline{Q}}[\text{conversion factor}] \tag{8.59}$$

With $\overline{C}$, $\overline{L}$, and $\overline{Q}$ in typical units of mg/l, kg/day, and m^3/s, the conversion factor is as follows:

$$\text{mg/l} = \left[\frac{\text{kg/day}}{\text{m}^3/\text{s}}\right]\left[\left(\frac{\text{day}}{86{,}400\ \text{s}}\right)\left(\frac{1{,}000{,}000\ \text{mg}}{\text{kg}}\right)\left(\frac{\text{m}^3}{1{,}000\ \text{l}}\right)\right] \tag{8.60}$$

$$\text{conversion factor} = 0.011574$$

With C, L, and Q in mg/l, lb/day, and ft³/s, respectively, the conversion factor is as follows:

$$\text{mg/l} = \left[\frac{\text{lb/day}}{\text{ft}^3/\text{s}}\right]\left[\left(\frac{\text{day}}{86{,}400\ \text{s}}\right)\left(\frac{453.6\ \text{g}}{\text{lb}}\right)\left(\frac{1{,}000\ \text{mg}}{\text{g}}\right)\left(\frac{\text{ft}^3}{28.316\ \text{l}}\right)\right] \tag{8.61}$$

$$\text{conversion factor} = 0.1854$$

The event mean concentration (EMC) is the average concentration that occurs in the runoff from a specific rainfall event.

$$\text{EMC} = \frac{M_P}{V_R}(\text{conversion factor}) \tag{8.62}$$

where M_P is the mass of a specific pollutant and V_R is the runoff volume. With EMC, M_P, and V_R in units of mg/l, mg, and liters, respectively, the conversion factor is 1.0. For a gaged watershed, with flow samples taken periodically during a rainfall event, the EMC is determined as a flow rate-weighted average of the measured concentrations of the samples.

$$\text{EMC} = \frac{\sum C_i Q_i}{\sum Q_i} \tag{8.63}$$

where C_i is the concentration of sample i and Q_i is the flow rate at the time sample i was taken.

8.8.2 Water, Sediment, and Other Quality Constituents

Predicting pollutant loads from a watershed involves estimates of (1) water quantities, (2) sediment quantities, and (3) concentrations of the water quality constituents of concern in the watershed runoff and sediment load. The water quantity modeling methods previously covered in this chapter provide an essential foundation for water quality modeling. Likewise, the sediment yield methods of the previous section are incorporated in water quality models. However, water quality modeling also requires procedures for estimating concentrations of dissolved and sediment-bound substances. In simpler models, concentrations are based on available empirical information. In more complex models, concentrations may be obtained from more mechanistic representations of physical, chemical, and biological processes that still require empirical data.

Most chemicals of concern in watershed modeling studies have both solid and dissolved phases, but often one phase will dominate. Chemicals may be divided into three categories based on their main transport phase in stormwater runoff.

- solid phase: chemicals that are strongly associated with sediment
- dissolved phase: chemicals that are dissolved in runoff
- distributed phase: significant chemical quantities are transported in both solid-phase and dissolved forms

Loading functions are relationships between dissolved and solid-phase pollutant concentrations and water flow rates or sediment mass fluxes. Loading functions for the first two categories are more straightforward than for the third. Runoff of distributed-phase chemicals is more difficult to model since dissolved and solid-phase concentrations are influenced by adsorption equilibrium phenomena (Mills *et al.*, 1985).

Solid-phase chemicals include organic nitrogen, particulate phosphorus, and heavy metals. Dissolved-phase heavy metals also occur in runoff under acidic conditions. Dissolved-phase chemicals include inorganic nitrogen and soluble phosphorus. Inorganic nitrogen in runoff is mostly nitrate-nitrogen, and this ion does not adsorb to soil particles. Although most phosphorus in runoff is solid phase, smaller amounts of dissolved phosphorus may also be important in certain situations, such as eutrophication studies, because the dissolved phosphorus is directly available to plants and algae. Loading functions for solid-phase and dissolved phosphorus are operational means of approximating complex soil chemistry. Modeling reactions that relate fixed, adsorbed, and soluble forms of phosphorus is complex and not highly accurate. Distributed-phase chemicals include most organic pesticides. Models for runoff of these chemicals are considerably more complex than the solid-phase and dissolved chemical loading functions.

8.8.3 Types of Water Quality Models

Various methods have been adopted for modeling water quality constituents in watershed runoff (Donigian, Imhoff, and Ambrose, 1995; Trudgill, 1995). The following approaches are representative of the array of modeling capabilities: (1) constant mean concentration, (2) loading functions, (3) regression equations, (4) urban buildup and washoff loading functions, and (5) comprehensive computer simulation models.

The first approach is based on estimating the mean concentration of specific pollutants that are assumed to be constant for different rainfall events. The most accurate way to determine a representative concentration is to actually measure water quality at the site of concern. Without field measurements for the watershed of interest, approximate estimates may be based on published data from measurements of concentrations in runoff from similar watersheds with the same land use and other characteristics. Runoff volumes and flow rates determined using the methods previously outlined in this chapter are combined with the concentration value to obtain the pollutant load. An annual runoff load may be determined by multiplying an annual runoff volume by a mean annual concentration. Likewise, the pollutant load for a given rainfall event is estimated by multiplying its runoff volume by a mean concentration. Applications of this simple constant mean concentration approach range from simple manual computational methods to incorporation as a simple component of a large computer simulation model. Because concentrations are known to vary greatly between rainfall events at the same location, the approach is considered to be very approximate.

The EPA screening procedures are among the more widely used of the array of methods reported in the literature based on manual (noncomputer) applications of loading functions. An EPA manual (Mills *et al.*, 1985) documents a set of screening procedures for assessing the loading and fate of conventional and toxic pollutants in streams, impoundments, estuaries, and groundwaters. The manual contains most of the required empirical data, as well as detailed instructions for applying the techniques. The portion of the manual covering modeling of pollutant loads from watersheds provides methods for both rural and urban runoff loads. The procedures for predicting loads from rural watershed are briefly discussed in Section 8.8.4.

Regression equations represent a purely empirical approach to watershed modeling. Pollutant loads are expressed as a function of selected watershed parameters that can be determined from readily available sources of information. A database of measured pollutant loads at many locations is required. Statistical regression analysis techniques are applied to relate the observed pollutant loads to selected watershed characteristics. The resulting regression equations provide predictive models that can be applied to other watersheds. A variety of regression models for predicting loads and concentrations are reported in the literature. The USGS urban regression equations discussed in Section 8.8.5 are probably the most significant of such models.

Buildup and washoff loading functions are incorporated in the EPA screening procedure for urban runoff (Mills *et al.*, 1985) and in several computer models including SWMM described in Section 8.9.2. The pollutant load from an urban watershed to result from a storm event is conceptualized as being dependent on the buildup (accumulation) of sediment (dirt, dust, and other solids) on streets and other surfaces during dry periods and the washoff of this sediment during rainfall events. Empirical relationships predict buildup of sediment and associated pollutants as a function of land use, street cleaning practices, and elapsed time since the last storm.

As discussed in Section 8.9, federal agencies and other entities have developed a number of generalized watershed simulation computer models with water quality analysis capabilities. The ARS-SWAT, EPA-SWMM, and EPA BASINS are described in Section 8.9.

8.8.4 EPA Screening Procedure for Rural Watersheds

The Environmental Protection Agency screening procedure for solid-phase and dissolved-phase pollutants from rural watersheds is based on the following loading functions (Mills et al., 1985).

$$L_S = 0.001\, C_S S_L \qquad \textbf{(8.64)}$$

$$L_D = 0.1\, C_D V_R \qquad \textbf{(8.65)}$$

where L_S is the solid-phase chemical load in kg/ha, C_g is the concentration of the chemical in the sediment in mg/kg, and S_L is the sediment load in metric tons/ha. L_D is the dissolved chemical load in kg/ha, C_D is the concentration of dissolved chemical in mg/l, and V_R is the runoff depth in cm. Equation 8.64 is applicable for solid-phase chemicals, such as organic nitrogen, particulate phosphorus, and heavy metals. Equation 8.65 is for dissolved substances, such as inorganic nitrogen and soluble phosphorus.

EPA recommends that V_R in Eq. 8.65 be determined using the NRCS CN method (Section 8.5.1) and that S_L in Eq. 8.64 be determined with the USLE (Section 8.7.1). Equation 8.64 can be applied with either single event, annual, or seasonal values of S_L from the USLE (Eq. 8.54), MUSLE (Eq. 8.57), or related methods to determine the corresponding pollutant load L_D. Equation 8.65 is applied with V_R from the NRCS CN method (Eqs. 8.15–8.18) to estimate L_D for individual rainfall events. Equation 8.65 can also be used to estimate an annual L_D corresponding to an annual V_R determined by other methods.

The concentrations C_S and C_D in Eqs. 8.64 and 8.65 are best determined by direct measurement through field sampling. The procedures outlined as follows provide approximate concentrations when sampling is not feasible (Mills *et al.*, 1985).

Soil nutrient (nitrogen and phosphorus) concentrations are available from soil surveys and the literature. Nitrogen concentrations may be inferred from soil organic matter percentages based on the premise that organic matter is about 5 percent nitrogen. Representative concentrations of dissolved nutrients are provided in Table 8.8. Locally observed concentrations should be used for C_D in Eq. 8.65 whenever possible. However, Table 8.8 provides C_D values for the common situation in which local water quality data are not available.

Nitrogen and phosphorus concentrations in eroded soil are generally larger than comparable concentrations in uneroded or *in situ* soil. Lighter organic matter and clay particles are more readily eroded than heavier sand and silt. Because nutrients tend to be associated with these light particles, sediment is enriched with nutrients compared with the soil from which it originates. A sediment nutrient concentration C_S can be related to the comparable *in situ* concentration C_I in soil by an enrichment ratio *en*.

$$C_S = enC_I \tag{8.66}$$

TABLE 8.8 REPRESENTATIVE DISSOLVED NUTRIENT CONCENTRATIONS IN RURAL RUNOFF (MILLS *et al.*, 1985)

Soil cover	Nitrogen (mg/l)	Phosphorus (mg/l)
Fallow	2.6	0.10
Corn	2.9	0.26
Small grains	1.8	0.30
Hay	2.8	0.15
Pasture	3.0	0.27
Forest		
Eastern U.S.	0.19	0.006
Midwest	0.06	0.009
West	0.07	0.012

The nutrient enrichment ratio *en* depends on the degree of erosion during the rainfall event. With very small storms, only the finest soil particles are eroded, and *en* is high. Conversely, large storms erode all size soil particles, and *en* approaches one. The following relationship is based on analyses of many field studies of nutrient transport.

$$en = \frac{7.39}{S_L^{0.2}} \tag{8.67}$$

The sediment load S_L during the storm event is in kg/ha. For very large soil losses ($S_l > 22{,}040$ kg/ha), Eq. 8.67 indicates that *en* is less than 1.0, in which case *en* should be set equal to 1.0. Equation 8.67 is for individual storm events. For annual loads, a midrange *en* value of about 2.0 is appropriate.

Runoff of pesticides can be described by the same general loading functions used for nutrients (Eqs. 8.64 and 8.65). However, the estimation of dissolved- and solid-phase concentrations is more difficult for pesticides. All pesticides are adsorbed to some extent by soil particles, and hence dissolved- and solid-phase concentrations cannot be determined independently. Also, these concentrations are dynamic, because pesticides are decomposed or decayed by photochemical, chemical, and microbiological processes. Decay rates are often sufficiently high that most of a pesticide will have decomposed within several weeks of application. A final complicating factor is the large number of pesticide compounds currently in use, each with its own properties and characteristic behavior in the soil.

Because pesticide concentrations in runoff depend on the relative timing of applications and storm events, and the specific adsorption and degradation properties of the pesticide, modeling is more complex. However, relatively simple equations can be used to describe the adsorption and decay phenomena, and calculations can be made for each storm event following a pesticide application. Mills *et al.* (1985) describe such a model and provide model parameters for a large number of pesticides. However, this analysis procedure is outside of the scope of our limited coverage.

8.8.5 USGS Regression Models for Urban Watersheds

In support of urban stormwater management efforts nationwide, the USGS has developed sets of regression equations for estimating constituent loads and concentrations for storm events, and mean seasonal and annual loads based on climatic, land use, and physical watershed characteristics (Driver and Tasker, 1990). These regression models provide capabilities for estimating loads and concentrations for urban watersheds for which direct water quality measurements are not available.

Databases compiled by the USGS and EPA were combined for use in developing the regression models. The national urban water quality database included observations of 2,813 storms at 173 gaging stations in 30 metropolitan areas. Multiple regression analyses, including ordinary least squares and generalized least squares, were used to determine the optimum linear regression models. Driver and

Tasker (1990) document the regression analyses in detail and present the resulting models for predicting, as functions of urban physical, land use, and climatic characteristics, the following: (1) discharge-weighted storm runoff constituent loads, (2) storm runoff volumes, (3) mean seasonal or annual constituent loads, and (4) mean concentrations of constituents during storm runoff events. Only the models for predicting mean concentrations during storm runoff events are reproduced here.

The following form of regression model was adopted.

$$\hat{Y} = \hat{\beta}_0 x_1^{\hat{\beta}1} x_2^{\hat{\beta}2} \ldots x_n^{\hat{\beta}n} \, \mathrm{BCF} \tag{8.68}$$

where $\hat{Y}$ is the response variable (constituent load or concentration or runoff volume); $x_1, x_2, \ldots, x_n$ are explanatory variables (watershed and climatic characteristics); n is the number of such characteristics included in the regression model; and $\hat{\beta}_0, \hat{\beta}_1, \hat{\beta}_2, \ldots, \hat{\beta}_n$ are regression coefficients. Without the bias-correction factor (BCF), the model provides an estimator of median response. The BCF converts the model to an estimator of mean response. Thus, Eq. 8.68 is applied to estimate the expected value or mean $\hat{Y}$. A median value of $\hat{Y}$ is obtained by omitting the BCF.

After analyzing alternative regionalization approaches, the United States was divided into three regions based on mean annual rainfall (MAR) as follows:

Region I	MAR $<$ 20 inches
Region II	20 inches $\leq$ MAR $<$ 40 inches
Region III	MAR $\geq$ 40 inches

The three regions are denoted by labeling variables with the numerals I, II, or III.

The models for predicting storm runoff mean concentrations consist of Eq. 8.68 with the coefficient values reproduced in Table 8.9. The response variables are mean concentrations of storm runoff, expressed in either milligrams per liter (mg/l) or micrograms per liter (μg/l), for 11 constituents. The mean concentration for a storm event is the constituent load divided by runoff volume converted to appropriate units. The water quality constituents modeled by the regression equations of Eq. 8.68 and Table 8.9 are as follows:

COD	chemical oxygen demand in mg/l
SS	suspended solids in mg/l
DS	dissolved solids in mg/l
TN	total nitrogen in mg/l
TKN	total ammonia plus organic nitrogen as nitrogen in mg/l
TP	total phosphorus in mg/l
DP	dissolved phosphorus in mg/l
CD	total recoverable cadmium in μg/l
CU	total recoverable copper in μg/l
PB	total recoverable lead in μg/l
ZN	total recoverable zinc in μg/l

TABLE 8.9 REGRESSION COEFFICIENTS FOR STORM RUNOFF MEAN CONCENTRATION MODELS

	$\hat{\beta}_0$	TRN	DA	IA +1	LUI +1	LUC +1	LUR +1	LUN +2	PD	DRN	INT	MAR	MNL	MJT	BCF	R^2
COD I	5.035	−0.473	−0.087	—	0.388	0.012	—	0.048	—	—	—	0.855	—	—	1.163	0.52
COD II	0.254	−0.259	−0.054	—	0.0003	0.025	—	−0.033	—	—	—	1.566	—	—	1.299	0.20
COD III	46.9	−0.179	−0.047	—	0.320	0.031	—	−0.169	—	—	—	—	—	—	1.270	0.18
SS I	2,041	0.143	0.108	—	—	—	—	—	—	−0.370	—	—	—	—	1.543	0.13
SS II	734	0.132	−0.342	−0.329	—	—	—	—	0.041	—	—	—	—	−0.519	1.650	0.19
SS III	176	0.054	0.286	—	0.168	0.072	—	−0.295	—	—	—	—	—	—	1.928	0.14
DS I	0.333	−0.402	0.469	0.445	—	—	—	—	—	—	—	1.497	—	—	1.352	0.68
DS II	2,398	−0.112	0.519	0.468	—	—	—	—	—	—	—	—	—	−1.373	1.179	0.66
TN I	3.52	−0.285	0.033	—	0.512	0.017	—	0.012	—	—	—	−0.129	—	—	1.096	0.54
TN II	1.65	−0.204	0.065	0.176	—	—	—	—	—	—	—	—	−0.296	—	1.256	0.10
TN III	26,915	−0.253	−0.169	0.057	—	—	—	—	—	—	—	−2.737	—	—	1.308	0.37
TKN I	1.282	−0.449	0.022	—	0.426	−0.016	—	−0.012	—	—	—	—	0.347	—	1.167	0.59
TKN II	0.830	−0.224	−0.066	0.039	—	—	—	—	—	—	—	—	0.106	—	1.321	0.12
TKN III	9,549	−0.157	−0.159	—	—	—	—	−0.086	—	—	—	−2.447	—	—	1.326	0.31
TP I	0.085	−0.232	−0.012	—	0.552	−0.080	—	0.038	—	—	—	0.530	—	—	1.261	0.51
TP II	0.022	−0.177	−0.133	0.006	—	—	—	—	—	—	2.019	—	—	—	1.521	0.15
TP III	2.630	−0.016	−0.107	—	—	0.053	0.184	−0.168	—	—	—	—	—	−0.710	1.365	0.29
DP I	0.352	−0.294	−0.013	—	0.629	−0.136	—	−0.046	—	—	—	−0.297	—	—	1.266	0.56
DP II	0.003	−0.209	−0.174	0.245	—	—	—	—	—	—	1.514	—	—	—	1.567	0.31
DP III	0.060	0.189	−0.076	—	—	—	—	0.358	—	—	—	—	—	—	1.341	0.24
CD I	0.338	−0.256	0.025	—	0.090	0.033	—	−0.110	—	—	—	0.481	—	—	1.166	0.20
CD II	0.851	0.223	0.189	—	—	—	—	—	—	—	—	—	—	0.394	1.284	0.10
CU I	11.3	−0.327	0.066	—	0.237	0.048	—	0.155	—	—	0.406	—	—	—	1.297	0.34
CU II	9.683	−0.298	−0.151	0.157	—	—	—	—	—	—	—	—	—	—	1.473	0.14
CU III	1,774	−0.104	−0.077	—	0.446	−0.78	—	−0.204	—	—	−3.247	—	—	—	1.348	0.67
PB I	141	−0.347	0.145	—	−0.109	0.034	—	−0.086	—	—	—	0.046	—	—	1.304	0.19
PB II	0.487	−0.268	−0.359	—	—	0.099	0.152	−0.008	—	—	—	1.088	—	—	1.433	0.41
PB III	39.8	−0.196	0.123	0.404	—	—	—	—	—	—	—	—	—	—	1.510	0.37
ZN I	199	−0.338	0.070	—	—	−0.029	0.114	0.068	—	—	—	−0.004	—	—	1.242	0.22
ZN II	0.149	−0.238	−0.201	0.278	—	—	—	—	—	—	—	—	—	1.961	1.650	0.15
ZN III	1,879	−0.149	−0.061	—	0.285	0.146	−0.078	—	—	—	—	—	—	−0.916	1.322	0.37

These constituents were selected based on their importance in urban stormwater management and availability of measured data. The regression models for dissolved solids (DS) and cadmium (CD) for region III (MAR > 40 inches) are omitted due to insufficient observed data.

The explanatory (independent) variables are as follows:

TRN	total storm rainfall in inches
DA	drainage area in square miles
IA	impervious area as a percent of total drainage area
LUI	industrial land use as a percent of total drainage area
LUC	commercial land use as a percent of total drainage area
LUR	residential land use as a percent of total drainage area
LUN	nonurban land use as a percent of total drainage area
PD	population density in people per square mile
DRN	storm duration in minutes
INT	maximum 24-hour precipitation intensity that has a 2-year recurrence interval in inches
MAR	mean annual rainfall in inches
MNL	mean annual nitrogen load in precipitation in pounds of nitrogen per acre
MJT	mean January temperature, in degrees Fahrenheit

Explanatory variables were selected on the basis of their frequency of availability in the USGS/EPA database, ease of measurement by practitioners in applying the models, and achievement of physically logical combinations of variables. Highly correlated explanatory variables were not combined in the same model. For example, percent impervious (IA) is not included in the same model with land use (LUI, LUC, and LUR). Explanatory variables for each regression model were selected using stepwise regression procedures available in a widely used software package called the Statistical Analysis System.

The coefficients $\hat{\beta}_0, \hat{\beta}_1, \hat{\beta}_2, \ldots, \hat{\beta}_n$ and bias correction factor BCF in Eq. 8.68 are tabulated in Table 8.9. For example, the regression equation for predicting the chemical oxygen demand (COD), in mg/l, for a storm in region III (MAR > 40 inches) is

$$\text{COD III} = 46.9\,\text{TRN}^{-0.179}\,\text{DA}^{-0.047}\,(\text{LUI}+1)^{0.320}\,(\text{LUC}+1)^{0.031}\,(\text{LUN}+2)^{-0.169}\,1.270$$

Event mean concentrations are determined from Eq. 8.68 with coefficients from Table 8.9. Equation 8.62 expresses the relationship between event mean concentration, load, and volume. Storm runoff loads may be determined by combining mean concentrations estimated using Eq. 8.68 and Table 8.9 with runoff volume determined with any of the methods previously covered in this chapter.

Driver and Tasker (1990) present the results of statistical correlation and error analyses for the models. The most accurate models are generally those for DS, TN, and TKN. Models for SS tend to be least accurate. The most accurate models were those for application in the more arid western states, and the least accurate are those for areas with large mean annual rainfall. The coefficient of multiple determination (R^2) tabulated in the last column of Table 8.9 measures the proportion of total variation about the mean explained by the regression.

Example 8.16

The climate of a particular city is characterized by a mean annual rainfall (MAR) of 32 inches/year, mean January temperature of 39°F, and a 2-year recurrence interval, 24-hour precipitation depth (INT) of 2.3 inches. A watershed in the city has a drainage area (DA) of 0.75 mi.^{2} and land use characterized by 30 percent impervious (IA) and a population density (PD) of 3,800 people per mi.^{2} Estimate the mean concentrations of suspended solids (SS), dissolved solids (DS), and total phosphorus (TP) for a 2.6-inch rainfall event.

Solution With a MAR of 32 inches, the city falls with region II. Models for estimating the mean concentration of the watershed runoff from the 2.6 inches of rainfall are obtained by substituting coefficients from Table 8.9 into Eq. 8.68.

$$\text{SSII} = 734\ \text{TRN}^{0.132}\ \text{DA}^{-0.342}\ (\text{IA} + 1)^{-0.329}\ \text{PD}^{0.041}\ \text{MJT}^{-0.519}\ 1.650$$

$$= 734(2.6)^{0.132}\ (0.75)^{-0.342}\ (31)^{-0.329}\ (3{,}800)^{0.041}\ (39)^{-0.519}\ 1.650 = 103\ \text{mg}/\ell$$

$$\text{DSII} = 2{,}398\ \text{TRN}^{-0.112}\ \text{DA}^{0.519}\ (\text{IA} + 1)^{0.468}\ \text{MJT}^{-1.373}\ 1.179$$

$$= 2{,}398(2.6)^{-0.112}\ (0.75)^{0.519}\ (31)^{0.468}\ (39)^{-1.373}\ 1.179 = 71\ \text{mg}/\ell$$

$$\text{TPII} = 0.022\ \text{TRN}^{-0.177}\ \text{DA}^{-0.133}\ (\text{IA} + 1)^{0.006}\ \text{INT}^{2.019}\ 1.521$$

$$= 0.022(2.6)^{-0.177}\ (0.75)^{-0.133}\ (31)^{0.006}\ (2.3)^{2.019}\ 1.521 = 0.16\ \text{mg}/\ell$$

8.9 GENERALIZED WATERSHED SIMULATION MODELS

Many available computer models for simulating watershed hydrology are generalized for broad applicability with values for parameters being input by the model user to describe a particular watershed. The methods covered in this chapter and other similar methods are incorporated into the computer programs. The models allow a watershed to be divided into any number of subbasins, as illustrated by Fig. 8.3, with the methods of Section 7.11 and Chapter 8 being applied to each individual subbasin. The routing methods of Chapter 6 are incorporated in the watershed models to route the runoff hydrographs through stream reaches and reservoirs. Thus, complex multiple-subbasin models may be constructed by combining and repeating the techniques presented in Chapters 6–8.

About 40 generalized watershed models are described by Wurbs (1995), Singh (1995), Donigian, Imhoff, and Ambrose (1995), and Bevin (2000). The remainder of this chapter focuses on the following models: Hydrologic Modeling System

(HMS) developed by the Hydrologic Engineering Center (HEC), Stormwater Management Model (SWMM) developed by the Environmental Protection Agency (EPA), and Soil and Water Assessment Tool (SWAT) developed by the Agricultural Research Service (ARS). HMS, SWMM, and SWAT provide a broad array of modeling capabilities that include most of the capabilities reflected in the many other available watershed models. HMS, SWMM, and SWAT:

- are readily available public domain models maintained by federal agencies
- have evolved through various versions over many years and continue to be improved and expanded
- are flexible packages of modeling options that may be used to construct models for particular watersheds and applications ranging in complexity from quite simple to very complex
- are extensively applied by agencies and consulting firms throughout the U.S. and other countries to a broad range of water resources engineering problems
- should be applied only by water resources engineers having a thorough knowledge of the methods incorporated in the models, including the material presented in this book

HEC-HMS simulates single precipitation events using a time step typically in the range of a few minutes to several hours. SWAT uses a daily time interval with typical simulation periods of several months to many years containing multiple precipitation events separated by extended dry periods. SWMM may be used for either single-event or continuous long-period simulations. SWMM and SWAT provide water quality, as well as quantity capabilities. HMS has no water quality features. SWMM is designed primarily for urban watersheds. SWAT is oriented toward agricultural and other rural land uses. HMS is applied to both urban and rural watersheds. All three models have options to use either metric or U.S. customary units. The public domain models are available from the sponsoring agencies and others either free-of-charge or for a nominal handling fee. Software, user documentation, and related information are available at the following web sites:

HEC-HMS	http://www.hec.usace.army.mil/
EPA-SWMM	http://www.epa.gov/ceampubl/ceamhome.htm
ARS-SWAT	http://www.brc.tamus.edu/swat/

Each of the methods covered in this chapter is itself a model. These models are combined to construct larger models. These models are incorporated in generalized models, such as HMS, SWMM, and SWAT. A model for a particular watershed may consist of combining parameter values and other input for that particular watershed with one of the generalized models. Although HMS, SWAT, and

SWMM are modeling systems themselves combining multiple models, they may be incorporated into larger modeling systems.

EPA's BASINS (Better Assessment Science Integrating Point and Nonpoint Sources) is an example of an integrated modeling framework that combines several generalized models. BASINS is a system of models and databases incorporated into an ArcView-based geographical information system environment. BASINS is a water quality modeling system that accesses SWAT and other models, including the EPA's Hydrologic Simulation Program-FORTRAN (HSPF), which is a watershed model somewhat similar to SWAT, and QUAL2E, which is a one-dimensional, steady-state stream water quality model. SWAT, HSPF, and QUAL2E have been extensively applied as individual models for many years prior to being combined within the geographical information system-based BASINS environment. BASINS integrates these models with databases and data management tools. BASINS has evolved through several improved and expanded versions since its initial release in 1996 (http://www.epa.gov/ostwater/BASINS/).

8.9.1 Hydrologic Modeling System (HEC-HMS)

HEC-HMS (HEC, 2001) and its predecessor, the HEC-1 Flood Hydrograph Package (HEC, 1998), are probably the most extensively applied of all watershed computer models. The model has been applied to watersheds ranging from small urban areas of less than 1 km^2 to large river basins of several hundred thousand km^2. Design flood hydrographs developed with HEC-1 have been used to size numerous dams and spillways, flood control improvements, and other hydraulic structures throughout the U.S. and abroad. The model has been applied extensively in floodplain management activities. Other common applications include economic evaluation of flood mitigation plans, dam safety studies, and real-time flood forecasting.

HEC-HMS, first released in 1997, is an object-oriented, menu-driven update of HEC-1 that has evolved through various versions dating back to 1968. The models compute discharge hydrographs at pertinent locations to result from observed precipitation events or synthetic design storms. A watershed consists of any number of subbasins, stream reaches, reservoirs, and other components (Fig. 8.3). Constructing a model for a particular watershed with HEC-HMS involves the selection of methods for representing each watershed component and providing parameter values and other input required for each method. The generalized modeling package provides the following capabilities.

- Rainfall depths for each time increment of a storm can be input in various formats.
- The model will develop a design storm for a specified exceedance frequency from an input IDF relationship using the approach outlined in Section 7.11.
- Snowfall and snowmelt are simulated using either degree-day or energy-budget methods.

- Precipitation is separated into losses and runoff using either of the following options: NRCS CN method (Section 8.5.1), Green and Ampt (Section 8.5.3), Holtan (Section 8.5.4), exponential, and initial/uniform.
- Runoff hydrographs for each subbasin are developed with either the unit hydrograph (Section 8.6) or kinematic wave (Section 6.2) approach as outlined in Fig. 8.10.
- A unit hydrograph can be either input or synthesized by the model using the NRCS dimensionless (Section 8.6.4), Snyder (Section 8.6.5), Clark (Section 8.6.6), or modified Clark methods (Section 8.6.6).
- The storage-outflow method is used for reservoir routing (Section 6.1). Stream routing options include storage-outflow (Section 6.1), Muskingum (Section 6.1), Muskingum-Cunge, and kinematic (Section 6.2).
- An optimization routine facilitates parameter calibration.

8.9.2 Storm Water Management Model (SWMM)

The EPA's SWMM has been widely applied throughout the U.S. and Canada, and elsewhere as well. SWMM is a comprehensive hydrologic, hydraulic, and water quality simulation model developed primarily for urban areas (Huber and Dickinson, 1988). Applications include various aspects of planning and design of urban drainage and storm water management facilities and studies addressing nonpoint pollution and related issues. The model was originally developed in 1969–1971 and continues to be periodically updated and expanded.

SWMM is a large, relatively complex software package capable of simulating the movement of precipitation and pollutants from the ground surface, through storm sewer networks, channels, and storage/treatment facilities, to receiving waters. The model can be used to simulate a single rainfall event or a long continuous period with multiple storms separated by dry periods. Snowmelt may also be simulated. SWMM computes runoff volumes, flow hydrographs, pollutant loads, and pollutographs.

A watershed is divided into a number of subwatersheds. Each subwatershed is modeled as a nonlinear reservoir with precipitation as input and infiltration, evaporation, and surface runoff as outflows. Infiltration is modeled with either the Green and Ampt or Horton equations. Infiltrated water is routed through upper and lower subsurface zones and may contribute to runoff. Flows are routed through channel and pipe systems using either kinematic or dynamic routing.

Quality constituents modeled by SWMM include suspended solids, settleable solids, biochemical oxygen demand, nitrogen, phosphorus, and grease. A variety of options are provided for determining surface runoff loads, including constant concentration, regression relationships of load versus flow, and buildup-washoff. The buildup-washoff option uses empirical relationships to express the accumulation (buildup) of pollutants on the subwatershed surface as a function of elapsed time since the last cleaning by street sweeping or rain. Other empirical relationships represent the washoff of the accumulated pollutants during a rainfall event. Dry-weather sanitary sewer flows and loads are generated based on land use, population density, and other factors. Pollutants are routed through the sewer system assuming complete mixing at inlets. Routines are provided to modify hydrographs and pollutographs at appropriate locations to represent storage and/or treatment.

8.9.3 Soil and Water Assessment Tool (SWAT)

The Agricultural Research Service (ARS) of the U.S. Department of Agriculture (USDA) has developed several watershed models that share certain computational methods. The Simulator for Water Resources in Rural Basins (SWRRB) was developed by modifying and expanding the Chemicals, Runoff, and Erosion from Agricultural Management Systems (CREAMS) model that simulates hydrology, erosion, nutrients, and pesticides from field-size areas. SWRRB expands CREAMS for applicability to larger more complex watersheds. The SWAT model was developed during the 1990's based largely on expanding SWRRB. SWAT is supported by the USDA ARS Grassland, Soil, and Water Research Laboratory in Temple, Texas.

SWAT is a daily time step continuous model designed to predict the impact of land management activities on water, sediment, and agricultural chemical yields in large ungaged watersheds. A watershed is divided into subbasins and stream reaches with yields for each subbasin being computed, routed through stream reaches and reservoirs, and combined as appropriate.

Daily precipitation may be provided as input. Alternatively, the model can synthetically generate sequences of daily precipitation based on preserving statistical characteristics determined from gage records. The continuous sequences of daily precipitation, including dry periods of zero precipitation, may be many years long. Precipitation is partitioned between rainfall and snowfall based on daily air temperature. Snow is melted as a function of temperature.

The NRCS CN or Green and Ampt methods are used to convert precipitation to runoff. A percolation component uses storage routing to predict flow through specified soil layers. Channel seepage losses are determined as a function of channel width, length, and flow duration. ET estimation options include the Hargreaves, Priestley-Taylor, and Penman-Monteith methods. Sediment yield is estimated for each subbasin with the modified USLE. The runoff volume and peak flow in the MUSLE (Eq. 8.57) are determined in SWAT using the NRCS CN method and a version of the rational method.

SWAT simulates crop growth and crop yield based on climatic variables. A harvest index increases as a function of heat units and is reduced by water stress during critical crop stages. Above-ground biomass at harvest is partitioned between harvest yield, biomass incorporated into the soil, and residue left on the soil surface.

Nutrient loadings are determined along with water and sediment yields. The amount of nitrate in the surface flow, lateral subsurface flow, and percolation is estimated as the products of water volumes and mean concentrations. An empirical loading function is used to estimate the daily organic nitrogen load in the runoff based on the concentration of organic nitrogen in the top soil layer, the sediment yield, and the enrichment ratio. Estimates of phosphorus in surface runoff is based on partitioning into the solution and sediment phases. Crop use of nitrogen and phosphorus is also modeled.

Empirical relationships are also used to simulate pesticide transport by runoff, percolation, evaporation, and sediment. Each pesticide is characterized by parameters that include solubility, half-life in soil and on foliage, washoff fraction, and organic carbon adsorption coefficient. The proportion of the applied pesticide that reaches the soil surface depends on a plant leaf-area index.

Hydrologic routing techniques are applied to route the runoff through stream reaches and reservoirs. Methods are included in SWAT to transport sediment, nutrient, and pesticide loadings along with the water. The water, sediment, nutrient, and pesticide yields determined by the daily computations are summed to obtain monthly and annual yields.

PROBLEMS

8.1. Use the NRCS lag equation to estimate the lag time for a watershed characterized by a curve number (CN) of 75 and an average watershed slope of 2.8 percent. The distance along the flow path from the outlet to the furthest point on the watershed divide measured from a topographic map is 2,100 m (6,900 ft).

8.2. Estimate the time of concentration for the watershed of Problem 8.1.

8.3. Precipitation drains from a park area into a street gutter to a storm sewer inlet. The flow path from the sewer inlet to the most hydraulically distant point in its watershed consists of 175 ft of overload flow through Bermuda grass on a 2.2 percent slope and gutter flow in a 260 ft length of paved gutter on a longitudinal slope of 1.5 percent. Use Eqs. 8.8–8.10 and Table 8.1 to estimate the time of concentration.

8.4. Estimate the 50-year recurrence interval peak discharge for a 560-acre watershed in Dallas characterized by a time of concentration of 20 minutes and a runoff coefficient C of 0.75. An IDF relationship for Dallas is provided by Eq. 7.50 and Table 7.6.

8.5. Determine the risk that the discharge determined in Problem 8.4 will be exceeded during any (a) 20-year period and (b) 50-year period.

8.6. Estimate the peak discharge for a watershed in Miami, Florida, that has a 0.10 probability of being equaled or exceeded in any year. The watershed is characterized by a drainage area of 1.0 mi.^{2}, a time of concentration of 25 minutes, and a runoff coefficient of 0.65. An IDF relationship for Miami is provided by Eq. 7.50 and Table 7.7.

8.7. What are the probabilities that the discharge determined in Problem 8.6 will be equaled or exceeded at least once during the next (a) 5 years and (b) 10 years?

8.8. Use the rational formula to compute the peak discharge from a 1.5 km^2 watershed for a rainfall intensity of 25 cm/hr. The runoff coefficient C is 0.75.

8.9. Apply the rational method to develop a discharge-frequency relationship for a selected small watershed located in your community.

8.10. A 200 m × 200 m square parking lot slopes toward one of its edges. The parking lot drains into a gutter along the downstream edge. Adjacent land drains away from the parking lot, such that the parking lot represents the entire watershed being considered. The gutter drains into a storm sewer inlet at its center. The longest flow path (corner to inlet) consists of 200 m of overland flow across the parking lot and 100 m of gutter. Velocities are 10 m/minute for overland and 20 m/minute for gutter flow. A cumulative rainfall depth versus duration relationship for a 25-year recurrence interval is provided in the table. C is 0.95

Time (min)	Depth (mm)
5	25
10	39
15	49
20	57
25	63
30	67
60	86

Use the rational method to estimate the 25-year recurrence interval peak inflow to the storm sewer inlet.

8.11. Use the rational formula to determine the peak inflow to the storm sewer inlet of Problem 8.10 for a design storm characterized by a 25-year recurrence interval, uniform temporal distribution, and rainfall duration of alternatively (a) 15 minutes and (b) 60 minutes. Explain why the rainfall duration is set equal to the time of concentration in the conventional rational method.

8.12. Assume the parking lot in Problem 8.10 is located in Dallas, Texas. Determine the 25-year peak inflow to the storm sewer inlet using the rainfall IDF relationship provided by Eq. 7.50 and Table 7.6.

8.13. The runoff coefficient is 0.90, and the time of concentration is 20 minutes for a 10-acre watershed. Which of the following design rainfall storms, associated with the same annual exceedance frequency, results in the greatest peak discharge by the rational formula: (a) 2.2 inches uniformly distributed over a duration of 20 minutes or (b) 2.6 inches uniformly distributed over a duration of 30 minutes. Explain why the rainfall duration is set equal to the time of concentration in the rational method.

8.14. Estimate the runoff volume from a 95-acre watershed that would result from 3.2 inches of rainfall. The watershed CN is 78. Express your answer alternatively in inches, acre-feet, and ft^3.

8.15. Use the NRCS CN method to determine the runoff volume to result from a storm with a rainfall depth of 98 mm falling on a watershed with a drainage area of 1.75 km^2 and a CN of 70. Express your answer alternatively in mm and m^3.

8.16. Use the NRCS CN method to determine the rainfall excess (runoff) volume resulting from each 1-hour increment of rainfall for the synthetic design storm of Problem 7.38 falling on a watershed with a CN of 86.

8.17. Use the NRCS CN method to determine the rainfall excess (runoff volume) resulting from each 1-hour increment of rainfall for the synthetic design storm of Problem 7.40 falling on a watershed with a CN of 82.

8.18. A sandy clay soil has an initial effective saturation S_e of 25 percent. Estimate the cumulative infiltration and infiltration rate after 30 minutes of rainfall at a constant intensity of 12 mm/hr.

8.19. A loam soil has an initial effective saturation S_e of 30 percent. Estimate the cumulative infiltration and infiltration rate after 1.0 hour of rainfall at a uniform intensity of 15 mm/hr.

8.20. A synthetic design storm was developed in Example 7.11 for 24-hour rainfall duration and 2-hour computational time step. This design storm will be applied to a watershed characterized by a drainage area of 45 acres, sandy loam soil, and an initial effective saturation S_e of 32 percent. Use the Green and Ampt model with parameter values from Table 8.4 to determine the runoff volume. Perform the computations in English units and express your final results alternatively in inches and acre-feet.

8.21. Repeat Problem 8.20 with the computations performed using metric units. Express your final results alternatively in mm and m^3.

8.22. A storm with a duration of about 24 hours resulted in the following hydrograph at a gaging station on a river. The flow was 52 m^3/s before the rain began. The drainage area above the gaging station is 1,450 km^2. Use the observed hydrograph to develop a 24-hour rainfall duration unit hydrograph for this watershed.

Time (hrs)	Flow (m^3/s)	Time (hrs)	Flow (m^3/s)	Time (hrs)	Flow (m^3/s)
0	52	60	328	120	124
6	52	66	307	126	114
12	55	72	280	132	107
18	66	78	259	138	97
24	97	84	238	144	86
30	176	90	214	150	79
36	349	96	193	156	66
42	450	102	173	162	62
48	442	108	145	168	58
54	370	114	135	174	52

8.23. Develop the flood hydrograph that would result from a spatial average of 15 cm of rain falling on the 1,450 km^2 watershed of Problem 8.22 during a 24-hour period. The watershed is characterized by a CN of 78. The streamflow just before the rain storm was 70 m^3/s. Assume a constant baseflow of 70 m^3/s during the flood event.

8.24. Develop the flood hydrograph that would result from 28 cm of rain falling on the 1,450 km^2 watershed of Problem 8.22 during a 72-hour period. The spatially averaged rainfall is 10 cm during the first 24 hours, 12 cm during the second 24 hours, and 6 cm during the third 24 hours, for a total of 28 cm falling over 72 hours. The watershed runoff characteristics are represented by a CN of 78. The streamflow just before the rain storm was 60 m^3/s. Assume a constant baseflow of 60 m^3/s during the flood.

8.25. The following unit hydrograph is for a rainfall duration of 30 minutes. Develop a unit hydrograph for this watershed for a rainfall duration of 90 minutes.

Time (min)	Flow (m^3/s)	Time (min)	Flow (m^3/s)
0	0	90	54
15	60	105	32
30	160	120	20
45	200	135	12
60	175	150	6
75	100	165	0

8.26. The triangular-shaped hydrograph given below is representative of a certain 64.8 km^2 watershed. The hydrograph resulted from a rainfall with an effective duration of 1.0 hour. This is not a unit hydrograph, but can be used to develop a unit hydrograph for this watershed for a rainfall duration of 1.0 hour. The watershed is characterized by a CN of 86.

Time (hrs)	0	1	2	3	4	5	6
Discharge (m^3/s)	0	150	300	225	150	75	0

Assume that a rainfall with a depth of 6.0 cm and duration of 3.0 hours falls over the watershed (2.0 cm/hour for 3 hours). The rainfall distribution is uniform over time as well as spatially. Compute the resulting runoff hydrograph using a computational time interval of 1.0 hour.

8.27. The triangular-shaped hydrograph given below is representative of a certain 1,190-acre watershed. The hydrograph resulted from a rainfall with an effective duration of 1.0 hour. This is not a unit hydrograph, but it can be used to develop a unit hydrograph for this watershed for a rainfall duration of 1.0 hour. The watershed is characterized by a CN of 80.

Time (hrs)	0	1	2	3	4	5	6
Discharge (cfs)	0	300	600	450	300	150	0

Assume that a rainfall with a depth of 3.0 inches and duration of 3.0 hours falls over the watershed (1.0 in./hour for 3 hours). The rainfall distribution is uniform over time as well as spatially. Compute the resulting runoff hydrograph using a computational time interval of 1.0 hour.

8.28. Observed flows at a stream gaging station resulting from a 3-hour rain storm are tabulated on the next page. The rain began at 3:00 p.m. (15:00 hours) on day 1 and continued for 3 hours. The stream gaging station has a drainage area of 146 mi.^{2}. Use this observed hydrograph to develop a unit hydrograph.

Day	Time (hrs)	Flow (cfs)	Day	Time (hrs)	Flow (cfs)
Day 1	15:00	720	Day 3	3:00	2,880
	18:00	720		6:00	2,520
	21:00	5,800		12:00	2,280
	24:00	9,600		15:00	2,040
Day 2	3:00	11,400		18:00	1,800
	6:00	8,400		21:00	1,560
	9:00	6,360		24:00	1,320
	12:00	5,520	Day 4	3:00	1,080
	15:00	4,800		6:00	960
	18:00	4,200		9:00	840
	21:00	3,750		12:00	720
	24:00	3,240		15:00	720

8.29. The runoff characteristics of the watershed above the stream gage of Problem 8.28 are represented by a CN of 76 and the unit hydrograph developed in Problem 8.28. Determine the hydrograph resulting from a storm with the cumulative rain depths tabulated below. The flow just before the rain storm was 650 cfs.

Time (hrs)	Cumulative rainfall depth (in.)
0	0
3	0.9
6	1.4
9	2.6
12	3.2

8.30. Develop a 1.0-hour rainfall duration unit hydrograph for a 6.8 mi.^{2} watershed characterized by a lag time of 4.5 hours. Use a 1.0-hour time step.

8.31. Problems 8.16 and 8.30 refer to a 6.8 mi.^{2} watershed in Harris County, Texas, characterized by a CN of 86 and a lag time of 4.5 hours. Develop a 100-year recurrence interval, 24-hour rainfall duration flood hydrograph for this watershed using a 1-hour computational time step.

8.32. Solve Problems 8.16, 8.30, and 8.31 using the Hydrologic Engineering Center (HEC)-Hydrologic Modeling System (HMS) using a 1.0-hour time step. Repeat the simulation using a 10-minute time step and compare results.

8.33. A 3.4 mi.^{2} watershed in Brazos County, Texas, has a CN of 82 and lag time of 2.25 hours. Develop a 50-year recurrence interval, 6-hour rainfall duration hydrograph for this watershed using a computational time step of 30 minutes.

8.34. Solve Problem 8.33 using HEC-HMS. Repeat the simulation for computational time intervals of 30 minutes and 5 minutes and compare results. Repeat the simulation for rainfall durations of 6 hours and 24 hours and compare results.

8.35. The triangular-shaped hydrograph given below is representative of a certain 297.52-acre watershed. The hydrograph resulted from a rainfall with an effective duration of 1.0 hour. This is not a unit hydrograph, but may be used to develop a unit hydrograph. The watershed is characterized by a CN of 86. A small detention basin (flood retarding reservoir) is located at the watershed outlet. The detention basin is characterized by the following table of water surface elevation versus storage volume in the reservoir and outflow through the outlet structures.

Streamflow hydrograph		Reservoir storage–outflow relationships		
Time (hrs)	Discharge (cfs)	Elevation (feet m.s.l.)	Storage (ft^3)	Outflow (ft^3/s)
0	0	212	0	0
1	150	216	1,296,000	80
2	300	220	3,240,000	200
3	225	224	5,184,000	320
4	150			
5	75			
6	0			

Assume that a rainfall with a depth of 6.0 inches and duration of 3.0 hours falls over the watershed (2.0 in./hr for 3 hours). The rainfall distribution is uniform over time as well as spatially. The base flow is zero. Compute the resulting outflow hydrograph from the detention basin. Use a computational time interval of 1.0 hour.

8.36. Note that the detention basin in Problem 8.35 has a linear relationship between storage and outflow. Thus, the flood routing part of Problem 8.35 could be performed using either the storage-outflow or Muskingum methods (Section 6.1) or a linear storage–outflow relationship may be substituted directly into the continuity equation (Eq. 6.3) to derive a routing equation. Route the inflow hydrograph (watershed runoff) in Problem 8.35 through the detention basin alternatively using (1) the storage-outflow method, (2) Muskingum method, and (3) a routing equation you develop by substituting the linear storage–outflow ($S = KO$) relationship into the continuity equation. These three approaches result in exactly the same outflow hydrograph.

8.37. Certain pasture land in the northwestern corner of Alabama is represented by a cover management factor C of 0.013. The soil is characterized as follows: 60 percent silt and very fine sand, no clay, 3 percent organic matter, and fine granular structure with slow permeability. The average slope is 5 percent and slope length is 400 feet.

a. Determine the average annual erosion rate in tons/acre/year.

b. Determine the erosion rate assuming the vegetative cover is stripped away leaving a freshly-tilled bare soil.

8.38. Estimate the average annual sediment yield for a 0.5 mile^2 watershed at the site described in Problem 8.37 for both conditions of undisturbed pasture and bare tilled soil.

8.39. Estimate the sediment yield for both conditions of land cover for the watershed described in Problems 8.37 and 8.39 for a flood with 4.0 inches of rain falling in 6 hours. Assume the rainfall is uniformly distributed spatially and temporally. The watershed is characterized as follows: A = 0.5 mi.^{2}, CN = 78, time of concentration (t_C) = 1.8 hours, and rational method runoff coefficient (C) = 0.6.

8.40. A watershed is characterized by the following parameters:
mean annual precipitation = 18 inches
drainage area = 1.7 mi.^{2}
NRCS curve number CN = 88
percent of watershed impervious = 50%
commercial land use as a percent of watershed area = 10%
industrial land use as a percent of watershed area = 15%
residential land use as a percent of watershed area = 63%
nonurban land use as a percent of watershed area = 12%
A particular storm resulted in 2.6 inches of rain falling on the watershed. Estimate the mean concentration, in mg/l, and load, in kg, in the storm runoff for the following water quality constituents: chemical oxygen demand (COD), total nitrogen (TN), dissolved phosphorus (DP), and recoverable lead (PB).

REFERENCES

American Society of Civil Engineers and Water Environment Federation, *Design and Construction of Urban Stormwater Management Systems,* ASCE Manual of Practice No. 77 and WEF Manual of Practice No. FDO-20, New York, NY, 1992.

American Society of Civil Engineers, *Hydrology Handbook,* 2nd Ed., ASCE Manuals and Reports on Engineering Practice No. 28, New York, NY, 1996.

Bevin, K. J., *Rainfall Runoff Modelling,* John Wiley & Sons, West Sussex, England, 2000.

Clark, C. O., "Storage and the Unit Hydrograph," *Transactions of the American Society of Civil Engineers,* 110, 1419–1446, 1945.

Donigian, A. S., J. C. Imhoff, and R. B. Ambrose, "Chapter 11 Modeling Watershed Water Quality," *Environmental Hydrology* (V. P. Singh, Ed.), Kluwer, Boston, MA, 1995.

Driver, N. E., and G. D. Tasker, *Techniques for Estimation of Storm-Runoff Loads, Volumes, and Selected Constituent Concentrations in Urban Watersheds in the United States,* Water-Supply Paper 2363, U.S. Geological Survey, Denver, CO, 1990.

Green, W. H., and G. A. Ampt, "Studies on Soil Physics, Part I, The Flow of Air and Water through Soils," *Journal of Agricultural Science,* Vol. 4, No. 1, 1911.

Haan, C. T., B. J. Barfield, and J. C. Hayes, *Design Hydrology and Sedimentology for Small Catchments,* 2nd Ed., Academic Press, San Diego, CA, 1994.

Holtan, H. N., "A Model for Computing Watershed Retention from Soil Parameters," *Journal of Soil and Water Conservation,* 20(3), 91–94, 1965.

Holtan, H. N., G. J. Stiltner, W. H. Henson, and N. C. Lopez, *USDAHL-74 Revised Model of Watershed Hydrology,* ARS Bulletin No. 1518, U.S. Department of Agriculture, Washington, D.C., 1975.

HUBER, W. C., and R. E. DICKINSON, *Storm Water Management Model User's Manual,* Version 4, EPA/600/3-88/001, Environmental Protection Agency, Athens, GA, 1988.

Hydrologic Engineering Center, *HEC-1 Flood Hydrograph Package Users Manual,* U.S. Army Corps of Engineers, Davis, CA, June 1998.

Hydrologic Engineering Center, *HMS Hydrologic Modeling System Users Manual,* U.S. Army Corps of Engineers, Davis, CA, January 2001.

Hydrologic Engineering Center, *HMS Hydrologic Modeling System Technical Reference Manual,* U.S. Army Corps of Engineers, Davis, CA, March 2000.

JENNINGS, M. E., W. O. THOMAS, and H. C. RIGGS, *Nationwide Summary of U.S. Geological Survey Regional Regression Equations for Estimating Magnitude and Frequency of Floods for Ungaged Sites,* Water Resources Investigations Report 94-4002, Reston, VA, 1994.

KUICHLING, E., "The Relationship between the Rainfall and Discharge of Sewers in Populous Districts," *Transactions of the American Society of Civil Engineers,* Vol. 20, 1889.

MCCUEN, R. H., *Hydrologic Analysis and Design,* 2nd Ed., Prentice Hall, Upper Saddle River, NJ, 1998.

MILLS, W. B., D. B. PORCELLA, M. J. UNGS, S. A. GHERINI, K. V. SUMMERS, L. MOK, G. L. RUPP, G. L. BOWIE, and D. A. HAITH, *Water Quality Assessment: A Screening Procedure for Toxic and Conventional Pollutants and Surface and Ground Water,* EPA/600/6-85/002a, U.S. Environmental Protection Agency, Environmental Research Laboratory, Athens, GA, September 1985.

Natural Resource Conservation Service, *National Engineering Handbook, Section 3: Sedimentation,* U.S. Department of Agriculture, Washington, D.C., 1983.

Natural Resource Conservation Service, *National Engineering Handbook, Section 4: Hydrology,* U.S. Department of Agriculture, Washington, D.C., 1985.

Natural Resource Conservation Service, *Urban Hydrology for Small Watersheds,* Technical Release 55, Department of Agriculture, Washington, D.C., 1986.

PONCE, V. M., and R. H. HAWKINS, "Runoff Curve Number: Has It Reached Maturity?," *Journal of Hydrologic Engineering,* American Society of Civil Engineers, Vol. 1, No. 1, January 1996.

RAWLS, W. J., D. L. BRACKENSIEK, and N. MILLER, "Green-Ampt Infiltration Parameters from Soils Data," *Journal of the Hydraulics Division,* American Society of Civil Engineers, Vol. 109, No. 1, 1983.

RENARD, K. G., G. R. FOSTER, G. A. WEESIES, D. K. MCCOOL, and D. C. YODER, *Predicting Soil Erosion by Water: A Guide to Conservation Planning with the Revised Universal Soil Loss Equation (RUSLE),* Agricultural Handbook 703, U.S. Department of Agriculture, Washington, D.C., 1997.

SHERMAN, L. K., "Streamflow from Rainfall by the Unit Hydrograph Method," *Engineering News Record,* 108, 501–505, April 1932.

SINGH, V. P., *Elementary Hydrology,* Prentice Hall, Upper Saddle River, NJ, 1992.

SINGH, V. P., *Computer Models of Watershed Hydrology,* Water Resources Publications, Highland Ranch, CO, 1995.

SNYDER, F. F., "Synthetic Unit Hydrographs," *Transactions of the American Geophysical Union,* Vol. 19:447, 1938.

TALLAKSEN, L. M., "A Review of Baseflow Recession Analysis," *Journal of Hydrology,* Vol. 165, 1995.

TRUDGILL, S. T. (Editor), *Solute Modeling in Catchment Systems,* John Wiley & Sons, New York, NY, 1995.

Transportation Research Board, *Erosion Control During Highway Construction,* Report 221, National Cooperative Research Program, Transportation Research Board and National Research Council, Washington, D.C., 1980.

U.S. Bureau of Reclamation, *Design of Small Dams,* 3rd Ed., U.S. Government Printing Office, Denver, CO, 1987.

WARD, A. D., and W. J. ELLIOT, *Environmental Hydrology,* Lewis Publishers, New York, NY, 1995.

WILLIAMS, J. R., and H. D. BERNDT, "Sediment Yield Prediction Based on Watershed Hydrology," *Transactions of the American Society of Agricultural Engineers,* 20(6), 1100–1104, 1977.

WISCHMEIER, W. H., and D. D. SMITH, *Predicting Rainfall Erosion Losses—A Guide to Conservation Planning,* Agricultural Handbook 537, U.S. Department of Agriculture, Washington, D.C., December 1978.

WURBS, R. A., *Water Management Models: A Guide to Software,* Prentice Hall, Upper Saddle River, NJ, 1995.

9

Groundwater Engineering

Groundwater is a major source of water supply, especially in arid or semiarid areas where surface water is limited. Because groundwater is filtered by flow through the formation, it generally requires little treatment for use as a water supply. Groundwater can be considered as subsurface storage of water with limited evaporation. However, if the water table is within the root zone of surface vegetation, groundwater can be lost by evapotranspiration.

The total dissolved solids in groundwater is typically greater than that of surface water. The dissolved constituents in groundwater partly depend on the formations through which it has passed. Water in a confined aquifer may have traveled many kilometers through the formation and have taken hundreds or thousands of years to travel from the recharge zone (Fig. 9.1). Water in an unconfined aquifer is usually within a short distance and travel time of the recharge zone. The water quality of an unconfined aquifer is generally dependent on the quality of the recharge water and is easily contaminated by man's activities on the ground surface.

9.1 SUBSURFACE WATER

As shown in Fig. 9.1, the vadose zone is the shallow partially saturated zone above the water table. Because the zone is partially saturated, capillary forces play a significant role in the movement of water in the vadose zone. The capillary rise above the water table may extend only a few centimeters for coarse porous media to several meters for a fine silt (Table 9.1). The lower portion of the capillary fringe may

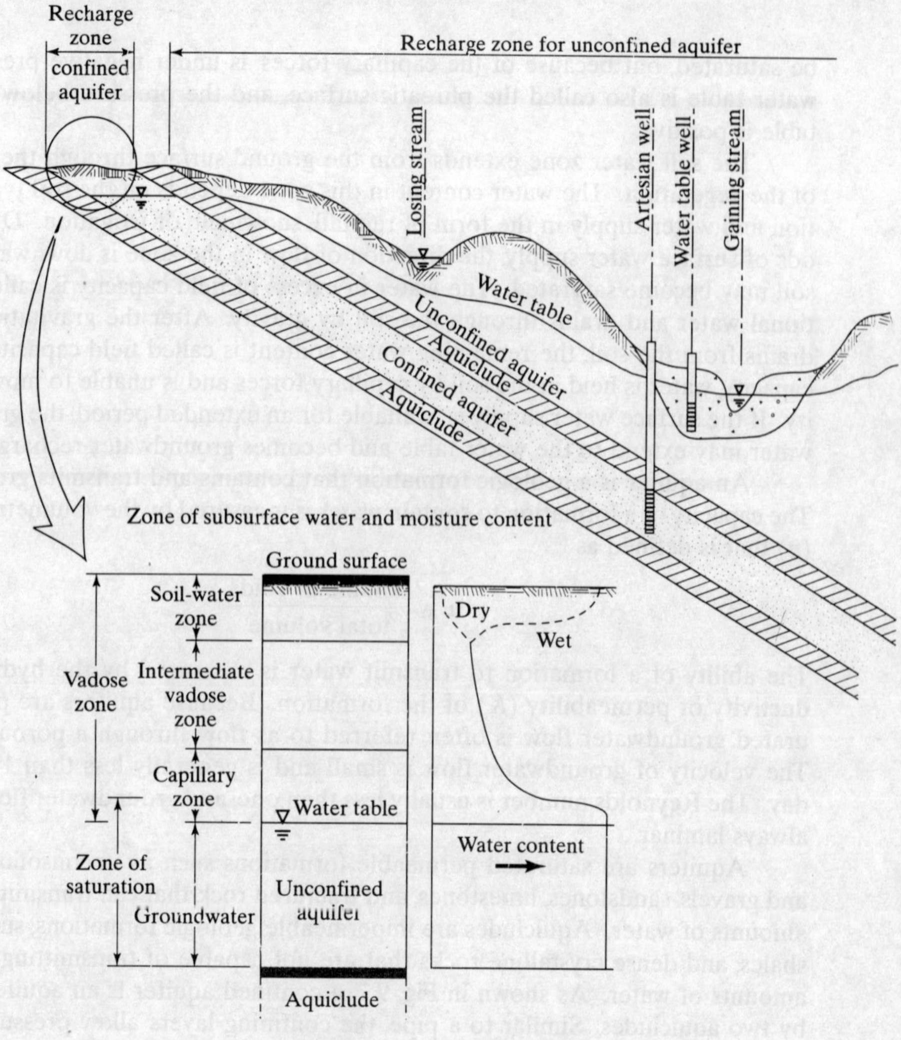

Figure 9.1 Subsurface water.

TABLE 9.1 CAPILLARY RISE IN POROUS MEDIA (LOHMAN, 1972)

Material	Grain size (mm)	Capillary rise (cm)
Fine gravel	2–5	2.5
Very coarse sand	1–2	6.5
Coarse sand	0.5–1	13.5
Medium sand	0.2–0.5	24.6
Fine sand	0.1–0.2	42.8
Silt	0.05–0.1	105.5
Fine silt	0.02–0.05	200

be saturated, but because of the capillary forces, is under negative pressure. The water table is also called the phreatic surface, and the pressure below the water table is positive.

The soil water zone extends from the ground surface through the root zone of the vegetation. The water content in this zone depends on the soil type, vegetation and water supply in the form of rainfall, snowmelt, or irrigation. During periods of surface water supply, the direction of flow in the zone is downward and the soil may become saturated. The water in excess of field capacity is called gravitational water and drains through the soil by gravity. After the gravitational water drains from the soil, the remaining water content is called field capacity. At field capacity, water is held in the soil by capillary forces and is unable to move by gravity. If the surface water supply is available for an extended period, the gravitational water may extend to the water table and becomes groundwater recharge.

An aquifer is a geologic formation that contains and transmits groundwater. The capacity of a formation to contain water is measured by the volumetric porosity (n) that is defined as

$$n = \frac{\text{volume of voids}}{\text{total volume}} \tag{9.1}$$

The ability of a formation to transmit water is measured by the hydraulic conductivity or permeability (K) of the formation. Because aquifers are porous, saturated groundwater flow is often referred to as flow through a porous medium. The velocity of groundwater flow is small and is generally less than 1 meter per day. The Reynolds number is usually less than one, and groundwater flow is nearly always laminar.

Aquifers are saturated permeable formations such as unconsolidated sands and gravels, sandstones, limestones, and fractured rock that can transmit significant amounts of water. Aquicludes are impermeable geologic formations, such as clays, shales, and dense crystalline rocks that are not capable of transmitting significant amounts of water. As shown in Fig. 9.2, a confined aquifer is an aquifer confined by two aquicludes. Similar to a pipe, the confining layers allow pressure to build up in the confined aquifer. The piezometric surface for a confined aquifer is analogous to the hydraulic grade line for pipe flow and represents the height that water would rise in a well (or piezometer) installed in the aquifer. The recharge zone for a confined aquifer is generally a narrow band where the aquifer formation outcrops with the ground surface.

An unconfined aquifer is an aquifer with a free water surface for an upper boundary. Above the water table is the unsaturated vadose zone and below the water table is the saturated aquifer. The recharge area for the unconfined aquifer is usually the ground surface above the aquifer. The unconfined aquifer will normally be hydraulically connected to stream channels that cross the aquifer. If the water table elevation is below that of the stream, the stream will recharge the aquifer, and if the water table elevation is above that of the stream, groundwater will discharge into the stream in the form of springs along the channel.

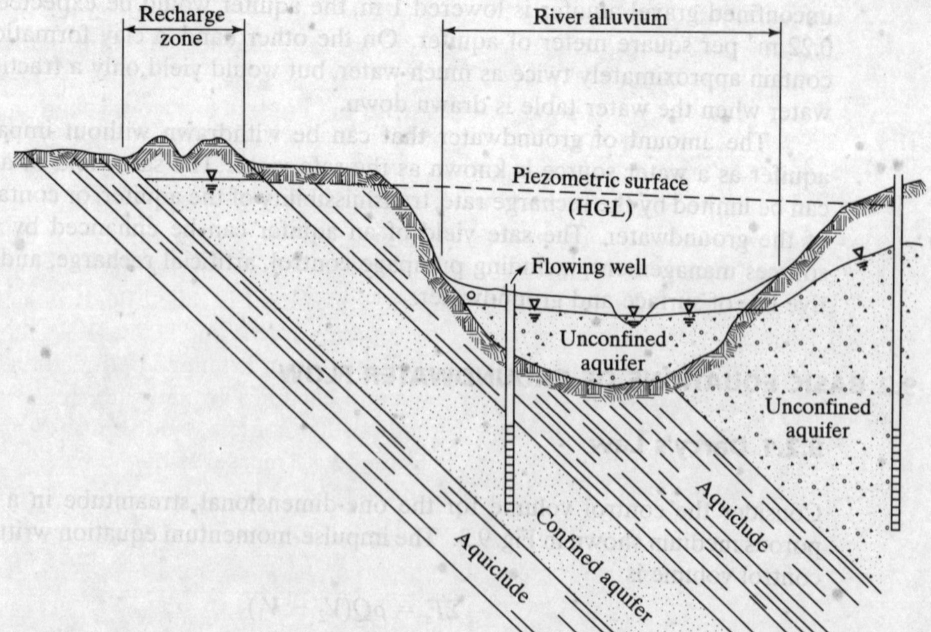

Figure 9.2 Confined and unconfined aquifers.

Typical values of porosity, specific yield, hydraulic conductivity, and intrinsic permeability are listed in Table 9.2 for several natural materials. The specific yield of a formation is the fractional volume of water that will drain freely by gravity from a unit volume of the formation. For example, 1 m^3 of saturated gravel aquifer would contain approximately 0.25 m^3 of water. If the water table elevation for an

TABLE 9.2 TYPICAL VALUES OF POROSITY, SPECIFIC YIELD, AND PERMEABILITY

Material	Volumetric porosity (n) %	Specific yield (S_s) %	Hydraulic conductivity (K) m/d	Intrinsic permeability (k) darcys
Clay	45	3	0.0004	0.0005
Sand	34	25	40	50
Gravel	25	22	4,000	5,000
Sand and gravel	20	16	400	500
Sandstone	15	8	4	5
Limestone	5	2	0.04	0.05
Shale	5	2	0.04	0.05
Granite	1	0.5	0.0004	0.0005

unconfined gravel aquifer is lowered 1 m, the aquifer would be expected to yield 0.22 m^3 per square meter of aquifer. On the other hand, a clay formation would contain approximately twice as much water, but would yield only a fraction of the water when the water table is drawn down.

The amount of groundwater that can be withdrawn without impairing the aquifer as a water source is known as the safe yield. The safe yield of an aquifer can be limited by the recharge rate, transmissibility of the aquifer, or contamination of the groundwater. The safe yield of an aquifer can be enhanced by water resources management, including pumpage control, artificial recharge, and conjunctive use of surface and groundwater.

9.2 BASIC EQUATIONS OF GROUNDWATER FLOW

9.2.1 Darcy's Law

Consider the control volume for the one-dimensional streamtube in a saturated porous medium shown in Fig. 9.3. The impulse-momentum equation written for the control volume is

$$\Sigma F_s = \rho Q(V_2 - V_1) \tag{9.2}$$

Because the groundwater velocity is small, the change in momentum is essentially zero. The primary forces acting along the streamtube are the pressure forces acting on the ends of the streamtube, the component of the weight of the water in the streamtube parallel to the direction of motion, and the shear force of resistance with the porous medium. Because the flow is saturated, the surface tension forces are not considered.

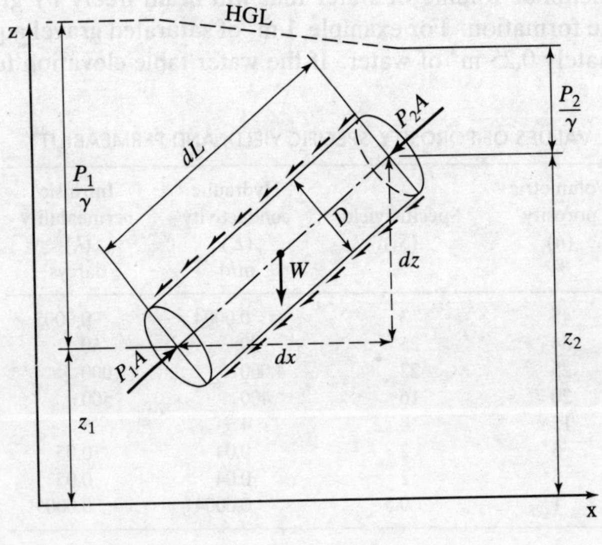

Figure 9.3 Control volume for a one-dimensional streamtube in a saturated porous medium.

Equation 9.2 can be rewritten as

$$PA - (P + dP)A - \gamma(AdL)\frac{dz}{dL} - \pi DdL\tau = 0 \tag{9.3}$$

Because groundwater flow is laminar, the shear stress can be estimated as

$$\tau = C'\mu\frac{V}{D} \tag{9.4}$$

Based on the assumption that the diameter of the streamtube (D) is proportional to the particle diameter (d) of the porous medium, the impulse-momentum equation for flow in a porous medium reduces to

$$\frac{dP}{\gamma} + dz = dh = -\frac{\mu VdL}{\gamma Cd^2}$$

or

$$V = -K\frac{dh}{dL} \tag{9.5}$$

where the hydraulic conductivity (K) is

$$K = \frac{\gamma}{\mu}Cd^2$$

or

$$K = \frac{\gamma}{\mu}k \tag{9.6}$$

The intrinsic permeability (k) of a formation is equal to Cd^2, where C is a dimensionless coefficient that depends on properties of the medium such as porosity, particle shape and particle size distribution, and d is a representative particle diameter of the formation. Intrinsic permeability has units of area (L^2). The Kozeny–Carman equation for estimating intrinsic permeability is

$$k = \frac{n^3}{180(1 - n)^2}d^2 \tag{9.7}$$

The darcy is the standard unit of intrinsic permeability where

$$1 \text{ darcy} = 0.99 \times 10^{-12}\ \text{m}^2$$

Equation 9.5 is commonly referred to as Darcy's law. The hydraulic conductivity (K) is a function of properties of both the fluid and the formation and has units of velocity (L/t). dh/dL represents the slope of the hydraulic grade line. For an unconfined aquifer, the slope of the hydraulic grade line is the slope of the water table and for a confined aquifer it is the slope of the piezometric surface.

TABLE 9.3 RANGE OF VALUES OF HYDRAULIC CONDUCTIVITY AND PERMEABILITY

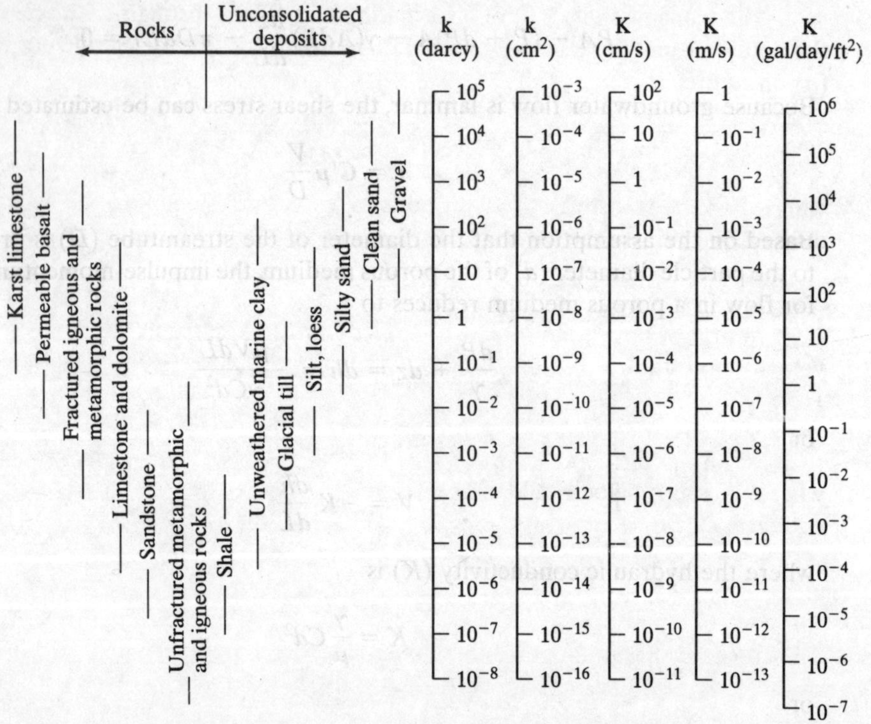

Source: Freeze and Cherry (1979)

Typical hydraulic conductivity and intrinsic permeability values are listed in Tables 9.2 and 9.3 for a range of rocks and unconsolidated deposits. Karst limestone, permeable basalt, gravel, and sand are the most permeable, whereas shale and unfractured igneous and metamorphic rocks are the least permeable.

The discharge rate (Q) is

$$Q = -AK\frac{dh}{dL} \tag{9.8}$$

where A is the total area perpendicular to the direction of flow. The velocity in Eq. 9.5 represents the specific discharge (Q/A) and not the velocity in the pore space. The average velocity in the pore space is

$$\overline{V} = \frac{V}{n_e} \tag{9.9}$$

where n_e is the effective porosity. Contaminated groundwater will travel at an average velocity ($\overline{V}$) through the formation. Because not all the pore space in the formation is available for fluid flow, the effective porosity (n_e) is less than the volumetric porosity (n).

The transmissivity (T) of a formation is the flow through a vertical section of the aquifer 1 unit wide with a hydraulic gradient of 1. The discharge per unit width (q) of aquifer is

$$q = -t\frac{dh}{dL} \tag{9.10}$$

where T is the transmissivity with units of L^2/T and is equal to the hydraulic conductivity (K) times the saturated thickness (m) of the aquifer or

$$T = Km$$

Writing Darcy's law for a homogeneous and isotropic medium in terms of velocity potential

$$V = -\frac{d\phi_v}{dL} \tag{9.11}$$

where V is the velocity along the streamtube, ϕ_v is the velocity potential and is equal to

$$\phi_v = K\left(\frac{P}{\gamma} + z\right) + C \tag{9.12}$$

The constant C was added to the velocity potential because of the arbitrary choice of the elevation datum. The units of ϕ_v are L^2/t.

Example 9.1 Groundwater Flow in a Confined Aquifer

As shown in the sketch below, two piezometers were installed in a confined aquifer. The piezometers were installed 1,000 m apart (dL). The head at piezometer A was measured as 42.1 m (h_a) and the head at piezometer B was measured as 38.3 m (h_b). The aquifer has a saturated thickness (m) of 10.5 m, an intrinsic permeability (k) of 100 darcys, a temperature of 20°C, and an effective porosity (n_e) of 0.20. Determine the discharge rate through the aquifer in m^3/day/m of aquifer width, the specific discharges, the pore velocity, and the time for water to travel from piezometer A to piezometer B.

The hydraulic conductivity of the aquifer (K) can be computed based on Eq. 9.6 for a temperature of 20°C as

$$K = \frac{\gamma}{\mu}k$$

$$1 \text{ darcy} = 0.99 \times 10^{-12} \text{ m}^2$$

$$K = \frac{9{,}790 \text{ N/m}^3 \times 100 \times 0.99 \times 10^{-12} \text{ m}^2}{0.001 \text{ Ns/m}^2} = 9.69 \times 10^{-4} \text{ m/s} = 83.7 \text{ m/day}$$

The transmissivity of the formation is

$$T = Km = 83.7 \times 10.5 = 879 \text{ m}^2\text{/day}$$

The discharge per unit width of the aquifer (q) is

$$q = -T\frac{dh}{dL} = -879 \times \frac{(38.3 - 42.1)}{1{,}000} = 3.34 \text{ m}^2/\text{day}$$

The specific discharge (Q/A) of the aquifer is

$$V = \frac{Q}{A} = -K\frac{dh}{dL} = -83.7(-0.0038) = 0.32 \text{ m/day}$$

The pore velocity

$$\bar{V} = \frac{V}{n_e} = \frac{0.32}{0.2} = 1.6 \text{ m/day}$$

Water travel time from piezometer A to B is

$$\text{Time} = \frac{L}{\bar{V}} = \frac{1{,}000}{1.6} = 625 \text{ days} = 1.7 \text{ years}$$

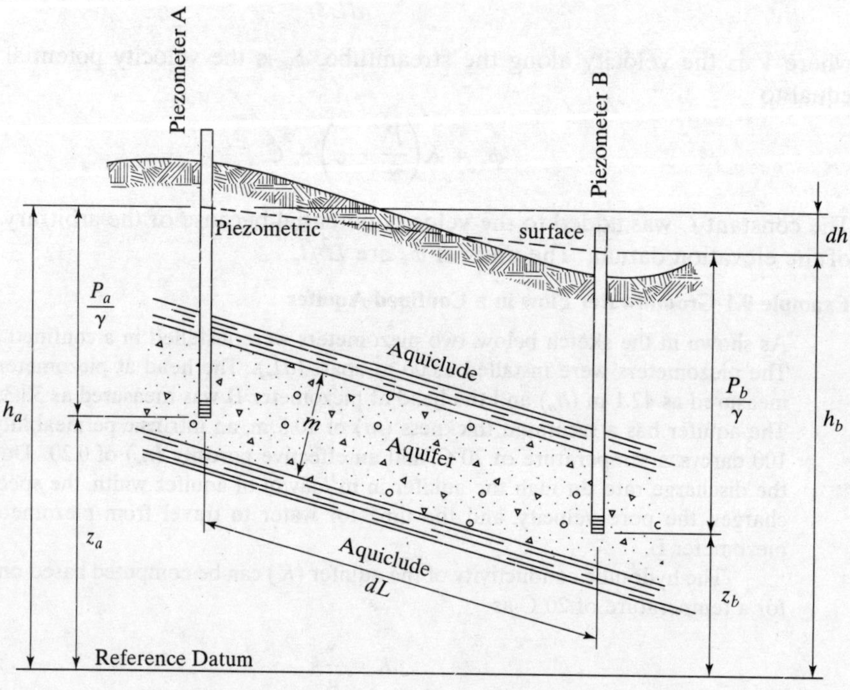

Example 9.1 Groundwater flow in a confined aquifer.

Example 9.2 Vertical Groundwater Flow

To measure the recharge rate at A and the discharge rate at B in Ex. 9.1, two additional piezometers were installed above and below the upper aquiclude at each location. The readings on the piezometers are shown in the sketch below. The intrinsic permeability of the aquiclude is 0.01 darcy.

From Example 9.1

$$1 \text{ darcy} = 0.837 \text{ m/day}$$

Hydraulic conductivity of the aquiclude

$$K = 0.01 \times 0.837 = 0.00837 \text{ m/day}$$

Recharge rate at A for an area of 1 km^2

$$Q = -KA\frac{dh}{dL} = -0.00837 \times 10^6 \times \left(-\frac{0.5}{10.0}\right) = 420 \text{ m}^3\text{/day/km}^2$$

Discharge rate at B for an area of 1 km^2

$$Q = -KA\frac{dh}{dL} = -0.00837 \times 10^6 \times \left(-\frac{0.7}{10.0}\right) = 590 \text{ m}^3\text{/day/km}^2$$

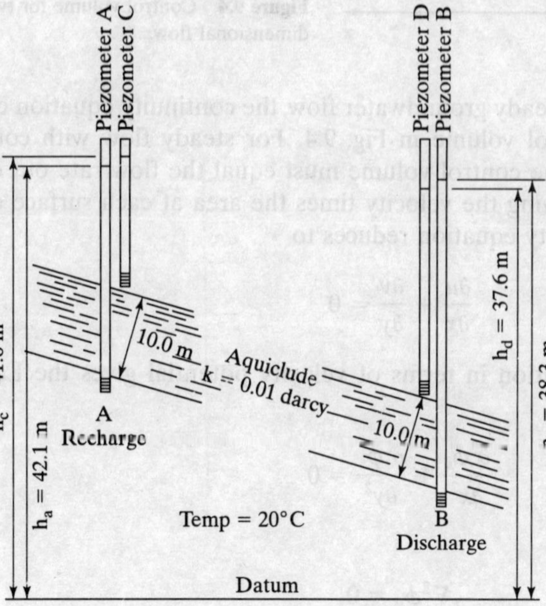

Example 9.2 Vertical groundwater flow.

9.2.2 Steady Two-Dimensional Flow

Equation 9.11 can be extended to two-dimensional flow or

$$u = -\frac{\partial \phi_v}{\partial x} \tag{9.13}$$

$$v = -\frac{\partial \phi_v}{\partial y}$$

where u is the horizontal component of the velocity in the x direction, and v is the horizontal component of the velocity in the y direction.

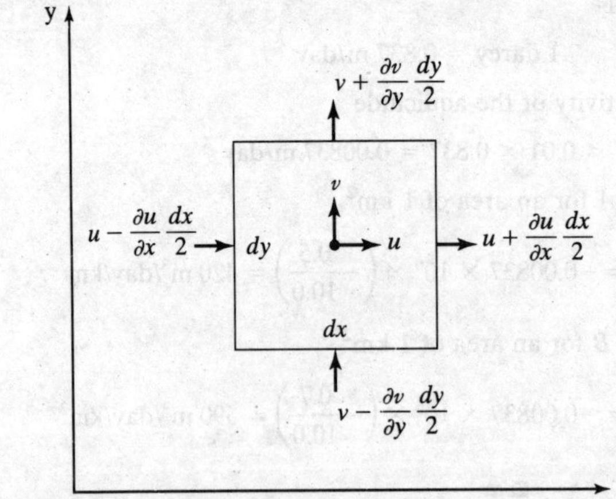

Figure 9.4 Control volume for two-dimensional flow.

For two-dimensional steady groundwater flow, the continuity equation can be derived based on the control volume in Fig. 9.4. For steady flow with constant density, the flow rate into the control volume must equal the flow rate out of the control volume. By multiplying the velocity times the area at each surface of the control volume, the continuity equation reduces to

$$\frac{\partial u}{\partial x} + \frac{\partial v}{\partial y} = 0 \tag{9.14}$$

Writing the continuity equation in terms of velocity potential gives the Laplace equation

$$\frac{\partial^2 \phi_v}{\partial x^2} + \frac{\partial^2 \phi_v}{\partial y^2} = 0 \tag{9.15}$$

or

$$\nabla^2 \phi_v = 0$$

Two-dimensional steady groundwater flow in a homogeneous, isotropic aquifer is also known as potential flow.

Defining the total derivative of the velocity potential as

$$d\phi_v = \frac{\partial \phi_v}{\partial x} dx + \frac{\partial \phi_v}{\partial y} dy \tag{9.16}$$

Along a velocity potential line $d\phi_v$ is equal to zero, and the direction of the velocity potential line is determined from Eq. 9.16 as

$$\left.\frac{dy}{dx}\right|_{\phi_v = c} = -\frac{u}{v} \tag{9.17}$$

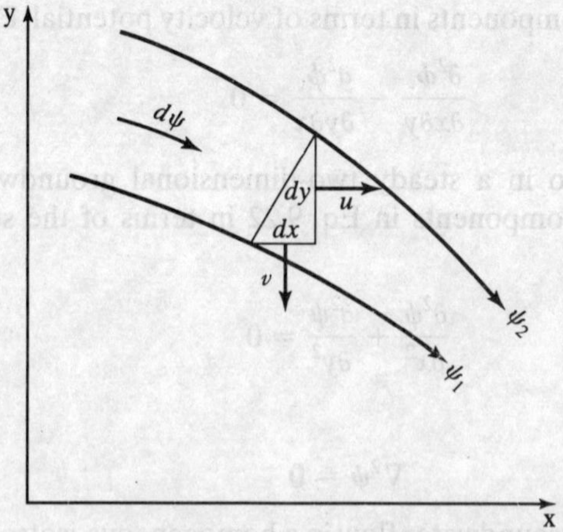

Figure 9.5 Stream functions.

Streamlines are tangent to the velocity vectors throughout the flow field. Consider the two streamlines ψ_1 and ψ_2 in Fig. 9.5 to form a streamtube. By definition, there can be no flow across the streamlines, and the flow rate between streamlines is constant and equal to $d\psi$. Based on continuity, the flow rate into and out of the triangle in Fig. 9.5 are equal

$$d\psi = -v dx + u dy \tag{9.18}$$

Defining the total derivative of ψ as

$$d\psi = \frac{\partial \psi}{\partial x} dx + \frac{\partial \psi}{\partial y} dy \tag{9.19}$$

By comparing the two equations (9.18 and 9.19), it is obvious that

$$u = \frac{\partial \psi}{\partial y}$$

$$v = -\frac{\partial \psi}{\partial x} \tag{9.20}$$

The direction of a streamline ($d\psi = 0$) can be obtained by setting Eq. 9.18 equal to zero and

$$\left.\frac{dy}{dx}\right|_{\psi = c} = \frac{v}{u} \tag{9.21}$$

By comparing the direction of a streamline (Eq. 9.21) with the direction of a velocity potential line (Eq. 9.17), it is apparent that the streamlines must be perpendicular to the velocity potential lines throughout the flow field.

Irrotational flow occurs when the vorticity is equal to zero or

$$\frac{\partial v}{\partial x} - \frac{\partial u}{\partial y} = 0 \tag{9.22}$$

Expressing the velocity components in terms of velocity potential, Eq. 9.22 becomes

$$\frac{\partial^2 \phi_v}{\partial x \partial y} - \frac{\partial^2 \phi_v}{\partial y \partial x} = 0 \tag{9.23}$$

Thus the vorticity is zero in a steady two-dimensional groundwater flow field. Expressing the velocity components in Eq. 9.22 in terms of the stream functions gives

$$\frac{\partial^2 \psi}{\partial x^2} + \frac{\partial^2 \psi}{\partial y^2} = 0 \tag{9.24}$$

or

$$\nabla^2 \psi = 0$$

Steady, two-dimensional groundwater flow in a homogeneous, isotropic aquifer must satisfy the Laplace equation in terms of both the velocity potential (ϕ_v) and the stream function (ψ).

9.2.3 Plane Potential Flow

In many cases, it is adequate to consider groundwater as uniform flow where the streamlines are all straight and parallel. An example of horizontal uniform flow is shown in Fig. 9.6. For the example, $u = U$ and $v = 0$ and in terms of velocity potential

$$\frac{\partial \phi_v}{\partial x} = -U \quad \text{and} \quad \frac{\partial \phi_v}{\partial y} = 0 \tag{9.25}$$

These two equations can be integrated to yield

$$\phi_v = -Ux + c \tag{9.26}$$

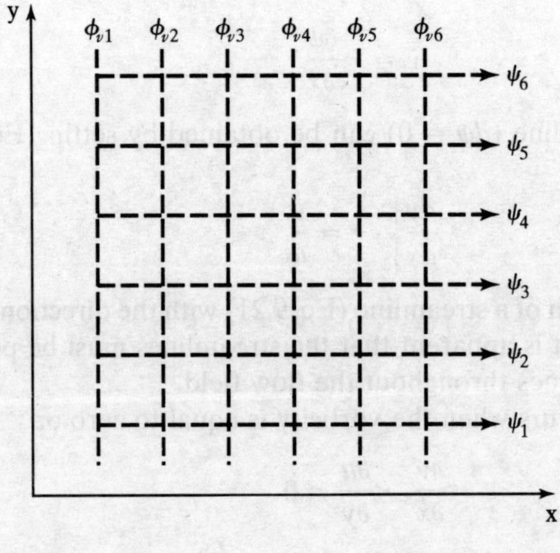

Figure 9.6 Uniform potential flow.

in rectangular coordinates or

$$\phi_v = -Ur\cos\theta + c \tag{9.27}$$

in polar coordinates (where $x = r\cos\theta$).

The corresponding stream functions for the flow field area

$$\frac{\partial\psi}{\partial y} = U \quad \text{and} \quad \frac{\partial\psi}{\partial x} = 0 \tag{9.28}$$

that yields

$$\psi = Uy + c \tag{9.29}$$

in rectangular coordinates or

$$\psi = Ur\sin\theta + c \tag{9.30}$$

In polar coordinates ($y = r\sin\theta$).

A pumped well can be considered a point sink, and a recharge well can be considered a point source in a potential flow field. Consider a source where radial streamlines are directed outward from a point (Fig. 9.7). Potential lines are circles with a radius r and continuity requires that the radial velocity V_r is

$$V_r = \frac{q}{2\pi r} \tag{9.31}$$

where q is the flow rate per unit thickness of aquifer. q is positive for a recharge well (source) and negative for a pumped well (sink). Because the flow is radial, it follows that

$$V_r = -\frac{\partial\phi_v}{\partial r} = \frac{q}{2\pi r} \tag{9.32}$$

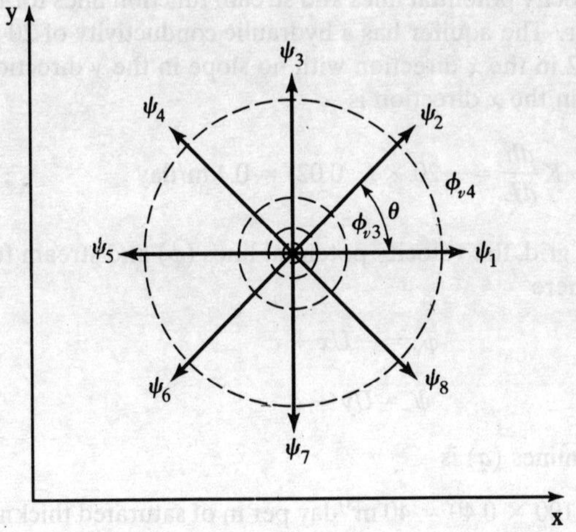

Figure 9.7 Streamlines and potential lines for a source.

and

$$V_\theta = -\frac{1}{r}\frac{\partial \phi_v}{\partial \theta} = 0 \quad \textbf{(9.33)}$$

in terms of the velocity potential and

$$V_r = \frac{1}{r}\frac{\partial \psi}{\partial \theta} = \frac{q}{2\pi r} \quad \textbf{(9.34)}$$

and

$$V_\theta = -\frac{\partial \psi}{\partial r} = 0 \quad \textbf{(9.35)}$$

in terms of the stream function.

Equations for the velocity potential and stream function can be obtained by integrating Eq. 9.32 and Eq. 9.34, respectively

$$\phi_v = -\frac{q}{2\pi}\ln r + c \quad \textbf{(9.36)}$$

and

$$\psi = \frac{q}{2\pi}\theta + c \quad \textbf{(9.37)}$$

If a well is represented as a source or a sink in a two-dimensional flow field, the smallest value of r is the radius of the well (V becomes infinite when $r = 0$).

Example 9.3 Uniform Potential Flow

Sketch and label the velocity potential lines and stream function lines for uniform flow in an unconfined aquifer. The aquifer has a hydraulic conductivity of 20 m/day and a water table slope of 0.02 in the x direction with no slope in the y direction.

The velocity (U) in the x direction is

$$U = -K\frac{dh}{dL} = -20 \times (-0.02) = 0.4 \text{ m/day}$$

Selecting a 100-m grid, the velocity potential lines (ϕ) and stream function lines (ψ) are shown below where

$$\phi_v = -Ux + c$$

$$\psi = Uy + c$$

the flow between streamlines (q) is

$$q = A \times U = 100 \times 0.40 = 40 \text{ m}^3\text{/day per m of saturated thickness}$$

$$q = d\psi = -d\phi_v = 40 \text{ m}^2\text{/day}$$

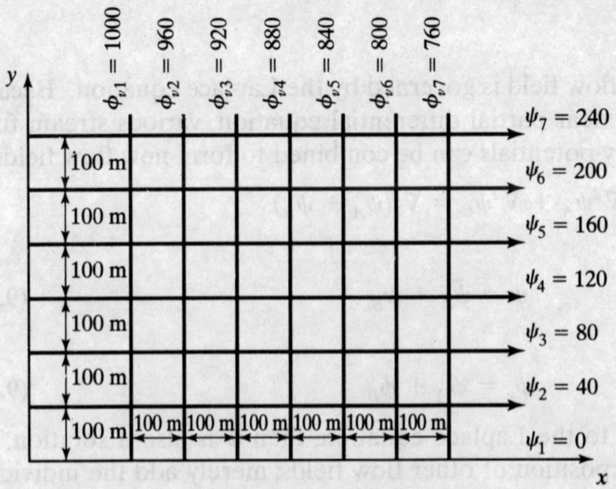

Example 9.3 Streamlines and potential lines for 100-m grid.

Example 9.4 Radial Potential Flow

A well has been installed in an aquifer with a saturated thickness (m) of 50 ft and is pumped at a rate (Q) of 100 gpm (0.22 cfs). Draw and label the velocity potential and stream function for the flow field.

$$q = \frac{Q}{m} = \frac{0.22}{50} \frac{\text{ft}^3}{\text{s} \times \text{ft}} = -0.0044 \text{ ft}^2/\text{s} = -380 \text{ ft}^2/\text{day}$$

$$\phi_v = -\frac{q}{2\pi} \ln r + 0$$

$$\psi = \frac{q}{2\pi} \theta + 380$$

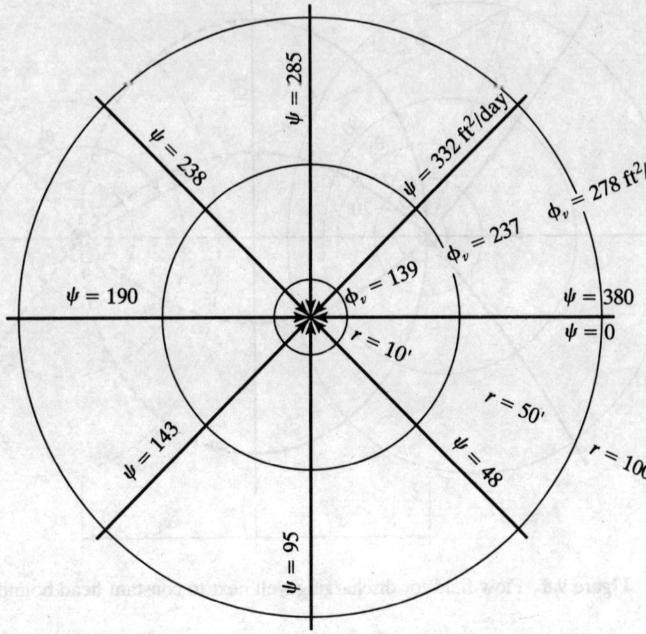

9.2.4 Superposition

Two-dimensional potential flow field is governed by the Laplace equation. Because the Laplace equation is a linear partial differential equation, various stream functions (A and B) and velocity potentials can be combined to form new flow fields or

$$\nabla^2\psi_A + \nabla^2\psi_B = \nabla^2(\psi_A + \psi_B)$$

and

$$\psi = \psi_A + \psi_B \tag{9.38}$$

Similarly

$$\phi_v = \phi_A + \phi_B \tag{9.39}$$

If ψ_A and ψ_B are solutions to the Laplace equation, then ψ is also a solution. To obtain a flow field by superposition of other flow fields, merely add the individual stream functions (or velocity potentials) to obtain the combined flow field.

In groundwater flow, a source and a sink of equal strength can be combined to represent a constant head boundary. As shown in Fig. 9.8, the source and sink are spaced a distance $2x_1$ apart and the y axis represents the constant head boundary. If a well is installed in an unconfined aquifer next to a river (constant head boundary), the y axis represents the edge of the river, the sink is the discharging well, and the source is the image well. The real domain exits where $x \leq 0$ and the imaginary domain is where $x > 0$.

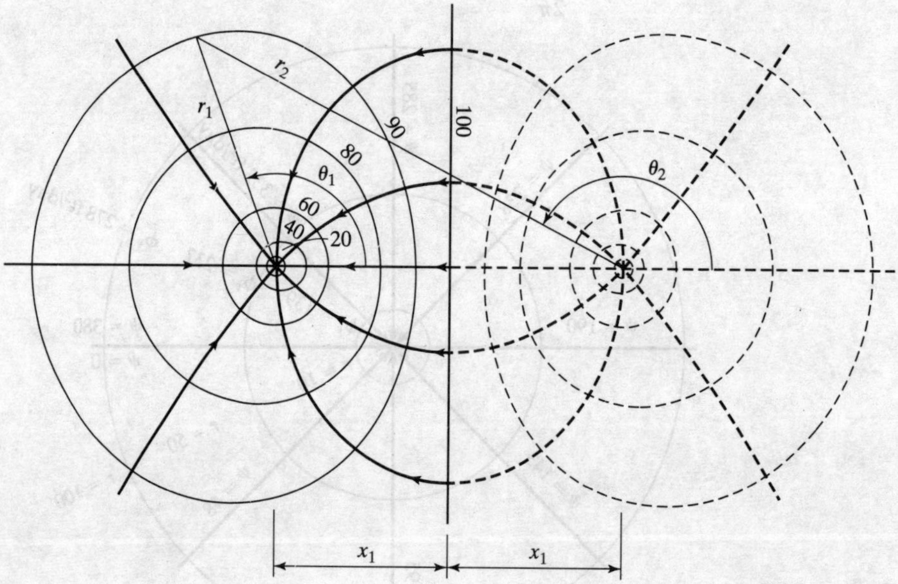

Figure 9.8 Flow field for discharging well next to constant head boundary (y axis).

The stream functions and velocity potentials for the combined flow are

$$\psi = \frac{q}{2\pi}\theta_2 - \frac{q}{2\pi}\theta_1 = \frac{q}{2\pi}(\theta_2 - \theta_1) + C \tag{9.40}$$

and

$$\phi_v = -\frac{q}{2\pi}(\ln r_2 - \ln r_1) = \frac{q}{2\pi}\ln\left(\frac{r_1}{r_2}\right) + C \tag{9.41}$$

For any location within the flow field, the values of r_1, r_2, θ_1, and θ_2 can be determined from geometry and the values of ψ and ϕ_v computed. The stream functions and velocity potentials can be mapped throughout the flow field.

The flow field for a well located next to a barrier (no flow) boundary can be simulated by adding an image well with the same sign and strength. For example, a pumped well in an unconfined aquifer located a distance x_1 from the valley wall (barrier boundary) can be simulated by adding a sink image well of the same strength a distance x_1 on the other side of the barrier boundary. The stream functions and velocity potentials for the combined flow are

$$\psi = \frac{q}{2\pi}(\theta_1 + \theta_2) + C \tag{9.42}$$

and

$$\phi_v = -\frac{q}{2\pi}\ln(r_1 \times r_2) + C \tag{9.43}$$

As shown in Fig. 9.9, the real ($x \leq 0$) and imaginary ($x > 0$) flow fields are mirror images of each other.

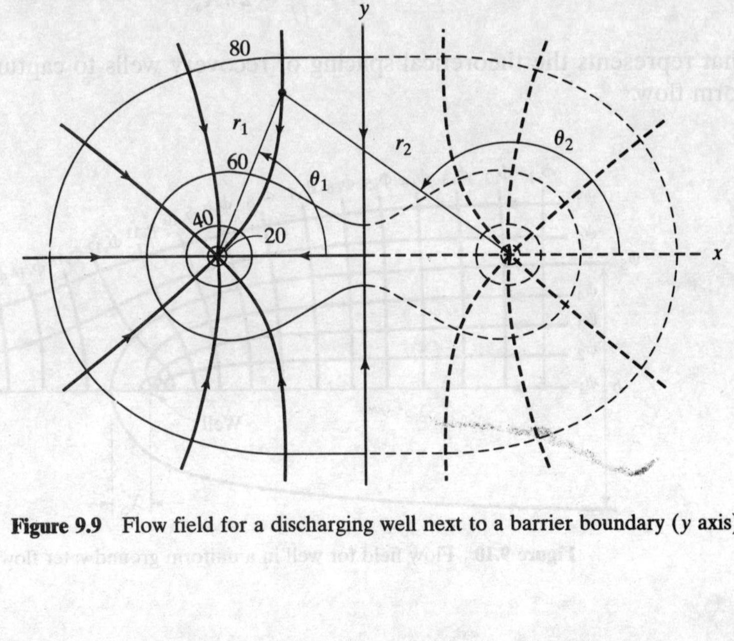

Figure 9.9 Flow field for a discharging well next to a barrier boundary (y axis)

The method of superposition is not limited to a single well, but can consider any number of wells. When the method of superposition is applied to an unconfined aquifer, it is assumed that the drawdown is small compared with the thickness of the aquifer so that the transmissivity of the aquifer remains essentially unchanged. There may be times when it is interesting to develop the flow field for a well in a uniform flow. For example, the water in the uniform flow may be contaminated and the contaminated groundwater is to be removed using a series of recovery wells located in a line perpendicular to the uniform flow. The stream function and velocity potential for a well in a uniform flow can be simulated by

$$\psi = Ur\sin\theta - \frac{q\theta}{2\pi} + C \qquad \textbf{(9.44)}$$

and

$$\phi_v = -Ur\cos\theta + \frac{q}{2\pi}\ln r + C \qquad \textbf{(9.45)}$$

The flow field for a well in a uniform stream is shown in Fig. 9.10. It is apparent that at some point along the x axis the velocity due to the well will just equal the velocity in the uniform flow and a stagnation point will occur. The stagnation point will occur at $X = X_s$ where

$$U = \frac{q}{2\pi X_s} \qquad \textbf{(9.46)}$$

The width of uniform groundwater flow entering the well (b) is

$$b = \frac{q}{U} = 2\pi X_s \qquad \textbf{(9.47)}$$

that represents the theoretical spacing of recovery wells to capture all of the uniform flow.

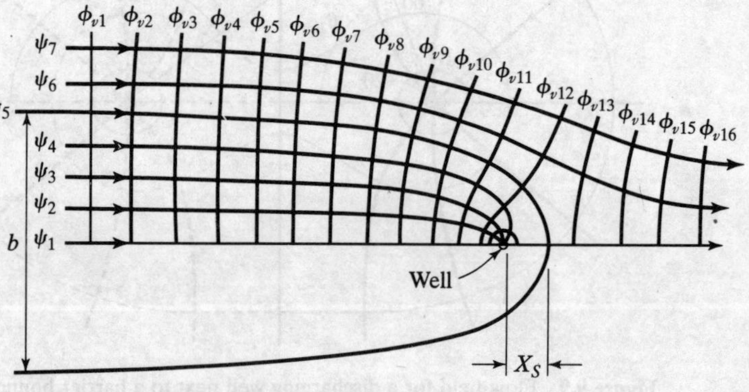

Figure 9.10 Flow field for well in a uniform groundwater flow.

Example 9.5 Well in Uniform Flow

Assume that the flow field in Example 9.3 is contaminated and that a series of wells from Example 9.4 are to be installed to capture all the flow. Determine the well spacing with (a) no overlap in capture zone and (b) 100 percent overlap in capture zone.

From Example 9.4, the well flow rate

$$q = -380 \text{ ft}^2/\text{day} = -35 \text{ m}^2/\text{day}$$

(a) Well spacing with no overlap

$$b = \frac{q}{U} = \frac{35}{0.4} = 88 \text{ m}$$

(b) Well spacing with 100 percent overlap spacing $= b/2 = 44$ m

9.2.5 Unsteady Flow

Consider an unconfined aquifer in a floodplain of a river as shown in Fig. 9.11. The aquifer is hydraulically connected to the river. During low river stage, the aquifer discharges into the river and during high river stage, the river recharges the aquifer. A cross-section of the aquifer taken parallel to the streamlines represents a one-dimensional aquifer segment of width b. The water table elevation changes with time. The expression for the conservation of mass for the control volume is

$$\rho\left(u - \frac{\partial u}{\partial x}\frac{\Delta x}{2}\right)(h - z)b - \rho\left(u + \frac{\partial u}{\partial x}\frac{\Delta x}{2}\right)(h - z)b = \rho b \Delta x S_y \frac{\Delta h}{\Delta t} \tag{9.48}$$

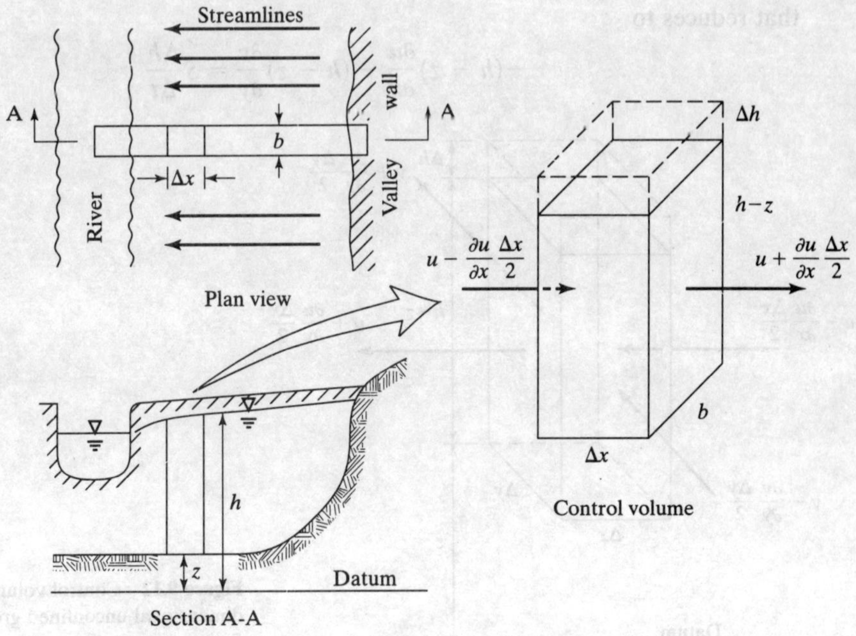

Figure 9.11 Control volume for one-dimensional unconfined groundwater flow.

where h is the elevation of the water table, z is the elevation of the base of the aquifer, u is the velocity, and S_y is the specific yield of the aquifer. For constant density groundwater, the equation reduces to

$$-(h - z)\frac{\partial u}{\partial x} = S_y \frac{\Delta h}{\Delta t} \tag{9.49}$$

Combining Eqs. 9.5 and 9.49 gives

$$\frac{\partial^2 h}{\partial x^2} = \frac{S}{T}\frac{\partial h}{\partial t} \tag{9.50}$$

where T is the transmissivity and S is the storage coefficient of the aquifer. For an unconfined aquifer, the storage coefficient is equal to the specific yield. Equation 9.50 assumes that the transmissivity is constant for the aquifer.

A control volume for a two-dimensional unconfined groundwater flow is shown in Fig. 9.12. Assuming a homogeneous and isotropic aquifer, the expression for the conservation of constant density water is

$$\begin{aligned}\rho\left(u - \frac{\partial u}{\partial x}\frac{\Delta x}{2}\right)\Delta y(h - z) - \rho\left(u + \frac{\partial u}{\partial x}\frac{\Delta x}{2}\right)\Delta y(h - z)\\ + \rho\left(v - \frac{\partial v}{\partial y}\frac{\Delta y}{2}\right)\Delta x(h - z)\\ - \rho\left(V + \frac{\partial V}{\partial y}\frac{\Delta y}{2}\right)\Delta x(h - z) = \rho S \Delta x \Delta y \frac{\Delta h}{\Delta t}\end{aligned} \tag{9.51}$$

that reduces to

$$-(h - z)\frac{\partial u}{\partial x} - (h - z)\frac{\partial v}{\partial y} = S\frac{\Delta h}{\Delta t} \tag{9.52}$$

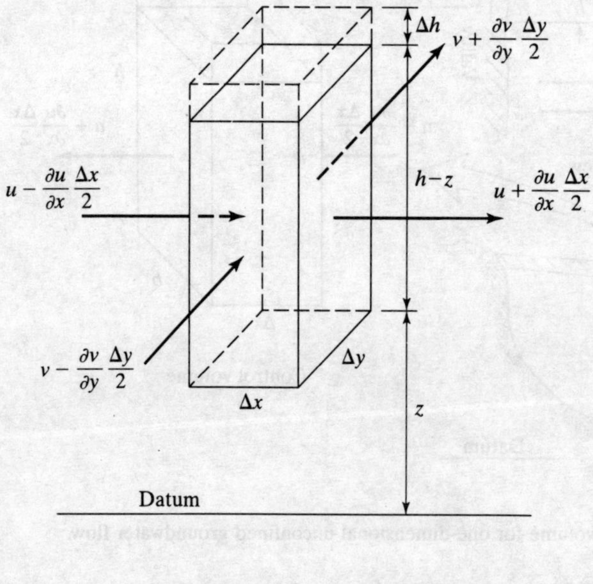

Figure 9.12 Control volume for two-dimensional unconfined groundwater flow.

Combining Eqs. 9.15 and 9.52 gives

$$\frac{\partial^2 h}{\partial x^2} + \frac{\partial^2 h}{\partial y^2} = \frac{S}{T}\frac{\partial h}{\partial t} \tag{9.53}$$

for two-dimensional groundwater flow.

Equations 9.50 and 9.53 were developed for unconfined aquifers, but the equations also apply to confined aquifers. A control volume for two-dimensional flow in a homogeneous, isotropic, confined aquifer is shown in Fig. 9.13. For a confined aquifer, h is the elevation of the piezometric surface, and m is the thickness of the aquifer. Specific storage (S_s) is defined as S/m, and Eq. 9.53 can be written as

$$\frac{\partial^2 h}{\partial x^2} + \frac{\partial^2 h}{\partial y^2} = \frac{S_s}{K}\frac{\partial h}{\partial t} \tag{9.54}$$

For an unconfined aquifer, it is easy to visualize the change in water storage resulting from a change in saturated thickness of the aquifer (Fig. 9.14). For a confined aquifer, the change in the elevation of the piezometric surface causes a change in pressure which results in a change in storage of water in the aquifer (Fig. 9.14). A decrease in the elevations of the piezometric surface results in an expansion of the water and a compression of the aquifer structure (subsidence of the ground surface). The storage coefficient of an aquifer is the volume of water released from storage from a unit area of the aquifer per unit decline in elevation of the piezometric surface. The water released from storage in a confined aquifer is

(1) from the expansion of the water = $\rho g m n \beta$, and

(2) from the compression of the aquifer structure = $\rho g m \alpha$

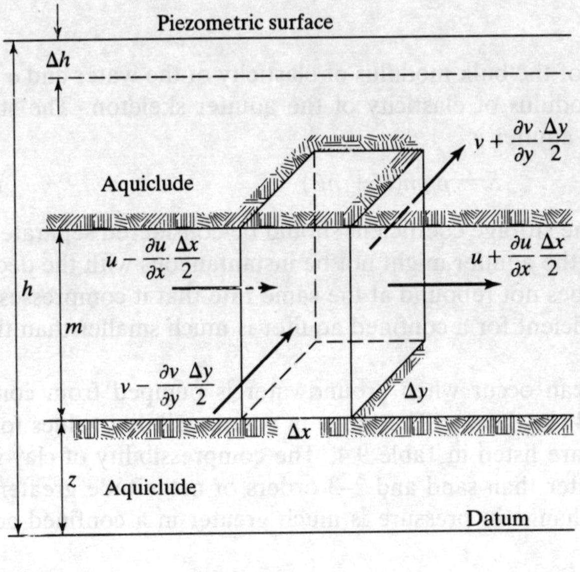

Figure 9.13 Control volume for a confined aquifer.

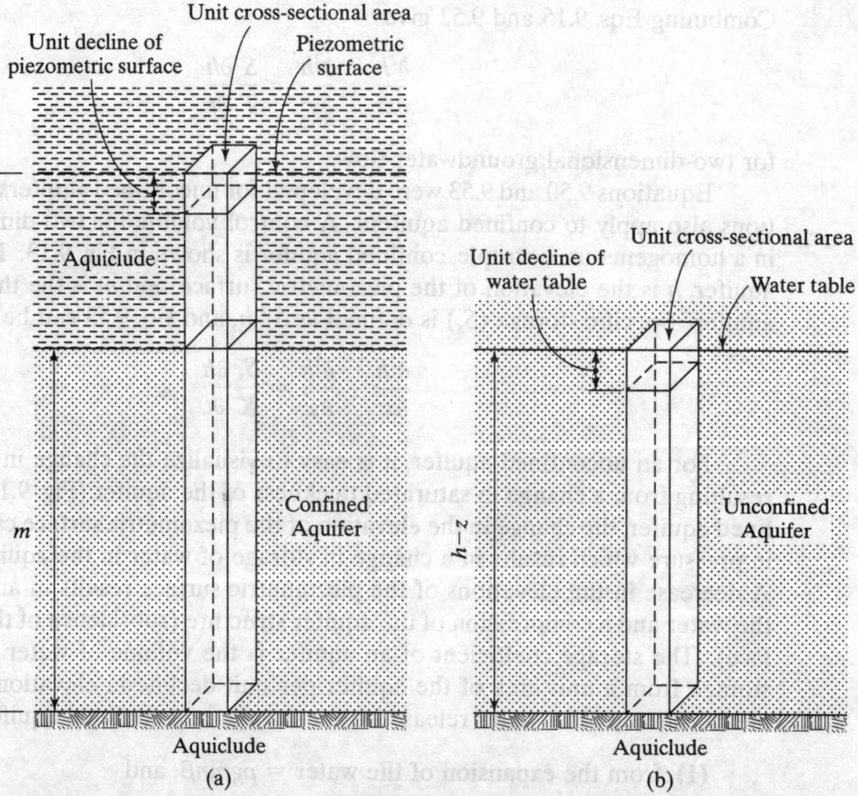

Figure 9.14 Sketches defining storage coefficients for (a) confined and (b) unconfined aquifers.

where β is the reciprocal of the bulk modulus of elasticity of the water and α is the reciprocal of the bulk modulus of elasticity of the aquifer skeleton. The storage coefficient for a confined aquifer is

$$S = \rho g m(\alpha + n\beta) \tag{9.55}$$

The two terms for the storage coefficient should be considered separately because the compression of the aquifer might not be instantaneous with the decrease in head, and an aquifer does not rebound at the same rate that it compresses. The value of the storage coefficient for a confined aquifer is much smaller than that of an unconfined aquifer.

Severe subsidence can occur when groundwater is pumped from confined aquifers with interbedded clay layers. The range in compressibility values for several formation materials are listed in Table 9.4. The compressibility of clay is 1–2 orders of magnitude greater than sand and 2–3 orders of magnitude greater than gravel. In addition, the change in pressure is much greater in a confined aquifer

TABLE 9.4 COMPRESSIBILITY VALUES FOR FORMATION MATERIALS

Formation material	Compressibility (α)* m^2/N
Sound rock	10^{-10}
Jointed rock	10^{-9}
Gravel	10^{-9}
Dense sand and gravel	10^{-9}
Dense sand	10^{-8}
Loose sand	10^{-7}
Dense clay	10^{-8}
Medium clay	10^{-7}
Loose clay	10^{-6}
Peat	10^{-5}

*1 psi = 6895 N/m^2.

than in an unconfined aquifer for the same pumping rate because of the magnitude of the storage coefficient.

Example 9.6 Subsidence Caused by Groundwater Pumping

The piezometric surface for a confined aquifer has been lowered 150 m due to pumping of groundwater. The aquifer is 100 m thick, which includes 25 m of interbedded clay layers. Compute the storage coefficient of the aquifer and the subsidence caused by pumping of groundwater. Use a compressibility (α) of 10^{-7} m^2/N for the clay and 10^{-9} m^2/N for the sand/gravel aquifer. The porosity of the clay is 0.40 and the sand/gravel aquifer is 0.25, and the compressibility of the water (β) is 4.4×10^{-10} m^2/N.

Storage coefficient

$$S = \rho g m(\alpha + n\beta) = 998 \times 9.81 \times 75(10^{-9} + 0.25 \times 4.4 \times 10^{-10}) + 998 \times 9.81 \times 25(10^{-7} + 0.40 \times 4.4 \times 10^{-10}) = 0.0008 + 0.0245 = 0.0253$$

Subsidence (dz)

$$\alpha = \frac{1}{E_\alpha} = \frac{\frac{d\forall}{\forall}}{dP} = \frac{\frac{dz}{z}}{dP}$$

$$dz_i = dP\alpha_i z_i$$

where subscript i refers to the formation layer and

$$dP = \gamma \Delta H = 998 \times 9.81 \times 150 = 1.469 \times 10^6 \text{N/m}^2$$

$$dz = 1.469 \times 10^6 (10^{-9} \times 75 + 10^{-7} \times 25) = 0.11 + 3.67 = 3.78 \text{ m}$$

9.3 WELLS

Water wells are installed to provide water supply, recharge, or observation of the aquifer. As shown in Fig. 9.15, a water well consists of a casing and well screen.

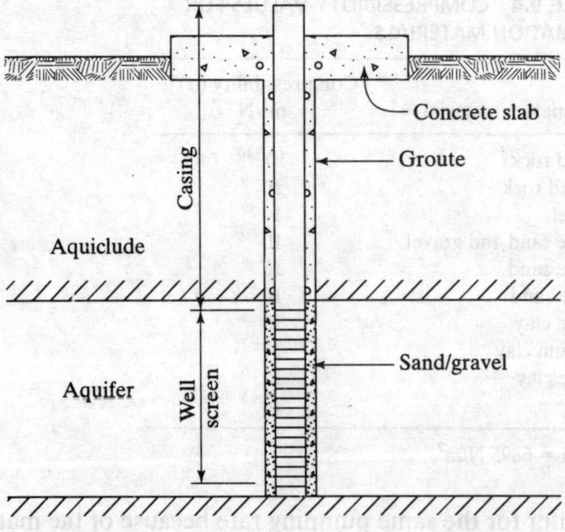

Figure 9.15 Water well.

The casing is usually grouted in the borehole to prevent contaminated water from flowing along the outside of the casing into the aquifer. A concrete slab is normally placed around the casing to prevent contaminated surface water from ponding around the well. The well screen provides the hydraulic connection with the aquifer and prevents formation sand from entering the well.

An observation well usually has a short well screen and is used to collect water quality samples and/or monitor water levels at a specific elevation in the aquifer. An observation well used to monitor water levels (or piezometric surface in a confined aquifer) is called a piezometer. Piezometers generally have a small diameter and are often nested to monitor the piezometric surface at several levels in an aquifer.

9.3.1 Hydraulics of Wells

When a water supply well is pumped, a cone of depression is formed around the well, and the drop in water level (or piezometric surface) is called drawdown. Well hydraulics involve the computation of drawdown around a well. The hydraulics of a recharge well is basically the same as that of a pumped well, except the discharge rate and drawdown are negative.

9.3.1.1 Steady-state equations. After a well has been pumped for an extended period, steady-state conditions are approached. Figure 9.16 shows a well that is completed in a confined aquifer with a thickness (m) and is pumped at a constant rate (Q). Based on the Darcy equation for an isotropic, homogeneous aquifer, the hydraulic equation for flow through a cylinder with radius (r) and height (m) is

$$Q = -KAS = K(2\pi rm)\frac{dh}{dr} \tag{9.56}$$

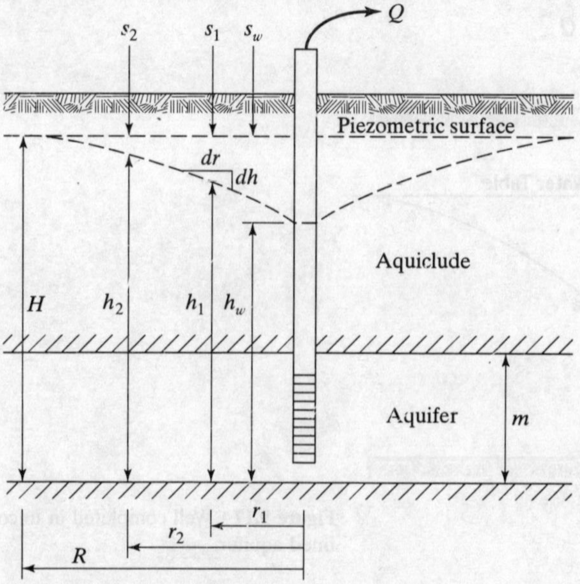

Figure 9.16 Well in confined aquifer.

where K is the hydraulic conductivity, $2\pi rm$ is the flow area, and $-dh/dr$ is the slope of the piezometric surface. Equation 9.56 can be integrated from r_1 to r_2

$$Q\int_{r_1}^{r_2}\frac{dr}{r} = 2\pi(mK)\int_{h_1}^{h_2} dh$$

to yield

$$Q = \frac{2\pi T(h_2 - h_1)}{\ln(r_2/r_1)} \qquad \textbf{(9.57)}$$

where the transmissivity (T) is equal to mK. Equation 9.57 is known as the Thiem equation.

If r_w is the radius of the well and R is the radius of influence of the well, then the drawdown at the well (s_w) is

$$s_w = H - h_w = \frac{Q}{2\pi T}\ln(R/r_w) \qquad \textbf{(9.58)}$$

Figure 9.17 is a pumped well installed in an isotropic, homogeneous unconfined aquifer. At any radial distance (r) from the well, the Darcy equation for flow toward the well is

$$Q = K(2\pi rh)\frac{dh}{dr}$$

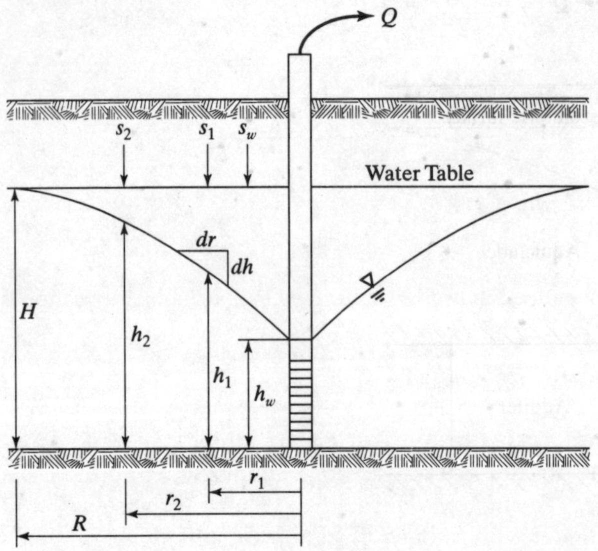

Figure 9.17 Well completed in unconfined aquifer.

The equation can be integrated from r_1 to r_2 to yield

$$Q = \frac{\pi K(h_2^2 - h_1^2)}{\ln(r_2/r_1)} \tag{9.59}$$

Equation 9.59 is the steady-state well equation for an unconfined aquifer. If the drawdown for an unconfined aquifer is small compared with the thickness of the aquifer, then $(h_1 + h_2)/2$ is approximately equal to H and Eq. 9.59 reduces to the steady-state well equation for a confined aquifer (Eq. 9.57). Equation 9.57 is often used for both confined and unconfined aquifers; however, the radius of influence for the two types of aquifers is very different.

Example 9.7 Steady-State Drawdown in a Confined Aquifer

A 24-inch diameter well is installed in a confined aquifer using a 6-inch gravel pack ($r_w = 1.5$ ft). The aquifer has a transmissivity of 40,000 gpd/ft, and the well is pumped at a rate of 2,000 gpm. If the radius of influence (R) of the well is 100,000 ft, determine the steady-state drawdown at the well.

$$s_w = \frac{Q}{2\pi T}\ln\left(\frac{R}{r_w}\right) = \frac{2{,}000 \times 1{,}440}{2\pi \times 40{,}000}\ln\left(\frac{100{,}000}{1.5}\right) = 127 \text{ ft}$$

Example 9.8 Steady-State Drawdown in an Unconfined Aquifer

A 12-inch diameter well is installed in an unconfined aquifer with a saturated thickness of 100 ft. The aquifer has a transmissivity of 40,000 gpd/ft, and the well is pumped at a rate of 600 gpm. If the radius of influence (R) of the well is 4,000 ft, determine the steady-state drawdown at the well.

$$h_2^2 - h_1^2 = \frac{Q}{\pi K}\ln\left(\frac{r_2}{r_1}\right)$$

$$100^2 - h_w^2 = \frac{600 \times 1{,}440}{3.1416 \times 400}\ln\left(\frac{4{,}000}{0.5}\right)$$

$$h_w^2 = 10{,}000 - 6{,}179$$

$$h_2 = 61.8 \text{ ft}$$

$$s_w = h_2 - h_w = 38.2 \text{ ft}$$

For comparison, the computed drawdown using the confined aquifer equation is

$$s_w = \frac{Q}{2\pi T}\ln\left(\frac{R}{r_w}\right) = \frac{600 \times 1{,}440}{2\pi \times 40{,}000}\ln\left(\frac{4{,}000}{0.5}\right) = 31.0 \text{ ft}$$

9.3.1.2 Unsteady-state equation. During the initial stages of pumping from a well, the flow in the aquifer has not reached equilibrium and the cone of depression continues to grow. The incremental control volume for unsteady well flow shown in Fig. 9.18 is a cylinder with radius r, thickness dr, and height m. During an incremental time step ($\Delta t = t_2 - t_1$), the change in drawdown for the control volume is Δh. The storage coefficient (S) is defined as the volume of water per unit area of aquifer that the confined (unconfined) aquifer will yield per unit drop in piezometric surface (water table). The yield of water from aquifer storage during the incremental time step (Δt) for the control volume is

$$\text{yield} = -S(2\pi r dr)\Delta h \tag{9.60}$$

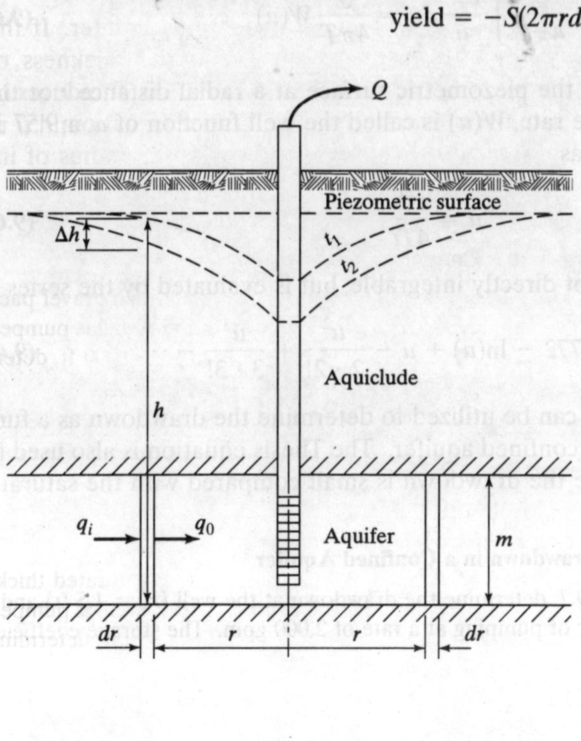

Figure 9.18 Unsteady state flow for well in a confined aquifer.

The continuity equation written for the incremental control volume is

$$q_o = q_i - S(2\pi r dr)\frac{\Delta h}{\Delta t} \quad \textbf{(9.61)}$$

where q_i is the flow into the incremental control volume from the aquifer and q_o is the flow from the control volume toward the well. Based on the Darcy equation, the aquifer flow rate q is

$$q = K(2\pi rm)\frac{\partial h}{\partial r} \quad \textbf{(9.62)}$$

The change in aquifer flow rate in the incremental control volume is

$$q_i - q_o = \frac{\partial q}{\partial r}dr = 2\pi Km\left(r\frac{\partial^2 h}{\partial r^2} + \frac{\partial h}{\partial r}\right)dr \quad \textbf{(9.63)}$$

Substituting Eq. 9.63 into Eq. 9.61 and replacing Km with the transmissivity (T) yields

$$\frac{\partial^2 h}{\partial r^2} + \frac{1}{r}\frac{\partial h}{\partial r} = \frac{S}{T}\frac{\partial h}{\partial t} \quad \textbf{(9.64)}$$

Solution to Eq. 9.64 was first presented by Theis in 1935 for well flow in an isotropic and homogeneous confined aquifer. The Theis equation is

$$s = \frac{Q}{4\pi T}\int_u^\infty \frac{e^{-u}}{u}du = \frac{Q}{4\pi T}W(u) \quad \textbf{(9.65)}$$

where s is the drawdown of the piezometric surface at a radial distance r at time t, Q is the constant discharge rate, $W(u)$ is called the well function of u, and u is a dimensionless term defined as

$$u = \frac{r^2 S}{4Tt} \quad \textbf{(9.66)}$$

The integral in Eq. 9.65 is not directly integrable, but is evaluated by the series

$$W(u) = -0.5772 - \ln(u) + u - \frac{u^2}{2 \cdot 2!} + \frac{u^3}{3 \cdot 3!} - \cdots \quad \textbf{(9.67)}$$

Equations 9.65 through 9.67 can be utilized to determine the drawdown as a function of r and t for a well in a confined aquifer. The Thesis equation is also used for an unconfined aquifer where the drawdown is small compared with the saturated thickness of the aquifer.

Example 9.9 Unsteady-State Drawdown in a Confined Aquifer

For the well in Example 9.7, determine the drawdown at the well (r_w = 1.5 ft) and at r = 100,000 ft after 1 year of pumping at a rate of 2,000 gpm. The storage coefficient of the aquifer is 0.00035.

$$s = \frac{Q}{4\pi T} W(u)$$

$$u = \frac{r^2 S}{4Tt}$$

Drawdown at the well

$$u = \frac{1.5^2(0.00035)7.48}{4 \times 40{,}000 \times 365} = 1.01 \times 10^{-10}$$

$$W(u) = -0.5772 - \ln u + u = -0.5772 + 23.02 + 0.0 = 22.44$$

$$s_w = \frac{2{,}000 \times 1{,}440}{4\pi\ 40{,}000} \times 22.44 = 128.6 \text{ ft}$$

Drawdown at $r = 100{,}000$ ft

$$u = \frac{(100{,}000)^2(0.00035)7.48}{4 \times 40{,}000 \times 365} = 0.45$$

$$W(u) = -0.577 - \ln(0.45) + 0.45 - \frac{0.45^2}{4} = 0.62$$

$$s = \frac{2{,}000 \times 1{,}440}{4\pi\ 40{,}000} \times 0.62 = 3.6 \text{ ft}$$

Example 9.10 Unsteady-State Drawdown in an Unconfined Aquifer

(a) The unconfined aquifer in Example 9.8 has a storage coefficient of 0.15. Compute the drawdown for $r = 0.5$ ft and 4,000 ft after 1 year of pumping at a rate of 600 gpm using the Thesis equation

$$s = \frac{Q}{4\pi T} W(u)$$

Drawdown for $r = 0.5$ ft

$$u = \frac{r^2 S}{4Tt} = \frac{0.5^2 \times 0.15 \times 7.48}{4 \times 40{,}000 \times 365} = 4.8 \times 10^{-9}$$

$$W(u) = -0.577 - \ln(u) + u = -0.577 - \ln(4.8 \times 10^{-9}) + 0 = 18.6$$

$$s = \frac{600 \times 1{,}440}{4\pi\ 40{,}000} \times 18.6 = 31.9 \text{ ft}$$

Drawdown at $r = 4{,}000$ ft

$$u = \frac{4{,}000^2 \times 0.15 \times 7.48}{4 \times 40{,}000 \times 365} = 0.307$$

$$W(u) = -0.577 - \ln(0.307) + 0.307 - \frac{(0.307)^2}{4} = 0.88$$

$$s = \frac{600 \times 1{,}440}{4\pi\ 40{,}000} \times 0.88 = 1.5 \text{ ft}$$

(b) Compute a table of values for $W(u)$ for values of u ranging from 0.9 to 10^{-15}. Code the equation

$$W(u) = -0.5772 - \ln(u) + u - \frac{u^2}{2 \cdot 2!} + \frac{u^3}{3 \cdot 3!} - \frac{u^4}{4 \cdot 4!}$$

and print values of $W(u)$ in tabular form. Output is listed below.

VALUES OF W(*u*) (WELL FUNCTION OF *u*)

	u	1	2	3	4	5	6	7	8	9
X	.1E + 00	1.82	1.22	.91	.70	.56	.45	.37	.31	.26
X	.1E − 01	4.04	3.35	2.96	2.68	2.47	2.30	2.15	2.03	1.92
X	.1E − 02	6.33	5.64	5.23	4.95	4.73	4.54	4.39	4.26	4.14
X	.1E − 03	8.63	7.94	7.53	7.25	7.02	6.84	6.69	6.55	6.44
X	.1E − 04	10.94	10.24	9.84	9.55	9.33	9.14	8.99	8.86	8.74
X	.1E − 05	13.24	12.55	12.14	11.85	11.63	11.45	11.29	11.16	11.04
X	.1E − 06	15.54	14.85	14.44	14.15	13.93	13.75	13.59	13.46	13.34
X	.1E − 07	17.84	17.15	16.74	16.46	16.23	16.05	15.90	15.76	15.65
X	.1E − 08	20.15	19.45	19.05	18.76	18.54	18.35	18.20	18.07	17.95
X	.1E − 09	22.45	21.76	21.35	21.06	20.84	20.66	20.50	20.37	20.25
X	.1E − 10	24.75	24.06	23.65	23.36	23.14	22.96	22.81	22.67	22.55
X	.1E − 11	27.05	26.36	25.96	25.67	25.44	25.26	25.11	24.97	24.86
X	.1E − 12	29.36	28.66	28.26	27.97	27.75	27.56	27.41	27.28	27.16
X	.1E − 13	31.66	30.97	30.56	30.27	30.05	29.87	29.71	29.58	29.46
X	.1E − 14	33.96	33.27	32.86	32.58	32.35	32.17	32.02	31.88	31.76

9.3.2 Aquifer Testing

The transmissivity and storage coefficient are hydraulic properties of an aquifer that are essential for groundwater management, including water supply yield, recharge evaluation, and contamination cleanup studies. Pump tests are conducted for the purpose of determining the hydraulic properties of the aquifer. The test is conducted by pumping the well at a constant rate and observing the drawdown in the piezometric surface or water table. To allow the cone of depression time to develop, the pump tests are generally conducted for approximately 24 hours in a confined aquifer and 72 hours in an unconfined aquifer. The drawdown should be measured at an observation well located approximately one to two times the saturated thickness of the aquifer away from the pumped well. Although one or more observation wells are desirable, it is often necessary to measure the drawdown at the pumped well because of cost constraints.

Figure 9.19 is a semi-log plot of time (t) on the log scale versus drawdown (s) on the arithmetic scale for a pump test. The plot becomes linear as time increases. Writing the Theis equation

$$s = \frac{Q}{4\pi T}\left[-0.577 - \ln(u) + u - \frac{u^2}{2 \cdot 2!} + \cdots\right] \tag{9.68}$$

and

$$u = \frac{r^2 S}{4Tt}$$

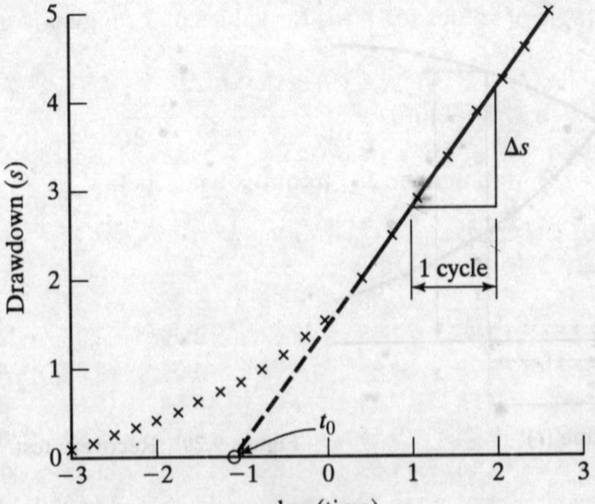

Figure 9.19 Semi-log plot of pump test data.

where r is the radial distance from the pumped well to the observation well, and T and S are the transmissivity and storage coefficient to be determined by the pump test. The value of u becomes small as time (t) increases. If u is small (<0.01), then Eq. 9.68 can be written

$$s = \frac{Q}{4\pi T}[-0.577 - \ln u]$$

$$= \frac{Q}{4\pi T}\ln\left[\frac{2.25Tt}{r^2S}\right] \quad \textbf{(9.69a)}$$

$$= \frac{2.3Q}{4\pi T}\log\left[\frac{2.25Tt}{r^2S}\right] \quad \textbf{(9.69b)}$$

Equations 9.69(a) (ln base e) and 9.69(b) (log base 10) are known as the Cooper–Jacob equation and show the drawdown to be a linear function of log t. Referring to the plot in Fig. 9.19, the slope (over one log cycle) of the semi-log plot is

$$\Delta s = \frac{2.3Q}{4\pi T}$$

or

$$T = \frac{2.3Q}{4\pi \Delta s} \quad \textbf{(9.70)}$$

In order for $s = 0$ when time $= t_o$, it is necessary that

$$\frac{2.25Tt_o}{r^2S} = 1$$

or

$$S = \frac{2.25Tt_o}{r^2} \quad \textbf{(9.71)}$$

Equations 9.70 and 9.71 are used to estimate values of T and S for the aquifer.

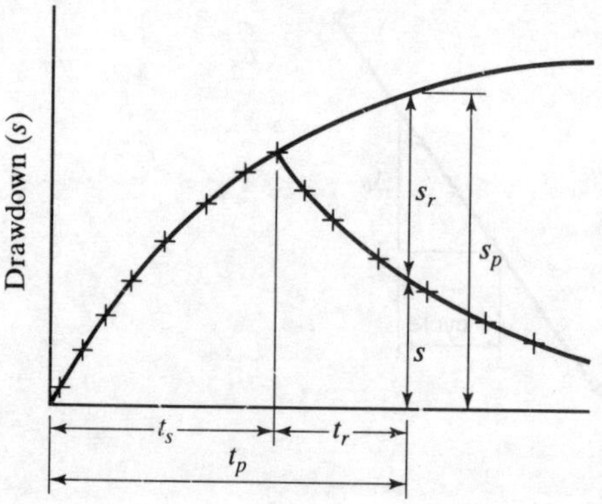

Figure 9.20 Recovery test.

The hydraulic properties of the aquifer determined from the pump test are representative of the aquifer from the pumped well to beyond the observation well. If after extended pumping the drawdown curve (Fig. 9.19) should deviate from a straight line, it would indicate that drawdown has been influenced by a boundary. If the plot in Fig. 9.19 deviates to the right, it would indicate recharge boundary, and if it deviates to the left, it would indicate a barrier boundary.

Measurements of the piezometric surface at the observation well are usually continued after the pump has been turned off. The recovery test can be utilized to provide a second estimate of the transmissivity. In the arithmetic drawdown plot (Fig. 9.20), the pump was turned off at time t_s and the recovery phase begins. The recovery phase of the test is simulated numerically by continuing the pumped well and adding a recharge well starting at time t_s. Utilizing the Cooper–Jacob equation for a pumped well and a recharge well at the same location, the drawdown during the recovery phase is

$$s = s_p - s_r = \frac{2.3Q}{4\pi T}\left[\log\frac{2.25T}{r^2S} + \log t_p\right] - \frac{2.3Q}{4\pi T}\left[\log\frac{2.25T}{r^2S} + \log t_r\right]$$

$$s = \frac{2.3Q}{4\pi T}\log\left(\frac{t_p}{t_r}\right) \tag{9.72}$$

A plot of drawdown versus log (t_p/t_r) is shown in Fig. 9.21 for the recovery phase of the pump test. The slope of the plot gives an estimate of the transmissivity using Eq. 9.70.

The pump test presented in this section to estimate the hydraulic properties of the aquifer was developed for a homogeneous, isotropic confined aquifer. The procedure assumes that the flow in the aquifer is horizontal and that the well is screen for the full length of the aquifer. The method is also applicable to unconfined aquifers where the drawdown is small, compared with the saturated thickness of the aquifer.

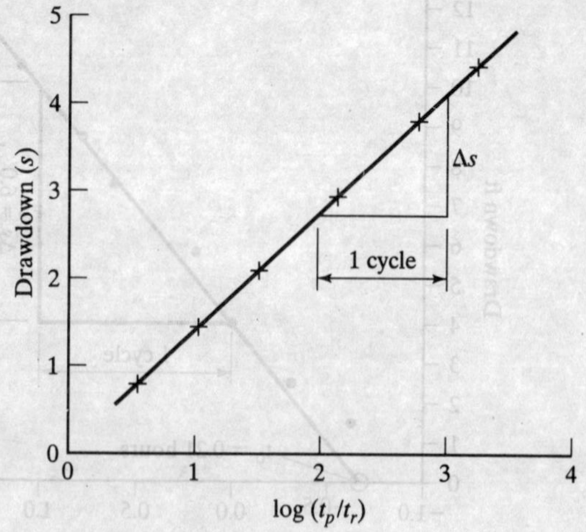

Figure 9.21 Semi-log plot of recovery test data.

Rewriting the Thiem equation (Eq. 9.57) as

$$s = H - h = \frac{Q}{4\pi T} \ln\left(\frac{R^2}{r^2}\right)$$

Comparing the above equation with the Cooper–Jacob equation [Eq. 9.69(a)], it is apparent that the radius of influence of the well (R) can be estimated as

$$R = 1.5\sqrt{\frac{Tt}{S}} \tag{9.73}$$

Example 9.11 Aquifer Pump Test

A 12-inch diameter well was pumped at a rate of 500 gpm. Water levels were observed at an observation well located 100 ft from the pumped well. Determine the transmissivity and storage coefficient of the aquifer for the data listed below using the Cooper–Jacob method. Solve the test both graphically and numerically.

Time hrs	Drawdown ft
0.25	1.5
0.5	2.5
1.0	4.0
2.0	5.8
4.0	7.5
6.0	8.7
12.0	10.5
18.0	11.4
24.0	12.2

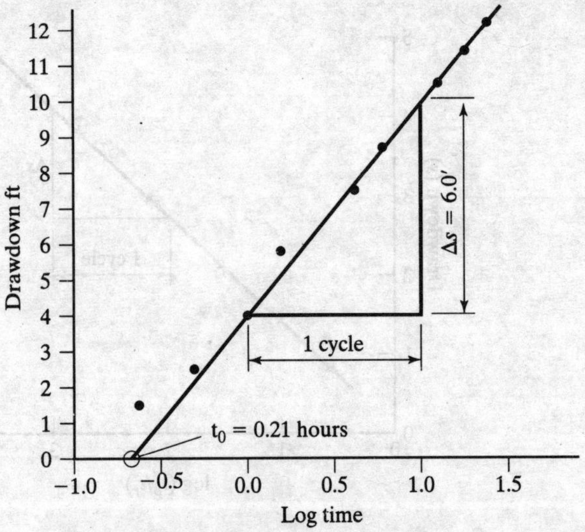

Example 9.11 Aquifer Pump Test.

Graphic Solution

Plot (s) versus log t and draw a line through the data points neglecting the first two points because t is small and u is large for these points.

$$T = \frac{2.3Q}{4\pi \Delta s} = \frac{2.3 \times 500 \times 1{,}440}{4\pi \times 6.0} = 22{,}000 \text{ gpd/ft}$$

$$S = \frac{2.25Tt_o}{r^2} = \frac{2.25 \times 22{,}000 \times 0.21}{10{,}000 \times 7.48 \times 24} = 0.0058$$

Numerical Solution

$$s = \frac{Q}{4\pi T} \ln\left[\frac{2.25Tt}{r^2 S}\right] = \frac{Q}{4\pi T}\left[\ln\left[\frac{2.25T}{r^2 S}\right] + \ln(t)\right]$$

or

$$s = B_1 + B_2 X$$

where

$$B_1 = \frac{Q}{4\pi T} \ln\left[\frac{2.25T}{r^2 S}\right]$$

$$B_2 = \frac{Q}{4\pi T}$$

$$X = \ln(t)$$

in matrix notation

$$[s] = [X][B]$$

$$[s] = \begin{bmatrix} 4.0 \\ 5.8 \\ 7.5 \\ 8.7 \\ 10.5 \\ 11.4 \\ 12.2 \end{bmatrix} \qquad [X] = \begin{bmatrix} 1 & \ln(1.0) \\ 1 & \ln(2.0) \\ 1 & \ln(4.0) \\ 1 & \ln(6.0) \\ 1 & \ln(12.0) \\ 1 & \ln(18.0) \\ 1 & \ln(24.0) \end{bmatrix} \qquad [B] = \begin{bmatrix} B_1 \\ B_2 \end{bmatrix}$$

Because there are more equations (7) than unknowns (2), a least-squares solution will be utilized to solve for $[B]$

$$[X]'[s] = [X]'[X][B]$$

$$[B] = ([X]'[X])^{-1}[X]'[s]$$

where the prime indicates the transpose of the $[X]$ array and the minus 1 exponent indicates the inverse of the square $([X]'[X])$ array.

A program (CJAQTEST.FOR) was written using CJAQTEST.DAT as the input file and CJAQTEST.OUT as the output file. The output file is listed below.

```
EXAMPLE 9.11 COOPER-JACOB METHOD AQUIFER TEST
    GROUNDWATER
     PUMPING RATE     =       500.000000 GPM
  RADIAL DISTANCE     =       100.000000 FT
     NO =           1          B =         4.002338
     NO =           2          B =         2.582281
          TRANSMISSIVITY =   22188.000000 GPD/FT
     STORAGE COEFFICIENT =   5.902888E-03

                   RELATION BETWEEN TIME AND DRAWDOWN
                               OBSERVED      COMPUTED
                   TIME       DRAWDOWN      DRAWDOWN       ERROR
         NO.      HOURS         FEET          FEET          FEET
          1        1.00         4.00          4.00           .00
          2        2.00         5.80          5.79          -.01
          3        4.00         7.50          7.58           .08
          4        6.00         8.70          8.63          -.07
          5       12.00        10.50         10.42          -.08
          6       18.00        11.40         11.47           .07
          7       24.00        12.20         12.21           .01
```

9.3.3 Superposition and Image Wells

Multiple wells with overlapping cones of depression are often installed in aquifers. If the water is pumped from several wells in an aquifer, the drawdown at any point

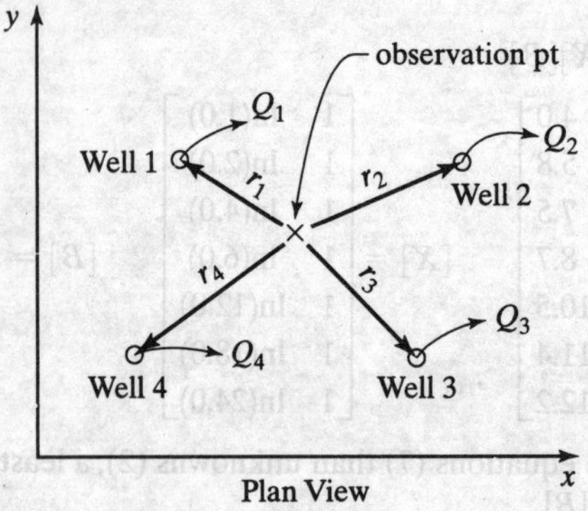

Figure 9.22 Multiple wells completed in the same aquifer.

is computed as the sum of the drawdown caused by each well. This principle is known as superposition. Figure 9.22 shows several pumped wells completed in the same aquifer. The drawdown at any point x is

$$s_x = \frac{1}{4\pi T} \sum_{i=1}^{n} Q_i W(u_i) \tag{9.74}$$

where

$$u_i = \frac{r_i^2 S}{4Tt_i}$$

and n is the number of wells in a confined aquifer.

An image well is an imaginary well used to reproduce the flow pattern representative of an aquifer boundary. If a real well is located near a boundary (Fig. 9.23), the image well is located on the opposite side and the same distance (a) from the boundary. To represent a barrier boundary, the image well has the same sign as the real well and to represent a recharge boundary, the image well has the opposite sign as the real well. The resulting drawdown for a well located near a boundary is found by superposition of the real and image well.

Based on the Cooper–Jacob equation (Eq. 9.69), the unsteady state drawdown at any point x in Fig. 9.23 for a well completed in a confined aquifer near a barrier boundary can be estimated from

$$s = \frac{Q}{2\pi T} \ln\left[\frac{2.25Tt}{r_1 r_2 S}\right] \tag{9.75}$$

For a well completed in an unconfined aquifer near a recharge boundary, the steady-state saturated thickness (h) may be computed from

$$H^2 - h^2 = \frac{Q}{\pi K} \ln\left(\frac{r_2}{r_1}\right) \tag{9.76}$$

where H is the initial saturated thickness of the aquifer.

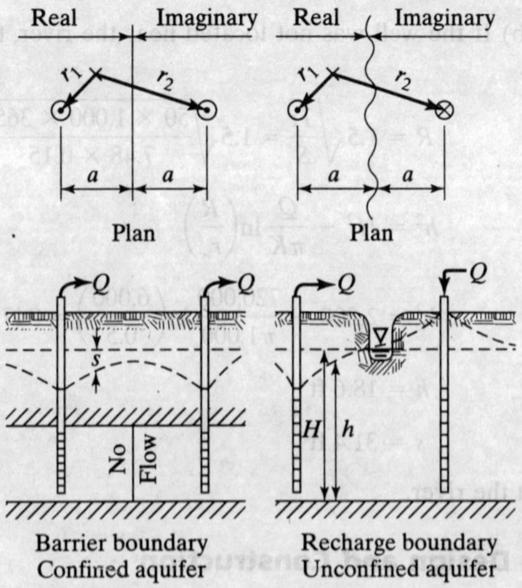

Figure 9.23 Image wells representing boundaries.

Example 9.12 Well Near Recharge Boundary

A 12-inch diameter water supply well is located in the Brazos River floodplain 300 ft from the river. The unconfined aquifer has a saturated thickness of 50 ft, a hydraulic conductivity of 1,000 gpd/ft^2, and a storage coefficient of 0.15. If the well is pumped at a rate of 500 gpm for an extended period, compute the drawdown at the well with and without the recharge boundary.

(a) With the recharge boundary

$$H^2 - h^2 = \frac{Q}{\pi K} \ln\left(\frac{r_2}{r_1}\right)$$

where

$$r_1 = 0.5 \text{ ft}$$

$$r_2 = 600 \text{ ft}$$

$$Q = 720{,}000 \text{ gpd}$$

$$K = 1{,}000 \text{ gpd/ft}^2$$

$$H = 50 \text{ ft}$$

$$h^2 = 2{,}500 - \frac{720{,}000}{\pi 1{,}000} \ln\left(\frac{600}{0.5}\right) = 875 \text{ ft}^2$$

$$h = 29.6 \text{ ft}$$

$$s = H - h = 20.4 \text{ ft}$$

(b) If the well was not located near the river, the drawdown would have been greater.

$$R = 1.5\sqrt{\frac{Tt}{S}} = 1.5\sqrt{\frac{50 \times 1{,}000 \times 365}{7.48 \times 0.15}} = 6{,}000 \text{ ft}$$

$$h^2 = H^2 - \frac{Q}{\pi K}\ln\left(\frac{R}{r_w}\right)$$

$$h^2 = 2{,}500 - \frac{720{,}000}{\pi 1{,}000}\ln\left(\frac{6{,}000}{0.5}\right) = 347 \text{ ft}^2$$

$$h = 18.6 \text{ ft}$$

$$s = 31.4 \text{ ft}$$

without the river.

9.3.4 Well Design and Construction

Wells are constructed for various purposes, including water supply, dewatering construction sites, artificial recharge, wastewater disposal, and groundwater contamination cleanup. Depending on the purpose of the well, depth of the aquifer, and geological material, the well may be constructed using a driven or jetted well point, auger bore hole, or drilled bore hole. Well points are limited to small diameter, shallow wells in unconsolidated material. They are often connected to a vacuum manifold pipe where a single pump is used for several wells. Major water supply wells are typically constructed using rotary drilling methods. This section includes general guidelines for the design and construction of water supply wells.

The diameter of the well depends on the size of the equipment to be placed in the well. For a water supply well, the diameter of the casing is generally selected as two nominal sizes (2 inches, 50 mm) larger than the bowl of the pump to be installed in the well. Typical casing sizes for various pump discharge rates are listed in Table 9.5.

TABLE 9.5 WELL CASING SIZE

Discharge rate		Casing diameter	
GPM	LPS	inches	mm
<100	<6.3	6	152
200	13	8	203
400	25	10	254
600	38	12	305
900	57	14	356
1,300	82	16	406
1,800	110	20	508
3,000	190	24	610

To prevent contamination of the aquifer, the well casing is typically grouted in the hole. Depending on the method used for grouting, the hole is drilled 2–6 inches (50–150 mm) larger than the outside diameter of the casing. After the casing is set in position, the annular space between the casing and hole is filled with grout.

For maximum yield from a confined aquifer, 70–80 percent of the aquifer thickness is screened. If the aquifer is homogeneous, the well screen is centered in the aquifer; otherwise, the well screen is set in the coarser part of the aquifer. For an unconfined aquifer, typically the lower one-third of the aquifer is screened and the upper two-thirds of the aquifer is reserved for drawdown. In a confined aquifer, the pump is usually set in the casing above the well screen, whereas in an unconfined aquifer, the pump is usually set in the well screen below the casing.

The well screen opening size is selected to ensure that formation sand does not enter the well during normal operation. The opening size is based on a grain size analysis of a sample from the aquifer formation. Figure 9.24 is a plot of an aquifer grain size analysis. The uniformity coefficient (C_u) is defined as the diameter with 40 percent larger than (d_{40}) divided by the diameter with 90 percent larger than (d_{90}) or

$$C_u = \frac{d_{40}}{d_{90}} \tag{9.77}$$

If the uniformity coefficient is less than three, indicating a uniform aquifer, the well screen opening size is usually selected as the d_{40} of the formation. If the uniformity coefficient is greater than 6, indicating a nonuniform aquifer, the well screen opening size is usually selected as the d_{30} of the formation. Some fine material from the aquifer will pass through the well screen during well development. This results in a graded filter about the well screen with an increased hydraulic conductivity near the screen.

The maximum desirable velocity through the opening in the well screen depends on the hydraulic conductivity of the aquifer and is listed in Table 9.6. The diameter of the well screen must be large enough to provide the required open area

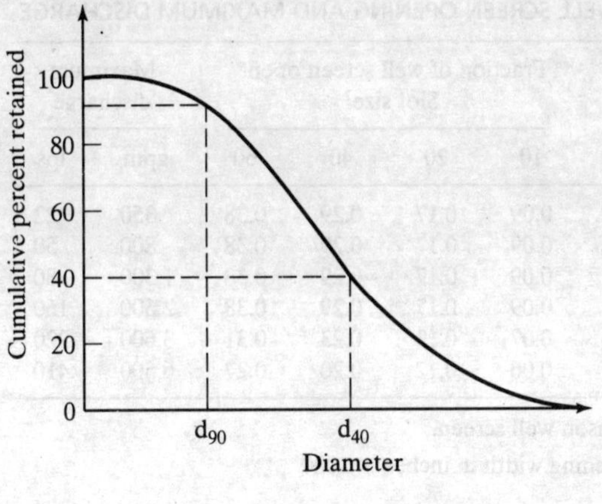

Figure 9.24 Aquifer grain size.

TABLE 9.6 MAXIMUM VELOCITY THROUGH WELL SCREEN OPENING

Aquifer hydraulic conductivity		Maximum velocity	
ft/day	m/day	ft/min	m/min
>800	>250	12	3.6
400	120	8	2.4
130	40	4	1.2
60	20	3	0.9

in the well screen. The amount of open area varies with manufacturer and depends on the well screen design and material. Table 9.7 lists the fraction of a typical well screen that is open for flow. The maximum recommended discharge rate for various diameter well screens is listed in Table 9.7.

A gravel pack is used for a fine grain aquifer when the d_{40} of the formation is less than 0.010 inches (0.25 mm). The gravel pack is placed between the well screen and the formation to provide a filter that prevents the fines in the formation from entering the well. To provide space for the gravel pack, the diameter of the hole for the well screen is made approximately 12 inches (305 mm) larger than the well screen to a height of approximately 10 ft (3 m) above the well screen. The gradation of the gravel pack is based on the finest layer in the aquifer. The d_{70} of the gravel pack is approximately four times the d_{70} of the aquifer for a uniform aquifer and six times the d_{70} of the aquifer for a nonuniform aquifer. The uniformity coefficient of the gravel pack should be about 2.5. The well screen slot size is selected as the d_{90} of the gravel pack.

The purpose of well development is to remove mud cake from the walls of the borehole and to remove fine material from the aquifer near the well screen. This is usually accomplished by mechanical surging, compressed air surging, or overpumping. A water supply well is also disinfected before being placed into operation.

TABLE 9.7 WELL SCREEN OPENING AND MAXIMUM DISCHARGE

Well screen diameter		Fraction of well screen open* Slot size[†]				Maximum discharge	
inches	mm	10	20	40	60	gpm	lps
6	152	0.09	0.17	0.29	0.38	350	22
8	203	0.09	0.17	0.29	0.38	800	50
10	254	0.09	0.17	0.29	0.38	1,300	80
12	305	0.09	0.17	0.29	0.38	2,500	160
14	356	0.07	0.13	0.23	0.31	3,600	220
18	457	0.06	0.12	0.20	0.27	6,500	410

*Based on Johnson well screen.

[†]Slot size = opening width in inches × 1000.

Example 9.13 Well Design

Design a well for a pumping rate of 1,000 gpm from a confined aquifer 200 ft thick with a hydraulic conductivity of 500 gpd/ft^2 and a storage coefficient of 0.0003. The base of the aquifer is 700 ft below the ground surface. Samples collected from the aquifer indicate a laminated formation of thin alternating layers of fine and coarse sand. Analyses of the formation samples for the fine and coarse sand are listed below.

	Fine sand inches	Coarse sand inches
d_{30}	0.010	0.030
d_{40}	0.008	0.025
d_{50}	0.007	0.021
d_{70}	0.005	0.012
d_{90}	0.003	0.010

Compute the drawdown at the well and at 10 miles after 1 year of continuous pumping.

- Well diameter from Table 9.5 is 16 inches.
- Well screen length is 80 percent of the aquifer thickness or 160 ft, which is to be centered in the aquifer. Depth of the well is 680 ft.
- The well screen is to be designed for the fine sand. The uniformity coefficient (C_u) is

$$C_u = \frac{d_{40}}{d_{90}} = \frac{0.008}{0.003} = 2.7 \text{ (uniform)}$$

- Gravel pack is required because the d_{40} of the fine sand is less than 0.010 inches. The d_{70} of the gravel pack is

$$(d_{70})_{\text{pk}} = 4 \times (d_{70})_{\text{aq}} = 4 \times 0.005 = 0.020 \text{ inches}$$

Gravel pack with $C_u = 2.5$

$$d_{30} = 0.035 \text{ inches}$$

$$d_{40} = 0.030$$

$$d_{50} = 0.027$$

$$d_{70} = 0.020$$

$$d_{90} = 0.012$$

- Well screen slot size is based on the d_{90} of the gravel pack slot size = 0.01 inches
- Based on the hydraulic conductivity of the aquifer (70 ft/day), the maximum velocity through the well screen is 3 ft/min (Table 9.6).
- The area (A) of the well screen opening is

$$A = \frac{Q}{V} = \frac{1{,}000/7.48}{3} = 44.6 \text{ ft}^2 = 40.1 \text{ in.}^2/\text{ft}$$

- A 12-inch diameter well screen has an area of (see Table 9.7)

$$= \pi \times 12 \times 12 \times 0.09 = 40.7 \text{ in.}^2/\text{ft}$$

- A 24-inch diameter hole will be required for the lower 170 ft of the well to accommodate the 12-inch diameter well screen and the 6-inch gravel pack. Volume of gravel pack ($\forall$) in yd^3 is

$$\forall = \frac{\pi}{4}(2^2 - 1^2)170/27 = 15 \text{ yd}^3$$

- The drawdown after 1 year of pumping is

$$u = \frac{r^2 S}{4Tt}$$

$$S = 0.0003$$

$$T = \frac{500 \times 200}{7.48} = 13{,}400 \text{ ft}^2/\text{day}$$

$r = 1.0$ ft	$r = 52{,}800$ ft
$u = 1.5 \times 10^{-11}$	$u = 4.3 \times 10^{-2}$
$W(u) = 24.4$	$W(u) = 2.6$
$s = \frac{Q}{4\pi T} W(u) = 28.0$ ft	$s = \frac{1{,}000 \times 1{,}440}{4\pi 500 \times 200} W(u) = 3.0$ ft

9.4 FLOW NET ANALYSIS

Flow net analysis is a useful tool for groundwater evaluation. It provides information on the direction and velocity of flow without the use of numerical models. Flow net analysis is based on steady flow in a piecewise homogeneous medium with a fluid of constant density and viscosity. Darcy's law must be valid for the medium and the Laplace equation must be satisfied.

Constructing a flow net is somewhat of an art and is usually done on a trial-and-error basis. For demonstration purposes, a number of flow nets are included in this section.

9.4.1 Flow in a Horizontal Plane

Flow nets to define horizontal flow in an aquifer are usually constructed based on observed water levels in existing water wells and/or installed observation wells. To define uniform flow in a horizontal plane, a minimum of three observation points (not in a straight line) of water levels are required. The equipotential (ϕ) lines represent contours of water levels in an unconfined aquifer or contours of piezometric levels in a confined aquifer. As shown in Fig. 9.25, the flow lines or streamlines (ψ) are drawn perpendicular to the equipotential lines. For uniform flow, the streamlines (and equipotential) lines are parallel. If the equipotential lines (ϕ) are

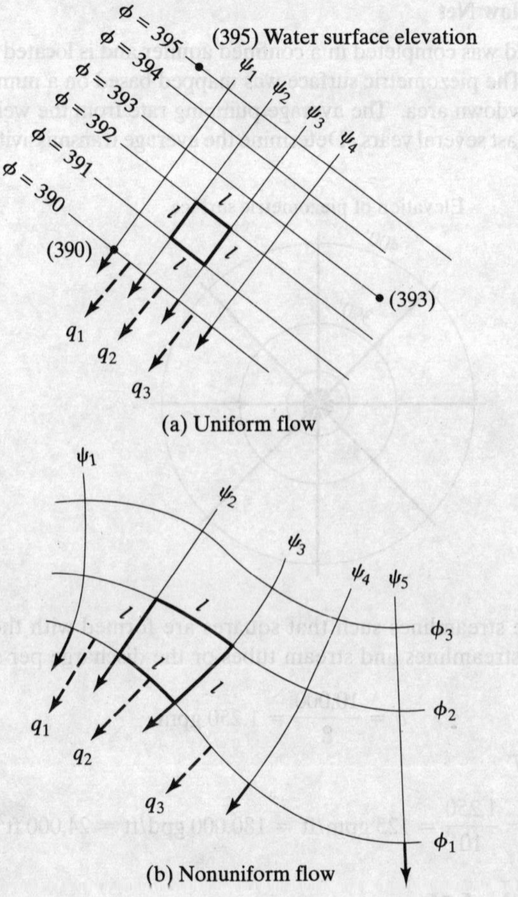

Figure 9.25 Horizontal flow net.

defined at equal intervals and the streamlines are drawn normal to the equipotential lines such that a series of squares (or curvilinear squares) are formed, then the rate of flow (q) between each pair of adjacent streamlines is equal. Based on the Darcy equation, the flow rate is (Fig. 9.25)

$$q_1 = q_2 = q_3 = KAS = Tl\frac{d\phi}{l} = Td\phi \tag{9.78}$$

where ℓ is the size of the square element, ϕ is the head potential, and T is the transmissivity.

Because the flow rate is constant between adjacent streamlines, the velocity is inversely proportional to the size of the flow net element. Because there is no flow across either a barrier boundary or a streamline, a barrier boundary can be represented as a streamline. Flow net analysis does not require defined boundaries on all sides of the region to be evaluated because any streamline can be replaced with an imaginary barrier boundary. A potential boundary is represented as equipotential line(s).

Example 9.14 Regional Flow Net

A municipal well field was completed in a confined aquifer and is located at the center of the diagram below. The piezometric surface was mapped based on a number of observation wells in the drawdown area. The average pumping rate from the well field has been 10,000 gpm over the last several years. Determine the average transmissivity of the aquifer.

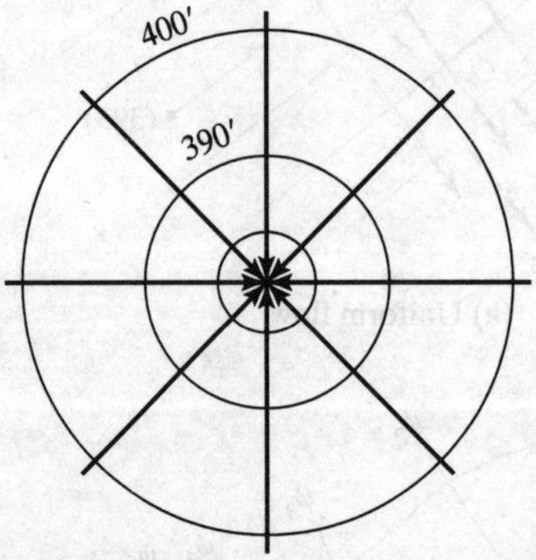

Solution Draw the streamlines such that squares are formed with the equipotential lines. This yields 8 streamlines and stream tubes or the discharge per stream tube is

$$q = \frac{10{,}000}{8} = 1{,}250 \text{ gpm}$$

The transmissivity T

$$T = \frac{q}{d\phi} = \frac{1{,}250}{10} = 125 \text{ gpm/ft} = 180{,}000 \text{ gpd/ft} = 24{,}000 \text{ ft}^2/\text{day}$$

9.4.2 Flow in a Vertical Plane

Equipotential lines are lines of constant energy where the sum of the elevation (z) and pressure head (p/γ) is constant.

9.4.2.1 Isotropic medium. Figure 9.26 is a flow net for seepage under a dam through a homogeneous, isotropic formation bounded below by an impermeable boundary. Lines A–B and C–D represent equipotential lines ($\phi = 90$ and $\phi = 10$, respectively), and lines B–C and E–F represent streamlines. Singularity points occur in a flow net where streamlines and equipotential lines do not intersect at right angles or are discontinuous. The velocity at singularity points either becomes large or approaches zero. Points B and C in Fig. 9.26 represent singularity points where the velocity becomes large. The flow net consists of curvilinear squares where each square can enclose a circle that is tangent to all four sides. The seepage (q) per unit length of the dam is

$$q = -KAS = K(l\text{N}_t)\frac{(El_1 - El_2)}{\ell\text{N}_d} = K(El_1 - El_2)(N_t/N_d) \qquad \textbf{(9.79)}$$

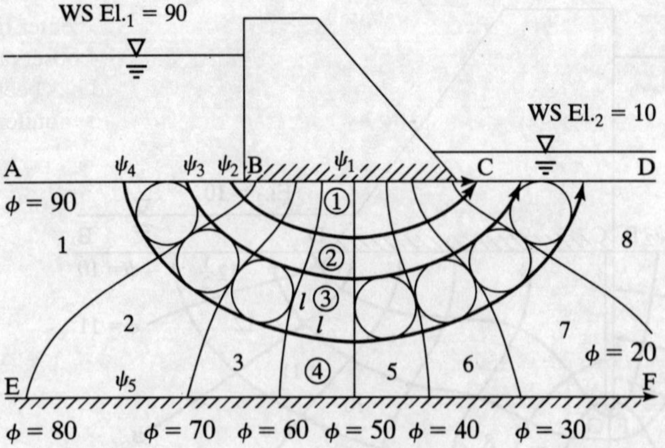

Figure 9.26 Seepage under a dam.

where N_t is the number of streamtubes (4), N_d is the number of equipotential drops (8), and l is the size of the flow net element. Flow nets are dimensionless, and the seepage rate is independent of the scale. If a number of people draw flow nets for the same problem, all the flow nets will be different, but the seepage rate computed from Eq. 9.79 will be essentially the same.

Figure 9.27 is a flow net for sheet piling. Lines A–B and C–D are equipotential lines, whereas lines B–E–F–C and G–H are streamlines. Points E and F are singularity points, where the seepage velocity is high.

Figure 9.28 is a flow net for seepage under a dam with a cutoff wall. The cutoff wall reduces both the seepage and the uplift pressure. The seepage rate under

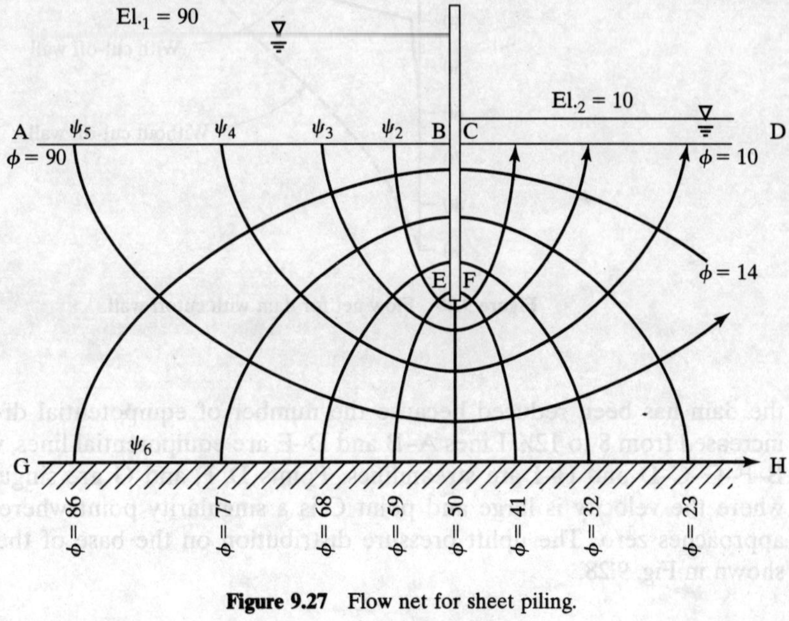

Figure 9.27 Flow net for sheet piling.

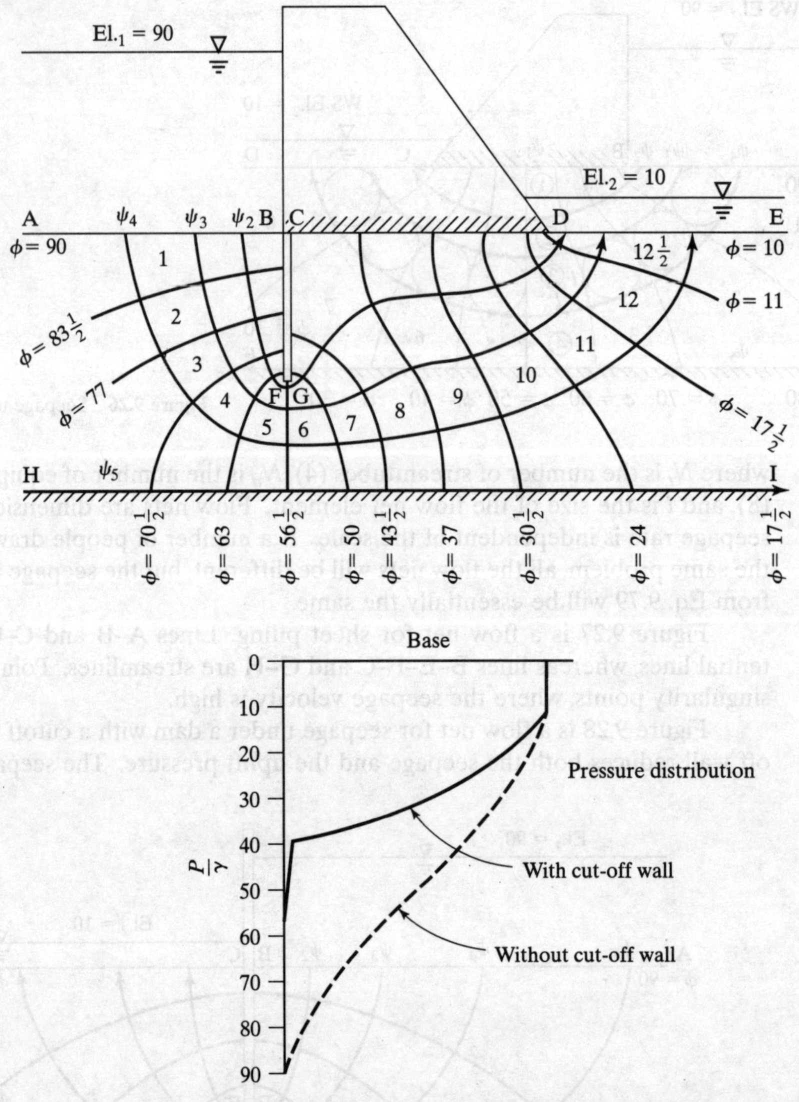

Figure 9.28 Flow net for dam with cutoff wall.

the dam has been reduced because the number of equipotential drops (N_d) has increased from 8 to 12½. Lines A–B and D–E are equipotential lines, whereas lines B–F–G–C–D and H–I are streamlines. Points D, F, and G are singularity points, where the velocity is large and point C is a singularity point where the velocity approaches zero. The uplift pressure distribution on the base of the dam is also shown in Fig. 9.28.

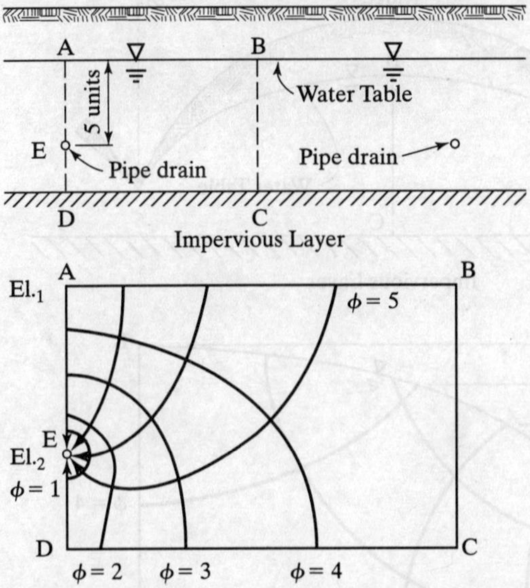

Figure 9.29 Flow net for pipe drains.

A flow net for parallel pipe field drains in a saturated porous medium with a horizontal water table is shown in Fig. 9.29. Because the flow net for the drain is symmetrical, only half the flow net (A–B–C–D) is drawn. The pipe drain represents a sink in the flow net with an equipotential value of zero. The water table (A–B) represents an equipotential line with a relative value of 5 units relative to the tile drain. Lines A–E and B–C–D–E represent streamlines. Points C, D, and E represent singularity points, where the seepage velocity at points C and D approach zero, and the velocity at E becomes large. The discharge rate computed from Eq. 9.79 represents half the seepage rate into the drain per unit length.

The flow net in Fig. 9.30 represents seepage for a gaining stream where groundwater discharges into the stream in the form of springs along the channel. The ridge between the parallel streams is a recharge area, and the stream is represented in the flow net as a sink. In this example, the water table is not horizontal, and equipotential lines are spaced along the water table at equal changes in elevation (Δz). The shape of the water table between A and B has considerable influence on the location and size of the recharge area and shape of the flow net. Line B–C–D–A represents a streamline with singularity points at C, D, and A. The discharge rate computed from Eq. 9.79 represents half the seepage into the stream.

Figure 9.31 represents the flow net for seepage from a stream where the hydraulic conductivity of the upper aquifer is much greater than the lower layer. The water table elevation adjacent to the stream is less than the water elevation in the stream and the stream is recharging the adjacent aquifer. The perimeter of the channel (A–B–C) is an equipotential lines and the streamlines originate perpendicular to the perimeter. Line C–D represents a free surface streamline and line A–E–F represents a boundary streamline.

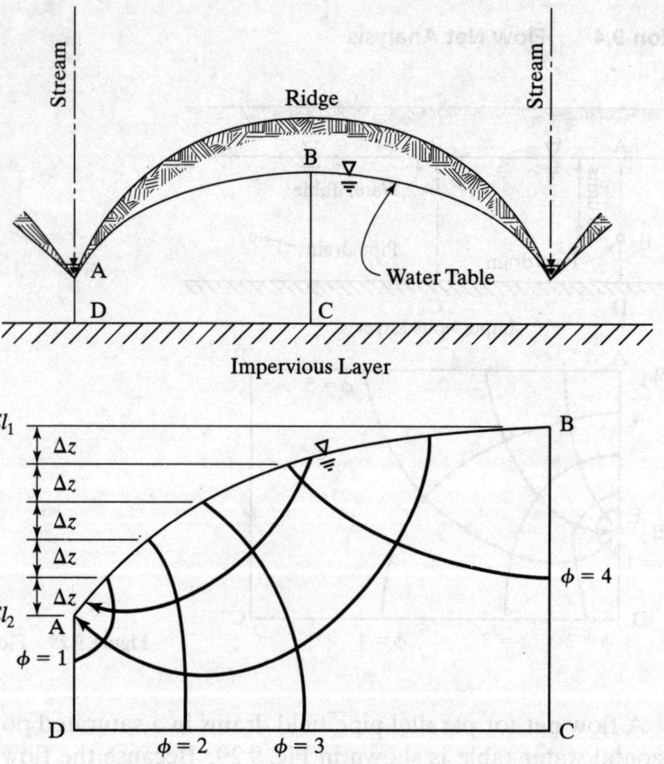

Figure 9.30 Flow net for gaining stream.

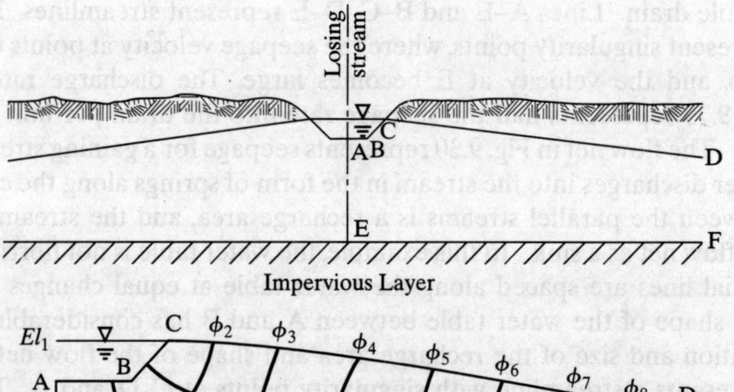

Figure 9.31 Flow net for losing stream.

9.4.2.2 Anisotropic medium. Fluvial formations are often formed in layers such that the horizontal hydraulic conductivity (K_x) is much greater than the vertical hydraulic conductivity (K_z). The flow net can be constructed for a homogeneous, anisotropic medium by expanding the vertical dimension of the flow region. If z is the original vertical dimension and Z is the transformed vertical dimension such that

$$Z = z\sqrt{\frac{K_x}{K_z}} \tag{9.80}$$

then the flow net can be constructed in the transformed domain using the same procedures presented for an isotropic medium.

The validity of this transformation is based on the steady state flow equations for an anisotropic medium which is

$$\frac{\partial}{\partial x}\left(K_x\frac{\partial\phi}{\partial x}\right) + \frac{\partial}{\partial z}\left(K_z\frac{\partial\phi}{\partial z}\right) = 0 \tag{9.81}$$

or

$$\frac{\partial^2\phi}{\partial x^2} + \frac{\partial}{\partial z}\left(\frac{K_z}{K_x}\frac{\partial\phi}{\partial z}\right) = 0 \tag{9.82}$$

Using the transformation in Eq. 9.80 and letting $X = x$, Eq. 9.82 reduces to the Laplace equation for a homogeneous, isotropic medium or

$$\frac{\partial^2\phi}{\partial X^2} + \frac{\partial^2\phi}{\partial Z^2} = 0 \tag{9.83}$$

where

$$\frac{\partial}{\partial z} = \sqrt{\frac{K_x}{K_z}}\frac{\partial}{\partial Z}$$

and

$$\frac{\partial\phi}{\partial z} = \sqrt{\frac{K_x}{K_z}}\frac{\partial\phi}{\partial Z}$$

The hydraulic conductivity of the transformed isotropic medium is

$$K = \sqrt{K_x \cdot K_z} \tag{9.84}$$

Figure 9.32 shows the flow net constructed in the transformed isotropic medium. The original vertical dimensions were multiplied by $\sqrt{K_x/K_z}$ to give the transformed section. The flow net was then developed for the transformed section using the basic procedures for flow net construction in an isotropic medium. Half the seepage rate is computed using Eq. 9.79, where El_1 and El_2 refer to the original heads and K is computed using Eq. 9.84.

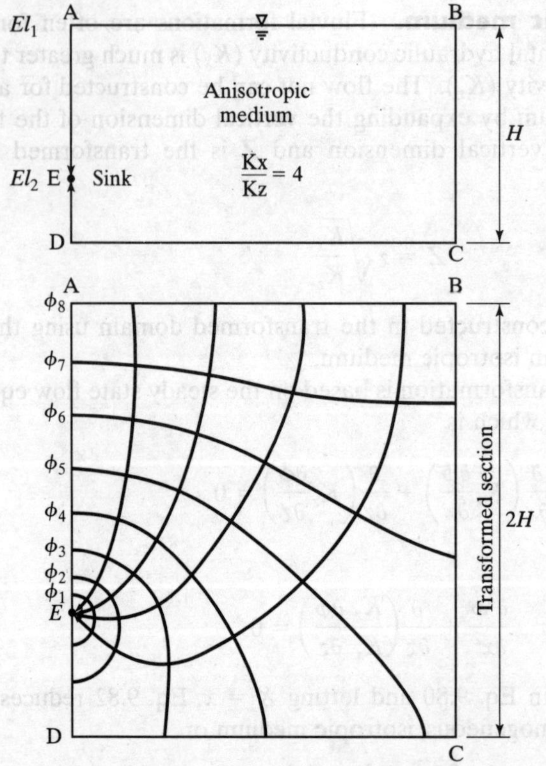

Figure 9.32 Flow net for anisotropic medium.

9.5 NUMERICAL METHODS

Computer simulation of groundwater flow is often used for complex aquifers. Flow nets can easily be constructed for homogeneous, isotropic aquifers, but construction of flow nets becomes more difficult for nonhomogeneous, anisotropic formations. Numerical methods can provide the detailed groundwater flow and head information for complex geology and boundary conditions. Numerical procedures require that the aquifer to be modeled be divided into grid elements. In the development of the model, either the grid elements or grid lines can be numbered. Because each grid element represents a portion of the aquifer, the grid elements have been numbered in Fig. 9.33. Nonhomogeneous aquifers can be represented numerically by assigning aquifer parameters to each grid element. The flow rates q_A through q_D shown in Fig. 9.33 represent grid-to-grid water transfer per unit area of aquifer, while q_F represents the net groundwater withdrawal rate (pumpage + evapotranspiration − recharge) per unit area of aquifer.

The continuity condition requires that the flow into a grid element minus the flow out of the grid element equals the change in storage in the grid element

$$\Delta x_i \Delta y_j \left[q_C - q_D - q_A + q_B - q_F = S\frac{\partial \phi}{\partial t} \right] \tag{9.85}$$

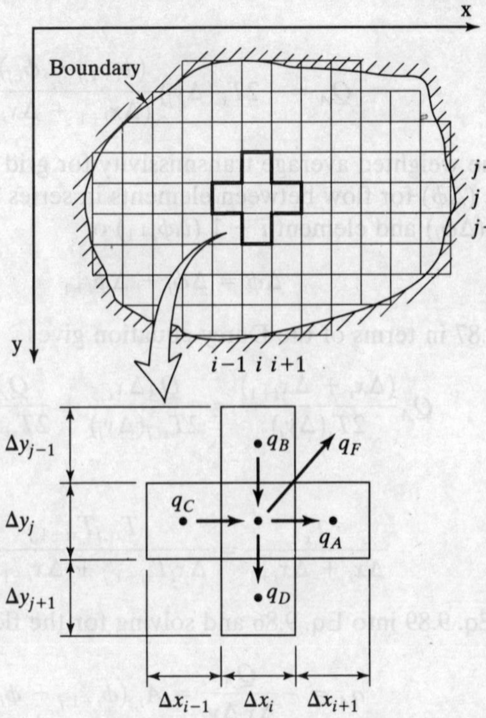

Figure 9.33 Finite difference grid.

where S is the storage coefficient, ϕ is the potential head (elevation plus pressure head with units of length), and t is time. Anisotropic medium can be represented by assigning different transmissivity values in the x (T_x) and y (T_y) directions.

The grid-to-grid water transfer from grid i to grid $i + 1$ is shown in Fig. 9.34 as Q_A. Based on the Darcy equation

$$Q_A = -KA\frac{\partial \phi}{\partial x}$$

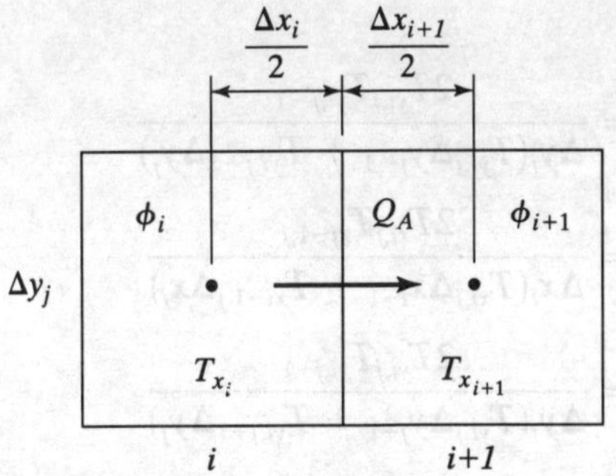

Figure 9.34 Grid to grid water transfer.

or

$$Q_A = -2T_x(\Delta y_j)\frac{(\phi_{i+1,j} - \phi_{i,j})}{(\Delta x_{i+1} + \Delta x_i)} \tag{9.86}$$

where T_x is the weighted average transmissivity for grid elements i and $i + 1$. The total headloss ($\Delta\phi$) for flow between elements in series is the sum of the headloss for element i ($\Delta\phi_i$) and element $i + 1$ ($\Delta\phi_{i+1}$) or

$$\Delta\phi = \Delta\phi_i + \Delta\phi_{i+1} \tag{9.87}$$

Writing Eq. 9.87 in terms of the Darcy equation gives

$$Q_A\frac{(\Delta x_i + \Delta x_{i+1})}{2T_x(\Delta y_j)} = \frac{Q_A\Delta x_i}{2T_{xi,j}(\Delta y_j)} + \frac{Q_A\Delta x_{i+1}}{2T_{xi+1,j}(\Delta y_j)} \tag{9.88}$$

or

$$\frac{T_x}{\Delta x_i + \Delta x_{i+1}} = \frac{T_{xi,j}T_{xi+1,j}}{\Delta x_i T_{xi+1,j} + \Delta x_{i+1}T_{xi,j}} \tag{9.89}$$

Substituting Eq. 9.89 into Eq. 9.86 and solving for the flow rate per unit area

$$q_A = -\frac{Q_A}{\Delta x_i \Delta y_j} = A_{i,j}(\phi_{i+1,j} - \phi_{i,j}) \tag{9.90}$$

where $A_{i,j}$ is the flow coefficient between elements i and $i + 1$ and is equal to

$$A_{i,j} = \frac{2T_{xi,j}T_{xi+1,j}}{\Delta x_i(T_{xi,j}\Delta x_{i+1} + T_{xi+1,j}\Delta x_i)} \tag{9.91}$$

The continuity equation (Eq. 9.85) can be written in terms of flow coefficients as

$$C_{i,j}(\phi_{i-1,j} - \phi_{i,j}) + D_{i,j}(\phi_{i,j+1} - \phi_{i,j}) + A_{i,j}(\phi_{i+1,j} - \phi_{i,j}) + B_{i,j}(\phi_{i,j-1} - \phi_{i,j}) - q_F = S\frac{\partial\phi}{\partial t} \tag{9.92}$$

where

$$B_{i,j} = \frac{2T_{yi,j}T_{yi,j-1}}{\Delta y_j(T_{yi,j}\Delta y_{j-1} + T_{yi,j-1}\Delta y_j)} \tag{9.93}$$

$$C_{i,j} = \frac{2T_{xi,j}T_{xi-1,j}}{\Delta x_i(T_{xi,j}\Delta x_{i-1} + T_{xi-1,j}\Delta x_i)} \tag{9.94}$$

$$D_{i,j} = \frac{2T_{yi,j}T_{yi,j+1}}{\Delta y_j(T_{yi,j}\Delta y_{j+1} + T_{yi,j+1}\Delta y_j)} \tag{9.95}$$

Dropping the subscripts on the flow coefficients, the continuity equation for element i, j reduces to

$$A\phi_{i+1,j} + B\phi_{i,j-1} + C\phi_{i-1,j} + D\phi_{i,j+1} - E\phi_{i,j} = S\frac{\partial\phi}{\partial t} + q_F \tag{9.96}$$

where

$$E = A + B + C + D$$

If $\Delta x = \Delta y$ and the aquifer is homogeneous and isotropic ($T_x = T_y = T$), then the continuity equation reduces to

$$\phi_{i+1,j} + \phi_{i,j-1} + \phi_{i-1,j} + \phi_{i,j+1} - 4\phi_{i,j} = \frac{\Delta x^2}{T}\left(S\frac{\partial\phi}{\partial t} + q_F\right) \tag{9.97}$$

Modification to the continuity equation (Eq. 9.96) is required at boundaries. Figure 9.35 shows an aquifer next to a river. The aquifer is bounded on the top, left and bottom with a barrier boundary and on the right with a potential boundary. For grid elements adjacent to a barrier boundary, the flow coefficient from the boundary side is equal to zero. For grid elements along the top row in Fig. 9.35 ($j = 1$), the B flow coefficient is equal to zero, and for grid elements along the bottom row ($j = 11$), the D coefficient is equal to zero.

For grid element (i,j) adjacent to a potential boundary (right boundary, Fig. 9.35), the flow coefficient from the boundary is

$$A = 2T_{xi,j}/\Delta x_i^2 \tag{9.98}$$

and the potential for the boundary ($\phi_{11,j}$) is specified as a boundary condition.

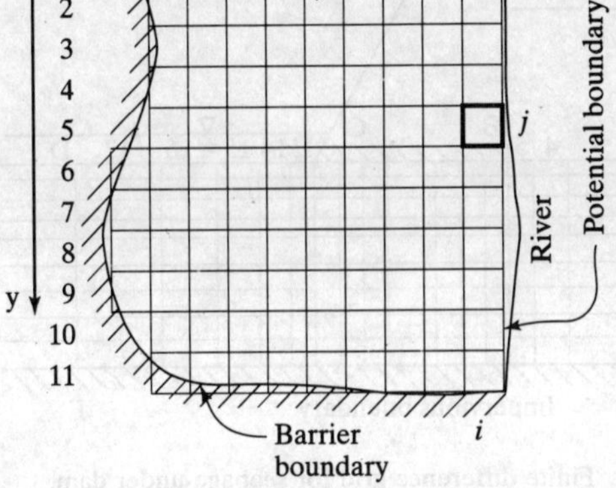

Figure 9.35 Boundaries.

9.5.1 Steady-State Flow Nets

Steady-state, two-dimensional groundwater flow problems can often be evaluated using a graphical solution to flow nets. However, for anisotropic, nonhomogeneous aquifers, a numerical solution will generally be preferred. The equation for steady-state, two-dimensional groundwater flow is

$$\frac{\partial}{\partial x}\left(T_x\frac{\partial \phi}{\partial x}\right) + \frac{\partial}{\partial y}\left(T_y\frac{\partial \phi}{\partial y}\right) = 0 \qquad \textbf{(9.99)}$$

where T_x and T_y represent the transmissivity of the formation in the x and y directions, respectively, and is the head potential.

Figure 9.36 represents the seepage under a dam where AB and CD are equipotential boundaries and BC and EF are barrier boundaries. It is assumed that the grid extends far enough upstream and downstream of the dam so that there is little flow across lines A–E and D–F and that line A–E–F–D is a barrier boundary. The continuity equation written for an interior grid element i,j reduces to

$$\phi^+_{i,j} = (A\phi_{i+1,j} + B\phi_{i,j-1} + C\phi_{i-1,j} + D\phi_{i,j+1})/E \qquad \textbf{(9.100)}$$

where the superscript + for the head potential refers to present iteration, and no superscript refers to previous iteration. The flow coefficients (A–E) in Eq. 9.100 have been defined in Section 9.5. For seepage in a vertical plane, the transmissivity is equal to the hydraulic conductivity since the thickness of the plane section is 1 unit.

An iterative procedure is used to compute the potential at a grid element and is based on the potentials at the adjacent grid elements. The potentials at the equipotential boundaries remain constant throughout the iterative procedure. For initial conditions, the potentials at the grid elements throughout the flow net may be either estimated or taken as zero. The iterative procedure continues until the

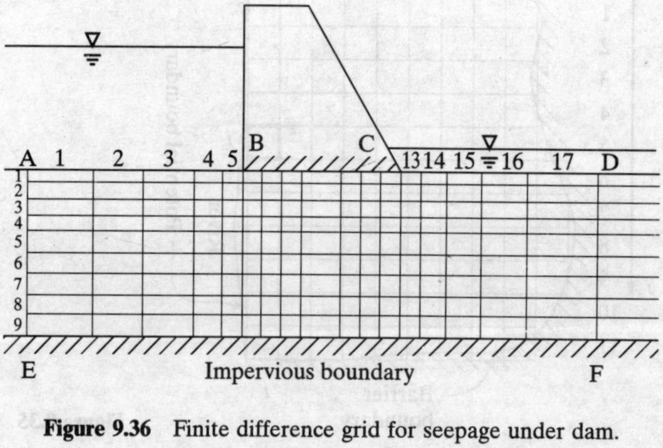

Figure 9.36 Finite difference grid for seepage under dam.

maximum change in potential is less than a specified value

$$\Delta\phi_{\max} = |\phi^{+}_{i,j} - \phi_{i,j}| \tag{9.101}$$

where $\Delta\phi_{\max}$ is the maximum change in the potential during an iteration.

After the procedure converges to a solution and the potentials have been computed at each grid element, the horizontal and vertical seepage rates can be computed between grid elements using the Darcy equation or

$$Q^{H}_{i,j} = C_{i,j}\Delta x_i \Delta y_j(\phi_{i-1,j} - \phi_{i,j}) \tag{9.102}$$

for the horizontal seepage from the $i - 1$ grid and

$$Q^{V}_{i,j} = B_{i,j}\Delta x_i \Delta y_j(\phi_{i,j-1} - \phi_{i,j}) \tag{9.103}$$

for the vertical seepage from the $j - 1$ grid.

Example 9.15 Seepage under a Dam

A dam is to be constructed on a semi-permeable foundation with an impermeable boundary located 60 ft deep. The hydraulic conductivity of each 10-ft layer of formation is shown on the sketch below. The base of the dam is 250 ft wide with a 25-ft wide cutoff wall 40 ft deep located near the upstream toe. Compute the seepage under the dam for a headwater depth of 45 ft and a tailwater depth of 5 ft.

Extend the computational grid 300 ft upstream of the dam and 400 ft downstream of the dam. The grid columns are numbered 1 through 24 in the horizontal

Layer(s)	Hydraulic conductivity K_x ft/day	K_z ft/day
1	100	10
2 & 3	50	5
4	10	1
5 & 6	5	0.5

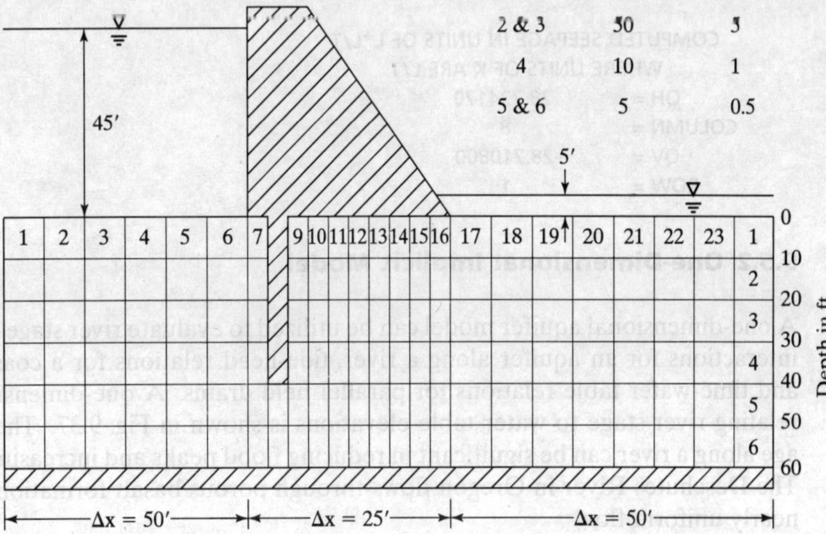

direction, and grid rows are numbered 1 through 6 in the vertical direction. Program (SEEPAG2D.FOR) was written to solve for steady-state seepage (Eq. 9.100). The program was dimensioned for 50 horizontal grid elements and 20 layers vertically. Both horizontal and vertical seepage are computed using Eqs. 9.102 and 9.103. The output file is listed below. Seepage under the dam is 28.7 ft^2/day.

```
EXAMPLE 9.15 SEEPAGE UNDER A DAM
 &GRID
 NX =          24
 NZ =           6
 DXD =     50.000000
 IDXC =         1
 DZD =     10.000000
 JDZC =         0
 NCHB =         2
 JR =           1
 IRS =         17
 IRE =         24
 IC =           8
 JCS =          5
 JCE =          6
 /
 0 2 2 2 2 2 2 0 0 0 0 0 0 0 0 0 0 3 3 3 3 3 3 3 3 0
 0 1 1 1 1 1 1 1 0 1 1 1 1 1 1 1 1 1 1 1 1 1 1 1 1 0
 0 1 1 1 1 1 1 1 0 1 1 1 1 1 1 1 1 1 1 1 1 1 1 1 1 0
 0 1 1 1 1 1 1 1 0 1 1 1 1 1 1 1 1 1 1 1 1 1 1 1 1 0
 0 1 1 1 1 1 1 1 0 1 1 1 1 1 1 1 1 1 1 1 1 1 1 1 1 0
 0 1 1 1 1 1 1 1 1 1 1 1 1 1 1 1 1 1 1 1 1 1 1 1 1 0
 0 1 1 1 1 1 1 1 1 1 1 1 1 1 1 1 1 1 1 1 1 1 1 1 1 0
 0 0 0 0 0 0 0 0 0 0 0 0 0 0 0 0 0 0 0 0 0 0 0 0 0 0

      COMPUTED SEEPAGE IN UNITS OF L*L/T
           WHERE UNITS OF K ARE L/T
          QH =      28.724170
      COLUMN =          8
          QV =     -28.710800
         ROW =          1
```

9.5.2 One-Dimensional Implicit Model

A one-dimensional aquifer model can be utilized to evaluate river stage–water table interactions for an aquifer along a river, tide-head relations for a coastal aquifer, and time-water table relations for parallel field drains. A one-dimensional model relating river stage to water table elevations is shown in Fig. 9.37. The bank storage along a river can be significant in reducing flood peaks and increasing low flows. The Deschutes River in Oregon flows through porous basalt formations and has a nearly uniform flow.

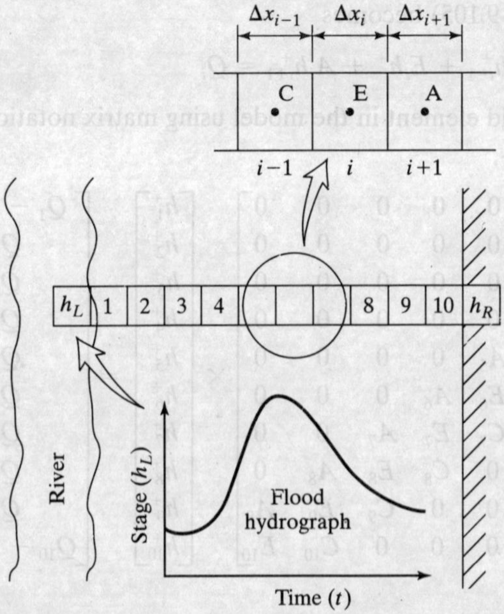

Figure 9.37 One-dimensional finite difference grid.

The partial differential equation for one-dimensional unsteady groundwater flow is

$$\frac{\partial}{\partial x}\left(T_x \frac{\partial \phi}{\partial x}\right) = S\frac{\partial \phi}{\partial t} + q_F \tag{9.104}$$

where ϕ is the water table elevation (h) for an unconfined aquifer or the elevation of the piezometric surface for a confined aquifer, S is the storage coefficient, and q_F is the net groundwater withdrawal rate per unit area.

The continuity equation written for the ith grid element of a water table aquifer is

$$A_i(h^+_{i+1} - h^+_i) + C_i(h^+_{i-1} - h^+_i) = \frac{S_i}{\Delta t}(h^+_i - h_i) + q_F \tag{9.105}$$

where h and h^+ are the water table elevations at the beginning and end of the time step (Δt), and A and C are flow coefficients defined by Eqs. 9.91 and 9.94, respectively.

Defining

$$E_i = -\left(A_i + C_i + \frac{S_i}{\Delta t}\right)$$

and

$$Q_i = -\frac{S_i}{\Delta t}h_i + q_F$$

The continuity equation (Eq. 9.105) becomes

$$C_i h_{i-1}^+ + E_i h_i^+ + A_i h_{i+1}^+ = Q_i \tag{9.106}$$

Writing Eq. 9.106 for each grid element in the model using matrix notation gives a tridiagonal coefficient matrix.

$$\begin{bmatrix} E_1 & A_1 & 0 & 0 & 0 & 0 & 0 & 0 & 0 & 0 \\ C_2 & E_2 & A_2 & 0 & 0 & 0 & 0 & 0 & 0 & 0 \\ 0 & C_3 & E_3 & A_3 & 0 & 0 & 0 & 0 & 0 & 0 \\ 0 & 0 & C_4 & E_4 & A_4 & 0 & 0 & 0 & 0 & 0 \\ 0 & 0 & 0 & C_5 & E_5 & A_5 & 0 & 0 & 0 & 0 \\ 0 & 0 & 0 & 0 & C_6 & E_6 & A_6 & 0 & 0 & 0 \\ 0 & 0 & 0 & 0 & 0 & C_7 & E_7 & A_7 & 0 & 0 \\ 0 & 0 & 0 & 0 & 0 & 0 & C_8 & E_8 & A_8 & 0 \\ 0 & 0 & 0 & 0 & 0 & 0 & 0 & C_9 & E_9 & A_9 \\ 0 & 0 & 0 & 0 & 0 & 0 & 0 & 0 & C_{10} & E_{10} \end{bmatrix} \times \begin{bmatrix} h_1^+ \\ h_2^+ \\ h_3^+ \\ h_4^+ \\ h_5^+ \\ h_6^+ \\ h_7^+ \\ h_8^+ \\ h_9^+ \\ h_{10}^+ \end{bmatrix} = \begin{bmatrix} Q_1 - C_1 h_L \\ Q_2 \\ Q_3 \\ Q_4 \\ Q_5 \\ Q_6 \\ Q_7 \\ Q_8 \\ Q_9 \\ Q_{10} - A_{10} h_R \end{bmatrix} \tag{9.107}$$

In the example (Fig. 9.37), h_L is the river stage at the potential boundary. Because the right boundary is a barrier boundary, A_{10} is equal to zero. The coefficient matrix is a tridiagonal matrix and the Thomas algorithm can be used to solve the system of linear equations. In the example, the head at the river (h_L) can be varied as a function of time to represent a flood wave and the heads in the aquifer computed for each time step using Eq. 9.107. Because the fully implicit approximation was used in the formulation, it is not necessary to limit the size of Δt except to define the river stage accurately. If the change in aquifer saturated thickness is large, then the transmissivity changes with time, and the flow coefficients may have to be computed for each time step.

After the water table elevations have been computed for a time step, the seepage rate between grid elements can be computed using Eq. 9.102, with Δy_j equal to 1 unit. In this example, the loss or gain at the river is the seepage at grid element 1 and can be considered when routing the flood hydrograph through the river channel.

Example 9.16 River-Groundwater System

As shown in Figure 9.37, a water table aquifer is located next to the river. On each side of the river, the distance to the barrier boundary is 2,000 ft. This is representative of a 10-mile reach of the river. The steady-state water table elevation is 350 ft. The base elevation of the aquifer at the river is 300 ft and increases to 310 ft at the barrier boundary. The aquifer has a hydraulic conductivity of 60 ft/day and a storage coefficient of 0.15. A flood occurs in the river with the river stage listed below. Model the system for a 30-day period using Δx = 200 ft and Δt = 1.0 day. Compute the daily loss (or gain) of streamflow for the 10-mile reach of the river.

Time days	River stage ft
0	350
1	365
2	370
3	369
4	368
5	366
6	363
7	360
8	358
9	356
10	354
11	353
12	353
13	352
30	350

A FORTRAN program (THOM1DU.FOR) was written to solve Eq. 9.107 using the Thomas algorithm. The input file (THOM1DU.DAT) is listed below. Comment lines in the program listing explain the items in the INPUT file. The maximum seepage from the river occurred on day 2 of 196 ft^2/day. For the 10-mile reach of the river, the loss would be

$$Q = \frac{196 \times 2 \times 10 \times 5{,}280}{86{,}400} = 240 \text{ cfs}$$

The seepage loss or gain in ft^2/day for the river is shown in the plot below.

Seepage from the river into the flood plain water table aquifer during a flood event.

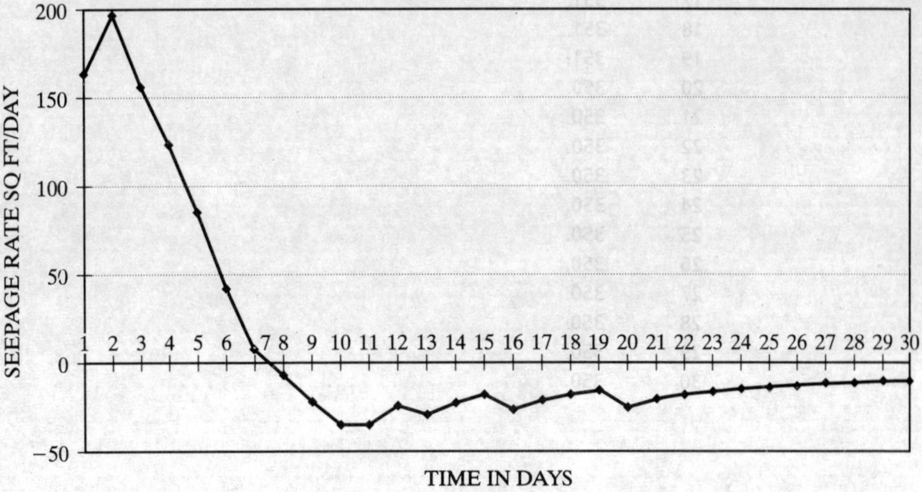

```
EXAMPLE 9.16 STREAM GROUNDWATER SYSTEM
&GRID  NX = 10,    DT = 1.,   DUR = 30.,   DXD = 200., IDXC = 0,   NCHB = 1,    NVHB = 1,
       DKX = 60.,  IDKC = 0,  DS = 0.15,   IDS = 0,      IC = 1,     DBZ = 300.,  IDBZ = 9/
              1        350.0
2  1  1  1  1  1  1  1  1  1  1  0
       2      2        301.0
       3      3        302.0
       4      4        303.0
       5      5        304.0
       6      6        305.0
       7      7        306.0
       8      8        307.0
       9      9        308.0
      10     10        309.0
              0        350.0
              1        365.0
              2        370.0
              3        369.0
              4        368.0
              5        366.0
              6        363.
              7        360.
              8        358.
              9        356.
             10        354.
             11        353.
             12        353.
             13        352.
             14        352.
             15        352.
             16        351.
             17        351.
             18        351.
             19        351.
             20        350.
             21        350.
             22        350.
             23        350.
             24        350.
             25        350.
             26        350.
             27        350.
             28        350.
             29        350.
             30        350.
```

9.5.3 Two-Dimensional Aquifer Model

For most groundwater applications, a two-dimensional model is adequate to simulate flow through the aquifer. The partial differential equation governing the nonsteady-state, two-dimensional flow in a nonhomogeneous, anisotropic aquifer is

$$\frac{\partial}{\partial x}\left(T_x \frac{\partial \phi}{\partial x}\right) + \frac{\partial}{\partial y}\left(T_y \frac{\partial \phi}{\partial y}\right) = S\frac{\partial \phi}{\partial t} + q_F \tag{9.108}$$

where the terms have been defined previously. The equivalent numerical model for Eq. 9.108 can be obtained by writing the continuity equation in terms of the unknown head potential (ϕ^+) for an interior grid element shown in Fig. 9.38 yielding

$$A_{i,j}(\phi^+_{i+1,j} - \phi^+_{i,j}) + B_{i,j}(\phi^+_{i,j-1} - \phi^+_{i,j}) + C_{i,j}(\phi^+_{i-1,j} - \phi^+_{i,j}) + D_{i,j}(\phi^+_{i,j+1} - \phi^+_{i,j}) = \frac{S_{i,j}}{\Delta t}(\phi^+_{i,j} - \phi_{i,j}) + q_{Fi,j} \tag{9.109}$$

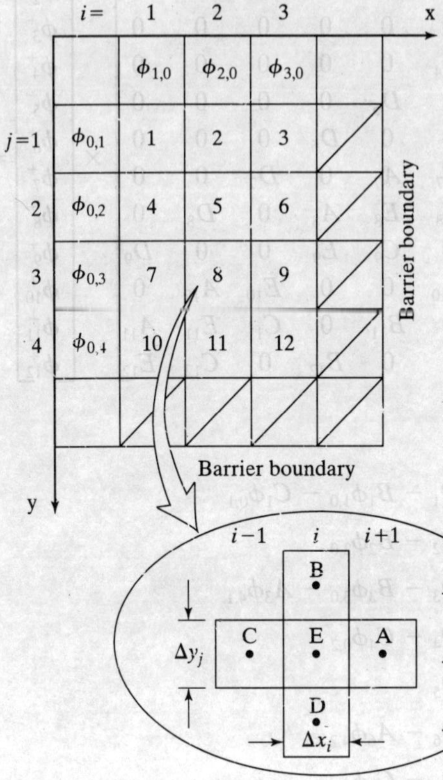

Figure 9.38 Two-dimensional model.

where the flow coefficients (A–D) have been defined by Eqs. 9.91, 9.93, 9.94, and 9.95, respectively. By letting

$$E_{i,j} = -\left(A_{i,j} + B_{i,j} + C_{i,j} + D_{i,j} + \frac{S_{i,j}}{\Delta t}\right)$$

and

$$Q_{i,j} = -\frac{S_{i,j}}{\Delta t}\phi_{i,j} + q_{Fi,j}$$

the implicit continuity Eq. 9.109 becomes

$$A_{i,j}\phi^{+}_{i+1,j} + B_{i,j}\phi^{+}_{i,j-1} + C_{i,j}\phi^{+}_{i-1,j} + D_{i,j}\phi^{+}_{i,j+1} + E_{i,j}\phi^{+}_{i,j} = Q_{i,j} \quad \textbf{(9.110)}$$

The aquifer in Fig. 9.38 consists of 12 grid elements with a potential boundary on the top and left where ϕ is specified and a barrier boundary on the right and bottom where the transmissivity is zero. Writing Eq. 9.110 for each grid element in matrix notation with grid number subscripts gives

$$\begin{bmatrix} E_1 & A_1 & 0 & D_1 & 0 & 0 & 0 & 0 & 0 & 0 & 0 & 0 \\ C_2 & E_2 & A_2 & 0 & D_2 & 0 & 0 & 0 & 0 & 0 & 0 & 0 \\ 0 & C_3 & E_3 & 0 & 0 & D_3 & 0 & 0 & 0 & 0 & 0 & 0 \\ B_4 & 0 & 0 & E_4 & A_4 & 0 & D_4 & 0 & 0 & 0 & 0 & 0 \\ 0 & B_5 & 0 & C_5 & E_5 & A_5 & 0 & D_5 & 0 & 0 & 0 & 0 \\ 0 & 0 & B_6 & 0 & C_6 & E_6 & 0 & 0 & D_6 & 0 & 0 & 0 \\ 0 & 0 & 0 & B_7 & 0 & 0 & E_7 & A_7 & 0 & D_7 & 0 & 0 \\ 0 & 0 & 0 & 0 & B_8 & 0 & C_8 & E_8 & A_8 & 0 & D_8 & 0 \\ 0 & 0 & 0 & 0 & 0 & B_9 & 0 & C_9 & E_9 & 0 & 0 & D_9 \\ 0 & 0 & 0 & 0 & 0 & 0 & B_{10} & 0 & 0 & E_{10} & A_{10} & 0 \\ 0 & 0 & 0 & 0 & 0 & 0 & 0 & B_{11} & 0 & C_{11} & E_{11} & A_{11} \\ 0 & 0 & 0 & 0 & 0 & 0 & 0 & 0 & B_{12} & 0 & C_{12} & E_{12} \end{bmatrix} \times \begin{bmatrix} \phi^{+}_1 \\ \phi^{+}_2 \\ \phi^{+}_3 \\ \phi^{+}_4 \\ \phi^{+}_5 \\ \phi^{+}_6 \\ \phi^{+}_7 \\ \phi^{+}_8 \\ \phi^{+}_9 \\ \phi^{+}_{10} \\ \phi^{+}_{11} \\ \phi^{+}_{12} \end{bmatrix} = \begin{bmatrix} G_1 \\ G_2 \\ G_3 \\ G_4 \\ G_5 \\ G_6 \\ G_7 \\ G_8 \\ G_9 \\ G_{10} \\ G_{11} \\ G_{12} \end{bmatrix}$$

(9.111)

where

$$G_1 = Q_1 - B_1\phi_{1,0} - C_1\phi_{0,1}$$

$$G_2 = Q_2 - B_2\phi_{2,0}$$

$$G_3 = Q_3 - B_3\phi_{3,0} - A_3\phi_{4,1}$$

$$G_4 = Q_4 - C_4\phi_{0,2}$$

$$G_5 = Q_5$$

$$G_6 = Q_6 - A_6\phi_{4,2}$$

$$G_7 = Q_7 - C_7\phi_{0,3}$$

$$G_8 = Q_8$$

$$G_9 = Q_9 - A_3\phi_{4,3}$$

$$G_{10} = Q_{10} - C_{10}\phi_{0,4} - D_{10}\phi_{1,5}$$

$$G_{11} = Q_{11} - D_{11}\phi_{2,5}$$

$$G_{12} = Q_{12} - A_{12}\phi_{4,4} - D_{12}\phi_{3,5}$$

The coefficient matrix in Eq. 9.111 is a banded matrix with values on five diagonals. Flow coefficients A_3, A_6, A_9, A_{12}, D_{10}, D_{11}, and D_{12} are zero because the adjacent elements are a barrier boundary. The potentials on the potential boundary ($\phi_{1,0}$, $\phi_{2,0}$, $\phi_{3,0}$, $\phi_{0,1}$, $\phi_{0,2}$, $\phi_{0,3}$, and $\phi_{0,4}$) are specified as either constant or a function of time.

The alternating direction implicit method is an iterative procedure for solving the continuity equations for a large aquifer. This method can be used with a small computer because the equations are solved one row or one column at a time. When solving for the potentials in a row (or column), the potentials in the adjacent rows (or columns) are assumed to be known. This results in a tridiagonal coefficient matrix that can be solved using the Thomas algorithm. One iteration consists of four computational sweeps through the grid. The equations are solved one row at a time moving forward through the grid (increasing j values). Then the equations are solved one column at a time moving forward through the grid (increasing i values). The process is repeated in the reverse direction row by row with decreasing j values and column by column with decreasing i values. The computations proceed both forward and backward through the grid, hence the name alternating direction implicit method. The process is repeated until convergence is achieved.

The row formulation of the linear equation in the forward direction for the jth interior row in Fig. 9.38 is

$$\begin{bmatrix} E_{1,j} & A_{1,j} & 0 \\ C_{2,j} & E_{2,j} & A_{2,j} \\ 0 & C_{3,j} & E_{3,j} \end{bmatrix} \begin{bmatrix} \phi^+_{1,j} \\ \phi^+_{2,j} \\ \phi^+_{3,j} \end{bmatrix} = \begin{bmatrix} G_{1,j} \\ G_{2,j} \\ G_{3,j} \end{bmatrix} \tag{9.112}$$

where

$$G_{1,j} = Q_{1,j} - B_{1,j}\phi^+_{1,j-1} - D_{1,j}\phi_{1,j+1} - C_{1,j}\phi_{0,j}$$

$$G_{2,j} = Q_{2,j} - B_{2,j}\phi^+_{2,j-1} - D_{2,j}\phi_{2,j+1}$$

$$G_{3,j} = Q_{3,j} - B_{3,j}\phi^+_{3,j-1} - D_{3,j}\phi_{3,j+1} - A_{3,j}\phi_{4,j}$$

In this example, $\phi_{0,j}$ is specified as a potential boundary and $A_{3,j}$ is zero because the barrier boundary is on the right.

The column formulation of the linear equations in the forward direction for the ith interior column is

$$\begin{bmatrix} E_{i,1} & D_{i,1} & 0 & 0 \\ B_{i,2} & E_{i,2} & D_{i,2} & 0 \\ 0 & B_{i,3} & E_{i,3} & D_{i,3} \\ 0 & 0 & B_{i,4} & E_{i,4} \end{bmatrix} \begin{bmatrix} \phi_{i,1}^{+} \\ \phi_{i,2}^{+} \\ \phi_{i,3}^{+} \\ \phi_{i,4}^{+} \end{bmatrix} = \begin{bmatrix} G_{i,1} \\ G_{i,2} \\ G_{i,3} \\ G_{i,4} \end{bmatrix} \tag{9.113}$$

where

$$G_{i,1} = Q_{i,1} - A_{i,1}\phi_{i+1,1} - C_{i,1}\phi_{i-1,1}^{+} - B_{i,1}\phi_{i,0}$$

$$G_{i,2} = Q_{i,2} - A_{i,2}\phi_{i+1,2} - C_{i,2}\phi_{i-1,2}^{+}$$

$$G_{i,3} = Q_{i,3} - A_{i,3}\phi_{i+1,3} - C_{i,3}\phi_{i-1,3}^{+}$$

$$G_{i,4} = Q_{i,4} - A_{i,4}\phi_{i+1,4} - C_{i,4}\phi_{i-1,4}^{+} - D_{i,4}\phi_{i,5}$$

Because of the barrier boundary $D_{i,4}$ is equal to zero and $\phi_{i,0}$ is specified as a potential boundary for this example.

Example 9.17 Recovery Wells and Water Barrier

The aquifer in the sketch below is bounded on three sides by a barrier boundary and on the fourth side by a recharge boundary. A tank farm is located over the water table aquifer. In the event of a spill, gasoline will infiltrate through the soil to the water table and then flow on the water table into the river. Install a water barrier system using a series of five recharge wells along the river in row 9. Four recovery wells located in row 5 will be utilized to recover the gasoline from the water table. The recovery wells are 24 inches in diameter and are equipped with a 50-gpm pump and a

$\Delta x = 50$ ft, $\Delta y = 25$ ft, $K_x = K_y = 1$ ft/hr, S = 0.15

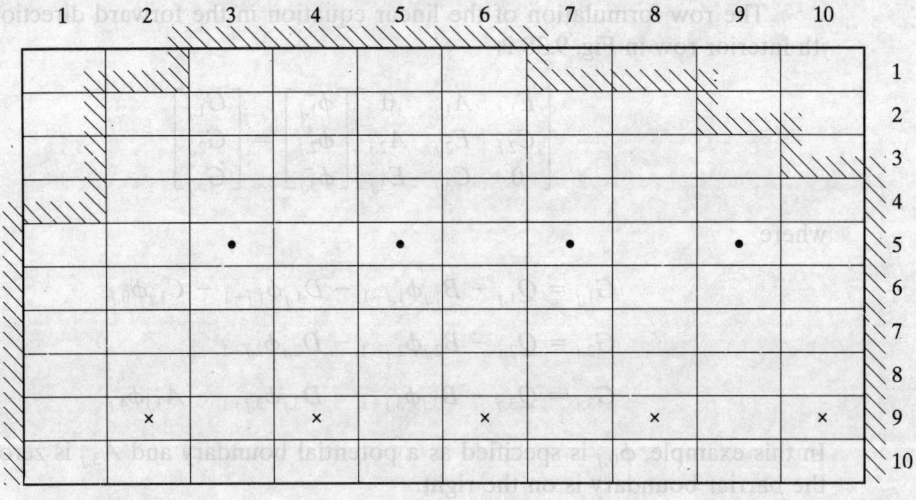

River

skimmer to remove the gasoline. In the event of a spill, the system is activated. Determine the time required to form an effective water barrier.

The wells are equipped with water level switch and will only operate when the water level in the grid is between 307 and 313 ft. The initial water table evaluation is 310.0 ft and the aquifer base elevation is 300.0 ft. Simulate a 6-hour period using a time step of 1 hour. Use a recharge rate of 40 gpm for each of the injection wells.

Solution (a) A FORTRAN program (ADI2DU.FOR) was written to solve Eqs. 9.112 and 9.113. The program was dimensioned for 40 grid elements in the X direction and 20 grid elements in the Y direction. A copy of the input file (ADI2DU.DAT) for this problem is included. The items in the input file are explained in the program listing. Part of the output file is attached. The output shows an effective barrier is formed after 1 hour, the recharge wells begin cycling off/on after 3 hours, and the pumped wells begin cycling off/on after 4 hours.

Solution (b) Because the coefficient matrix is a banded matrix with values on 5 diagonals, Eq. 9.111 can be written in a more compact form

$$AF = \begin{bmatrix} 0 & 0 & E_1 & A_1 & D_1 & G_1 \\ 0 & C_2 & E_2 & A_2 & D_2 & G_2 \\ 0 & C_3 & E_3 & 0 & D_3 & G_3 \\ B_4 & 0 & E_4 & A_4 & D_4 & G_4 \\ B_5 & C_5 & E_5 & A_5 & D_5 & G_5 \\ B_6 & C_6 & E_6 & 0 & D_6 & G_6 \\ B_7 & 0 & E_7 & A_7 & D_7 & G_7 \\ B_8 & C_8 & E_8 & A_8 & D_8 & G_8 \\ B_9 & C_9 & E_9 & 0 & D_9 & G_9 \\ B_{10} & 0 & E_{10} & A_{10} & 0 & G_{10} \\ B_{11} & C_{11} & E_{11} & A_{11} & 0 & G_{11} \\ B_{12} & C_{12} & E_{12} & 0 & 0 & G_{12} \end{bmatrix} \quad IAF = \begin{bmatrix} 0 & 4 \\ 0 & 5 \\ 0 & 6 \\ 1 & 7 \\ 2 & 8 \\ 3 & 9 \\ 4 & 10 \\ 5 & 11 \\ 6 & 12 \\ 7 & 0 \\ 8 & 0 \\ 9 & 0 \end{bmatrix}$$

Because flow coefficients C, E, and A are centered along the main diagonal, the IAF pointer array need only include the positions of B and D flow coefficients in the coefficients matrix. Because the example is for a rectangular-shaped aquifer, each row has the same number of aquifer elements and in this example problem, the B and D flow coefficients are located at $I - 3$ and $I + 3$ from the diagonal.

The Gauss–Seidel method to solve simultaneous linear equations use an iterative procedure where the next estimate of the potentials are

$$\phi_1^+ = (G_1 - \phi_2 A_1 - \phi_4 D_1)/E_1$$

$$\phi_2^+ = (G_2 - \phi_1 C_2 - \phi_3 A_2 - \phi_5 D_2)/E_2$$

$$\phi_3^+ = (G_3 - \phi_2 C_3 - \phi_6 D_3)/E_3$$

$$\phi_4^+ = (G_4 - \phi_1 B_4 - \phi_5 A_4 - \phi_7 D_4)/E_4$$

$$\phi_5^+ = (G_5 - \phi_2 B_5 - \phi_4 C_5 - \phi_6 A_5 - \phi_8 D_5)/E_5$$

•

•

•

$$\phi_{12}^+ = (G_{12} - \phi_9 B_{12} - \phi_{11} C_{12})/E_{12}$$

The procedure is repeated until the largest change in the potentials is within some limit.

A FORTRAN program (GS2DU.FOR) was written to solve Eq. 9.111 using the Gauss–Seidel method. Program was dimensioned to solve up to 500 linear equations simultaneously. The input file (GS2DU.DAT) is identical to the input file for the alternating direction implicit (ADI) method. The two output files are nearly identical.

```
EXAMPLE 9.17(a) TWO-DIMENSIONAL GROUNDWATER FLOW USING THE ADI METHOD
&GRID  NX = 10,   NY = 10,  DXD = 50.,  IDXC = 0,  DYD = 25.,  JDYC = 0,
       DKX = 1.0, DKY = 1.0, IJDK = 0,  DZ = 300.0, IDZ = 0, DS = 0.15, IDS = 0,
       NCHB = 1,  NW = 9,   DT = 1.,  DUR = 6./
        2      310.0
   3    5      401.0      307.0     313.0
   5    5      401.0      307.0     313.0
   7    5      401.0      307.0     313.0
   9    5      401.0      307.0     313.0
   2    9     -321.0      307.0     313.0
   4    9     -321.0      307.0     313.0
   6    9     -321.0      307.0     313.0
   8    9     -321.0      307.0     313.0
  10    9     -321.0      307.0     313.0
 0 0 0 0 0 0 0 0 0 0 0 0
 0 0 0 1 1 1 1 0 0 0 0 0
 0 0 1 1 1 1 1 1 1 0 0 0
 0 0 1 1 1 1 1 1 1 1 0 0
 0 0 1 1 1 1 1 1 1 1 1 0
 0 1 1 1 1 1 1 1 1 1 1 0
 0 1 1 1 1 1 1 1 1 1 1 0
 0 1 1 1 1 1 1 1 1 1 1 0
 0 1 1 1 1 1 1 1 1 1 1 0
 0 1 1 1 1 1 1 1 1 1 1 0
 0 1 1 1 1 1 1 1 1 1 1 0
 0 2 2 2 2 2 2 2 2 2 2 0
```

```
EXAMPLE 9.17(b) TWO-DIMENSIONAL GROUNDWATER FLOW USING THE ADI METHOD
    TIME =        1.000000   ITER =        4
 310.00  310.00  310.00  310.00  310.00  310.00  310.00  310.00  310.00  310.00
 310.00  310.00  310.00  310.00  310.00  310.00  310.00  310.00  310.00  310.00
 310.00  310.00  309.99  310.00  309.99  310.00  309.99  310.00  309.99  310.00
 310.00  309.99  309.85  309.99  309.85  309.99  309.85  309.99  309.85  309.99
 310.00  309.96  308.28  309.93  308.28  309.93  308.28  309.93  308.28  309.96
 310.00  309.99  309.85  309.99  309.85  309.99  309.85  309.99  309.85  309.99
 310.00  310.01  309.99  310.01  309.99  310.01  309.99  310.01  309.99  310.01
```

310.01	310.12	310.01	310.12	310.01	310.12	310.01	310.12	310.01	310.12
310.03	311.37	310.06	311.37	310.06	311.37	310.06	311.37	310.06	311.40
310.01	310.12	310.01	310.12	310.01	310.12	310.01	310.12	310.01	310.12
TIME =		2.000000	ITER =		4				
310.00	310.00	310.00	310.00	310.00	310.00	310.00	310.00	310.00	310.00
310.00	310.00	310.00	310.00	310.00	310.00	310.00	310.00	310.00	310.00
310.00	310.00	309.96	309.99	309.96	309.99	309.96	309.99	309.95	310.00
309.99	309.98	309.64	309.96	309.64	309.96	309.64	309.96	309.64	309.98
310.00	309.91	306.83	309.81	306.83	309.81	306.83	309.81	306.83	309.90
310.00	309.98	309.64	309.96	309.64	309.96	309.64	309.96	309.64	309.98
310.00	310.03	309.96	310.03	309.96	310.03	309.96	310.03	309.96	310.03
310.02	310.32	310.03	310.32	310.03	310.32	310.03	310.32	310.03	310.33
310.08	312.46	310.16	312.46	310.16	312.46	310.16	312.46	310.17	312.54
310.02	310.31	310.03	310.31	310.03	310.31	310.03	310.31	310.04	310.33
TIME =		3.000000	ITER =		4				
310.00	310.00	310.00	310.00	310.00	310.00	310.00	310.00	310.00	310.00
310.00	310.00	309.99	310.00	309.99	310.00	309.99	310.00	310.00	310.00
309.99	309.99	309.93	309.99	309.93	309.99	309.93	309.98	309.91	310.00
309.97	309.96	309.51	309.92	309.51	309.92	309.51	309.92	309.51	309.96
309.99	309.87	307.31	309.75	307.31	309.75	307.31	309.75	307.31	309.87
310.00	309.97	309.52	309.93	309.51	309.93	309.51	309.93	309.52	309.97
310.01	310.07	309.94	310.06	309.94	310.06	309.94	310.06	309.94	310.08
310.04	310.57	310.07	310.57	310.07	310.57	310.07	310.57	310.07	310.61
310.15	313.32	310.30	313.33	310.30	313.33	310.30	313.33	310.30	313.47
310.04	310.56	310.08	310.56	310.08	310.56	310.08	310.56	310.08	310.60
TIME =		4.000000	ITER =		4				
310.00	310.00	310.00	310.00	310.00	310.00	309.99	309.99	309.99	309.99
310.00	310.00	309.98	310.00	309.98	309.99	309.98	309.99	309.99	310.00
309.99	309.99	309.88	309.97	309.88	309.97	309.88	309.97	309.86	310.00
309.95	309.94	309.31	309.88	309.30	309.88	309.30	309.88	309.30	309.93
309.99	309.82	305.94	309.64	305.93	309.64	305.93	309.64	305.94	309.81
310.00	309.96	309.31	309.90	309.31	309.90	309.31	309.90	309.31	309.96
310.01	310.11	309.91	310.10	309.91	310.10	309.91	310.10	309.91	310.13
310.07	310.71	310.11	310.71	310.11	310.71	310.11	310.71	310.11	310.78
310.20	312.68	310.38	312.69	310.38	312.69	310.38	312.69	310.39	312.86
310.07	310.70	310.12	310.70	310.12	310.70	310.12	310.70	310.13	310.76
TIME =		5.000000	ITER =		4				
310.00	310.00	309.99	310.00	309.99	310.00	309.99	309.99	309.99	309.99
309.99	309.99	309.97	309.99	309.97	309.99	309.97	309.99	309.99	309.99
309.98	309.98	309.84	309.96	309.84	309.96	309.84	309.96	309.80	309.99
309.93	309.91	309.20	309.83	309.19	309.83	309.19	309.83	309.19	309.90
309.99	309.79	306.49	309.57	306.47	309.57	306.47	309.57	306.48	309.78
310.00	309.95	309.21	309.86	309.20	309.86	309.20	309.86	309.21	309.95
310.02	310.16	309.88	310.14	309.88	310.14	309.88	310.14	309.89	310.18
310.10	310.91	310.16	310.91	310.16	310.91	310.16	310.91	310.17	311.01
310.26	313.54	310.50	313.56	310.50	313.56	310.50	313.56	310.51	313.78

9.6 GROUNDWATER QUALITY

The quality of groundwater depends on many factors, including the quality of groundwater recharge water, the soil and plant effects in the shallow root zone in the recharge area, the natural reactions between the water, and the formations through which the groundwater flows and man's activities.

Groundwater is typically higher in total dissolved solids (TDS) than surface water and is classified as fresh if the TDS is less than 1,000 mg/l, brackish 3,000–10,000 mg/l, saline 10,000–50,000 mg/l, and brine 50,000–300,000 mg/l. Generally, the recommended maximum TDS for drinking water is 500 mg/l, whereas the maximum limit for irrigation is slightly higher. The major cations in groundwater are typically Na^+, Ca^{++}, Mg^{++}, and K^+. The major anions are Cl^-, HCO_3^-, and $SO_4^=$.

The hardness of water is defined as the concentration of ions that react with soap to produce scum and cause scale to be produced when evaporated in boilers. The total hardness (TH) is

$$TH = 2.5[Ca^{++}] + 4.1[Mg^{++}] \tag{9.114}$$

where the concentrations are in mg/l. Soft water has a TH value less than 60 mg/l, whereas very hard water has a TH value greater than 150 mg/l.

A high concentration (>45 mg/l) of nitrate (NO_3^-) in drinking water can be potentially harmful to infants and can cause blue baby disease. Common sources of nitrate contamination include septic tanks, landfills, cemeteries, agricultural fertilizers, and animal wastes.

Groundwater quality is affected by the quality of the recharge water that may be direct infiltration of precipitation, snowmelt, or seepage of surface water. A schematic sketch of precipitation recharging a confined aquifer is shown in Fig. 9.39. The precipitation water quality is affected by the air quality. In coastal areas, the concentration of Cl^- in the precipitation may be as high as 8 mg/l. The precipitation in industrial areas will often have a high concentration of $SO_4^=$ and low pH.

Evaporation from the ground surface and transpiration by plants in the recharge zone can cause a buildup of salts in the surface soils. The salts are leached by deep percolation to the groundwater. Evapotranspiration by deep-rooted phreatophytes in arid regions not only cause increased salt concentration in shallow aquifers, but can also cause a significant loss of groundwater in floodplains and stream valleys.

The schematic in Fig. 9.39 shows the changing water quality as the groundwater flows through the aquifer. Aquifer effects include filtration, decay, ion exchange, and adsorption. The shallow aquifer in the recharge zone is subject to man-made and natural contamination. As the water travels from the recharge zone, some of the contaminants will be removed by the aquifer. Generally, as groundwater travels through the aquifer, the TDS will increase.

Point source pollution of aquifers include leakage from petroleum storage tanks, service station fuel storage tanks, and buried pipelines. Generally, the leakage rate is small, but can occur over an extended time period. Except for the fraction that dissolves in the water, most of the petroleum products will float on the water table or are trapped in the capillary fringe zone above the water table.

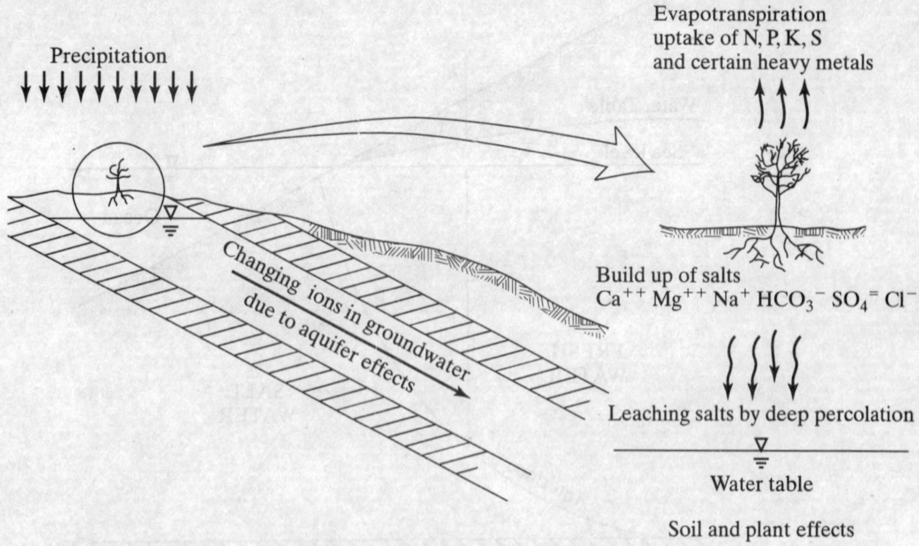

Figure 9.39 Schematic of factors affecting groundwater quality.

Point source pollution of pathogenic bacteria include septic tanks, landfills, cemeteries, and animal wastes. Bacteria and viruses are usually filtered by fine porous media and are a problem in very shallow aquifers near the pollutant source. Exceptions include coarse porous media, fracture formations, or solution channels in limestone aquifer where bacteria and viruses can travel much farther. Waste disposal pits, improperly plugged abandoned oil wells, waste disposal wells, and mines are other forms of point source pollution.

9.6.1 Saltwater Intrusion

Saltwater intrusion is an example where the concentration of the contaminant affects the groundwater flow pattern and there is little mixing of water across the saltwater–freshwater interface. Saltwater intrusion commonly occurs in coastal areas because of overpumping of freshwater aquifers. Because saltwater is heavier than freshwater, saltwater will intrude into coastal freshwater aquifers. A general sketch of saltwater intrusion is shown in Fig. 9.40. The pressure on each side of the interface is equal or

$$\rho_f(h + z) = \rho_s z \qquad \textbf{(9.115)}$$

where

ρ_f = density of freshwater
ρ_s = density of saltwater
h = height of freshwater table is above sea level and
z = distance the interface is below sea level.

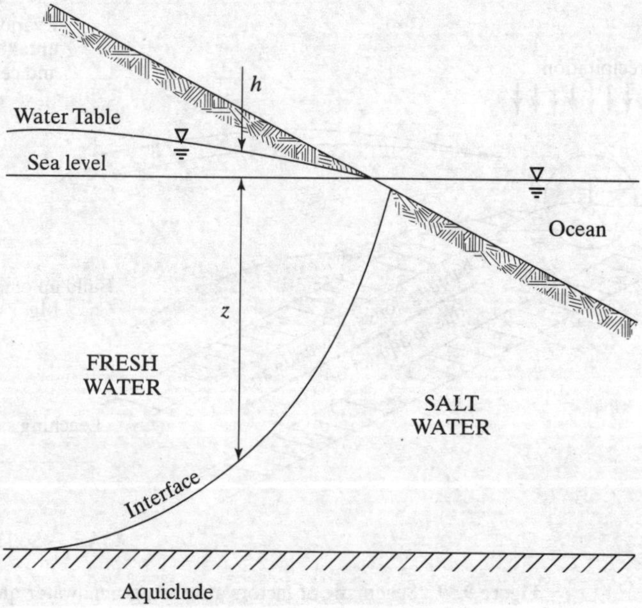

Figure 9.40 Saltwater intrusion.

For a coastal aquifer, $\rho_s = 1.025$ and $z = 40$ h. For every unit that the water table is above sea level, freshwater will extend 40 units below sea level. Groundwater withdrawal from coastal aquifers will lower the water table and increase saltwater intrusion. Pumpage from coastal aquifers must be carefully managed to avoid excessive saltwater intrusion. Recharge wells might be required to create a freshwater barrier between the seawater and the water supply well field.

9.6.2 Solute Transport Equations

The concentration of the contamination (solute) varies as the groundwater moves from the pollution source. It is generally assumed that the change in concentration of the solute is not sufficient to affect the groundwater density, viscosity, or flow characteristics. The velocity distribution in the aquifer can therefore be computed independent of the solute concentration. Many groundwater contamination problems are analyzed by combining a groundwater flow model with a solute transport model.

The partial differential equation representing advection, two-dimensional dispersion, and losses is

$$\frac{\partial c}{\partial t} = -\frac{\partial(\overline{U}c)}{\partial x} + \frac{\partial}{\partial x}\left(D_x \frac{\partial c}{\partial x}\right) + \frac{\partial}{\partial y}\left(D_y \frac{\partial c}{\partial y}\right) - \text{losses} \tag{9.116}$$

where c is the solute concentration, t is the time, $\overline{U}$ is the pore velocity of the groundwater in the x direction, x is the coordinate direction along the streamline, D_x is the longitudinal dispersion coefficient, D_y is the transverse dispersion coefficient, and y is the coordinate direction perpendicular to the streamline. The first term to the right of the equal sign is the advection term, whereas the following two terms represent longitudinal and transverse dispersion, respectively. The loss term depends on the type of solute. For a solute that has a constant decay rate

$$\text{losses} = KC \tag{9.117}$$

or if the losses are caused by adsorption

$$\text{losses} = \frac{\rho_b}{n} K_d \frac{\partial c}{\partial t} \tag{9.118}$$

where K is the constant decay rate, ρ_b and n are the bulk density and porosity of the aquifer, and K_d is the distribution coefficient. Because the solute migrates slower than water, Eq. 9.116 is often written

$$R\frac{\partial c}{\partial t} = -\frac{\partial(\overline{U}c)}{\partial x} + \frac{\partial}{\partial x}\left(D_x \frac{\partial c}{\partial x}\right) + \frac{\partial}{\partial y}\left(D_y \frac{\partial c}{\partial y}\right) - KRc \tag{9.119}$$

where $R = 1 + (\rho_b/n)K_d$ is called the retardation factor. K_d has units of ml/g. A distribution coefficient of 0 indicates no absorption of the solute, whereas a distribution coefficient greater than 1 indicates the solute is nearly immobile. If K_d is equal to 1, the solute would be retarded relative to the groundwater flow by a factor between 5 and 10.

The dispersion coefficient has units of L^2/t and is a function of the heterogeneity of the aquifer, the groundwater velocity, and the scale of the dispersion or mixing process. The dispersion coefficients are often written as

$$D_x = \overline{U}\alpha_x \tag{9.120a}$$

and

$$D_y = \overline{U}\alpha_y \tag{9.120b}$$

where α_x and α_y are the longitudinal and transverse dispersivity, respectively. Because the dispersion coefficients are a function of scale, physical laboratory model studies have limited value in evaluating groundwater contamination problems.

9.6.3 Analytical Models

Direct solutions to the partial differential equation describing dispersion in a porous media are available for only a few simple cases. The contaminated plume shown in Fig. 9.41(a) is for a continuous point source. The partial differential equation for steady-state, continuous source, one-dimensional advection and two-dimensional dispersion with no losses is

$$0 = -\overline{U}\frac{\partial c}{\partial x} + D_x \frac{\partial^2 c}{\partial x^2} + D_y \frac{\partial^2 c}{\partial y^2} \tag{9.121}$$

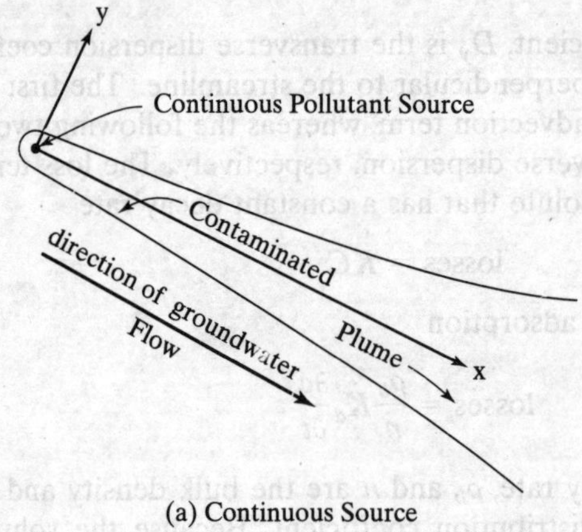

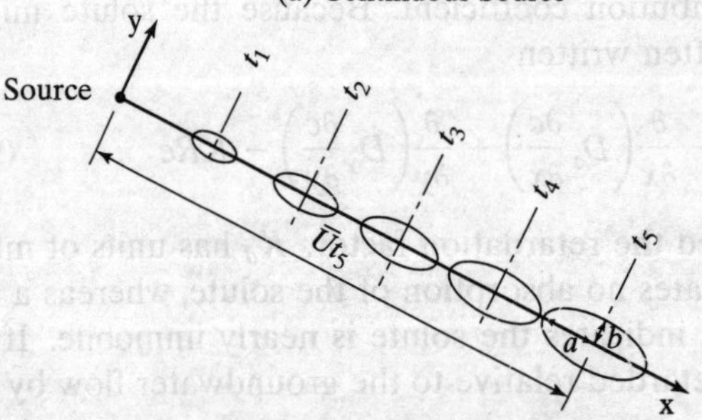

Figure 9.41 Pollutant sources.

Solution to Eq. 9.121 for a homogeneous, isotropic aquifer is

$$C_{x,y} = \frac{C_o q}{2\pi\sqrt{D_y D_x}} \exp\left[\frac{x\overline{U}}{2D_x}\right] K_o[G] \tag{9.122}$$

where

- $C_{x,y}$ is the concentration of the contaminant at position x, y
- C_o is the concentration of the contaminant at the source
- q is the pollutant discharge rate per unit thickness of contaminated aquifer (L^2/t)
- D_x, D_y are constant longitudinal and lateral dispersion coefficients, respectively (L^2/t)
- $\overline{U}$ is the groundwater pore velocity along the flow line and is equal to the Darcy velocity divided by the porosity
- x is the distance along the flow line from the point source
- y is the lateral distance from the flow line
- K_o is the modified Bessel function of second kind and zero order and

$$G = \left[\frac{x^2\overline{U}^2}{4D_x^2} + \frac{y^2\overline{U}^2}{4D_x D_y}\right]^{1/2}$$

The Bessel function can be evaluated using published tables or computer software such as MathCad.

Figure 9.41(b) represents an instantaneous point source. The pollution source represents a spill or slug loading that is carried with the groundwater in the x direction. In the three-dimensional dispersion model, the contaminant plume is described as an ellipsoid with semi-axes a, b, and c. The semi-axes are equal to

$$a = 3\sigma_x = \sqrt{2D_x t} \tag{9.123a}$$

$$b = 3\sigma_y = \sqrt{2D_y t} \tag{9.123b}$$

$$c = 3\sigma_z = \sqrt{2D_z t} \tag{9.123c}$$

where σ is the standard deviation of the concentration distribution, and t is the time elapsed since slug loading with a mass (M) occurred. Theoretically, the ellipsoid will contain 99.7 percent of contaminant mass. The center of the contaminant mass will travel along the flow line a distance $X = \overline{U}t$ from the instantaneous point source. The maximum concentration (C_M) at the center of the ellipsoid is

$$C_M = \frac{M}{8(\pi t)^{3/2}\sqrt{D_x D_y D_z}} \tag{9.124}$$

The concentration at a local coordinate point x, y, z relative to the center of the ellipsoid at time t is

$$C_{(x,y,z,t)} = C_M \exp - \left[\frac{x^2}{2a^2} + \frac{y^2}{2b^2} + \frac{z^2}{2c^2}\right] \tag{9.125}$$

9.6.4 Numerical Models

Numerical models can be used to simulate the transport of solutes in porous media. A major concern with most numerical models is numerical dispersion. Because the average concentration is often used in a grid element, the contaminant is spread to adjacent grid elements during each computational time step. If the computations were repeated 50 times, then the contaminant would be spread numerically a distance of 50 grid elements. Although beyond the scope of this text, particle tracking has been used to reduce numerical dispersion.

The partial differential equation describing unsteady transport of solute in a two-dimensional, isotropic homogeneous porous media is

$$\frac{\partial c}{\partial t} = D_x \frac{\partial^2 c}{\partial x^2} + D_y \frac{\partial^2 c}{\partial y^2} - \overline{U}\frac{\partial c}{\partial x} \tag{9.126}$$

Bear (1979) developed a solution to this partial differential equation in the form of an integral equation. The equation can be solved numerically to compute the concentration of solute in the plume from a continuous point source at any location (x,y) and time (t).

$$C_{(x,y,t)} = \frac{C_o q}{4\pi\sqrt{D_x D_y}} \int_0^t \frac{1}{t-\tau} \exp(B)\, d\tau \tag{9.127}$$

where

$$B = -\frac{1}{t - \tau}\left[\frac{x^2}{4D_x} + \frac{y^2}{4D_y}\right] + \frac{2x\overline{U}}{4D_x} - \frac{(t - \tau)\overline{U}^2}{4D_x}$$

The terms in Eq. 9.127 have been defined in Section 9.6.3.

Example 9.18 Continuous Point Source of Pollution

(a) A continuous point source of pollution is contaminating the aquifer. Observation wells were installed, and the water table slope is 0.0010. Pump test was conducted and hydraulic conductivity is estimated as 13.4 ft/day. The porosity of the formation is 0.25. Estimate the distribution of contaminated groundwater after 10 years of leakage. Use a longitudinal dispersivity of 30 ft and transverse dispersivity of 2 ft. The contaminant is not absorbed by the formation (distribution coefficient is equal to zero).

Solution A basic computer program (PLUME2D.BAS) was written to evaluate Eq. 9.127. Remark lines are included in the program listing to explain the coding and input data. Input data is entered from the keyboard when the model is run. The output file (PLUME.OUT) indicates the pollution would have traveled 450 ft longitudinally and 70 ft laterally.

RELATIVE CONCENTRATION FROM POINT SOURCE

C = 1.00 AT X = 0 AND Y = 0

TIME 3650 DAYS, VELOCITY .0536 L/DAY

DL = 1.608 L*L/DAY, DT = .1072 L*L/DAY

KX = 13.4 L/DAY, SLOPE = .001

LENGTH UNIT L

DIST	17	50	83	116	149	182	215	248	281	314	347	380	413	446	479
101	.000	.000	.000	.000	.000	.000	.000	.000	.000	.000	.000	.000	.000	.000	.000
68	.000	.001	.001	.001	.001	.001	.001	.000	.000	.000	.000	.000	.000	.000	.000
51	.003	.004	.005	.006	.006	.005	.005	.003	.002	.002	.001	.000	.000	.000	.000
35	.013	.019	.022	.024	.022	.019	.015	.011	.007	.005	.003	.001	.001	.000	.000
18	.061	.075	.075	.068	.057	.045	.033	.023	.015	.009	.005	.003	.001	.001	.000
10	.135	.135	.115	.095	.075	.057	.041	.028	.018	.011	.006	.003	.001	.001	.000
3	.261	.182	.139	.108	.083	.062	.045	.030	.019	.012	.006	.003	.002	.001	.000
−0	.310	.190	.142	.110	.085	.063	.045	.031	.020	.012	.007	.003	.002	.001	.000
−3	.261	.182	.139	.108	.083	.062	.045	.030	.019	.012	.006	.003	.002	.001	.000
−10	.135	.135	.115	.095	.075	.057	.041	.028	.018	.011	.006	.003	.001	.001	.000
−18	.061	.075	.075	.068	.057	.045	.033	.023	.015	.009	.005	.003	.001	.001	.000
−35	.013	.019	.022	.024	.022	.019	.015	.011	.007	.005	.003	.001	.001	.000	.000
−51	.003	.004	.005	.006	.006	.005	.005	.003	.002	.002	.001	.000	.000	.000	.000
−68	.000	.001	.001	.001	.001	.001	.001	.000	.000	.000	.000	.000	.000	.000	.000
−101	.000	.000	.000	.000	.000	.000	.000	.000	.000	.000	.000	.000	.000	.000	.000
RELATIVE MASS	.530	.498	.449	.390	.323	.255	.190	.130	.085	.052	.029	.014	.007	.002	.000

(b) A continuous point source discharges at a rate of 0.38 sq ft per day at a concentration of 1.0 mg/l into the aquifer described in part (a). Compute the steady state concentration along the x axis at distances of 0.1, 17, 50, 83, 116, 149, 182, and 215 ft from the source using Eq. 9.122. Compare the concentrations with those computed in Part (a).

Solution

$$\overline{U} = 13.4 \times 0.001/0.25 = 0.0535 \text{ ft/day}$$

$$D_x = \overline{U} \times \alpha_x = 0.0535 \times 30 = 1.60 \text{ ft}^2\text{/day}$$

$$D_y = \overline{U} \times \alpha_y = 0.0535 \times 2 = 0.11 \text{ ft}^2\text{/day}$$

$$C_{x,o} = \frac{C_o q}{2\pi\sqrt{D_x D_y}} \exp\left[\frac{X\overline{U}}{2D_x}\right] K_o\left[\frac{X\overline{U}}{2D_x}\right]$$

$$= 0.14 \exp[0.0167\, X]\, K_o[0.0167\, X]$$

LIST OF CONCENTRATION VALUES

X ft	$C_{x,o}$ Part b	C Part a
0.1	0.98	1.0
17	0.27	0.31
50	0.17	0.19
83	0.14	0.14
116	0.12	0.11
149	0.11	0.08
182	0.10	0.06
215	0.09	0.04

9.7 GENERALIZED GROUNDWATER MODELS

The International Ground Water Modeling Center (IGWMC) is an information, education, and research center for groundwater modeling. IGWMC distributes software, organizes short courses and workshops, provides technical assistance, and conducts applied research on groundwater modeling. Numerous models are distributed through the IGWMC (Table 1.7).

MODFLOW developed by the U.S. Geological Survey is probably the most widely used of the many groundwater flow models. The public domain model may be obtained through the U.S. Geological Survey or IGWMC. The modular structure of the model consists of a main program and series of highly independent modules grouped into packages, which deal with specific features of the hydrologic system or specific methods for solving the governing equations. MODFLOW simulates two- or three-dimensional, steady or transient, saturated flow in anisotropic, heterogeneous, layered aquifer systems. Layers may be confined, unconfined, or convertible between the two conditions. The model allows for analysis of external influences, such as wells, areal recharge, drains, evapotranspiration, and streams. A block-centered, finite-difference approach is used to solve the governing equations.

PROBLEMS

9.1. An undisturbed core sample was obtained from an unsaturated aquifer 1.0 m (3.28 ft) above the water table. The sample is 0.150 m (0.492 ft) high and 0.050 m (0.164 ft) in diameter. The solids in the aquifer have a specific gravity of 2.65. The weight of the sample was 630 grams (1.39 lbs) before drying and 570 grams (1.26 lbs) after drying. Calculate the porosity, volumetric water content, and bulk density of the aquifer.

9.2. A confined aquifer has a transmissivity of 100 m^2/day (1,076 ft^2/day). If the slope of the piezometric surface is 0.0005, compute the flow rate of water through the aquifer per km (mile) of width.

9.3. A water table aquifer has a hydraulic conductivity of 10.0 m/day (32.8 ft/day). Compute the velocity of water through the aquifer if the slope of the water table is 0.001 and the porosity is 0.2. How far will the ground water travel in 1.0 year?

9.4. An unconfined aquifer with a specific yield of 0.20 is used as a water supply for the irrigation of farm land. The recharge area of the aquifer is the same area as the irrigated area. The recharge is limited to 76 mm (3.0 in.) per year. The saturated thickness of the aquifer is 15.2 m (50 ft). How many years will the water supply last if 254 mm (10.0 in.) of water per year is pumped from the aquifer for irrigation?

9.5. Two piezometers were installed at the same location: one installed above the aquitard and the other installed below the aquitard. Well information is listed below. If the intrinsic permeability of the aquitard is 0.005 darcys, determine the seepage rate m^3/day · per km^2 (ft^3/day/mi.^{2}). Is the seepage upward or downward?

	Observation well	
	A	B
Depth to bottom of well		
m	120.6	130.8
ft	395.6	429.0
Depth to water surface		
m	50.2	53.1
ft	164.7	174.2

9.6. How much water can be removed from an unconfined aquifer with a specific yield of 0.2 when the water table is lowered 1.0 m (3.28 ft)? How much water can be removed from a confined aquifer with storage coefficient of 0.0005 when the piezometric surface is lowered 1.0 m (3.28 ft)? Express your answers in m^3/km^2 (ft^3/mi.^{2}).

9.7. Two observation wells were installed along a streamline in an unconfined aquifer 1,000 m (3,280 ft) apart. Compute the time for water to travel from well A to well B if the hydraulic conductivity is 10.0 m/day (32.8 ft/day) and the effective porosity of the aquifer is 0.2. The well information is listed at the top of the next page.

	Observation well	
	A	B
Top of casing elevation,		
m	252.0	251.0
ft	826.8	823.5
Depth of water,		
m	8.0	7.5
ft	26.2	24.6
Depth of well,		
m	25.5	31.5
ft	83.6	103.3

9.8. Three observation wells were installed in an unconfined aquifer with a saturated thickness of 15.2 m (49.9 ft) and a hydraulic conductivity of 10.0 m/day (32.8 ft/day). Determine the magnitude and direction of the hydraulic gradient and the total flow in the aquifer per unit width. Sketch the equal potential lines and streamlines. The well information is listed below.

	Observation well		
	A	B	C
x Coordinate	0	500(1,640)	0
y Coordinate	0	0	500(1,640)
Water elevation,			
m	128.5	129.2	130.0
ft	421.5	423.9	426.5

9.9. As shown in the sketch below, an unconfined aquifer on a hillside drains into a stream. The aquifer has an average saturated thickness of 3.0 m (9.8 ft), has a hydraulic conductivity of 0.30 m/day (1.0 ft/day), an effective porosity of 0.20, and a slope of 0.03. Compute the maximum seepage into the stream per km (mi.) considering seepage is from both sides of the channel. Compute the travel time for water to flow from the ridge to the channel 300 m (984 ft).

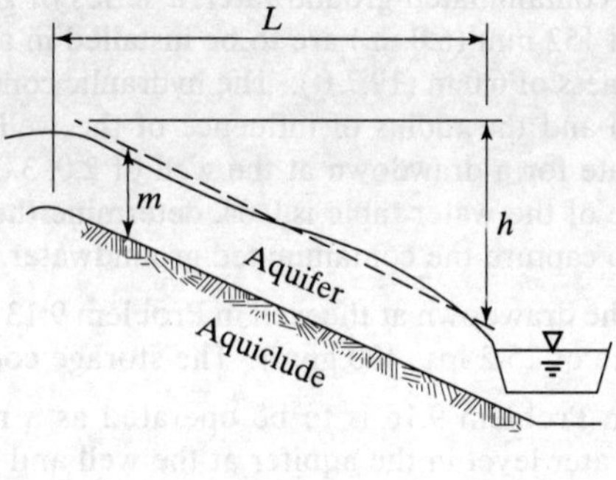

9.10. An industrial storage site was located near the ridge in Problem 9.9 and has contaminated the aquifer. To prevent the contaminated groundwater from reaching the stream, a line (parallel to the stream) of pumped wells will be installed between the ridge and the stream. Pump tests indicate that each well will produce 0.032 lps (0.5 gpm). Compute the maximum spacing of the wells to capture the contaminated groundwater.

9.11. An unconfined aquifer is used as a municipal water supply. A railroad car is derailed and contaminates the aquifer 3.0 km (9,840 ft) up gradient (Slope = 0.005). Estimate the travel time for the contaminated groundwater to reach the water supply if the intrinsic permeability of the aquifer is 100 darcys and the effective porosity is 0.15.

9.12. A series of wells are to be used to dewater a construction site. An aquifer 30 m (98 ft) thick is located below the site. The aquifer consists of alternating layers of loose sand and loose clay (see Table 9.4). The total thickness of the sand layers is 10.0 m (32.8 ft), and the total thickness of the clay layers is 20.0 m (65.6 ft). During construction, the piezometric surface of the aquifer will be lowered 7.6 m (25.0 ft). Estimate the subsidence caused by the project.

9.13. Determine the maximum steady-state pumping rate in lps (gpm) for a 305-mm (12 in.) diameter well completed in a confined aquifer with a transmissivity of 124 m^2/day (10,000 gpd/ft). The maximum allowable drawdown at the well is limited to 61.0 m (200 ft). The radius of influence for the well is 18,300 m (60,000 ft).

9.14. Determine the maximum discharge rate for Problem 9.13 if the diameter of the well is increased to 610 mm (24 in.).

9.15. Determine the drawdown at the well in Problem 9.13 for a pumping rate of 25.2 lps (400 gpm).

9.16. Determine the maximum steady-state pumping rate in lps (gpm) for a 305-mm (12-in.) diameter well installed in an unconfined aquifer if the allowable drawdown at the well is limited to 8.0 m (26.2 ft). The aquifer has a saturated thickness of 12.0 m (39.4 ft) and a hydraulic conductivity of 40 m/day (131.2 ft/day). The radius of influence for the well is 2,000 m (6,560 ft).

9.17. Determine the maximum steady-state pumping rate for the well in Problem 9.16 if the allowable drawdown at the well is limited to 6.0 m (19.7 ft).

9.18. Determine the drawdown at the well and 30 m (98.4 ft) from the well in Problem 9.16 for a steady-state discharge of 18.9 lps (300 gpm).

9.19. To recover contaminated groundwater, a series of groundwater recovery wells with a diameter of 152 mm (6.0 in.) are to be installed in an unconfined aquifer with a saturated thickness of 6.0 m (19.7 ft). The hydraulic conductivity of the aquifer is 2 m/day (6.6 ft/day) and the radius of influence of the well is 50 m (164 ft). Determine the pumping rate for a drawdown at the well of 2.0, 3.0, and 4.0 m (6.6, 9.8, and 13.1 ft). If the slope of the water table is 0.04, determine the theoretical maximum spacing of the wells to capture the contaminated groundwater.

9.20. Compute the drawdown at the well in Problem 9.13 after 1, 30, and 365 days of pumping at a rate of 25.2 lps (400 gpm). The storage coefficient for the aquifer is 0.001.

9.21. The well in Problem 9.16 is to be operated as a recharge well. Determine the increase in water level in the aquifer at the well and 30 m (98.4 ft) from the well after

24 hrs of recharging at a rate of 20 lps (317 gpm). The storage coefficient for the aquifer is 0.15.

9.22. Determine the drawdown at a 305-mm (12-in.) diameter well installed in an unconfined aquifer after pumping for 90 days at a rate of 18.9 lps (300 gpm). The aquifer has a saturated thickness of 12.2 m (40.0 ft), a storage coefficient of 0.12, and a hydraulic conductivity of 45.7 m/day (150 ft/day).

9.23. If the well in Problem 9.22 is located 61.0 m (200 ft) from a river (recharge boundary), determine the drawdown at the well after pumping 18.9 lps (300 gpm) for 90 days.

9.24. If the well in Problem 9.22 is located 61.0 m (200 ft) from a barrier boundary, determine the drawdown at the well after pumping 18.9 lps (300 gpm) for 90 days.

9.25. An irrigation well is located in the floodplain of a major river. The well is 305 mm (12 in.) in diameter and is pumped at a rate of 30 lps (476 gpm). The transmissivity of the aquifer is 800 m^2/day (8,607 ft^2/day), and the storage coefficient is 0.15. Determine the drawdown(s) in the well at times (t) of 1, 2, 6, 12, 24, and 48 hrs. Plot s versus log(t).

9.26. Compute the drawdown at the well in Problem 9.25 for $t = 24$ hrs if the well is located 100 m (328 ft) from the river (recharge boundary).

9.27. Compute the drawdown at the well in Problem 9.25 for $t = 24$ hrs if the well is located 100 m (328 ft) from the valley wall (barrier boundary).

9.28. Four water supply wells are located at the corners of a square 1.61 km (1.0 mile) on each side. The wells are 610 mm (24 in.) in diameter and are pumped at a rate of 126.2 lps (2,000 gpm). The confined aquifer has a transmissivity of 1,243 m^2/day, 100,000 gpd/ft, and a storage coefficient of 0.00035. Determine the drawdown in a well after 365 days of pumping.

9.29. The wells in Problem 9.28 are located 24.4 km (15.2 mi) from the formation outcrop (recharge boundary). Determine the drawdown in a well after 365 days of pumping.

9.30. The drawdowns (s) listed below were measured at an observation well located at 15.2 m (50.0 ft) from the pumped well. Determine the transmissivity and storage coefficient of the aquifer. The pumping rate was 31.5 lps (500 gpm).

t	hrs	0.5	1.0	2.0	4.0	6.0	12.0	24.0	48.0
s	m	0.15	0.30	0.46	0.76	0.98	1.31	1.65	1.95
	ft	0.5	1.0	1.5	2.5	3.2	4.3	5.4	6.4

9.31. A 305-mm (12-in.) diameter well was pumped at a rate of 31.5 lps (500 gpm) for 48 hrs. The following drawdowns (s) were measured at an observation well located 30.5 m (100 ft) from the pumped well. Determine the transmissivity and storage coefficient of the aquifer and compute the drawdown at the pumped well after 24 hrs of pumping.

t	hrs	0.5	1.0	2.0	4.0	6.0	12.0	24.0	48.0
s	m	0.03	0.15	0.46	0.91	1.22	1.98	2.90	3.66
	ft	0.1	0.5	1.5	3.0	4.0	6.5	9.5	12.0

9.32. A 610-mm (24-in.) diameter well was pumped at a rate of 126.2 lps (2,000 gpm). Determine the transmissivity and storage coefficient of the aquifer. The following drawdowns (s) were measured at the pumped well.

t	hrs	0.5	1.0	2.0	4.0	6.0	12.0	24.0
s	m	8.54	8.84	9.15	9.76	10.06	10.37	10.98
	ft	28.0	29.0	30.0	32.0	33.0	34.0	36.0

9.33. Design a water supply well for a pumping rate of 126.2 lps (2,000 gpm) from a confined aquifer 122.0 m (400 ft) thick with a hydraulic conductivity of 10.2 m/day (250 gpd/ft^2) and a storage coefficient of 0.0003. The base of the aquifer is 610 m (2,000 ft) below the ground surface and the piezometric surface is 100 m (328 ft) below the ground surface. Samples of the aquifer collected while drilling indicate a uniform formation with the following grain sizes.

	Grain size	
	mm	inch
d_{30}	0.76	0.030
d_{40}	0.64	0.025
d_{50}	0.53	0.021
d_{70}	0.30	0.012
d_{90}	0.25	0.010

Compute the drawdown at the well after 1 year of continuous pumping.

9.34. Design a water supply well for a pumping rate of 95 lps (1,506 gpm) from a confined aquifer 100 m (328 ft) thick with a hydraulic conductivity of 8.0 m/day (26.2 ft/day) and a storage coefficient of 0.001. The base of the aquifer is 450 m (1,476 ft) below the ground surface and the piezometric surface is 10 m (32.8 ft) below the ground surface. Samples collected from the aquifer indicate a uniform formation with the following grain sizes.

	Grain size	
	mm	inch
d_{30}	0.25	0.010
d_{40}	0.20	0.008
d_{50}	0.18	0.007
d_{70}	0.13	0.005
d_{90}	0.08	0.003

Compute the drawdown in the well after 1 year of continuous pumping.

9.35. Design a well for a pumping rate of 31.5 lps (500 gpm) from an unconfined aquifer with a saturated thickness of 18.3 m (60.0 ft), a hydraulic conductivity of 30.6 m/day (750 gpd/ft^2), and a storage coefficient of 0.15. The base of the aquifer is 25.9 m (85 ft) below the ground surface. Aquifer samples indicate a uniform formation with the following grain size.

	Grain size	
	mm	inch
d_{30}	1.52	0.060
d_{40}	0.69	0.027
d_{50}	0.30	0.012
d_{70}	0.13	0.005
d_{90}	0.03	0.001

Compute the drawdown at the well after 90 days of continuous pumping at a rate of 31.5 lps (500 gpm).

9.36. Design a well for a pumping rate of 13.0 lps (206.1 gpm) from an unconfined aquifer with a saturated thickness of 15.0 m (49.2 ft), a hydraulic conductivity of 15.0 m/day (368 gpd/ft^2), and a storage coefficient of 0.15. The base of the aquifer is 30 m (98.4 ft) below the ground surface. Aquifer samples collected while drilling indicate the following grain size to be representative of the formation.

	Grain size	
	mm	inch
d_{30}	0.81	0.032
d_{40}	0.66	0.026
d_{50}	0.53	0.021
d_{70}	0.25	0.010
d_{90}	0.10	0.004

Compute the drawdown at the well after 90 days of continuous pumping at the design rate.

9.37. Observation wells were used to map the piezometric surface around an industrial well field completed in a confined aquifer with a transmissivity of 1,243 m^2/day (100,000 gpd/ft). Estimate the average pumping rate from the well field.

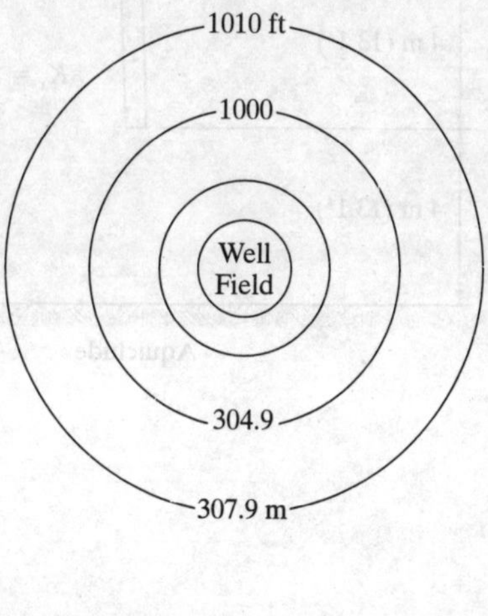

9.38. The piezometric surface map of the confined aquifer includes two pumping centers (A and B). Determine the amount of water being withdrawn from the aquifer at pumping centers A and B. The transmissivity of the aquifer is 500 m^2/day (40,000 gpd/ft).

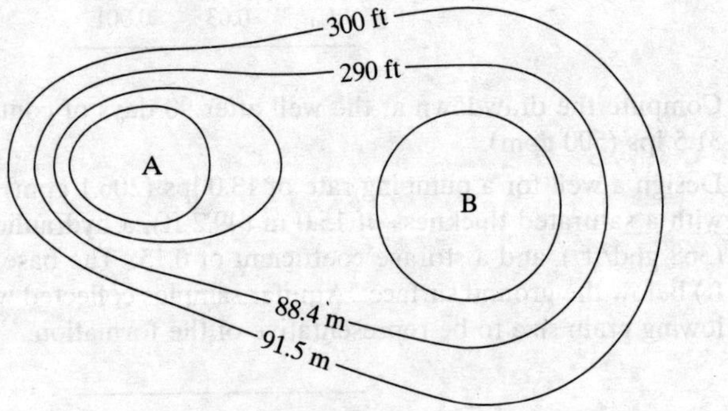

9.39. A vertical cutoff wall is used to confine flood flows and reduce subsurface seepage. Compute the seepage rate under the cutoff wall.

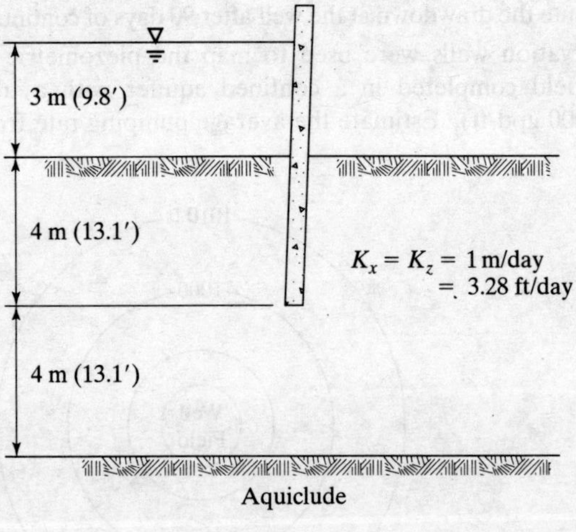

9.40. A dam is to be constructed on a semi-permeable foundation as shown below. The base of the dam is 60 m (196.8 ft) wide. The aquifer is 10.0 m (32.8 ft) thick and has a hydraulic conductivity of 0.5 m/day (1.64 ft/day). Compute the seepage rate under the dam for a headwater depth of 10.0 m (32.8 ft) and a tailwater depth of 2.0 m (6.56 ft).

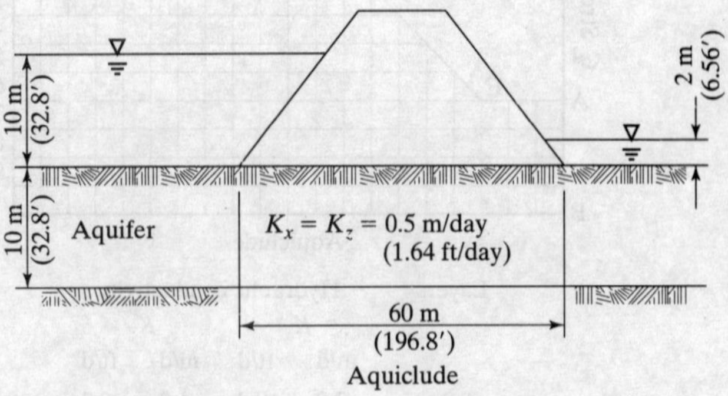

9.41. A cutoff wall is added to the dam in Problem 9.40. The cutoff wall is located in the center of the dam, is 5.0 m (16.4 ft) wide, and extends 5.0 m (16.4 ft) into the aquifer. Compute the seepage under the dam.

9.42. Repeat Problem 9.40 for an anisotropic, layered aquifer. The aquifer consists of 5 layers, each 2.0 m (6.56 ft) thick. The hydraulic conductivity of each layer is listed below with the top layer as 1.

	Hydraulic conductivity			
	K_x		K_z	
Layer	m/day	ft/day	m/day	ft/day
1	10.0	32.8	1.0	3.3
2	5.0	16.4	0.5	1.6
3	0.5	1.6	0.05	0.2
4	1.0	3.3	0.1	0.3
5	10.0	32.8	1.0	3.3

9.43. Repeat Problem 9.42 with the cutoff wall described in Problem 9.41.

9.44. Compute the regional seepage into the stream for the saturated formation shown below. ABCD represents a barrier boundary, whereas AED represents a constant head boundary with the head equal to the elevation.

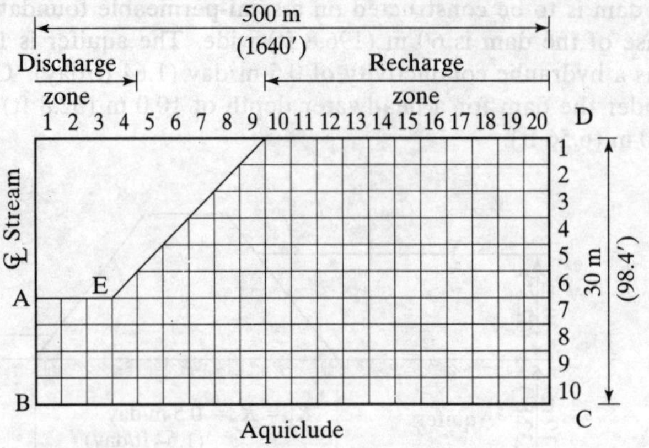

Layers	Hydraulic conductivity			
	K_x		K_z	
	m/d	ft/d	m/d	ft/d
1-8	3.0	9.8	1.0	3.3
9-10	30.0	98.4	10.0	32.8

9.45. Parallel field drains are spaced 61.0 m (200 ft) apart and placed in a depth of 2.48 m (8.0 ft). The depth to the aquiclude is 3.05 m (10.0 ft). Determine the maximum discharge into the drain if the water table is at the ground surface. AB and DEFG are barrier boundaries, and AG and C are constant head boundaries.

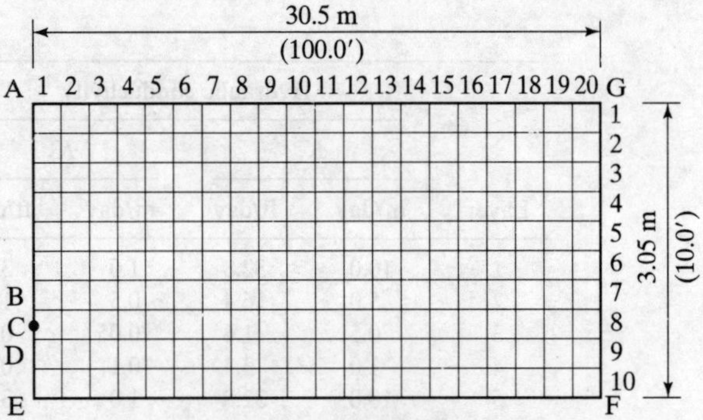

Hydraulic conductivity			
K_x		K_z	
m/d	ft/d	m/d	ft/d
6.10	20.0	1.22	4.0

9.46. The floodplain of a river extends 2.0 km (1.24 mi.) on each side of the river. The aquifer adjacent to the river has a static water level of 100.0 m (328.0 ft), a base elevation of 80.0 m (262.4 ft), a hydraulic conductivity of 40 m/day (131.2 ft/day), and a storage coefficient of 0.15. Compute the daily loss or gain in streamflow for a 15-km (9.32-mi.) reach of the river for river stages listed below.

Time, days	Stage m	Stage ft	Time, days	Stage m	Stage ft
0	100.0	328.0	20	107.5	352.6
1	102.1	334.9	21	107.1	351.3
2	104.3	342.1	22	106.3	348.7
3	108.7	356.5	23	106.8	350.3
4	110.8	363.4	24	108.1	354.6
5	111.2	364.7	25	107.9	353.9
6	111.8	366.7	26	106.5	349.3
7	112.3	368.3	27	105.8	347.0
8	112.8	370.0	28	104.9	344.1
9	113.2	371.3	29	104.2	341.8
10	112.8	370.0	30	103.7	340.1
11	112.3	368.3	31	103.1	338.2
12	111.7	366.4	32	102.8	337.2
13	111.2	364.7	33	102.5	336.2
14	110.6	362.8	34	102.2	335.2
15	110.0	360.8	35	101.9	334.2
16	109.5	359.2	36	101.3	332.3
17	109.2	358.2	37	100.8	330.6
18	108.7	356.5	38	100.5	329.6
19	108.1	354.6	39	100.0	328.0

9.47. The sketch below represents an industrial site where there is a high potential for a hydrocarbon product spill to occur over an aquifer. The sand and gravel aquifer is very permeable and a spill would result in the product reaching the water table aquifer and

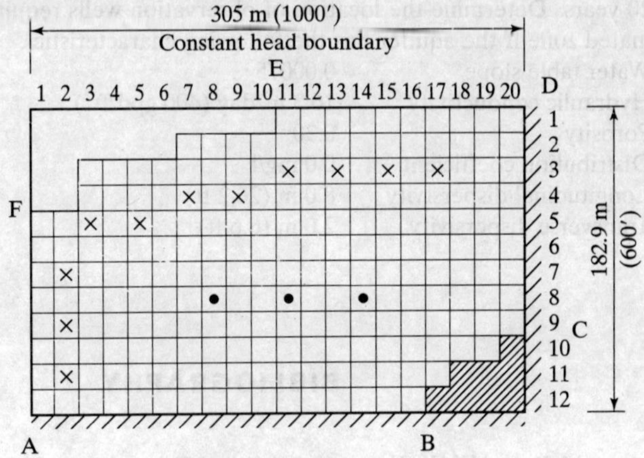

$K_x = K_y = 40.8$ m/d (1000 gpd/ft^2)

S = 0.15

• Pumped well 6.3 lps (100 gpm)

× Recharge well 1.6 lps (25 gpm)

seeping into the bay. AFED represents a constant head boundary (bay) with an elevation of 96.0 m (315.0 ft). ABCD represents a barrier boundary. The bottom of the aquifer has an elevation of 91.5 m (300.0 ft). The wells are equipped with switches so that the pumped wells turn off when the water level drops below 93.0 m (305.0 ft), and the recharge wells turn off when the water level exceeds 99.1 m (325.0 ft). Determine the time for an effective barrier to form.

9.48. A storage tank with 25 m^3 (6,600 gal) of liquid contains 200 kg (441 lbs) of contaminant. The tank ruptured and contaminated a shallow aquifer with a hydraulic conductivity of 5.0 m/day (16.4 ft/day), porosity of 0.20, a saturated thickness of 10 m (32.8 ft), and a water table slope of 0.005. What will be the location, maximum concentration, and approximate dimensions of the contaminated area in 5 years? Use a longitudinal dispersivity of 3 m (10.0 ft) and a transverse dispersivity of 1 m (3.3 ft) and a vertical dispersivity of 0.1 m (0.33 ft). Assume the contaminant does not decay and is not absorbed by the formation.

9.49. Repeat Problem 9.48 for a continuous leak where the tank drained at a uniform rate over a 5-year period. Assume that the contaminant is uniformly mixed in the upper 5 m of the aquifer and that the two-dimensional steady state model is applicable near the site. Compute the concentration at $y = 0$ and $x = 1, 10$, and 100 m (3.3, 33, 328 ft).

9.50. A continuous point source of pollution is contaminating the aquifer. Observation wells were installed, and the water table slope is 0.0010. Pump test was conducted, and hydraulic conductivity is estimated as 4.07 m/day (100 gpd/ft^2). The porosity of the formation is 0.25. Estimate the distribution of contaminated groundwater after 10 years of leakage. Use a longitudinal dispersivity of 10 m (32.8 ft) and transverse dispersivity of 1 m (3.28 ft). The contaminant is not absorbed by the formation (distribution coefficient is equal to zero).

9.51. Repeat Problem 9.50 with a distribution coefficient = 0.1 ml/g.

9.52. A contaminant has been leaking into the groundwater aquifer for approximately 20 years. Determine the location of observation wells required to define the contaminated zone if the aquifer has the following characteristics.

Water table slope	0.00075
Hydraulic conductivity	16.3 m/day (400 gpd/ft^2)
Porosity	0.30
Distribution coefficient	0.0 mg/l
Longitudinal dispersivity	8.0 m (26.2 ft)
Transverse dispersivity	2.0 m (6.6 ft)

BIBLIOGRAPHY

ANDERSON, M. P., and W. W. WOESSNER, *Applied Groundwater Modelling, Simulation of Flow and Transport,* Academic Press, 1992.

BEAR, J., *Hydraulics of Groundwater,* McGraw-Hill, New York, NY, 1979.

BEDIENT, P. B., H. S. RIFAI, and C. J. NEWELL, *Groundwater Contamination: Transport and Remediation,* 2nd Ed., Prentice Hall, Upper Saddle River, NJ, 1999.

BOUWER, H., *Groundwater Hydrology,* McGraw-Hill, New York, NY, 1978.

CHARBENEAU, R. J., *Groundwater Hydraulics and Pollutant Transport,* Prentice Hall, Upper Saddle River, NJ, 2000.

DEWIEST, R. J. M., *Geohydrology,* John Wiley & Sons, New York, NY, 1965.

DOMENICO, P. A., and F. W. SCHWARTZ, *Physical and Chemical Hydrogeology,* John Wiley & Sons, New York, NY, 1990.

FETTER, C. W., *Applied Hydrogeology,* 4th Ed., Prentice Hall, Upper Saddle River, NJ, 2001.

FREEZE, R. A., and J. A. CHERRY, *Groundwater,* Prentice Hall, Upper Saddle River, NJ, 1979.

Johnson Division of Universal Oil Products Co., *Ground Water and Wells,* St. Paul, MN, 1975.

LEHR, J., S. HURLBURT, B. GALLAGHER, and J. VOYTEK, *Design and Construction of Water Wells, A Guide for Engineers,* Van Nostrand Reinhold, 1988.

LOHMAN, S. W., *Ground-Water Hydraulics,* Professional Paper 708, U.S. Geological Survey, Washington, D.C., 1972.

TODD, D. K., *Ground Water Hydrology,* John Wiley & Sons, New York, NY, 1980.

WALTON, W. C., *Groundwater Resource Evaluation,* McGraw-Hill, New York, NY, 1970.

10

Urban Stormwater Management

The purpose of urban stormwater management is to enhance the quality of life in urban areas by (1) protecting human life and reducing flood safety risks, (2) preventing damage to private and public property, (3) minimizing the disruption of normal urban activities caused by storm runoff, and (4) protecting water quality. Urban stormwater management includes the development of structural and nonstructural measures necessary to achieve these objectives, and requires a basic knowledge of hydrology and open channel hydraulics.

The National Flood Insurance Program was established in 1968 to restrict development in flood-prone areas and provide federally-backed flood insurance. The program is administered through the Federal Insurance Administration and Mitigation Directorate of the Federal Emergency Management Agency. Over 19,000 communities, representing most of the urban areas in the U.S., participate in the program. Participating communities must enforce mandatory land use regulations for flood prone areas. A flood insurance study is required to delineate the 10-, 50-, 100-, and 500-year recurrence interval flood-prone areas. The flood insurance study is used to establish flood insurance premiums and land use restrictions for the 100-year floodplain. The 100-year return period flood has become widely accepted as the standard for flood damage mitigation.

Because of the flood safety risks, low water crossing structures on major streams should be avoided in urban areas. A low water crossing structure (sometimes referred to as dips) is when the road cross drainage structure (such as a culvert) is undersized, and during a major flood event, water flows over the roadway. A vehicle might be stalled in the roadway during a flood, and if the water gets high enough, the vehicle can be washed off the roadway into the downstream channel. The roadway becomes a broad-crested weir, and with a low tailwater elevation, the velocity of the water on the roadway will be near critical velocity.

Example 10.1 Low Water Crossing Structure

Assume a vehicle was stalled at a low water crossing structure, and the water depth on the roadway was 2.0 ft and rising. The person in the vehicle can either stay in the vehicle and hope that vehicle remains on the roadway or try and walk to safety. Compute the force on the vehicle for a water depth of 2.0 ft and a drag coefficient of 2.0. The vehicle is 16 ft long, with 1.0 ft of clearance under the vehicle and at a water depth of 2.0 ft has an area of 19.0 ft^2 exposed to the flow.

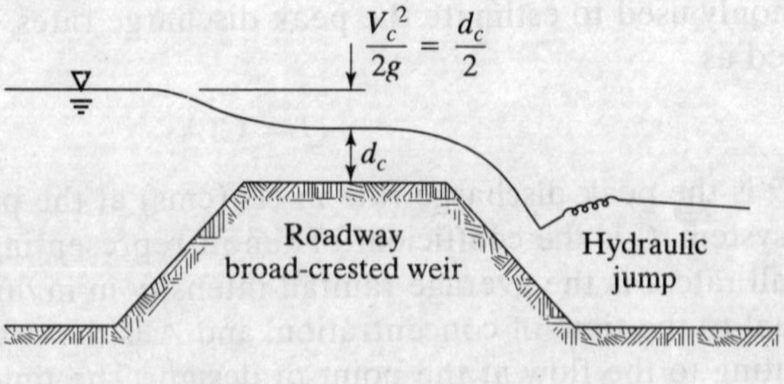

Velocity

$$V_c = \sqrt{2gd_c/2} = \sqrt{2g(2.0)/2} = 8.0 \text{ fps}$$

Force

$$F_D = \frac{C_D \rho A V_c^2}{2} = 2.0 \times 1.94 \times 19.0 \times \frac{8.0^2}{2.0} = 2{,}360 \text{ lbs}$$

Compute the drag force on a person who is standing sideways to the flow and 0.75 wide, with a drag coefficient of 1.2.

$$F_D = 1.2 \times 1.94 \times (0.75 \times 2.0) \times \frac{8.0^2}{2.0} = 112 \text{ lbs}$$

Because of buoyancy, the vehicle will probably not remain on the roadway, and it will be extremely difficult to walk to safety.

10.1 STORMWATER COLLECTION SYSTEMS

In urban areas, stormwater runoff generally occurs as sheet flow to the streets, conveyed as gutter flow along the streets to drain inlets; collected by storm sewers in buried conduits; and discharged into streams, lakes, or ponds. Because of the adverse effect that urbanization has on water quality and the peak discharge rate, a detention and/or sedimentation basin may be required before the effluent from the storm sewer is discharged into the receiving stream or lake. A schematic of a stormwater collection system is shown in Fig. 10.1. The stormwater collection system is a combination of streets and storm sewers. The lots are graded so that the runoff flows toward the street, and the street conveys the runoff to the storm sewer drain inlets. The storm sewers are generally designed for a 5-, 10-, or 25-year return period storm. Some street flooding can be expected during a major flood event. Because the stormwater collection system consists of both the streets and storm sewers, the system is designed for at least a 100-year return period storm. The streets must be able to carry flood flow that is in excess of the storm sewer capacity.

The system of streets and storm sewers used to collect stormwater runoff is often referred to as the minor urban drainage system, and the rational method is commonly used to estimate the peak discharge rates. The rational equation is expressed as

$$Q = \mathrm{CIA} \tag{8.11}$$

where Q is the peak discharge rate in cfs (cms) at the point of design in the collection system, C is the coefficient of runoff representing the ratio of runoff rate to rainfall rate, I is the average rainfall intensity in in./hr (m/s) during a time period equal to the time of concentration, and A is the drainage area in acres (m^2) contributing to the flow at the point of design. The time of concentration (t_c) is the time required for water to flow from the most hydraulically remote point in the drainage area to the point of design. t_c can also be defined as the time for the runoff from the drainage area to reach equilibrium under a steady rainfall rate.

Although the total storm duration may be considerably longer than t_c, the rainfall intensity (I) is the maximum average rainfall intensity that occurs in the storm for a period of time equal to t_c. The rainfall intensity for the design of the stormwater collection system can be given by the intensity-duration frequency (IDF) equation such as

$$I = \frac{a}{(t_c + b)^c} \tag{7.50}$$

where I is rainfall intensity in in./hr (mm/hr); t_c is time of concentration in minutes; and a, b, and c are constants for the location and storm frequency.

The area contributing to the flow at design point 1 in Fig. 10.2 is ABGF, whereas the area contributing to the flow at design point 5 is AELK. As the drainage area becomes larger, the time of concentration increases and the design rainfall intensity decreases. The time of concentration for the storm sewer system consists of overland flow time plus gutter flow time plus conduit flow time. The gutter flow time and the conduit flow time can be estimated by dividing the channel length by the average water velocity.

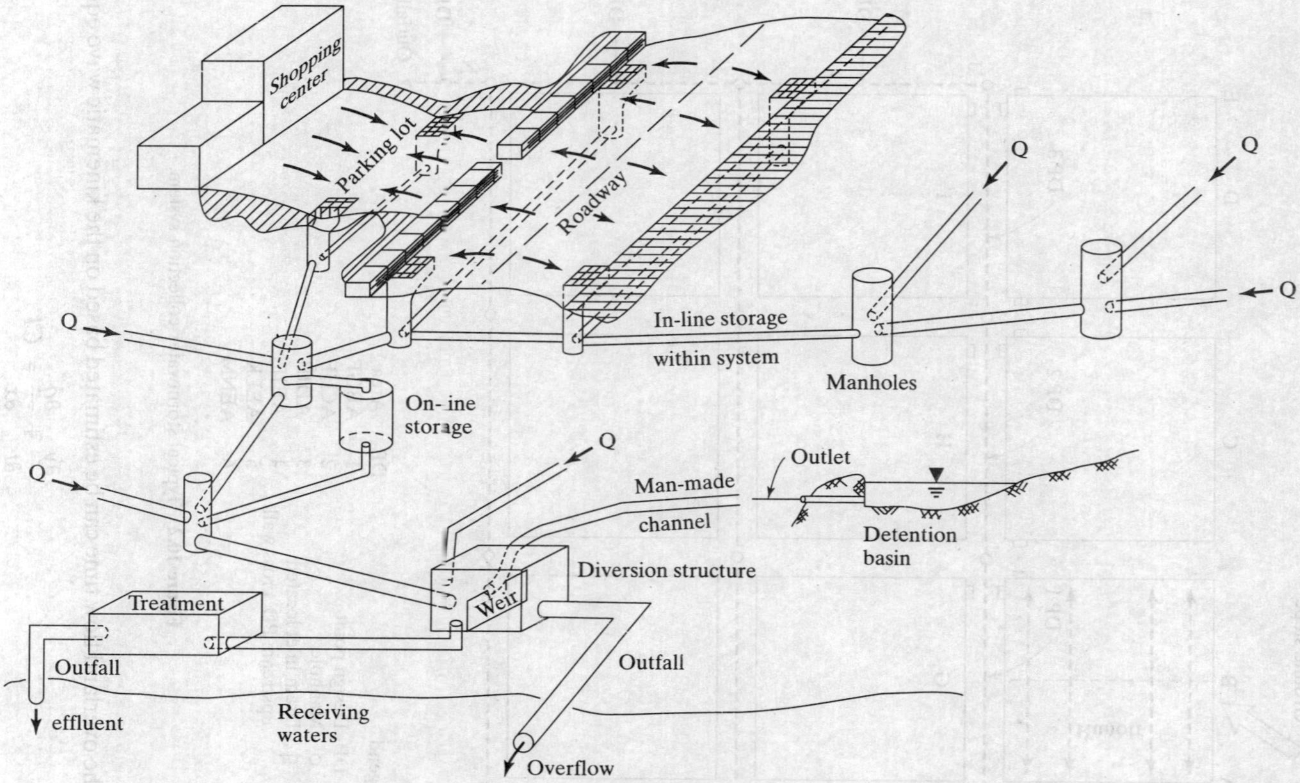

Figure 10.1 Principal elements in urban stormwater collection system. (American Society of Civil Engineers and Copyright Water Environment Federation, Alexandria, Va. 1992; reprinted with permission.)

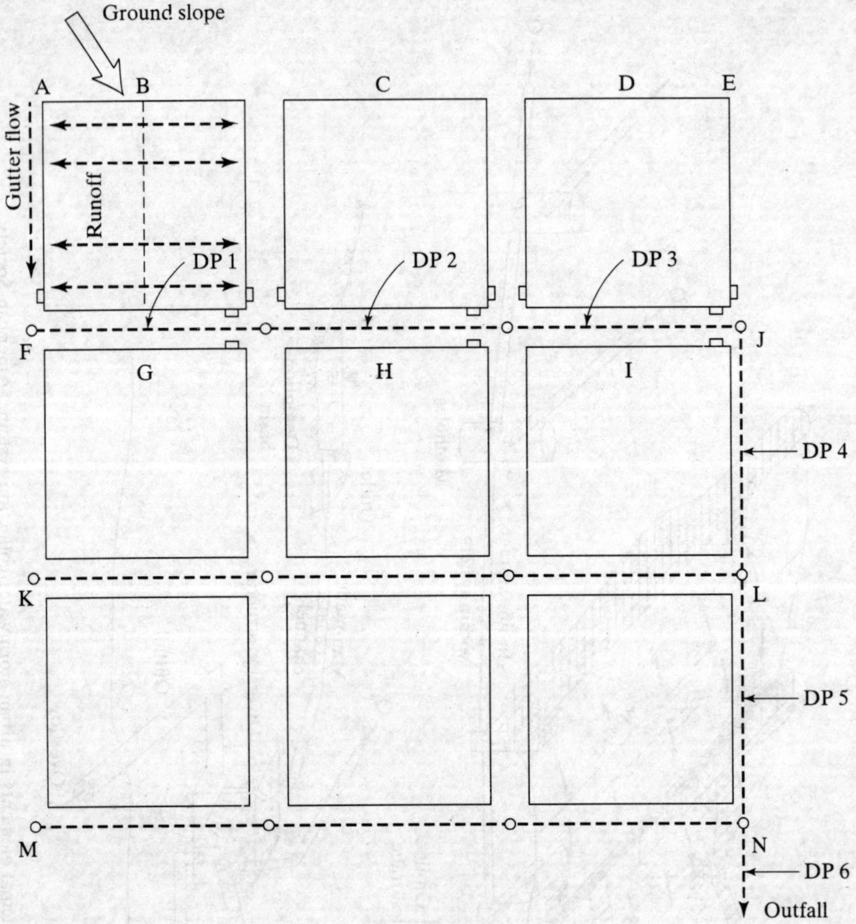

Legend

DP - Design point
○ - Manhole
▯ - Drain inlet located upstream PD cross walk

DP	Area
1	ABGF
2	ACHF
3	ADIF
4	AEJF
5	AELK
6	AENM

Figure 10.2 Typical stormwater collection system.

The overland flow time can be estimated based on the kinematic wave equation

$$\frac{\partial y}{\partial t} + \frac{\partial q}{\partial x} = CI \qquad \textbf{(6.13)}$$

Using Manning's equation

$$q = \alpha y^m$$

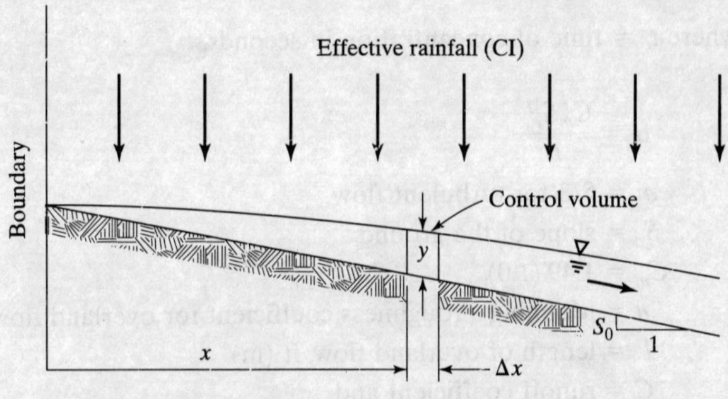

Figure 10.3 Kinematic wave overland flow travel time.

where

$$\alpha = \frac{C_m S_o^{1/2}}{n}$$

and

$$m = 5/3$$

Combining the kinematic wave and the Manning's equations gives

$$\frac{\partial y}{\partial t} + \alpha m y^{m-1} \frac{\partial y}{\partial x} = CI \quad \textbf{(10.1)}$$

The control volume in Fig. 10.3 is moving at a velocity of dx/dt. If t is the time for the control volume to travel from the boundary to a distance x, it follows that

$$\frac{dy}{dt} = CI = \frac{\partial y}{\partial t} + \frac{dx}{dt}\frac{\partial y}{\partial x} \quad \textbf{(10.2)}$$

Comparing Eqs. 10.1 and 10.2 gives

$$\frac{dx}{dt} = \alpha m y^{m-1} \quad \textbf{(10.3)}$$

Integrating Eq. 10.2 gives

$$y = CIt \quad \textbf{(10.4)}$$

Combining Eqs. 10.3 and 10.4 and integrating gives

$$x = \alpha (CI)^{m-1} t^m \quad \textbf{(10.5)}$$

The travel time (t_c) for $x = L$ is

$$t_c = \left(\frac{L}{\alpha (CI)^{m-1}}\right)^{1/m} \quad \textbf{(10.6)}$$

where t_c = time of concentration in seconds,

$$\alpha = \frac{C_m S_o^{1/2}}{n}$$

m = 5/3 for turbulent flow

S_o = slope of the ground

C_m = 1.49 (1.0)

n = Manning's roughness coefficient for overland flow (Table 6.1)

L = length of overland flow, ft (m)

C = runoff coefficient and

I = average rainfall intensity, fps (mps).

Table 10.1 can be used to estimate the runoff coefficient (C) based on land use, soil type, ground slope, and rainfall intensity. Two sets of runoff coefficients

TABLE 10.1 RUNOFF COEFFICIENTS FOR USE IN THE RATIONAL FORMULA

	NRCS Hydrologic soil group and ground surface slope range											
Land use	A			B			C			D		
	0–2%	2–6%	6%+	0–2%	2–6%	6%+	0–2%	2–6%	6%+	0–2%	2–6%	6%+
Impervious areas	0.90*	0.90	0.90	0.90	0.90	0.90	0.90	0.90	0.90	0.90	0.90	0.90
	0.95†	0.95	0.95	0.95	0.95	0.95	0.95	0.95	0.95	0.95	0.95	0.95
Commercial	0.71	0.71	0.72	0.71	0.72	0.72	0.72	0.72	0.72	0.72	0.72	0.72
	0.88	0.89	0.89	0.89	0.89	0.89	0.89	0.89	0.90	0.89	0.89	0.90
Industrial	0.67	0.68	0.68	0.68	0.68	0.69	0.68	0.69	0.69	0.69	0.69	0.70
	0.85	0.85	0.86	0.85	0.86	0.86	0.86	0.86	0.87	0.86	0.86	0.88
Freeways and	0.57	0.59	0.60	0.58	0.60	0.61	0.59	0.61	0.63	0.60	0.62	0.64
expressways	0.70	0.71	0.72	0.71	0.72	0.74	0.72	0.73	0.76	0.73	0.75	0.78
High density‡	0.47	0.49	0.50	0.48	0.50	0.52	0.49	0.51	0.54	0.51	0.53	0.56
residential	0.58	0.60	0.61	0.59	0.61	0.64	0.60	0.62	0.66	0.62	0.64	0.69
Medium density§	0.25	0.28	0.31	0.27	0.30	0.35	0.30	0.33	0.38	0.33	0.36	0.42
residential	0.33	0.37	0.40	0.35	0.39	0.44	0.38	0.42	0.49	0.41	0.45	0.54
Low density¶	0.14	0.19	0.22	0.17	0.21	0.26	0.20	0.25	0.31	0.25	0.28	0.35
residential	0.22	0.26	0.29	0.24	0.28	0.34	0.28	0.32	0.40	0.31	0.35	0.46
Agricultural	0.08	0.13	0.16	0.11	0.15	0.21	0.14	0.19	0.26	0.18	0.23	0.31
	0.14	0.18	0.22	0.16	0.21	0.28	0.20	0.25	0.34	0.24	0.29	0.41
Open space	0.05	0.10	0.14	0.08	0.13	0.19	0.12	0.17	0.24	0.16	0.21	0.28
	0.11	0.16	0.20	0.14	0.19	0.26	0.18	0.23	0.32	0.22	0.27	0.39

*Smaller runoff coefficients for use with storm recurrence intervals less than 25 years.

†Larger runoff coefficients for use with storm recurrence intervals of 25 years or more.

‡High density residential = greater than 15 dwelling units per acre.

§Medium density residential = 4–15 dwelling units per acre.

¶Low density residential = 1–4 dwelling units per acre.

are listed in Table 10.1 for each land use category. The smaller values of runoff coefficients are to be used with storm recurrence intervals of less than 25 years, and the larger values are to be used for storm recurrence intervals of 25 years or more.

Example 10.2 Time of Concentration

Determine the time of concentration and peak discharge rate for design point 1 in Fig. 10.2. The stormwater collection system is for a high-density residential area with NRCS (Natural Resource Conservation Service) hydrologic soil group C. The lots have been graded to drain toward the street at a slope of 0.01 and have a Manning "n" value for overland flow of 0.2. The street gutter (A–F) has a longitudinal slope of 0.015, a Manning "n" value of 0.017, and a transverse side slope of 30:1.

The blocks in Fig. 10.2 are 350 feet square, which includes a 50-ft wide street. The storm sewers are to be designed for a 25-year return period storm. The design IDF (intensity-duration-frequency) equation for the system is

$$I = \frac{89}{(t_c + 8.5)^{0.754}}$$

where I is rainfall intensity in in./hr.

Solution The residential area contributing to the flow is

$$A_1 = \frac{300 \times 150}{43{,}560} = 1.03 \text{ ac}$$

with a runoff coefficient of 0.60 (Table 10.1). The area of the street contributing to the flow is

$$A_2 = \frac{325 \times 25}{43{,}560} = 0.19 \text{ ac}$$

with a runoff coefficient of 0.95.

Equation 10.6 is used to compute the overland flow travel time. The time of concentration (t_c) is

$$t_c = t_{\text{overland}} + t_{\text{gutter}}$$

For BG units, Eq. 10.6 can be written as

$$t_c = \frac{1}{60}\left(\frac{L}{\alpha(CI)^{m-1}}\right)^{1/m}$$

where

$$\alpha = \frac{C_m S_o^{1/2}}{n}$$

$$C_m = 1.49$$

$$S_o = 0.01$$

$$n = 0.2$$

$$\alpha = 0.74$$

$$L = 150 \text{ ft}$$

$$C = 0.60$$

$$m = 5/3$$

and

$$I = \frac{89}{(t_c + 8.5)^{0.754}} \text{ in./hr} = \frac{0.00206}{(t_c + 8.5)^{0.754}} \text{ fps}$$

Using an assumed value of 1.5 minutes for gutter flow time, the above equation becomes

$$t_c = \frac{1}{60} \frac{150^{0.6}(t_c + 8.5)^{0.30}}{0.74^{0.6}(0.6 \times 0.00206)^{0.4}} + 1.5 = 5.88(t_c + 8.5)^{0.30} + 1.5$$

$$t_c = 17.2 \text{ min}$$

$$I = \frac{89}{(17.2 + 8.5)^{0.754}} = 7.7 \text{ in./hr}$$

$$Q_{\text{peak}} \text{ at design point 1} = \text{CIA} = (0.60 \times 1.03 + 0.95 \times 0.19)7.7 = 6.1 \text{ cfs}$$

Check gutter travel time based on an average discharge rate of 3.1 cfs.

$$Q = \frac{1.49}{n} \frac{SS}{3.2} d^{8/3} S_o^{1/2}$$

$$3.1 = \frac{1.49}{0.017} \frac{30}{3.2} d^{8/3}(0.015)^{1/2} = 101\, d^{8/3}$$

$$d = 0.27 \text{ ft}$$

$$A_g = \frac{SS}{2} d^2 = 15(0.27)^2 = 1.09 \text{ ft}^2$$

$$V_g = \frac{Q}{A_g} = \frac{3.1}{1.09} = 2.8 \text{ fps}$$

$$t_{\text{gutter}} = \frac{300}{2.8 \times 60} = 1.8 \text{ min}$$

Correcting for gutter travel time

$$t_c = 17.3 \text{ min}$$

$$I = 7.7 \text{ in./hr}$$

$$Q = 6.1 \text{ cfs}$$

10.1.1 Design Criteria

The design capacity of the combined street and storm sewer collection system must equal or exceed the runoff rate for the 100-year return period storm. During a major flood event, street flooding may occur, but residential, public, commercial, and industrial buildings should not flood (Fig. 10.4). The capacity of the storm sewer collection system is generally selected to handle a 5-, 10-, or 25-year return period storm. Streets and open channels are required to carry the flood flows that are in excess of the storm sewer capacity.

The design of the stormwater collection system is based on the principles of steady flow in open channels, and Manning's equation is typically used to compute

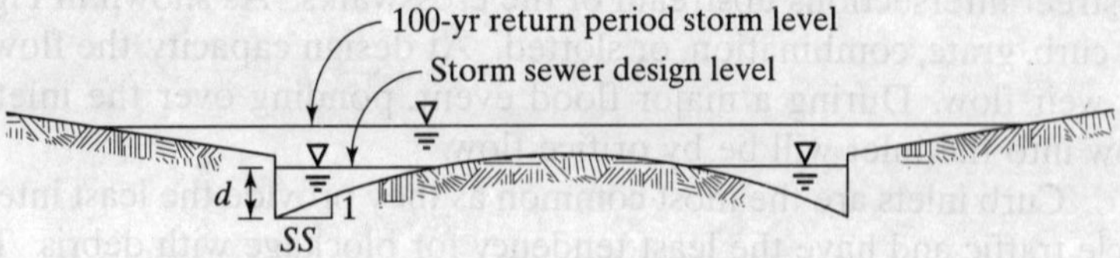

Figure 10.4 Water levels in street.

the water velocity and depth in gutters and the velocity and pipe diameter in the storm sewers.

10.1.2 Street Gutters

The discharge rate in a triangular curb gutter based on Manning's equation is

$$Q = \frac{C_m}{n}\frac{SS}{3.2}d^{8/3}S_o^{1/2} \qquad \textbf{(5.9)}$$

where Q is the discharge rate in cfs (cms), C_m is 1.49 for British Gravitational (BG) units and 1.0 for the International System (SI) of units, n is Manning roughness coefficient, SS is the street cross slope (ranges from 10 to 50), d is the normal depth in ft (m), and S_o is the longitudinal slope of the gutter.

A Manning "n" value of 0.017 is recommended for gutter flow. Because of the shallow depth of flow, the recommended "n" value is greater than that normally used for smooth concrete or asphalt pavement. The curb height should be at least 0.5 ft (0.15 m). No overtopping of the curb is allowed for the design storm. For the safety of children, the drain inlets should be spaced so that the maximum water depth in the gutter does not exceed 0.5 ft (0.15 m), and the water velocity does not exceed 10 fps (3.0 mps). A minimum longitudinal slope of 0.004 will generally ensure a low flow cleansing velocity in the gutter.

The top width of flow (T) is the spread of water on the street and is equal to

$$T = SS \times d \qquad \textbf{(10.7)}$$

For local streets, the water can spread to the crown of the street during the design storm. Collector streets are usually designed so that one traffic lane is free of water, and arterial streets are usually designed so that one traffic lane in each direction is free of water during the design storm. Because of the potential for hydroplaning, no encroachment of water is allowed on any traffic lane for high speed roads.

10.1.3 Drain Inlets

Inlets are used in the drainage system to intercept the surface runoff (primarily in the street gutter) and direct the flow into the storm sewer. They are often located at street intersections upstream of the crosswalks. As shown in Fig. 10.5, inlets can be curb, grate, combination, or slotted. At design capacity, the flow into the inlet is by weir flow. During a major flood event, ponding over the inlet may occur and flow into the inlet will be by orifice flow.

Curb inlets are the most common as they provide the least interference with bicycle traffic and have the least tendency for blockage with debris. The height of the

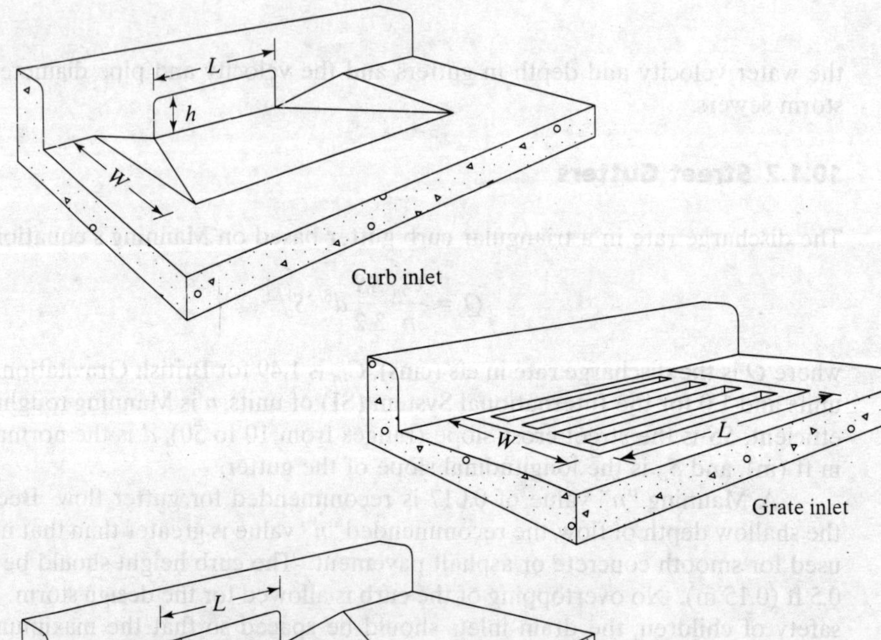

Figure 10.5 Perspective views of curb, grate, combination, and slotted drain inlets.

curb opening should not exceed 0.5 ft (0.15 m) because of the risk to small children. To increase the capacity of the inlet, the invert of the inlet is depressed below the invert of the street gutter. Typically the inlet is depressed 0.25–0.50 ft (0.07–0.15 m) with W (Fig. 10.5) ranging from 1.0–2.0 ft (0.3–0.6 m). The weir flow equation gives the discharge into the curb inlet on a mild slope (Q_I)

$$Q_I = C_w(L + 1.8W)d^{3/2} \tag{10.8}$$

where C_w is 2.3 for BG units and 1.25 for SI units, L and W are defined in Fig. 10.5, and d is the normal depth of the water in the approach gutter. Because some flow may bypass the curb inlet located on a gutter with a steep slope, drainage manuals should be consulted for sizing drain inlets located on a gutter with a steep slope.

Inlets in sumps are located at the low point in a roadway (such as the sag in a vertical curve) and are located to prevent ponding of water on the roadway. If clogging by debris is not anticipated, grate inlets work well when located in a sump. A grate inlet will operate as a weir at shallow water depths (d) less than 0.5 ft (0.12 m) and the grate inlet capacity (Q_I) is

$$Q_I = C_w P d^{3/2} \tag{10.9}$$

where C_w is 3.0 for BG units and 1.7 for SI units, P is the perimeter of the grate opening in ft (m), not including the side against the curb, and d is the depth of flow in ft (m).

A slotted drain inlet will function as a weir at shallow depths $d < 0.2$ ft (0.06 m) and operate as an orifice when submerged with a depth > 0.4 ft (0.12 m). The capacity (Q_I) of a slotted drain inlet with a width (W) > 0.15 ft (0.04 m) is

$$Q_I = C_w L d^{3/2} \tag{10.10}$$

for weir flow and

$$Q_I = 0.6A\sqrt{2gd} \tag{10.11}$$

for orifice flow. C_w is equal to 2.3 for BG units and 1.25 for SI units, and A is the area of the open space in the grate.

Example 10.3 Gutter Flow and Drain Inlet

Determine the maximum water depth and velocity in gutter and the length of a curb inlet at point F in Fig. 10.2 ($W = 1.5$ ft).

From Ex. 10.2

$$Q = 6.1 \text{ cfs} = 101d^{8/3}$$

Maximum depth in gutter

$$d = 0.35 \text{ ft}$$

$$A_g = 15d^2 = 1.84 \text{ ft}^2$$

Maximum velocity in gutter

$$V_g = \frac{Q}{A_g} = \frac{6.1}{1.84} = 3.3 \text{ fps}$$

Length of curb opening

$$Q_I = C_w(L + 1.8W)d^{3/2}$$

$$6.1 = 2.3(L + 1.8 \times 1.5)0.35^{3/2}$$

$$L = 12.8 - 2.7 = 10.1 \text{ ft}$$

10.1.4 Storm Sewers

The design of storm sewers is based on flow in open channels under steady flow conditions. Some general guidelines are as follows:

(1) The storm sewer pipe is assumed to flow full, and the pipe size is computed using the Manning equation.

(2) The minimum pipe size considered for storm sewers ranges from 12 to 18 inches (0.30–0.46 m) in diameter.

(3) To prevent sedimentation, the minimum water velocity for the pipe flowing full is 2.5 fps (0.75 mps).

(4) To avoid clogging with debris in the storm sewer, the pipe size should not decrease downstream.

(5) Pipe grades are typically specified in terms of elevation of the invert at manholes.

(6) At changes in pipe size, the crown of the two pipes are at the same elevation.

(7) Pipe slopes should conform with the ground slope where possible, and there should be at least 3.0 ft (0.9 m) of cover over the pipe to prevent excessive wheel loading on the pipe.

(8) Manholes are used for maintenance access and are located where two or more pipes join, where there is a change in pipe size or grade, and at sharp bends. The maximum spacing between manholes depends on the pipe size and may range from 500 ft (150 m) to over 1,000 ft (300 m).

Headlosses in manholes are computed as minor losses and are included in the design by lowering the invert elevation of outlet pipe at the manhole. The typical form of the minor loss equation is

$$H_M = K\frac{V^2}{2g} \quad \textbf{(4.11)}$$

The minor loss includes junction loss, bend loss, and transition loss. For inflow and outflow pipes aligned with each other, K is between 0.1 and 0.3. As the amount of turbulence in the manhole increases, the value for K will also increase.

Example 10.4 Pipe Size

Determine the discharge rates and pipe sizes for design points 1, 2, and 3 in lateral F–J and design points 4, 5, and 6 in main J–N. The lateral has a length of 1,050 ft and a slope of 0.01. The main has a length of 700 ft and slope of 0.02. Use a concrete pipe with a Manning's "n" of 0.015. The minimum pipe size is 18 inches in diameter. Use an overland flow time of concentration of 16 minutes and a runoff coefficient of 0.65.

Design point 1 (pipe 1)

From Example 10.2, $Q = 6.1$ cfs

$$Q = \frac{C_m}{n}\frac{\pi D^2}{4}\left(\frac{D}{4}\right)^{2/3} S_o^{1/2}$$

$$D = \left[\frac{3.21}{1.49}\frac{Qn}{S_o^{1/2}}\right]^{3/8} = \left[\frac{3.21}{1.49} \times \frac{Q}{(0.01)^{1/2}} \times 0.015\right]^{3/8}$$

$$D_1 = 0.65\,Q^{3/8} = 1.28 \text{ ft}$$

Compute travel time in pipe 1 (T_1) using a minimum pipe size of 1.5 ft.

$$V_1 = \frac{1.49}{n}R^{2/3}S_o^{1/2} = \frac{1.49}{0.015}\left(\frac{1.5}{4}\right)^{2/3}(0.01)^{1/2} = 5.2 \text{ fps}$$

$$T_1 = \frac{350}{5.2 \times 60} = 1.1 \text{ min}$$

Design point 2 (pipe 2)

$$t_c = 16 + 1.8 + 1.1 = 18.9 \text{ min}$$

$$I = \frac{89}{(18.9 + 8.5)^{0.754}} = 7.3 \text{ in./hr}$$

Area contributing to flow

$$A = \frac{350 \times 525}{43{,}560} = 4.2 \text{ ac}$$

$$Q_2 = \text{CIA} = 0.65 \times 7.3 \times 4.2 = 19.9 \text{ cfs}$$

$$D_2 = 0.65(19.9)^{0.375} = 1.99 \text{ ft}$$

Use $D_2 = 2.0$ ft.

Travel time, pipe 2 (T_2)

$$V_2 = \frac{1.49}{0.015}\left(\frac{2}{4}\right)^{2/3}(0.01)^{1/2} = 6.3 \text{ fps}$$

$$T_2 = \frac{350}{6.3 \times 60} = 0.9 \text{ min}$$

Design point 3 (pipe 3)

$$t_c = 18.9 + 0.9 = 19.8 \text{ min}$$

$$I = \frac{89}{(19.8 + 8.5)^{0.754}} = 7.2 \text{ in./hr}$$

$$A = \frac{350 \times 875}{43{,}560} = 7.0 \text{ ac}$$

$$Q_3 = \text{CIA} = 0.65 \times 7.2 \times 7.0 = 32.8 \text{ cfs}$$

$$D_3 = 0.65(32.8)^{0.375} = 2.4 \text{ ft}$$

Use $D_3 = 2.5$ ft.

Travel time in pipe 3 (T_3)

$$V_3 = \frac{1.49}{0.015}\left(\frac{2.5}{4}\right)^{2/3}(0.01)^{1/2} = 7.3 \text{ fps}$$

$$T_3 = \frac{350}{7.3 \times 60} = 0.8 \text{ min}$$

Design point 4 (pipe 4)

$$t_c = 19.8 + 0.8 = 20.6 \text{ min}$$

$$I = \frac{89}{(20.6 + 8.5)^{0.754}} = 7.0 \text{ in./hr}$$

$$A = \frac{350 \times 1{,}050}{43{,}560} = 8.4 \text{ ac}$$

$$Q_4 = 0.65 \times 7.0 \times 8.4 = 38.2 \text{ cfs}$$

$$D_4 = \left[\frac{3.21 \times 0.015}{1.49 \times (0.02)^{1/2}}\right]^{3/8} Q^{3/8} = 0.57 \times Q^{3/8} = 2.25 \text{ ft}$$

Pipe diameter cannot be smaller than upstream pipe. Use $D_4 = 2.5$ ft.

Travel time in pipe 4 (T_4)

$$V_4 = \frac{1.49}{0.015}\left(\frac{2.5}{4}\right)^{2/3}(0.02)^{1/2} = 10.3 \text{ fps}$$

$$T_4 = \frac{350}{10.3 \times 60} = 0.6 \text{ min}$$

Design point 5 (pipe 5)

$$t_c = 20.6 + 0.6 = 21.2 \text{ min}$$

$$I = \frac{89}{(21.2 + 8.5)^{0.754}} = 6.9 \text{ in./hr}$$

$$A = \frac{700 \times 1{,}050}{43{,}560} = 16.9 \text{ ac}$$

$$Q_5 = \text{CIA} = 0.65 \times 6.9 \times 16.9 = 75.8 \text{ cfs}$$

$$D_5 = 0.57(75.8)^{3/8} = 2.9 \text{ ft}$$

Use $D_5 = 3.0$ ft.

$$V_5 = \frac{1.49}{0.015}\left(\frac{3.0}{4}\right)^{2/3}(0.02)^{1/2} = 11.6 \text{ fps}$$

$$T_5 = \frac{350}{11.6 \times 60} = 0.5 \text{ min}$$

Design point 6 (pipe 6)

$$t_c = 21.2 + 0.5 = 21.7 \text{ min}$$

$$I = \frac{89}{(21.7 + 8.5)^{0.754}} = 6.8 \text{ in./hr}$$

$$A = \frac{1{,}050 \times 1{,}050}{43{,}560} = 25.3 \text{ ac}$$

$$Q_6 = \text{CIA} = 0.65 \times 6.8 \times 25.3 = 111.8 \text{ cfs}$$

$$D_6 = 0.57(111.8)^{3/8} = 3.35$$

Use $D_6 = 3.5$ ft.

Example 10.5 Pipe Invert Elevations

If the upstream invert elevation of pipe 1 of Example 10.4 is 98.35 ft, determine the remaining invert elevations of pipes 1, 2, and 3. The manholes are 4.0 ft in diameter and have headloss coefficients (Eq. 4.11) of 0.3. Plot profile of lateral 1.

Length of line 1

$$L_1 = 350 - 4.0 = 346 \text{ ft}$$

Downstream invert elevation pipe 1

$$El = 98.35 - L_1 \times S_1 = 98.35 - 346 \times 0.01 = 94.89 \text{ ft}$$

$$H_L = 0.3\frac{V^2}{2g} = 0.3 \times \left(\frac{6.3^2}{64.4}\right) = 0.18 \text{ ft}$$

Upstream invert elevation pipe 2

$$El = 94.89 - H_L - \Delta D = 94.89 - 0.18 - 0.50 = 94.21 \text{ ft}$$

Downstream invert elevation pipe 2

$$El = 94.21 - 346 \times 0.01 = 90.75 \text{ ft}$$

Headloss in manhole

$$H_L = 0.3 \times \frac{7.3^2}{64.4} = 0.25 \text{ ft}$$

Upstream invert elevation pipe 3

$$El = 90.75 - H_L - \Delta D = 90.75 - 0.25 - 0.50 = 90.00 \text{ ft}$$

Downstream invert elevation pipe 3

$$El = 90.00 - 346 \times 0.01 = 86.54 \text{ ft}$$

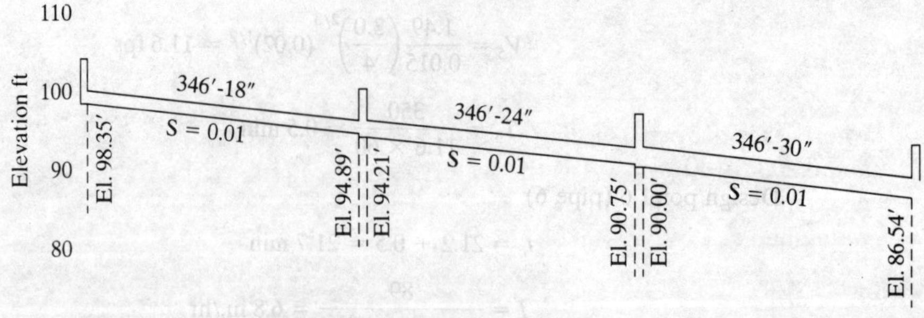

Example 10.5 Profile of lateral 1.

10.2 ON-SITE DETENTION BASINS

As the drainage area is urbanized, the peak discharge rate increases. Because the runoff channels have been hydraulically improved, the runoff occurs faster after urbanization. Because of increased impervious areas caused by pavements, sidewalks, and roofs, the amount of runoff increases. In addition, grading and compacting the landscape and brush and forest litter removal can result in less interception and depression storage of precipitation after urbanization.

Most stormwater management plans require that the peak runoff discharge rate after development be no greater than the peak discharge rate before development. Detention basins can be used to reduce the peak discharge rate after urbanization. Normally, the detention basin is designed to reduce peak discharge rate to predevelopment levels for flood events ranging from a 2-year to a 100-year return period flood level. Generally, roof-top storage should be avoided; particularly for large buildings where wind set-up might cause design depth exceedance and failure of the structure. Because detention basin design requires knowledge of both discharge rates and runoff volumes, the unit hydrograph methodology will be used in the following discussion.

10.2.1 Unit Hydrograph

The Natural Resource Conservation Service (NRCS), formerly the Soil Conservation Service (SCS), developed a dimensionless unit hydrograph based on instrumentation of a large number of natural watersheds representing a wide range of sizes and geographical areas. Although it was developed from natural watersheds, it is frequently applied to urban watersheds. The NRCS dimensionless unit hydrograph is tabulated in Table 8.6 and plotted in Fig. 8.11. The normalized runoff (q/q_p)

is a function of the normalized time (t/t_p), where q_p is the unit hydrograph peak discharge and t_p is the time to peak. The unit hydrograph is computed by multiplying t/t_p in Table 8.6 by t_p and q/q_p in Table 8.6 by q_p. Because the unit hydrograph has a fixed shape as specified in Table 8.6 and because the area under the unit hydrograph in Fig. 8.11 represents one unit of runoff from the watershed, there is a direct relation between q_p and t_p that is

$$q_p = K_u \frac{A}{t_p} \tag{10.12}$$

where q_p is the peak discharge rate in cfs/in. (cms/mm), K_u equals 484 for BG units and 0.208 for SI units, A is the watershed area in mi.^{2} (sq km), and t_p is the time to peak in hours. The time to peak is related to the time of concentration of the watershed (t_c) by

$$t_p = \frac{dt}{2} + 0.6 \times t_c \tag{10.13}$$

where dt is the duration of the rainfall excess and is equal to the computational time step of the runoff model. Because of variation in the rainfall intensity during a storm, the storm is divided into a number of short duration storms, each with a duration dt. To define the rising limb of the unit hydrograph in the runoff model, dt is typically taken as t_p divided by 5 and Eq. 10.13 becomes

$$t_p = 2/3\, t_c \tag{10.14}$$

For hand computations, the curvilinear unit hydrograph is often approximated with a triangular unit hydrograph with a height q_p and a base $2.67t_p$ (Fig. 8.11). Equations 10.12 and 10.14 can also be used with the triangular unit hydrograph to compute the peak discharge rate and time to peak.

Example 10.6 Unit Hydrograph

Plot the before and after development triangular unit hydrograph for a 30-acre watershed if the time of concentration before development was 60 minutes and after development is 40 minutes.

Before development unit hydrograph

$$Q_p = 484 \times \frac{A}{t_p}$$

$$A = \frac{30}{640} = 0.0469 \text{ mi.}^2$$

$$t_p = 0.667 \times t_c = 0.667 \times \frac{60}{60} = 0.667 \text{ hr}$$

$$q_p = 484 \times \frac{0.0469}{0.667} = 34.0 \text{ cfs}$$

Base of unit hydrograph = 2.67 × 0.667 = 1.78 hrs

After development unit hydrograph

$$t_p = 0.667 \times \frac{40}{60} = 0.44 \text{ hr}$$

$$q_p = 484 \times \frac{0.0469}{0.44} = 51.6 \text{ cfs}$$

$$\text{Base of unit hydrograph} = 2.67 \times 0.44 = 1.17 \text{ hrs}$$

Check to see if area of unit hydrograph represents 1 inch of runoff over watershed.

$$\text{Volume of 1 inch of runoff for 30 acres} = \frac{1}{12} \times 30 \times 43{,}560 = 108{,}900 \text{ ft}^3$$

$$\text{Before development U.H. area} = 34.0 \times 1.78 \times \frac{3{,}600}{2} = 108{,}900 \text{ ft}^3$$

$$\text{After development U.H. area} = 51.6 \times 1.17 \times \frac{3{,}600}{2} = 108{,}700 \text{ ft}^3$$

The unit hydrographs are plotted below.

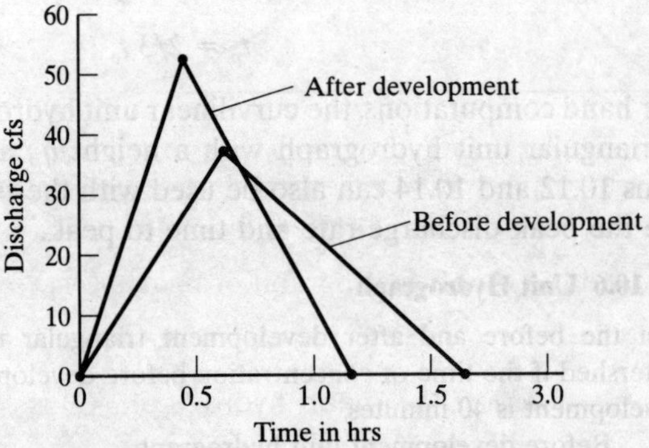

Example 10.6 Unit hydrograph for 30 ac watershed.

10.2.2 Runoff Hydrograph

The runoff hydrograph from the watershed is computed by multiplying the rainfall excess times the unit hydrograph. As shown in Fig. 10.6, the design storm has been divided into *Nr* (four) parts, each with a duration *dt* and a precipitation excess of r_1, r_2, r_3, and r_4, respectively. The unit hydrograph has been approximated with *Nu* (five) ordinates (u_1, u_2, u_3, u_4, and u_5) each spaced *dt* apart along the time axis. The relation between the rainfall excess ordinates (*Nr*), the unit hydrograph ordinates

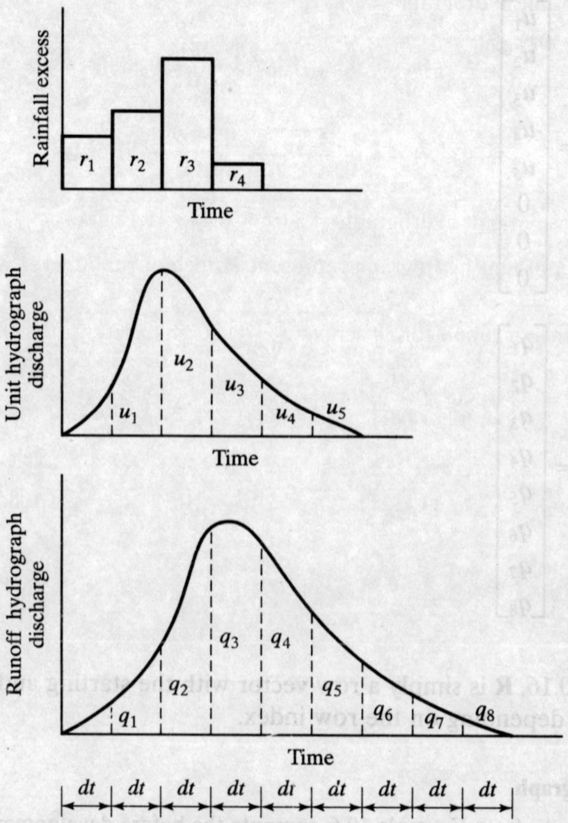

Figure 10.6 Rainfall excess, unit hydrograph and runoff hydrograph.

(Nu), and the runoff hydrograph ordinates (Nq) is

$$Nq = Nr + Nu - 1 \qquad \textbf{(10.15)}$$

In matrix notation, the runoff hydrograph ($\mathbf{Q}$) is computed

$$\mathbf{R}\,\mathbf{U} = \mathbf{Q} \qquad \textbf{(10.16)}$$

where

$$\mathbf{R} = \begin{bmatrix} r_1 & 0 & 0 & 0 & 0 & 0 & 0 & 0 \\ r_2 & r_1 & 0 & 0 & 0 & 0 & 0 & 0 \\ r_3 & r_2 & r_1 & 0 & 0 & 0 & 0 & 0 \\ r_4 & r_3 & r_2 & r_1 & 0 & 0 & 0 & 0 \\ 0 & r_4 & r_3 & r_2 & r_1 & 0 & 0 & 0 \\ 0 & 0 & r_4 & r_3 & r_2 & 0 & 0 & 0 \\ 0 & 0 & 0 & r_4 & r_3 & 0 & 0 & 0 \\ 0 & 0 & 0 & 0 & r_4 & 0 & 0 & 0 \end{bmatrix}$$

$$\mathbf{U} = \begin{bmatrix} u_1 \\ u_2 \\ u_3 \\ u_4 \\ u_5 \\ 0 \\ 0 \\ 0 \end{bmatrix}$$

$$\mathbf{Q} = \begin{bmatrix} q_1 \\ q_2 \\ q_3 \\ q_4 \\ q_5 \\ q_6 \\ q_7 \\ q_8 \end{bmatrix}$$

In the application of Eq. 10.16, **R** is simply a row vector with the starting and ending column index changed depending on the row index.

Example 10.7 Runoff Hydrograph

Using the unit hydrographs from Example 10.6, compute the before development and after development runoff hydrographs for a 60-minute duration storm. The storm was divided into six parts, each with a duration of 0.167 hours. The incremental precipitation excess is listed below for before and after development.

	Precipitation excess	
Time hrs	Before in.	After in.
0	0	0
0.167	0.1	0.3
0.33	0.3	0.5
0.50	0.5	0.8
0.667	0.4	0.6
0.83	0.2	0.4
1.00	0.1	0.2
Total	1.6	2.8

Solution The unit hydrograph (U.H.) values are tabulated below for before development condition and the runoff hydrograph computed.

Time hrs	U.H. cfs	1 0.1	2 0.3	3 0.5	4 0.4	5 0.2	6 0.1	Total cfs
0	0	0	0	0	0	0	0	0
0.17	8	0.8	0	0	0	0	0	0.8
0.33	17	1.7	2.4	0	0	0	0	4.1
0.5	25	2.5	5.1	4.0	0	0	0	11.6
0.67	34	3.4	7.5	8.5	3.2	0	0	22.6
0.83	29	2.9	10.2	12.5	6.8	1.6	0	34.0
1.00	23	2.3	8.7	17.0	10.0	3.4	0.8	42.2
1.17	17	1.7	6.9	14.5	13.6	5.0	1.7	43.4
1.33	11	1.1	5.1	11.5	11.6	6.8	2.5	38.6
1.50	5	0.5	3.3	8.5	9.2	5.8	3.4	30.7
1.67	0	0	1.5	5.5	6.8	4.6	2.9	21.3
1.83			0	2.5	4.4	3.4	2.3	12.6
2.00				0	2.0	2.2	1.7	5.9
2.17					0	1.0	1.1	2.1
2.33						0	0.5	0.5
2.50							0	0

Similarly, the unit hydrograph for the after development watershed is tabulated below and the runoff hydrograph computed.

Time hrs	U.H. cfs	1 0.3	2 0.5	3 0.8	4 0.6	5 0.4	6 0.2	Total cfs
0	0	0.0	0.0	0.0	0.0	0.0	0.0	0.0
0.17	17	5.1	0.0	0.0	0.0	0.0	0.0	5.1
0.33	35	10.5	8.5	0.0	0.0	0.0	0.0	19.0
0.50	52	15.6	17.5	13.6	0.0	0.0	0.0	46.7
0.67	39	11.7	26.0	28.0	10.2	0.0	0.0	75.9
0.83	26	7.8	19.5	41.6	21.0	6.8	0.0	96.7
1.00	13	3.9	13.0	31.2	31.2	14.0	3.4	96.7
1.17	0	0.0	6.5	20.8	23.4	20.8	7.0	78.5
1.33			0.0	10.4	15.6	15.6	10.4	52.0
1.50				0.0	7.8	10.4	7.8	26.0
1.67					0.0	5.2	5.2	10.4
1.83						0.0	2.6	2.6
2.00							0.0	0.0

The two runoff hydrographs are plotted below and show that the peak discharge rate after development was more than twice the peak discharge rate before development.

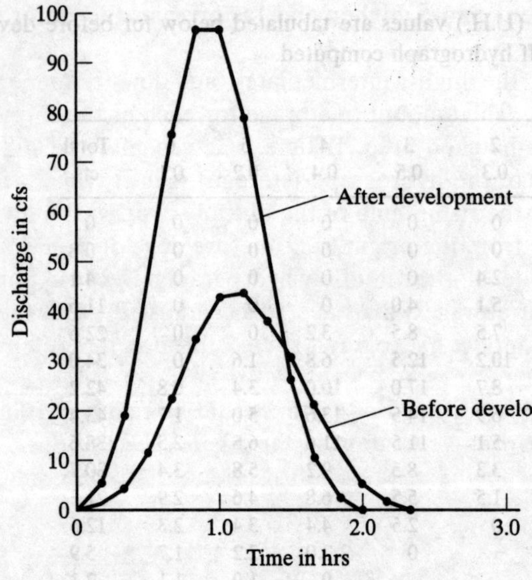

Example 10.7 Runoff hydrograph.

10.2.3 Detention Basin Storage

If all the runoff from the watershed flows through the detention basin, it is called an in-line detention basin. The detention basin is designed so that the peak discharge rate after development is no greater than it was before development for a full range of flood flows. This requires that runoff hydrographs be computed for before development conditions and after development conditions for flood flows ranging from 2- to 100-year return periods. Typically, the 2-, 10-, and 100-year return period precipitation events are used to compute the storage requirements and size the outlet works. The schematic diagram in Fig. 10.7 shows the before and after runoff hydrographs for a specific return period storm. The

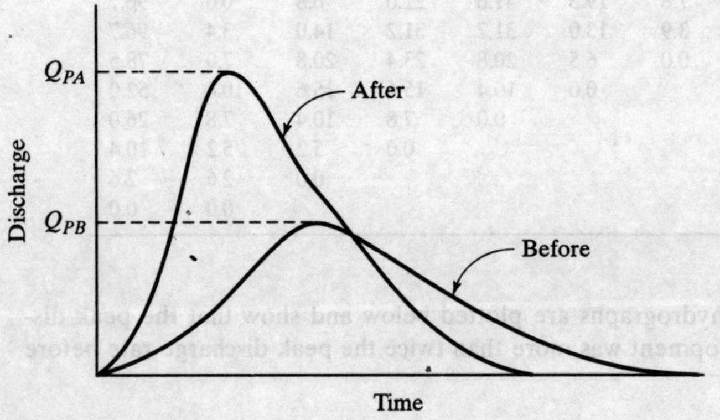

Figure 10.7 Runoff hydrographs before and after development.

runoff hydrograph after development peaks higher and faster than the before development hydrograph.

The storage requirements for the high-, intermediate-, and low-frequency design storms are shown in Fig. 10.8. The amount of storage for each of the three design storms is shown as the cross-hatched area. If time is in seconds and discharge is in cfs (cms), then the area of the hydrograph (storage) is in ft^3 (m^3). In estimating the storage requirements, the rising limb of the outflow hydrograph has been approximated as a straight line from the origin to Q_p before development located on the falling limb of the after development inflow hydrograph. Because this is an approximation, the amount of storage may have to be adjusted after the detention basin is designed and the inflow hydrograph routed through the detention basin.

The side slopes of the basin, if constructed of earth, should be relatively flat (4:1 or flatter) for safety and for mowing vegetation in the basin. The bottom of the basin should be sloped to prevent stagnant water pools and mosquito breeding.

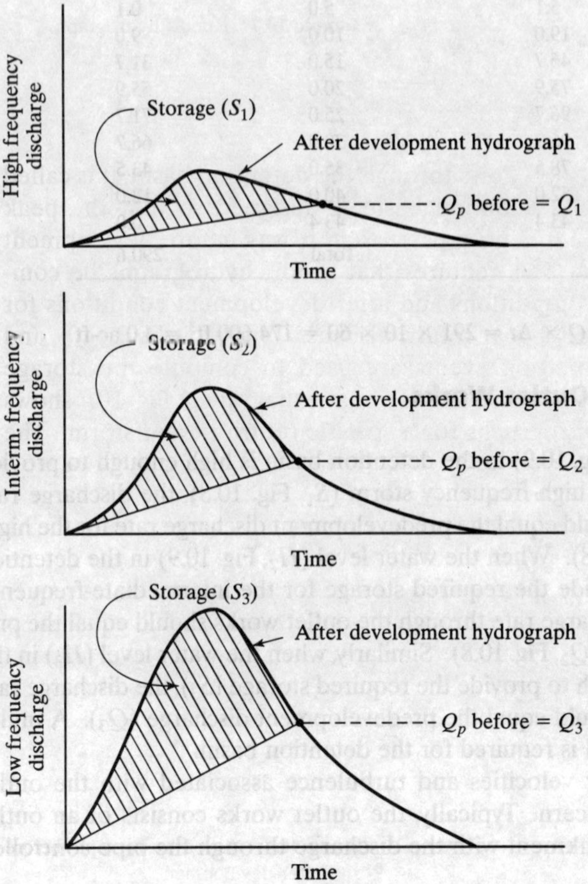

Figure 10.8 Detention basin storage requirements.

Maintenance access to the basin is necessary for trash and sediment removal. The basin need not be a rectangular box, but should be shaped and landscaped to be an attractive open space. Because of the increased frequency of flooding and increased water depth, in-line detention basins (except in arid regions) have limited value as parks or playgrounds.

Example 10.8 Detention Storage

Determine the amount of storage required in a detention basin to reduce the peak discharge rate after development to the before development rate in Example 10.7.

Solution The inflow hydrograph after development is tabulated below, along with the rising limb of the outflow hydrograph that is assumed to be a straight line.

Time hrs	Q_1 Inflow hydrograph cfs	Q_2 Outflow hydrograph cfs	ΔQ Difference cfs
0	0.0	0.0	0.0
0.17	5.1	5.0	0.1
0.33	19.0	10.0	9.0
0.50	46.7	15.0	31.7
0.67	75.9	20.0	55.9
0.83	96.7	25.0	71.7
1.0	96.7	30.0	66.7
1.17	78.5	35.0	43.5
1.33	52.0	40.0	12.0
1.50	43.4	43.4	0.0
		Total	290.6

Storage volume required $= \Sigma \Delta Q \times \Delta t = 291 \times 10 \times 60 = 174{,}600 \text{ ft}^3 = 4.0$ ac-ft.

10.2.4 Detention Basin Outlet Works

When the water level (H_1, Fig. 10.9) in the detention basin is high enough to provide the required storage for the high-frequency storm (S_1, Fig. 10.8), the discharge rate through the outlet works should equal the predevelopment discharge rate for the high-frequency storm (Q_1, Fig. 10.8). When the water level (H_2, Fig. 10.9) in the detention basin is high enough to provide the required storage for the intermediate-frequency storm (S_2, Fig. 10.8), the discharge rate through the outlet works should equal the predevelopment discharge rate (Q_2, Fig. 10.8). Similarly, when the water level (H_3) in the detention basin is high enough to provide the required storage (S_3), the discharge rate through the outlet works should equal the predevelopment discharge (Q_3). A multiple-level outlet works control is required for the detention basin.

Because of high water velocities and turbulence associated with the outlet works, safety is a major concern. Typically, the outlet works consists of an outlet pipe through an earth embankment with the discharge through the pipe controlled

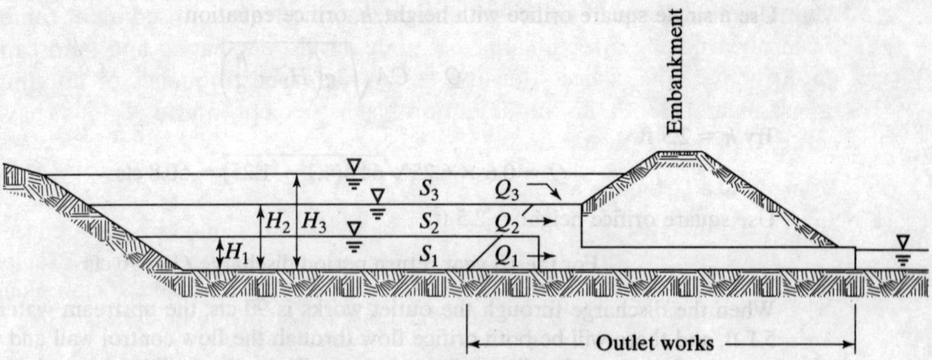

Figure 10.9 Detention basin outlet works.

by an inlet structure at the upstream end of the pipe. The discharge through the outlet works can usually be computed using the orifice, weir, or culvert equations. The trash rack for debris removal can often be combined with safety bars so that children can climb out of the basin. For safety and maintenance, the side slopes on the embankment should be 4:1 or flatter. Because of the high velocity of water through the outlet works, erosion may be a problem in the downstream channel and riprap may be required.

Example 10.9 Outlet Works Design

Design the outlet works of a detention basin so that the peak discharge in the downstream channel after development is no greater than before development. The storage (S) and discharge (Q) requirements for the 2-, 10-, and 100-year return period storms are listed below

Return period yrs	Storage (S) ac-ft	Discharge (Q) cfs	Height* (H) ft
2	2.6	50	4.1
10	3.7	90	5.1
100	5.0	125	6.2

*Height determined from detention basin capacity table.

The outlet pipe is a 120-ft-long concrete pipe ($n = 0.015$) and is set on a grade of 1 percent. The tailwater depth in the downstream channel is 3.0 ft when the discharge through the outlet works is 125 cfs.

For the 2-year return period discharge $Q = 50$ cfs

Use a single square orifice with height, h, orifice equation

$$Q = CA\sqrt{2g\left(H_1 - \frac{h}{2}\right)}$$

Try $h = 2.5$ ft

$$Q = 0.6 \times 6.25\sqrt{64.4(4.1 - 1.25)} = 50.8 \text{ cfs}$$

Use square orifice height = 2.5 ft.

For the 10-year return period discharge $Q = 90$ cfs

When the discharge through the outlet works is 90 cfs, the upstream water depth is 5.1 ft, and there will be both orifice flow through the flow control wall and weir flow over the flow control wall (see figure below). The orifice will become submerged, and the head on the orifice will be the difference in water surface elevation upstream and downstream of the wall. Because the water surface elevation downstream of the wall depends on the size of pipe, it is assumed that the discharge through the orifice remains at 50 cfs. The length of the weir (L) required to discharge 40 cfs is computed from the weir equation

$$Q = K_w L(2g)^{1/2}(H_2 - H_1)^{3/2}$$

where

$$K_w = 0.40$$

$$H_2 = 5.1 \text{ ft}$$

$$H_1 = 4.1 \text{ ft}$$

$$40 = 0.4L(64.4)^{1/2}(1.0)^{3/2}$$

$$L = 12.5 \text{ ft}$$

For the 100-year return period discharge $Q = 125$ cfs

Size the outlet pipe to carry 125 cfs when the upstream water depth is 6.2 ft and the downstream tailwater depth is 3.0 ft. The allowable headloss through the pipe is

$$H_L = H_3 + LS_o - y_{tw} = 6.2 + 120 \times 0.01 - 3.0 = 4.4 \text{ ft}$$

$$H_L = \frac{V^2}{2g}\left(K_e + \frac{29n^2L}{R^{4/3}} + 1.0\right)$$

Considering the trash rack and grated inlet, assume an entrance loss coefficient of 1.0

$$4.4 = \frac{16 \times 125^2}{\pi^2 D^4 2g}\left(2.0 + \frac{29(0.015)^2 120}{(D/4)^{4/3}}\right)$$

$$4.4 = \frac{393}{D^4}\left(2.0 + \frac{4.97}{D^{4/3}}\right)$$

$$D = 4.0 \text{ ft}$$

Drop the invert of the pipe 1.0 ft so that the crown of the pipe is at the downstream tailwater elevation.

Check length of weir when Q = 90 cfs. Because the tailwater depth was not specified for a discharge of 90 cfs, normal depth at the pipe entrance will be assumed.

$$Q_{\text{full}} = \frac{1.49}{n} A R^{2/3} S_o^{1/2} = \frac{1.49}{0.015} 12.57(1)^{2/3}(0.01)^{1/2} = 125 \text{ cfs}$$

$$\frac{Q}{Q_{\text{full}}} = \frac{90}{125} = 0.72$$

From Fig. 5.3

$$y_n = 4.0 \times 0.68 = 2.72 \text{ ft}$$

Referring to Fig. 5.2(b), the area of flow

$$A = \frac{D^2}{4}\left(\frac{\theta}{2} + \sin\alpha\cos\alpha\right)$$

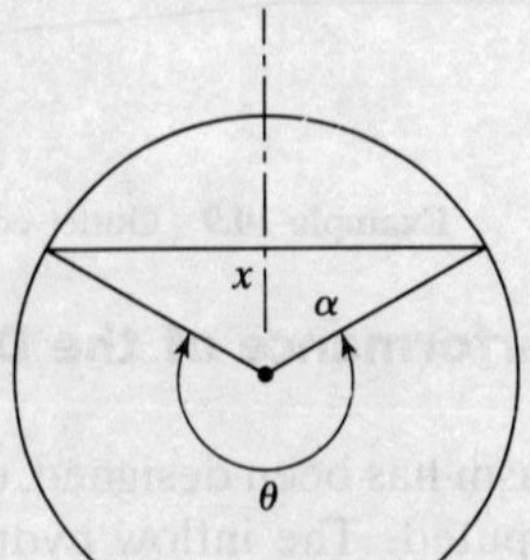

$$A = \frac{4^2}{4}\left(\frac{3.88}{2} + 0.93 \times 0.36\right) = 9.10 \text{ ft}^2$$

$$V_p = \frac{90}{9.10} = 9.9 \text{ fps}$$

Depth at entrance to pipe (just downstream of flow control wall)

$$H = y_n + \frac{V_p^2}{2g} + K_e\frac{V_p^2}{2g} - \Delta z = 2.72 + 1.5 \times \frac{(9.9)^2}{2g} - 1.0 = 4.0 \text{ ft}$$

Check for entrance control using Eq. 4.33

$$Q = A_e C_o \sqrt{2g(H - D/2)}$$

For concrete pipe with the bell end upstream, C_o is assumed to be 0.8.

$$Q = 12.57 \quad 0.8\sqrt{64.4(5 - 2)} = 140 \text{ cfs}$$

Pipe flow is not under entrance control.

The discharge through the orifice in the flood control wall is

$$Q = 0.6 \times A \times \sqrt{2g(5.1 - 4.0)} = 0.6 \times 6.25 \times \sqrt{64.4 \times 1.1} = 32 \text{ cfs}$$

Length of weir based on $Q = 58$ cfs

$$58 = 0.4L(64.4)^{1/2}(1.0)^{3/2}$$

$$L = 18 \text{ ft}$$

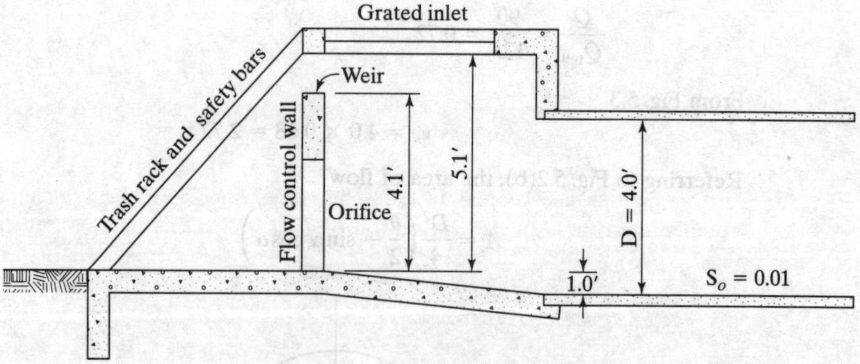

Example 10.9 Outlet control structure.

10.2.5 Evaluate Performance of the Detention Basin

After the detention basin has been designed, elevation versus storage and discharge tables should be computed. The inflow hydrograph after development should be routed through the detention basin for the high-, intermediate-, and low-frequency storm events using Eq. 6.4. The peak discharge on the outflow hydrographs should be compared with the before development peak discharge rates. Some adjustment in the basin size or outlet works may be required to ensure that the peak discharge after development is not greater than it was before development.

Example 10.10 Detention Basin Evaluation

Develop a computer model to evaluate a detention basin design. The program is to use the SCS dimensionless unit hydrograph to compute runoff hydrographs and level pool storage routing to route hydrographs through the detention basin. Evaluate the detention basin design listed in Example 10.9. The watershed area for the example is 30 acres and is located in Harris County, Texas. The soils in the watershed are sandy clay loam.

Solution The program uses the following code to identify model components:

100–299 Detention basins or reservoirs
300–499 Subbasins or watersheds
500–699 Precipitation equations

The rainfall intensity equations used in the program are in the form of

$$I = \frac{a}{(t + b)^c}$$

where I is the rainfall intensity in./hr; t is time in minutes; and a, b, and c are coefficients for the location and rainfall return period. For this problem in Harris County, Texas

ID no.	a	b	c	Return period
500	68	7.9	0.800	2
501	81	7.7	0.753	10
502	91	7.9	0.706	100

A balance triangular rainfall distribution is used in the model to compute incremental precipitation.

The subbasin file includes information needed to compute the unit hydrograph and precipitation excess. It includes the subbasin ID number, precipitation file ID number, area (ac), time of concentration (hr), soil ID number, volumetric soil moisture (VSM) content, surface retention (SR; in.), and percent watershed impervious (IMP). The subbasin file for this example problem is listed below.

ID	PRECIP	AREA	T_p	Soil ID	VSM	SR	IMP
300	500	30.0	1.0	6	0.16	1.0	0
301	501	30.0	1.0	6	0.16	1.0	0
302	502	30.0	1.0	6	0.16	1.0	0
303	500	30.0	0.67	7	0.16	0.5	35
304	501	30.0	0.67	7	0.16	0.5	35
305	502	30.0	0.67	7	0.16	0.5	35

Subbasin ID numbers 300–302 represent predevelopment conditions for the 2-, 10-, and 100-year storm events, whereas subbasin ID numbers 303–305 represent developed conditions for the 2-, 10-, and 100-year storm events.

The model uses the Green and Ampt infiltration equation to compute the precipitation excess where the potential infiltration (f) is

$$f = K_s\left(1 + \frac{(n - \text{VSM})\text{WFS}}{F}\right)$$

K_s is the hydraulic conductivity of the soil (in./hr), n is the soil porosity, VSM is the volumetric soil moisture, WFS is the wetting front suction, and F is the cumulated infiltration. The Green and Ampt parameters (K_s, n, WFS) are read from a file (SOILS.DAT), and the user only needs to specify soil type (see program listing for soil code).

The Green and Ampt parameters depend on the grain size classification shown in the triangle of soil textures developed by the NRCS shown below. The triangle shows the fraction of clay, silt, and sand for each soil texture class. For example, loam would have approximately 40 percent sand, 40 percent silt, and 20 percent clay.

Soils that develop in the watershed depend upon the geology, geological processes, and climate of the area. In the development of the soil profile over many years, the surface layer (called the A horizon) tends to accumulate organic material and tends to be loamy (leached of clay particles). The A horizon generally varies in thickness from less than 0.5 ft (0.15 m) to over 1.0 ft (0.30 m). The second layer (B horizon) tends to accumulate the fine particles leached from the A horizon. As a result, the hydraulic characteristics of the two layers tend to be very different. For a minor storm event, the infiltration tends to be controlled by the characteristics of the A horizon, whereas for a major storm event, the infiltration tends to be controlled by the characteristics of the B horizon. A two-layer Green and Ampt infiltration model can be used to more accurately simulate the infiltration process (James, Warriner, and Reedy, 1992). The hydraulic properties of the soil layers can be estimated from the NRCS soil surveys available for most counties in the United States.

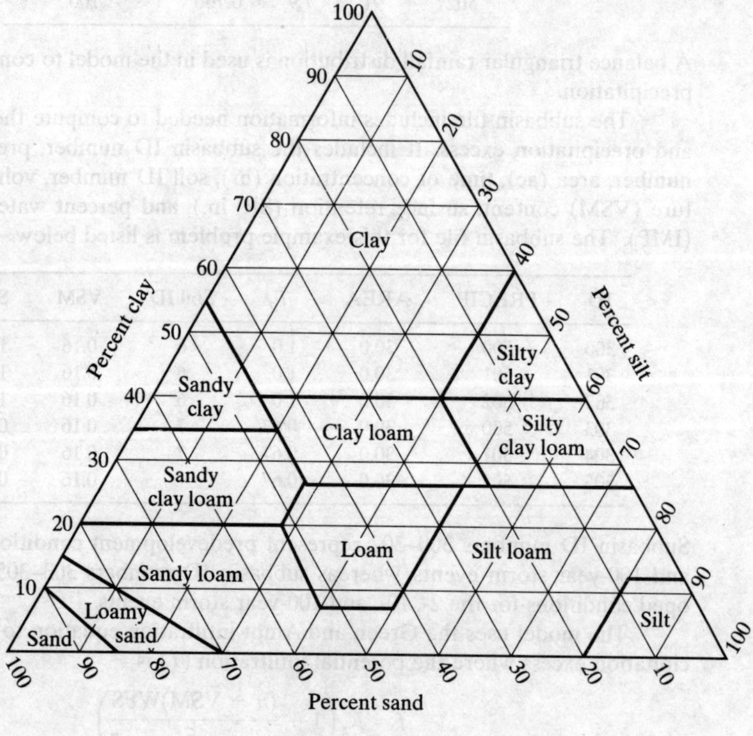

Triangle of soil textures (NRCS)

The detention basin file (RES.DAT) contains the information required to route a hydrograph through the basin. For this example problem, the detention basin was labeled 200, and the file is listed below for a starting water surface elevation of 300.0 ft and 5 lines of data (elevation, storage, and discharge).

200	300.0	5
300.0	0.0	0.0
304.1	2.6	50.0
305.1	3.7	90.0
306.2	5.0	125.0
308.2	12.0	200.0

The model does only three hydrologic operations (compute hydrograph, store hydrograph, and route hydrograph). The command file (COMMAND.DAT) is a listing of the hydrologic operations in the sequence they are to be performed using the following code

C 301	Compute hydrograph for subbasin 301
S 301	Store hydrograph and zero array in memory
S 1301	Store hydrograph, but don't zero array in memory
R 200	Route hydrograph in memory through reservoir 200

For the example problem, dt = 0.0833 hrs and the storm duration is 12 hrs.

Results

The program was run with the following results (DETENT.OUT)

Return period	PRE DEV Q_p cfs	DEV INFLOW Q_p cfs	DEV OUTFLOW Q_p cfs	MAX WSEL RES ft
2 yr	55	88	50	303.8
10 yr	94	130	90	305.1
100 yr	135	175	125	306.2

The peak discharge rates after development are less than predevelopment peak discharge rates, and the design is acceptable. The computed hydrographs are shown in the plots below.

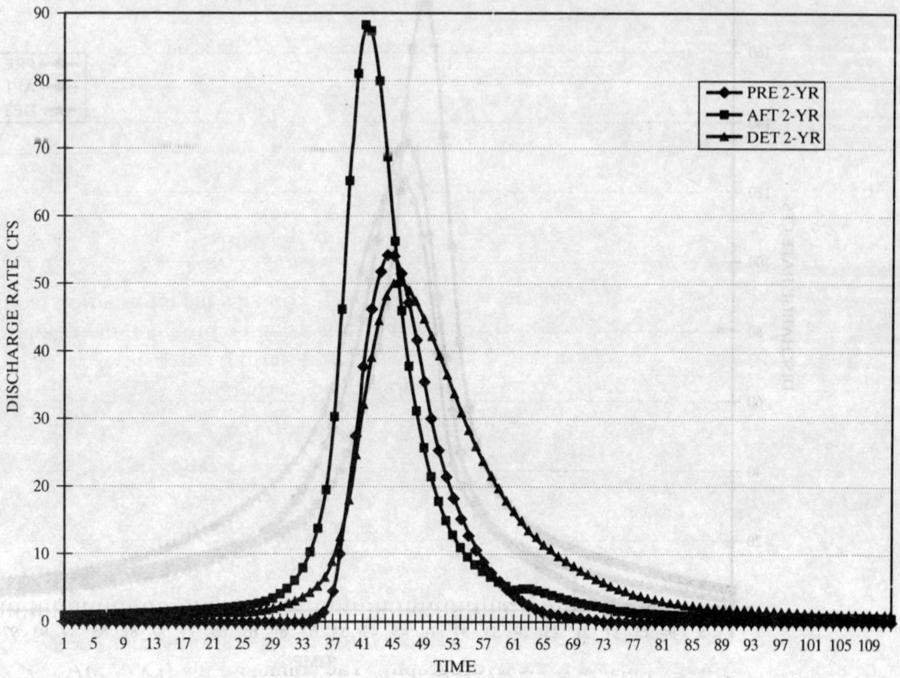

2-year return period detention basin design hydrographs.

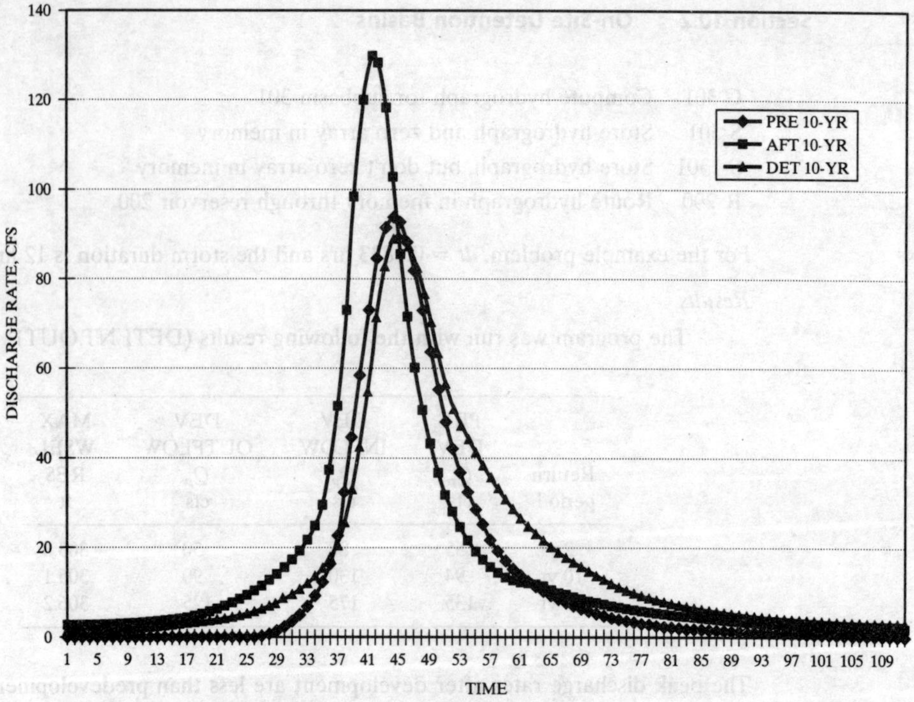

10-year return period detention basin design hydrographs.

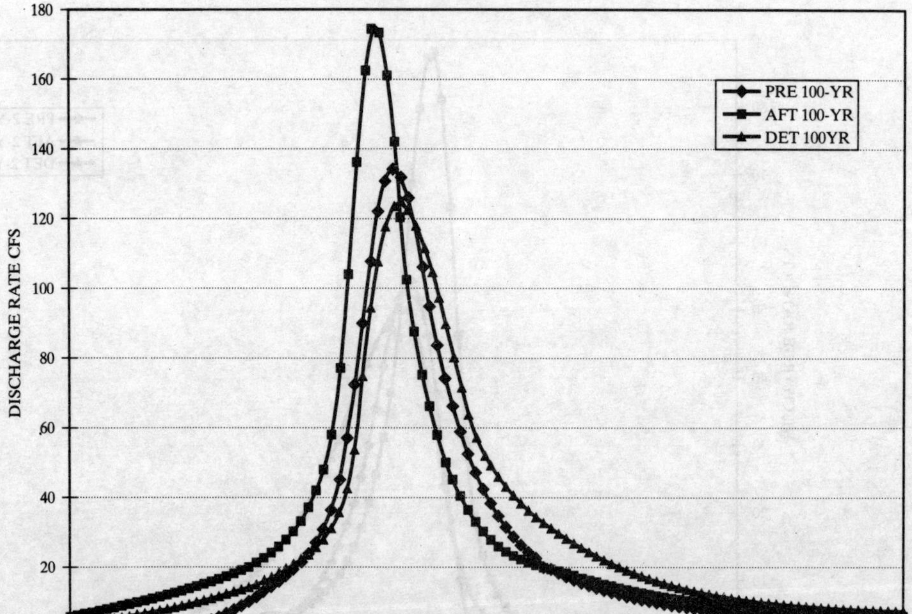

100-year return period detention basin design hydrographs.

10.3 REGIONAL DETENTION FACILITIES

Figure 10.10(a) shows on-site detention basins located on first-order stream channels that discharge into a second-order channel. On-site detention basins are designed to ensure that the after development peak discharge is no greater than the predevelopment peak discharge in the channel immediately downstream of the basin. As the distance downstream from the on-site detention increases, the effectiveness in reducing the peak discharge rate decreases. If the objective is to reduce the peak discharge at the outlet of the second-order watershed in Fig. 10.10(a), then on-site detention in the lower part of the watershed (basins E and F) should not be constructed, because they will tend to increase the peak discharge at the outlet. Because of the timing of the runoff, runoff from the lower part of the watershed tends to occur before the runoff from the upper part of the watershed. In general, for a given amount of storage, the maximum reduction in the peak discharge rate at the watershed outlet will occur when detention facilities are located in the upper part of the watershed to delay the runoff and the channels in the lower part of the watershed are hydraulically improved to enhance the runoff from the area.

Figure 10.10(b) shows on-site detention facilities located on first-order stream channels in a third-order watershed. If the objectives of the detention facilities is to reduce the peak discharge rate at the outlet of a third-order watershed, the detention facilities on tributaries 5 and 6 should not be constructed, because they will delay the runoff from the lower watershed and increase the peak discharge rate at the outlet. As the order of the stream channel increases, the effectiveness of on-site detention to reduce the peak discharge rate becomes less.

To reduce the flood peak on the main channel in a large urbanized watershed, a regional detention facility may be required. If the downstream channel has a limited capacity and the objective of the regional detention facility is to limit the

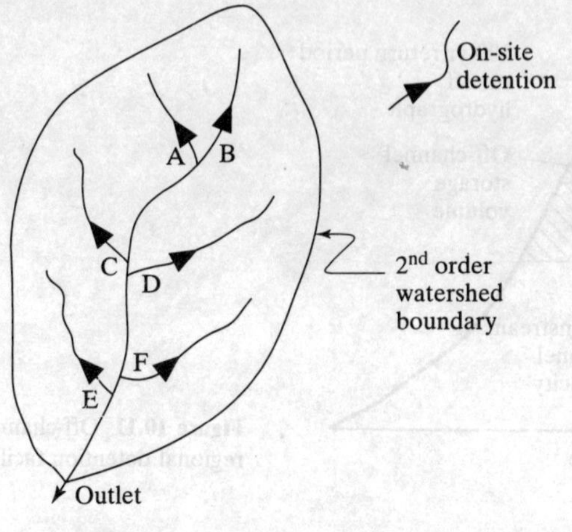

Figure 10.10(a) On-site detention facilities located on first-order channels in a second-order watershed.

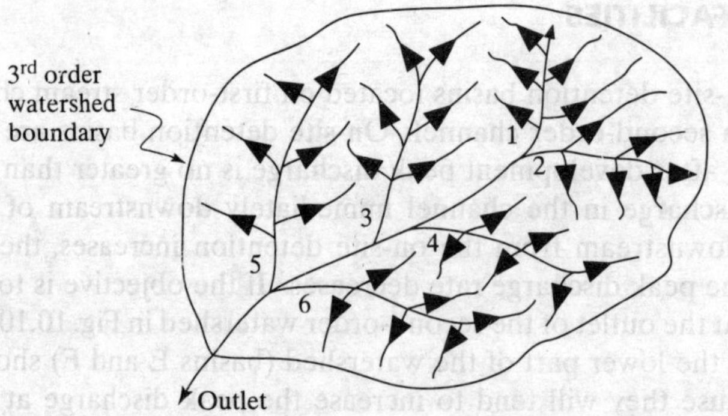

Figure 10.10(b) On-site detention facilities located on first-order channels in a third-order watershed.

discharge in the downstream channel to that capacity, then off-channel detention should be considered. With off-channel detention, only the runoff that is in excess of the downstream channel capacity is diverted into the detention basin and stored (Fig. 10.11).

Off-channel detention has several advantages over on-line detention, including: (1) less storage volume, (2) less frequent flooding of the basin area that can be used for athletic fields and playground, and (3) less prone to filling with sediments, reducing the associated cost of sediment removal.

If the downstream channel has a capacity to handle a 10-year return period discharge rate, then the detention basin should be used to store floodwater on the average of only once in 10 years and should not create a problem for scheduled athletic activities. The detention storage basin must be graded so that the water depth gradually decreases away from the river and there is no danger of someone being trapped in the basin during filling.

As shown in Fig. 10.12, a side-channel weir can be used to control the discharge into the off-channel storage basin. Because the highest concentration of suspended sediments in the runoff is in the main channel near the bottom, most of the suspended sediments will bypass the detention basin. The water stored in the detention basin

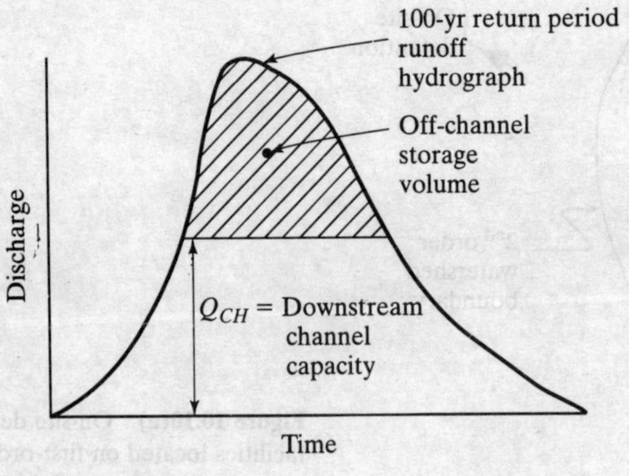

Figure 10.11 Off-channel storage for regional detention facility.

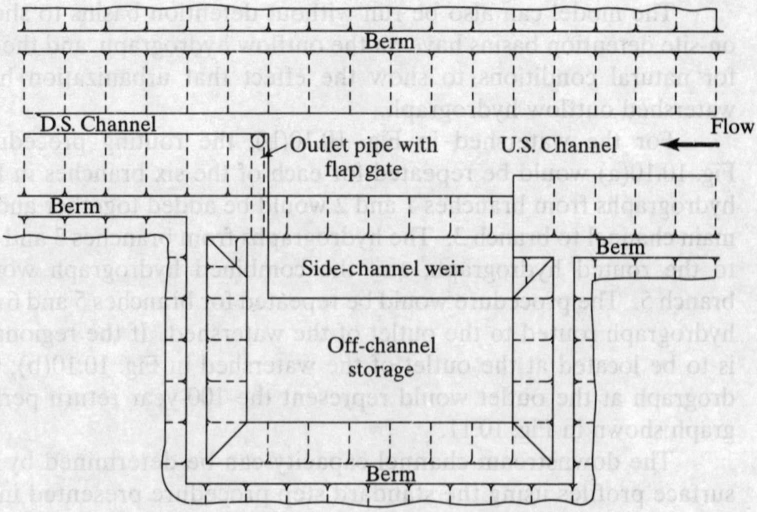

Figure 10.12 Plan view of off-channel detention basin.

can be drained back into the channel after the water level in the channel is less than the water level in the detention basin through a conduit with a flap gate at the channel end (Fig. 10.12).

10.3.1 Watershed Modeling

Watershed modeling will be necessary to compute the 100-year return period hydrograph required to design the regional detention facility. The watershed is divided into subbasins. To show the effect the on-site detention basins have on the outflow hydrographs, subbasins used in the model should be delineated so that the drainage areas of the on-site detention basins are represented.

For the watershed in Fig. 10.10(a), the runoff hydrograph for subbasin A would be computed (Eq. 10.16) based on a 100-year return period rainfall event and routed through the detention basin (Eq. 6.4). The procedure would be repeated for subbasin B. The two hydrographs would be added together and routed along the main channel to the next major downstream tributary using hydraulic routing presented in Section 6.3. The procedure of computing the runoff hydrographs and routing through the detention basins would be repeated for branches C and D. The two hydrographs would then be added to the routed hydrograph and the combined hydrograph would be routed along the main channel to branch E. The procedure of computing hydrographs, reservoir routing, and combining hydrographs would be repeated for branches E and F. The combined hydrograph would be routed along the main channel to the outlet. Runoff from other areas within the watershed that were not included in the six subbasins (if any) can be computed as necessary and added to the hydrograph as appropriate.

The model can also be run without detention basins to show the effect that on-site detention basins have on the outflow hydrograph, and the model can be run for natural conditions to show the effect that urbanization has on the runoff watershed outflow hydrograph.

For the watershed in Fig. 10.10(b), the routing procedure described for Fig. 10.10(a) would be repeated for each of the six branches in Fig. 10.10(b). The hydrographs from branches 1 and 2 would be added together and routed along the main channel to branch 3. The hydrographs from branches 3 and 4 would be added to the routed hydrograph, and the combined hydrograph would be routed to branch 5. The procedure would be repeated for branches 5 and 6 and the combined hydrograph routed to the outlet of the watershed. If the regional detention basin is to be located at the outlet of the watershed in Fig. 10.10(b), the combined hydrograph at the outlet would represent the 100-year return period runoff hydrograph shown in Fig. 10.11.

The downstream channel capacity can be determined by computing water surface profiles using the standard step procedure presented in Section 5.7.3 for a range of discharge rates. The downstream channel capacity is the largest discharge rate that results in acceptable water levels in the channel. The volume of off-channel storage required can be computed as the cross-hatched area in Fig. 10.11.

Example 10.11 Watershed Modeling

Assume the watershed in Fig. 10.10(a) is divided into six identical subbasins each 30 acres in size with the same subbasin characteristics as Example 10.10. Determine the peak discharge rate at the outlet of the watershed for a 100-year return period, 12-hour duration precipitation event for the following conditions: (1) predevelopment, (2) developed without detention, (3) developed with detention, and (4) developed with detention in subbasins A through D.

The channel in Fig. 10.10(a) is divided into three routing reaches, each 2,650 ft long with a slope of 0.001. The upper reach is labeled 710 and 700, the middle reach is 711 and 701, and the lower reach is 712 and 702 for predeveloped and developed conditions, respectively. For predeveloped conditions, the Manning "*n*" value for the channel is 0.06 and for the floodplain is 0.15. For after development conditions, the Manning "*n*" value is 0.03 for both the channel and floodplain. The cross-sections for each routing reach are listed in input file ROUTE.FIL.

Solution The model developed for Example 10.10 was modified by adding two additional commands

A Add a hydrograph that has been previously stored

R Route a hydrograph through a stream channel

Stream channel routing reaches are identified with ID numbers ranging from 700 to 999. The command *R* 700 will cause the computer to route the hydrograph in memory through stream routing reach 700 using the hydraulic routing procedure presented in Chapter 6.

ROUTE.FIL (predevelopment)

710	10	2.000	0.15	0.06	0.15	2,650.	0.001
345.	335.	334.	330.	330.	334.	335.	345.
0.	10.	20.	27.	28.	35.	45.	55.
711	10	2.000	0.15	0.06	0.15	2,650.	0.001
342.	334.	332.	328.	325.	332.	333.	342.
0.	10.	25.	30.	33.	38.	50.	60.
712	10	2.000	0.15	0.06	0.15	2,650.	0.001
340.	330.	327.	324.	320.	328.	329.	340.
0.	50.	60.	65.	70.	75.	90.	100.

ROUTE.FIL (after development)

700	10	2.000	0.03	0.03	0.03	2,650.	0.001
345.	335.	334.	330.	330.	334.	335.	345.
0.	10.	20.	27.	28.	35.	45.	55.
701	10	2.000	0.03	0.03	0.03	2,650.	0.001
342.	334.	332.	328.	325.	332.	333.	342.
0.	10.	25.	30.	33.	38.	50.	60.
702	10	2.000	0.03	0.03	0.03	2,650.	0.001
340.	330.	327.	324.	320.	328.	329.	340.
0.	50.	60.	65.	70.	75.	90.	100.

PRECIP.FIL

500	68.000	7.900	0.800
501	81.000	7.700	0.753
502	91.000	7.900	0.706

RES.FIL

200	300.0	5
300.0	0.0	0.0
304.1	2.6	50.0
305.1	3.7	90.0
306.2	5.0	125.0
308.2	10.0	200.0

WATSHD.FIL (predevelopment)

310	502	30.000	1.000	6	0.16	1.0	0.0
311	502	30.000	1.000	6	0.16	1.0	0.0
312	502	30.000	1.000	6	0.16	1.0	0.0
313	502	30.000	1.000	6	0.16	1.0	0.0
314	502	30.000	1.000	6	0.16	1.0	0.0
315	502	30.000	1.000	6	0.16	1.0	0.0

WATSHD.FIL (after development)

300	502	30.00	0.667	7	0.16	0.5	35.0
301	502	30.00	0.667	7	0.16	0.5	35.0
302	502	30.00	0.667	7	0.16	0.5	35.0
303	502	30.00	0.667	7	0.16	0.5	35.0
304	502	30.00	0.667	7	0.16	0.5	35.0
305	502	30.00	0.667	7	0.16	0.5	35.0

Results

Peak discharge at the outlet of the watershed [Fig. 10.10(a)] are

Part 1 Predevelopment

$$Q = 475 \text{ cfs}$$

Part 2 After development without detention basins

$$Q = 860 \text{ cfs}$$

Part 3 After development with detention basins in all subbasins

$$Q = 665 \text{ cfs}$$

Part 4 After development with detention basins in subbasins A through D

$$Q = 615 \text{ cfs}$$

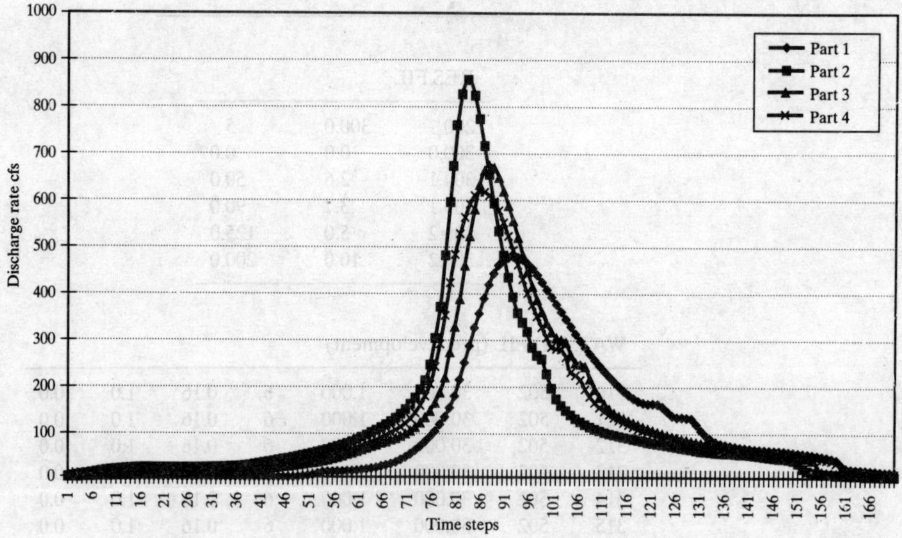

Example 10.11 Watershed modeling hydrographs..

Discussion

The detention basin was designed to decrease the peak discharge from the basin to below the predevelopment peak discharge rate (Example 10.10). However, the peak discharge from the watershed in Fig. 10.10(a) after development and with detention basins in all subbasins was greater than the predevelopment peak discharge. When the detention basins were removed from subbasins E and F, the peak discharge from the watershed after development was less than the after development conditions with detention basins in all subbasins.

10.3.2 Side-Channel Weir

The initial estimate of the height of the side-channel weir above the channel bottom (H_w) can be established as the water depth in the channel when the discharge rate is equal to the downstream channel capacity. To reduce the length of the side-channel weir, the height of the weir may be adjusted. The crest of the weir is normally placed parallel to the bottom of the channel. If the discharge in the channel is greater than the downstream channel capacity, the discharge (q_ℓ) over the weir per incremental length (Δx) can be estimated as

$$q_\ell = -K_w \Delta x \sqrt{2g}(y - H_w)^{3/2} \tag{10.17}$$

where K_w is the weir discharge coefficient and y is the water depth in the channel. Flow over the side-channel weir is shown in Fig. 10.13.

During a flood event, the flow in the channel is spatially varied unsteady flow. The hydraulics of a side-channel weir is complex, and the required length of the weir cannot be determined with a simple formula. Because of the drawdown effect of the weir, the depth of flow upstream (y_1) of the weir is less than the normal depth of flow in the upstream channel. To determine the water depth in the channel (y) and the weir discharge (q_ℓ), the water surface profile in the channel must be computed using the unsteady flow equations. The weir length can be determined by trial and error. The weir design can be evaluated using the hydraulic channel routing procedure presented in Section 6.3, except the lateral inflow (q_ℓ) in Eq. 5.88 is computed using Eq. 10.17, with Δx equal to one. The upstream boundary condition will be the inflow hydrograph, and the downstream boundary condition will be a

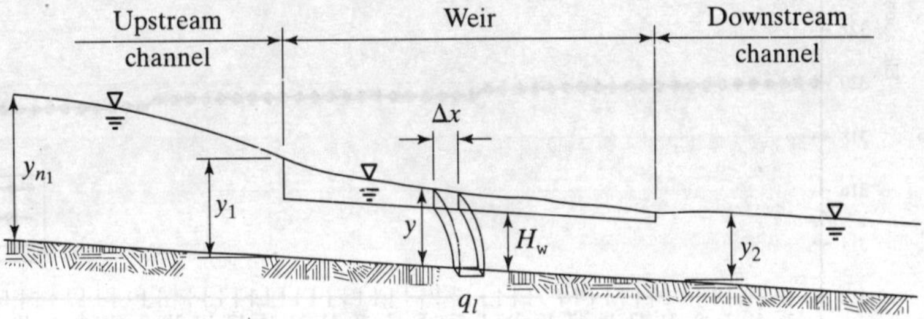

Figure 10.13 Flow over side-channel weir.

rating table based on the Manning equation. The design hydrograph for the 100-year return period storm can be routed through the channel and the downstream hydrograph computed. The side-channel weir design can be modified as necessary until the downstream discharge is within the channel capacity.

Example 10.12 Side-Channel Weir

A side-channel weir is 250 ft long with a crest height above the bottom of the channel of 6.0 ft and a discharge coefficient of 0.40. There are three channels in series. Channel 1 is the upstream channel, channel 2 is the weir section, and channel 3 is the downstream channel. The channel size, length, roughness, and slope are listed below.

Channel	Bottom width ft	Side slope	Length ft	Manning "n"	Channel slope
1	8.0	3.0	250	0.03	0.001
2	8.0	2.0	250	0.03	0.0005
3	4.0	3.0	250	0.03	0.0005

The inflow hydrograph was computed in part 2 of the previous example. Route the inflow hydrograph through the three reaches, and compute the outflow hydrograph for reach 3 and the hydrograph for the flow over the weir.

Solution The FORTRAN Program HYRT2.FOR was modified to route an inflow hydrograph through three channels in series with the side channel weir as the center channel. The modified program is called HYWEIR.FOR.

Results

The inflow, outflow, and weir flow hydrographs are plotted in the attached figure. The water surface profile at peak discharge rate is also plotted and shown in the attached figure.

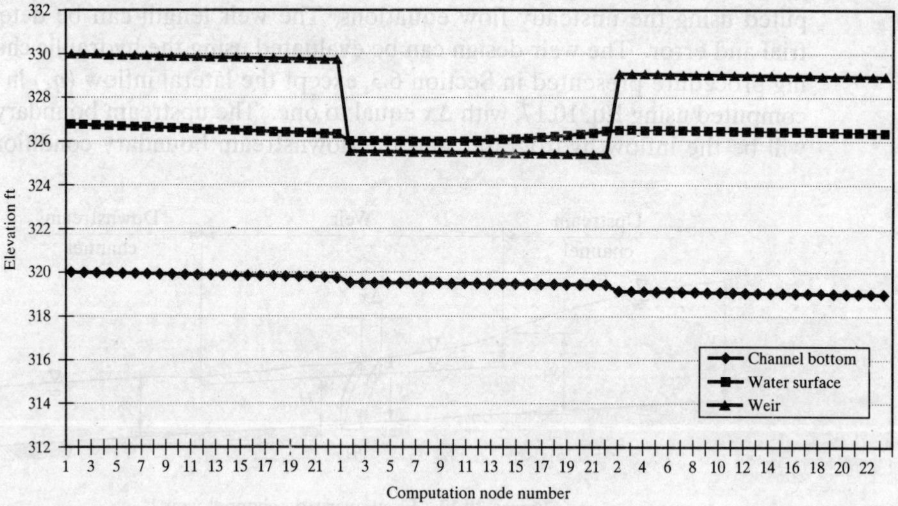

Water surface profile at peak discharge.

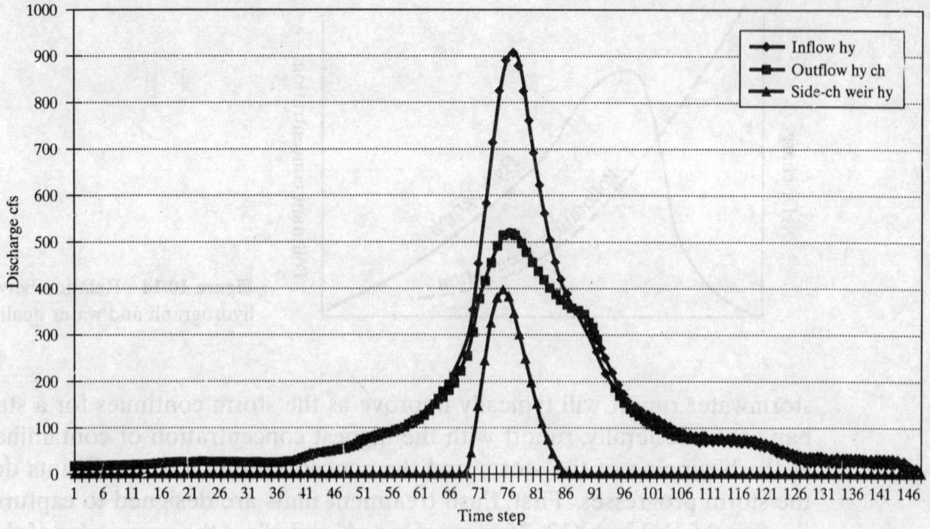

Hydrographs for side-channel weir.

10.4 WATER QUALITY

Prior to urbanization, runoff flows into channels as sheet or shallow swale flow where the land surface is typically covered with vegetation. The vegetation controls erosion and acts as a biofilter to trap sediment in the runoff. Urbanization will adversely affect the quality of stormwater runoff because of change in land cover, urban activities, and air pollution. Rain tends to wash out the pollutants in the atmosphere and the water quality of the rain that falls in the urban area is often reduced. After an area is urbanized, much of the runoff is from lawns, often with excess application of fertilizers and chemicals and from impervious surfaces with a buildup of atmospheric pollution fallout, human litter, tire and brake wear, oil drippings, and exhaust from vehicles.

There are typically three phases of urbanization that affect the water quality of stormwater runoff. During the initial phase of urbanization, the dominant source of pollution will be sediments from bare soil areas at construction sites. During the intermediate phase of urbanization, sediments from construction sites will decline, but sediments from stream bank erosion will increase because of the increased runoff rate and volume. During the mature phase of urbanization (when the stream channels have stabilized and there is limited new construction), the primary source of pollution will be from washoff of accumulated deposit on impervious surfaces.

The buildup of pollutants on the surface occurs between rainfall events, and the pollutants are washed off during rainfall events. In general, the greater the percent impervious and the longer the time since the last rainfall, the higher the concentration of pollutants in the runoff. As shown in Fig. 10.14, the quality of the

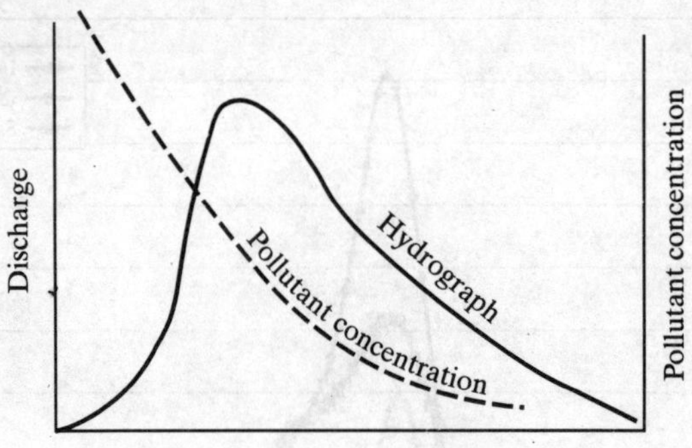

Figure 10.14 Relation between runoff hydrograph and water quality.

stormwater runoff will typically improve as the storm continues for a stabilized urban area. Generally, runoff with the highest concentration of contaminants occurs at the beginning of the storm and the concentration of contaminants decreases as the storm progresses. First, flush treatment units are designed to capture and treat the first 0.5–1.0 inch (13–25 mm) of runoff and allow the remainder of the runoff to bypass the treatment unit.

The total rainfall amount has little effect on the amount of sediment yield that occurs from impervious surfaces. Once the sediment on an impervious surface has been detached and transported away, it does not matter how much additional rainfall occurs. The amount of sediment yield during a rainfall event depends primarily on the antecedent time since the last rainfall event.

Sedimentation is the primary pollutant removal process used in stormwater quality management. Typical removal rates of pollutants in urban stormwater runoff by sedimentation are listed in Table 10.2.

The design of detention facilities for flood control is different from that for water quality control. Flood control detention facilities are designed to reduce the peak discharge rate for large, less frequent runoff events, whereas water quality

TABLE 10.2 TYPICAL REMOVAL RATES OF POLLUTANTS IN URBAN STORMWATER RUNOFF BY SEDIMENTATION

Pollutant	Removal rate %
Total suspended solids	50–70
Total phosphates	10–20
Nitrogen	10–20
Organic matter	20–40
Pb	75–90
Zn	30–60
Hydrocarbons	50–70
Bacteria	50–90

Adapted from Urbonas and Stahre (1993).

control detention facilities are designed to improve water quality based primarily on sedimentation for the small, more frequent runoff events. For a stabilized urban area, the runoff from the smaller storms generally will have a higher concentration of pollutants than the runoff from larger storms. The mean storm precipitation depth in the United States ranges from 0.4 to 0.8 inches (10–20 mm). A water quality control detention facility sized to capture the first flush runoff volume will typically be able to treat 80–90 percent of the annual runoff. A detention time of approximately 24 hours will be required to achieve good removal efficiencies by sedimentation.

10.4.1 Empirical Equations

Empirical equations used to predict pollutant loadings are presented in this section to provide an understanding of best management practices for improving water quality. The equations are not necessarily accurate for a specific area, but provide a better understanding of the factors that affect the quality of the runoff.

10.4.1.1 Unit loading. The water quality of urban runoff varies with land use. Annual unit loads for TSS (total suspended solids), BOD (biochemical oxygen demand), total N (nitrogen), and total P (phosphorus) are listed in Table 10.3 for residential, commercial, industrial, transportation, recreational, and construction land use. The table was adapted from the Southeastern Wisconsin Regional Planning Commission in 1978 and shows that the runoff from construction sites and transportation areas tend to have the highest concentration of pollutants, while runoff from recreational areas tend to have the lowest concentration of pollutants. The table indicates that urban runoff from different land use areas will not have the same water quality and perhaps should be treated differently. Runoff from bare soil areas and construction sites should be treated on site and the sediments not allowed to enter the stormwater drainage system. Runoff associated with impervious surfaces, high traffic areas, and high population density areas may require treatment before being allowed to mix with other less polluted runoff.

TABLE 10.3 ANNUAL UNIT POLLUTION LOADS IN THE KINNICKINNIC RIVER WATERSHED

Land use	TSS lbs/ac	BOD lbs/ac	N lbs/ac	P lbs/ac
Residential	550	25	4	0.3
Commercial	750	100	9	0.8
Industrial	1,000	40	8	0.7
Transportation	3,000–40,000	20–160	10–20	1–3
Recreation	400	1	3	0.1
Construction	150,000	120	60	45

10.4.1.2 Preliminary screening procedure. Heaney, Haber, and Nix (1976) developed the preliminary screening procedure for estimating the unit loading of pollutants based on data from the Environmental Protection Agency Nationwide Urban Runoff Program. The basic equation is

$$U = u(i,j) \times P_a \times \text{PDF} \times \text{SWF} \tag{10.18}$$

where

- U average annual amount of pollutant j generated per unit area of land use i
- $u(i,j)$ load of pollutant j generated per unit of precipitation per unit area of land use i (Table 10.4)
- P_a average annual precipitation
- PDF a dimensionless population density factor has a value of 1.0 for commercial and industrial areas; a value of 0.14 for parks, cemeteries, and schools; and a value of $0.14 + 0.218\ (\text{PD})^{0.54}$ for residential areas, where PD is the population density in persons per acre
- SWF a dimensionless street-sweeping factor has a value of 1.0 if the average time between street sweeping (SS) is greater than 20 days and a value SS/20 if SS is less than 20 days.

Equation 10.18 indicates that population density is a factor in estimating the amount of pollutant in the runoff from a residential area and the street sweeping can be an effective management practice for improving the water quality of urban runoff.

10.4.1.3 Universal soil loss equation. The Universal Soil Loss Equation (USLE) was developed to estimate annual sediment yield from agricultural land caused by sheet erosion. The USLE is included so that the relative effects of various erosion control measures can be evaluated as management practices for urban water quality improvement. The USLE is

$$E = A \times R \times K \times LS \times C \times P \tag{10.19}$$

TABLE 10.4 UNIT LOAD OF POLLUTANT *j* FOR LAND USE *i* *u(i,j)* (lbs PER ac-in. OF PRECIPITATION)

	Pollutant (*j*)			
Land use (*i*)	TSS	BOD	N	P
Residential	16	0.8	0.13	0.03
Commercial	22	3.2	0.30	0.07
Industrial	29	1.2	0.28	0.07
Other developed	2.7	0.1	0.06	0.01

Adapted from Heaney *et al.* (1976).

TABLE 10.5 USLE *K* FACTOR BASED ON SOIL TYPES*

Soil	Soil erodibility (*K*)[†]
Sand and gravel	0.10
Loamy coarse sand, sand and fine sand	0.15
Loamy fine sand and loamy sand	0.17
Fine sandy loam and sandy loam	0.22
Loam, clay loam and sandy clay loam	0.30
Silt-loam and silty clay loam	0.34
subsoil	0.43
Clay and silty clay	
<50% clay	0.32
>50% clay	0.28

*Adapted from Walesh (1989).

[†]Units are tons/acre per erosion index unit.

where

E the annual soil loss caused by water erosion in the rill and inter-rill areas (tons/yr)

A the area (acres)

R a rainfall factor that depends on geographic location and accounts for the erosion force of rainfall and runoff (Fig. 8.14)

K the soil erodibility factor that depends on soil type (Table 10.5 and Eq. 8.55)

LS a dimensionless length-slope factor reflecting the effects of length and steepness on soil erosion (Fig. 8.15)

C a dimensionless land cover and management factor (Table 8.7)

P_c a dimensionless erosion control practice factor (Table 10.6).

Table 8.7 indicates that straw mulch or other land surface treatments (hydromulch, surface fabric, etc.) can be very effective in reducing soil erosion from bare soil areas and that grass cover should be established as soon as possible. Table 10.6 indicates that contouring and terracing the landscape can be beneficial as a management practice in improving water quality of the runoff from pervious surfaces.

TABLE 10.6 EROSION CONTROL PRACTICE FACTOR FOR USLE

	1–12% Slope
No control practice	1.0
Contouring	0.6
Terracing	0.12

Adapted from Walesh (1989).

When the site landscaping is being planned, a long steep slope with a high *LS* factor should be avoided.

Example 10.13 Sediment Yield from Construction Site

A construction site is 1,000 ft long and 500 ft wide. The soils are silty clay loam and the site slopes 2 percent for a length of 500 ft. The site is located in Denver, Colorado, and no erosion control measures are used. Estimate the sediment yield from the construction site for bare soil.

Solution

$$E = A \times R \times K \times LS \times C \times P$$

$$A = 500 \times 1{,}000/43{,}560 = 11.5 \text{ ac}$$

$$R = 50 \text{ erosion index units/yr (Fig. 8.14)}$$

$$K = 0.34 \text{ tons/ac per erosion index unit (Table 10.5)}$$

$$LS = 0.3 \text{ (Fig. 8.15)}$$

$$C = 1.0 \text{ (Table 8.7)}$$

$$P_c = 1.0 \text{ (Table 10.6)}$$

$$E = 11.5 \times 50 \times 0.34 \times 0.3 \times 1.0 \times 1.0 = 59 \text{ tons/yr}$$

10.4.2 Stream Channel Mechanics

This section provides the student with a better understanding of the effects of channel modifications on the remainder of the stream. A channel whose width, depth, slope, roughness and alignment remains constant is called a rigid channel. However, most natural channels are constantly changing. A channel whose width, depth, slope, roughness and alignment are functions of flow rate and sediment load is called an alluvial channel. A stream is called a graded stream when it has developed the right bed slope and cross-section to transport the water and sediment from the watershed. If the slope of the channel is too flat for the sediment load, deposition will occur, which will steepen the grade; and if the slope of the channel is too steep, erosion will occur, which will flatten the grade. The sediment processes that occur along an alluvial channel system can be divided into production (erosion), transition (stable), and deposition (Figure 10.15).

The slope in the upper reach of a river tends to be steep, with high velocities and large sediment load. Braided channels are typically wide with multiple islands and poorly defined banks. They generally have steep slopes with excessive sediment load composed primarily of bed load material. As the large, angular particles are carried downstream, they become rounded and smaller. Downstream, the channel slope is flatter, the average discharge is larger, and the sediments are finer. When the fine particles in the channel bottom are washed away leaving only the larger particles on the surface, it is called channel armoring.

In the upper reaches, the sediments in the channel bed tend to be coarser than the sediments in the channel bank, and the channel tends to be wide and shallow.

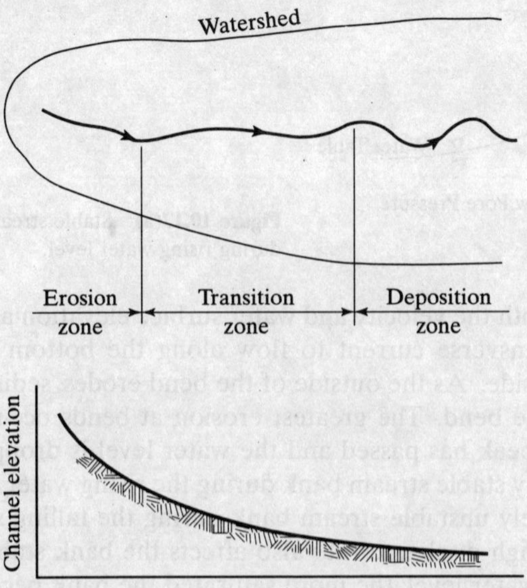

Figure 10.15 Sediment zones and stream profile.

Farther downstream, the banks are composed of fine cohesive material and are more resistant to erosion. Where the bank material is easy to erode, the channel tends to be wide and shallow, and where the bank material is difficult to erode, the channel tends to be narrow and deep. Bank erosion occurs at concentrated points primarily at the outside of bends.

Meandering channels (Fig. 10.16) have an S-shaped planform caused by multiple meanders. The thalweg is the deepest part of the channel and is located at the

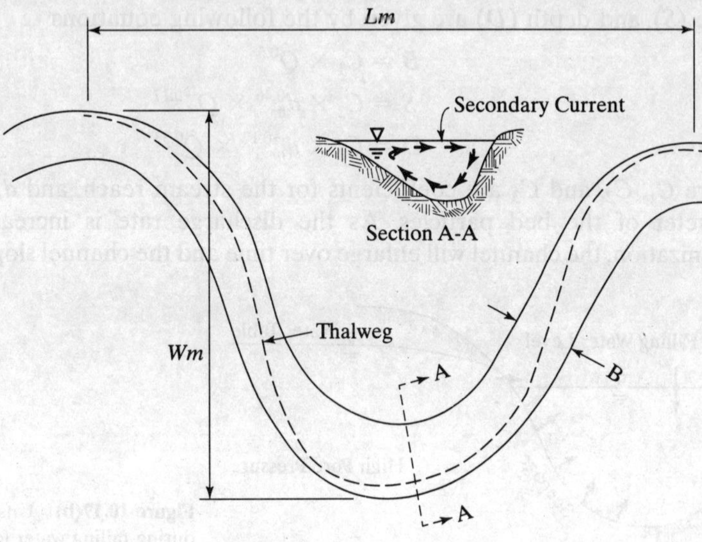

Figure 10.16 Channel meander.

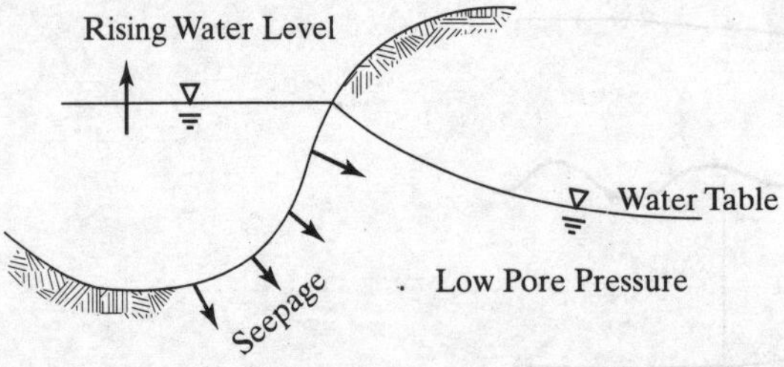

Figure 10.17(a) Stable stream bank during rising water level.

outside of the bend, where both the velocity and water surface elevation are higher. This causes a secondary, transverse current to flow along the bottom from the outside of the bend to the inside. As the outside of the bend erodes, sediments are deposited at the inside of the bend. The greatest erosion at bends occurs during flood stages after the flood peak has passed and the water level is dropping. Figure 10.17(a) shows a relatively stable stream bank during the rising water level, and Fig. 10.17(b) shows a relatively unstable stream bank during the falling stage of a flood. The duration of the high discharge rate also affects the bank stability. The longer the duration of high water level, the more saturated the bank becomes and the more unstable the bank will be during the falling stage. Although detention basins and reservoirs can be used to decrease the peak discharge rate, they also increase the duration of the high water level in the downstream channel.

As shown in Fig. 10.16, the channel length is greater than the floodplain length (Lm), and the slope of the channel is less than the slope of the floodplain. Sinuosity is defined as the length of the thalweg divided by the length of the floodplain. The sinuosity of a natural channel typically ranges from 1.2 to 4.0.

Observation of canals and rivers by researchers indicate that the size and slope of the channel are related to the discharge rate (Q). The width of the channel (B), slope (S), and depth (D) are given by the following equations

$$B = C_1 \times Q^{0.5} \quad \textbf{(10.20)}$$

$$S = C_2 \times d_m^{0.46} \times Q^{-0.17} \quad \textbf{(10.21)}$$

$$D = C_3 \times d_m^{-0.17} \times Q^{0.33} \quad \textbf{(10.22)}$$

where C_1, C_2, and C_3 are coefficients for the stream reach, and d_m is the median diameter of the bed particles. As the discharge rate is increased because of urbanization, the channel will enlarge over time and the channel slope will decrease.

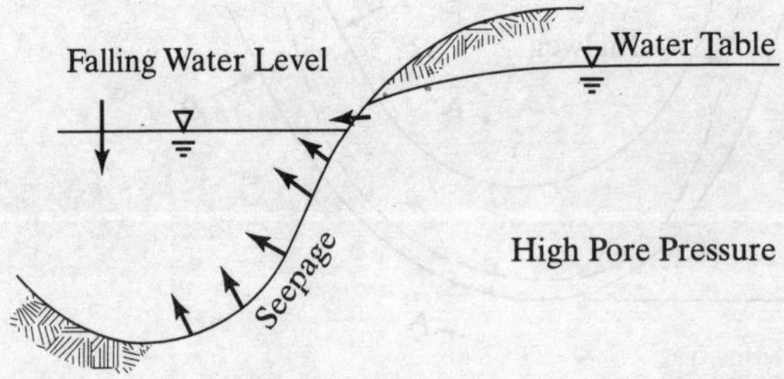

Figure 10.17(b) Unstable stream bank during falling water level.

TABLE 10.7 SCHUMM'S CLASSIFICATION OF ALLUVIAL CHANNELS

Mode of sediment transport	Silt-clay in channel perimeter (%)	Channel stability: Depositing (excess load)	Stable (graded)	Eroding (excess capacity)
Suspended 85–100%	100	Bank deposition	B/D < 7 Sinuosity > 2 Slope: flat	Streambed erosion
Mixed	30	Bank deposition, then streambed deposition	B/D 7–25 Sinuosity 1.5–2 Slope: moderate	Streambed erosion, then channel widening
Bed load 35–70%	0	Streambed deposition	B/D > 25 Sinuosity 1.0–1.5 Slope: steep	Channel widening

The Schumm (1963) classification of alluvial channels is listed in Table 10.7. Schumm used sediment load as the basis of the classification (suspended, mixed, and bed load) because of its relation to channel slope, shape, sinuosity, and perimeter material (percent silt and clay). For the three modes of sediment transport, the classification system uses three classes of channel stability (depositing, stable, and eroding). Reservoirs installed on a stream channel will cause a stable channel to become depositing upstream and eroding downstream.

Example 10.14 Channel Erosion

(a) A stream channel is 30 ft wide, 6 ft deep, and has a slope of 0.001. Because the watershed has been urbanized, the discharge in the channel has increased 30 percent. Estimate the eventual size and slope of the channel after urbanization.

$$\frac{B}{B_0} = \frac{C_1}{C_1}\left(\frac{Q}{Q_0}\right)^{1/2}$$

$$B = B_0(1.3)^{1/2} = 30 \times 1.14 = 34 \text{ ft}$$

$$S = S_0 \times \left(\frac{Q}{Q_0}\right)^{-0.17} = 0.001 \times (1.3)^{-0.17} = 0.00096$$

$$D = D_0\left(\frac{Q}{Q_0}\right)^{0.33} = 6(1.3)^{0.33} = 6 \times 1.09 = 6.5 \text{ ft}$$

The channel would erode approximately 4 ft wider and 0.5 ft deeper over time.

(b) If the B/d of a small channel (mean discharge of 100 cfs) is 3, estimate the B/d for a channel with a mean discharge of 1,000 cfs and 10,000 cfs.

$$B = C_1 Q^{0.5}$$

$$d = C_4 Q^{0.33}$$

$$\frac{B}{d} = 1.37 Q^{0.17}$$

Q (cfs)	B/d
100	3.0
1,000	4.4
10,000	6.7

10.4.3 Sediment Transport

Sediment in a stream may be transported by rolling, sliding, or bouncing along the channel bottom called bed load and by suspension in the water column called suspended load. If the sediment is coarse, most of the sediment transport will be bed load and if the sediment is fine, most of the sediment transport will be suspended load. In estimating the sediment transport capacity of a channel, the bed load and suspended load are usually considered separately. However, in this section we will only consider the total sediment load.

Studies by Laursen (1958) and Garde and Albertson (1958) indicated that the total sediment transport rate of a channel can be estimated with the following equation

$$g_s = C_4 \frac{V^4 n^3}{d_p^{1.5} d} \tag{10.23}$$

where g_s is the sediment transport rate per unit width of the channel, C_4 is a constant for the channel, V is the water velocity, n is the Manning roughness value for the channel, d_p is the sediment particle diameter, and d is the water depth. When the sediment transport capacity of the stream is greater than the available sediments in the stream, erosion will occur; and when the available sediments in the stream are greater than the sediment transport capacity, deposition will occur. If the velocity in a channel is increased by 10 percent, the sediment transport capacity will increase nearly 50 percent.

A constriction (such as levees) in an alluvial channel will usually cause upstream deposition where the velocity has decreased and erosion in the channel reach with increased velocity. A dam across a channel will cause deposition in the reservoir and erosion in the downstream channel.

Example 10.15 Channel Erosion and Deposition

A levee system has been installed along a stream channel. During a major flood, the levee system causes the water depth to increase 10 percent and the water velocity to decrease 15 percent in the upstream channel. If the sediment transport in the stream during the flood is 1,000 tons/day, estimate the rate of deposition upstream of the levees.

Solution

$$g_s = C_4 \frac{V^4 n^3}{d_p^{1.5} d}$$

Upstream sediment transport capacity

$$G_s = 1{,}000\left(\frac{V}{V_0}\right)^4\left(\frac{d_0}{d}\right) = 1{,}000\left(\frac{0.85}{1.0}\right)^4\left(\frac{1.0}{1.1}\right) = 475 \text{ tons/day}$$

Deposition rate upstream = 525 tons/day.

10.4.4 Sedimentation

Sediment includes individual particles, aggregates, chemicals, and organic materials. Size, shape, and density affect the settling velocity. As shown in Fig. 10.18, there are four types of particle settling based on concentrations and interaction between particles.

Discrete particle settling occurs when the sediment concentration is low and particles fall independently of each other. For a spherical particle of diameter (d_p) falling turbulent free water, the submerged weight of the particle is equal to the drag force of the water on the particle or

$$\frac{\pi d_p^3}{6}(\rho_s - \rho_w)g = C_D \frac{\pi d_p^2}{4}\rho_w \frac{V_s^2}{2} \tag{10.24}$$

where ρ_s is the density of the particle, ρ_w is the density of the water, V_s is the settling velocity, and C_D is the drag coefficient.

For laminar flow, the drag coefficient is given by

$$C_D = \frac{24}{\text{Re}} \tag{10.25}$$

where $R_e = V_s\, d_p/\nu$ and ν is the kinematic viscosity of water. Equation 10.24 can be simplified

$$V_s = \frac{g}{18\nu}(sg - 1)d_p^2 \tag{10.26}$$

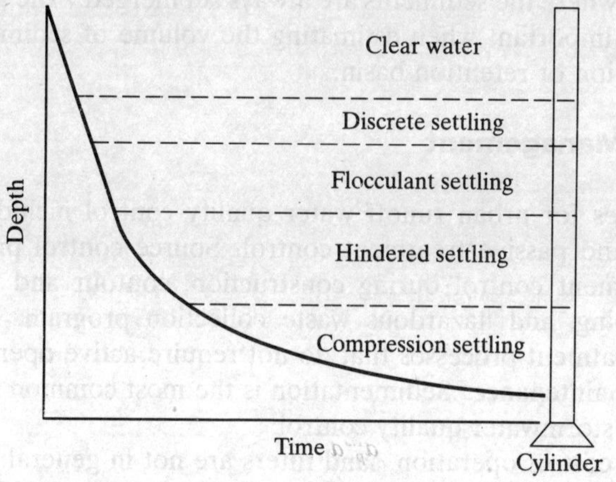

Figure 10.18 Four types of particle settling. (Adapted from Hann, Barfield and Hayes 1994)

TABLE 10.8 DISCRETE PARTICLE SETTLING VELOCITIES

Particle	Size mm	Settling velocity ft/hr	Settling velocity m/hr
Very fine sand	0.10–0.05	100–25	30–7
Silt	0.05–0.002	25–0.04	7–0.01
Clay	<0.002	<0.04	<0.01

where *sg* is the specific gravity of the particle. For water at 68°F (19°C) and a sand particle ($sg = 2.65$), Eq. 10.26 can be reduced to

$$V_s = K_p d_p^2 \tag{10.27}$$

where d_p is in mm, V_s is in ft/hr (m/hr), and K_p is equal to 10,000 for BG units and 3,000 for SI units. The settling velocities computed in Table 10.8 indicates that sedimentation basins with a 24-hour detention time will be effective in removing sand and larger silt-sized particles, but probably will be ineffective in removing small silt and clay size particles unless flocculent settling occurs.

Flocculent settling occurs when colloidal particles in a dilute suspension coalesce and form flocs that have sufficient mass to settle rapidly. Because flocculation is a naturally occurring process and depends on the chemistry of the water and colloids, the rate of flocculation in a sedimentation basin is difficult to predict. A settling tube analysis of the runoff may be required to estimate the rate of flocculation.

Hindered settling occurs when the particles are so concentrated that they interfere with the settling of surrounding particles. The settling of particles displaces water that flows upward through the openings between particles. As a result, there is uniform settling of all particles with a sharp liquid–solid interface.

Compression settling is when the sediment layer must be compressed for settling to occur. Sediments will consolidate faster in a detention basin (normally empty) than in a reservoir where the sediments are always submerged. The amount of compression settling is important when estimating the volume of sediments to be removed from a detention or retention basin.

10.4.5 Water Quality Management

Best management practices for urban runoff water quality control include both source control practices and passive treatment control. Source control practices include erosion and sediment control during construction, contour and terrace landscaping, street sweeping, and hazardous waste collection program. Passive treatment controls are treatment processes that do not require active operational control but only routine maintenance. Sedimentation is the most common passive treatment process use for stormwater quality control.

Because of the high cost of operation, sand filters are not in general use for urban stormwater quality control. The sediments trapped in the filter will have to be removed frequently to maintain the capacity of the filter. Typically, the sediment content in the effluent will be less than the sediment transport capacity of the downstream channel and the use of a sand filter may contribute to downstream channel erosion.

The use of infiltration basins, percolation trenches, and porous pavements as a passive treatment method for urban stormwater management is questionable. If the sediments have not been removed from the stormwater prior to being used for groundwater recharge, the system will typically seal and clog within a few years. If the stormwater passes through a sedimentation basin prior to being used for groundwater recharge, only about 60 percent of the sediments and 50 percent of pollutants (Table 10.2) will have been removed, and the life of the infiltration system will be extended a few years. An effective recharge system will normally require both sedimentation and filtration treatment before the stormwater is used for recharging the groundwater. If groundwater recharge is an objective of stormwater management, then the system should be designed and operated as a recharge project with active monitoring influent quality and not as a passive method of disposing of potentially contaminated surface water.

The drainage system has historically been used for waste disposal. In an urban area, there is a high probability of toxic chemicals, sewage overflow, household hazardous wastes, used motor oil, gas or oil spills, and winter road salt getting into the stormwater drainage system. It will usually be better to have the pollutants in the surface water where they can be detected and removed rather than to inject them underground where they might remain for years undetected.

10.4.5.1 Initial phase of urbanization. Landscaping should be planned to avoid long steep slopes and provide terracing, contouring, and swales as appropriate to control runoff and minimize soil erosion. Bare soil areas should be surface-treated with straw, hay, wood chips, or hydromulch and vegetation established as soon as possible to limit erosion. The ground cover intercepts the rainfall and absorbs the raindrop energy. The surface roughness improves infiltration into the soil and reduces the runoff velocity. The vegetation roots and soil mulch will tend to hold the soil in place during storm events.

Sediment traps may be required around bare soil areas to prevent eroded soil from entering the stormwater collection system. Straw bale check dams, rock-filled check dams, and filter fences are often used for this purpose. Straw bales can be effective in removing sediments from the runoff when installed correctly. As shown in Fig. 10.19, the straw bales should be placed in a shallow trench and secured in place with stakes. Rock-filled check dams can be used to trap the larger (bed load) particles in the runoff. Rock-filled check dams are simply constructed by dumping rock in the channel. As shown in Fig. 10.20, when placed in series along a channel, they can be used for grade stabilization. Filter fence is normally used to control sediments in overland flow and is typically installed on the contour to prevent concentrated flow at any location. As shown in Fig. 10.21, the filter fabric can be placed over a wire mesh fence with the toe of the fabric buried in a trench along the front of the fence.

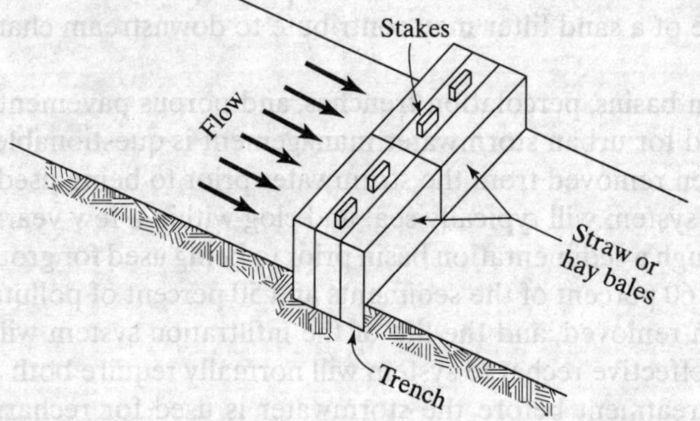

Figure 10.19 Straw bale check dam.

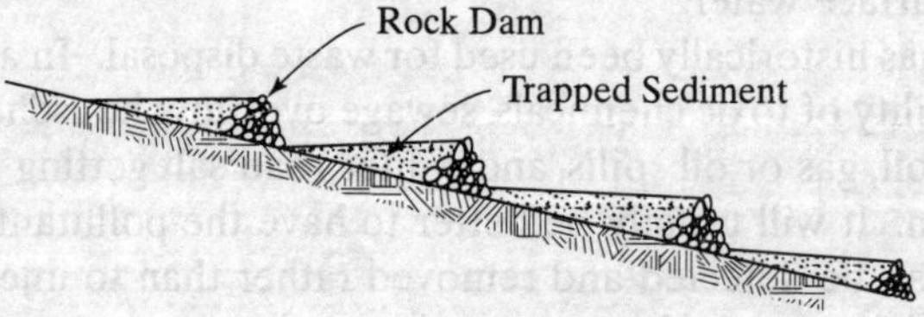

Figure 10.20 Rock check dams used for grade stabilization and sediment traps along the channel.

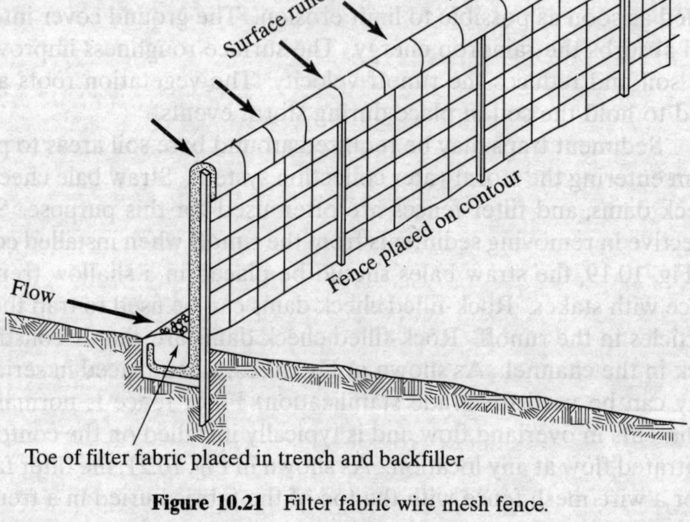

Figure 10.21 Filter fabric wire mesh fence.

10.4.5.2 Intermediate phase of urbanization. During the intermediate phase of urbanization, erosion from bare soil areas at construction sites will decrease, while channel erosion will tend to increase. Because of the increase of impervious area in the urbanized watershed, the volume of runoff will increase. The effect of urbanization on the runoff volume will generally be the greatest where the original soils were porous (sandy) and the land cover was forest. The increase in runoff will probably result in channel erosion. Rather than allowing the channels to erode to a graded condition where the stream bed slope and cross-section are in equilibrium with the water and sediments being transported from the watershed, it will be less damaging to the environment to design and construct modified channels capable of handling the discharge rate after urbanization.

It will be necessary to develop hydrologic models of the watershed and estimate the discharge rates both before and after the watershed has been urbanized. The two sets of watershed models (before and after urbanization) should be based on the same division of the watershed into subbasins and the same stream routing reaches. The models will need to be very detailed so that the area contributing to each on-site detention basin will be modeled as a subbasin. The after development hydrologic models will include changes in land use and ground cover as they affect surface retention and infiltration rates, and changes in stream channels as they affect the stream routing and all on-site and regional detention basins. These hydrologic models can be used to evaluate the system of detention basins, locate and design regional detention, design modified drainage channels, and determine the 100-year return period discharge rate.

For the purpose of this discussion, the drainage channels are divided into smaller channels where the watershed has been or will be completely urbanized and the larger drainage channels where the watershed will be only partially urbanized. The city or drainage district should purchase or obtain a drainage easement for the smaller channels to be modified because of urbanization of the watershed. The city is responsible for any damage to private property caused by urbanization, including channel erosion and increased flood levels caused by the increased runoff. Typically, the smaller drainage channels might be designed for a 25-year return period flood (with freeboard), but are required to contain the 100-year return period discharge rate (with reduced freeboard) within the drainage easement.

Normann (1975) presented a procedure for the design of open channels using vegetative liners. The hydraulic radius of flow for a stable channel design is

$$d_{max} = mS^n \tag{10.28}$$

where d_{max} is the maximum water depth in ft and S is the channel slope. The values of the coefficients m and n can be obtained from Table 10.9 for erodible and resistant soils. The values of K (Table 10.5) used in the USLE is a measure of erodibility and can be used to interpolate values of m and n from Table 10.9. A value of $K > 0.5$ is considered erodible, and a value of $K < 0.17$ is considered resistant.

The watershed of the larger streams in the area will probably be only partially urbanized. The impact of urbanization on the discharge rate will be less. Instead

TABLE 10.9 COEFFICIENTS FOR STABLE CHANNEL DESIGN WITH VEGETATIVE LINER

Height Bermuda grass liner inches	Erosion-resistant $K < 0.17$		Erodible $K > 0.5$	
	m	n	m	n
12	0.20	−0.60	0.21	−0.51
6	0.13	−0.62	0.11	−0.59
2.5	0.10	−0.67	0.08	−0.63
1.5	0.07	−0.65	0.05	−0.67

Adapted from Normann (1975).

of designing a new channel, only minor channel improvements may be required to control channel erosion.

The Federal Highway Administration procedure for estimating the size of riprap lining on the bed of a sloping channel uses a maximum stable depth of flow (d) in place of d_{max} in Eq. 10.28, with $n = -1.0$ and $m = 5\ D_{50}/\gamma$, where D_{50} is the riprap diameter in ft (50 percent riprap by weight smaller than D_{50}), and γ is the unit weight of water (62.4 lbs/ft^3).

$$D_{50} = \frac{\gamma d S}{5} \tag{10.29}$$

The design of a riprap-lined channel requires the selection of a rock size (D_{50}) large enough so that the force of the water will be less than the gravitational force holding the rock in place. The riprap is graded so that the spaces between the larger stones are filled with smaller stones. Typically, the riprap is graded in size from 0.2 D_{50} to 2.0 D_{50}.

The riprap is normally placed on a gravel bed so that the water will not erode the base material. The gravel bed should be approximately half the thickness of the riprap, but not less than 6 inches (152 mm). If the base material is primarily silt or fine sand, a filter cloth should be placed under the gravel bed.

Example 10.16 Stable Grass-Lined Channel

Determine the size of a grass-lined channel to carry 1,000 cfs. The channel will be lined with Bermuda grass cut to a height of 2.5 inches on erodible soils. The channel has 4:1 side slopes and a longitudinal slope of 0.001. The Mannings "n" value for the channel is 0.03.

Solution

$$d_{max} = mS^n$$

$$d_{max} = \text{maximum water depth, ft}$$

$$m = 0.08 \text{ (Table 10.9)}$$

$$n = -0.63 \text{ (Table 10.9)}$$

$$S = 0.001$$

$$d_{max} = 0.08(0.001)^{-0.63} = 6.2 \text{ ft}$$

$$Q = \frac{1.49}{n} AR^{2/3} S_o^{1/2}$$

$$AR^{2/3} = \frac{Qn}{1.49 S_o^{1/2}} = \frac{1{,}000 \times 0.03}{1.49\sqrt{0.001}} = 637$$

$$AR^{2/3} = \frac{(Bd + 4d^2)^{5/3}}{(B + 8.25d)^{2/3}}$$

$$637 = \frac{(6.2B + 154)^{5/3}}{(B + 51.2)^{2/3}}$$

Solving for B

$$B = 20 \text{ ft}$$

$$A = 20 \times 6.2 + 4 \times 6.2^2 = 278 \text{ ft}^2$$

$$P = 20 + 8.25 \times 6.2 = 71.2 \text{ ft}$$

$$R = \frac{278}{71.2} = 3.9 \text{ ft}$$

$$Q = \frac{1.49}{0.03} \times 278 \times 3.9^{2/3}\, 0.001^{1/2} = 1{,}080 \text{ cfs}$$

OK water depth will be slightly less than the maximum allowable according to Eq. 10.28.

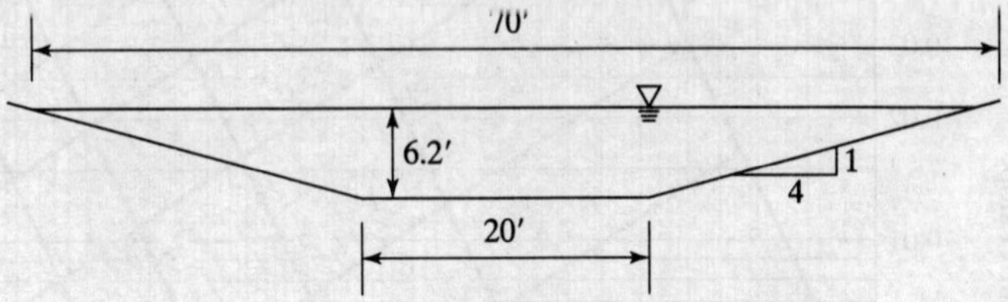

Stable grass-lined channel for a discharge of 1000 cfs

Example 10.17 Riprap Design

Design riprap for a channel in erodible soil on a slope of 0.01 and a maximum water depth of 5.0 ft. Determine the size of riprap required.

$$D_{50} = \frac{d}{5}\gamma S = \frac{5}{5} 62.4 \times 0.01$$

$$D_{50} = 0.62 \text{ ft}$$

Riprap graded in size from 0.12 to 1.24 ft.

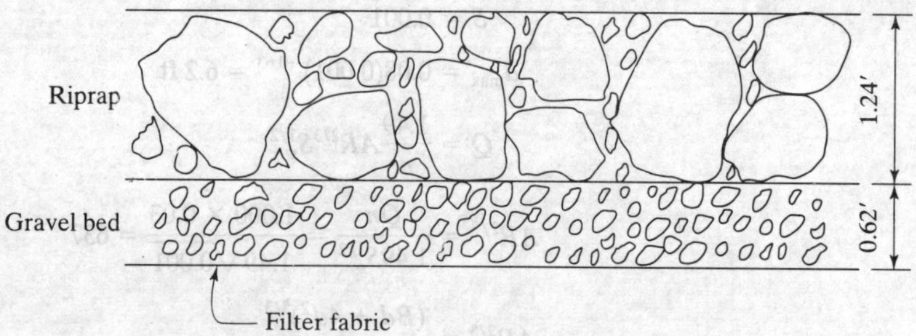

Example 10.18 Channel Design Charts

Develop channel design charts for the determination of channel size for urban runoff. Plot design charts on log-log graphs for triangular gutters, circular storm sewers, and rectangular channels.

A computer program called CHANNEL.FOR was written to generate the discharge and velocity data for a range of channel sizes and slopes. The output was plotted using log-log graph paper, and the design charts are as follows.

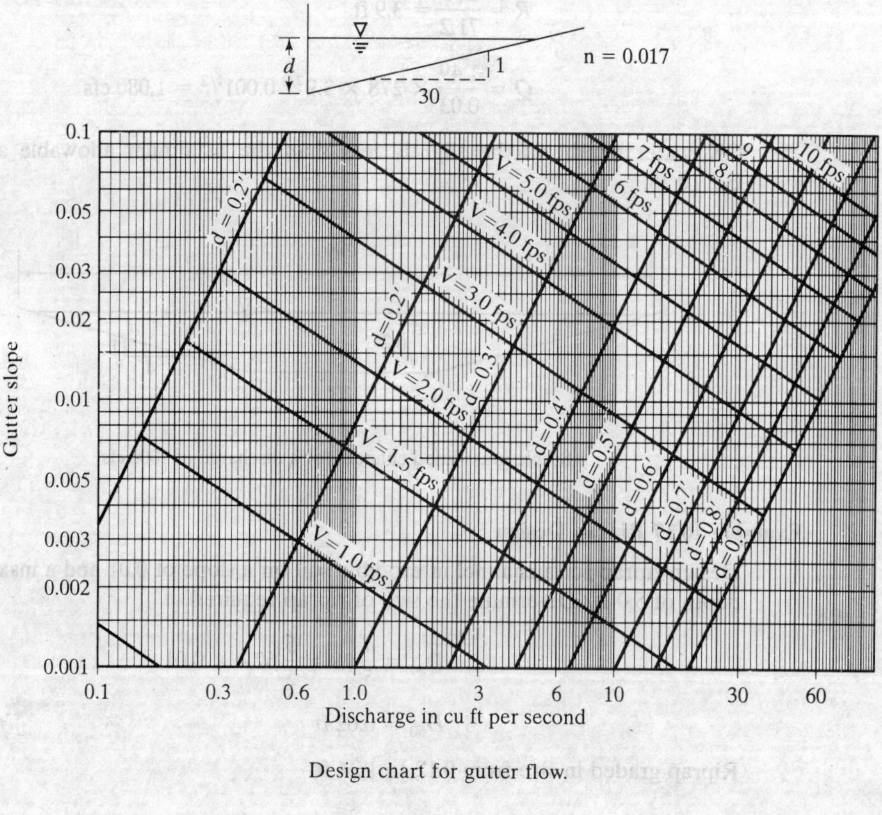

Design chart for gutter flow.

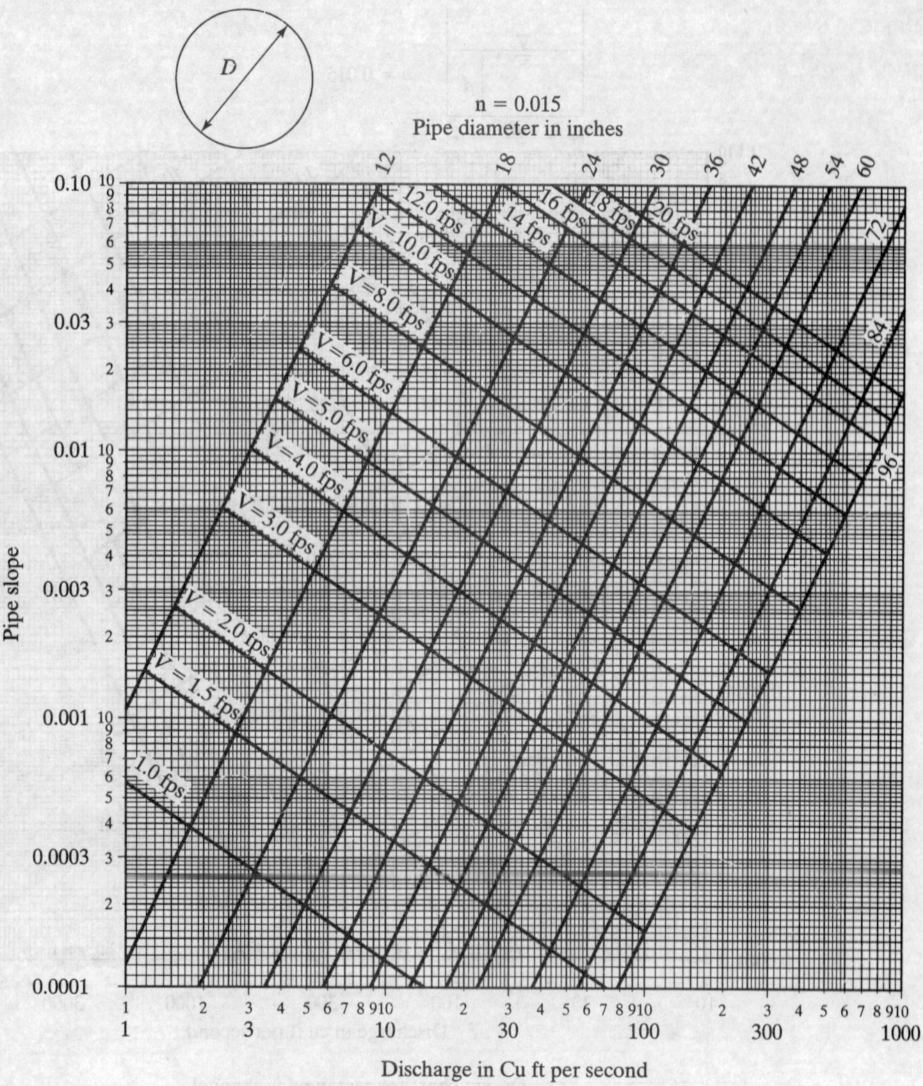

Design chart for circular pipe flowing full $n = 0.015$.

10.4.5.3 Mature phase of urbanization. During the mature phase of urbanization, most of the pollutants in the stormwater will be from washoff of accumulated deposits on impervious surfaces. Street sweeping can be effective in removal of solids from the surface of streets and parking lots prior to washoff by rainfall. Because a large amount of pollutants are associated with the smaller particles, it is important to use vacuum-type street sweepers to remove fine dust, as well as aesthetically objectionable debris and other material.

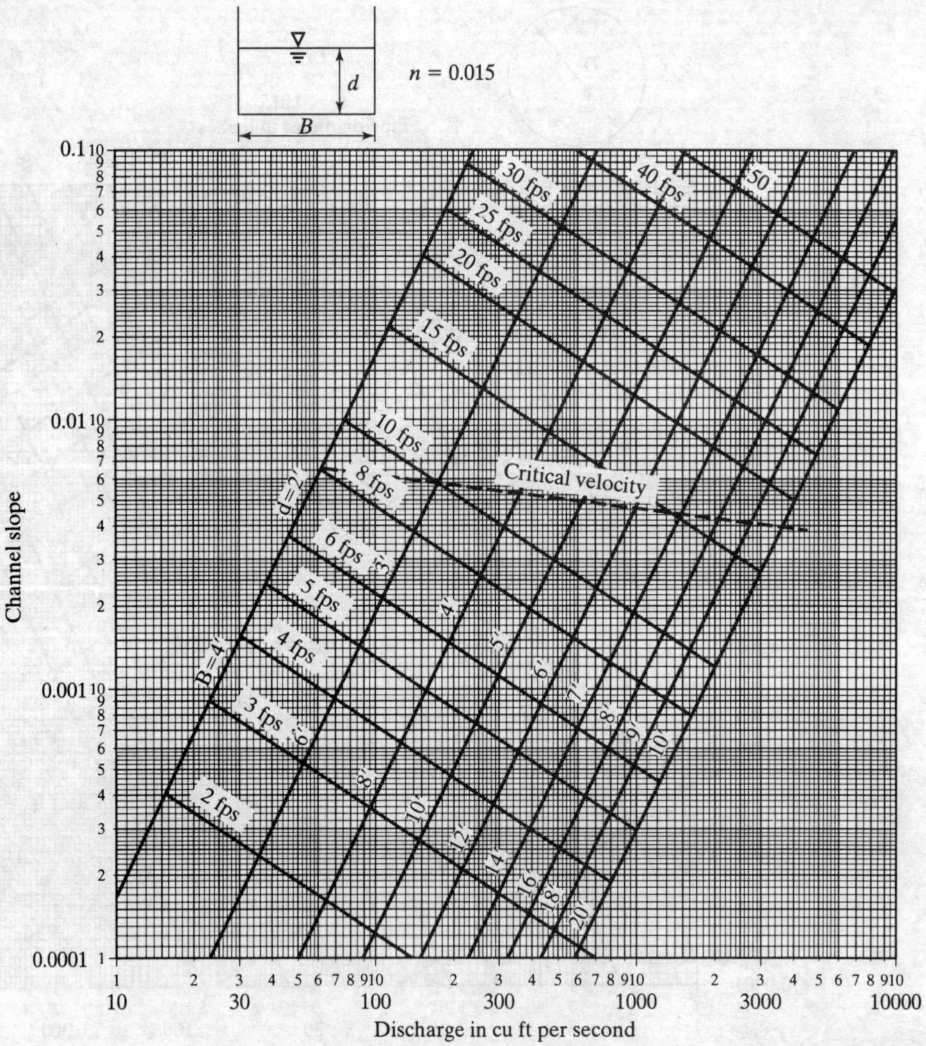

Design chart for rectangular channel.

Detention basins can be used as a sedimentation basin. A dual-purpose detention basin is a basin designed to control both water quality and peak discharge. Water quality control generally requires an average detention time of 24 hours, and if the basin is drained by gravity, the basin will require 48–72 hours to empty after a storm event. If a second storm occurs shortly after the first storm, storage would not be available for flood control. For water quality control, storage for the second storm is not critical because the pollutants have not had time to build up on the impervious surfaces, and treatment may not be necessary. Should the sediments not be removed in a dual-purpose detention basin and a major storm occurs, the settled solids in the basin could be resuspended and flushed downstream.

Vegetative filter strips are zones of vegetation that the stormwater runoff passes through as shallow sheet flow before it becomes concentrated channel flow. They are normally located on the contour perpendicular to the general direction of flow. Filter strips along a stream channel are often called riparian vegetative filter strips.

If the sediments in the stormwater runoff are contaminated, the use of vegetative filter strips may not be appropriate. The sediment trapped in the filter will be contaminated, and the exposure to humans might be increased in parks and playground areas. The sediment trapped by the filter will have to be removed and the vegetation reestablished on a continuous basis to ensure proper operation and prevent the trapped sediments from being washed away during extreme storm events.

For the filter strip to operate properly, the following design criteria are recommended (ASCE and WEF, 1998):

- Hydraulic retention time of not less than 5 minutes
- Average velocity of less than 0.9 ft/s (0.3 m/s)
- Manning "n" value
  0.20 for routinely mowed strips
  0.24 for infrequently mowed strips
- Average flow depth of less than 1.0 inch (25 mm)
- Select a vegetation cover suitable for the site

10.4.5.4 First flush treatment. First flush treatment units are designed to store and treat the first 0.5–1.0 inch (18–25 mm) of runoff with the remainder of the runoff bypassing the treatment unit. Figure 10.22(a) is a flow diagram for a first flush treatment system. The treatment unit is sized for a volume equal to the first flush. When the treatment unit is full, the water in the treatment unit is at the same level as the top of the weir in the bifurcation structure causing the remainder of the runoff to flow over the weir and bypass the treatment unit [Fig. 10.22(b)]. A typical sedimentation basin is shown in Fig. 10.23. The baffles at the entrance are used to distribute the inflow across the treatment unit and to confine the

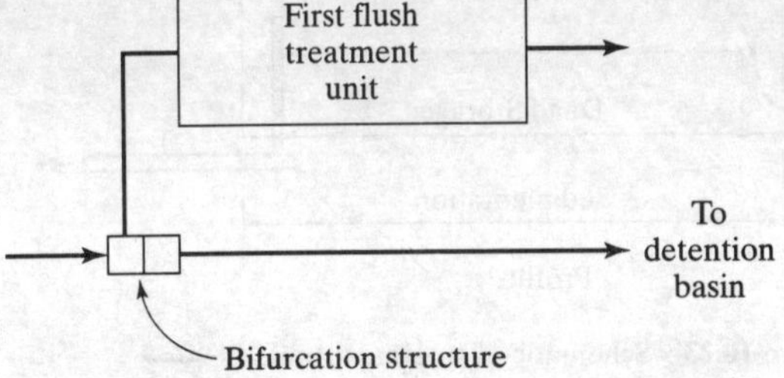

Figure 10.22(a) First flush treatment flow diagram.

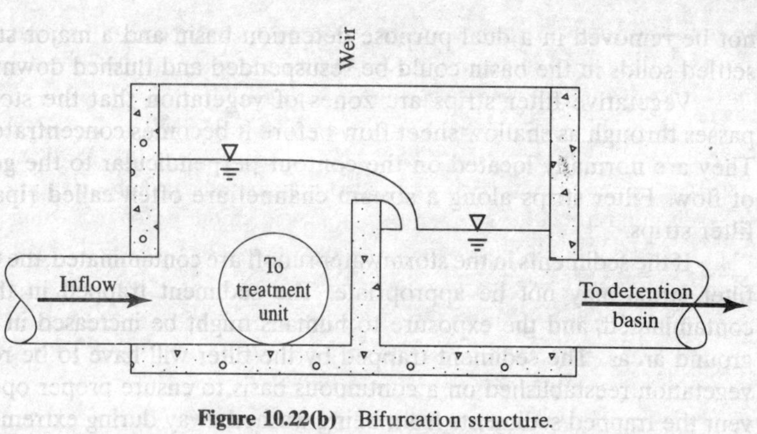

Figure 10.22(b) Bifurcation structure.

turbulence to the entrance. The oil, grease, and floating debris along with coarse sediments will be contained in the trap before the sedimentation basin.

A floating trough and weir can be utilized at the outlet to skim off the higher quality water at the surface near the end of the sedimentation basin. The dead storage shown in Fig. 10.23 is used to minimize resuspension of stored sediments during filling. By removing both floating material and suspended solids, the sedimentation unit will also remove the pollutants attached to them. Provisions must

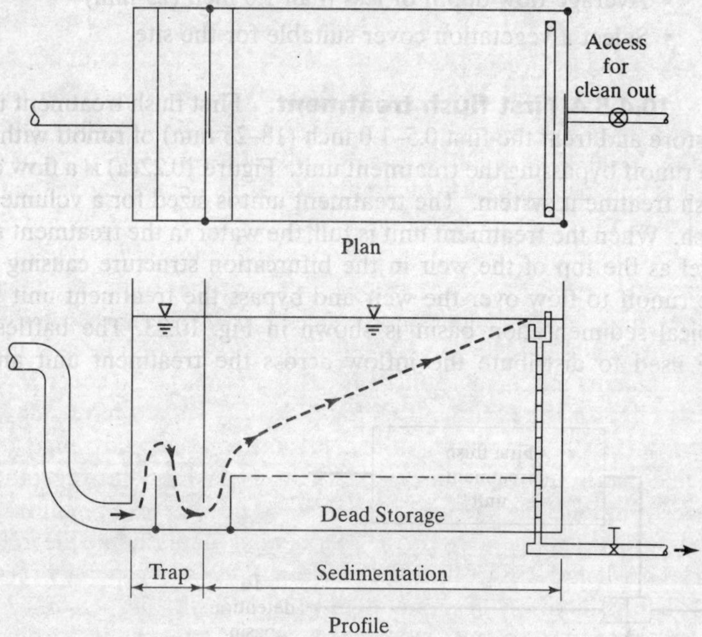

Figure 10.23 Schematic of treatment unit.

be made for easy clean out of the accumulated debris and sediments in the treatment unit. The site should be accessible by truck and the unit designed for cleaning with a front-end loader or similar equipment.

Example 10.19 First Flush Treatment Unit

Determine the dimensions of a first flush treatment unit to capture the first 1/2 inch of runoff from a 10-acre watershed. The operating depth of the unit is to be 8.0 ft, and the L/W ratio of the unit is to be approximately 4.

$$\text{Runoff volume} = d \times A = \frac{0.5}{12} \times 10 \times 43{,}560 = 18{,}150 \text{ ft}^3$$

$$\text{Surface area} = \frac{18{,}150}{8} = 2{,}269 \text{ ft}^2$$

$$4W^2 = 2{,}269$$

$$W = 24 \text{ ft}$$

$$L = 96 \text{ ft}$$

10.4.5.5 Retention ponds. A retention pond is a small artificial lake with wetland vegetation (littoral zone) around the perimeter. They are attractive and are considered to enhance neighborhood property values. They are often called amenity lakes, and can be included in the design of golf courses and the landscaping of parks and open spaces. They are more attractive than detention basins because the sediments and debris accumulate in the permanent pool and are not visible (Fig. 10.24).

Retention ponds are designed to remove pollutants from stormwater by physical, chemical, and biological processes. Sedimentation, chemical flocculation, and vegetative filtration occur in the pond to remove 70–90 percent of the suspended solids. Biological removal of pollutants is by uptake by wetland vegetation and metabolism by photoplankton and microorganisms (ASCE and WEF, 1998).

The depth of the deep pool should be greater than 6 ft (2 m) to prevent wind-generated waves from resuspending accumulated bottom sediments and to reduce bottom plant growth by limiting sunlight at the bottom. The pool depth should not be greater than about 10 ft (3 m) so that the water will be well mixed and the bottom sediments remain aerobic. The littoral zone should cover 25–50 percent of the surface area of the pond and for public safety should have side slopes no steeper than 6:1 (H:V). The depth of the littoral zone should range from 0.5 to 1.5 ft (0.25–0.45 m). Information on plants for use as wetland vegetation should be obtained from local agricultural agencies and nurseries. To minimize short-circuiting and improve sedimentation, the permanent pool should be relatively long and narrow. A length-to-width ratio between 3 and 4 is generally recommended.

The volume of the permanent pool ($\forall_p$) should be sufficient to provide an average detention time of 2–4 weeks. $\forall_p$ can be estimated from the mean runoff volume ($\forall_R$) of all runoff events during the year. The ratio $\forall_p/\forall_R$ should be

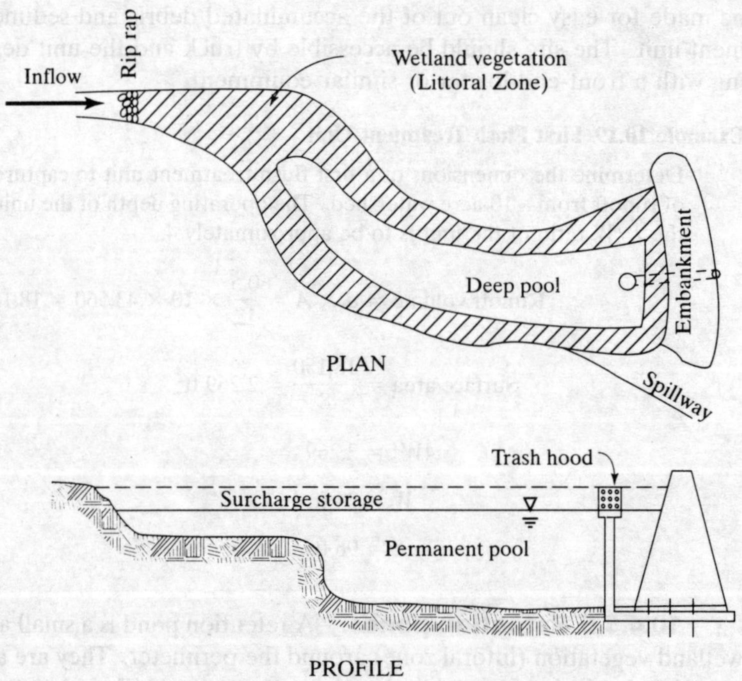

Figure 10.24 Retention pond plan and profile.

approximately 2.5. Figure 10.25 is a map of the United States showing the mean storm precipitation depth (P_m). The mean runoff volume is the mean storm precipitation from Fig. 10.25 times the average runoff coefficient (C) for the watershed (see Table 10.1) times the watershed area.

Additional storage is recommended to increase the removal of suspended solids and to reduce short-circuiting in the lake. The additional storage is called surcharge detention storage and should be approximately equal to the permanent pool volume. The orifice control on the outlet pipe in Fig. 10.24 should be sized so that the time to evacuate 95 percent of the surcharge storage volume is at least 12 hours. The trash hood should be perforated to draw off water into the outlet pipe from all levels in the surcharge detention storage and should have a grated top and a solid annular space at the bottom between the hood and the riser pipe.

Figure 10.26 shows the evacuation time as the time from the peak of the outflow hydrograph until the time that 95 percent of the surcharge storage has been evacuated. The detention time is usually estimated as the time to the centroid of the outflow hydrograph minus the time to the centroid of the inflow hydrograph. Plug flow computations are often used to compute the detention time for a retention basin. It is assumed that all the old water in the permanent pool is discharged from the basin before any of the new water is discharged. The time to the centroid

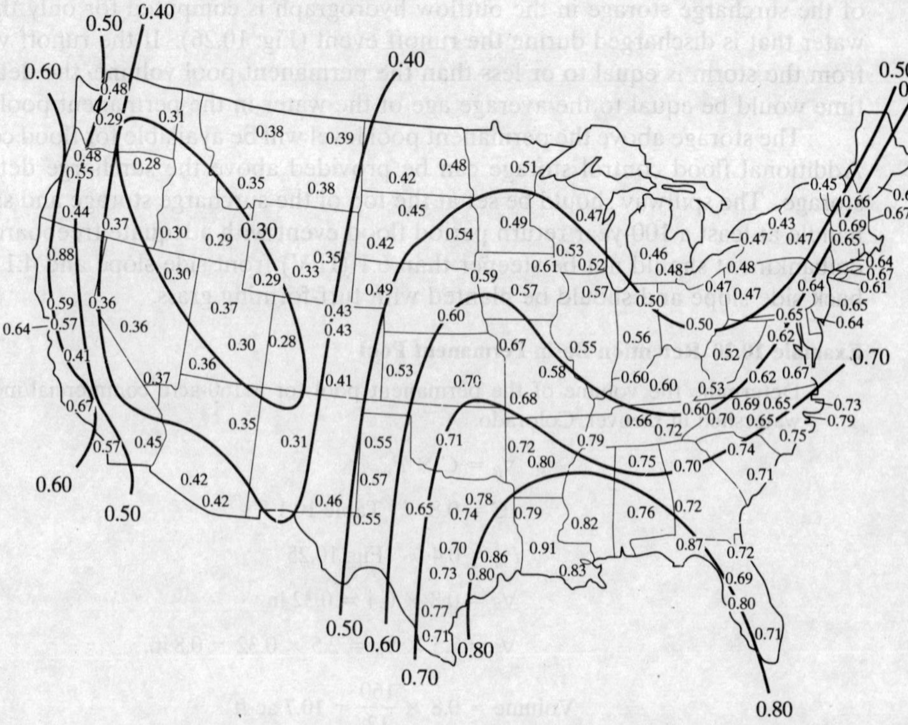

Figure 10.25 Mean storm precipitation depth (inches) (American Society of Civil Engineers and Water Environment Federation, 1998; reprinted with permission).

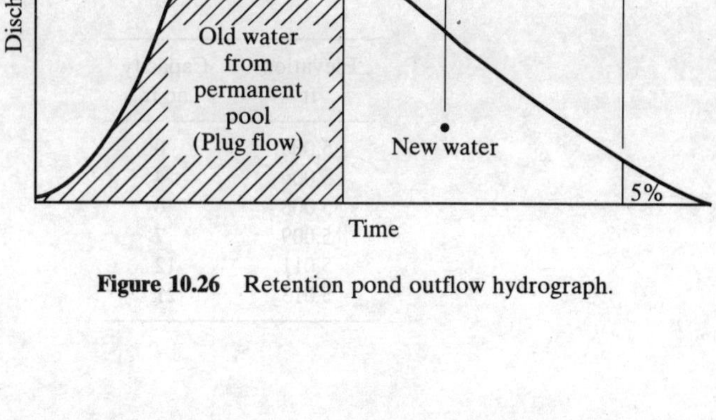

Figure 10.26 Retention pond outflow hydrograph.

of the surcharge storage in the outflow hydrograph is computed for only the new water that is discharged during the runoff event (Fig. 10.26). If the runoff volume from the storm is equal to or less than the permanent pool volume, the detention time would be equal to the average age of the water in the permanent pool.

The storage above the permanent pool level will be available for flood control. Additional flood control storage can be provided above the surcharge detention storage. The spillway should be set at the top of the surcharge storage and sized to handle at least a 100-year return period flood event with adequate freeboard. The embankment should not be steeper than 6:1 (H:V) front side slope and 4:1 (H:V) back side slope and should be planted with turf-forming grass.

Example 10.20 Retention Basin Permanent Pool

Determine the volume of the permanent pool for a 160-acre commercial/industrial watershed in Denver, Colorado.

$$\forall_R = C \times P_m$$

$$C = 0.8 \qquad \text{Table 10.1}$$

$$P_m = 0.4 \qquad \text{Fig. 10.25}$$

$$\forall_R = 0.8 \times 0.4 = 0.32 \text{ in.}$$

$$\forall_p = 2.5 \times \forall_R = 2.5 \times 0.32 = 0.8 \text{ in.}$$

$$\text{Volume} = 0.8 \times \frac{160}{12} = 10.7 \text{ ac-ft}$$

Example 10.21 Retention Basin Outlet Control

Determine the area (A) of the orifice in the riser pipe of a retention basin such that 95 percent of the surcharge storage is discharged from the reservoir in 48 hours. The orifice discharge coefficient (C_o) is 0.6.

The retention basin capacity table is listed below and plotted in the attached sketch. The retention basin permanent pool is 10.7 ac-ft, and the surcharge storage is 10.7 ac-ft. From the capacity curve, the normal pool elevation is 5,010.7 ft and the top of the surcharge storage is 5,013.1 ft. The water surface elevation in the reservoir when 95 percent of the surcharge storage has been released from the retention basin is 5,010.9 ft.

Elevation ft	Capacity ac-ft
5,000	0
5,006	4
5,008	6
5,009	7
5,011	12
5,013	21

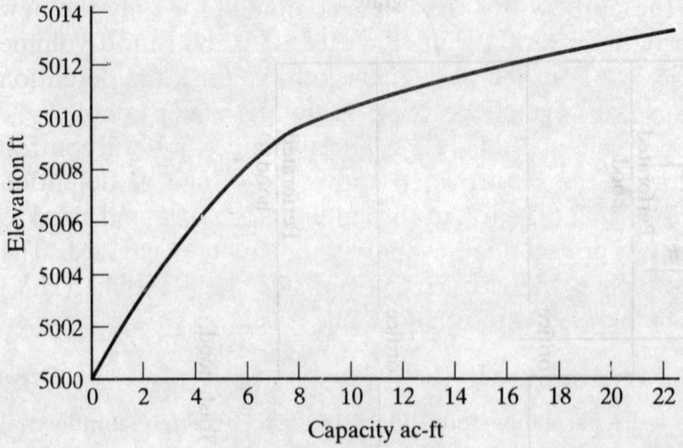

Capacity curve.

A sketch of the trash hood and riser pipe is shown. The discharge (Q_o) through the orifice in the riser pipe is

$$Q_o = C_o A \sqrt{2g}\, h^{1/2}$$

Using a Δh equal to 0.5 ft the evacuation time table is listed below.

Elevation ft	Surcharge storage ac-ft	ΔS ac-ft	h ft	$\frac{Q_o}{A}$* ft/s	$A\Delta t^\dagger$
5,013.1	21.4				
		3.1	3.15	8.54	4.39
5,012.6	18.3				
		2.5	2.65	7.84	3.86
5,012.1	15.8				
		2.2	2.15	7.06	3.77
5,011.6	13.6				
		1.6	1.65	6.18	3.13
5,011.1	12.0				
		0.8	1.30	5.49	1.76
5,010.9	11.2				
				Total	16.91

$$^*\frac{Q_o}{A} = C_o \sqrt{2g}\, h^{1/2}$$

$$^\dagger \Delta t = \frac{\Delta S}{Q_o} \times \frac{43{,}560}{3{,}600} = 12.1 \frac{\Delta S}{Q_o} \text{ hrs}$$

$$A \Sigma \Delta T = 16.91$$

$$\Sigma \Delta T = 48.0$$

$$A = \frac{16.91}{48.0} = 0.35 \text{ ft}^2 = 50 \text{ in}^2$$

Use four orifices each 4 inches high and 3 inches wide.

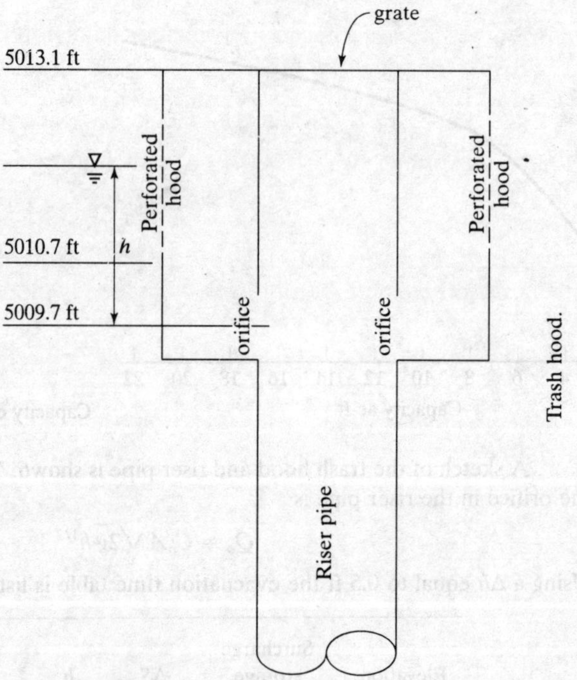

10.5 FLOOD-DAMAGE MITIGATION

Flood-damage mitigation attempts to reduce flood damages by both structural and nonstructural methods. Nonstructural methods include restricting development in flood-prone areas, flood-proofing existing structures in flood-prone areas, temporary evacuation of flood-threatened areas, and reduction of runoff by watershed management. Structural methods include reducing the peak discharge rate by reservoirs, increasing the capacity of streams by channel improvements, confining the flood flow using levees and flood walls, and diverting floodwater around flood-prone areas.

Many of the structural methods of flood control tend to increase development. A multipurpose reservoir may increase development both around the reservoir and along the downstream channel. Should the lake level of the reservoir exceed the design level or the release rate from the reservoir exceed the capacity of the downstream channel, flooding of structures may occur. Levees also encourage development behind the levee and when the level of the river exceeds the height of the levee, flooding of homes and businesses may occur.

10.5.1 Nonstructural Methods

Many of the larger communities and drainage districts have flood alert systems. Rain gages and stream gages are located at critical locations throughout the watersheds that contribute flow to flooding in the area. Data from the gages are transmitted to a central location for analysis, display, and storage. The flood alert system, along with the

National Weather Service forecast and weather radar, can be used to issue flood warnings and provide a basis for temporary evacuation of flood-threatened areas.

Flood warning might be issued on the bases of rainfall amounts or rates and water levels in streams. Real-time data, along with flood warnings, can be made available to the public using television, radio, and the internet. Flood forecasts might be made based on projected rainfall amounts from weather radar using relatively simple relationships developed from previous storms or using complex computer models of the watershed. The stored historical data are extremely valuable for engineering studies and are generally made available through the internet.

Communities participating in the National Flood Insurance Program are required to have mandatory land use regulations restricting development in flood-prone areas to receive federally-backed flood insurance. For the purpose of defining flood-prone areas, the 100-year return period flood is used. A cross-section of a natural stream channel is shown in Fig. 10.27(a). Typically, the capacity of the channel is exceeded about every 2 years. A floodplain delineated based on identifying the extent of alluvial soils will have a return period of approximately 10 years. The 100-year return period floodplain is shown in Fig. 10.27(b), along with the floodway and the floodway fringe.

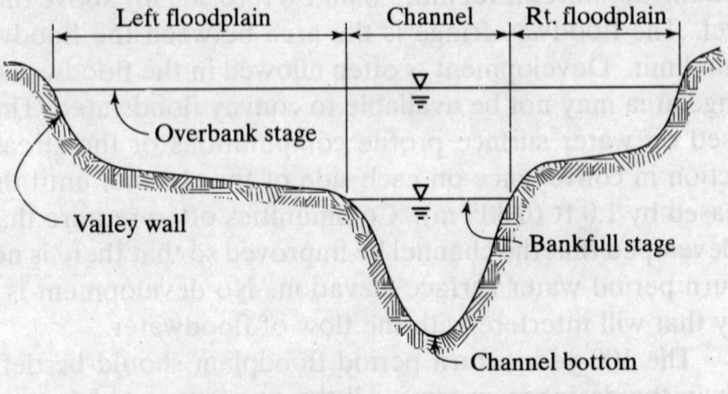

Figure 10.27(a) Typical stream cross-section.

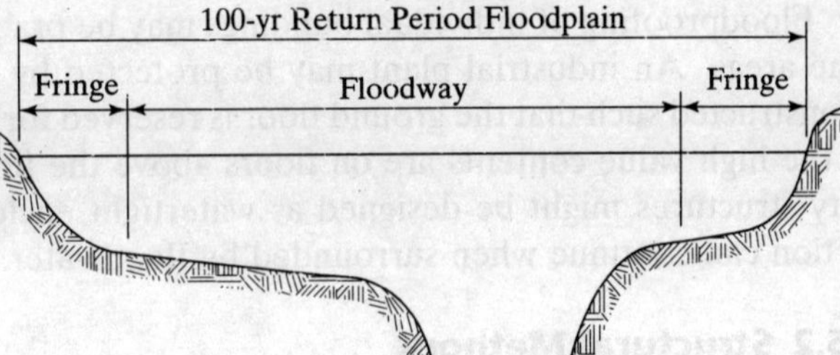

Figure 10.27(b) Floodway and floodway fringe.

To define the 100-year return period floodplain, the 100-year return period discharge is computed for existing watershed conditions using the appropriate hydrologic methods, depending on the availability of data. For larger streams, adequate historical stream gage data may be available, and the 100-year discharge rate can be computed using the log-Pearson type III distribution. For other channels in areas where adequate stream gage data are not available, the 100-year return period discharge rate can be computed using a detailed hydrologic model of the watershed. Typically, the 100-year return period precipitation with a balanced triangular distribution is used in the model. For the smaller channels, a storm duration of 24 hours is generally used. For the larger channels, the storm duration should be greater than or equal to the time to peak of the outflow hydrograph.

A water surface profile is computed for the channel using the standard step method based on surveyed cross-sections or cross-sections from detailed topographic maps and the computed 100-year return period discharge rate. The computed water surface elevations combined with topographic maps are used to delineate the 100-year return period floodplain, assuming a level water surface across the channel.

The floodway has been arbitrarily defined as that part of the cross-section that includes the channel and will pass the 100-year return period flood without increasing the water level more than 1.0 ft (0.305 m) above the existing 100-year flood level. The floodway fringe is the area between the floodway and 100-year floodplain limit. Development is often allowed in the floodway fringe, and the floodway fringe area may not be available to convey floodwater. The floodway is computed based on water surface profile computations of the stream reach with equal reduction in conveyance on each side of the channel until the water elevation is increased by 1.0 ft (0.305 m). Communities often require that if the floodway fringe is developed that the channel be improved so that there is no change in the 100-year return period water surface elevation. No development is permitted in the floodway that will interfere with the flow of floodwater.

The 100-year return period floodplain should be defined for all open channels in the drainage system. All the computer models used to establish discharge rates and water surface profiles should be maintained by the community or drainage district and made available to engineering firms working in the area.

Floodproofing of individual buildings may be practical when located in flood-prone areas. An industrial plant may be protected by a ring levee. Buildings can be constructed such that the ground floor is reserved for vehicles that can be moved, and the high value contents are on floors above the flood level. Important public utility structures might be designed as watertight enclosures so that their normal function can continue when surrounded by floodwater.

10.5.2 Structural Methods

Major reservoirs used to control flooding will generally be multipurpose reservoirs and are usually beyond the scope of urban stormwater management. Detention

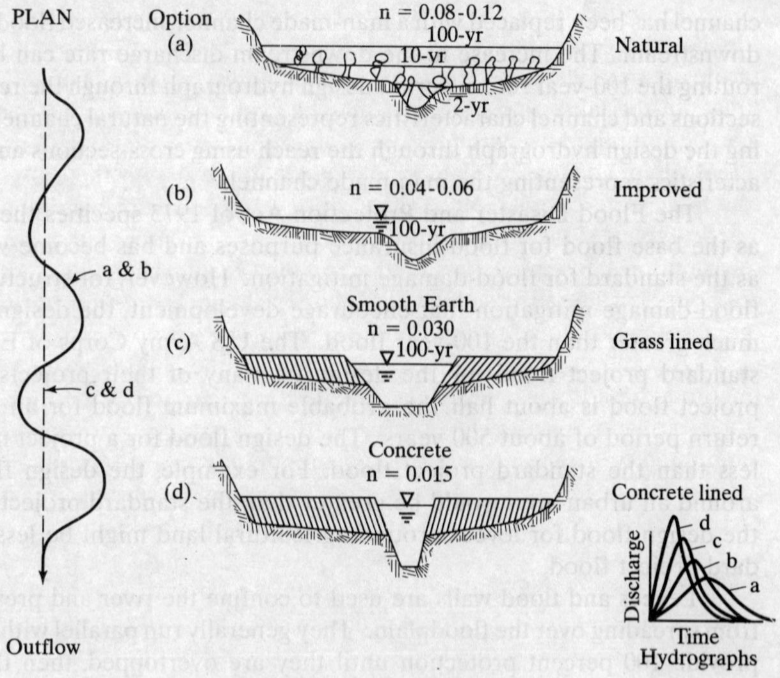

Figure 10.28 Channel improvementss.

basins are small flood control reservoirs used to control the runoff from small, local watersheds and were discussed in Section 10.2.

The capacity of existing channels can be increased using channel improvements. Figure 10.28(a) shows the cross-section of a natural channel and the 100-year flood level. Under natural conditions, the floodplain is usually uneven ground covered with brush and the main channel is nonuniform and meandering. A typical Manning "n" value for the channel might be around 0.10. If the brush is removed and the surface smoothed in the floodplain [Fig. 10.28(b)], the Manning "n" value might be reduced to 0.05. If the natural channel is replaced with a smooth earth, grass-lined channel [Fig. 10.28(c)], the Manning "n" value might be reduced to about 0.03. If the natural channel is replaced with a man-made channel, the alignment would probably be straightened and the slope of the channel increased. The increase in slope and the reduction in the Manning "n" value will increase the water velocity that might result in channel erosion. Figure 10.28(d) shows the natural channel being replaced with a concrete-lined channel with a Manning "n" value of approximately 0.015. The concrete-lined channel will usually be narrower and deeper than the grass-lined earth channel.

A natural channel will provide some natural flood storage. If the channel is hydraulically improved, the natural flood storage in the reach is reduced and the peak flow at the outlet of the reach will be increased. If a long reach of a natural channel has been replaced with a man-made channel, increased flooding might occur downstream. The increase in the downstream discharge rate can be evaluated by routing the 100-year return period design hydrograph through the reach using cross-sections and channel characteristics representing the natural channel, and then routing the design hydrograph through the reach using cross-sections and channel characteristics representing the man-made channel.

The Flood Disaster and Protection Act of 1973 specifies the 100-year flood as the base flood for flood insurance purposes and has become widely accepted as the standard for flood-damage mitigation. However, for structural methods of flood-damage mitigation that encourage development, the design flood must be much greater than the 100-year flood. The U.S Army Corps of Engineers uses a standard project flood for the design of many of their projects. The standard project flood is about half the probable maximum flood for an area and has a return period of about 500 years. The design flood for a project may be more or less than the standard project flood. For example, the design flood for levees around an urban area would be greater than the standard project flood, whereas the design flood for levees around agricultural land might be less than the standard project flood.

Levees and flood walls are used to confine the river and prevent floodwater from spreading over the floodplain. They generally run parallel with the river. They provide 100 percent protection until they are overtopped, then they provide no flood protection. The height of the levee must be high enough to transmit the design flow with adequate freeboard.

A levee is an earth dike and is usually constructed of material excavated near the levee. The side slopes of the levee are usually 4:1 or less. The slopes should be protected against erosion by sodding and where necessary by riprap. The topsoil should be removed before the levee fill is placed. A cutoff trench may be required to reduce seepage under the levee. If seepage through the foundation is a serious problem, a sheet pile cutoff wall might be used. The top width of the levee should be at least 10 ft (3 m) to allow access of maintenance equipment. Seepage through the embankment should be evaluated, and the back slope of the levee should be flat enough to contain the seepage line. A toe drain on the back slope may also be used to contain the seepage line within the levee.

A flood wall is usually constructed of concrete and requires less right of way than a levee. They are designed to withstand the hydrostatic pressure exerted by the water at the design flood level. They are designed similar to a low concrete dam and can fail by sliding, overturning, and foundation failure.

A flood bypass might be used where it is not feasible to enlarge the existing channel to carry the increased flow caused by urbanization. Because deposition or erosion may occur at the point of diversion, a bifurcation structure might be required to ensure correct division of flow. A side-channel weir might be considered if the bypass is only used to carry the floodwater that is in excess of this downstream channel capacity. The bypass channel is designed as an open channel. Opportunities for a flood bypass are limited by the topography of the area.

Example 10.22 Channel Modification

This example problem is intended to show the impact of channel modification on downstream flooding. A natural stream channel has been modified and replaced with a grass-lined channel. The original channel was 1.3 miles long, with a longitudinal slope of 0.0007. The Manning's "*n*" values were 0.06 for the channel and 0.15 in the overbank. An eight-point cross-section for the natural channel is listed below.

Elevation, ft	340.	330.	327.	324.	320.	328.	329.	340.
Distance, ft	0.	50.	60.	65.	70.	75.	90.	100.

The new channel is 0.91 miles long, with a longitudinal slope of 0.001. The channel has a bottom width of 10.0 ft, 4:1 side slopes, and a Manning's "*n*" value of 0.03. The following eight-point cross-section represents the modified channel.

Elevation, ft	340.	330.	329.	321.	321.	329.	330.	340.
Distance, ft	0.	15.	18.	50.	60.	92.	95.	100.

The inflow hydrographs are those computed in Ex. 10.11 Part I (preurbanization), Part II (urbanization without detention), and Part III (urbanization with detention). Route the three hydrographs through both the natural channel and the modified channel to show the impact of channel modification on the peak discharge rate.

Solution The routing file for Example 10.11 was modified by adding the natural channel cross-section as routing reach 703 and the modified channel as routing reach 704. The command file was modified by adding *R* 703 or *R* 704 as appropriate.

Results

The results of the routing are tabulated below and plotted on the attached graphs.

	Outflow hydrograph peak discharge rates		
Watershed conditions	Natural channel cfs	Modified channel cfs	Increase cfs
Natural	340	460	120
Developed without detention	460	785	325
Developed with detention	410	630	220

Conclusion

Modifying the channel will result in the peak discharge rate being 100–300 cfs higher, depending on the watershed condition.

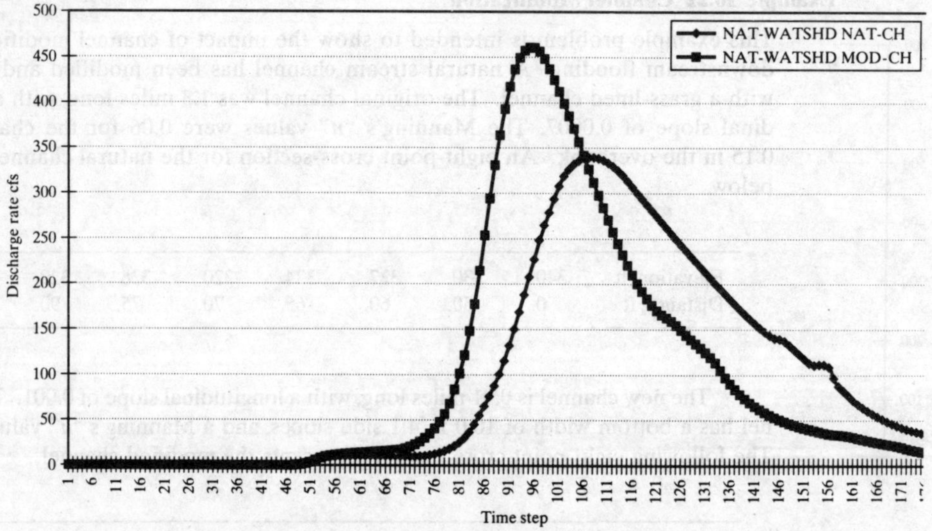

A comparison of hydrographs from an undeveloped watershed for a natural channel and a modified channel.

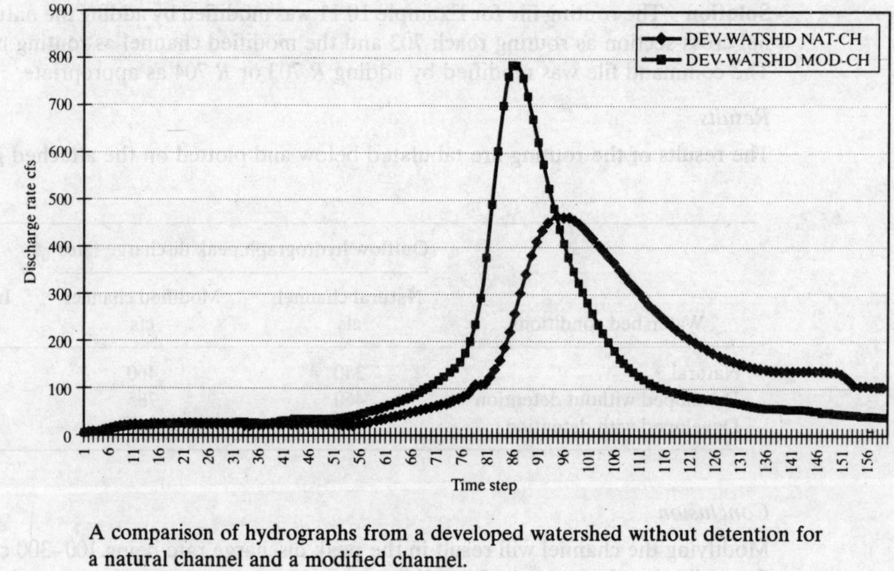

A comparison of hydrographs from a developed watershed without detention for a natural channel and a modified channel.

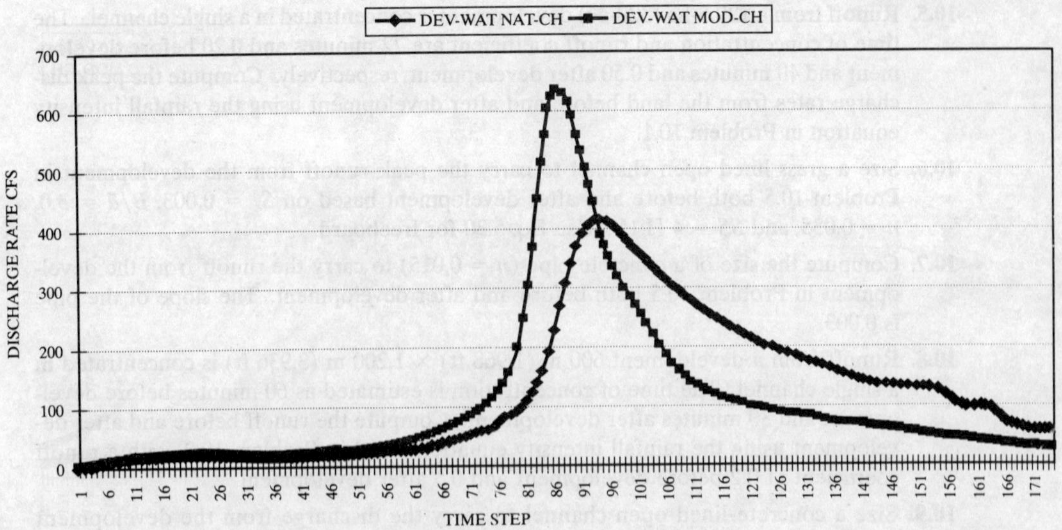

A comparison of hydrographs from a developed watershed with detention for a natural channel and a modified channel.

PROBLEMS

10.1. A 1.0-ha (2.47-ac) parcel of agricultural land is to be developed into medium density residential housing. The soils on the property are in hydrologic soil group C with an average slope of 3 percent. Compute the runoff for the parcel before and after development for a 25-year return period storm. The estimated time of concentration was 30 minutes before development and 20 minutes after development. Coefficients for the rainfall intensity equation are a = 2,261 for I in mm/hr (89 for I in in./hr), b = 8.5, and c = 0.754.

10.2. A 152 m × 152 m (500 ft × 500 ft) parcel of agricultural land is to be developed into high-density residential housing. Soils on the property are in hydrologic soil group B, with an average slope of 2 percent. Compute the runoff from the parcel before and after development for a 100-year return period storm. The estimated time of concentration was 40 minutes before development and 20 minutes after development. Coefficients for the rainfall intensity equation (10.2) are a = 2,692 for I in mm/hr (106 I in in./hr), b = 9.0, and c = 0.792.

10.3. A gutter with a Manning "n" value of 0.017 has a longitudinal slope of 2.0 percent and a cross slope of 30 H:1V. Determine the discharge rate and velocity in the gutter for a water depth of 0.15 m (0.5 ft).

10.4. Determine the water depth and velocity in a gutter (n = 0.017) when the discharge rate is 0.20 cms (7.0 cfs). The gutter has a longitudinal slope of 3 percent and a cross slope 20 H:1V.

10.5. Runoff from a 129.5-ha (320-ac) development is concentrated in a single channel. The time of concentration and runoff coefficient are 72 minutes and 0.20 before development and 40 minutes and 0.50 after development, respectively. Compute the peak discharge rates from the land before and after development using the rainfall intensity equation in Problem 10.1.

10.6. Size a grass-lined open channel to carry the peak runoff from the development in Problem 10.5 both before and after development based on $S_o = 0.003$, $B/d = 3.0$, $n = 0.035$, and $SS = 4$ H:1V. Use Eq. 5.30 for freeboard.

10.7. Compute the size of a concrete pipe ($n = 0.015$) to carry the runoff from the development in Problem 10.5 both before and after development. The slope of the pipe is 0.003.

10.8. Runoff from a development 600 m (1,968 ft) × 1,200 m (3,936 ft) is concentrated in a single channel. The time of concentration is estimated as 60 minutes before development and 30 minutes after development. Compute the runoff before and after development using the rainfall intensity equation listed in Problem 10.2, with a runoff coefficient of 0.2 before development and 0.5 after development.

10.9. Size a concrete-lined open channel to carry the discharge from the development in Problem 10.8 both before and after development based on $S_o = 0.003$, $B/d = 1.5$, $n = 0.015$, and $SS = 2$ H:1V. Use Eq. 5.30 for the freeboard.

10.10. Compute the time of concentration (t_c) for overland flow from a plane surface, with a length of 45.7 m (150 ft) and a slope of 0.5 percent. Use the rainfall intensity equation given in Problem 10.2, with a runoff coefficient of 0.4. Compute t_c for both a grass surface ($n = 0.2$) and a bare soil surface ($n = 0.1$).

10.11. Compute the time of concentration (t_c) for overland flow from a plane surface, with a length of 30 m (98.4 ft) and a slope of 2 percent. Use the rainfall intensity equation given in Problem 10.1, with a runoff coefficient of 0.6. Compute t_c for both a grass surface ($n = 0.2$) and a bare soil surface ($n = 0.1$).

10.12. Determine the capacity of a depressed curb inlet where the depth of flow in the gutter is 0.12 m (0.39 ft), the width of the depression is 0.40 m (1.31 ft), and the inlet length is 3.0 m (9.84 ft).

10.13. Determine the length of a curb inlet for a discharge rate 0.17 cms (6.0 cfs), a longitudinal gutter slope of 1.0 percent, a cross-slope 20 H:1V, $n = 0.017$, and a depression width of 0.5 m (1.64 ft).

10.14. Determine the length of a curb inlet for a discharge rate 0.25 cms (8.8 cfs), a longitudinal gutter slope of 2.5 percent, a cross-slope of 25 H:1V, $n = 0.017$, and a depression width 0.40 m (1.31 ft).

10.15. Size a concrete pipe ($n = 0.015$) to carry a discharge of 0.5 cms (17.6 cfs) on a slope of 0.005. What is the water depth in the pipe selected?

10.16. Size a corrugated metal pipe ($n = 0.025$) to carry a discharge of 3.0 cms (105.9 cfs) on a slope of 0.003. What is the water depth in the pipe selected? What size concrete pipe ($n = 0.015$) would be required?

10.17. The invert elevation of a 0.61-m (24-in.) diameter pipe flowing full into a manhole is 30.49 m (100.00 ft). What should the invert elevation be for a 0.61-m (24-in.) diameter pipe flowing out of the manhole with a velocity of 1.22 mps (4.0 fps). The minor loss coefficient for the manhole is 0.3.

10.18. The invert elevation of a 0.61-m (24-inch) diameter pipe flowing into a manhole is 30.49 m (100.00 ft). Determine the invert elevation of a 0.76-m (30-in.) diameter pipe flowing out of the manhole with a velocity of 1.22 mps (4.0 fps). The minor loss coefficient for the manhole is 0.3.

10.19. Compute the before and after development triangular-shaped unit hydrographs for 129.5-ha (320-ac) watershed if the time of concentration changes from 150 minutes before development to 90 minutes after development. Use $dt = 0.33$ hour.

10.20. Compute the runoff hydrographs before and after development for the watershed in Problem 10.19 based on the following incremental precipitation excess.

Time	Before		After	
hrs	mm	inches	mm	inches
0.00	0.0	0.0	0.0	0.0
0.33	5.1	0.2	15.2	0.6
0.67	15.2	0.6	25.4	1.0
1.00	25.4	1.0	40.6	1.6
1.33	20.3	0.8	30.5	1.2
1.67	10.2	0.4	20.3	0.8
2.00	5.1	0.2	10.2	0.4

10.21. Compute the storage in m^3 (ft^3) required in Problem 10.20 for a detention basin so that the peak discharge after development is equal to the peak discharge before development.

10.22. Compute the before and after development unit hydrographs for a 16.2-ha (40-ac) commercial development. The time of concentration before development is 60 minutes and after development is 30 minutes. Use $dt = 0.167$ hour.

10.23. Compute the runoff hydrographs from the watershed in Problem 10.22 based on the following values of incremental precipitation excess.

Time	Before		After	
hrs	mm	inches	mm	inches
0.00	0.0	0.0	0.0	0.0
0.17	2.5	0.1	5.1	0.2
0.33	7.6	0.3	15.2	0.6
0.50	12.7	0.5	25.4	1.0
0.67	10.2	0.4	20.3	0.8
0.83	5.1	0.2	10.2	0.4
1.00	2.5	0.1	5.1	0.2

10.24. Compute the storage required for a detention basin in Problem 10.23 such that the peak discharge rate after development is equal to the peak discharge rate before development.

10.25. Determine the size (h) of a square orifice in a detention basin flow control wall shown below to limit the discharge rate to 0.50 cms (17.6 cfs) under a head (H) of 1.52 m

(5.0 ft). Use an orifice coefficient of 0.6. What is the discharge rate through the orifice for an upstream water depth equal to the height of the orifice? Assume critical depth in the orifice and an entrance loss coefficient of 0.5.

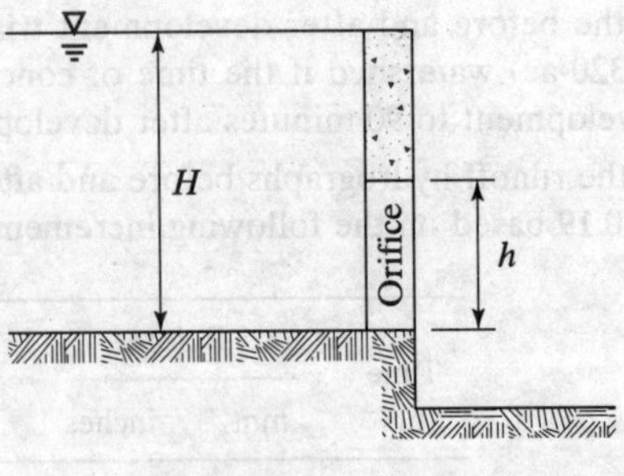

10.26. Determine the length of a weir in a detention basin to limit the discharge rate to 3.0 cms (105.9 cfs) under a head (H) of 0.76 m (2.49 ft). Use a weir discharge coefficient of 0.4.

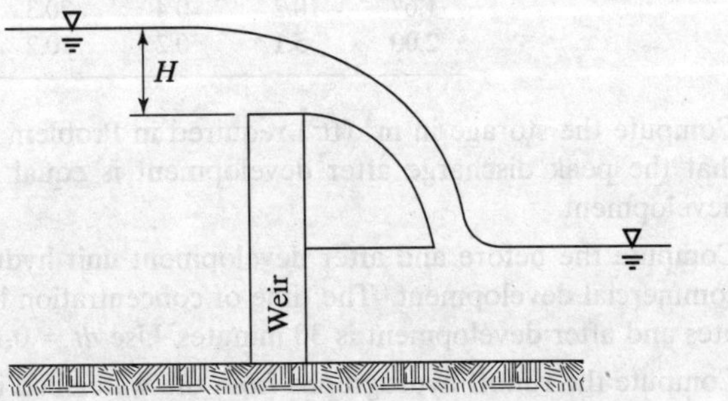

10.27. Determine the size of a concrete outlet pipe to limit the discharge from a detention basin to 5.7 cms (201 cfs) under a head of 5.0 m (16.4 ft). The outlet pipe has an entrance loss coefficient of 1.2, a Manning "n" value of 0.015, and an exit loss coefficient of 1.0.

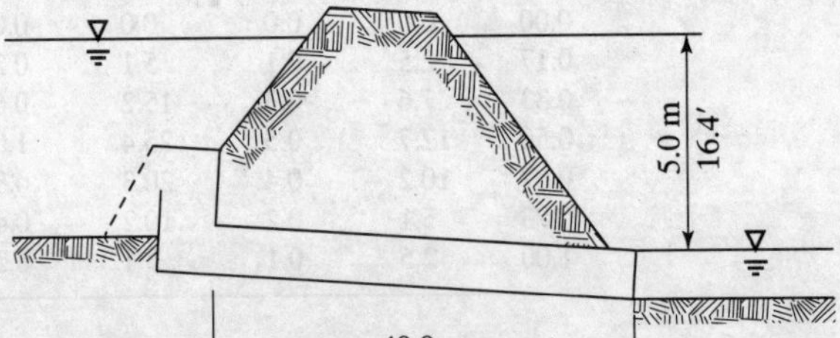

10.28. A weir shown below is used to limit the discharge from a detention basin. Compute the length of the weir (L_1 and L_2) using a weir coefficient of 0.4. Neglect end contractions.

Water depth H		Discharge rate	
ft	m	cfs	cms
5.0	1.52	—	—
8.0	2.44	70	2.0
10.0	3.05	247	7.0

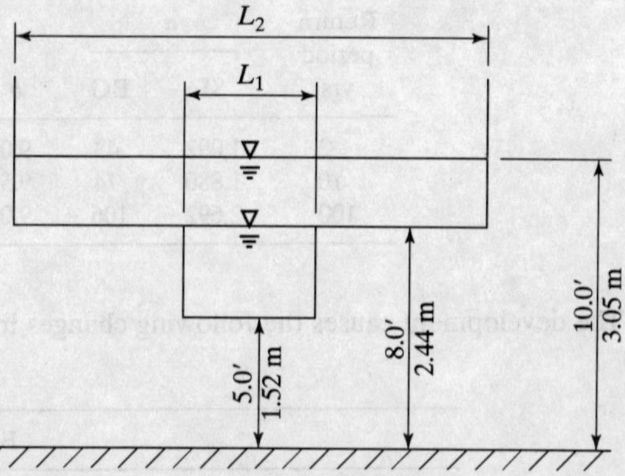

10.29. A detention basin discharge outlet control is a rectangular opening 0.91 m (3.0 ft) wide. Compute the discharge through the opening when the water depth (H) in the detention basin is 1.52 m (5.0 ft), 2.44 m (8.0 ft), and 3.05 m (10.0 ft). Assume a contraction loss coefficient of 0.5 and critical depth in the constriction.

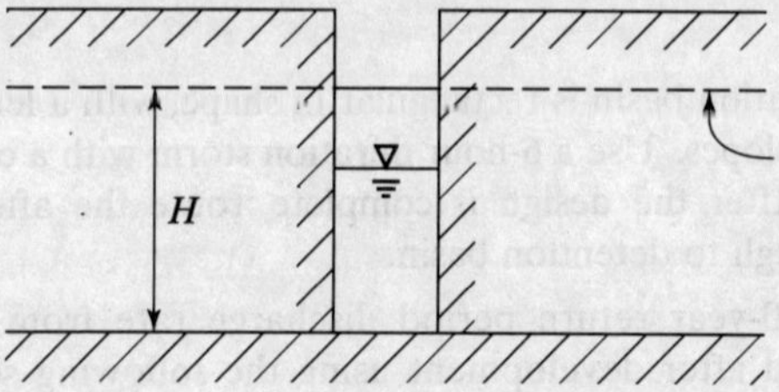

10.30. A triangular weir ($\theta = 20°$) is to be used to control the discharge from a detention basin. Compute the discharge through the weir ($K_T = 0.6$) for heads (H) of 1.52 m (5.0 ft), 2.44 m (8.0 ft), and 3.05 m (10.0 ft).

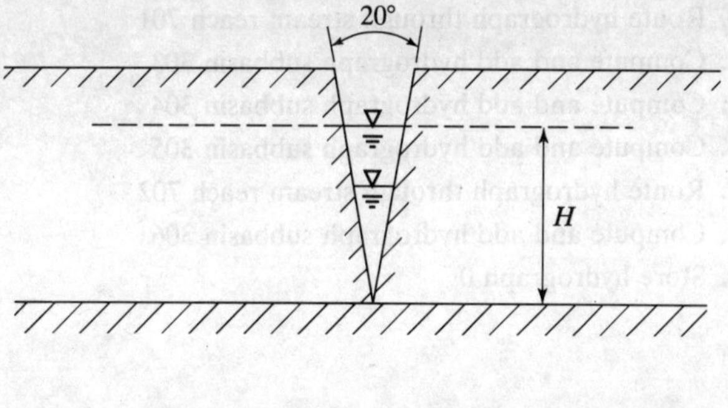

10.31. Design a detention basin for a 6.1-ha (15-ac) commercial development so that the peak discharge rates after development are no greater than the predevelopment peak discharge rates for the 2-hour, 10-year, and 100-year return period storms. Soils for the watershed are classified as silty clay loam, with an initial volumetric moisture content of 0.18. The coefficients for the rainfall intensity equation (Eq. 7.50) are:

Return period yrs	*a*		*b*	*c*
	SI	BG		
2	1,092	43	9.0	0.794
10	1,880	74	9.5	0.801
100	2,692	106	9.0	0.792

The development causes the following changes in the watershed:

	Before	After
Time of concentration (minutes)	60	30
Percent impervious	0	50
Initial precipitation abstraction		
mm	25.4	12.7
inches	1.0	0.5

Assume the detention basin is rectangular in shape, with a length to width ratio of 3 and 4 H:1V side slopes. Use a 6-hour duration storm with a computational time step of 10 minutes. After the design is complete, route the after development runoff hydrograph through to detention basin.

10.32. Compute the 100-year return period discharge rate from the watershed shown below before and after development using the following sequence of hydrologic operations:

1. Compute hydrograph from subbasin 301
2. Compute and add hydrograph subbasin 302
3. Route hydrograph through stream reach 701
4. Compute and add hydrograph subbasin 303
5. Compute and add hydrograph subbasin 304
6. Compute and add hydrograph subbasin 305
7. Route hydrograph through stream reach 702
8. Compute and add hydrograph subbasin 306
9. Store hydrograph 0

The 100-year rainfall intensity equation for the area is

$$I = \frac{95}{(t + 8.0)^{0.728}} \text{ in./hr}$$

$$= \frac{2{,}413}{(t + 8.0)^{0.728}} \text{ mm/hr}$$

Use a 24-hour duration storm with dt = 10 minutes and an initial rainfall abstraction of 1.0 inch (25.4 mm) before development and 0.5 inches (12.7 mm) after development. The volumetric soil moisture content of the soil is 0.25. Subbasin characteristics are listed below.

ID	Area		Time (hrs) concentration		Soil	Percent impervious	
no.	ac	ha	Before	After	type*	Before	After
301	380	154	1.9	1.1	4	0	35
302	220	89	1.8	1.0	5	0	35
303	210	85	1.0	0.6	6	0	40
304	150	61	1.6	0.9	5	0	30
305	140	57	0.9	0.5	6	0	40
306	200	81	1.5	0.8	5	0	30

*See program listing WATERSHD.FOR.

The channel has a slope of 0.003, with a channel Manning "n" value of 0.08 and an overbank "n" value of 0.12. After development, "n" values will be 0.04 and 0.06, respectively. Use the following cross-sections for routing with NSEC = 80 and PERQ = 2.

Reach 701 L = 5,800 ft (1,770 m)

BG units

Elevation	350.	345.	343.	340.	339.	342.	344.	350.
Distance	0	15.	40.	45.	55.	60.	85.	100.

SI units

Elevation	106.7	105.2	104.6	103.6	103.4	104.3	104.9	106.7
Distance	0	4.6	12.2	13.7	16.8	18.3	25.9	30.5

Reach 702 L = 4,100 ft (1,250 m)

BG units

Elevation	320.	315.	314.	310.	309.	313.	315.	320.
Distance	0	20.	50.	55.	65.	70.	100.	120.

SI units

Elevation	97.6	96.0	95.7	94.5	94.2	95.4	96.0	97.6
Distance	0	6.1	15.2	16.8	19.8	21.3	30.5	36.6

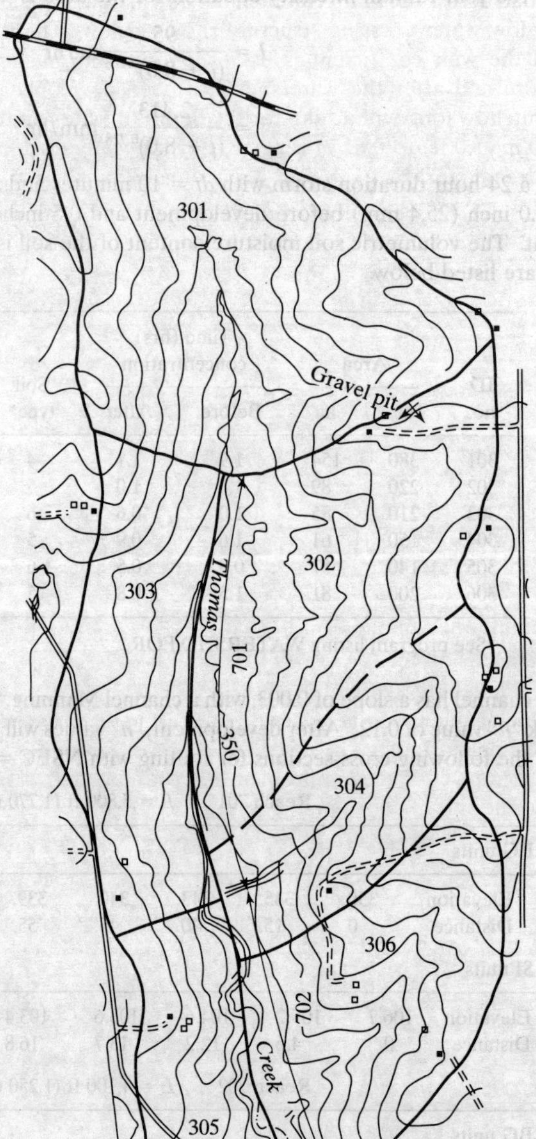

Problem 10.32 Watershed maps.

10.33. A road across the main channel just downstream of the watershed in Problem 10.32 is a low water crossing structure shown below. The weir length is 61.0 m (200 ft) and the weir coefficient is 0.35. Compute the maximum depth on the roadway before and after the watershed has been developed. During the 100-year flood event, how long will it take for the depth of water on the roadway to increase from 0.30 m (1.0 ft) to 0.61 m (2.0 ft) after the watershed has been developed? Neglect upstream storage.

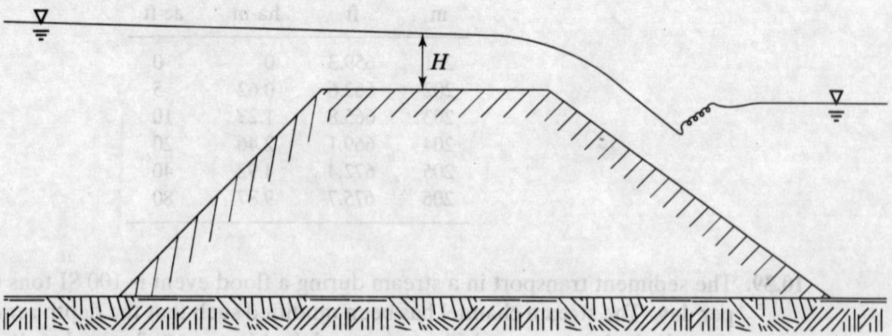

10.34. Compute the volume of off-channel storage required to reduce the 100-year return period peak discharge rate in the downstream channel after development to the pre-development discharge rate in Problem 10.32.

10.35. A side-channel weir is to be used to divert the water from the main channel to the off-channel storage in Problem 10.34. Route the after development outflow hydrograph computed in Problem 10.32 through the three channels in series listed below. The side-channel weir crest is 2.13 m (7.0 ft) above the channel bottom and has a discharge coefficient of 0.4. Use MNOS = 40.

Channel no.	Bottom width m	Bottom width ft	Side slopes	Length m	Length ft	Channel slope	Manning "n"
1	13.0	42.8	3	100	328	0.001	0.035
2	13.0	42.8	2	100	328	0.0005	0.035
3	10.0	32.8	2	100	328	0.001	0.035

10.36. A commercial development covers an area of 10 ha (24.7 ac). Determine the size of a first flush treatment unit designed to capture and treat the first 12.7 mm (0.5 in.) of runoff. Determine the dimensions of the unit if the maximum depth is 3.5 m (11.4 ft) and the L/W ratio is approximately 5. Allow 0.50 m (1.6 ft) for dead storage.

10.37. Determine the volume of a retention basin permanent pool for a 100-ha (247-ac)high-density residential development in Dallas, Texas. Use hydrologic soil group D with a 6 percent slope.

10.38. Determine the area of the orifice in the riser pipe in the retention basin in Problem 10.37 such that 95 percent of the surcharge storage is discharged from the reservoir in 48 hours. The retention basin capacity table is listed below. Use an orifice discharge coefficient of 0.6. Assume the surcharge storage volume is equal to the permanent pool storage volume.

Elevation		Storage	
m	ft	ha-m	ac-ft
201	659.3	0	0
202	662.6	0.62	5
203	665.8	1.23	10
204	669.1	2.46	20
205	672.4	4.92	40
206	675.7	9.87	80

10.39. The sediment transport in a stream during a flood event is 100 SI tons (110 BG tons) per day. The stream channel has been modified such that the depth remains the same, the velocity has increased 50 percent, and the Manning "n" value has decreased 30 percent. Estimate the rate of erosion or deposition in the channel.

10.40. Compute the 100-year return period floodplain limits for the downstream channel in Problem 10.32 before, Q_p = 69.0 cms (2,400 cfs), and after development, Q_p = 124.6 cms (4,400 cfs). Use Manning "n" values of 0.08 for the channel and 0.12 for the overbank. Start the water surface profile computations using normal depth (S_o = 0.003). Base the computations on the following cross-sections 30.5 m (100 ft) apart. Use a contraction loss coefficient of 0.2 and expansion loss coefficient of 0.4.

Cross-section 1

Elevation									
	m	94.48	92.96	91.44	89.91	89.91	91.44	92.96	94.48
	ft	310.0	305.0	300.0	295.0	295.0	300.0	305.0	310.0
Distance									
	m	0	60.96	76.20	85.34	97.53	106.67	121.91	182.87
	ft	0	200.0	250.0	280.0	320.0	350.0	400.0	600.0

Cross-section 2

Elevation									
	m	94.48	93.05	91.50	90.03	90.00	91.56	93.08	94.48
	ft	310.0	305.3	300.2	295.4	295.3	300.4	305.4	310.0
Distance									
	m	0	57.91	79.24	85.34	94.48	103.63	118.87	167.63
	ft	0	190.0	260.0	280.0	310.0	340.0	390.0	550.0

Cross-section 3

Elevation								
m	94.79	93.11	91.53	90.16	90.09	91.53	92.96	94.79
ft	311.0	305.5	300.3	295.8	295.6	300.3	305.0	311.0
Distance								
m	0	54.86	73.15	82.29	94.48	109.72	121.91	152.39
ft	0	180.0	240.0	270.0	310.0	360.0	400.0	500.0

Cross-section 4

Elevation								
m	94.79	93.26	91.74	90.22	90.19	91.71	93.23	94.79
ft	311.0	306.0	301.0	296.0	295.9	300.9	305.9	311.0
Distance								
m	0	45.72	79.24	91.44	103.63	109.72	121.91	167.63
ft	0	150.0	260.0	300.0	340.0	360.0	400.0	550.0

Cross-section 5

Elevation								
m	95.09	93.33	91.83	90.31	90.22	92.05	93.57	95.09
ft	312.0	306.2	301.3	296.3	296.0	302.0	307.0	312.0
Distance								
m	0	64.00	91.44	97.53	106.67	112.77	121.91	182.87
ft	0	210.0	300.0	320.0	350.0	370.0	400.0	600.0

BIBLIOGRAPHY

American Iron and Steel Institute, *Modern Sewer Design,* 4th Ed., American Iron and Steel Institute, Washington, D.C., 1999.

American Society of Civil Engineers and Water Environment Federation, *Design and Construction of Urban Stormwater Management Systems,* WEF Manual of Practice FD-20 and ASCE Manual and Report on Engineering Practice No. 77, New York, 1992.

American Society of Civil Engineers and Water Environment Federation, *Urban Runoff Quality Management,* WEF Manual of Practice No. 23 and ASCE Manual and Report on Engineering Practice No. 87, New York, NY, 1998.

BEDIENT, P. B., and W. C. HUBER, *Hydrology and Floodplain Analysis,* 2nd Ed., Addison-Wesley, Reading, MA, 1988.

GARDE, R. J., and M. L. ALBERTSON, "Discussion of Paper by Laursen," *Journal of Hydraulics Division,* ASCE, No. 1856, November 1958.

HANN, C. T., B. J. BARFIELD, and J. C. HAYES, *Design Hydrology and Sedimentology for Small Catchments,* Academic Press, San Diego, CA, 1994.

HEANEY, J. P., W. C. HUBER, and S. J. NIX, *Storm Water Management Model: Level I—Preliminary Screening Procedure,* U.S. Environmental Protection Agency, Washington, D.C., 1976.

HOGGAN, D. H., *Computer-Assisted Floodplain Hydrology and Hydraulics,* McGraw-Hill, New York, NY, 1989.

JAMES, W. P., J. WARINNER, and M. REEDY, "Application of the Green-Ampt Infiltration Equation to Watershed Modeling," Water Resources Bulletin, American Water Resources Association, Vol. 28, No. 3, 1992.

LAURSEN, E. M., "The Total Sediment Load in Streams," *Journal of Hydraulics Division,* ASCE, No. 1530, February 1958.

MASON, J. M., and B. E. MILLER, *Guidelines for Drainage Design,* Texas Engineering Extension Service, College Station, TX, 1986.

MCCUEN, R. H., *A Guide to Hydrologic Analysis Using SCS Methods,* Prentice Hall, Upper Saddle River, NJ, 1982.

MORRIS, H. M., and J. M. WIGGERT, *Applied Hydraulics in Engineering,* Ronald Press Company, New York, NY, 1972.

NORMANN, J. M., *Design of Stable Channels with Flexible Linings,* FHA Circular 15, U.S. Department of Transportation, Washington, D.C., 1975.

PETERSEN, M. S., *River Engineering,* Prentice Hall, Upper Saddle River, NJ, 1986.

SCHUMM, S. A., *A Tentative Classification of Alluvial River Channels,* Circular 477, U.S. Geological Survey, Reston, VA, 1963.

URBONAS, B., and P. STAHRE, *Stormwater: Best Management Practice and Detention for Water Quality, Drainage and CSO Management,* Prentice Hall, Upper Saddle River, NJ, 1993.

VIESSMAN, W., and G. L. LEWIS, *Introduction to Hydrology,* Harper Collins, New York, NY, 1996.

WALESH, S. G., *Urban Surface Water Management,* John Wiley & Sons, New York, NY, 1989.

WANIELISTA, M. P., and Y. A. YOUSEF, *Stormwater Management,* John Wiley & Sons, New York, NY, 1993.

11

Water Resources Systems Analysis

Deciding among alternative courses of action is fundamental to water resources engineering, as well as to many other aspects of our lives. A broad spectrum of analysis methodologies have been developed to provide a systematic, quantitative basis for decision-making. This chapter applies selected methods from the allied fields of systems engineering and engineering economics to water resources planning and management decisions. Three interrelated types of decision-support tools are introduced: benefit-cost analysis, simulation, and optimization. Economic evaluation methods are combined with hydrologic and hydraulic simulation models covered in previous chapters to determine economically optimal flood damage reduction plans. Linear programming is used to develop simulation and optimization models to support project selection, reservoir system operation, water allocation, pollution load allocation, and other water resources planning and management decisions.

11.1 THE SYSTEMS PHILOSOPHY

Water resources systems engineering may be defined as the art and science of formulating and evaluating alternative water management plans and selecting that particular set of actions that will best accomplish specified objectives, within the constraints of governing natural laws, engineering principles, economics, environmental protection requirements, social and political concerns, legal restrictions, and institutional and financial capabilities. Systems analysis principles and methods adopted by water resources engineers are also applied in a broad spectrum of other fields. The terms *systems engineering*, *operations research*, and *management science* are used in various fields of engineering and business management to refer to similar analysis strategies and techniques.

Systems analysis consists of both (1) a philosophy or general approach for analyzing problems and making decisions and (2) computer-based mathematical modeling tools. In general, systems analysis, whether applied in water resources planning and management or other totally different fields, has the following characteristics:

- Systematic quantitative approach to determining the optimum solutions to complex problems
- Decision-making support
- Comprehensive integrated systems focus
- Interdisciplinary aspects
- Reliance on mathematical models and computers

A *system* can be defined as a collection of components, connected by some type of interaction or interrelationship, that collectively responds to some stimulus or demand and fulfills some specific purpose or function. Examples of natural and/or man-made water resources systems include the hydrologic cycle, river basins, streams, aquifers, municipal water supply systems, wastewater management systems, irrigation systems, stormwater management systems, flood damage reduction systems, navigation systems, multiple-purpose reservoir systems, and conjunctive surface/ground water management systems. These are all physical systems. Systems may also consist partially or wholly of resource management policies and practices, such as environmental regulatory systems, water right allocation systems, water conservation programs, and floodplain management regulations.

A model of a system is a conceptualization of that system which preserves its essential characteristics. Models are used to study the performance of existing or proposed systems. A model is simpler, easier to understand, easier to construct, and easier to manipulate than the system it represents. The ultimate value of a model depends on its ability to aid in decision making. Models are used in all stages of the problem-solving process. Models never perfectly represent reality. They preserve the essential characteristics of a system without considering all of the system's complexities. You must refine the model sufficiently to serve your particular purposes.

Examples of types of models include thoughts, language, schematic representations, physical representations (scale and analog models), and mathematical equations. Any simplified representation of a real-world system is a model. Our focus is on mathematical models that simulate and optimize water resources systems.

The modeling methods outlined in this chapter are applicable to the full spectrum of water management activities, including various levels of planning, project design, construction management, system operations and maintenance, and administration of water rights and environmental regulatory programs. Regardless of the application, the general steps in the systems approach are as follows:

- *Define the problem*: Compile and analyze information to develop an understanding of relevant systems and the problems, needs, and opportunities being addressed.
- *Establish objectives*: Define broad goals, specific objectives, and quantitative measures to evaluate how well alternative solutions meet the objectives.
- *Formulate feasible alternatives*: Develop alternative water management strategies, plan configurations, and system designs that address the objectives while satisfying constraints.
- *Evaluate the alternatives*: Assess the consequences of each alternative and the tradeoffs between the alternatives.
- *Select the best alternative*: Decide on the optimum course of action.

Although occurring generally in sequential order, these tasks are overlapping and iteratively repeated throughout the decision-support process. For example, additional insight gained in formulating and evaluating alternative solutions often leads to a refined definition of problems and needs and modification of objectives.

Evaluating the feasibility of alternative plans is a key aspect of decision processes. Feasibility typically includes engineering, economic, financial, environmental, social, institutional, legal, and political dimensions. Engineering feasibility means that the plan will physically function as required from the perspective of hydrologic, hydraulic, geotechnical, structural, and other engineering specifications. Economic feasibility means the benefits exceed the costs. A plan is financially feasible if funds can be obtained to pay for it. Environmental protection laws and requirements must be met. Social feasibility relates to the plan having public support and not adversely impacting human welfare. Institutional feasibility implies that agencies and programs either exist or can be created to carry out all aspects of the plan. The plan must be consistent with federal and state law. Often the key element in plan implementation is political support from city councils, state legislatures, the U.S. Congress, and a myriad of public officials and their constituents.

11.2 ECONOMIC ANALYSIS

Economic feasibility and optimality represent just one of many considerations in the decision-making process. However, economic analysis does play an important central role in decision-making at many levels, in various settings. Economic evaluation of water resources management plans combines basic methods of engineering economics with benefit estimation procedures developed for specific water management sectors, such as flood mitigation, municipal and industrial water supply, irrigation, hydropower, navigation, and recreation. Analyses of economic costs and benefits provide important information for use, along with various other forms of information, in making a myriad of decisions in planning, design, operations, and other water resources engineering activities.

One of the major accomplishments of the federal programs for water resources development has been the introduction of economic criteria into government decision-making (James and Lee, 1971). In the Flood Control Act of 1936 (Public Law 74–738), the U.S. Congress stated that federal participation in flood control projects was warranted only if *"the benefits to whomsoever they may accrue are in excess of the estimated costs."* Economic evaluation of irrigation projects dates back to the Reclamation Act of 1902 (PL 57–161). Federal policies and procedures are outlined by the Water Resources Council (1983). Concepts of economic evaluation encompass the different water resources development and management purposes and are applied to nonfederal as well as federal projects (James and Lee, 1971; Goodman, 1984; American Water Works Association, 2001).

Economic evaluation provides both a feasibility constraint and an objective function for comparing alternatives. The economic feasibility criterion is that benefits must exceed costs. The economic objective function for comparing alternative plans may be in either of the following alternative forms:

- maximize net benefits, which are benefits less costs
- minimize cost required to provide a specified level of service
- maximize benefits derived from fixed resources

In many cases, both benefits and costs are relevant and are included in the analysis. For example, flood mitigation system evaluation procedures outlined later incorporate both the benefits of reducing flood damages and the costs of implementing and maintaining flood mitigation measures. Other decision problems may be based on minimizing cost while providing a specified level of service, without actually estimating benefits. Storm drainage systems are typically designed to provide a specified level of protection at minimum cost. Wastewater treatment facilities are based on meeting water quality standards at minimum cost. Water supply systems are designed to provide needed amounts of water at acceptable levels of reliability at minimum cost. On the other hand, a decision problem may involve maximization of benefits in situations where costs are essentially fixed. For example, operating plans for an existing multipurpose reservoir system may be optimized based on maximizing benefits.

11.2.1 Engineering Economics

Engineering economics is a set of principles applied in comparing alternative plans to determine the economically optimal (Newnam and Johnson, 1995; Revelle, Whitlach, and Wright, 1997). Equivalence of kind and equivalence of time are required so that all relevant costs and benefits of each alternative are comparable. Equivalence of kind is achieved by expressing all benefits and costs included in the analysis in dollars (or other monetary units in other countries). Equivalence of time is achieved through discounting techniques using compound interest formulas. Having a dollar today is worth more than obtaining a dollar at some future time, because the dollar in hand today can be invested and accrue interest.

TABLE 11.1 DISCOUNTING FORMULAS AND FACTORS

Equation	Notation form	Factor in parentheses	
$F = P(1+i)^N$	$F = P(F/P,i,N)$	Single payment present worth	**(11.1)**
$P = F\left(\frac{1}{(1+i)^N}\right)$	$P = F(P/F,i,N)$	Single payment compound amount	**(11.2)**
$P = A\left(\frac{(1+i)^N - 1}{i(1+i)^N}\right)$	$P = A(P/A,i,N)$	Series present worth	**(11.3)**
$A = P\left(\frac{i(1+i)^N}{(1+i)^N - 1}\right)$	$A = P(A/P,i,N)$	Capital recovery	**(11.4)**
$F = A\left(\frac{(1+i)^N - 1}{i}\right)$	$F = A(F/A,i,N)$	Series compound amount	**(11.5)**
$A = F\left(\frac{i}{(1+i)^N - 1}\right)$	$A = F(A/F,i,N)$	Sinking fund	**(11.6)**

Benefits and costs associated with water projects occur at various times. Initial investment costs occurring at the beginning of the project life are associated with construction or implementation. Operation and maintenance costs continue throughout the life of the project. Major replacement and rehabilitation costs may occur periodically. Benefits typically accrue over long periods of time. Time streams of benefits and costs may be converted to other equivalent cash flows for purposes of comparison using discounting formulas from Table 11.1 with a specified discount rate.

The discounting formulas presented in Table 11.1 convert cash flows between a present amount P, future amount F, and uniform annual series A, which are illustrated in Fig. 11.1. The factors within the parentheses are a function of the annual interest or discount rate i and number of compounding periods (years) N. In comparing alternatives, all cash flows are converted to the same time base and expressed as either an equivalent present worth, future worth, or uniform annual series.

In addition to adopting a consistent time base, all cash flows are also expressed at constant price levels. For example, all benefits and costs might be expressed in

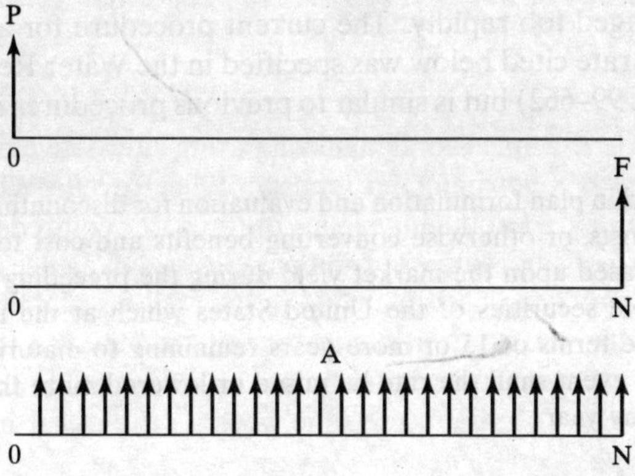

Figure 11.1 The equivalence formulas in Table 11.1 relate a present amount P at time 0, a future amount F at the end of year N, and a uniform annual series A.

constant year 2002 dollars, even though they are projected to occur at various times over a 50-year period of analysis with price levels varying with inflation and other economic factors. A consistent period of analysis and discount rate are also used for all alternatives.

The period of analysis is the length of time over which project benefits and costs are included in the comparative evaluation. The economic life ends when the incremental benefits from continued use no longer exceed the cost of continued operation. The period of analysis should not exceed the economic life, unless project replacement is considered, and may be shortened arbitrarily to exclude the highly uncertain events of the very distant future. The same period of analysis must be adopted for all alternatives even if their economic lives differ. As illustrated by Example 11.2, annual worth analysis conveniently handles varying lifetimes based on assuming an infinite analysis period with projects being replaced at the end of each economic life.

The discount rate is an expression of the time value of capital used in equivalence computations comparing alternatives. Although the arithmetic of the compound interest formulas of Table 11.1 is the same for banking or business transactions and comparing water management plans, the conceptual meaning of the interest rate differs. In business transactions, the interest rate is the fee a lender charges a borrower for the use of money and is determined by the capital market. In water resources engineering and other areas of public works, the ideal discount rate would achieve a rate of capital formation maximizing total social welfare. The rate is essentially a value judgment based on a compromise between present consumption and capital formation from the perspective of the general public. Construction projects, with high initial investment and benefits accruing in the future, are less likely to be economically feasible as the discount rate increases.

The discount rate is often linked to the concept of marginal internal rate of return in private industry. If funds were committed to the project yielding the highest return first, and then to subsequent projects in order of rate of return, the rate of return of the last project selected before funds ran out would be the marginal internal rate of return. The discount rate used by the federal water agencies in the U.S. is based on the market interest rate for risk free investment, with the limitation that the rate not be changed too rapidly. The current procedure for annually updating the federal discount rate cited below was specified in the Water Resources Development Act of 1986 (PL 99–662) but is similar to previous procedures in effect since the early 1960s.

> The interest rate to be used in plan formulation and evaluation for discounting future benefits and computing costs, or otherwise converting benefits and cost to a common time basis, shall be based upon the market yield during the preceding year on interest-bearing marketable securities of the United States which at the time the computation is made, have terms of 15 or more years remaining to maturity. Provided, however, that in no event shall the rate be raised or lowered more than one-quarter of 1 percent for any year.

Example 11.1

A person borrows $20,000 to be repaid in 10 annual payments at an annual interest rate of 8.0 percent. Compute the payments.

$$A = P\left(\frac{i(1+i)^N}{(1+i)^N - 1}\right) = \$20{,}000\left(\frac{0.08(1+0.08)^{10}}{(1+0.08)^{10} - 1}\right) = \$2{,}980.59/\text{year}$$

Example 11.2

A new pumping plant is to be constructed for a water supply system. Either alternative A or B will provide the required flow capacity. As indicated in Table 11.2, alternative A costs more to construct but lasts longer. Alternative B has a lower initial investment but higher operation and maintenance cost and a shorter economic life. The economically optimum plan is to be determined.

Solution Cash flow diagrams are presented in Fig. 11.2. The initial investment costs for construction and equipment purchase and installation are assumed to occur at time 0. Continuing operation and maintenance costs are treated as an annual aggregated average amount placed at the end of each year. The salvage value is an income treated as a negative cost. A common 50-year period of analysis is adopted for both alternatives based on the premise that pumping plant B can be replaced at the end of its 25-year economic life with identically the same facility and equipment.

TABLE 11.2 COST DATA FOR EXAMPLE 11.2

Alternative	Initial investment costs	Annual operation and maintenance	Salvage value	Life (yrs)
A	$525,000	$26,000	0	50
B	$312,000	$48,000	$50,000	25

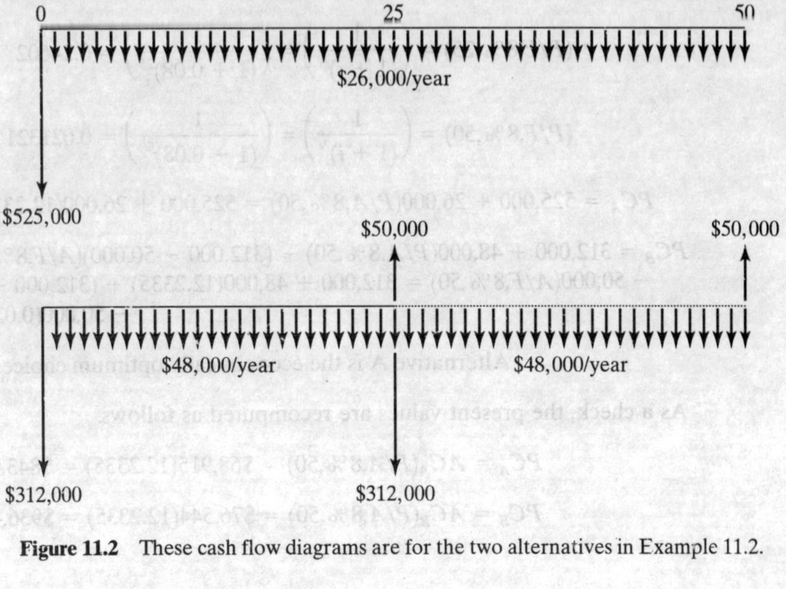

Figure 11.2 These cash flow diagrams are for the two alternatives in Example 11.2.

Both annual worth and present worth analysis solutions are presented. Annual worth analysis is more convenient because it automatically handles the different lives based on the inherent premise of equal replacement forever. A present worth analysis is also presented to demonstrate that it yields the same result.

In the first solution, all amounts are converted to equivalent annual values to determine which alternative has the lowest annual cost, AC.

$$(A/P,8\%,50) = \left(\frac{i(1+i)^N}{(1+i)^N - 1}\right) = \left(\frac{0.08(1+0.08)^{50}}{(1+0.08)^{50} - 1}\right) = 0.081743$$

$$(A/P,8\%,25) = \left(\frac{i(1+i)^N}{(1+i)^N - 1}\right) = \left(\frac{0.08(1+0.08)^{25}}{(1+0.08)^{25} - 1}\right) = 0.093679$$

$$(A/F,8\%,25) = \left(\frac{i}{(1+i)^N - 1}\right) = \left(\frac{0.08}{(1+0.08)^{25} - 1}\right) = 0.013679$$

$$AC_A = 525{,}000(A/P,8\%,50) + 26{,}000 = 525{,}000(0.081743) + 26{,}000 = \$68{,}915$$

$$AC_B = 312{,}000(A/P,8\%,25) + 48{,}000 - 50{,}000(A/F,8\%,25)$$
$$= 312{,}000(0.093679) + 48{,}000 - 50{,}000(0.013679) = \$76{,}544$$

Alternative A is the economically optimum choice.

The problem is solved again based on converting all amounts to equivalent present costs PC to demonstrate that the results are the same with either approach. Since the same period of analysis must be used for both alternatives, an equal cost replacement of alternative A after 25 years is assumed as shown in Fig. 11.2.

$$(P/A,8\%,50) = \left(\frac{(1+i)^N - 1}{i(1+i)^N}\right) = \left(\frac{(1+0.08)^{50} - 1}{0.08(1+0.08)^{50}}\right) = 12.2335$$

$$(P/F,8\%,25) = \left(\frac{1}{(1+i)^N}\right) = \left(\frac{1}{(1+0.08)^{25}}\right) = 0.14602$$

$$(P/F,8\%,50) = \left(\frac{1}{(1+i)^N}\right) = \left(\frac{1}{(1+0.08)^{50}}\right) = 0.021321$$

$$PC_A = 525{,}000 + 26{,}000(P/A,8\%,50) = 525{,}000 + 26{,}000(12.2335) = \$843{,}070$$

$$PC_B = 312{,}000 + 48{,}000(P/A,8\%,50) + (312{,}000 - 50{,}000)(A/F,8\%,25)$$
$$- 50{,}000(A/F,8\%,50) = 312{,}000 + 48{,}000(12.2335) + (312{,}000 - 50{,}000)(0.14602)$$
$$- 50{,}000(0.021321) = \$936{,}400$$

Alternative A is the economically optimum choice.

As a check, the present values are recomputed as follows.

$$PC_A = AC_A(P/A,8\%,50) = \$68{,}915(12.2335) = \$843{,}070$$

$$PC_B = AC_B(P/A,8\%,50) = \$76{,}544(12.2335) = \$936{,}400$$

Example 11.3

The following alternative plans for reducing flood damages along a reach of a river are being considered:

Plan 1—channel improvements consisting of widening and straightening the river reach
Plan 2—flood retarding dam A
Plan 3—flood retarding dam B, which is at the same site but is larger than dam A
Plan 4—both the channel improvements and dam A
Plan 5—both the channel improvements and dam B

The initial construction and related investment costs and average annual operation and maintenance costs for each plan are tabulated in columns 2 and 3 of Table 11.3. Without implementation of flood control improvements, average annual flood damages of \$525,000/year are expected to occur (Section 11.3). The channel improvements and dam will reduce the average annual flood damages to the estimated amounts in column 4. The discount rate is 7 percent. A 50-year period of analysis has been adopted.

(a) Select the optimum plan based on the objective of minimizing total annual cost.
(b) Select the optimum plan based on the objective of maximizing net benefits as discussed in the next section.
(c) Select the optimum plan based on an incremental benefit-cost ratio analysis.

These alternative analysis approaches always yield the same results. They are conceptually identical and simply represent alternative ways of organizing the comparative evaluation of alternative plans.

Solution The initial investment costs in column 2 of Table 11.3 are converted to equivalent annual costs shown in column 5 by multiplying by the following capital recovery factor.

$$(A/P,7\%,50) = \left(\frac{i(1+i)^N}{(1+i)^N - 1}\right) = \left(\frac{0.07(1+0.07)^{50}}{(1+0.07)^{50} - 1}\right) = 0.072460$$

TABLE 11.3 TOTAL ANNUAL COST COMPARISON FOR EXAMPLE 11.3

Flood control plan (1)	Initial investment (2)	Operation and maintenance (3)	Average annual damages (4)	Annual worth of investment (5)	Total annual cost (6)
	\$	\$	\$	\$	\$
No project	0	0	525,000	0	525,000
Plan 1	380,000	125,000	312,000	27,540	464,540
Plan 2	1,620,000	77,000	238,000	117,390	432,390
Plan 3	1,970,000	113,000	156,000	142,750	411,750
Plan 4	2,000,000	202,000	125,000	144,920	471,920
Plan 5	2,350,000	238,000	83,000	170,280	491,280

TABLE 11.4 NET ANNUAL BENEFIT COMPARISON FOR EXAMPLE 11.3

Plan (1)	Annual benefits (2)	Annual costs (3)	Net benefits (4)	BCR (5)	Incremental BCR (6)
	$	$	$		
1	213,000	152,540	60,460	1.40	1.40
2	287,000	194,390	92,610	1.48	1.77
3	369,000	255,750	113,250	1.44	1.34
4	400,000	346,920	53,080	1.15	0.34
5	442,000	408,280	33,720	1.08	0.68

The total annual costs in column 6 of Table 11.3 are computed as the sum of columns 3, 4, and 5. Plan 3 is the economically optimum plan because its total annual cost of $411,750 is less than any of the other plans, including the default no-action plan of constructing no flood control improvements at all.

An alternative comparison is presented in Table 11.4 based on computing benefits as the reduction in average annual damages achieved by each plan. Thus, the benefits tabulated in column 2 of Table 11.4 are computed as $525,000 less the annual damages from column 4 of Table 11.3. The annual costs in column 3 of Table 11.4 are the sum of columns 3 and 5 of Table 11.3. Net benefits in column 4 are benefits less costs. The benefit-to-cost ratio (BCR) in column 5 is column 2 divided by column 3. The five flood control plans are all economically feasible, because their benefits exceed costs (net benefits > 0 and BCR > 1.0). Plan 3 is optimum because its net annual benefits of $113,250 are greater than the other plans. Determining the optimum plan based on maximizing net annual benefits (Table 11.4) yields the same result as minimizing total annual costs (Table 11.3), because the benefits are computed as a cost reduction.

Maximizing net benefits is not the same as maximizing the benefit-to-cost ratio (BCR), unless the incremental BCR approach is used. For example, with a BCR of 1.48, plan 2 has the highest BCR but is not the economically optimal plan.

In an incremental analysis, plans ranked in order of increasing cost are compared two at a time. The alternative with higher cost is selected if and only if the incremental benefits exceed the incremental costs. Incremental BCRs are shown in column 6 of Table 11.4. In comparing each plan with the next more costly plan, the more costly plan is selected if the incremental BCR exceeds 1.0. Plan 2 is better than plan 1 because the incremental BCR of 1.77 is greater than 1.0. The annual cost for plan 3 is $61,360 greater than for plan 2 ($255,750 − $194,390 = $61,360). The corresponding incremental benefits are $82,000. Because the incremental BCR of 1.34 ($82,000/$61,360 = 1.34) is greater than 1.0, plan 3 is selected over plan 2. Plan 4 is not better than plan 3, because the incremental BCR of 0.34 is less than 1.0. Plan 3 is optimal.

11.2.2 Benefits and Costs

In many cases, all of the alternative plans being compared provide essentially the same service or benefit, and the service must be provided. The economic evaluation is simply a cost analysis without the necessity for estimating benefits. In other situations, alternative plans provide different levels of service or benefits. Benefits must be evaluated to determine the optimum alternative. Project justification may also require a demonstration that benefits exceed costs.

Cost estimation plays a central role in all fields of civil engineering, including water resources engineering. Agencies and consulting firms routinely prepare cost estimates for many different types of projects using similar procedures. However, benefit estimation procedures vary greatly, depending on the water management purpose. Different benefit estimating methods have been developed for flood mitigation, irrigation, municipal and industrial water supply, navigation, hydropower, recreation, water quality management, and other water management purposes.

Cost estimates include various items. Construction costs are based on unit costs for clearing land, excavating earth, placing concrete, and other materials and labor. Real estate costs are based on the market value of the lands and properties to be acquired for the project. Equipment manufacturers supply prices for their products. Engineering, design, construction supervision, and administration are included in project cost estimates. Interest during construction is included for multiple-year construction projects. Operation and maintenance costs include labor hours, materials, and equipment. A contingency factor is usually applied to cost estimates to cover unforeseen expenses.

Water resources planning textbooks, such as James and Lee (1971) and Goodman (1984), outline benefit estimation principles and methods. The Water Resources Council (1983) provides general guidelines followed by the federal water agencies. Detailed benefit evaluation procedures developed within the agencies are documented by agency manuals and reports. The Institute for Water Resources of the U.S. Army Corps of Engineers is one of the organizations that have played key roles in developing and disseminating benefit evaluation methods.

Methods for measuring benefits may be based on (1) market observations, (2) cost reductions or savings, (3) cost of the least costly alternative, or (4) market surrogates. After the following brief comments illustrating these concepts from the perspectives of the various water management sectors, we will limit our attention specifically to flood damage reduction benefits.

Two alternative approaches have been adopted in estimating hydroelectric power benefits: (1) revenues generated and (2) least-cost alternative analysis. The first approach is based on revenues generated from the sale of electrical energy. The second approach is based on the cost of alternative thermal plants that would be required to generate the electricity in the absence of the hydropower project. For existing systems, the benefits of hydropower generation are measured in terms of savings in fuel costs at the thermal plants in the system.

The primary benefits of irrigation projects are estimated as the net difference in agricultural income between with and without project conditions based on farm budget analyses. Secondary benefits may involve the impacts of increased farm production on related industries such as tractor and irrigation equipment manufacturers, fertilizer and pesticide suppliers, and the food and clothing industries. The

U.S. Bureau of Reclamation has developed detailed procedures for evaluating irrigation projects.

Municipal and industrial water supply planning is based on the concept that a reliable water supply is a necessity. Demand management should restrict water needs to reasonable per capita amounts. Water supply requirements are met at minimum cost. When municipal and industrial water supply is included in a federal multipurpose reservoir project, the water supply benefits are estimated as the cost of the least costly water supply project that would be implemented if the proposed multipurpose reservoir project is not constructed.

The Corps of Engineers has developed procedures for measuring the benefits of navigation projects based on savings in transportation costs. The commodities shipped on an existing navigation system and projected to be transported on a proposed system are identified. The cost of transporting the commodities by barge versus truck and train are estimated and compared.

Various methods have been developed by federal and state water agencies to estimate the benefits of water-based recreation. The oldest practice is to set dollar values per visitor day for various recreation activities, such as boating, swimming, and fishing, which are combined with projections of the number of visitor days. Questionnaires have been used to obtain information from recreationists to help planners assign the unit values. Another approach is to base the value of recreation on the amount of money people are willing to spend for travel and other expenses to visit the recreation site.

Flood damage reduction benefits are classified as location, intensification, and inundation reduction. Location benefits refer to facilitating a new economic use of floodplain land, such as shifting from agricultural to industrial use. Intensification benefits result from intensifying the use of floodplain land, such as shifting from lower value to higher value crops in response to implementation of a flood control project. Inundation reduction benefits are derived from reducing flood damages to floodplain occupants. Inundation reduction benefits are measured as the reduction in expected annual damages to result from a particular plan. Methodologies for estimating expected annual damages are outlined in the next section.

11.3 SIMULATION OF FLOOD DAMAGE REDUCTION SYSTEMS

Hydrologic and hydraulic simulation models covered in previous chapters are combined with economic evaluation methods to determine economically optimal flood damage reduction plans (Hansen, 1987; Davis *et al.*, 1988; U.S. Army Corps of Engineers, 1996a,b). Economic analyses focus specifically on monetary flood damages, which though important, are certainly not the only consideration in evaluating alternative plans. Loss of life, health and social welfare, and environmental impacts are also important considerations in flood studies. Economic benefits are evaluated as the reduction in annual losses that would result from a particular course of action.

Example 11.3 illustrates procedures for comparing alternative flood mitigation plans. The residual average annual damages occurring with each plan are given in column 4 of Table 11.3. The reductions in average annual damage are tabulated in column 2 of Table 11.4. Methodologies for estimating average annual damage (AAD) are outlined in this section. First, the basic relationships required to determine AAD are described. Evaluation of the impacts of reservoirs, channel improvements, levees, nonstructural measures, watershed development, and other modifications are then addressed. Finally, methods for dealing with uncertainties in the basic data are discussed.

11.3.1 Hydrologic, Hydraulic, and Economic Relationships

The stream-floodplain system is divided into reaches, and damage index locations are selected for each reach. For example, the floodplain in Fig. 11.3 is subdivided into four reaches. As illustrated in Fig. 11.4, the following relationships are developed for each location: annual exceedance frequency versus peak discharge, discharge versus stage, and stage versus damage. These basic hydrologic (frequency–discharge), hydraulic (discharge–stage), and economic (stage–damage) relationships are developed based on field data and computer modeling. The frequency–damage relationship is derived from these other three relationships. AAD is the integral of the frequency-damage function. AAD are computed for each reach and summed. The functional relationships are developed for each index location and represent variables for the entire reach.

Annual exceedance frequency versus peak discharge relationships are developed using standard hydrologic engineering techniques covered in Chapters 7 and 8. An example of a flow–frequency relationship developed from gaged streamflow data is presented in Fig. 7.2 and Table 7.4. Due to lack of gaged flow data, watershed

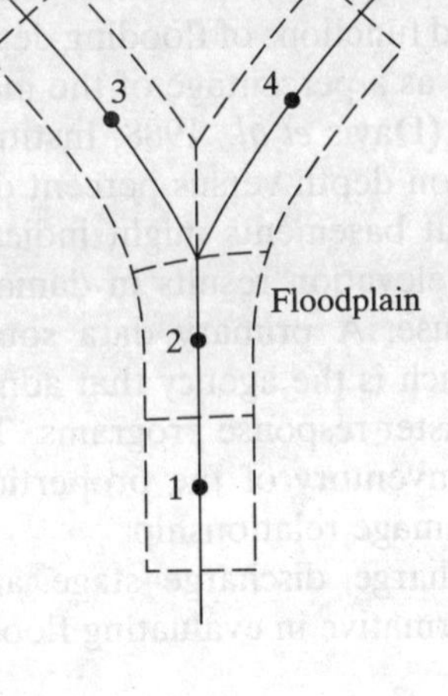

Figure 11.3 The four index locations represent stream reaches and floodplain sections.

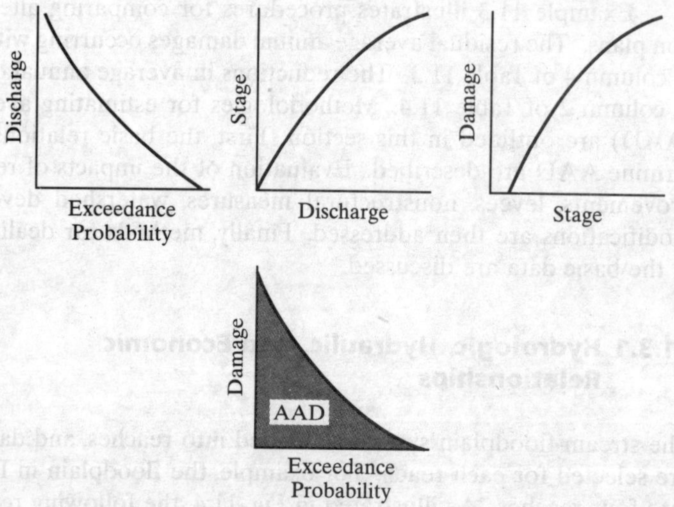

Figure 11.4 These relationships are fundamental to flood damage reduction studies.

models (Section 7.11 and Chapter 8) are usually used to develop frequency–discharge relationships.

The hydraulic relationship between water surface elevation (stage) and discharge is developed based on water surface profile computations that are covered in Chapter 5. A stage at an index location corresponds to a water surface profile along the river reach.

The stage versus damage relationship represents the damages in dollars that would occur along a river reach if flood waters reach various levels. Three alternative approaches for developing stage–damage relationships involve using: historical flood damage data for the study area, engineering cost estimates of damages for assumed flooding levels, and generalized inundation depth versus percent damage functions (James and Lee, 1971). The most common approach for urban floodplains is to use generalized functions of flooding depth relative to the floor elevation versus damage expressed as a percentage of the market value of different types of buildings and properties (Davis *et al.*, 1988; Institute for Water Resources, 1992). For example, an inundation depth versus percent damage function for single-story residential houses without basements might indicate that flooding to a depth of 4.0 feet above the floor elevation results in damages equal to 29 percent of the market value of the house. A primary data source is the Federal Emergency Management Agency, which is the agency that administers the National Flood Insurance Program and disaster response programs. The generic depth-damage functions are applied to an inventory of the properties in a particular floodplain to develop a stage versus damage relationship.

The frequency–discharge, discharge–stage, and stage–damage relationships are each individually informative in evaluating flooding problems. The frequency–

discharge and discharge–stage relationships are used to delineate floodplains and set insurance rates for the National Flood Insurance Program and associated local floodplain management activities. The frequency–damage relationship derived from these other basic relationships also provides meaningful information, defining the probability of various levels of damage occurring. However, concise indices are particularly useful in comparing alternatives in support of decision processes. Average annual damage provides such an index.

11.3.2 Average Annual Damage

AAD or the expected value of annual damages, in dollars, is a probability-weighted average of the full range of possible flood magnitudes and can be viewed as what might be expected to occur, on average, in any future year. The terms *expected* and *average* annual damage are used interchangeably. As discussed in statistics textbooks, the expected value of the random variable X is defined as

$$E[X] = \int_{-\infty}^{\infty} x f_x(x)dx \qquad \textbf{(11.7)}$$

where $f_x(x)$ is the probability density function. The exceedance probability can be expressed as

$$P(X) = \int_{x}^{\infty} f_x(x)dx \qquad \textbf{(11.8)}$$

Equations 11.7 and 11.8 are combined to obtain

$$E[X] = \int_{-\infty}^{\infty} x \frac{dP(x)}{dx} dx \qquad \textbf{(11.9)}$$

The expected value of annual damage is determined by integrating the exceedance frequency versus damage function. The integration is performed numerically as illustrated in Example 11.4.

Example 11.4

Expected annual damage is estimated for the floodplain along a reach of Chester Creek in Pennsylvania (U.S. Army Corps of Engineers, 1996a,b). The basic relationships are tabulated in Table 11.5. The frequency-discharge function is based on the log-Pearson type III probability distribution, with parameters estimated from gaged streamflow data (Chapter 7). Exceedance probabilities covering the full range from relatively frequent to extremely rare are tabulated. The stage–discharge relationship was developed from a hydraulic analysis of the river reach using the computer program HEC-2 Water Surface Profiles (Chapter 5). Water surface profiles are computed for a range of assumed flow rates and related to the stage at the index location. The

TABLE 11.5 HYDROLOGIC, HYDRAULIC, AND ECONOMIC DAMAGE RELATIONSHIPS FOR EXAMPLE 11.4

Frequency–discharge		Discharge–stage		Stage–damage	
Exceedance probability	Discharge (m^3/s)	Discharge (m^3/s)	Stage (m)	Stage (m)	Damage ($1,000)
0.002	899	84	1.97	3.35	0
0.005	676	100	2.39	4.27	19
0.01	539	168	3.39	4.57	26
0.02	423	228	4.07	5.18	339
0.05	299	278	4.58	5.49	525
0.10	223	384	5.50	6.10	1,100
0.20	158	606	7.13	6.71	2,150
0.50	87	652	7.47	8.23	5,130
0.80	51	722	7.75	8.53	5,650
0.90	39	838	8.10	9.14	6,420
0.95	32	1,031	8.79	9.45	6,590

stage–damage relationship was developed based on an inventory of property in the floodplain combined with generic depth-damage functions. The field surveys and computer modeling studies required to develop these basic relationships involve a significant amount of work.

The three relationships tabulated in Table 11.5 are combined to develop the annual exceedance probability versus damage relationship tabulated in Table 11.6. The

TABLE 11.6 NUMERICAL INTEGRATION OF FREQUENCY-DAMAGE FUNCTION TO DETERMINE AVERAGE ANNUAL DAMAGES IN EXAMPLE 11.4

Exceedance probability	Damage ($)	Probability increment	Mean damage for increment ($)	Weighted damage ($)
		0.002	5,290,000	10,580
0.002	5,290,000			
		0.003	4,560,000	13,680
0.005	3,830,000			
		0.005	2,938,000	14,690
0.01	2,045,000			
		0.01	1,426,000	14,260
0.02	808,000			
		0.03	525,000	15,750
0.05	242,000			
		0.05	127,800	6,390
0.10	13,600			
		0.10	6,800	680
0.20	0			
			Average Annual Damage =	76,030

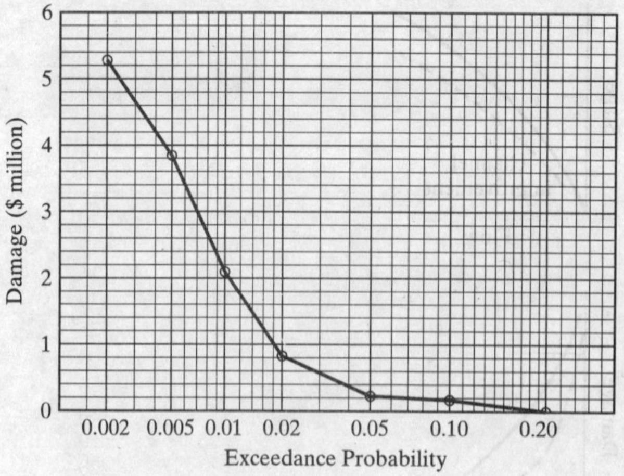

Figure 11.5 The AAD is computed in Example 11.4 as the area under the annual exceedance probability versus damage relationship.

discharge associated with each exceedance probability is obtained from the first relationship in Table 11.5. The stage–discharge tabulation is interpolated to obtain the corresponding stage. The stage–damage relationship is then interpolated to obtain the corresponding damage. The resulting exceedance probability versus damage relationship is numerically integrated in Table 11.6 to determine the AAD. This probability-weighted average damage may be visualized as the area under the frequency-damage curve plotted as Fig. 11.5. The computations in Table 11.6 approximate this area as the summation of the incremental rectangular areas computed by multiplying a probability interval (width) by the mean damage in the interval (height).

There is a probability of 0.002 that flood damages in any year will exceed $5,290,000. In a given year, the probability is 0.01 that damages will exceed $2,045,000 and 0.10 that damages will exceed $13,600. The expected value of annual damages or the AAD is $76,030.

11.3.3 Evaluation of Modified Conditions

Flood damage reduction measures and watershed modifications are modeled in terms of impacts on the basic relationships. Figure 11.6 illustrates the effects of various types of measures. Reservoir storage, reflected in routing computations, modifies the frequency–discharge relationships at downstream locations. Watershed modifications due to urbanization and other land use changes, reflected in precipitation-runoff computations, also result in corresponding changes in the frequency–discharge relationships at downstream locations. Channel improvements, reflected in water surface profile computations, modify the discharge–stage relationship. Levees and floodwalls are also reflected in the depth to which properties are inundated by a given river discharge. Nonstructural measures, such as floodplain regulation, floodproofing, and removal of buildings, reduce the susceptibility of properties to damage. Nonstructural measures are reflected in stage-damage functions.

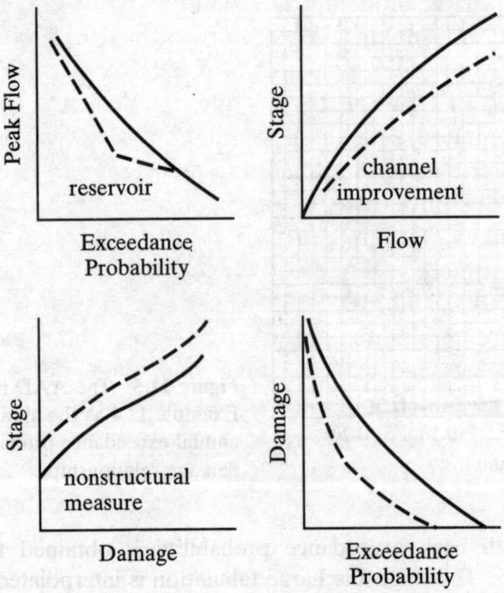

Figure 11.6 Various types of flood damage reduction measures are modeled based on their impacts on the basic relationships.

Any changes in the frequency–discharge, discharge–stage, and stage–damage relationships result in a corresponding change in the frequency-damage function and, thus, result in a change in AAD. As illustrated by Example 11.3, economic benefits associated with a flood damage reduction plan are estimated as the reduction in AAD achieved by the plan.

11.3.4 Modeling Uncertainties in Flood Damage Reduction Analyses

The concept of explicitly considering uncertainty, discussed here in the context of flood studies, is pertinent to many other areas as well. Uncertainty is inherent in all aspects of water resources engineering. Hydrologic phenomena are highly variable and random in nature. Data limitations and modeling approximations add to the uncertainties. Traditional approaches for dealing with uncertainty include conservative estimates of model parameters; safety factors, redundancy, and other forms of conservatism in designs; and sensitivity analyses. More recently, development of techniques that explicitly model uncertainty has become a major emphasis throughout the engineering profession.

The AAD evaluation procedures just outlined have been routinely applied for several decades by the Corps of Engineers, the Natural Resource Conservation Service, and other federal and nonfederal entities. During the 1990's, the Corps of Engineers established a policy of incorporating additional uncertainty considerations in the analysis procedures [U.S. Army Corps of Engineers, 1996a,b; Hydrologic Engineering Center (HEC), 1997]. Uncertainties in developing the hydrologic, hydraulic, and economic relationships are explicitly considered.

The reason for explicitly modeling uncertainties is to enhance information available to support decision-making. With conventional evaluation approaches, costs, net benefits, BCRs, and other measures of project performance are provided as single-value estimates. In Example 11.4, using conventional evaluation procedures, the AAD was estimated as a single number, \$76,030. By modeling uncertainties as probability distributions, the resulting measures of system performance are also expressed as probability distributions. For example, in Example 11.4, the explicit uncertainty-based evaluation approach results might show that the average annual damage has an estimated 85 percent, 60 percent, and 20 percent likelihood of exceeding \$42,000, \$75,000, and \$100,000, respectively. Thus, the uncertainties inherent in our estimates are highlighted in the information used to support decisions.

The frequency–discharge, discharge–stage, and stage–damage relationships developed in the conventional flood evaluation procedures continue to be used when modeling uncertainties. However, uncertainties inherent in the relationships are expressed as probability distributions. A Monte Carlo simulation algorithm incorporating random sampling replaces the conventional procedure of developing just one frequency-damage function by directly combining the other relationships. As illustrated by Fig. 11.7, probability distribution functions are superimposed on the basic relationships. The peak discharge, for a given annual exceedance probability, is treated as varying in accordance with a specified probability distribution, rather than being treated as a single point estimate. Likewise, the stage for a specified discharge is treated as a random variable. The damage to result from a given stage is handled similarly as a probability distribution. The U.S. Army Corps of Engineers (1996a,b) and the HEC (1997, 1998a–c) outline considerations in developing the frequency/discharge/stage/damage relationships and associated uncertainty descriptors.

Probability descriptors may be formulated in different ways, depending on data availability and choice of modeling methods. The confidence limit statistics described in Section 7.7.2 may be used to develop probability descriptors for frequency–discharge relationships. Uncertainties in frequency–discharge and discharge–stage relationships are due largely to approximations in design storms (Chapter 7) and estimation of model parameters, such as the watershed curve number (Chapter 8) and Manning *n*

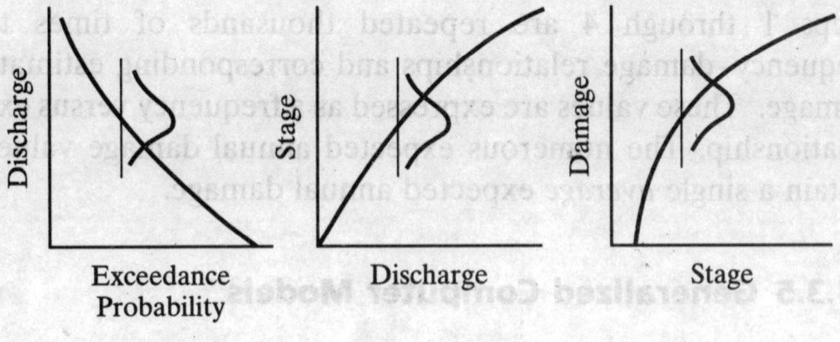

Figure 11.7 Uncertainty is modeled by describing the basic variables as probability distributions.

(Chapter 5). Possible blockages of culverts and bridges by debris during floods adds to the uncertainties in discharge–stage relationships. Inaccuracies in estimating market values and floor elevations of houses contribute to uncertainties in stage–damage relationships. Schemes for developing probability descriptors may be based on high and low estimates defining ranges for each of selected input data that must be estimated with a high degree of uncertainty.

The general procedure for evaluating expected annual damage includes the following steps:

A. Best estimates of frequency–discharge, discharge–stage, and stage–damage relationships are developed using previously discussed conventional modeling methods.

B. Probability descriptors are developed for each relationship that explicitly reflects the uncertainties involved in developing the relationships as discussed in the preceding paragraph.

C. The relationships are combined to develop frequency–damage relationships based on Monte Carlo simulations. This step is iteratively repeated to synthesize thousands of frequency–damage relationships.

D. The numerous frequency–damage relationships are numerically integrated to obtain numerous estimates of expected annual damage.

The Monte Carlo simulations of step C are performed as follows.

1. An annual exceedance probability is selected by random sampling from a standard uniform probability distribution.
2. For that exceedance probability, a discharge is selected by sampling the probability–discharge relationship with discharge treated as a probability distribution.
3. For that discharge, a stage is selected by sampling the discharge–stage relationship with stage treated as a probability distribution.
4. For that stage, a damage is selected by sampling the stage–damage relationship with damage treated as a probability distribution.

Steps 1 through 4 are repeated thousands of times to develop numerous frequency–damage relationships and corresponding estimates of expected annual damage. These values are expressed as a frequency versus expected annual damage relationship. The numerous expected annual damage values may be averaged to obtain a single *average* expected annual damage.

11.3.5 Generalized Computer Models

Computer simulation models are required for the various analyses included in the evaluation of flood mitigation plans. As previously discussed, Hydrologic Engineering

Center computer programs are widely used by agencies and consulting firms throughout the U.S. and abroad. The hydrologic, hydraulic, and economic analyses outlined in this section may be performed using a set of three software packages: (1) *HEC-FFA*, *HEC-1*, or *HEC-HMS*; (2) *HEC-2* or *HEC-RAS*; and (3) *HEC-FDA*. *HEC-5 Simulation of Flood Control and Conservation Systems* (HEC, 1998c) may also be used to model complex reservoir system operating policies. The *HEC-FFA Flood Frequency Analysis* (HEC, 1992b) is used to develop frequency–discharge relationships for gaged sites. For locations with inadequate or no streamflow records, watershed hydrology is modeled using either the *HEC-HMS Hydrologic Modeling System* (HEC, 2001a) or its predecessor the *HEC-1 Flood Hydrograph Package* (HEC, 1998a). River hydraulics is modeled with the *HEC-RAS River Analysis System* (HEC, 2001b) or its predecessor *HEC-2 Water Surface Profiles* (HEC, 1992b). Information from the hydrologic and hydraulic models are combined with economic data to compute average annual damages within the *HEC-FDA Flood Damage Analysis* program (HEC, 1998b). The procedures for explicit consideration of uncertainties are incorporated in *HEC-FDA*, as well as capabilities for managing large amounts of data on floodplain properties susceptible to damage. Conventional average annual damage computations, including development and integration of frequency–damage relationships, are also incorporated in *HEC-1* and *HEC-5*.

11.4 SIMULATION AND OPTIMIZATION

A simulation model is a representation of a system used to predict its behavior under a given set of conditions. Simulation is the process of experimenting with a simulation model to analyze the performance of the system under various conditions. Alternative executions of a simulation model are made to analyze alternative plans. As illustrated by the flood damage reduction system simulation methodology outlined in Section 11.3, a simulation study may involve combining several simulation models.

In a broad sense, optimization is the process of determining the best plan. Optimization includes human judgment, use of simulation and/or optimization models, and use of other decision support tools. An optimization strategy may involve numerous runs of one or several simulation models. A discrete number of plans may be evaluated based on the results of simulating each plan.

The term *optimization model* is also often used, synonymously with mathematical programming, to refer to a mathematical formulation in which a formal algorithm is used to compute a set of decision variable values that minimize or maximize an objective function subject to constraints. Optimization models automatically search for an optimal decision policy. The remainder of this chapter focuses on the particular optimization modeling approach of linear programming.

Optimization and simulation are two different but overlapping modeling approaches. Many models contain elements of both approaches. Optimization models also simulate the system. Mathematical programming algorithms are often used to perform computations within simulation models. Models based on mathematical programming are often used in combination with other types of simulation models.

Systems analysis models are also categorized as being either descriptive or prescriptive. Descriptive models demonstrate what will happen if a specified plan is adopted. Prescriptive models determine the plan that should be adopted to satisfy specified decision criteria. Although it is desirable for decision-support models to be as prescriptive as possible, the real-world complexities of water resources systems often necessitate model orientation toward the more descriptive end of the descriptive/prescriptive spectrum. In general, models should be as prescriptive as the scope of the analysis demands and the complexities of the application allow.

Because mathematical programming requires adherence to a certain mathematical format, other simulation methods may provide greater flexibility, permitting a more detailed and realistic representation of the complex characteristics of a system. The advantages of mathematical programming are related to facilitating a more prescriptive analysis and providing more systematic computational algorithms. Many different models, representing diverse applications in engineering, science, and business, can be developed based on the same standard mathematical programming algorithms. Useful capabilities are also provided for analyzing problems characterized by a need to consider an extremely large number of combinations of values for decision variables.

11.4.1 Mathematical Programming

Mathematical programming models are formulated in a specified format for solution with available standard methods. The objective function x_0 and constraints are represented by mathematical expressions as a function of n decision variables $x_1, x_2, x_3, \ldots, x_n$. The general form of the formulation is as follows.

Maximize or minimize an objective function

$$x_0 = f(x_1, x_2, x_3, \ \ldots, x_n) \tag{11.10}$$

Subject to constraints

$$\begin{aligned} G_1(x_1, x_2, x_3, \ldots, x_n) &= \text{ or } \leq \text{ or } \geq \ b_1 \\ G_2(x_1, x_2, x_3, \ldots, x_n) &= \text{ or } \leq \text{ or } \geq \ b_2 \\ G_3(x_1, x_2, x_3, \ldots, x_n) &= \text{ or } \leq \text{ or } \geq \ b_3 \\ \vdots \qquad & \qquad\qquad \vdots \\ G_m(x_1, x_2, x_3, \ldots, x_n) &= \text{ or } \leq \text{ or } \geq \ b_m \end{aligned} \tag{11.11}$$

The remainder of this chapter focuses on linear programming (LP) and its extensions that include network flow programming and zero-one integer LP. Other optimization techniques not covered in this book include quadratic programming (nonlinear objective function), dynamic programming, and search algorithms. Optimization techniques are covered by water resources systems engineering books (Loucks, Stedinger, and Haith, 1981; Mays and Tung, 1992; Mays, 1997), civil engineering systems books (Jewell, 1986; ReVelle, Whitlach, and Wright, 1997; ReVelle and McGarity, 1997), operations research books (Carter and Price, 2001), and numerous books focused specifically on mathematical programming (Jensen and Barnes, 1980; Phillips and Garcia-Diaz, 1981; Vanderbei, 1996).

11.4.2 Definition of Optimization Modeling Terms

Definitions of fundamental terms used in discussing mathematical programming are as follows:

optimization—finding the best (optimum) solution; in mathematical programming, a decision policy is determined that minimizes or maximizes an objective function subject to not violating constraints.

decision variables—variables that can be controlled; these are the variables for which optimum values are to be determined.

decision policy—each of the decision variables is assigned a value; a decision policy is a set of values for the decision variables.

constraints—limitations or restrictions on possible decision policies.

feasible policy—a decision policy that does not violate any constraints.

infeasible policy—a decision policy that violates one or more constraints.

objective function—a statement of the consequences of a decision policy; the objective function is a criterion by which *optimum* is defined and may also be called a criterion function.

optimum solution—a feasible decision policy that optimizes (minimizes or maximizes) the objective function.

11.5 LINEAR PROGRAMMING

Although other nonlinear programming techniques also have been adopted, most water resources engineering applications of mathematical programming involve LP or extensions thereof. The popularity of LP in water resources systems analysis, as well as in other operations research, management science, and systems engineering fields, is due to the following considerations. LP is applicable to a wide variety of types of problems. Efficient solution algorithms are available. Generalized computer software packages are available for applying the solution algorithms.

LP consists of finding values for a set of n decision variables $x_1, x_2, \ldots, x_n$ that minimize or maximize an objective or criterion function x_0 of the form:

$$x_0 = c_1x_1 + c_2x_2 + \cdots + c_nx_n \quad \textbf{(11.12)}$$

subject to a set of m constraint equations and/or inequalities of the form:

$$\begin{aligned} a_{11}x_{11} + a_{12}x_{12} + \cdots + a_{1n}x_{1n} &= \text{or} \leq \text{or} \geq b_1 \\ a_{21}x_{11} + a_{22}x_{22} + \cdots + a_{2n}x_{2n} &= \text{or} \leq \text{or} \geq b_2 \\ \vdots \qquad & \qquad \vdots \\ a_{m1}x_{m1} + a_{m2}x_{m2} + \cdots + a_{mn}x_{mn} &= \text{or} \leq \text{or} \geq b_m \end{aligned} \tag{11.13}$$

and a set of constraints requiring that the decision variables be nonnegative:

$$x_j \geq 0 \quad \text{for} \quad j = 1, 2, \ldots, n \tag{11.14}$$

where a_{ij}, b_i, and c_j are constants. Methods are available for circumventing the non-negativity constraints of Eqs. 11.14. The LP model is expressed in more concise notation as:

Minimize or maximize

$$x_0 = \sum_{j=1}^{n} c_j x_j \tag{11.15}$$

Subject to

$$\sum_{j=1}^{n} a_{ij}x_j \leq b_i \quad \text{for} \quad i = 1, 2, \ldots, m \tag{11.16}$$

and

$$x_j \geq 0 \quad \text{for} \quad j = 1, 2, \ldots, n \tag{11.17}$$

where x_0 is the objective function; x_j is the decision variables; c_j, a_{ij}, and b_i are constants; n is the number of decision variables; and m is the number of constraints. The "less than or equal" sign in the constraint inequalities may be replaced by "greater than or equal" or "equal" signs to suit the particular problem being modeled. Maximizing $-x_0$ is equivalent to minimizing x_0. The objective function and all constraints are linear functions of the decision variables. A set of values for the n variables is called a decision policy.

Considerable ingenuity and significant approximations may be required to formulate a real-world problem in the required mathematical format. However, if the problem can be properly formulated, standard LP algorithms and computer codes are available to perform the computations. Solution techniques are briefly mentioned in Section 11.5.2 and covered in depth by many other books. Our main concern is formulating water resources systems analysis problems as LP models. Several examples are presented to illustrate the basic concept of formulating a problem in the LP format of Eqs. 11.12–11.17. The examples develop relatively small LP models with relatively few decision variables that can be easily solved with a spreadsheet program. The basic water management concepts illustrated by the simplified examples are incorporated in actual models that often have hundreds or thousands of decision variables.

The examples presented in the remainder of this chapter illustrate the following common types of LP applications: allocation of water to competing uses/users; selection among alternative water supply sources; blending supplies with different water quality; optimizing reservoir operating policies; optimizing agricultural cropping patterns; and pollutant load allocation. The decision variables associated with these applications include:

- reservoir releases and storage levels
- amount of water allocated to each use/user
- amount of water supplied from each source
- reductions in pollutant loads
- amount of land planted with alternative crops

The objective functions represent the following objectives:

- minimize costs or maximize revenues or net benefits
- maximize reservoir yield or minimize storage capacity
- minimize water supply depletions
- adhere to relative priorities

Constraints involve mass balances, capacities, water availability, and demands. These represent just a few of the many types of applications of LP in water resources engineering.

11.5.1 Graphical Solution of Two-Variable LP Problems

A couple of simple two-variable problems are presented first so that the basics can be visualized graphically. The two-variable graphical solution serves to illustrate the terms defined in Section 11.4.2 and introduce fundamental concepts incorporated in standard n-variable solution algorithms.

Example 11.5

The decision problem is to determine the amount of land to be planted in crop A and crop B that will maximize income within the constraints of limited land and water resources. Twenty million m^3 of water is available in storage for irrigation of the two crops. Crop A requires 9,000 m^3 of water per hectare of irrigated land and produces a net income of $720 per hectare. Crop B requires 6,000 m^3 per hectare and produces a profit of $1,200 per hectare. Crop A is limited to 1,600 hectares. Up to 2,400 hectares of land is available for planting crop B. The linear programming (LP) model is formulated to determine the amount of land, in hectares, planted in crop A (x_1) and crop B (x_2) that will maximize income (x_0), in dollars.

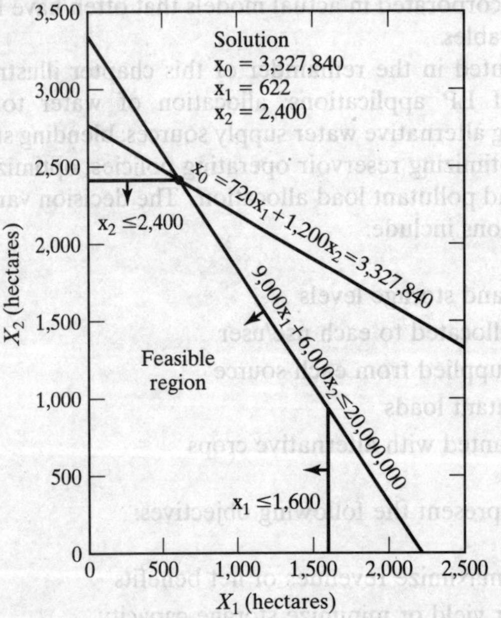

Figure 11.8 Example 11.5 is solved graphically to illustrate basic concepts of LP.

Maximize

$$x_0 = 720x_1 + 1{,}200x_2$$

Subject to

$$9{,}000x_1 + 6{,}000x_2 \leq 20{,}000{,}000$$

$$x_1 \leq 1{,}600$$

$$x_2 \leq 2{,}400$$

$$x_1 \text{ and } x_2 \geq 0$$

The LP problem is solved graphically in Fig. 11.8 to obtain the optimum solution of planting 622 and 2,400 hectares of crops A and B, respectively, resulting in a net income of $3,327,840.

Example 11.6

An industrial water demand of 10,000 m^3 during a particular time period is supplied by withdrawals from two sources: (1) a groundwater aquifer and (2) a reservoir on a river. The total dissolved solids (TDS) concentrations in the aquifer and reservoir are 980 and 100 mg/l (g/m^3), respectively. The maximum allowable TDS concentration is 500 mg/l (g/m^3) for the water use being supplied. The capacities of the well and reservoir diversion structure constrain withdrawals, during the time period, to not exceed 6,000 m^3 and 10,000 m^3, respectively, from the aquifer and reservoir. In the current time period, operating decisions are based on minimizing the amount of water withdrawn from the better quality reservoir (lower TDS) to maximize the amount of good quality water remaining for future use.

The decision problem is to determine the amount of water to withdraw from the aquifer (x_1) and reservoir (x_2) while meeting the total demand of 10,000 m^3 with a concentration of 500 g/m^3 or less. The amount of water withdrawn from the reservoir (x_2) is minimized.

Solution A TDS mass balance constraint is formulated that equates the TDS load in the water supply diversion to the sum of the TDS mass withdrawn from each of the two sources. The concentration of the water supply diversion is limited to 500 g/m^3.

$$980x_1 + 100x_2 \le 500(x_1 + x_2)$$

or

$$480x_1 - 400x_2 \le 0$$

The LP model is formulated as follows, with the decision variables x_1 and x_2 expressed in m^3.

Minimize

$$x_0 = x_2$$

Subject to

$$480x_1 - 400x_2 \le 0$$

$$x_1 + x_2 \ge 10{,}000$$

$$x_1 \le 6{,}000$$

$$x_2 \le 10{,}000$$

$$x_1 \quad \text{and} \quad x_2 \ge 0$$

The problem is solved graphically in Fig. 11.9 to obtain the optimum solution of withdrawing 4,545 m^3 and 5,455 m^3, respectively, from the aquifer and reservoir.

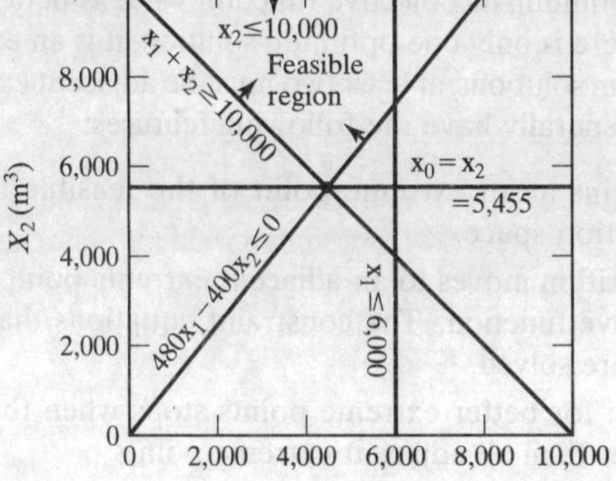

Figure 11.9 Example 11.6 is also solved graphically.

The plots of the constraints and objective function in Figs. 11.8 and 11.9 illustrate fundamental features of LP. The constraints bound a feasible region. Decision policies within this region are feasible in that all constraints are satisfied. Values lying outside of the feasible region violate one or more constraints. In the two-variable graphical solution, the objective function x_0 can be plotted as a function of x_1 and x_2 for a specified value of x_0. Plots for alternative values of x_0 are parallel. At the maximum or minimum value of x_0, the objective function plot passes through at least one corner of the feasible region. If the objective function is parallel to a controlling constraint, an infinite number of optimum solutions are represented by two corners of the feasible region and the boundary connecting the two corners. A problem formulation may involve a set of constraints that cannot all be satisfied simultaneously, and thus there is no feasible region and no solution.

Realistic LP problems typically involve hundreds or thousands of decision variables. The graphical approach is applicable only to problems involving two decision variables. For problems with more than two decision variables, the objective function and constraints can be visualized as hyperplanes in n-dimensional space. Mathematical solution algorithms solve the equations at the corners, called extreme points, of the planes enclosing the feasible space, in a systematic iterative search for the optimum.

11.5.2 Solution of LP Problems

The simplex algorithm, used in many LP computer codes, is explained in many textbooks, including Mays and Tung (1992), Vanderbei (1996), ReVelle, Whitlach, and Wright (1997), and Carter and Price (2001). Special computationally efficient algorithms are available for certain forms of LP problems, such as the network flow programming models discussed later. General characteristics of LP solution algorithms are briefly summarized as follows. The optimum set of values for the n decision variables occurs at a corner or extreme point of the feasible region in n-dimensional space. Algebraic solution algorithms are based on the following properties of extreme points. Only a finite number of extreme points exist for any problem. If an extreme point is a global optimum, its objective function value is better than all adjacent extreme points. If there is only one optimum solution, it is an extreme point. If there are multiple optimum solutions, at least two must be adjacent extreme points.

Solution algorithms generally have the following features:

- The procedure begins at an extreme point of the feasible region of the n-dimensional solution space.
- Iteratively, the algorithm moves to an adjacent extreme point with a better value of the objective function. The constraint equations that intersect at the extreme point are solved.
- The iterative search for better extreme points stops when the current extreme point is better than all adjacent extreme points.

At the beginning of the solution procedure, the inequality and equality constraints are converted to a set of m equations and n unknowns, with $n > m$. The n unknowns include the original decision variable plus slack and surplus variables used to convert inequalities to equalities. With more unknowns than equations, the system of equations is indeterminate. Thus, solution algorithms are based on letting $(n - m)$ unknowns equal zero and solving for the remaining m unknowns. In the iterative algorithm, the choice of variables to set equal to zero depends on the current extreme point.

LP formulations may have no feasible solution or, in other cases, an unbounded solution. Complexities arise when the constraints define either no feasible solution space or an unbounded feasible solution space. Constraints may be formulated such that they cannot all be satisfied simultaneously. No feasible solution exists if no set of values for the decision variables satisfies all constraints. An unbounded solution occurs when the objective function can be increased or decreased infinitely, and thus no solution is found. In this case, the problem must be reformulated.

Various generalized optimization computer programs are available. The user inputs values for the coefficients in the objective function and constraint equations. The optimizer program computes values for the decision variables. LP capabilities are included in popular spreadsheet programs, such as those noted in Table 1.5. However, the spreadsheet programs are designed for problems limited to several hundred decision variables and constraints. Water resources system analysis models often involve many thousand decision variables and constraints. A number of generalized optimization programs are available for solving linear and, in some cases, nonlinear programming problems, including very large problems. LINDO is an LP package available from LINDO Systems, Inc. (http://www.lindo.com). The General Algebraic Modeling System (GAMS) is a general-purpose optimization package designed for solving large linear, nonlinear, and mixed integer programming problems that is distributed by the GAMS Development Corporation (http://www.gams.com).

Optimization models are also sometimes coded in Fortran or other languages. Already-written subroutines for performing LP computations may be incorporated in model development. The same code for optimizer routines may be used in any number of different models.

11.5.3 Reservoir System Analysis Models

Reservoir system analysis is an example of the various types of applications of LP in water resources engineering. Numerous variations of LP formulations have been applied in analyzing an array of reservoir operations problems. Example 11.7 involves determining reservoir release decisions, for each time step of the analysis period, that maximize revenues in dollars. Example 11.8 deals with firm yield analysis.

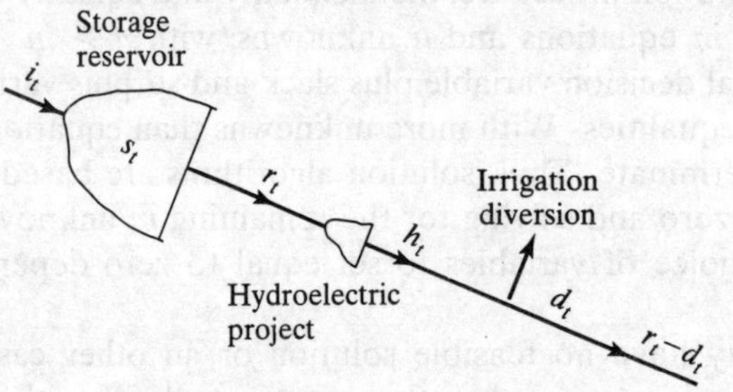

Figure 11.10 Multipurpose reservoir system operations are modeled in Example 11.7.

Example 11.7

As indicated in Fig. 11.10, releases from the storage reservoir flow through a hydroelectric power plant located downstream, supply water for an irrigation diversion further downstream, and maintain instream flows. Flows of up to 180 million m^3/month can be used to generate hydroelectric energy; flows in excess of this amount bypass the turbines. The demands shown in column 4 of Table 11.7 provide upper limits for the irrigation diversion. Irrigation diversions are limited to 6 months of the year. An instream flow requirement of 20 million m^3 per month must be maintained. The reservoir storage capacity is 600 million m^3. Revenues for supplying irrigation are \$900 per million m^3 of water diverted. Each million m^3 of water used to generate hydroelectric power results in revenues of \$400. The decision problem consists of determining the set of monthly reservoir releases that maximize revenues, given the inflows tabulated in column 3 of Table 11.7.

Solution The following 42 decision variables are incorporated in the LP formulation, with t denoting the monthly time interval.

TABLE 11.7 DATA (COLUMNS 1–4) AND RESULTS (COLUMNS 5–8) FOR EXAMPLE 11.7

Month (1)	t (2)	Inflow i_t (10^6 m^3) (3)	Irrigation demand (10^6 m^3) (4)	Storage s_t (10^6 m^3) (5)	Total release r_t (10^6 m^3) (6)	Hydropower release h_t (10^6 m^3) (7)	Irrigation diversion d_t (10^6 m^3) (8)
Jan	1	95	0	210	20	20	0
Feb	2	112	0	302	20	20	0
Mar	3	170	0	335	137	137	0
Apr	4	250	0	405	180	180	0
May	5	265	50	600	70	70	50
Jun	6	62	150	492	170	170	150
Jul	7	35	260	319	208	180	188
Aug	8	18	260	157	180	180	160
Sep	9	55	190	32	180	180	160
Oct	10	88	100	0	120	120	100
Nov	11	85	0	65	20	20	0
Dec	12	90	0	115	20	20	0

end-of-month storage	s_t	for	$t = 1, 2, \ldots, 12$
reservoir release	r_t	for	$t = 1, 2, \ldots, 12$
hydropower discharge	h_t	for	$t = 1, 2, \ldots, 12$
irrigation diversion	d_t	for	$t = 5, 6, 7, 8, 9, 10$

The objective is to maximize total annual revenues, in dollars.

$$\text{Revenues} = 400h_1 + 400h_2 + 400h_3 + 400h_4 + 400h_5 + 400h_6 + 400h_7 + 400h_8 + 400h_9 + 400h_{10} + 400h_{11} + 400h_{12} + 900d_5 + 900d_6 + 900d_7 + 900d_8 + 900d_9 + 900d_{10}$$

Note that, in the objective function, the coefficients are \$400 and \$900 per million m^3, respectively, for the 18 decision variables h_t and d_t and zero for the 24 decision variables s_t and r_t. Constraints are as follows. Monthly reservoir mass balances are maintained for each of the 12 months of the analysis for the inflows i_t tabulated in column 3 of Table 11.7. The storages at the beginning and end of the year are set equal ($s_0 = s_{12}$) based on the premise that the annual cycle is repeated.

$$s_t - s_{t-1} + r_t = i_t \quad \text{for} \quad t = 1, 2, \ldots, 12$$

End-of-month storages cannot exceed the reservoir capacity of 600 million m^3.

$$s_t \leq 600 \quad \text{for} \quad t = 1, 2, \ldots, 12$$

Diversions do not exceed the demands of column 4 of Table 11.7, and no more than 180 million m^3/month is used for hydroelectric power generation.

$$d_t \leq \text{demand} \quad \text{for} \quad t = 1, 2, \ldots, 12$$

$$h_t \leq 180 \quad \text{for} \quad t = 1, 2, \ldots, 12$$

The instream flow requirement is 20 million m^3/month. The flow used for hydropower cannot exceed the reservoir release. The irrigation diversion cannot exceed the reservoir release.

$$r_t - d_t \geq 20 \quad \text{for} \quad t = 1, 2, \ldots, 12$$

$$r_t - h_t \geq 0 \quad \text{for} \quad t = 1, 2, \ldots, 12$$

Nonnegativity constraints specify that the 42 decision variables have values that are zero or positive numbers.

$$s_t, r_t, h_t, d_t \geq 0$$

The complete formulation of the linear programming model is as follows.

Maximize

$$\text{Revenues} = 400h_1 + 400h_2 + 400h_3 + 400h_4 + 400h_5 + 400h_6 + 400h_7 + 400h_8 + 400h_9 + 400h_{10} + 400h_{11} + 400h_{12} + 900d_5 + 900d_6 + 900d_7 + 900d_8 + 900d_9 + 900d_{10}$$

Subject to

$$
\begin{array}{lll}
s_1 - s_{12} + r_1 = 95 & s_1 \le 600 & \\
s_2 - s_1 + r_2 = 112 & s_2 \le 600 & \\
s_3 - s_2 + r_3 = 170 & s_3 \le 600 & \\
s_4 - s_3 + r_4 = 250 & s_4 \le 600 & \\
s_5 - s_4 + r_5 = 265 & s_5 \le 600 & d_5 \le 50 \\
s_6 - s_5 + r_6 = 62 & s_6 \le 600 & d_6 \le 150 \\
s_7 - s_6 + r_7 = 35 & s_7 \le 600 & d_7 \le 260 \\
s_8 - s_7 + r_8 = 18 & s_8 \le 600 & d_8 \le 260 \\
s_9 - s_8 + r_9 = 55 & s_9 \le 600 & d_9 \le 190 \\
s_{10} - s_9 + r_{10} = 88 & s_{10} \le 600 & d_{10} \le 100 \\
s_{11} - s_{10} + r_{11} = 85 & s_{11} \le 600 & \\
s_{12} - s_{11} + r_{12} = 90 & s_{12} \le 600 & \\
r_1 \ge 20 & r_1 - h_1 \ge 0 & h_1 \le 180 \\
r_2 \ge 20 & r_2 - h_2 \ge 0 & h_2 \le 180 \\
r_3 \ge 20 & r_3 - h_3 \ge 0 & h_3 \le 180 \\
r_4 \ge 20 & r_4 - h_4 \ge 0 & h_4 \le 180 \\
r_5 - h_5 \ge 20 & r_5 - h_5 \ge 0 & h_5 \le 180 \\
r_6 - h_6 \ge 20 & r_6 - h_6 \ge 0 & h_6 \le 180 \\
r_7 - h_7 \ge 20 & r_7 - h_7 \ge 0 & h_7 \le 180 \\
r_8 - h_8 \ge 20 & r_8 - h_8 \ge 0 & h_8 \le 180 \\
r_9 - h_9 \ge 20 & r_9 - h_9 \ge 0 & h_9 \le 180 \\
r_{10} - h_{10} \ge 20 & r_{10} - h_{10} \ge 0 & h_{10} \le 180 \\
r_{11} \ge 20 & r_{11} - h_{11} \ge 0 & h_{11} \le 180 \\
r_{12} \ge 20 & r_{12} - h_{12} \ge 0 & h_{12} \le 180
\end{array}
$$

$$s_t \ge 0 \quad \text{for} \quad t = 1, 2, \ldots, 12$$

$$r_t \ge 0 \quad \text{for} \quad t = 1, 2, \ldots, 12$$

$$h_t \ge 0 \quad \text{for} \quad t = 1, 2, \ldots, 12$$

$$d_t \ge 0 \quad \text{for} \quad t = 5, 6, \ldots, 10$$

The model can be solved using the Microsoft Excel solver, other spreadsheet programs, or other LP software. The resulting values for the decision variables are tabulated in columns 5–8 of Table 11.7. The objective function is \$1,246,000 for this optimum decision policy. This is the optimum objective function value, but only one of multiple optimum decision policies. Other sets of values for the decision variables also result in the objective function being $1,246,000.

The model formulated in Example 11.7, based on 1 year of streamflows, has 42 decision variables, 42 nonnegativity constraints, and 66 other constraints. Reservoir/river system reliability analyses are often based on monthly streamflow sequences covering 50–100 years. Using monthly streamflows for the 50-year (600-month) period 1951–2000, instead of just 12 months, increases the number of decision variables from 42 to 2,100. Example 11.7 has only one reservoir and one hydropower plant. The same general formulation has been applied to systems of many reservoirs and hydropower plants, using hydrologic simulation periods of 50–100 years, resulting in LP models with many thousands of decision variables. Weekly or daily, rather than monthly, time steps are often adopted. As discussed later in conjunction with Examples 11.10 and 11.11, another common variation of this general type of model applies LP to operate multiple reservoirs and hydropower plants during one sequential time step (month, week, day) at a time. The LP computations are repeated within the model for each time step as discussed in Section 11.5.5. Reservoir evaporation is an important term that is omitted from Examples 11.7 and 11.8 to simplify the model but discussed later in Section 11.5.6.

In addition to economic measures, reliability indices such as volume and period reliabilities (Eqs. 7.48 and 7.49) are computed from the results of simulation models such as the LP formulations of Examples 11.7 and 11.10. Reliabilities may be computed for the irrigation and hydroelectric energy demands of Example 11.7. Reliability indices are more meaningful if computed from the results of a simulation with a long period of analysis, like 50 years, rather than the one-year period of analysis adopted in Example 11.7.

As discussed in Section 7.10, firm yield is the maximum demand with period and volume reliabilities of 100 percent. Firm yield is the maximum demand that can be met continuously during sequences of reservoir inflows representing historical hydrology. The relationship between storage capacity and firm yield is fundamental to the planning and design of reservoir projects. The firm yield for a site on a river for zero storage (no reservoir) is the lowest inflow during the simulation period. For the reservoir site of Example 11.8, this run-of-river firm yield is 19×10^6 m^3/month from column 3 of Table 11.8. The upper limit on firm yield for unlimited reservoir storage capacity is the mean streamflow at the site. Evaporation prevents this upper limit from actually being achieved. Between these lower and upper extremes, a storage capacity versus firm yield relationship is developed by determining firm yields achieved by specified storage capacities or vice versa.

LP is one of several alternative approaches for performing firm yield analyses. The LP formulation discussed next may be applied in developing storage capacity versus firm yield relationships for a single reservoir. Iterative executions of reservoir/river system simulation models (Section 12.8) are typically used to determine firm yields and other reliability measures for multiple-reservoir systems with complex configurations and operating policies, as well as for simpler systems.

The storage capacity C required to meet a specified set of water demands y_t representing a firm yield can be determined with the following LP formulation.

Minimize C **(11.18)**

Subject to

$$s_t = s_{t-1} + i_t - y_t - r_t \quad \text{for} \quad t = 1, 2, \ldots, T \qquad \textbf{(11.19)}$$

$$s_t \le C \quad \text{for} \quad t = 1, 2, \ldots, T \qquad \textbf{(11.20)}$$

$$s_t, y_t, r_t \ge 0 \quad \text{for} \quad t = 1, 2, \ldots, T \qquad \textbf{(11.21)}$$

where

C reservoir storage capacity
s_t storage content at the end of period t
i_t streamflow inflow to the reservoir during period t
y_t demand during period t representing the firm yield
r_t all spills and releases other than y_t during period t
T number of time periods in the analysis

Thus, the reservoir storage capacity C is minimized subject to constraints that include: the reservoir mass balances; not allowing storage content s_t to exceed storage capacity C; and not allowing the variables to have negative values. The constraints are repeated for each time period t. The known firm yield withdrawals y_t and reservoir inflows i_t are provided as input data. The model computes the value of the storage capacity C and also values of end-of-period storage s_t and other releases r_t for each period t.

Yield analyses require an assumption regarding starting and ending storage conditions. Specification of the storage at the beginning s_0 and end s_T of the overall analysis period could be added to the model formulation. Alternatively, the entire streamflow sequence representing reservoir inflows may be assumed to be repeated as necessary to achieve a repetitive cycle. This is reflected in the model by assuming that the first ($t = 1$) period of a T-period cycle follows the last period ($t = T$) of the prior cycle and specifying that the beginning ($t = 0$) and ending ($t = T$) storages be equal ($s_0 = s_T$).

With the above formulation, the reservoir storage capacity C is computed for a user-specified firm yield y_t. Alternatively, a model can be formulated to determine a constant firm yield Y provided by a specified storage capacity, by changing the objective function to:

$$\text{Maximize } Y \qquad \textbf{(11.22)}$$

subject to the same constraints as before. With this formulation, a storage capacity is specified as input, and the model computes the firm yield Y, along with values of s_t and r_t for each period.

Example 11.8

The reservoir storage capacity required to provide a firm yield of 50 m^3/s for a given sequence of inflows is determined. Monthly inflow volumes are tabulated in column 3 of Table 11.8. The monthly diversion volumes in million m^3 corresponding to a constant diversion of 50 m^3/s are shown in column 4. An LP model is formulated to determine the storage capacity required to meet the monthly demands for the 24-month sequence of reservoir inflows.

The LP problem is formulated with 49 decision variables consisting of the reservoir storage capacity C, end-of-period storage contents s_t, and spills r_t, for each of 24 months ($t = 1, 2, \ldots, 24$). The objective is to:

Minimize C

Note that, in the objective function, the coefficients are 1 for the decision variable C and zero for the 48 decision variables s_t and r_t. A set of 24 constraints are formulated to represent the reservoir mass balance for each of the 24 months of the analysis. The mass balance is expressed as:

$$s_t = s_{t-1} + i_t - y_t - r_t$$

TABLE 11.8 DATA (COLUMNS 1–5) AND RESULTS (COLUMNS 6 AND 7) FOR THE FIRM YIELD MODEL OF EXAMPLE 11.8

Month (1)	t (2)	Inflow i_t (10^6 m^3) (3)	Yield Y_t (10^6 m^3) (4)	$i_t - y_t$ (10^6 m^3) (5)	Storage s_t (10^6 m^3) (6)	Release r_t (10^6 m^3) (7)
Jan	1	123	134	−11	293	0
Feb	2	172	121	51	304	40
Mar	3	163	134	29	304	29
Apr	4	334	130	204	304	204
May	5	421	134	287	304	287
Jun	6	130	130	0	304	0
Jul	7	37	134	−97	207	0
Aug	8	19	134	−115	92	0
Sep	9	109	130	−21	71	0
Oct	10	88	134	−46	25	0
Nov	11	140	130	10	35	0
Dec	12	134	134	0	35	0
Jan	13	150	134	16	51	0
Feb	14	167	121	46	97	0
Mar	15	230	134	96	193	0
Apr	16	288	130	158	304	47
May	17	362	134	228	304	228
Jun	18	67	130	−63	241	0
Jul	19	32	134	−102	139	0
Aug	20	27	134	−107	32	0
Sep	21	98	130	−32	0	0
Oct	22	276	134	142	142	0
Nov	23	223	130	93	235	0
Dec	24	209	134	75	304	6

that is rearranged with the decision variables on the left of the equal sign:

$$s_t - s_{t-1} + r_t = i_t - y_t$$

The storage s_0 at the beginning of the first month is assumed to equal the storage s_T at the end of the last month. Thus, for the first month ($t = 1$), the mass balance constraint is:

$$s_1 - s_{24} + r_1 = i_1 - y_1$$

With values for i_1 and y_1 from Table 11.8:

$$s_1 - s_{24} + r_1 = 123 - 134 = -11$$

A set of 24 constraints specify that storage content cannot exceed capacity.

$$s_t - C \leq 0 \quad \text{for} \quad t = 1, 2, \ldots, 24$$

Nonnegativity constraints specify that the 49 decision variables have values that are zero or positive numbers.

The complete formulation of the LP model is as follows.

Minimize C

Subject to

$$s_1 - s_{24} + r_1 = -11 \qquad s_{13} - s_{12} + r_{13} = 16$$
$$s_2 - s_1 + r_2 = 51 \qquad s_{14} - s_{13} + r_{14} = 46$$
$$s_3 - s_2 + r_3 = 29 \qquad s_{15} - s_{14} + r_{15} = 96$$
$$s_4 - s_3 + r_4 = 204 \qquad s_{16} - s_{15} + r_{16} = 158$$
$$s_5 - s_4 + r_5 = 287 \qquad s_{17} - s_{16} + r_{17} = 228$$
$$s_6 - s_5 + r_6 = 0 \qquad s_{18} - s_{17} + r_{18} = -63$$
$$s_7 - s_6 + r_7 = -97 \qquad s_{19} - s_{18} + r_{19} = -102$$
$$s_8 - s_7 + r_8 = -115 \qquad s_{20} - s_{19} + r_{20} = -107$$
$$s_9 - s_8 + r_9 = -21 \qquad s_{21} - s_{20} + r_{21} = -32$$
$$s_{10} - s_9 + r_{10} = -46 \qquad s_{22} - s_{21} + r_{22} = 142$$
$$s_{11} - s_{10} + r_{11} = 10 \qquad s_{23} - s_{22} + r_{23} = 93$$
$$s_{12} - s_{11} + r_{12} = 0 \qquad s_{24} - s_{23} + r_{24} = 75$$
$$s_t - C \leq 0 \quad \text{for} \quad t = 1, 2, \ldots, 24$$
$$s_t \geq 0 \quad \text{for} \quad t = 1, 2, \ldots, 24$$
$$r_t \geq 0 \quad \text{for} \quad t = 1, 2, \ldots, 24$$
$$C \geq 0$$

The model can be solved using a spreadsheet solver or LP program. The results consist of a storage capacity C of 304 million m^3 and the values for s_t and r_t tabulated columns 6 and 7 of Table 11.8.

11.5.4 Pollution Load Allocation Models

LP models are applied in studies of nonpoint pollution loads from watersheds and point loads from wastewater treatment plants. Models are formulated for the particular problem of interest and may vary greatly in their complexity and approach. Example 11.9 provides a general idea of how LP models may be formulated to analyze pollutant load allocation problems.

Example 11.9

Regulatory agencies have imposed stream water quality criteria requiring that the concentration of a particular pollutant not exceed 120 mg/l (0.12 kg/m^3) at any point in the river system of Fig. 11.11. Four treatment plants discharge treated wastewater into the river. Pollutant loads from sources other than effluent from the four wastewater treatment plants are negligible. The treatment facilities are operated in coordination to meet the water quality standards, with the goal being to minimize the combined daily operating cost of all plants. As the water flows down the stream, natural processes reduce the pollutant load. The pollutant mass is reduced by the percentages $P_{i,j}$ by natural processes between site i and site j as follows.

$$P_{1,3} = 10\% \qquad P_{2,3} = 20\% \qquad P_{3,4} = 15\%$$

Treatment plant efficiency limits the percentage of the pollutant load that can be removed before discharging into the stream. For example, a maximum of 92 percent of the pollutant load processed through plant 1 can be removed. Pertinent data describing the system is provided in Table 11.9. The amount of the pollutant to remove at each treatment plant to minimize total costs is to be determined.

Solution The decision variables x_1, x_2, x_3, and x_4 are the amount of the pollutant in 1,000 kg/day to be removed at each treatment plant. The objective function x_0 is the

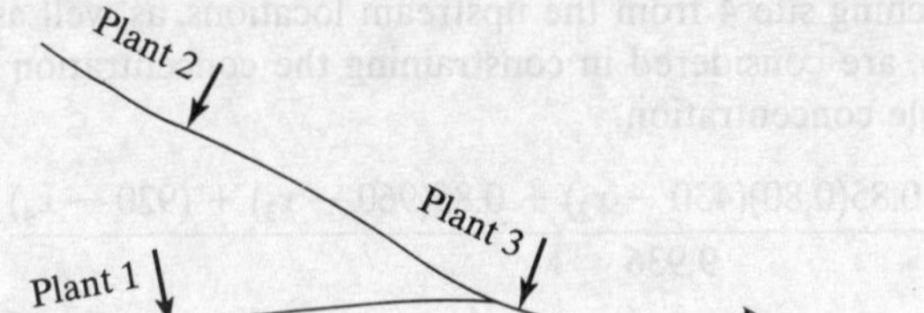

Figure 11.11 In Example 11.9, pollutant loads are discharged into the stream system at four locations.

TABLE 11.9 DATA FOR EXAMPLE 11.9

Wastewater treatment plant	1	2	3	4
Pollutant load generated by city (1,000 kg/day)	510	430	960	920
Treatment costs ($/1,000 kg removed)	$2.50	$1.80	$4.00	$3.50
Maximum plant efficiency	0.92	0.90	0.91	0.92
Total streamflow at city (m^3/s)	39.1	45.5	98.2	115
Total streamflow converted to 1,000 m^3/day	3,378	3,931	8,484	9,936

total cost in \$/day to be minimized. Constraints are formulated as follows. The decision variables must be nonnegative.

$$x_i \geq 0$$

The water quality requirement must be met at sites 1 and 2 considering only the pollutant load discharged into the river at these locations. The concentration in kg/m^3 at location 1 is the pollutant load $(510 - x_1)$ in thousand kg/day discharged into the river divided by the streamflow of 3,378 thousand m^3/day. This concentration cannot exceed the allowable concentration of 0.12 kg/m^3.

$$\frac{510 - x_1}{3{,}378} \leq 0.12 \text{ kg/m}^3$$

$$x_1 \geq 104.6$$

Likewise, a water quality constraint is formulated for location 2.

$$\frac{430 - x_2}{3{,}931} \leq 0.12 \text{ kg/m}^3$$

$$x_2 \geq -41.7 \qquad \text{(redundant since } x_2 \geq 0)$$

The water quality requirement must be met at plant 3 considering the portion of the pollutant load discharged upstream at sites 1 and 2, as well as the load that enters the river at site 3. The loads at locations 1 and 2 are reduced 10 and 20 percent by natural processes before reaching site 3.

$$\frac{0.90(510 - x_1) + 0.80(430 - x_2) + (960 - x_3)}{8{,}484} \leq 0.12 \text{ kg/m}^3$$

$$0.9x_1 + 0.8x_2 + x_3 \geq 744.9$$

The pollutant load reaching site 4 from the upstream locations, as well as the pollutant discharge at site 4, are considered in constraining the concentration at site 4 to not exceed the allowable concentration.

$$\frac{0.85(0.90)(510 - x_1) + 0.85(0.80)(430 - x_2) + 0.85(960 - x_3) + (920 - x_4)}{9{,}936} \leq 0.12 \text{ kg/m}^3$$

$$0.765x_1 + 0.68x_2 + 0.85x_3 + x_4 \geq 1{,}226$$

The amount removed at each treatment plant is limited by the removal efficiency of the plant.

$$x_1 \leq 0.92(510) = 469$$

$$x_2 \leq 0.90(430) = 387$$

$$x_3 \leq 0.91(960) = 874$$

$$x_4 \leq 0.92(920) = 846$$

The complete formulation is as follows:

Minimize

$$x_0 = 2.50x_1 + 1.80x_2 + 4.00x_3 + 3.50x_4$$

Subject to

$$x_1 \geq 104.6$$
$$0.9x_1 + 0.8x_2 + x_3 \geq 744.9$$
$$0.765x_1 + 0.68x_2 + 0.85x_3 + x_4 \geq 1{,}226$$
$$x_1 \leq 469$$
$$x_2 \leq 387$$
$$x_3 \leq 874$$
$$x_4 \leq 846$$
$$x_i \geq 0 \quad \text{for} \quad \text{all } x_i$$

The optimum decision policy is x_1, x_2, x_3, and x_4 values of 469, 387, 13.2, and 592.8 thousand kg/day, with a x_0 of \$3,997. This solution is obtained using software with mathematical programming capabilities such as the Excel solver.

11.5.5 Water Allocation Models

LP formulations have been incorporated into a number of river basin management models (Section 12.8) to allocate streamflow and reservoir storage contents among numerous water users in accordance with water rights priority systems. For example, a simulation model might allocate water to several hundred water users during each month of a 60-year (720-month) hydrologic period of analysis using a monthly computational time step. The simulation results are used to compute reliabilities (Section 7.10) for each of the water users. An LP formulation performs the water allocation during each month of the simulation. Thus, the LP algorithm is activated 720 times during a simulation. Examples 11.10 and 11.11 have only a few water users, but illustrate the general concept of formulating LP models to allocate water among users based on specified priorities.

Example 11.10

A schematic of a river/reservoir system is presented in Fig. 11.12. Reservoirs A and B, located at nodes 1 and 2, have storage capacities of 750×10^6 and 900×10^6 m^3, respectively. The initial storage in reservoirs A and B at the beginning of the time interval is 460×10^6 and 215×10^6 m^3, respectively. Releases are made as necessary to maintain instream flow requirements and then, to the extent possible, to meet water supply diversion targets. Instream flow requirements for the particular time period are as follows:

Reach	1–4	2–3	3–4	4–5	Below 5
Flow (10^6 m^3)	0	5	10	10	30

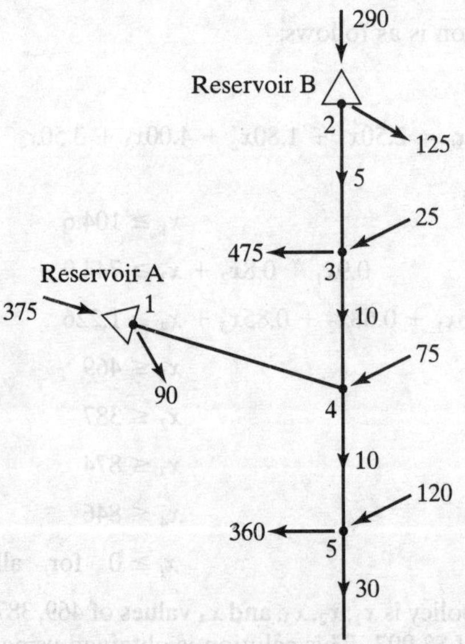

Figure 11.12 The water allocation problem represented by this system schematic is solved in both Examples 11.10 and 11.11.

The supply and demand for water are shown in Table 11.10 for each of the node locations shown in Fig. 11.12. The total supply at each node consists of reservoir storage at the beginning of the time interval and the local flow entering the river between the node and adjacent upstream node(s) during the time period. The demand in Table 11.10 is a target diversion at each node. If supplies are insufficient to meet all demands, allocations are based on the relative priorities tabulated in the last column. For example, with a relative priority of 5, the diversion of 360×10^6 m^3 at node 5 has the highest priority of the four diversions. Lesser priority diversions are met only to the extent that higher priority diversions are not adversely affected.

Solution The model incorporates the 11 decision variables defined in Table 11.11 The decision variables include instream flows in each of five river reaches (x_1, x_2, x_3, x_4, x_5), water supply diversions at four nodes (x_6, x_7, x_8, x_9), and the ending storage in the two reservoirs (x_{10}, x_{11}).

The objective function is formulated to reflect relative priorities between water users as follows.

$$\text{maximize } x_0 = 4x_6 + 3x_7 + 2x_8 + 5x_9 + x_{10} + x_{11}$$

TABLE 11.10 GIVEN INFORMATION FOR EXAMPLES 11.10 AND 11.11

Location or node	Initial storage (10^6 m^3)	Local inflow (10^6 m^3)	Total supply (10^6 m^3)	Demand target (10^6 m^3)	Relative priority
1	460	375	835	90	4
2	215	290	505	125	3
3	—	25	25	475	2
4	—	75	75	—	—
5	—	120	120	360	5

TABLE 11.11 DECISION VARIABLES FOR EXAMPLE 11.10

Decision variables	Definition of decision variables	Solution (10^6 m^3)
X_1	Instream flow from 1 to 4	195
X_2	Instream flow from 2 to 3	380
X_3	Instream flow from 3 to 4	10
X_4	Instream flow from 4 to 5	270
X_5	Instream flow below node 5	30
X_6	Diversion at node 1	90
X_7	Diversion at node 2	125
X_8	Diversion at node 3	395
X_9	Diversion at node 5	360
X_{10}	Reservoir A ending storage	550
X_{11}	Reservoir B ending storage	0

The objective function coefficients are used simply to assign relative priorities to guide the allocation of the limited water resources to competing uses. The absolute values of the coefficients are arbitrary. Only the relationship of the coefficient values relative to each other affect the results. Water supply diversion x_9 is assigned the highest priority of the four diversions, as reflected by its coefficient of five. Filling the two reservoirs (x_{10}, x_{11}) is assigned lower priority as reflected by coefficients of one. The diversion targets are fully met prior to filling the reservoirs.

The instream flow requirements (x_1, x_2, x_3, x_4, x_5) are assigned zero priority (coefficient values) in the objective function and are handled as constraints.

$$x_1 \geq 0 \qquad x_2 \geq 5 \qquad x_3 \geq 10 \qquad x_4 \geq 10 \qquad x_5 \geq 30$$

The constraints force the instream flow requirements to be met even if diversion targets cannot be fully met. Sufficient water is available to meet all instream flow requirements. Otherwise, there would be no feasible solution, and the model would have to be reformulated.

A volume balance constraint is written for each node. For example, at node 1, initial storage and inflows of 835 m^3 from Table 11.10 supply downstream flows (x_1), diversions (x_6), and reservoir filling (x_{10}).

$$x_1 + x_6 + x_{10} = 835$$

The complete LP model is formulated as follows

Maximize

$$x_0 = 4x_6 + 3x_7 + 2x_8 + 5x_9 + x_{10} + x_{11}$$

Subject to

$$x_1 \geq 0 \qquad x_2 \geq 5 \qquad x_3 \geq 10 \qquad x_4 \geq 10$$

$$x_5 \geq 30 \qquad x_6 \leq 90 \qquad x_7 \leq 125 \qquad x_8 \leq 475$$

$$x_9 \leq 360 \qquad x_{10} \leq 750 \qquad x_{12} \leq 900$$

$$x_1 + x_6 + x_{10} = 835$$

$$x_2 + x_7 + x_{11} = 505$$

$$-x_2 + x_3 + x_8 = 25$$

$$-x_1 - x_3 + x_4 = 75$$

$$-x_4 + x_5 + x_9 = 120$$

The model can be solved using any LP software. The resulting values for the decision variables are tabulated in the last column of Table 11.11.

11.5.6 Dealing with Nonlinearities in LP Models

Nonlinear relationships complicate LP models, but can be handled in many cases. Nonlinear equations can sometimes be adequately approximated by linear equations. In other cases, successive iterative solutions of the LP model are used to handle nonlinearities.

Examples 11.7, 11.8, 11.10, and 11.11 neglect reservoir evaporation in the volume accounting computations. Evaporation volumes are typically computed as a net evaporation rate multiplied by the mean water surface area during the computational time step. Water surface area is a nonlinear function of storage. The mean storage or area during a time interval, such as a month, is typically approximated as the average of the values at the beginning and ending of the time interval. Thus, end-of-period storage is computed as a function of evaporation volume, which in turn is computed as a function of end-of-period storage. Likewise, hydroelectric power is a function of both head and discharge. Head is a nonlinear function of storage. End-of-period reservoir storage is computed as a function of releases required to meet hydroelectric energy requirements, which in turn depend on the available head provided by the reservoir storage. Various approaches are adopted to deal with these nonlinearities. The simplest approach is to treat reservoir evaporation and hydropower as functions of beginning-of-period storage only, rather than both beginning and ending period storage and to approximate storage–area and storage–head relationships as linear equations. However, the approach outlined in the next paragraph is more accurate.

The common approach of successive approximations consists of iteratively executing an LP routine, with successive approximations of nonlinear terms computed outside of the LP routine. For example, for reservoir evaporation computations, an initial estimate of evaporation volume based on the known beginning-of-month storage and nonlinear storage–area relationship is input to the LP algorithm. The LP algorithm computes the end-of-month storage, which is then used to develop an improved estimate of the evaporation volume. The LP algorithm is iteratively executed with improved estimates of evaporation inputted until a specified stop criterion is met. Models often contain algorithms that automatically execute an LP submodel numerous times each time the model is run.

Objective functions may also be nonlinear. For example, benefits and costs may be nonlinear functions of decision variables. A separable function of n variables can be written as the sum of n individual functions

$$f(x_1, x_2, x_3, \ldots, x_n) = f_1(x_1) + f_2(x_2) + f_3(x_3) + \cdots + f_n(x_n)$$

each of which depends on only one of the variables. Loucks, Stedinger, and Haith (1981), Mays and Tung (1992), and others present methods for approximating separable nonlinear objective functions as piecewise linear functions.

11.6 NETWORK FLOW PROGRAMMING

Network flow programming is a computationally efficient form of LP that can be applied to problems that can be formulated in a specified format representing a system as a network of nodes and arcs having certain characteristics. Network flow programming is addressed in detail by Jensen and Barnes (1980), Phillips and Garcia-Diaz (1981), and others. Wurbs (1996) reviews reservoir/river system management models based on network flow programming.

In a network flow model, the system is represented as a collection of nodes and arcs. For a reservoir/river system, the nodes are locations of reservoirs, diversions, stream tributary confluences, and other pertinent system features. Nodes are connected by arcs representing the way "*flow*" is conveyed. For a reservoir/river system, flow represents either a discharge rate, such as instream flows and diversions, or a change in storage per unit of time.

The general form of the network flow programming problem is as follows:

Minimize

$$\sum\sum c_{ij}\, q_{ij} \qquad \text{for all arcs} \qquad (11.23)$$

Subject to

$$\sum q_{ij} - \sum q_{ji} = 0 \qquad \text{for all nodes} \qquad (11.24)$$

$$l_{ij} \le q_{ij} \le u_{ij} \qquad \text{for all arcs} \qquad (11.25)$$

where q_{ij} is the flow rate in the arc connecting node i to node j; c_{ij} is a penalty or weighting factor for q_{ij}; l_{ij} is a lower bound on q_{ij}; and u_{ij} is an upper bound on q_{ij}. In Eq. 11.24, q_{ij} is flow from node i to node j, and q_{ji} is flow from node j to node i.

A solution algorithm computes the values of the flows q_{ij} in each of n arcs (node i to node j), which minimizes an objective function consisting of the sum of the flows multiplied by corresponding weighting factors, subject to constraints, including maintaining a mass balance at each node and not violating user-specified upper and lower bounds on the flows. Each arc has three parameters: a weighting, penalty, or unit cost factor c_{ij} associated with q_{ij}; lower bound l_{ij} on q_{ij}; and an upper bound u_{ij} on q_{ij}. The requirement for lower and upper bounds results in the term capacitated flow networks. Network flow programming provides considerable flexibility for formulating a particular application. The weighting factors c_{ij} in the objective function are defined in various ways. The c_{ij} may be unit costs in dollars or penalty or utility terms that provide mechanisms for expressing relative priorities. A penalty weighting factor is the same as a negative utility weighting factor.

Network flow programming problems can be solved using conventional LP algorithms. However, the network flow format facilitates the use of much more computationally efficient algorithms that save computer time and allow analysis of larger problems with numerous variables and constraints.

Example 11.11

This example consists of repeating the previous Example 11.10 using the network flow format of Eqs. 11.23–11.25. Flows q_{ij} are computed for the arcs connecting the set of five nodes shown in Fig. 11.12 along with a source node and a sink node. The source node represents the source of water entering the stream/reservoir system. Water leaving the system flows to the sink node.

The decision variables q_{ij}, lower and upper bounds l_{ij} and u_{ij} on q_{ij}, and objective function coefficients c_{ij} are shown in Table 11.12. The c_{ij} are the relative priorities, which for the diversions are given in Table 11.12. Five constraints in the form of Eq. 11.24 represent the volume balance at each of the five node locations shown in Fig. 11.12. Constraints in the form of Eq. 11.25 place upper and lower bounds on the decision variables. In order to fit the capacitated network flow format, the given initial storage and inflows are treated as decision variables with both lower and upper bounds set at the values specified in Table 11.12. In reality,

TABLE 11.12 TERMS IN NETWORK FLOW FORMULATION OF EXAMPLE 11.11

Decision q_{ij}	Definition of q_{ij}	Nodes i	Nodes j	Lower bound l_{ij}	Upper bound u_{ij}	Priority c_{ij}	Solution (10^6 m^3) q_{ij}
$q_{1,4}$	Instream flow from 1 to 4	1	4	0	999	0	195
$q_{2,3}$	Instream flow from 2 to 3	2	3	5	999	0	380
$q_{3,4}$	Instream flow from 3 to 4	3	4	10	999	0	10
$q_{4,5}$	Instream flow from 4 to 5	4	5	10	999	0	270
$q_{5,S}$	Instream flow below 5	5	Sink	30	999	0	30
$q_{S,1}$	Inflow + initial storage	Source	1	835	835	0	835
$q_{S,2}$	Inflow + initial storage	Source	2	505	505	0	505
$q_{S,3}$	Inflow + initial storage	Source	3	25	25	0	25
$q_{S,4}$	Inflow + initial storage	Source	4	75	75	0	75
$q_{S,5}$	Inflow + initial storage	Source	5	120	120	0	120
$q_{1,D}$	Diversion at node 1	1	Sink	0	90	4	90
$q_{2,D}$	Diversion at node 2	2	Sink	0	125	3	125
$q_{3,D}$	Diversion at node 3	3	Sink	0	475	2	395
$q_{5,D}$	Diversion at node 5	5	Sink	0	360	5	360
$q_{1,S}$	Reservoir A storage	1	Sink	0	750	1	550
$q_{2,S}$	Reservoir B storage	2	Sink	0	900	1	0

streamflows do not have known upper bounds, but Eq. 11.25 requires that bounds be specified for all q_{ij}. Consequently, arbitrarily large values of 999 are assigned to u_{ij} for the instream flows.

The network flow model is formulated as follows:

Maximize

$$4q_{1,D} + 3q_{2,D} + 2q_{3,D} + 5q_{5,D} + q_{1,S} + q_{2,S}$$

Subject to

$$q_{S,1} - q_{1,D} - q_{1,4} - q_{1,S} = 0$$

$$q_{S,2} - q_{2,D} - q_{2,3} - q_{2,S} = 0$$

$$q_{S,3} + q_{2,3} - q_{3,D} - q_{3,4} = 0$$

$$q_{S,4} + q_{1,4} - q_{4,5} = 0$$

$$q_{S,5} + q_{4,5} - q_{5,D} - q_{5,S} = 0$$

$$0 \le q_{1,4} \le 999 \qquad 835 \le q_{S,1} \le 835 \qquad 0 \le q_{1,D} \le 90$$

$$5 \le q_{2,3} \le 999 \qquad 505 \le q_{S,2} \le 505 \qquad 0 \le q_{2,D} \le 125$$

$$10 \le q_{3,4} \le 999 \qquad 25 \le q_{S,3} \le 25 \qquad 0 \le q_{3,D} \le 475$$

$$10 \le q_{4,5} \le 999 \qquad 75 \le q_{S,4} \le 75 \qquad 0 \le q_{5,D} \le 360$$

$$30 \le q_{5,S} \le 999 \qquad 120 \le q_{S,5} \le 120 \qquad 0 \le q_{1,S} \le 750$$

$$0 \le q_{2,S} \le 900$$

Because the model is formulated in the format of Eqs. 11.23–11.25, either regular LP or network flow programming algorithms can be applied. The solution is tabulated in the last column of Table 11.12.

11.7 ZERO-ONE INTEGER PROGRAMMING

In *integer LP*, one or more decision variables are constrained to be integers (0, 1, 2, 3, ...). *Mixed integer* refers to models in which some variables are constrained to be integer, and others are not. Zero-one or binary LP is a special case of integer LP in which one or more decision variables are constrained to be either zero or one. Zero-one LP is useful for problems in which the solution is a set of *yes* or *no* decisions. Actions are either selected ($x_i = 1$) or not selected ($x_i = 0$) as illustrated by Examples 11.12 and 11.13.

Special algorithms are required to solve integer programming models. Microsoft Excel, GAMS, and other software packages include capabilities for integer and zero-one integer programming.

Example 11.12

A water district has four new customers (1–4) that may be either cities, companies, and/or agricultural irrigators. The district has three possible sources (A–C) from which to supply its four new customers. Water supply source A costs \$14,200,000 to develop. Sources B and C cost \$8,500,000 and \$17,100,000 to develop. One, two, or three water supply sources must be selected. A pipeline must be constructed from each of the four customers to at least one of the three water supply sources. Costs for constructing the four pipelines to the customers are provided in Table 11.13. The demands for each customer and available supply from each source are also shown in the table. A zero-one integer LP model is formulated to determine which water supply sources to develop and which pipelines to construct to minimize total costs.

Solution Fifteen decision variables are defined as follows.

A = 1 if source A is developed and 0 otherwise

B = 1 if source B is developed and 0 otherwise

C = 1 if source C is developed and 0 otherwise

$Ai = 1$ if customer i is served by source A and 0 otherwise, for $i = 1, 2, 3, 4$

$Bi = 1$ if customer i is served by source B and 0 otherwise, for $i = 1, 2, 3, 4$

$Ci = 1$ if customer i is served by source C and 0 otherwise, for $i = 1, 2, 3, 4$

The zero-one LP model is formulated as follows:

Minimize

$$\text{Total Cost} = 14.2\,A + 8.5\,B + 17.1\,C + 2.2\,A1 + 8.3\,A2 + 8.2\,A3 + 5.2\,A4 + 4.6\,B1 + 6.1\,B2 + 2.4\,B3 + 9.8\,B4 + 1.7\,C1 + 5.8\,C2 + 1.3\,C3 + 3.6\,C4$$

Subject to 10 constraints:

Each customer must be connected to at least one source.

$$A1 + B1 + C1 \geq 1 \quad A2 + B2 + C2 \geq 1 \quad A3 + B3 + C3 \geq 1 \quad A4 + B4 + C4 \geq 1$$

If a pipeline is constructed to a source, the source must be developed.

$$4A - A1 - A2 - A3 - A4 \geq 0 \qquad 4B - B1 - B2 - B3 - B4 \geq 0$$
$$4C - C1 - C2 - C3 - C4 \geq 0$$

TABLE 11.13 DATA FOR EXAMPLE 11.12

Supply source	Pipeline Costs (\$) Customer 1	Customer 2	Customer 3	Customer 4	Capacity (m^3/day)
A	\$2,200,000	\$8,300,000	\$8,200,000	\$5,200,000	340,000
B	\$4,600,000	\$6,100,000	\$2,400,000	\$9,800,000	235,000
C	\$1,700,000	\$5,800,000	\$1,300,000	\$3,600,000	480,000
Demand (m^3/day)	135,000	98,000	64,000	192,000	

The demands cannot exceed the amount of water available from each source.

$$135\text{ A1} + 98\text{ A2} + 64\text{ A3} + 192\text{ A4} \le 340$$

$$135\text{ B1} + 98\text{ B2} + 64\text{ B3} + 192\text{ B4} \le 235$$

$$135\text{ C1} + 98\text{ C2} + 64\text{ C3} + 192\text{ C4} \le 480$$

Excel or other software can be used to obtain the optimum solution consisting of B, C, B2, C1, C3, and C4 equalling 1 and the other decision variables equalling zero. The total cost is $38,300,000.

Example 11.13

The 10 possible projects that are being considered as components of a river basin development plan include reservoirs, conveyance/pumping facilities, flood control improvements, and hydroelectric power plants. Investment costs and net annual benefits (annual benefits less annual costs) for each project are tabulated below. This problem consists of formulating a zero-one LP model to select the set of projects that will maximize net benefits subject to constraints described as follows.

Project	Investment cost ($ million)	Net annual benefits ($ million)
1	120	4
2	150	3
3	90	5
4	210	9
5	160	7
6	180	6
7	90	4
8	100	5
9	80	2
10	140	7

Due to budget constraints, the total of the investment costs for the selected projects cannot exceed $750,000,000. Projects 2 and 3 can be implemented only if project 6 is implemented. (*Projects 2 and 3 are each contingent on project 6.*) Project 10 can be implemented only if both projects 8 and 9 are implemented. (*Project 10 is contingent on projects 8 and 9*). Project 4 cannot be selected if project 3 is selected. (*Projects 4 and 3 are mutually exclusive.*) At least one, but no more than three, of the following projects must be selected: projects 5, 6, 7, and/or 8.

Solution Ten decision variables x_i are defined as follows: $x_i = 1$ if the project is selected and zero otherwise

Maximize

$$x_0 = 4x_1 + 3x_2 + 5x_3 + 9x_4 + 7x_5 + 6x_6 + 4x_7 + 5x_8 + 2x_9 + 7x_{10}$$

Subject to: Budget constraint

$$120x_1 + 150x_2 + 90x_3 + 210x_4 + 160x_5 + 180x_6 + 90x_7 + 100x_8 + 80x_9 + 140x_{10} < 750$$

Contingency constraints

$$x_2 - x_6 \le 0 \quad \text{alternatively} \quad x_2 + x_3 - 2x_6 \le 0$$

$$x_3 - x_6 \le 0$$

$$x_{10} - x_8 \le 0 \quad \text{alternatively} \quad 2x_{10} - x_8 - x_9 \le 0$$

$$x_{10} - x_9 \le 0$$

Mutual exclusivity constraint

$$x_3 + x_4 \le 1$$

Constraint specifying one to three of these projects

$$x_5 + x_6 + x_7 + x_8 \ge 1$$

$$x_5 + x_6 + x_7 + x_8 \le 3$$

All x_i are either zero or one.

The optimum solution consists of selecting projects 3, 5, 6, 8, 9, and 10 ($x_3 = x_5 = x_6 = x_8 = x_9 = x_{10} = 1$ and $x_1 = x_2 = x_4 = x_7 = 0$). The total net annual benefits (x_0) are $32,000,000.

PROBLEMS

11.1. Three alternative plans for expanding a municipal water distribution system will each meet demands for water and maintain required pressures. Each plan has a project life of 30 years. The initial investment cost and annual operation and maintenance cost for each plan are as follows:

Plan	Initial investment cost ($)	Annual operation and maintenance cost ($)
A	120,000	28,000
B	160,000	24,000
C	290,000	10,000

Select the economically optimum plan based on a discount rate of 8 percent. Perform the evaluation alternatively based on present worth and annual worth.

11.2. Refer to the three plans in Problem 11.1. We are somewhat uncertain regarding our selection of discount rate. For what ranges of values for the discount rate will each plan be economically optimal?

11.3. Cost and benefit estimates for two alternative hydroelectric power projects are as follows:

Project	A	B
Initial investment cost	$25,000,000	$42,000,000
Annual operation and maintenance cost	$1,200,000/yr	$2,000,000/yr
Annual benefits	$3,800,000/yr	6,200,000/yr

Both projects have an expected life of about 100 years, with no salvage value. Determine the economically optimum project using a discount rate of 8 percent and the following evaluation methods: (a) present worth analysis, (b) annual worth analysis, and (c) incremental benefit-cost analysis.

11.4. A municipal water district needs to replace a pump. Two alternatives are being considered. Pump A will cost $42,400 installed. It is expected to be replaced in 20 years. The estimated salvage value at the end of 20 years is $3,200. Annual maintenance and electric power costs for operating pump A are estimated to be $3,600/year. The other alternative (pump B) will cost $16,000 installed and is expected to have a salvage value of $2,400 at the end of its useful life of 10 years. Annual operation and maintenance will cost $6,000/year for pump B. Which of the two alternatives is economically optimum at a discount rate of (a) 6 percent and (b) 10 percent?

11.5. Rework Example 11.2 in Section 11.2.1 using a discount rate of 10 percent.

11.6. An economic evaluation is being performed for a proposed multiple-purpose reservoir project. Would you expect the benefit-to-cost ratio to be higher for a discount rate of 4 percent or 8 percent? Explain why.

11.7. An irrigation system is to be installed for a public golf course. Each of three alternative designs consists of pipes and a pump. The piping system is expected to last 30 years, with zero salvage value. The pump is expected to last 10 years and have zero salvage value. Replacement pumps of the same design and cost will be used to extend the operation to 30 years. Cost estimates for the three alternative system designs are provided below.

Alternative	A	B	C
Initial pump cost	$7,500	$15,000	$30,000
Initial piping cost	$15,000	$15,000	$22,500
Annual maintenance	$6,000/yr	$4,500/yr	$3,000/yr
Pumping cost per 1,000 gallons	$4.50	$3.75	$3.00

The optimum alternative depends on the amount of water pumped each year to irrigate the golf course. Determine the range in the amount of water pumped for which each alternative would be economically optimum based on a discount rate of 8 percent.

11.8. An agricultural irrigation project is being designed. Two alternatives involve either (1) manually operated gates or (2) motor-operated gates that can be controlled electronically from headquarters. The remotely controlled gates cost more initially, but result in significant savings in operation cost over the life of the project. Economic justification for installing the more expensive gates depends on the life of the project. The remotely controlled gates cost $160,000 installed. Annual operation and power costs are $5,500/year. Manually operated gates cost $82,000 installed. These gates will require daily adjustments on each of the 250 days of the irrigation season. A person will spend about 2 hours traveling to the intake structure, adjusting the gates, and returning to the headquarters each day. Transportation costs for the 30-mile round trip (both ways) is $0.30/mile. Personnel cost including salary and fringe benefits is $20/hour. What project life would be required for the two alternatives to break even based on a discount rate of 8 percent and zero salvage value?

11.9. Project costs and residual flood damages for two alternative flood mitigation plans are provided below, along with damages that will occur if no project is implemented. Compute net benefits and benefit-cost ratios and determine the economically optimum plan using a 100-year period of analysis and 6 percent discount rate.

Plan	Average annual damage ($)	Initial investment ($)	Annual operation and maintenance ($)
No project	100,000	0	0
Alternative 1	40,000	600,000	5,000
Alternative 2	10,000	1,000,000	10,000

11.10. Estimates of project costs and residual flood damages for six flood damage reduction plans are as follows:

Flood mitigation plan	Initial investment cost ($)	Operation and maintenance ($/yr)	Average annual damages ($/yr)
No action	0	0	1,450,000
Plan A	5,640,000	572,000	210,000
Plan B	4,820,000	485,000	300,000
Plan C	4,730,000	271,000	375,000
Plan D	3,890,000	185,000	570,000
Plan E	2,250,000	216,000	725,000
Plan F	910,000	300,000	750,000

Determine the benefit-to-cost ratio and net benefits for each plan for a 100-year period of analysis and discount rate of 8 percent. From an economic perspective, which plan is best?

11.11. A flood control project consists of purchasing and moving selected residential houses from a floodplain. The houses can be purchased and relocated at a total investment cost of $5.0 million. Compute the benefit-to-cost ratio for this project based on a 100-year period of analysis, 6 percent discount rate, and data provided in the table below. A discharge–frequency relationship is provided in the first two columns. Damages resulting from the various discharge levels are tabulated for with and without project conditions in the last two columns.

Recurrence interval (yrs)	Peak flows (cfs)	Damages ($ million)	
		Without project	With project
5	2,100	0	0
10	3,600	2	0
50	5,200	8	3
100	9,700	12	5
200	17,500	18	10
500	24,200	20	12

11.12. Exceedance probability versus discharge relationships are given below alternatively for conditions with and without a proposed flood control reservoir project. A discharge versus damage relationship is also tabulated. Determine the economic benefits that would be provided by construction of the reservoir project.

	Discharge (cfs)			
Exceedance probability	Without project	With project	Discharge (cfs)	Damage ($ million)
0.25	10,000	10,000	10,000	0
0.20	14,000	10,000	15,000	5
0.15	18,000	10,000	20,000	10
0.10	25,000	14,000	30,000	20
0.05	35,000	20,000	40,000	30
0.01	50,000	40,000	50,000	40
0.005	60,000	58,000	60,000	50

11.13. A flood control reservoir project and channel improvement project are proposed for reducing flood damages along a stream in an urban area. Estimated costs are as follows:

	Initial investment	Annual maintenance
Dam and reservoir project	$1,500,000	$15,000/yr
Channel improvement project	$1,000,000	$20,000/yr

Hydrologic, hydraulic, and economic studies have been conducted to develop the following relationships for the river reach and floodplain: (a) annual exceedance probability versus peak discharge with and without the reservoir, (b) discharge versus stage with and without the channel improvement, and (c) stage versus damage.

	Peak discharge (cfs)			Stage (ft)			
Annual exceedance probability	Without reservoir	With reservoir	Discharge (cfs)	Without project	With project	Stage (ft)	Damage ($ million)
0.50	3,000	2,700	0	0	0	9	0.0
0.20	5,400	3,500	5,000	10	7	10	0.3
0.10	7,600	4,200	10,000	15	11	15	1.3
0.05	10,400	5,500	15,000	20	14	20	2.4
0.02	15,000	8,600	20,000	23	17	25	4.2
0.01	19,000	14,000	25,000	25	19	28	5.4
0.005	24,000	20,000	30,000	27	22	30	7.0
0.002	31,500	29,500	35,000	28	24		

Perform an economic evaluation of the following alternative plans using a 50-year period of analysis and 8 percent discount rate. Determine the net benefits and benefit-to-cost ratio for each plan. Select the economically optimum plan.

Plan 1—no action
Plan 2—construct the reservoir project
Plan 3—construct the channel improvement project
Plan 4—construct both the reservoir and channel improvement projects

11.14. A farming operation is considering two types of crops. Each crop requires the amounts of land, labor, fertilizer, and water tabulated below per unit of crop produced. The maximum available units of each resource are also shown. Net income in \$ per unit of crop produced is \$60/unit and \$50/unit for crops A and B, respectively. Formulate and graphically solve an LP (linear programming) model to determine the number of units of each crop to plant to maximize net income. You may use Excel or another software package to check your solution.

	Units of resource per unit of crop		Limit on available
Resource	Crop A	Crop B	resource
Land	5	2	80
Labor	1	2	40
Fertilizer	3	2	60
Water	2	3	60

11.15. A farm has 1,800 acre-feet of water available annually. Two crops are considered for which annual irrigation water requirements are 3 acre-feet/acre and 2 acre-feet/acre, respectively. For various reasons, no more than 400 acres can be planted in crop 1, and no more than 600 acres can be allocated to crop 2. Estimated profits are \$300 per acre planted in crop 1 and \$500 per acre planted in crop 2. Determine how many acres to plant in each crop to maximize profits. Formulate and graphically solve an LP model. You may use Excel or another LP program to check your solution.

11.16. A company produces two products, A and B. Tabulated below are: the amount of raw materials and water required to produce each unit of products A and B, production rates (units of A or B produced per hour of production time), and profit per unit produced for each product.

Product	A	B
Profit (\$/unit)	24	20
Production rate (units/hr)	30	15
Raw materials required (lbs/unit)	10	6
Water required (gallons/unit)	8	10

The two products must share the total raw material, water, and production time available each day. The total amount of raw material available per day to share between the two products is 3,150 pounds. A maximum of 14 hours/day of production time is available to allocate between the two products. A total of 3,000 gallons/day of water is available.

a. Formulate an LP model to determine the daily production rate for each product that will maximize profits.

b. Solve the model graphically. You may use Excel or another program to check your solution.

c. Is there one or more nonbinding constraints in your solution? If so, what does the non-binding constraint physically mean in this particular problem? If not, what does having no non-binding constraints mean in this problem?

d. How many feasible decision policies does the problem have?

e. How many optimum decision policies does the problem have?

f. Assume that the unit profits for both products A and B are reduced 50 percent to $12 and $10. How will this affect your solution?

g. Assume the unit profit for product A is fixed at $24 but the unit profit for product B is subject to change. How much would the $20 unit profit for B have to change to change the number of units of product B produced each day?

11.17. An industrial water treatment plant requires 300,000 m^3 of water per day. The water must be chlorinated and softened before use. The treatment process requires 100 units of a chlorination chemical and 150 units of a particular softening agent. Two alternative types of water additive contain the chlorination chemical and softening agent. Additive A contains 3 units of the chlorination chemical and 8 units of the softening agent per package of water additive. Additive B contains 9 and 4 units of the chlorination chemical and softening agent per package. Water additive options A and B cost $8 and $10 per package, respectively. Formulate an LP model to determine the combination of treatment additives A and B that will minimize cost. Solve the LP problem graphically.

11.18. Water is conveyed from alternative sources 1 3 to alternative users A–E. The demands for any of the five users (A–E) can be met by any combinations of supply from the three sources (1–3). The water demands for Users A–E are 1, 2, 3, 8, and 10 million cubic meters (Mm^3), respectively. The amount of water available from sources 1–3 are 4, 8, and 12 Mm^3, respectively. The unit costs in dollars per million m^3 ($/Mm^3) for supplying water from each source to each user are provided below.

Destination	A	B	C	D	E
Source 1	$7/Mm^3	$10/Mm^3	$5/Mm^3	$4/Mm^3	$12/Mm^3
Source 2	$3/Mm^3	$2/Mm^3	0	$9/Mm^3	$1/Mm^3
Source 3	$8/Mm^3	$13/Mm^3	$11/Mm^3	$6/Mm^3	$14/Mm^3

Use LP to determine the allocation of water from each source to each user that meets demands at minimum cost.

11.19. A city must increase its water supply by 150 million gallons/day (mgd) in response to population growth. Three alternative sources of supply are available. A river flowing through the city provides a low-cost, good-quality source of supply. However, because the city is already obtaining most of its existing supply from this source, the maximum available additional supply is limited to 25 mgd. Up to 120 mgd may be obtained from a groundwater aquifer, but the quality is not as good as surface water. The hardness of the ground water is 2,300 pounds/million gallons (lbs/mgal). A river/reservoir system located a significant distance from the city provides a third alternative source, which is very expensive. The maximum amount of water available, unit cost, and hardness for each of the three alternative sources are provided below. Hardness consists primarily of calcium and magnesium. Water quality standards limit the maximum allowable hardness to 1,200 lbs/mgal.

Source	1	1	3
Supply limit (mgd)	25	120	100
Hardness(lbs/mgal)	200	2,300	700
Cost ($/mgd)	$750	$1,500	$3,000

Formulate an LP model to determine the amount of water to obtain from each of the three alternative sources that minimizes costs while limiting the hardness of the combined supply to not exceed 1,200 lbs/mgal. Use Excel or another software package to obtain the optimum solution.

11.20. A municipal water district supplies water to three cities from the following three sources. Source A is a well with relatively low pumping cost in an aquifer with relatively high salinity. Source B is a well with higher pumping cost in an aquifer with better quality water. Source C is a reservoir on a river with good quality (low salinity) and no pumping cost. Each of the three cities may be supplied from combinations of the three sources. The unit cost per m^3 for pumping water from each source to each city is tabulated below. The daily water demand for each city is shown in the last row of the table.

	City 1	City 2	City 3
Source A	$0.15/m^3	$0.20/m^3	$0.10/m^3
Source B	$0.20/m^3	$0.40/m^3	$0.30/m^3
Source C	No cost	No cost	No cost
Demand	1,000 m^3/day	2,000 m^3/day	3,000 m^3/day

The total dissolved solids (TDS) concentration in grams per cubic meter (g/m^3) for each source is shown below. The amount of water available from each source (capacity) is also tabulated.

Source	TDS concentration	Capacity
Source A	800 g/m^3	4,000 m^3/day
Source B	400 g/m^3	2,500 m^3/day
Source C	100 g/m^3	5,000 m^3/day

The TDS concentration of the municipal water supply cannot exceed 500 g/m^3. Water from alternative sources can be blended to achieve this water quality standard. Source C is a multiple-purpose reservoir from which water can be withdrawn for municipal water supply only if the water demands and quality standards cannot be met in any other way. Water supply diversions adversely affect hydropower, recreation, and other reservoir purposes. The amount of water supplied from source C must be minimized, subject to meeting water supply demands and the 500 g/m^3 maximum allowable TDS limit. Formulate and solve an LP model to determine the amount of water supplied from each source for each city that minimizes pumping costs while meeting water demands and water quality standards.

11.21. In this problem, the inflow in column 3 of Table 11.8, that is used in Example 11.8, represents streamflows at the site of the reservoir system of Example 11.7 shown in Fig. 11.10. Develop an LP model for the reservoir system of Example 11.7, except use the 24-month streamflow sequence from Table 11.8 as the inflows to the reservoir. The reservoir storage content at the end of the 24-month simulation period should be the same as at the beginning. Use this model to test the impacts of environmental instream flow requirements on irrigation and hydropower generation. First run the model with the instream flow requirement of 20 million m^3/month. Rerun the model with no instream flow requirement. Rerun the model again with the instream flow requirement doubled to 40 million m^3/month. Compare the irrigation and hydropower revenues generated under these three alternative scenarios. In addition to economic revenues, volume reliability defined by Eq. 7.48 is a useful measure of water supply capabilities. Determine the volume reliability for the irrigation diversion from the results of the three simulations.

11.22. Reformulate the LP model of Problem 11.21 so that the environmental instream flow requirement is given higher priority than any other water use, but is not treated as an absolute constraint. The model should run even if there is not enough water to meet the instream flow requirement. Run the model for alternative instream flow requirements of zero, 20, 40, 60, 80, 100, 150, 200, and 300 million m^3/month. Determine the period reliability for the instream flow requirements, as defined by Eq. 7.49, from the results of each simulation. Develop a table comparing hydropower and irrigation revenues and period reliability for the instream flow requirement for each of the alternative scenarios of specified instream flow requirements.

11.23. This problem consists of developing a firm yield versus storage capacity relationship for a proposed water supply reservoir. The monthly streamflows (inflows) at the site of the proposed reservoir are provided below for a critical 2-year drought period. The firm yield provided by a reservoir at this site is the maximum constant water supply diversion rate that can be maintained continuously if the sequence of 24 monthly inflows are repeated cyclically forever. Firm yield is a function of reservoir storage capacity, as well as inflows. Compute at least six points on the firm yield versus storage capacity curve. Firm yields for the two extremes of zero storage capacity and an infinitely large reservoir can be easily determined without using an LP model. Determine at least four other additional points on the yield-storage curve by developing an LP model. Plot the storage capacity versus firm yield curve.

Month	Inflows (ac-ft/month) First year	Second year
Jan	76,000	93,000
Feb	110,000	103,000
Mar	100,000	142,000
Apr	206,000	178,000
May	260,000	223,000
Jun	80,000	41,000
Jul	23,000	20,000
Aug	9,000	17,000
Sep	67,000	61,000
Oct	54,000	170,000
Nov	87,000	138,000
Dec	83,000	129,000

11.24. A firm yield versus storage capacity relationship for a proposed reservoir is to be developed based on the monthly streamflows (inflows) at the site during a critical 2-year drought period. The firm yield is the maximum constant water supply diversion rate from the reservoir that can be maintained continuously if the sequence of 24 monthly inflows tabulated below are repeated cyclically forever. Firm yield is a function of reservoir storage capacity, as well as inflows. Compute at least six points on the firm yield versus storage capacity curve. Firm yields for the two extremes of (1) zero storage capacity and (2) an infinitely large reservoir can be easily determined without using an LP model. Determine at least four other additional points on the yield-storage curve by developing an LP model. Plot the storage capacity versus firm yield curve.

Month	Inflows (million m^3/month) First year	Second year
Jan	204	94
Feb	170	176
Mar	210	123
Apr	75	254
May	68	321
Jun	12	99
Jul	22	28
Aug	72	10
Sep	220	83
Oct	175	146
Nov	127	158
Dec	115	153

11.25. Regulatory agencies have imposed stream water quality criteria which require that the load of a particular pollutant not exceed 0.00120 units per cubic meter (units/m^3) of streamflow at any point in the river system. Five cities discharge wastewater into the river. Loads of the pollutant from sources other than the five cities is negligible. The cities cooperate to meet the water quality standards, with their goal being to minimize the combined daily operating cost of all treatment plants. As the water flows down the stream, natural processes reduce the pollution level. The pollutant load is reduced by the following percentages ($P_{i,j}$) by natural processes between city i and city j as follows.

$$P_{1,2} = 20\% \qquad P_{2,4} = 10\% \qquad P_{3,4} = 10\% \qquad P_{4,5} = 15\%$$

Treatment plant efficiency limits the percentage of the pollutant load that can be removed before discharging into the stream. For example, a maximum of 95 percent of the pollutant load generated by city 1 can be removed by the treatment plant at city 1. Pertinent data describing the system is as follows:

City	1	2	3	4	5
Pollutant load generated by city (units/day)	1,900	3,400	4,000	7,800	6,600
Treatment costs ($ per unit removed)	0.17	0.23	0.19	0.40	0.30
Maximum plant efficiency	0.95	0.92	0.90	0.90	0.92
Total streamflow at city (m^3/s)	34.0	41.0	43.0	96.0	112.0

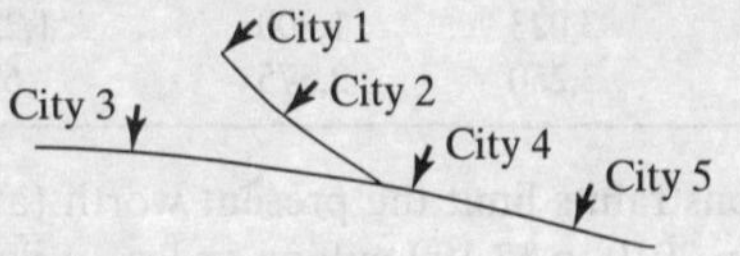

Use LP to determine the amount of pollutant that should be removed at the treatment plant at each city to minimize daily operating costs for the entire system while meeting the water quality criterion.

11.26. Four water use entities (users 1–4) are to be supplied from one or more of four possible sources (A–D). Sources A, B, and C cost $21,000,000, $38,000,000, and $30,000,000, respectively, to develop. Source D costs only $12,000,000 to develop, but can only supply customer 3. Sources C and D are mutually exclusive, meaning either, but not both, may be selected. One, two, or three water supply sources may be developed and used. A pipeline must be constructed from each of the four customers to at least one of the four water supply sources. A source is developed only if at least one customer's pipeline is constructed to it. Costs for constructing the four pipelines to the customers are provided below.

Supply source	Customer 1	Customer 2	Customer 3	Customer 4
A	$6,200,000	$8,800,000	$72,000,000	$25,000,000
B	$2,500,000	$7,500,000	$30,000,000	$ 7,700,000
C	$5,100,000	$25,000,000	$24,000,000	$7,500,000
D	Not applicable	Not applicable	$10,000,000	Not applicable

Each of the four customers are to be supplied from one of the four possible sources. The problem is to decide: (1) which source(s) to develop and (2) if more than one source is developed, which source will supply each customer. Use zero-one LP to determine which water supply sources to develop and which pipelines to construct in order to minimize total costs (costs of developing sources plus cost of constructing 4 pipelines).

11.27. Five water resources development projects are being considered for implementation. Construction staging (timing) is also being considered. Each project can be implemented in either the year 2010 or 2020, except project 3 is being considered for construction in 2020 only. The problem is to determine which of the five projects to implement and at what time (2010 or 2020). The objective is to maximize the present worth of net benefits. Present worths of benefits and costs associated with each project are shown below. The net benefits are equivalent present worth in 2010. A discount rate of 6 percent [$(P/F,6\%,10 \text{ years}) = 0.5584$] was used to convert the 2020 present worth of net benefits to equivalent 2010 values.

Project	Benefits ($ million)	Costs ($ million)	2010 Projects net benefits ($ million)	2020 Projects net benefits ($ million)
1	3,353	1,885	1,468	820
2	1,600	1,200	400	223
3	1,725	1,045	680	380
4	3,023	1,800	1,223	683
5	3,250	2,575	675	377

Budgetary constraints limit the present worth (at year 2010) of costs for all projects constructed in 2010 to $3,100 million or less. Likewise, present worths (at year 2020) of costs of all projects constructed in 2020 cannot exceed $4,400 million. Projects 2, 4, or 5 cannot be constructed unless project 1 is constructed at the same time or earlier. Project 3 is contingent on project 2 being constructed during the previous time period. Projects 2 and 4 are mutually exclusive, meaning project 2 cannot be constructed if project 4 is constructed and vice versa. Likewise, projects 2 and 5 are mutually exclusive. Project 3 can only be constructed in year 2020, not 2010. Obviously, projects constructed in 2010 cannot be constructed again in 2020. Formulate a zero-one integer programming model to determine which projects to construct and when. Define the decision variables as follows, where $P_{i,j}$ is one if the project is selected for implementation and zero otherwise.

$P_{1,1}$ = implement project 1 in year 2010
$P_{1,2}$ = implement project 1 in year 2020
$P_{2,1}$ = implement project 2 in year 2010
$P_{2,2}$ = implement project 2 in year 2020
$P_{3,2}$ = implement project 3 in year 2020
$P_{4,1}$ = implement project 4 in year 2010
$P_{4,2}$ = implement project 4 in year 2020
$P_{5,1}$ = implement project 5 in year 2010
$P_{5,2}$ = implement project 5 in year 2020

11.28. A water resources development agency is considering nine projects, called projects A–I. The agency cannot implement more than 4 projects, except a total of 5 projects can be implemented if both A and H are included. Implementation of A and B are contingent on implementation of C. Also, D cannot be selected unless both F and G are selected. A and D cannot both be adopted; if either A or D is implemented, the other cannot be implemented. If F is selected, neither B nor H can be selected. Each project, if implemented, would result in the discounted annual net benefits indicated by the table below.

Project	A	B	C	D	E	F	G	H	I
Net annual benefits ($1,000)	43	68	25	94	57	72	12	35	80

Formulate and solve a zero-one LP model to select a set of projects that maximize net benefits while meeting all constraints.

REFERENCES

American Water Works Association, *Water Resources Planning,* Manual M50, Denver, CO, 2001.

CARTER, M. W., and C. C. PRICE, *Operations Research: A Practical Introduction,* CRC Press, Boca Raton, FL, 2001.

DAVIS, S. A., N. B. JOHNSON, W. J. HANSEN, A. WARREN, F. R. REYNOLDS, C. O. FOLEY, and R. L. FULTON, *National Economic Development Procedures Manual: Urban Flood Damage,* Institute for Water Resources, U.S. Army Corps of Engineers, Alexandria, VA, 1988

GOODMAN, A. S., *Principles of Water Resources Planning,* Prentice Hall, Upper Saddle River, NJ, 1984.

HANSEN, W. J., *National Economic Development Procedures Manual: Agricultural Flood Damage,* Institute for Water Resources, U.S. Army Corps of Engineers, Alexandria, VA, 1987.

Hydrologic Engineering Center, *HEC-2 Water Surface Profiles, User's Manual,* CPD-2, U.S. Army Corps of Engineers, Davis, CA, 1992a.

Hydrologic Engineering Center, *HEC-1 Flood Hydrograph Package, User's Manual,* CPD-1, U.S. Army Corps of Engineers, Davis, CA, 1998a.

Hydrologic Engineering Center, *HEC-FFA Flood Frequency Analysis, User's Manual,* CPD-13, U.S. Army Corps of Engineers, Davis, CA, 1992b.

Hydrologic Engineering Center, *Proceedings of a Hydrology and Hydraulics Workshop on Risk-Based Analysis for Flood Damage Reduction Studies,* SP-28, U.S. Army Corps of Engineers, Davis, CA, 1997.

Hydrologic Engineering Center, *HEC-HMS Hydrologic Modeling System, User's Manual,* CPD-74, USACE, Davis, CA, 2001a.

Hydrologic Engineering Center, *HEC-RAS River Analysis System, User's Manual,* CPD-68, U.S. Army Corps of Engineers, Davis, CA, 2001b.

Hydrologic Engineering Center, *HEC-FDA Flood Damage Analysis, User's Manual,* CPD-72, U.S. Army Corps of Engineers, Davis, CA, 1998b.

Hydrologic Engineering Center, *HEC-5 Simulation of Flood Control and Conservation Systems, User's Manual,* CPD-5, U.S. Army Corps of Engineers, Davis, CA, 1998c.

Institute for Water Resources, *Catalog of Residential Depth-Damage Functions Used by the Army Corps of Engineers in Flood Damage Estimation,* IWR Report 92-R-3, U.S. Army Corps of Engineers, Alexandria, VA, 1992.

JAMES, L. D., and R. R. LEE, *Economics of Water Resources Planning,* McGraw-Hill, New York, NY, 1971.

JENSEN, P. A., and J. W. BARNES, *Network Flow Programming,* John Wiley & Sons, New York, NY, 1980.

JEWELL, T. K, *A Systems Approach to Civil Engineering Planning and Design,* Harper & Row, New York, NY, 1986.

LOUCKS, D. P., J. R. STEDINGER, and D. A. HAITH, *Water Resource Systems Planning and Analysis,* Prentice Hall, Upper Saddle River, NJ, 1981.

MAYS, L. W., *Optimal Control of Hydrosystems,* Marcel Dekker, New York, NY, 1997.

Mays, L. W., and Y.-K. TUNG, *Hydrosystems Engineering and Management,* McGraw-Hill, New York, NY, 1992.

NEWNAM, D. G., and B. JOHNSON, *Engineering Economic Analysis,* 5th Ed., Engineering Press, San Jose, CA, 1995.

PHILLIPS, D. T., and A. GARCIA-DIAZ, *Fundamentals of Network Flow Programming,* Prentice Hall, Upper Saddle River, NJ, 1981.

REVELLE, C. S., and A. E. MCGARITY (Eds.), *Design and Operation of Civil and Environmental Engineering Systems,* John Wiley & Sons, New York, NY, 1997.

REVELLE, C. S., E. E. WHITLACH, and J. R. WRIGHT, *Civil and Environmental Systems Engineering,* Prentice Hall, Upper Saddle River, NJ, 1997.

U.S Army Corps of Engineers, *Risk-Based Analysis for Evaluation of Hydrology/Hydraulics, Geotechnical Stability, and Economics in Flood Damage Reduction Studies,* Engineering Regulation 1105-2-101, Washington, D.C., 1996a.

U.S. Army Corps of Engineers, *Risk-Based Analysis for Flood-Damage-Reduction Studies,* Engineering Manual 1110-2-1619, USACE, Washington, D.C., 1996b.

VANDERBEI, R. J., *Linear Programming: Foundations and Extensions,* International Series in Operations Research and Management Science, Vol. 4, Kluwer Academic Publishers, 1996.

Water Resources Council, *Principles and Guidelines for Water and Related Land Resources Planning,* Government Printing Office, Washington, D.C., March 1983.

WURBS, R. A., "Application of Network Flow Programming Models in Managing Reservoir/River Systems," *Applications of Management Science, Vol. 9: Engineering Applications* (K. D. Lawrence and G. R Reeves, Ed.), JAI Press, Greenwich, CT, 1996.

12

River Basin Management

River basin management involves the development, conservation, control, regulation, protection, allocation, and beneficial use of water in streams, rivers, lakes, and reservoirs. Management of the water and related land and environmental resources of a river basin integrates natural and man-made systems. Systems of institutional resource management programs and constructed facilities are also integrally connected. This chapter focuses on the following aspects of river basin management: (1) reservoir systems for storing and regulating streamflows, (2) water rights systems for allocating water resources among users, and (3) environmental policies and practices for protecting ecological systems and water quality. The chapter ends with a brief review of computer modeling capabilities for supporting river basin management.

The chapter is organized based on using actual river basins and reservoir projects to illustrate water resources development and management practices and issues. Well-known larger river/reservoir systems have been selected as illustrative examples. However, basic concepts and methods are applicable, regardless of the size of the system. Water resources engineers deal with the full spectrum ranging from small watersheds with one or two reservoirs to extremely large complex river basins. The numerous smaller systems are just as important as the larger river basins highlighted in this chapter. For the people served by the professional water management community, the most important water resources in the world are those regional and local systems that meet their own needs.

12.1 RIVER BASIN SYSTEMS

The water resources of river basins are managed for the purposes outlined in Table 1.1 and discussed in Chapter 1. Water is withdrawn from river/reservoir systems for domestic, municipal, industrial, and agricultural supplies. Instream uses include hydroelectric power generation, navigation, and recreation. Protection and restoration of wetlands, fisheries, wildlife habitat, and other biological resources are important components of river basin management. Water quality management deals with pollutants stemming from human activities, as well as naturally occurring constituents. Erosion and sedimentation are major concerns. Periodic overflows of stream channels into floodplains are a natural characteristic of river systems. Because floodplain lands near rivers and lakes offer significant advantages for urban and agricultural development, flood mitigation is a key aspect of river basin management. Although individual projects may serve a single purpose, multiple-purpose development and management are common.

As discussed in Section 2.7 and Chapter 8, river basins are composed of subbasins, each of which may be further subdivided into more subwatersheds. Most streams and rivers are components of tributary systems discharging into rivers that flow into an ocean or other body of water connected to an ocean. Exceptions include closed basins of inland seas and salt lakes that have no outlets to the oceans. Such drainage basins include those of the Caspian Sea in Russia and Iran, the Aral Sea in Kazakhstan and Uzbekistan, Lake Balkhash in Russia, the Dead Sea in Israel and Jordan, the Great Salt Lake in Utah, and Salton Sea in California. Various rivers flow into these seas and lakes.

Although groundwater engineering (Chapter 9) and river basin management (Chapter 12) are covered in separate chapters, the importance of the interactions between surface water and groundwater are certainly recognized. Groundwater provides base flows for surface streams. Seepage from streams contributes to groundwater. Because they are hydrologically connected, water withdrawn from one source may affect the amount of water available from the other. Water quality constituents in one affects the water quality of the other. Water pumped from groundwater aquifers for water supply is discharged to streams as municipal wastewater treatment plant effluent and irrigation return flows. Conjunctive management of surface water and groundwater supplies can be very beneficial. Supplies may be withdrawn from reservoir/river systems during periods of plentiful streamflow, switching to pumping from groundwater reserves during drought.

Comprehensive river basin planning and management integrates the myriad of considerations involved in solving water-related problems and meeting needs associated with population growth, economic development, and environmental protection. The fundamental importance of comprehensive water resources planning and management has been recognized for decades. Varying degrees of success have been achieved in actually incorporating the holistic systems concept in planning studies and implementation of management strategies. The concept emphasizes comprehensive integration of

- multiple purposes (water supply, hydropower, flood mitigation, etc.)
- economic development, social welfare, and environmental protection
- water supply augmentation and demand management
- structural and nonstructural flood damage reduction strategies
- human and ecosystem needs for water
- water quantity and quality considerations
- conjunctive management of surface water and groundwater resources
- management of water, land, energy, and biological resources

12.2 DAMS, RESERVOIRS, AND ASSOCIATED FACILITIES

Reservoir systems are essential to river basin management. The hydrographs of Figs. 2.14, 2.15, and 7.6 illustrate the tremendous temporal variability of streamflows that characterize most rivers. In addition to cyclic seasonal fluctuations within each year and random year-to-year variations, severe droughts with durations of several years and extreme flood events are major concerns. Reservoir storage is necessary to regulate streamflow fluctuations, maintain instream flow requirements, develop reliable water supplies, and reduce damages caused by floods. Multiple-purpose reservoir system operating policies and procedures are described by Wurbs (1996).

Reservoir projects include dams and appurtenant outlet structures, pumping plants, pipelines, canals, channel improvements, hydroelectric power plants and transmission facilities, navigation locks, fish ladders, recreation facilities, and various other structures. The extensive literature on design, maintenance, operation, and rehabilitation of dams, spillways, outlet works, gates, energy dissipators, and related hydraulic structures includes books by the U.S. Bureau of Reclamation (1976, 1977, 1987), Jansen (1983), Kollgaard and Chadwick (1988), Senturk (1994), Singh and Varshney (1995), Kutzner (1997), Vischer and Hager (1998), and Herzog (1999).

12.2.1 Reservoir Pools

Reservoir operating policies typically involve dividing the storage capacity into designated pools. A typical reservoir consists of one or more of the vertical zones or pools illustrated by Fig. 12.1. The allocation of storage capacity between pools may be constant or may vary seasonally.

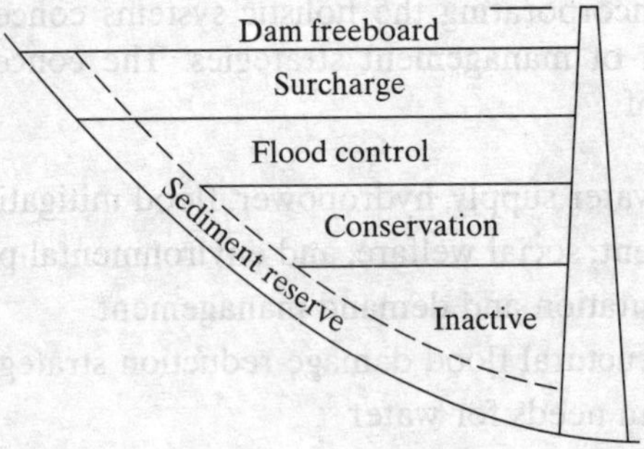

Figure 12.1 The storage capacity of a multiple-purpose reservoir is divided into zones or pools.

Water is not withdrawn from the inactive pool, except through the natural processes of evaporation and seepage. The top of inactive pool elevation may be fixed by the invert of the lowest outlet or by conditions of operating efficiency for hydroelectric turbines. An inactive pool also may be contractually set to facilitate withdrawals from outlet structures that are significantly higher than the invert of the lowest outlet at the project. The inactive pool is sometimes called dead storage. It may provide a portion of the sediment reserve, head for hydroelectric power, and water for recreation and fish habitat.

Conservation storage purposes, such as municipal and industrial water supply, irrigation, navigation, hydroelectric power, and instream flow maintenance, involve storing water during periods of high streamflow and/or low demand for later beneficial use as needed. Conservation storage also provides opportunities for recreation. The reservoir water surface is maintained at or as near the designated top of the conservation pool level as streamflows and water demands allow. Drawdowns are made as required to meet the various needs for water.

The flood control pool remains empty, except during and immediately following a flood event. The top of flood control pool elevation is often set by the crest of an uncontrolled emergency spillway, with normal flood releases being made through other outlet structures. Gated spillways allow the top of flood control pool elevation to exceed the spillway crest elevation. To be prepared for the next flood, operating procedures include emptying of flood control pools as quickly as possible after a flood event without making releases that contribute to downstream flooding.

The surcharge pool is essentially uncontrolled storage capacity above the flood control pool (or conservation pool if there is no designated flood control storage capacity) and below the maximum design water surface. The top of the surcharge storage (maximum design water surface) is established during project design from the perspective of dam safety. Reservoir design and operation are based on ensuring that this designated level is never exceeded under any conditions. For most dams, particularly earthfill embankments, the top of dam elevation includes a freeboard allowance above the top of the surcharge pool to account for wave action and provide an additional safety factor against overtopping.

12.2.2 Sedimentation

Reservoir storage capacity is lost over time due to sedimentation. The rate of sediment deposition varies tremendously between reservoir sites, depending on flow rates and sediment loads in the rivers flowing into the reservoirs and the trap efficiencies of the reservoirs (Section 8.7.4). Because sediment transport increases greatly during flood events, reservoir sedimentation also varies greatly over time with the random occurrence of floods. Sediment deposits occur throughout the reservoir in each of the designated pools, as illustrated in Fig. 12.1. As streamflow velocities decrease in the upper reaches of a reservoir, sediments are deposited forming deltas. Smaller particles will move further into the reservoir before depositing. Reservoir sediment surveys may be performed periodically to determine current bottom topography and resulting storage capacities.

For many smaller reservoirs constructed by local entities, no special provisions are made to allow for sedimentation. However, for most federal projects and other large reservoirs, sediment reserve storage capacity is provided to accommodate sediment deposition expected to occur over a specified design life, typically 50–100 years. The volume and location of the sediment deposits and resulting changes in reservoir topography are predicted using methods outlined by the U.S. Bureau of Reclamation (1987) and the U.S. Army Corps of Engineers (1989). Storage capacity reserved for future sediment accumulation is reflected in water supply contracts and other planning and management actions.

12.2.3 Dams

Dams are classified based on the type and materials of construction as embankment, gravity, arch, and buttress. More than one type of dam may be included in a single structure. Dams are often earth embankments for most of their length, combined with a concrete section containing outlet conduits and spillways. Curved dams may combine both gravity and arch effects to achieve stability.

The *World Register of Dams* maintained by the International Commission on Large Dams (ICOLD; 1998) is discussed in Section 12.6. The number of large dams in the worldwide inventory in 1984 is categorized in Table 12.1 by type and height. Most dams are of the embankment type; most are earthfill, but some are rockfill. For heights greater than 60 m, there are more concrete gravity and arch dams than embankment dams. Most dams are less than 30 m in height. The ICOLD register does not include hundreds of thousands of dams with heights of less than 10 m. The number of dams decrease with height. Most dams are small, but most reservoir storage capacity is provided by the relatively fewer larger dams. Discrepancies in comparing data on dam heights may occur because measurements may be from either the dam foundation or river bed.

Earthfill embankments are the most common type of dam, because materials are usually available at the construction site, and foundation requirements are usually less stringent than for other types of dams. Earthfill embankments are of

TABLE 12.1 DISTRIBUTION OF DAMS BY TYPE AND HEIGHT

Dam height (m)	Type of dam					
	Embankment	Gravity	Arch	Buttress	Multi-arch	Total
10–30	24,567	2,222	775	175	74	27,813
30–60	3,657	1,294	428	110	48	5,537
60–100	477	361	204	40	13	1,095
100–150	116	65	83	12	—	276
150–200	21	8	24	—	—	53
Over 200	6	4	13	—	1	24
Total	28,844	3,954	1,527	337	136	34,798

three types: homogeneous, zoned, and diaphragm. A homogeneous dam is composed of essentially the same material throughout. The material must be sufficiently impervious to provide a water barrier, and the slopes must be relatively flat for stability. Although small dams have often been constructed in this manner, few large dams are homogeneous. Embankment dams are commonly zoned to have a central impervious core, a transition zone along the faces of the core, and outer zones of more porous material for stability and protection of the impervious core. The impervious zone may consist of clay or a mixture of clay, silt, and sand. The pervious zones may consist of sand, gravel, cobbles, or rock, or mixtures of these materials. Diaphragm-type dams have a thin diaphragm of concrete, steel, timber, or impervious earth, which serves as a water barrier with the bulk of the embankment constructed of pervious material such as sand, gravel, or rock. The position of the diaphragm may vary from a blanket on the upstream face to a central vertical core. Earthfill embankments typically have a rock layer protecting the slopes from erosion.

The bulk of a rockfill embankment dam is composed of rocks of various sizes that provide the stability for the structure. An impervious membrane is required to make the dam watertight. This membrane may be an upstream face of concrete, asphalt, impervious earth, or other material, or a core of impervious soil near the center of the dam. There is not a well-defined distinction between earthfill and rockfill dams. Some dams are composed of a combination of both types of material. Typically, rockfill dams are designed with steeper slopes than earthfill dams.

Gravity, arch, and buttress dams are usually constructed of concrete, though a number of old dams are built of stone blocks or masonry. A gravity dam depends on its own weight for stability. Gravity dams are usually roughly triangular in cross-section, with the base width so related to height as to ensure stability against overturning, sliding, and foundation crushing. Most gravity dams are straight in plan, but some are slightly curved.

The curved shape of arch dams results in most of the horizontal load of the water being transmitted to the abutments. Arch dams have thinner sections than comparable gravity dams. Arch dams are feasible only in canyons, with walls capable of withstanding the thrust produced by the arch action.

A buttress dam consists of a sloping membrane that transmits the water load to a series of supporting members, called buttresses, at right angles to the axis of the dam. Typical configurations for buttress dams include the flat slab and multiple arch. The face of a flat slab buttress dam is a series of flat reinforced concrete slabs. The face of a multiple-arch dam consists of a series of arches that permit wider spacing of the buttresses. Buttress dams usually require only one-third to one-half as much concrete as gravity dams of similar height. Consequently, buttress dams may be used on foundations that are too weak to support a gravity dam.

Dams may also be classified as overflow or nonoverflow. Overflow dams are concrete structures designed for water to flow over their crests. Nonoverflow dams have spillways to prevent overtopping. Earthfill and rockfill embankments are damaged by erosive action of overflowing water and thus must be designed as nonoverflow dams.

12.2.4 Spillways and Outlet Works

Reservoir releases to the river below a dam are made through spillways and outlet works. Spillways provide the capability to release high flow rates during major floods without damage to the dam and appurtenant structures. Spillways are required to allow flood inflows to safely flow over or through the dam, regardless of whether the reservoir contains flood control storage capacity. Spillways may be controlled or uncontrolled. A controlled spillway is provided with crest gates that allow the outflow rate to be adjusted. For an uncontrolled spillway, the outflow rate is simply a function of the head or height of the water surface above the spillway crest. Because spillway flows involve extremely high velocities, stilling basins or other types of energy dissipation structures are required to prevent catastrophic erosion damage to the downstream river channel and dam. For many reservoir projects, a full range of outflow rates are discharged through a single spillway. Some reservoirs have more than one spillway. A service spillway conveys smaller, more frequently occurring release rates, and an emergency spillway is used only rarely during extreme floods.

A variety of configurations have been adopted in spillway design. Spillways may be classified based on the path the water takes over, through, or around the dam. Types include overflow, chute, side-channel, and shaft. An overflow spillway is a section of dam designed to permit water to flow over the crest. A chute spillway is a channel located along a dam abutment or some distance from the dam. With a side-channel spillway, water flows over the dam into a channel running parallel to the dam. In a shaft spillway, the water drops through a vertical shaft to a horizontal conduit through the dam.

The major portion of the storage volume in most reservoirs is located below the spillway crest. Flows over the spillway can occur only when the storage level is above the spillway crest. Outlet works are used for releases from storage both below and above the spillway crest. Discharge capacities for outlet works are typically much smaller than for spillways.

Outlet works are used to release water for the various beneficial uses, such as water supply diversions and maintenance of instream flows. Flood control releases may also be made through outlet works. The components of an outlet works facility include an intake structure in the reservoir, one or more conduits or sluices through the dam, gates located either in the intake structure or conduits, and a stilling basin or other energy dissipation structure at the downstream end.

12.2.5 Other Facilities

Water supply diversions may be either lakeside or downstream. Lakeside withdrawals involve intake structures, along with pumps and pipeline or canal conveyance facilities. Downstream releases through an outlet works may be diverted from the river at locations that may be great distances below the dam. Downstream releases may be made through hydroelectric power penstocks, navigation locks, or other structures, as well as outlet works and spillways. Releases from multiple reservoirs may be diverted at a common downstream location. The term *barrage* is used in some countries to refer to relatively low-head diversion dams often associated with irrigation. The function of a barrage is to raise the river level sufficiently to divert flow into a water supply canal.

Each hydroelectric power project has its own unique layout and design. Hydropower facilities typically include, in some form, a diversion and intake structure, a penstock or conduit to convey the water from the reservoir to the turbines, the turbines and governors, housing for the equipment, transformers, and transmission lines to distribution centers. A forebay or surge tank regulates the head. Trash-racks and gates are typically provided in the intake structure. A draft tube delivers the water from the turbines to the tailrace, through which it is returned to the river. The powerhouse may be located at one end of the dam, directly downstream from the dam, or between buttresses in a buttress dam. In some cases, water is conveyed through a penstock to a powerhouse located some distance below the dam. With favorable topography, a high head can be achieved in this manner even with a low dam. A re-regulating dam is often provided below the hydroelectric plant.

Dams on rivers used for navigation often include locks. A navigation lock is a rectangular box-like structure with gates at either end that allows vessels to move upstream or downstream through a dam. Lockage occurs as follows, assuming a vessel is traveling upstream. The lock chamber is emptied. The downstream gate is opened and the vessel enters the lock. The chamber is filled, with the water lifting the vessel to the level of the reservoir. The upstream gate is opened and the vessel departs. The highest lock in the U.S. is the John Day lock on the Columbia River with a lift of 34.5 m.

12.2.6 Examples of Reservoir Projects

The reservoir projects shown in Figs. 1.9 and 1.10 and in Figs. 12.2–12.6 illustrate a few of the many types and configurations of water control facilities. Lakes Mead, Roosevelt, Shasta, Folsom, and Somerville are federal multiple-purpose reservoirs with storage capacities of 36,700, 6,400, 5,610, 1,250, and 626 million m^3, respectively. Lake Mead is the largest reservoir in the United States. Roosevelt and Shasta Reservoirs are also extremely large. Folsom and Somerville Reservoirs are representative of medium-sized projects. The outlet facilities at Somerville Reservoir are much simpler than those at the other projects.

Hoover Dam (Figs. 1.10 and 12.2), impounding Lake Mead on the Colorado River, is operated by the U.S. Bureau of Reclamation (USBR) primarily for water supply, hydropower, and flood control. At the time of its construction in 1931–1936, Hoover Dam was over twice as high and twice as voluminous as any other concrete

Figure 12.2 Lake Mead, impounded by the Hoover Dam on the Colorado River, is the largest reservoir in the United States. (Courtesy U.S. Bureau of Reclamation)

Figure 12.3 Grand Coulee Dam on the Columbia River is the most massive concrete structure and largest hydropower project in the United States. Franklin D. Roosevelt Reservoir extends 240 km upstream to the Canadian border. (Courtesy U.S. Bureau of Reclamation)

dam previously constructed (Kollgaard and Chadwick, 1988). The concrete arch dam has a height of 221 m (726 ft) above its foundation. The crest is 379 m (1,244 ft) long. Two concrete-lined, side-channel, drum gate-controlled spillways have a maximum discharge capacity of 11,300 m^3/s. Four tunnels branching into 16 power penstocks and 12 outlet pipes have a maximum discharge capacity of 1,270 m^3/s. Nineteen hydropower turbines provide an installed capacity of 1,344,800 kW. Only a small portion of the 658 km^2 (254 $mi.^2$) water surface area at the top of conservation pool is seen in Fig. 12.2.

Grand Coulee Dam and Franklin D. Roosevelt Reservoir on the Columbia River, constructed in 1933–1942, are also operated by the USBR primarily for water

Figure 12.4 Shasta Dam and Lake serve to control floodwater; store surplus winter runoff for irrigation use in the Sacramento and San Joaquin Valleys; maintain flows in the Sacramento River for navigation, fish conservation, and protection of the Sacramento-San Joaquin Delta from intrusion of saline ocean water; generate hydroelectric power; and supply water for municipal and industrial use. (Courtesy U.S. Bureau of Reclamation)

supply, hydropower, and flood control. The concrete gravity dam has a height of 168 m (550 ft) above its foundation. The crest is 1,730 m (5,670 ft) long. A 503-m-long overflow section at the center of the dam, controlled with 11 drum gates, has a maximum discharge capacity of 28,300 m^3/s. Forty 2.6-m-diameter steel conduits through the dam provide maximum discharge capacities of 7,930 m^3/s to the 27 hydropower turbines and 7,500 m^3/s directly to the river. The installed electric energy capacity is 6,480,000 kW.

Shasta Dam and Reservoir on the Sacramento River is a component of the California Central Valley Project. It was constructed by the USBR in the early 1940's. The curved concrete gravity dam is 184 m high, with a crest length of 1,060 m. The 539,000-kW powerplant is located just below the dam. Water is released from

Figure 12.5 Folsom Dam and Lake on the American River in California are operated by the USACE for flood control and by the USBR for hydroelectric power, irrigation, municipal and industrial water supply, fish and wildlife, and recreation. (Courtesy U.S. Bureau of Reclamation)

the reservoir through five 4.6-m-diameter penstocks to the five main generating units and two station service units. Keswick Dam, located 14 km below Shasta Dam, stabilizes the uneven hydropower releases. The overflow section near the center of Shasta Dam is controlled by three 34-m × 8.5-m drum gates. The outlet works also includes eighteen 2.6-m-diameter conduits through the dam controlled by fourteen 2.4-m wheel-type gates and four 2.6-m tube valves.

Folsom Dam was constructed by the U.S. Army Corps of Engineers (USACE) and transferred to the USBR for coordinated operation as an integral part of the Central Valley Project. The dam was constructed during the period from 1948 to 1956. The dam has a concrete river section with a height of 104 m and length of 427 m, flanked by long earthfill wing dams extending from the ends of the concrete section on both abutments. The powerplant, constructed and operated by the USBR, is located at the foot of the dam. The plant has a total capacity of

Figure 12.6 Somerville Reservoir on Yequa Creek in Texas is a component of a 9-reservoir system operated by the USACE for flood control and a 13-reservoir system operated by the Brazos River Authority for water supply. (Photo by R. A. Wurbs, May 1996)

198,720 kW. Water from the reservoir is released through three 4.7-m-diameter penstocks to three generating units. The much smaller Nimbus Dam, located 11 km below Folsom Dam, re-regulates hydroelectric power releases. Releases from Folsom Reservoir to the river are also made through eight 1.5-m × 2.7-m outlet conduits controlled by two slide gates. A 2.1-m-diameter conduit conveys water from the reservoir to a pumping plant. Flood flows over the two adjacent overflow spillway sections of the dam are controlled by five 13-m × 15-m and three 13-m × 16-m radial gates.

Somerville Dam and Reservoir are located in Texas on Yequa Creek, a tributary of the Brazos River. The dam is an earthfill embankment with a height of 24 m and crest length of 6,160 m. The conservation pool and flood control pool have storage capacities of 197 and 428 million m^3, respectively. If the flood control storage capacity is ever exceeded, excess flows will pass over an uncontrolled ogee-shaped overflow spillway with a length of 381 m. The spillway crest elevation coincides with the top of the flood control pool. The outlet works intake structure shown in Fig. 12.6 contains two 1.5-m × 3-m gates that control releases through a 3-m-diameter conduit through the dam that discharges to a stilling basin and then to Yequa Creek. Releases are made through this outlet works from either the conservation or flood control pools. Water is supplied via pipeline to the city of Brenham through another intake structure located in the conservation pool. Somerville is a component of a system of several reservoirs that release for downstream diversions from the lower Brazos River.

12.3 WATER RIGHTS AND WATER ALLOCATION SYSTEMS

Streamflow and reservoir storage capacity in major river basins are typically shared by many water users who use the water for a variety of purposes. Water rights systems provide a basis to (1) allocate resources among users, (2) protect existing users from having their supplies diminished by new users, and (3) govern the sharing of limited streamflow and water in storage during droughts when supplies are inadequate to meet all needs. Allocation of water resources among different entities is a key aspect of river basin management that becomes particularly important as demands approach and exceed supplies.

The institutional framework for river basin management involves a hierarchy of water allocation systems. The water resources of international river basins may be allocated between nations by treaties and other agreements. In the U.S., water is allocated among states through river basin compacts and other means. Within individual states, water is shared by river authorities, municipal water districts, cities, irrigation districts, individual farmers, industries, and private citizens through water rights systems. A water district or river authority distributes water to its customers in accordance with contractual commitments.

12.3.1 International River Basins

Principles and rules of international water law are found in treaties, international custom, general principals of law, and writings of international institutions (Wouters, 2000). Treaties and other agreements between the United States and Mexico and Canada regarding the sharing of the waters of the Rio Grande, Colorado, and Columbia River Basins are discussed later in this chapter. Other nations have also negotiated successful arrangements for sharing water resources. However, in many international river basins, little progress has been made in developing effective water allocation systems.

Wolf, Natharius, Danielson, Ward, and Pender (1999) list 261 international river basins that cover about 45 percent of the world's land area, excluding Antarctica. In this inventory, *river basin* is defined as the area that contributes hydrologically (including both surface and groundwater) to a first-order stream, which is defined as a river that flows to the ocean or to a terminal lake or inland sea.

International refers to a river basin with at least one perennial tributary that crosses or forms political boundaries of two or more nations.

Effective joint multiple-nation river basin management will be a major determinant in achieving stability, peace, and prosperity in many regions of the world in the 21st century. Gleick (1998) reviews experiences throughout the world regarding cooperation and conflicts between nations over shared water resources. A few of the many regions with great potential for either cooperation or conflict include the following. The Jordan River—shared by Israel, Jordan, Syria, and the Palestinians—is a small stream with remarkably great historical, hydrological, and political importance. The Euphrates and Tigris Rivers flow from Turkey through Syria, Iraq, and Iran. Most of the flow of the Euphrates and Tigris Rivers originates in their upper watersheds in Turkey and is controlled by the GAP Project described in Section 12.6. The Ganges and Brahmaputra River Basins in Nepal, China, India, Bhutan, and Bangladesh, with a history of centuries of water conflicts, contained an estimated 400 million people in 2000 living at an impoverished standard of living. In the Southern African Region encompassing Angola, Botswana, Lesotho, Malawi, Mozambique, Namibia, South Africa, Swaziland, Tanzania, Zambia, and Zimbabwe, every major river is shared by two or more nations. Population growth and economic development are resulting in intensified demands on limited water resources with a long history of controversy.

12.3.2 Water Rights in the United States

A water right is the legal right for an entity to store, regulate, and/or divert water for beneficial use (Rice and White, 1991; Wolfe, 1995; Getches, 1997). Water law is the creation, allocation, and administration of water rights. Water is a renewable resource owned by the state and used by the public. Water right systems in the U.S. are established primarily at the state level. However, federal laws govern water rights for military installations, other federal lands such as national parks, and Indian reservations. Many river basins encompass portions of multiple states. Water resources are allocated between states based on interstate compacts developed by the states and approved by the U.S. Congress. Disputes may arise in implementing interstate compacts, particularly during droughts. Water rights systems within each state govern the distribution of the water resources of that state. These systems vary between states. Water rights for groundwater are also different than for surface water.

12.3.2.1 Surface water rights. Legal rights to the use of streamflow are generally based on two alternative doctrines: riparian and prior appropriation. The basic concept of the riparian doctrine is that water rights are incidental to the ownership of land adjacent to a stream. The prior appropriation doctrine is based on the concept of protecting senior water users from having their supplies diminished by newcomers developing water supplies later in time. In a prior appropriation system, water rights are not inherent in land ownership, and priorities are established based on dates that water is appropriated.

The doctrine of riparian rights common in the eastern U.S. is based on English common law. Under the strictest interpretation of the riparian doctrine, the owner of land adjacent to a stream (riparian land) is entitled to receive the full natural flow of the stream without change in quantity or quality. Because a strict interpretation imposes impractical constraints on water use, the riparian doctrine is normally interpreted to allow riparian landowners to divert reasonable amounts of streamflow for beneficial purposes.

The doctrine of prior appropriation is associated with settling the American West. As settlers moved from the eastern states to the West in the 1800's, farmers and ranchers claimed land, and miners claimed gold and other minerals. Likewise, water was appropriated by the first to arrive and claim the resources for beneficial use. In developing their farms and communities, people needed protection from having their water supplies diminished with later population growth and economic development.

Most of the western states have established water allocation systems in which a state agency issues permits to water right holders specifying amounts and conditions of water use. Riparian and/or appropriative rights may be incorporated into the original development of the permit system, with additional new permits being issued based on prior appropriation. With growing demands on limited water resources, permit systems will likely continue to be developed in the eastern states, which have more abundant streamflows, similar to those already in place in the drier western states.

Water rights in the eight driest states (Nevada, Arizona, Utah, Idaho, Montana, Wyoming, Colorado, and New Mexico) are based purely on the prior appropriation doctrine. Alaska is also a prior appropriation state. Ten western states that have hybrid systems merging both riparian and appropriative rights into permit systems include California, Oregon, Washington, Texas, Oklahoma, Kansas, Nebraska, South Dakota, North Dakota, and Idaho. Hawaii has a unique hybrid system. Water rights in 30 eastern states are based primarily on the riparian doctrine.

12.3.2.2 Groundwater rights. The rights and obligations for groundwater use are generally tied to two legal principles: (1) property ownership and (2) shared ownership of a common public resource. A variety of state approaches to groundwater rights has evolved from these concepts. State groundwater law is mixtures of the following doctrines.

- *Absolute Ownership Doctrine*—Landowners own the groundwater under their land and may drill wells and pump as much water as they wish. Texas and several other states have historically adhered to this doctrine but are slowly changing.
- *Reasonable Use Doctrine*—Landowners own groundwater, but their pumping is limited to reasonable use, which has been defined in a variety of ways. This doctrine is common in the eastern states.

- *Correlative Rights Doctrine*—In times of shortage, groundwater is shared by overlying landowners in proportion to the amount of land they own. This extension of the reasonable use rule is primarily associated with California.
- *Prior Appropriation Doctrine*—Groundwater is allocated similarly to surface water with priorities assigned based on the dates that users first appropriate the water for beneficial use. This doctrine is common in the western states.
- *Permit Systems*—Systems in which state agencies issue permits specifying the amounts and conditions of water use have been adopted in a number of both western and eastern states. The other doctrines may be reflected in the water rights documented by the permits.

Some states divide groundwater into categories with different water right rules applied to each classification. Percolating groundwater may be legally differentiated from underground streams with definable flow paths. Underground streams are sometimes treated as being similar to surface streams. The issue of the impacts of groundwater pumping on surface streamflow has been addressed to varying extents in different states. In some states, groundwater is classified as either tributary or nontributary. Tributary groundwater hydrologically contributes to surface streamflow. Nontributary groundwater does not. Water right rules and management strategies for tributary groundwater are based on protecting surface water rights.

12.3.2.3 Components of water allocation systems. Each state has developed its own set of rules and practices governing water rights. These water allocation systems have evolved historically and continue to change. State water rights systems generally have the following components or features.

- State-negotiated compacts approved by the federal government allocate waters of interstate river basins between states. Some states are also affected by federal agreements with Canada or Mexico for sharing international waters or by rights reserved for Indian reservations, military installations, or other federal lands.
- A legally established priority system based generally on variations of the riparian and/or prior appropriation concepts guides the allocation of the waters within a state among numerous water management entities and water users.
- An administrative system is needed to grant, limit, and modify water rights and to enforce the allocation of water resources, particularly during droughts and times of insufficient supply. These systems may or may not include formal issuance of written permits to water right holders.
- Increasingly more states are adopting computer modeling systems to support administration of their water allocation systems and associated water resources planning and management activities.

These features may vary greatly between groundwater and surface water. From a water law perspective, most states treat groundwater and surface water as separate resources. The extent to which the important hydrologic and water management interconnections are recognized varies between the states.

States in the western and eastern halves of the U.S. have generally adopted different approaches to water rights due largely to the western states having much drier climates. Water allocation and accounting systems tend to be more rigorous in regions where demands approach or exceed supplies. The experience of the state of Texas is discussed in the next section to illustrate key aspects of developing and administering water right systems. Regions of Texas are representative of both western and eastern states from various perspectives, including climate. Mean annual precipitation varies from 20 cm (8 in.) in west Texas to 142 cm (56 in.) in the eastern extreme of the state. Texas actually provides two case studies, since water allocation in the Lower Rio Grande Valley has distinct differences from the remainder of the state.

12.3.3 Water Rights in Texas

Texas is divided into the 15 major river basins shown in Fig. 12.7 and eight coastal basins located between the lower reaches of the major river basins. Interstate compacts govern sharing of the interstate rivers with neighboring states. Two similar, but yet distinctly different, water rights systems have been developed: one in the Lower Rio Grande Valley and another for the remainder of the state (Wurbs, 1995, 2002). Water rights are a major consideration in river basin management statewide.

12.3.3.1 Historical development of the Texas water rights system.

Rights to use streamflow in Texas have been granted over several centuries under Spanish, Mexican, Republic of Texas, and State of Texas laws. Early water rights were based on various versions of the riparian doctrine. The prior appropriation doctrine was adopted in the 1890's, while still maintaining existing riparian rights. An essentially unmanageable system evolved, with various types of water rights existing simultaneously, with many rights being unrecorded. The drought of 1950–1957 motivated a 20-plus-year effort to merge the numerous varied rights into a unified permit system. A 1995–1996 drought provided an impetus for further refinements.

The Texas share of the waters of the Rio Grande below Fort Quitman was allocated among numerous water rights holders in conjunction with a massive lawsuit commonly called the Lower Rio Grande Valley Water Case. The litigants included 42 water districts and 2,500 individuals. The lawsuit was filed in 1956, the trial was held in 1964–1966, the final judgment was filed in 1969, and regulations implementing the court decision were adopted in 1971. The assorted versions of riparian and appropriative rights were combined into a permit system. The expense and effort involved in this lawsuit demonstrated the impracticality

Figure 12.7 Texas is divided into 15 major river basins and eight coastal basins.

of a purely judicial determination of water rights for the entire state and led to enactment of the Water Rights Adjudication Act passed by the Texas Legislature in 1967.

The stated purpose of the 1967 Act was to require a recording of all claims for water rights that were not already recorded, to limit the exercise of those claims to actual use, and to provide for the adjudication and administration of water rights. The adjudication process required to merge all existing rights into a permit system was initiated in 1968 and completed in the late 1980's.

Some centralized agency has administered some type of water rights system statewide since 1913. However, the agencies and water rights system have changed over time. The Texas Natural Resource Conservation Commission (TNRCC) was created in 1993 by merging programs of predecessor agencies. The TNRCC consists of three full-time commissioners appointed by the governor and a professional and administrative staff of over 3,000 employees. Water rights represent just one of many statewide regulatory responsibilities of the TNRCC. The National Pollutant Discharge Elimination System (NPDES) discussed in Section 12.4.1 and a dam safety program are two other responsibilities.

12.3.3.2 Allocation of the waters of the Rio Grande. The Rio Grande Basin shown in Fig. 12.8 is shared by Mexico and three states in the U.S. Water allocation is governed by two international treaties and two interstate compacts. Allocation of the Texas share of the water to irrigators, cities, and other users is based on state law. Fort Quitman, located 140 km downstream of the City of El Paso, is a key location in both the international and state water allocation systems. Fort Quitman is 1,850 river kilometers above the Gulf of Mexico.

A 1906 treaty between the U.S. and Mexico provides for delivery of 60,000 acre-feet/year (74 million m^3/yr) of Rio Grande water to Mexico in the El Paso-Juarez Valley above Fort Quitman. Elephant Butte Reservoir in New Mexico, operated by the USBR, and the American and International diversion dams near El Paso, operated by the International Boundary and Water Commission (Section 1.3.6), are used to implement the water allocation provisions of the treaty.

The Rio Grande Compact approved by the legislatures of Colorado, New Mexico, and Texas in 1939 allocates the uncommitted waters of the Rio Grande above Fort Quitman. The Pecos River Compact adopted in 1949 allocates the waters of that tributary between Texas and New Mexico.

The Water Treaty of 1944 expanded the International Boundary Commission to the International Boundary and Water Commission (IBWC) described in

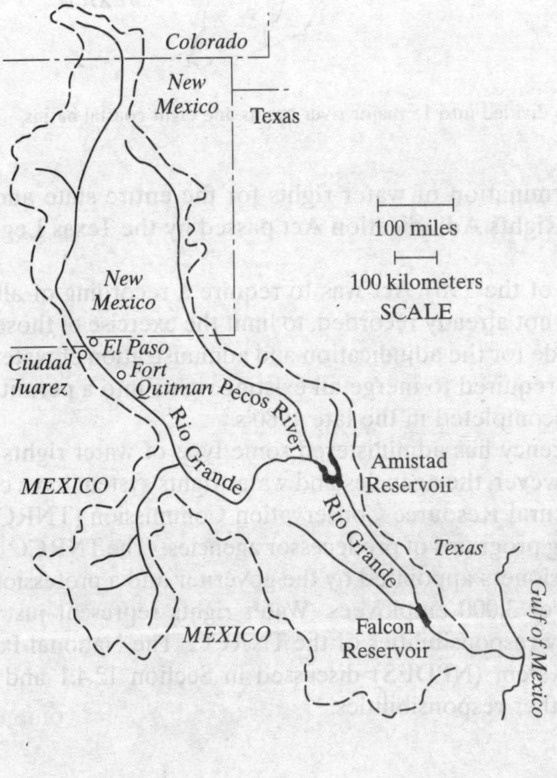

Figure 12.8 The Rio Grande is an international and interstate river basin.

Section 1.3.6, provided for the distribution of waters of the Rio Grande from Fort Quitman to the Gulf of Mexico between the two nations, and authorized construction of Amistad and Falcon Reservoirs. The 1944 treaty also includes provisions for allocation of the waters of the Colorado River (Section 12.7.7).

The International Amistad and Falcon Reservoirs are operated by the IBWC primarily for flood control and water supply for the Lower Rio Grande Valley. They also provide hydroelectric power and recreation. Amistad Reservoir contains 4.17 billion m^3 of conservation storage and 2.15 billion m^3 of flood control storage. Falcon Reservoir contains 3.29 billion m^3 of conservation storage and 0.63 billion m^3 of flood control storage. In accordance with the 1944 treaty, the U.S. has 56.2 percent and 58.6 percent of the conservation storage capacity of Amistad and Falcon, respectively, with Mexico owning the remainder. The IBWC operates Anzaldus and Retamal Dams on the lower reach of the Rio Grande to facilitate diversions. The travel time for releases from Falcon Reservoir to reach the most downstream diversion locations is about 1 week.

Streamflows into Falcon and Amistad Reservoirs are allocated between the two countries. Flows on a number of major tributaries named in the treaty are gaged and allocated as specified by the treaty. All other flows not otherwise allocated are divided equally between the two countries. Computations are performed weekly to allocate the reservoir inflow and evaporation volumes that are combined with recorded releases to determine the amount of water that each country has in storage.

The IBWC is responsible for flood control operations. Hydroelectric power generation is essentially limited to using water released anyway for other purposes. The U.S. share of the water supply storage in Amistad and Falcon Reservoirs is used to meet demands in the lower basin administered by the TNRCC in accordance with the state water rights system. Irrigation districts, individual farmers, and cities communicate their water needs directly to the TNRCC water master office, which in turn schedules releases from Falcon and Amistad Reservoirs with the IBWC The water master office maintains a weekly accounting of the amount of water used and the amount of water in reservoir storage allocated to each of about 1,600 water rights accounts. Most of the water is used in the very productive agricultural region below Falcon Reservoir.

12.3.3.3 Statewide water rights permit system. Water rights are granted by a state license, or permit, which allows the holder to divert a specified amount of water annually at a specific location, for a specific purpose, and to store water in reservoirs of specified capacity. As of 2002, the TNRCC was administering about 7,000 active water rights permits for use of surface water. The water rights are held by river authorities, cities, municipal water districts, irrigation districts, individual farmers, companies, and private citizens.

Anyone may submit an application to the TNRCC for a new water right or to change an existing water right at any time. The TNRCC will approve the application if unappropriated water is available, a beneficial use of the water is contemplated, water conservation will be practiced, existing water rights are not impaired, and the water use is not detrimental to the public welfare. After approval of an application, the TNRCC issues a permit giving the applicant the right to use a stated amount of water in a prescribed manner. Once the right to the use of water has been perfected by the issuance of a permit by the TNRCC and the subsequent beneficial use of the water by the permit holder, the water authorized to be appropriated under the terms of the particular permit is not subject to further appropriation unless the permit is canceled. A permit may be canceled if water is not used during a 10-year period. Special term permits may also be issued allowing water use for specified periods of time. The Rio Grande and segments of other rivers are overappropriated with no new rights for additional water use being granted.

A permit holder has no actual title to the water but only a right to use the water. However, a water right can be sold, leased, or transferred to another person. The Lower Rio Grande Valley has been the only region of Texas with an active water market historically. In 1993, the legislature established a statewide water bank to be administered by the Texas Water Development Board. Although transfers can be accomplished independently of the water bank, the program was created to encourage and facilitate water marketing, transfer, and reallocation.

The Texas Water Code requires that the TNRCC consider environmental instream flow needs in the water rights permitting process. Such needs include maintenance of aquatic habitat and species, water quality, public recreation, wetlands, and freshwater inflows to bays and estuaries. Instream flow uses have become a major consideration in issuing permits since 1985. However, most water rights in the state were granted earlier without specifying instream flow requirements. Developing methodologies for establishing instream flow criteria and incorporating them into the water rights system continues to be an important issue.

Although water-master operations are common in some western states, the Rio Grande and South Texas water-master offices, which are components of the TNRCC, are the only such programs in Texas. A water-master office has administered water rights and accounted for water use in the Rio Grande Basin since the 1960's. The South Texas water-master was established in the late 1980's to administer water rights allocations in the Guadalupe, Nueces, and San Antonio River Basins. Plans during the 1980's to establish water-master programs throughout the state were later abandoned due to political considerations. The TNRCC responds to reports of illegal water use anywhere in the state. However, with the exception of the Lower Rio Grande Valley, water withdrawals are not routinely monitored.

The Texas Water Code is based on the prior appropriation doctrine. In general, senior water users are legally protected from more junior appropriators taking their water. However, water marketing is encouraged; rights may be bought and sold. In emergency drought situations, nonmunicipal water users may be forced to sell rights to municipalities. For permits issued statewide during the adjudication of existing rights pursuant to the Water Rights Adjudication Act of 1967, priority dates were established based on historical legal rights and actual water use.

Since completion of the adjudication process, priorities for additional new rights are based on the dates that the permit applications are filed.

For the Lower Rio Grande, priorities were set in conjunction with the previously discussed lawsuit. Water rights are divided into three categories. Municipal rights have the highest priority. Irrigation rights are divided into class A and class B rights, with class A rights receiving more storage in Falcon and Amistad Reservoirs storage accounts in the allocation procedure. In water-short years, this weighted priority system for irrigation rights results in greater shortages occurring on lands with class B water rights.

12.3.3.4 Groundwater rights. Unlike many states with groundwater permit systems, groundwater rights in Texas have historically been based on the common law rule allowing landowners to pump as much water as they wish from under their land. The water rights permit system applies only to surface water. However, increased state regulation of groundwater is evolving over time.

The Texas Water Code allows local groundwater conservation districts to be created "in order to provide for the conservation, preservation, protection, recharging, and prevention of waste of groundwater." Groundwater conservation districts have been created with voter approval at the county level since 1949. As of 2001, forty-three groundwater districts covered all or portions of 80 of the 254 counties in Texas. Most of the 43 districts were created since 1980. Most of the districts have assumed little if any regulatory authority. Texans are reluctant to have anyone tell them how much water they can pump from their own wells. Governmental regulation of pumping has been driven by necessity as depleting aquifers resulted in major problems. The Harris-Galveston Coastal Subsidence District and Edwards Underground Water Authority have developed the strongest regulatory programs.

The Harris-Galveston Coastal Subsidence District was created in 1957 in response to severe subsidence in the vicinity of the cities of Houston and Galveston. Due to decades of overdrafting groundwater, the ground surface has been lowered over 3 m in places in this low-lying, heavily urbanized coastal region.

The Edwards Aquifer Authority was created in 1993 largely due to a federal court ruling related to protection of endangered species under the Endangered Species Act. The Edwards is a limestone aquifer shared by San Antonio, several smaller cities, and extensive irrigated farming interests. San Antonio is the largest city in the nation that relies solely on groundwater for its water supply. Springs fed by the Edwards Aquifer maintain the flow of several rivers and support ecosystems that include several endangered species. During the late 1990's and 2000's, the Edwards Aquifer Authority developed a water rights permitting system to facilitate management of the aquifer.

12.3.3.5 Water availability modeling. The Texas Legislature in 1997 passed a comprehensive water management legislative package labeled Senate Bill 1. The designation *"Senate Bill 1"* is traditionally reserved each legislative session for legislation of the greatest importance. The 1997 Senate Bill 1 addressed a wide range of water management issues, including expanding statewide water availability modeling capabilities in support of regulatory and planning activities. The TNRCC, its partner agencies, and contractors developed a Water Availability Modeling (WAM) System pursuant to the 1997 Senate Bill 1 (Wurbs, 2001).

The Water Rights Analysis Package (WRAP) is the river basin water management simulation model incorporated in the WAM System. The generalized WRAP model provides capabilities for assessing water availability and reliability within the framework of a priority-based water allocation system. Complex river/reservoir system management facilities and practices may be simulated. WRAP hydrology and water rights input files developed by the TNRCC and its contractors for each river basin are publicly available to the water management community. The WAM system is used by water management entities and their consultants both in planning studies and in preparing permit applications and is used by the TNRCC in evaluating permit applications. Permit applications and are approved only if certain conditions are met, including demonstration that adequate water supply reliability (Section 7.10) can be ensured for the intended use and that other existing water rights will not be adversely affected.

12.4 WATER QUALITY MANAGEMENT

Water quality encompasses the physical, chemical, and biological characteristics of water. Both natural water quality and man-induced changes in quality are important considerations in river basin management. This section highlights three aspects of water quality management: (1) the NPDES regulating the discharge of pollutants into stream systems, (2) reservoir management problems and practices, and (3) salinity problems and control measures.

12.4.1 National Pollutant Discharge Elimination System

The NPDES is the cornerstone of the nation's water pollution control efforts (Greenway, 2000). This regulatory permit program implements the prohibition by the Clean Water Act of discharging pollutants from a point source to stream systems. The NPDES was established pursuant to the Water Pollution Control Act Amendments of 1972 (Public Law 92-500), which has been amended by several subsequent congressional acts, including the Clean Water Act of 1977 (PL 95-217) and the Water Quality Act of 1987 (PL 100-4). In Public Law 92-500, the Congress articulated goals for cleaning up the nation's streams and created a basic framework for achieving these goals, which included the NPDES and a wastewater treatment facility construction grants program. Later provisions in Public Law 100-4 converted the wastewater treatment works grant program to a state revolving fund loan program and incorporated urban stormwater runoff into the NPDES permit program.

NPDES permits are required for point sources, such as municipal and industrial wastewater treatment plants, stormwater sewer systems, construction activities, animal feedlots, and mining operations. The program is administered by the Environmental Protection Agency (EPA) in collaboration with state environmental management agencies. By 2001, EPA had authorized 42 states and territories to issue and enforce NPDES permits. For the remaining unauthorized states, EPA regional offices administer the program.

NPDES permits are designed to limit pollutants entering stream systems. Permits typically specify limits on allowable ranges for specified chemicals, monitoring and reporting requirements, operating and maintenance procedures, spill response procedures, and methods for handling emergencies. Best management practices may be required to minimize the amount of contaminant being released. Monitoring is a self-policing activity, but reports on discharges must be submitted to the EPA or state regulatory agency in a prescribed manner.

All discharges must be chemically or biologically treated to meet levels that are prescribed by EPA for each specific industry or specific discharge conditions. This limitation ensuring the quality of effluent is based on available technology. The discharge must meet the level of quality required by the receiving water body. Another set of limits, called national effluent guidelines, placed on discharges under the NPDES, is technology-based for over 50 industrial facility classifications.

12.4.2 Water Quality Management Aspects of Reservoir System Operations

Water quality and the aquatic environment may be significantly impacted by reservoir management practices. Water quality requirements for reservoir releases may involve both flow rates and quality parameters. Low-flow augmentation, or maintenance of minimum streamflow rates at downstream locations, is a primary water quality operating objective at many reservoir projects. The quality of the releases is controlled at many projects through multiple-level selective withdrawals.

Common reservoir water quality problems include turbidity, suspended solids, and associated impacts on fisheries, algae, and water quality. Pollution from watershed activities, such as acid mine drainage, oil field operations, agricultural activities, and municipal and industrial wastewater effluents, are problems in many areas. Problems are often related to eutrophication. Eutrophication is the process of excessive addition of organic matter, plant nutrients, and silt to reservoirs at rates sufficient to cause increased production of algae and rooted plants. Symptoms of eutrophication include algae blooms, weed-choked shallow areas, low dissolved oxygen, and accumulation of bottom sediments. Resulting problems include elimination of reservoir fisheries, adverse impacts on downstream ecosystems, degradation of water supplies, and reduced storage capacity.

Reservoir water quality problems may also be related to seasonal stratification. As illustrated by Fig. 12.9, in a stratified lake, the well-mixed surface layer,

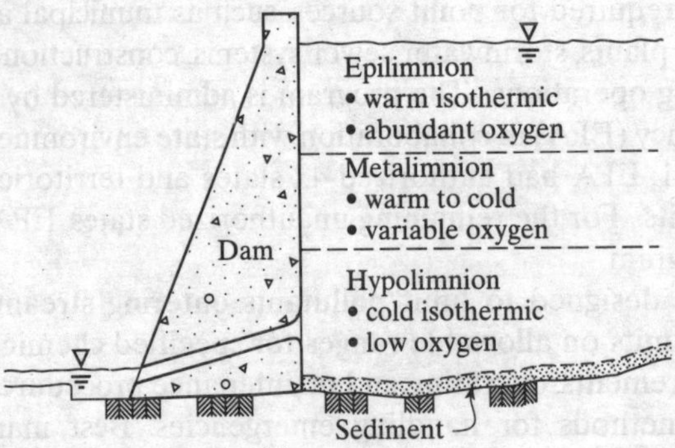

Figure 12.9 Reservoirs may seasonally stratify.

called the *epilimnion,* and the colder bottom layer, called the *hypolimnion*, are separated by a layer of sharp temperature gradient, called the *metalimnion.* Most impoundments exhibit some degree of temperature stratification. In general, deeper lakes are more likely to become highly stratified each summer and are not as likely to become mixed by wind or short-term temperature changes. When the surface of the lake begins to receive a greater amount of heat from the sun and air than is lost, it becomes warmer and less dense, whereas the colder, more dense water remains on the bottom. In the layer of colder water near the bottom, little if any oxygen is transferred from the air to replace that depleted by oxidation of organic substances, and, eventually anoxia may develop. Under this condition, a reducing environment is created, resulting in elevated levels of parameters such as iron, manganese, ammonia, and hydrogen sulfide. Changes such as these may result in water that is degraded and toxic to aquatic life.

A primary means of managing the water quality of reservoir releases is to control the vertical levels at which water is withdrawn from the reservoir. Many reservoir projects include outlet works intake structures providing multilevel withdrawal capabilities. The reservoir operating decision problem involves establishing the desired temperature, dissolved oxygen, and other water quality criteria and selecting the elevations at which to make releases to meet the criteria. Water from different levels may have to be mixed to meet the different water quality criteria. Management of water quality in the reservoir pool may also be a consideration in selective withdrawals from multilevel intake structures. Good and poor quality water can be blended to meet the release criteria with a minimum of good and maximum of poor quality water. This type of release policy will help to prevent a deterioration of quality in the reservoir that could lead to an eventual inability to meet the release criteria.

12.4.3 Salinity

Dissolved solids or salts are the inorganic solutes that occur in all natural waters because of weathering of rocks and soils. Total dissolved solids or salinity increases as waters move over the land surface and through soils and aquifers. Evaporation and transpiration increase concentrations. Human activities, such as irrigated agriculture and construction of reservoirs, increase evapotranspiration and the salinity of land and water resources. Groundwater pumping, oil field operations, and municipal use and wastewater disposal activities may also increase salinity. The ocean is a major source of salt in coastal areas. Salinity plays an important role in water resources development and management throughout the world, particularly in more arid regions. In the U.S., salinity is a major concern in river basins in the West and Southwest.

Natural salt pollution is particularly notable in the region shown in Fig. 12.10. During the Permian age about 230 million years ago, this region was covered by a large inland sea. Thick deposits of halite were formed as evaporating seawater precipitated salts. Most of the salt loads in the rivers originate from formations at shallow depths within the Permian Basin geologic region delineated in Fig. 12.10. This semiarid region consists of gypsum and salt-encrusted rolling plains containing numerous salt springs and seeps that contribute large salt loads to the rivers. The mineral pollutants consist largely of sodium chloride, with moderate amounts of calcium sulfate and other dissolved solids. Salt concentrations are

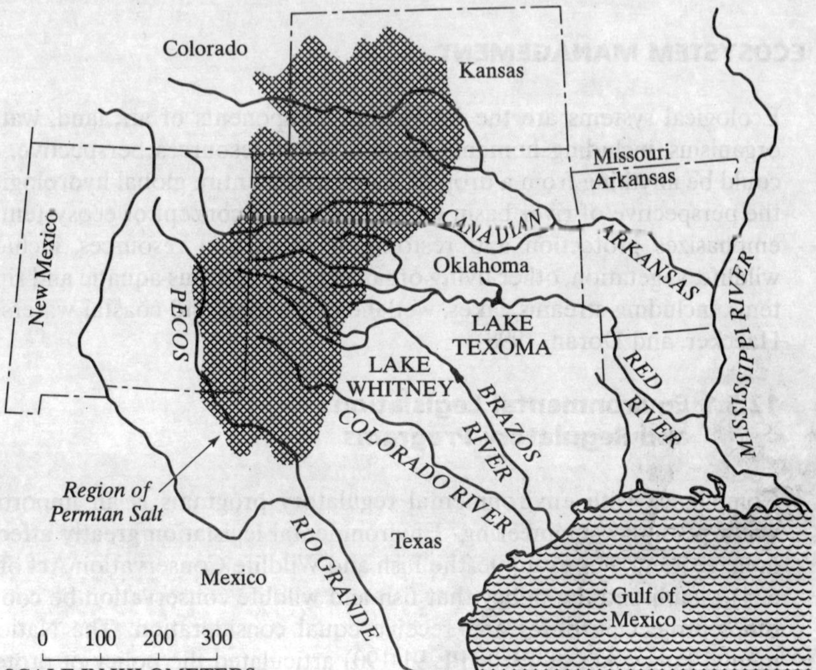

Figure 12.10 Geologic formations in the Permian Basin Region contribute large salt loads to these rivers and their upper watershed tributaries.

extremely high in streams in the upper Arkansas, Canadian, Red, Brazos, Colorado, and Pecos River Basins, exceeding seawater at some locations. Salt concentrations in the downstream reaches of the rivers decrease with dilution from low-salinity tributary inflows.

Population and economic growth combined with depleting groundwater reserves have greatly increased demands on the water resources of these river basins. Salinity severely limits use of large amounts of streamflow and reservoir storage in areas where water demands are surpassing supplies. Many studies have been conducted to develop strategies for dealing with the salinity. Several of many proposed salt control plans have been implemented. Economic feasibility, institutional difficulties, environmental concerns, and lack of funding have prevented or delayed construction of most proposed projects. Measures for dealing with salinity that have been implemented or seriously considered fall into three categories: (1) dilution of high-salinity water with better quality water, (2) desalination plants for treating municipal and industrial supplies, and (3) collection and disposal of brine in primary source areas. Brine collection facilities include both shallow well systems and surface impoundments. Disposal facilities include both deep-well injection and surface storage.

12.5 ECOSYSTEM MANAGEMENT

Ecological systems are the interacting components of air, land, water, and living organisms, including humans. From a water resources perspective, an ecosystem could be anything from a drop of water to the entire global hydrologic cycle. From the perspective of river basin management, the concept of ecosystem management emphasizes protection and restoration of natural resources, including fish and wildlife, vegetation, other living organisms, and various aquatic and riparian ecosystems, including streams, lakes, wetlands, estuaries, and coastal waters (Mac, Opler, Haecker, and Doran, 1998).

12.5.1 Environmental Legislation and Regulatory Programs

Compliance with environmental regulatory programs is an important aspect of water resources engineering. Environmental legislation greatly affects river basin management. For example, the Fish and Wildlife Conservation Act of 1958 (PL 85-624) established the policy that fish and wildlife conservation be coordinated with other project purposes and receive equal consideration. The National Environmental Policy Act of 1970 (PL 91-190) articulated the policy of protecting the environment and established requirements for evaluating the environmental impacts of federal actions. The following discussion highlights two other efforts that significantly affect river basin management: the Section 404 regulatory program and the endangered species protection program.

The Water Pollution Control Act Amendments of 1972 (PL 92-500), as further amended by the Clean Water Act of 1977 (PL 95-217), was cited in Section 12.4.1 from the perspective of the NPDES. Section 404 of Public Laws 92-500 and 95-217 established the dredge and fill permit program administered by the USACE. Construction, land development, and other activities involving placing of dredge or fill materials into streams, rivers, lakes, or wetlands requires USACE approval of a permit application. The objective is to ensure that every reasonable effort is made to minimize adverse impacts of development activities on the environment. The Corps of Engineers seeks public input and collaborates with the EPA and Fish and Wildlife Service in determining whether the permit application should be approved and, if so, what modifications to a proposed project may be required.

Protection of wetlands is an important part of the Section 404 regulatory program. The term *wetland* includes swamps, marshes, bogs, bottomlands, sloughs, and wet meadows (Mitsch and Gosselink, 2000). Wetlands have water at or near the ground surface for much of the year. Wetland ecosystems represent the transition between aquatic and terrestrial ecosystems. Prior to the 1970's, drainage of wetlands for agricultural and urban development was accepted practice. Wetlands disappeared at alarming rates. The importance of preserving wetland ecosystems became widely recognized. Restoration of the Florida Everglades during the 1990's–2000's is a particularly notable example of efforts to protect and preserve wetlands. Preventing further loss of wetlands nationwide is a major objective addressed through the Section 404 permit program.

Requirements for conservation of endangered species, pursuant to the 1973 Endangered Species Act (PL 93-205) as amended by the Endangered Species Act Amendments of 1978 and 1979 (PL 95-632 and PL 96-159) and other legislation, are administered by the Fish and Wildlife Service in coordination with other agencies. Many species of fish, wildlife, and plants have been rendered extinct as a consequence of economic development. The objective of the Endangered Species Act is to prevent loss of additional species. Endangered species are officially identified, and they and their habitat are protected from actions that could cause their destruction.

Endangered species have significantly impacted river basin management nationwide, including operations of several major reservoir systems. For example, although salmon migration had been for decades an important consideration in reservoir system management in the Columbia River Basin, the 1992 listing of certain types of salmon as endangered resulted in intensified fish protection efforts. Programs have been implemented in the Susquehanna River to restore populations of American Chad. Reservoir operations on the Missouri River have been modified to prevent inundation of sandbars that serve as nesting habitat for the least tern and piping plover, which are endangered birds.

12.5.2 Environmental Management Aspects of Reservoir System Operations

Environmental resources management opportunities and problems associated with reservoir operations vary widely between regions and between reservoirs.

Reservoir operations influence fish, wildlife, and ecological systems both in the reservoir pool and in the river downstream.

Reservoir releases contribute to maintenance of instream flows necessary for the support of aquatic habitat and species, protection or enhancement of water quality, preservation of wetlands, and provision of freshwater inflows to bays and estuaries. Reservoir operating plans may include maintenance of specified minimum flow rates at downstream locations. Periodic flooding, as well as low-flow augmentation, may be important for certain ecosystems. The required flow rates may be specified as a function of season, reservoir storage, reservoir inflows, and other factors.

Reservoir releases for downstream fishery management depend on water quality characteristics and water control capabilities. Achieving optimal temperatures for either cold water or warm water fisheries through selective multilevel releases may be an operating objective. Maintenance of dissolved oxygen levels may be an operating objective. Releases can be beneficial for maintaining gravel beds for certain fish species. Dramatic changes in release rates, typically associated with hydropower and flood control operations, can be detrimental to downstream fisheries.

Migration of anadromous fish, such as salmon in the Pacific Northwest and striped bass in the Northeast, is a concern in some regions. Declines in anadromous fish populations have been attributed to dams due to blockage of migration, alteration of normal streamflow patterns, habitat modification, blockage of access to spawning and rearing areas, and changes in water quality. Regulation for anadromous fish is particularly important during certain seasons of the year.

Project regulation can influence fisheries in the pool as well as downstream. Water surface level fluctuations is one of the most apparent influences of reservoir operation. Periodic fluctuations in water levels present both problems and opportunities in regard to reservoir fisheries. The seasonal fluctuations that occur at many flood control projects and daily fluctuations at hydropower projects often eliminate shoreline vegetation and cause subsequent shoreline erosion, water quality degradation, and loss of habitat. Adverse impacts of water level fluctuations also include loss of shoreline shelter and physical disruption of spawning and nests. Beneficial fisheries management techniques include: pool level management for weed control; forcing forage fish out of shallow cover areas, making them more susceptible to predation; and maintaining appropriate pool levels during spawning.

12.6 RIVERS AND RESERVOIRS OF THE WORLD

Twenty-five of the largest river basins in the world are listed in Table 12.2 with their approximate drainage area, length of the main stem of the river system, and mean discharge at the outlet. Most of these river basins are shared by multiple countries. The country with the greatest proportion of the watershed area is listed

TABLE 12.2 LARGEST RIVER BASINS IN THE WORLD

River	Country with greatest area	Drainage area (km^2)	Length (km)	Mean flow (1,000 m^3/s)
Amazon	Brazil	5,870,000	6,400	180
Congo	Congo	3,700,000	4,700	41
Mississippi	United States	3,230,000	6,020	17
Nile	Sudan	3,040,000	6,650	3
Parana	Brazil	2,800,000	4,880	22
Ob-Irtysh	Russia	2,730,000	5,410	15
Yenisey	Russia	2,500,000	5,540	19
Lena	Russia	2,490,000	4,400	16
Niger	Nigeria	2,120,000	4,200	6
Amur	Russia	1,880,000	2,820	12
Mackenzie	Canada	1,801,000	4,240	11
Yangtze	China	1,680,000	6,300	34
Ganges-Brahmaputra	India	1,680,000	2,900	38
Volga	Russia	1,550,000	3,530	8
Zambezi	Zambezi	1,390,000	3,500	7
Nelson	Canada	1,150,000	2,680	2
Indus	Pakistan	1,090,000	2,900	8
St. Lawrence	Canada	1,060,000	4,000	10
Tocantis	Brazil	906,000	2,700	10
Yukon	United States, Alaska	830,000	3,190	6
Tigris-Euphrates	Iraq	794,000	2,800	1
Danube	Romania	780,000	2,850	7
Columbia	United States	668,000	2,000	7
Colorado	United States	651,000	2,400	0.6
Rio Grande	United States	549,000	1,360	0.08

in Table 12.2. A few other river basins not listed are larger than several of the smallest basins included in the table.

The Amazon in South America is the largest river basin in the world in terms of both drainage area and flow. Sixty-three percent of the Amazon River Basin lies within Brazil, with the remainder in Peru, Bolivia, Colombia, Ecuador, Venezuela, and Guyana. The Amazon is second only to the Nile River in length. The 25 river basins listed in Table 12.2 cover about 35 percent of the world's land surface, excluding Antarctica. Most but not all of the basins listed in Table 12.2 are among the 261 international river basins noted in Section 12.3.1 that cover about 45 percent of the world's land area, excluding Antarctica.

12.6.1 Worldwide Inventory of Large Dams

History does not record exactly when dams were first constructed. However, reservoir projects have served people for at least 5,000 years, beginning in the cradles of civilization of Babylonia, Egypt, India, Persia, and the Far East. The history of dams closely follows the rise and decline of civilizations, especially in cultures highly dependent on irrigation (Jansen, 1983; Schnitter, 1994). Most dams were constructed during the 20th century.

The International Commission on Large Dams (ICOLD), which was founded in France in 1928, is a primary source of data covering the worldwide inventory of dams. The ICOLD developed and maintains the *World Register of Dams,* which is a listing of large dams with information on type, dimensions, and ownership. The register is a compilation of data developed by national committees in the participating countries. The U.S. Society on Dams, which prior to 2000 was called the U.S. Committee on Large Dams, represents the United States. For purposes of inclusion in the register, the ICOLD has defined a large dam as either: (1) greater than 15 m in height, or (2) between 10 and 15 m in height, with a storage volume of at least 3 million m^3. The World Commission on Dams (2000) summarizes the ICOLD worldwide inventory of dams and discusses key issues related to reservoir project construction and operation.

The number of large dams in the register increased from 5,196 in the year 1950, to 34,798 in 1982, to 47,425 in 1999. An additional 1,648 dams were reported as being under construction in 1999. Data collection constraints caused some large dams to not be counted in these tallies. Hundreds of thousands of smaller water supply reservoirs, barrages, farm ponds, and flood detention structures do not meet the criteria cited in the previous paragraph for being classified as a large dam and thus are not included in the count. At least 140 countries have at least one large dam. The nine countries listed in Table 12.3 have over 90 percent of the large dams

TABLE 12.3 NINE COUNTRIES WITH THE GREATEST NUMBER OF LARGE DAMS IN 1999

	Number of Large Dams			1999 Under
Country	in 1950	in 1982	in 1999	construction
China	8	19,595	26,094	330
United States	1,543	5,338	6,775	42
India	202	1,085	3,796	650
Japan	1,173	2,142	2,560	100
Spain	205	690	1,191	31
South Korea	116	628	805	133
Canada	189	580	797	0
South Africa	79	342	789	7
Mexico	109	487	615	2
Totals for:				
9 countries	3,624	30,887	43,422	1,295
Worldwide	5,196	34,798	47,425	1,648

in the world. Over half of the dams are in China. The U.S. places a distant second in number of dams. For the nine countries having the largest number of large dams in 1999, Table 12.3 shows their tally of dams in 1950, 1982, and 1999, and the additional dams under construction in 1999. Other countries, not listed in Table 12.3, that have between 500 and 600 large dams include Turkey, Brazil, France, Italy, and Great Britain.

The extremely large dams are not all located in the countries with the most numerous projects. The five highest dams in the world are the Rogun (335 m), Nurek (300 m), Grand Dixence (285 m), Inguri (272 m), and Vaiont (860 m). The Grand Dixence and Vaiont are in Switzerland and Italy. The other three are in the Commonwealth of Independent States (former USSR). With a height of 235 m, Oroville Dam on the Feather River in California is the highest dam in the U.S. and 14th highest in the world.

12.6.2 River Basin Development

Rivers in the U.S. and many other developed nations are highly regulated by dams and appurtenant structures. A massive infrastructure of constructed facilities is in operation. The pre-1980's construction era transitioned to an emphasis on optimal operations, maintenance, and rehabilitation of existing facilities. However, major projects are still being constructed in various countries.

Turkey's development of its portion of the Tigris-Euphrates River Basin is a notable example of recent major construction projects. The Southeast Anatolia Development Project (acronym GAP from its Turkish name) is a comprehensive large-scale water management plan implemented during the 1980's–2000's for irrigation and hydroelectric power for arid southeastern Turkey (Kolars and Mitchell, 1991). GAP includes 22 dams, 19 hydroelectric power plants, and many irrigation distribution networks. The Ataturk Dam on the Euphrates River is the largest dam in the GAP. Construction of the 180-m-high and 1,820-m-long rock-fill dam and appurtenant structures was accomplished during the 1980's and early 1990's. The Ataturk Dam storage capacity of 49 billion m^3 is the largest of any reservoir in Turkey, among the largest in the world, and larger than any reservoir in the U.S.

The Three Gorges Project on the Yangtze (or Chang Jiang) River in China is being constructed during the period 1997–2009. The 185-m-tall and 2,000-m-long dam will impound a reservoir that will be 600 km long, with a storage capacity of almost 40 billion m^3. The reservoir will extend through scenic canyons formed by immense limestone cliffs known as the Three Gorges. The project will provide critically needed flood control, hydroelectric energy, and water supply. It has generated enormous international controversy because of its massive size, serious environmental impacts, and displacement of the homes of over a million people.

12.7 MAJOR RIVER/RESERVOIR SYSTEMS IN THE UNITED STATES

Most of the major reservoir projects in the U.S. were constructed during the period from 1900 through the 1970's, which has been called the construction era of water resources development. Although additional new reservoir projects are needed and continue to be developed, most of the major reservoir systems required to manage our rivers are in place. Economic, environmental, and institutional considerations constrain construction of water resources development projects. Since the 1970's, water resources management policy and practice have shifted to a greater reliance on improving water use efficiency, managing floodplain land use, optimizing the operation of existing facilities, developing water allocation systems, and environmental regulation.

Public needs and objectives and numerous factors affecting reservoir/river system management change over time. Population and economic growth in various regions of the nation are accompanied by increased needs for flood control, water supply, energy, recreation, and the other services provided by water resources development. Depleting groundwater reserves have resulted in an increased reliance on surface water in many areas. Concerns have grown for maintenance of instream flows for preservation of riverine habitat and species, wetlands, and freshwater inflows to bays and estuaries. With an aging inventory of numerous dams and reservoirs being operated in an environment of change and intensifying demands on limited resources, operational improvements are being considered increasingly more frequently.

12.7.1 Institutional Framework for Reservoir System Management

Numerous reservoir projects, located throughout the U.S., are operated by federal, state, and regional agencies, local water districts, cities, and private industry. As discussed in Chapter 1, the water management community consists of water users, concerned citizens, public officials, professional engineers and scientists, special interest groups, businesses, utilities, cities, and local, state, regional, federal, and international agencies. In addition to the entities that construct and operate a reservoir system, numerous other public agencies, project beneficiaries, and interest groups play significant roles in developing new projects and determining operating policies for existing systems. Within this institutional framework, a number of organizations are directly responsible for developing and managing reservoir projects. Most reservoirs are owned and operated by private electrical and water utilities, cities, water districts, and other local entities. However, the majority of the storage capacity is contained in federal reservoirs. Most, though certainly not all, of the very large reservoir systems in the U.S. are operated by the federal water agencies. The much more numerous nonfederal reservoirs tend to be much smaller in size than the federal projects.

The USACE is the largest reservoir management agency in the nation, with over 500 reservoirs in operation. The Corps of Engineers is unique in having nationwide responsibilities for construction and operation of large-scale multiple-purpose reservoir projects. The USBR operates about 130 reservoirs in the 17 western states and has constructed numerous other projects that have been turned over to local interests for operation. The Tennessee Valley Authority (TVA) operates a system of about 50 reservoirs in the seven-state Tennessee River Basin. The NRCS has constructed many thousands of small flood retarding dams throughout the U.S.

The responsibilities of organizations involved in operating reservoir systems are based on project purposes. The USACE has played a clearly dominant role nationwide in constructing and operating major reservoir systems for navigation and flood control. The USBR water resources development program was founded on facilitating development of the arid West by constructing irrigation projects. The activities of the federal water resources development agencies have evolved over time to emphasize comprehensive multiple-purpose water resources management. Hydroelectric power, municipal and industrial water supply, recreation, and fish and wildlife conservation are major purposes of USACE and USBR projects.

Municipal and industrial (M&I) water supply has been primarily a nonfederal responsibility, though significant municipal and industrial storage capacity has been included in federal reservoirs for the use of nonfederal project sponsors. All costs allocated to M&I water supply are reimbursed by nonfederal sponsors that contract for storage capacity in accordance with the Water Supply Act of 1958, as amended by the Water Resources Development Act of 1986 and other legislative acts. Nonfederal sponsors for federal projects are often regional water authorities that sell water to municipalities, industries, and other water users, under various contractual arrangements. Numerous cities, municipal water districts, and other local agencies operate their own reservoir projects. Private companies, as well as governmental entities, play key roles in hydroelectric power generation, thermal-electric cooling water projects, and industrial water supply.

The Reclamation Acts of 1902 and 1939 and other legislation dictate the policy that costs allocated to irrigation in federal projects be reimbursed by the project beneficiaries. The details of repayment requirements for irrigation projects have varied over the years with changes in reclamation law. Congressional acts authorizing specific USBR projects have often included repayment provisions tailored to the circumstances of the individual project.

Water supply operations are controlled by agency responsibilities, contractual commitments, and legal systems for allocating and administering water rights. Water allocation and use are regulated by state water rights systems (Section 12.3). Operations of reservoir systems in international and interstate river basins are controlled by agreements between states and/or nations that were negotiated over many years.

Hydroelectric power generated at USACE and USBR reservoirs is marketed to electric utilities by the five regional power administrations of the Department of Energy. The power administrations operate through contracts and agreements with the electric cooperatives, municipalities, and utility companies that buy and distribute the power. The Tennessee Valley Authority (TVA) is directly responsible for marketing, dispatching, and transmission of power generated at its plants. Many private and public electrical power companies operate their own reservoirs and hydropower plants. Several large hydroelectric power systems are composed of multiple-storage and generating components owned and operated by federal, state, or local public agencies and private companies. Hydroelectric power facilities are typically components of systems that rely primarily on thermal plants for the base load, with hydropower supplying peak loads.

12.7.2 Largest River Basins in the United States

The Mississippi River is the largest river in the U.S. in terms of discharge, watershed area, or length. With a length of 6,020 km (3,740 miles), the Mississippi is the fourth longest river in the world. It is the seventh largest river in the world ranked in terms of the mean discharge at its mouth of 17,300 m^3/s (611,000 ft^3/s). The Mississippi River has a drainage area shown in Fig. 2.3 of 3,230,000 km^2 (1,250,000 mi^2), which encompasses 41 percent of the conterminous United States and a small portion of Canada. The largest subwatersheds of the Mississippi River Basin are the Missouri River Basin (1,370,000 km^2), Ohio River Basin (528,000 m^2), Arkansas River Basin (416,000 km^2), and Red River Basin (93,200 km^2).

Outside of the Mississippi and its subbasins, the next largest river basins in the U.S. are the Yukon, St. Lawrence, Columbia, Colorado, and Rio Grande, which are listed in Table 12.2. The Yukon River Basin encompasses more than half of Alaska and a small portion of Canada. The St. Lawrence River Basin lies primarily in Canada. The Mississippi, Columbia, Colorado, and Rio Grande Basins—which are the largest river basins in the conterminous U.S.—encompass about 62 percent of the land area of the 48 states. These four largest basins are delineated in Fig. 12.11. Major river basins along the East Coast, such as the Hudson, Delaware, Susquehanna, and Potomac, and in the Southeast, such as the Appalachicola, Alabama, and Tombigbee, not shown in Fig. 12.11, have smaller drainage areas but relatively large flows and supply large urban population centers. As discussed in Chapter 2, the largest rivers, in terms of mean flow rates, in the U.S. are shown in Fig. 2.13.

12.7.3 Major Reservoirs in the United States

Numerous dam and reservoir projects regulate streams ranging from small creeks to major rivers. These projects range in size from many thousands of small farm ponds owned by individual farmers, with storage capacities of several hundred m^3,

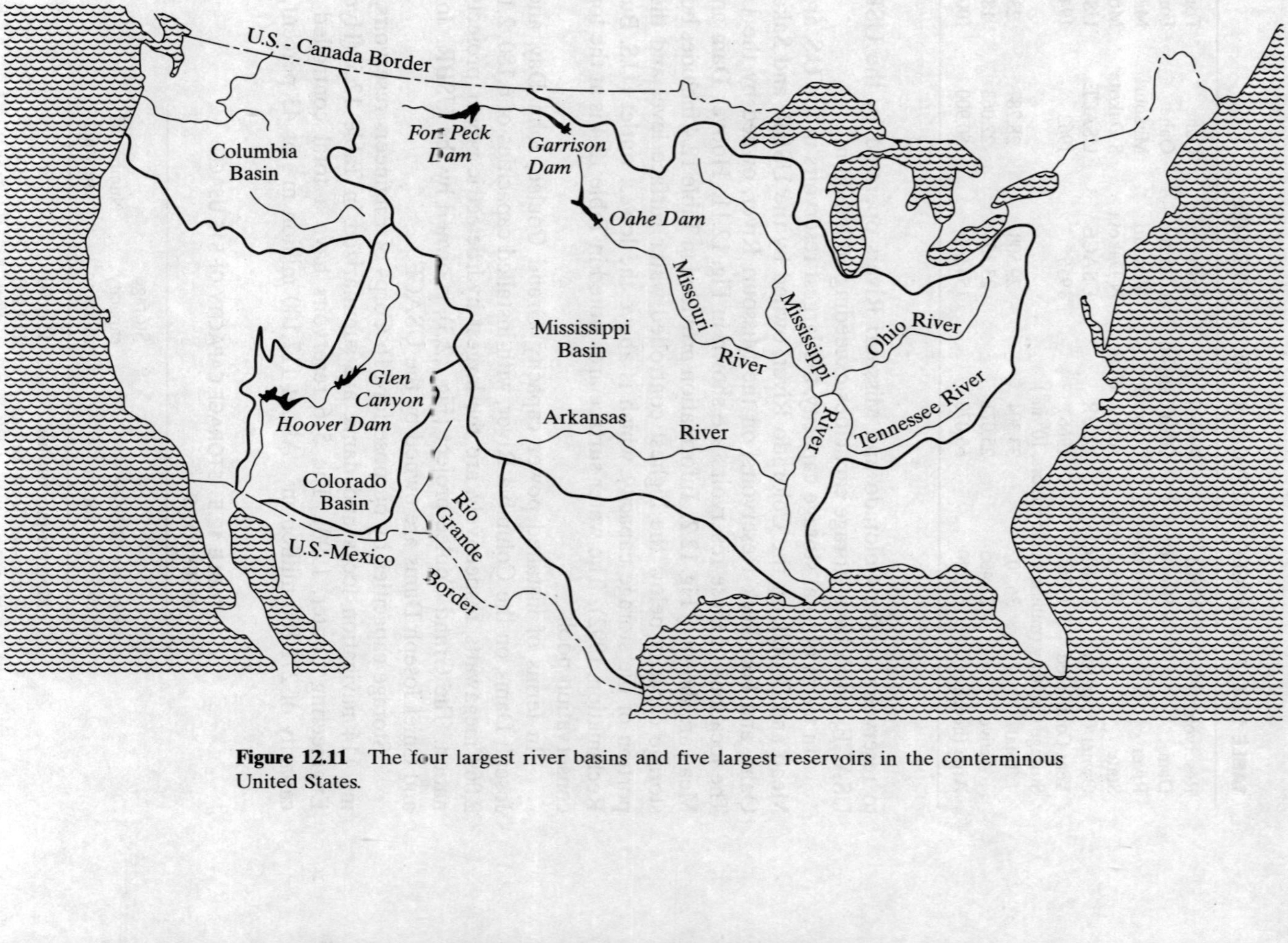

Figure 12.11 The four largest river basins and five largest reservoirs in the conterminous United States.

TABLE 12.4 FIVE LARGEST RESERVOIRS IN THE UNITED STATES

Reservoir	Mead	Powell	Sakakawea	Oahe	Fort Peck
Dam	Hoover	Glen Canyon	Garrison	Oahe	Fort Peck
River	Colorado	Colorado	Missouri	Missouri	Missouri
State	Nev & Ariz	Ariz & Utah	N Dakota	S Dakota	Montana
Owner/Operator	USBR	USBR	USACE	USACE	USACE
Year Completed	1936	1966	1956	1962	1940
Storage Capacity (million cubic meters, 10^6 m^3)					
Total	36,702	33,304	29,508	28,787	23,324
Active	21,405	25,071	23,354	22,063	18,041
Area (hectares)	65,840	65,070	154,600	150,900	100,800

to reservoirs on the Colorado and Missouri Rivers operated by the USBR and USACE, which have storage capacities exceeding 30 billion m^3.

In terms of total storage capacity, the largest reservoirs in the U.S. are Lakes Mead and Powell, on the Colorado River, owned by the USBR, and Sakakawea, Oahe, and Fort Peck Reservoirs, on the Missouri River, owned by the USACE. The locations of these reservoirs are shown in Fig. 12.11. Hoover Dam and Lake Mead are shown in Fig. 12.2. Information provided in Table 12.4 includes both total storage capacity below the highest controlled water surface level and the active portion of the storage capacity, which is above the lowest outlet (U.S. Bureau of Reclamation, 1992). The water surface area noted in Table 12.4 is at the top of the conservation pool.

In terms of installed power capacity, Grand Coulee, John Day, and Chief Joseph Dams, on the Columbia River, with installed capacities of 6,180, 2,160, and 2,069 megawatts, respectively, are the largest hydroelectric power projects in the nation. The Grand Coulee project (Fig. 12.3), is owned by the USBR. John Day and Chief Joseph Dams are owned by the USACE.

Storage capacities by purpose for 516 Corps of Engineers reservoirs, including 114 navigation locks and dams, are summarized in Table 12.5 (Hydrologic Engineering Center, 1990). The 516 reservoirs have a total controlled storage capacity of 272,100 million m^3. About 117,100 million m^3, or 43 percent, of this

TABLE 12.5 STORAGE CAPACITY OF 516 USACE RESERVOIRS

Storage allocation	Storage capacity (million m^3)	Number of reservoirs
Multiple-purpose use	151,600	385
Exclusive flood control	117,100	330
Exclusive navigation	2,900	135
Exclusive hydropower	475	5
Total storage capacity	272,100	516

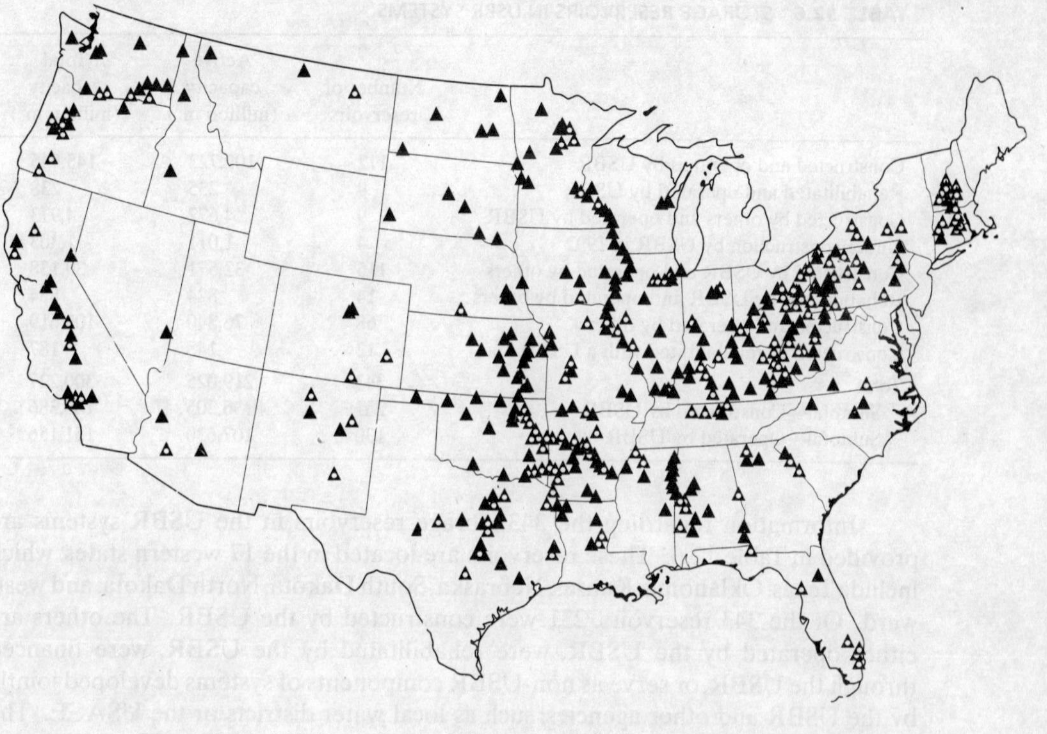

Figure 12.12 U.S. Army Corps of Engineers reservoirs are located throughout the nation.

capacity is designated for use exclusively for flood control. At least a portion of the storage capacity in each of 330 reservoirs is designated for use exclusively for flood control. An additional 1.1 percent and 0.2 percent of the total capacity is used exclusively for navigation and hydroelectric power, respectively. The remaining 151,600 million m^3, or 56 percent, of the storage capacity provides multiple-purpose use in 385 reservoirs. The multiple-use storage capacity in each individual reservoir serves two or more purposes, which may include municipal and industrial water supply, irrigation, recreation, fish and wildlife enhancement, low-flow augmentation, flood control, navigation, or hydropower. Figure 12.12 shows the geographic distribution of the larger Corps of Engineers reservoirs. The reservoirs are operated by USACE district offices.

The 516 reservoirs reflected in Table 12.5 are owned and operated by the USACE. There are about 88 other reservoirs owned by other agencies for which the Corps of Engineers shares operational responsibilities. Most of these projects are included in the USBR database described next. The USACE is responsible for flood control operations at a number of projects constructed by the USBR and has also constructed or rehabilitated a number of additional projects which are operated by other agencies.

TABLE 12.6 STORAGE RESERVOIRS IN USBR SYSTEMS

	Number of reservoirs	Active capacity (million m^3)	Total capacity (million m^3)
Constructed and operated by USBR	112	102,723	145,945
Rehabilitated and operated by USBR	9	235	238
Constructed by others and operated by USBR	9	4,672	4,973
Under construction by USBR in 1992	4	1,011	1,303
Constructed by USBR and operated by others	115	32,571	39,138
Rehabilitated by USBR and operated by others	14	824	824
Constructed and operated by others	68	76,840	109,619
Constructed or rehabilitated with a USBR loan	12	148	187
Total	343	219,025	300,227
Subtotal—Constructed by USBR	231	136,305	186,386
Subtotal—Operated by USBR	130	107,630	151,156

Information regarding the 343 storage reservoirs in the USBR systems are provided in Table 12.6. These reservoirs are located in the 17 western states, which include Texas, Oklahoma, Kansas, Nebraska, South Dakota, North Dakota, and westward. Of the 343 reservoirs, 231 were constructed by the USBR. The others are either operated by the USBR, were rehabilitated by the USBR, were financed through the USBR, or serve as non-USBR components of systems developed jointly by the USBR and other agencies, such as local water districts or the USACE. The Bureau operates 130 reservoirs. Many reservoirs constructed by the USBR are turned over to local irrigation districts or other entities for operation.

Almost all of the 343 reservoirs in Table 12.6 include irrigation as a primary purpose. The majority serve multiple purposes. About 28 percent are single-purpose irrigation reservoirs. Municipal and industrial water supply storage is included in 87 reservoirs. Eighty-five are operated for hydroelectric power, including a few single-purpose hydropower reservoirs. Eleven of the multiple-purpose reservoirs include navigation. Other purposes include fish and wildlife, recreation, river regulation, and sediment control.

The active storage capacity above the lowest outlet and total capacity below the highest controlled water surface are summarized in Table 12.6. The total storage capacity of individual reservoirs range from less than 100,000 m^3 at several reservoirs to 36,702,000,000 m^3 in Lake Mead behind Hoover Dam. The five largest reservoirs in the nation, listed in Table 12.4, account for half of the total storage capacity of the 343 reservoirs in Table 12.6.

12.7.4 Tennessee Valley Authority System

As shown in Figs. 2.3 and 12.11, the Tennessee River flows into the Ohio River just above its confluence with the Mississippi River. The Tennessee River Basin encompasses about 106,000 km^2 of seven states. The mean annual rainfall over the basin is about 132 cm. Annual snowfall averages about 20 cm, but does not create a snowpack. The mean flow of the Tennessee River at its mouth is about 1,870 m^3/s. Terrain in the basin ranges from mountains and forests in the east to hills and open fields in the west.

The TVA system is operated in accordance with operating priorities mandated by the 1933 Congressional act that created the TVA. This act specified that the TVA system be used to regulate streamflow primarily for purposes of promoting navigation and controlling floods and, so far as may be consistent with such purposes, for generation of electric energy. In addition to these primary purposes, recreation, fish and wildlife, water quality, water supply, and vector control are also important aspects of system management.

The TVA system includes 50 dams, with a total storage capacity of about 16.9 million m^3, which is about 30 percent of the mean annual runoff. Some of the dams were constructed by the TVA. Others were constructed by other entities, but are operated by the TVA. Norris Dam, shown in Fig. 1.9, was the first dam constructed by the TVA. The multiple-purpose Kentucky Reservoir, located on the Tennessee River near its confluence with the Ohio River, is the largest reservoir in the system. The 10 dams located on the main stem of the Tennessee River have navigation locks permitting barge traffic to be maintained on the 1,000-km (625-mi.) reach from the Ohio River to the city of Knoxville. Streamflows are highest during the winter months and at minimum levels during the summer and fall. The reservoir system is operated to provide flood control, particularly in the winter months, and to augment streamflows in the summer and fall months for navigation, power generation, and other purposes.

The TVA operates 36 hydroelectric power plants at 22 multiple-purpose reservoir projects and 14 single-purpose power dams. The TVA hydroelectric facilities were originally designed to provide most of the system power demand. However, TVA electrical energy generation has evolved into a thermal-based power system, with the hydroelectric plants being used primarily for carrying intermediate and peak loads. A major objective in developing operating plans is to minimize system fuel costs. An operating plan is developed each year based on the expected value of power generation. Decisions to draft storage at any point in time are based on the amount of reservoir storage available and the cost of the thermal generation that would otherwise be required. Hydropower release decisions are also influenced by minimum flow requirements for nonpower purposes and the desirability of maintaining reservoirs as high as possible during the summer months in the interest of reservoir recreation. The operating plans include seasonal rule curves for allocating storage capacity between flood control and conservation purposes.

12.7.5 Missouri River Basin

The Missouri River Basin, above its confluence with the Mississippi River, encompasses 1,350,000 km^2 of 10 midwestern states and about 26,000 km^2 of Canada. Average annual precipitation over the basin ranges from 20 cm just east of the Rocky

Mountains to about 100 cm in the southeastern portion of the basin. April through June is typically the wettest season of the year. Snowfall in northern and central portions of the basin ranges from 50 cm in the lower basin to more than 250 cm in high elevation Rocky Mountain locations. High streamflows on the Missouri River are caused by snowmelt and rainfall on the plains during March and April and by mountain snowmelt and rainfall during the period May through July.

Numerous reservoirs located on tributaries throughout the basin are operated by various entities. The USACE and USBR have constructed 44 and 22 reservoirs, respectively, on tributaries. Six other USACE reservoirs on the mainstem Missouri River are the cornerstone for water management in the basin. The three largest reservoirs are included in Table 12.4 and Fig. 12.11. The six mainstream reservoirs are operated for flood control, hydropower, water supply, water quality, irrigation, navigation, recreation, and preservation of environmental resources. The USACE system also includes navigation and bank stabilization improvements along the Missouri River from Sioux City, Iowa, to the mouth at St. Louis, Missouri.

Navigation releases are made to maintain depth requirements in the Missouri River between Sioux City and the confluence with the Mississippi River from early April to early December. Ice forms on the river during the winter, restricting navigation. However, a minimum discharge is normally maintained throughout the winter months for water quality and power production.

The hydroelectric power plants at the six reservoirs have a total installed capacity of 2,048 megawatts. The firm power output of the projects is used to meet a portion of the power requirements of the Western Area Power Administration's customers. This power is a mix of base load, intermediate, and peaking power. When secondary energy is available, it is marketed by the Western Area Power Administration to area utilities at a cost based on the value of the thermal plant fuel saved. The peak power demand occurs between mid-December and mid-February in the north due to home heating and between mid-June and mid-August in the south due to air conditioning loads.

Impacts on the environment are a major consideration in managing the Missouri River system. For example, the endangered least tern and piping plover have had major impacts on reservoir system operations. These bird species are dependent on nesting habitat that consists of sparsely vegetated sandbars on the rivers. River/reservoir system management includes protection of this habitat.

12.7.6 Columbia River Basin

From its source in Columbia Lake in Canada's Selkirk Mountains, the Columbia River flows northwestward and then southward through British Columbia to the U.S. It crosses the international border north of Spokane, Washington, then flows southward across central Washington where it is joined by the Snake River. The Columbia then turns westward to form the border between Washington and Oregon and enters the Pacific Ocean at Astoria, Oregon. The climate of the region ranges from arid in parts of the interior to very wet in coastal areas. In parts of eastern Washington and Oregon and the Snake River basin, the mean annual precipitation is less than 20 cm. In contrast, some of the coastal mountain rain forests receive more than 510 cm annual precipitation.

The first navigation locks in the region were constructed during the 1870's on the Willamette River and at the cascades area near the present location of Bonneville Dam. Hydroelectric development began in the late 1880's when electric "dynamos," now called generators, were installed on the Spokane River in Spokane, on the Willamette River at Oregon City, and on the Snoqualmie River at Snoqualmie Falls, east of Seattle. Harnessing the energy of the mainstem Columbia River began with the construction of Rock Island Dam near Wenatchee in 1932 and initiation of construction on Grand Coulee (Fig. 12.3) and Bonneville Dams in the late 1930's. Bonneville Dam was constructed with state-of-the-art fish ladders to facilitate the passage of returning salmon. There were two other main periods of federal dam construction in the basin: in the 1950's when Hungry Horse, Chief Joseph, The Dalles, McNary, and Albeni Falls were built and in the mid-1970's with construction of the Columbia River Treaty projects, Dworshak, and the lower Snake projects. Most construction of privately owned dams was also during these periods.

Figure 12.13 The Feeder Canal with a capacity of 450 m^3/s is a component of the USBR Columbia Basin Irrigation Project. Water pumped from Lake Roosevelt (Fig. 12.3) is transported through this canal to farms in central and eastern Washington. (Photo by W. P. James, August 1998)

The Columbia River Treaty between the United States and Canada, adopted in 1964, provided for the construction and operation of Mica, Arrow, and Duncan Dams in Canada, and Libby Dam in the United States. Under the terms of the treaty, each nation has designated an operating entity. The Canadian entity is the British Columbia Hydro and Power Authority. The Bonneville Power Administration and USACE North Pacific Division represent the U.S. These entities appoint representatives to two committees, the Operating Committee and Hydrometeorological Committee, which are charged with carrying out the operating arrangements necessary to implement the treaty.

Over 250 major reservoir projects on the Columbia River and its tributaries are operated for flood control, hydroelectric power, irrigation, navigation, fish and wildlife enhancement, recreation, low-flow augmentation, and municipal and industrial water supply. The larger storage projects are generally filled and emptied in an annual cycle. Some storage reservoirs may not necessarily refill each year if all the active storage is withdrawn. Many of the reservoirs are run-of-river or pondage hydroelectric power projects with little or no active storage capacity relative to mean annual streamflow. At-site power generating facilities are located at some of the reservoirs. Other storage reservoirs make releases that pass through many powerhouses located different distances downstream. Some projects are designed specifically for daily re-regulation of outflows from an upstream reservoir.

Fishery considerations are important, particularly in regard to migration, habitat, and production of salmon. Operations include moving fish past the dams with minimal losses and augmenting instream flows to better carry juveniles downstream to the ocean.

Operation of a number of hydroelectric power projects owned by a variety of public and private entities are coordinated in accordance with the Pacific Northwest Coordination Agreement. The seasonal operation of the system reservoirs are coordinated for the optimal use of their collective storage capacity. The Coordination Agreement provides that prior to the start of each operating year, an operating plan is developed to provide the optimum firm energy load-carrying capacity for each reservoir in the coordinated system.

12.7.7 Colorado River Basin

The Colorado River and its major tributaries originate as snowmelt-fed streams high in the Rocky Mountains. The river drains about 629,000 km^2 in seven states as it meanders southward to Mexico and the Gulf of California. For purposes of water management activities, the Colorado River Basin is divided into the Upper and Lower Basins. The dividing point is at Lee Ferry, Arizona, which is located below Glen Canyon Dam near the Arizona-Utah border. Flows vary greatly from year to year, as well as seasonally. Snowmelt in the upper basin results in flows generally being highest from April to July. Mean annual precipitation in the arid lower basin is about 13 cm. Demands exceed supplies. In an average year, all available flow is diverted for consumptive use, and the river is dry before reaching the Gulf of California.

Management and use of the waters of the Colorado River Basin are governed by a series of international and interstate negotiations, legislative acts, U.S. Supreme Court decisions, compacts, and treaties, which were developed historically over several decades and are collectively called the *Law of the River.* Early major agreements include a 1922 compact between the several states that share the basin and the 1944 treaty between the U.S. and Mexico discussed in Section 12.3.3.2. From the perspective of legal agreements, the Colorado River is perhaps the most regulated river in the world. The water allocation system is complex. Water management is complicated by the fact that the amount of water that is legally allocated among the two nations and seven states is greater than the estimated mean annual flow of the river.

Nine USBR reservoirs have a total active storage capacity of about four times the mean annual runoff. Lake Mead impounded by Hoover Dam and Lake Powell behind Glen Canyon Dam, the two largest reservoirs in the U.S., are included in Table 12.4 and Fig. 12.11. System operations are based on maintaining flows to Mexico in accordance with the Water Treaty of 1944 and other agreements and meeting obligations to agricultural, industrial, and municipal water users, such as the cities of Las Vegas and Los Angeles, in the seven states within the U.S. as required by compacts and other allocation commitments. The reservoirs are also operated for flood control, water quality control, recreation, and hydroelectric power. Hydroelectric power generation is limited essentially to water supply releases for downstream users. Releases, including controlled flooding, have been made specifically for protection and restoration of ecosystems, including endangered species.

Salinity is a major problem in the Colorado River Basin, as well as in other western river basins. Naturally high salt concentrations are further increased by evaporation associated with river basin development projects and irrigated agriculture. In 1973 the U.S. committed to limiting salinity in the flow to Mexico. Salinity control measures that have been implemented in the basin include irrigation management programs, desalting plants, and construction of facilities, such as wells, dikes, pumps, and evaporation ponds to collect and dispose of saline water.

12.7.8 California Central Valley

The Sacramento River and San Joaquin River watersheds comprise the Central Valley Basin shown in Fig. 12.14. The Sacramento River and its tributaries flow southward, draining the northern part of the basin. The San Joaquin River and its tributaries flow northward, draining the southern portion. The two river systems join at the Sacramento-San Joaquin Delta, flow through Suisun Bay and Carquinez Straits into San Francisco Bay, and then out the Golden Gate to the Pacific Ocean.

Water resources development and management in the Central Valley have been driven by California's pattern of precipitation and water demands. Most snow

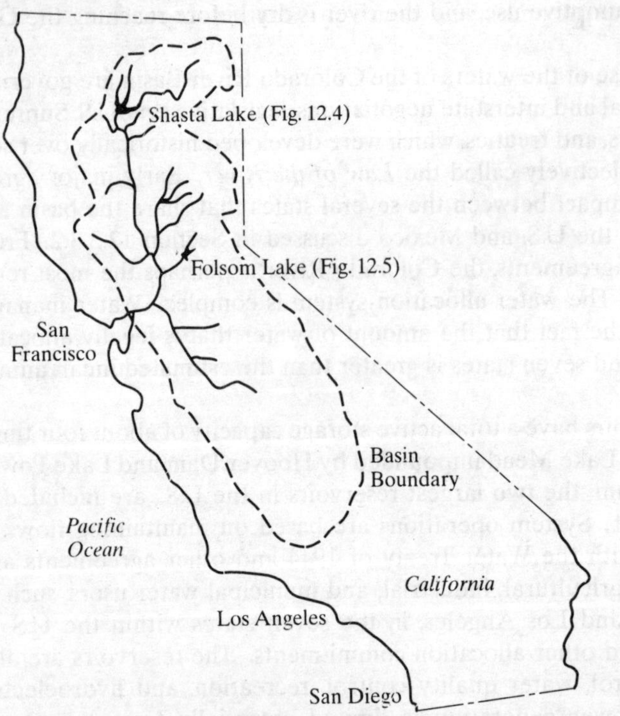

Figure 12.14 The California Central Valley is among the most productive agricultural regions in the world.

and rain occur in the north during winter, while water demands are concentrated in the south during the summer. The two largest water systems in California are the interrelated State Water Project, developed by the State of California, and the federal Central Valley Project. Major features of the State Water Project include: Oroville Dam and Reservoir on the Feather River; Harvey O. Banks Pumping Plant in the Sacramento-San Joaquin Delta; California Aqueduct delivery to the Southern San Joaquin Valley and Southern California; South Bay Aqueduct delivery to the San Francisco Bay area; and San Luis Reservoir, which stores water on the west side of the San Joaquin Valley for later release into the California Aqueduct. The following discussion focuses on the Central Valley Project.

The USBR's Central Valley Project consists of six major storage reservoirs, 12 other smaller reservoirs for flow regulation and power generation, 39 pumping plants, and more than 800 km of canals. Most of the reservoirs and other project elements were constructed by the USBR, but the Folsom (Fig. 12.5) and New Melones Reservoirs were constructed by the USACE. The Bureau is responsible for operating the project.

The primary purpose of the Central Valley Project is to provide a reliable water supply for the rich agricultural lands of the semi-arid Sacramento and San Joaquin Valleys. Precipitation ranges from 76 cm annually in the north to 13 cm in the south. Most rainfall occurs in December–March, during the nonirrigation season. Agriculture is highly dependent on irrigation, which is supplied from both surface water and groundwater. Water is stored in the high runoff winter and spring months to meet irrigation requirements, which are greatest during the summer months. The extensive system of canals and pumping plants is used to transfer water from the water-rich Sacramento River Basin in the north to the water-poor, but intensively cultivated, San Joaquin Valley in the south.

Flood control and hydroelectric power generation are also important functions. A substantial amount of power generation is required to meet pumping requirements for the water supply conveyance components of the system, and revenues from generation above these requirements help repay the cost of reservoirs and other facilities. Reservoir recreation, navigation on the Sacramento River, municipal and industrial water supply, fish and wildlife, and control of salinity intrusion in the Sacramento-San Joaquin River delta are secondary functions, but they have an important influence on how the system is operated.

Clair Engle (Trinity Dam), Shasta, Folsom, New Melones, and San Luis Reservoirs are key storage projects. San Luis is a large seasonal pumped-storage reservoir that is filled during the winter and used to provide irrigation and hydroelectric power requirements during the summer and fall. It is one of only a few seasonal pumped-storage projects in the U.S. About 30 percent of the usable storage space in the major reservoirs is allocated to seasonal joint use storage, and the remaining 70 percent is allocated to carryover storage used to meet water and energy needs during extended droughts.

Much of the irrigated land in the San Joaquin Valley is served by the Delta-Mendota Canal. The canal originates in the Sacramento River delta area and extends in a southeasterly direction, generally parallel to the San Joaquin River, for about 185 km, terminating about 50 km west of Fresno. Although the irrigation demand occurs primarily in the summer months, water is pumped into the canal from the Sacramento River throughout the year. Water excess to irrigation needs is pumped into the San Luis Reservoir, to be held until the peak irrigation demand season, when it is released back into the Delta-Mendota Canal. A portion of the San Luis Storage is also allocated to the state-operated California Water Project, with water being pumped from and discharged back into the California Aqueduct, which runs generally parallel to the Central Valley Project's Delta-Mendota Canal.

The hydropower plants of the Central Valley Project provide a dependable capacity of 880 megawatts to the Pacific Gas and Electric Company (PG&E). Contracts with PG&E specify minimum 12-month, 6-month, and monthly energy delivery and provide benefits for exceeding these levels. The USBR submits a daily generating schedule, which is based on Central Valley Project reservoir conditions, to PG&E that uses this energy on an hour-by-hour basis in such a way as to minimize fuel costs at the thermal-electric plants in the system.

The water pumped from the Sacramento River delta is a mix of natural runoff and releases from storage projects such as Clair Engle and Shasta Reservoirs. Minimum flows must be maintained in the delta below the pumping plants to prevent salt water intrusion. Protection of ecosystems in the delta and bay is a major concern. Delta flows are controlled by regulating both upstream reservoir releases and the amounts of water pumped for export from the delta by the Central Valley Project and State Water Project.

12.8 RIVER BASIN MANAGEMENT COMPUTER MODELS

River/reservoir system management and associated computer modeling capabilities are major concerns throughout the U.S. and world. The published literature and unpublished information available within the water agencies are extensive. Wurbs (1996) provides a detailed review of river basin management models.

12.8.1 Modeling Applications

River basin management system analysis models support:

- planning and design of new facilities
- reevaluation of the operating plans of existing systems
- administration of water allocation systems involving water rights, commitments between water suppliers and users, and international and interstate agreements
- operational planning for developing management strategies for the next year or irrigation season
- real-time decision support, including operations during droughts and floods, as well as during normal hydrologic conditions

In many countries, such as the U.S., the earlier construction era of water resources development evolved after the 1970's into a focus on better management of existing facilities, water resources allocation, and environmental protection/restoration. Thus, in developing and applying river basin system analysis models, the emphasis has shifted to evaluating the operations of existing systems, but planning and design of proposed new projects are also still important.

Reservoir/river system analysis models are used for various purposes in a variety of settings. Models are used in feasibility studies to aid in the formulation and evaluation of alternative plans for responding to water-related problems and needs. Feasibility studies range from broad comprehensive river basin planning to detailed project implementation studies. Feasibility studies may involve proposed construction projects, as well as reallocations of storage capacity or other operational modifications at completed projects. Other modeling applications involve studies performed specifically to reevaluate operating policies for existing reservoir systems. Periodic evaluations may be made to ensure system responsiveness to current conditions and objectives. Modeling studies are often accomplished in response to a particular perceived problem or need. Studies may be motivated by drought conditions, a major flood event, water quality problems, or environmental losses, such as fish kills. Water availability models support the administration of water rights permit programs. Operational planning studies are conducted annually for some reservoir systems to establish operating strategies for the next year. Execution of models during actual reservoir operations in support of real-time release decisions represents another major area of application.

A river basin management application often involves integration of several different types of models. Optimization and simulation models (Chapter 11) dealing with water quantities might be used along with models addressing water quality. Other models are used to establish diversion and instream flow requirements to be met by the reservoir releases. A watershed precipitation-runoff model (Chapter 8) is used to develop runoff hydrographs and pollutant loadings for input to the reservoir operation models, which in turn determine discharges and contaminant concentrations at pertinent locations in the river/reservoir system. The example modeling system could also include a river hydraulics model (Chapters 5 and 6) to compute flow depths and velocities. A groundwater model (Chapter 9) may be used to assess groundwater resources and interconnections between surface water and groundwater. A geographic information system and other data management programs are included in the modeling system to: (1) develop and manage voluminous input data, (2) perform statistical and graphical analyses of simulation output, and (3) display and communicate results.

12.8.2 Reservoir/River System Operations Models

Reservoir/river system operations models typically include features for representing the:

- spatial configuration of the river basin system
- basin hydrology
- physical characteristics of reservoirs, spillways and outlet works, hydroelectric power plants, conveyance facilities, and other water control structures
- system operating rules
- water use requirements
- effects of basinwide water management on the reservoir/river/use system of concern
- measures of system performance

River basin hydrology is represented in system operations models by sequences of streamflows at all pertinent locations representing a specified condition of development. Modeling flood control operations is typically based on gaged or synthesized streamflow sequences representing flood events. The computational time step is typically in the range of one to 24 hours. Studies of operations for conservation purposes such as water supply, hydropower, and environmental management are typically based on a simulation period of several decades with a monthly or weekly computational time step. For example, a planning study might be based on streamflows at relevant locations for each of the 732 months during a 1940–2002 simulation period. These flows represent historical hydrology adjusted to reflect a specified condition of river basin development, typically natural unregulated flows. Since the future is unknown, historical hydrology is adopted as being representative of the statistical characteristics of future inflows to the system. Physical characteristics of reservoirs and other facilities, operating rules, and water use requirements are combined with the natural streamflow sequences to assess capabilities for meeting water supply, environmental instream flow, hydropower generation, flood mitigation, and other objectives under various scenarios, with alternative water management strategies. Reliability indices, flow-frequency relationships, economic measures, and other forms of simulation results are used to evaluate alternative plans.

A computer model may be developed for a specific reservoir/river system or generalized for application to essentially any river basin. Many models of both types have been applied to each of the river basins cited in this chapter (Wurbs, 1996). Recent trends favor use of generalized simulation models that can be applied to a broad range of applications in any river basin. Several generalized models are listed in Table 12.7. Other similar models are also available. Application of these software packages involves developing input data sets for the particular system of concern.

The USBR maintains a Hydrologic Modeling Inventory (HMI) for the stated purpose of "providing water resources managers, public stakeholders, and model

TABLE 12.7 GENERALIZED RIVER BASIN MANAGEMENT MODELS

Common name	Descriptive name	Organizations that developed and maintains model
HEC-5	Simulation of Flood Control and Conservation Systems	USACE Hydrologic Engineering Center
HEC-ResSim	Reservoir System Simulation	USACE Hydrologic Engineering Center
HEC-PRM	Prescriptive Reservoir Model	USACE Hydrologic Engineering Center
SSARR	Streamflow Synthesis and Reservoir Regulation	USACE North Pacific Division
RiverWare	Reservoir and River Operations Model	U.S. Bureau of Reclamation, TVA, and Center for Advanced Decision Support for Water and Environmental Systems at University of Colorado
MODSIM	River Basin Network Flow Model	Colorado State University
WRAP	Water Rights Analysis Package	Texas A&M University, Texas Water Resources Institute, and TNRCC
IRIS and	Interactive River System Simulation	Cornell University and
IRAS	Interactive River-Aquifer Simulation	Resources Planning Associates
MIKE BASIN	Multiple Purpose River Network	Danish Hydraulic Institute

developers a forum in which the latest modeling and decision support tools may be shared." The generalized models listed in Table 12.7 are described in the USBR HMI (http://www.usbr.gov/hmi). The models have been extensively applied in many river basins. They are widely applied by water resources engineers working for various firms and agencies other than the original model development entity. Each of these generalized modeling systems was developed over several years, and continues to be improved and expanded with experience in river basin modeling and advances in computer technology.

All of the models in Table 12.7 provide capabilities for modeling multiple-reservoir systems operated for conservation purposes, including water supply, hydropower, and instream flow maintenance. HEC-5, HEC-ResSim, and SSARR also have detailed capabilities for simulating flood control operations. All of the models in Table 12.7 focus on water quantities, but several also have water quality simulation features. Other models not cited deal specifically with water quality (Wurbs, 1996). All of the models focus on surface water, but several include features for modeling interactions with groundwater. IRIS was originally developed to support multiple-nation water allocation negotiations for international river basins. MIKE BASIN is applied within an ArcView geographic information system environment. WRAP is briefly discussed in Section 12.3.3.5. HEC-PRM, RiverWare, and MODSIM are based on linear programming (Section 11.5). Examples 11.7, 11.10, and 11.11 illustrate basic computations performed by these LP models. The generalized models listed in Table 12.7 also incorporate a variety of other methods covered in the various chapters of this book.

REFERENCES

GETCHES, D. H., *Water Law in a Nutshell,* 3rd Ed., West Publishing, St. Paul, MN, 1997.

GLEICK, P. H., *The World's Water: The Biennial Report on Freshwater Resources 1998–1999,* Island Press, Washington, D.C., 1998.

GREENWAY, A. R. (Editor), *Environmental Permitting Handbook,* McGraw-Hill, New York, NY, 2000.

HERZOG, M. A. M., *Practical Dam Analysis,* Thomas Telford Publishing, London, UK, 1999.

Hydrologic Engineering Center, *A Preliminary Assessment of Corps of Engineer Reservoirs, Their Purposes and Susceptibility to Drought,* RD-33, USACE, Davis, CA, December 1990.

International Commission on Large Dams, *World Register of Dams,* Paris, France, 1998.

JANSEN, R. B., *Dams and Public Safety,* U.S. Bureau of Reclamation, Denver, CO, 1983.

KOLARS, J. F., and W. A. MITCHELL, *The Euphrates River and the Southeast Anatolia Development Project,* Southern Illinois University Press, Carbondale, IL, 1991.

KOLLGAARD, E. B., and W. L. CHADWICK (Editors), *Development of Dam Engineering in the United States,* Pergamon Press, Elmsford, NY, 1988.

KUTZNER, C., *Earth and Rockfill Dams: Principles of Design and Construction,* A. A. Balkema, Rotterdam, The Netherlands, 1997.

MAC, M. J., P. A. OPLER, C. E. HAECKER, and P. D. DORAN, *Status and Trends of the Nation's Biological Resources,* U.S. Geological Survey, Reston, VA, 1998.

MITSCH, W. J., and J. G. GOSSELINK, *Wetlands,* 3rd Ed., John Wiley, New York, NY, 2000.

RICE, L., and M. D. WHITE, *Engineering Aspects of Water Law,* Krieger Publishing, Malabar, FL, 1991.

SCHNITTER, N. J., *A History of Dams, the Useful Pyramids,* A. A. Balkema, Rotterdam, The Netherlands, 1994.

SENTURK, F., *Hydraulics of Dams and Reservoirs,* Water Resources Publications, Highlands Ranch, CO, 1994.

SINGH, B., and R. S. VARSHNEY, *Engineering for Embankment Dams,* A. A. Balkema, Rotterdam, The Netherlands, 1995.

U.S. Army Corps of Engineers, *Management of Water Control Systems,* Engineering Manual 1110-2-3600, Washington, D.C., November 1987.

U.S. Army Corps of Engineers, *Sedimentation Investigations of Rivers and Reservoirs,* Engineering Manual 1110-2-4000, Washington, D.C., December 1989.

U.S. Bureau of Reclamation, *Design of Gravity Dams,* Denver, CO, 1976.

U.S. Bureau of Reclamation, *Design of Arch Dams,* Denver, CO, 1977.

U.S. Bureau of Reclamation, *Design of Small Dams,* Denver, CO, 1987.

U.S. Bureau of Reclamation, *Statistical Compilation of Engineering Features of Bureau of Reclamation Projects*, Denver, CO, 1992.

VISCHER, D. L., and W. H. HAGER, *Dam Hydraulics,* John Wiley, New York, NY, 1998.

WOLF, A. T., J. A. NATHARIUS, J. J. DANIELSON, B. S. WARD, and J. K. PENDER, "International River Basins of the World," *International Journal of Water Resources Development,* Vol. 14, No. 4, pp. 387–427, 1999.

WOLFE, M. E., *A Landowners Guide to Western Water Rights,* Roberts Rinehart Publishers, Boulder, CO, 1995.

World Commission on Dams, *Dams and Development: A New Dimension for Decision-Making,* Earthscan Publishers, London, UK, November 2000.

WOUTERS, P., "National and International Water Law: Achieving Equitable and Sustainable Use of Water Resources," *Water International,* International Water Resources Association, Vol. 25, No. 4, pp. 499–512, December 2000.

WURBS, R. A., "Water Rights in Texas," *Journal of Water Resources Planning and Management,* American Society of Civil Engineers, Vol. 121, No. 6, pp. 447–454, December 1995.

WURBS, R. A., *Modeling and Analysis of Reservoir System Operations,* Prentice Hall, Upper Saddle River, NJ, 1996.

WURBS, R. A., "Assessing Water Availability Under a Water Rights Priority System," *Journal of Water Resources Planning and Management,* American Society of Civil Engineers, Vol. 127, No. 4, July/August 2001.

WURBS, R. A., "Administration and Modeling of the Texas Water Rights System," *Water for Texas: 2000 and Beyond,* Texas A&M University Press, 2002.

Index